Acta Numerica 2010

Acta Numerica

Volume 19 2010

CAMBRIDGE UNIVERSITY PRESS
Cambridge, New York, Melbourne, Madrid, Cape Town,
Singapore, São Paulo, Delhi, Tokyo, Mexico City

Cambridge University Press
The Edinburgh Building, Cambridge CB2 8RU, UK

Published in the United States of America by Cambridge University Press, New York

www.cambridge.org
Information on this title: www.cambridge.org/9780521290494

First published 2010
First paperback edition 2011

A catalogue record for this publication is available from the British Library

ISBN 978-0-521-19284-2 Hardback
ISBN 978-0-521-29049-4 Paperback

Contents

Acta Numerica (2010), pp. 1–120
doi:10.1017/S0962492910000012

Finite element approximation of eigenvalue problems

Daniele Boffi
Dipartimento di Matematica, Università di Pavia,
via Ferrata 1, 27100 Pavia, Italy
E-mail: daniele.boffi@unipv.it

We discuss the finite element approximation of eigenvalue problems associated with compact operators. While the main emphasis is on symmetric problems, some comments are present for non-self-adjoint operators as well. The topics covered include standard Galerkin approximations, non-conforming approximations, and approximation of eigenvalue problems in mixed form. Some applications of the theory are presented and, in particular, the approximation of the Maxwell eigenvalue problem is discussed in detail. The final part tries to introduce the reader to the fascinating setting of differential forms and homological techniques with the description of the Hodge–Laplace eigenvalue problem and its mixed equivalent formulations. Several examples and numerical computations complete the paper, ranging from very basic exercises to more significant applications of the developed theory.

CONTENTS

1. Introduction

The aim of this paper is to provide the reader with an overview of the state of the art in the numerical analysis of the finite element approximation of eigenvalue problems arising from partial differential equations.

The work consists of four parts, which are ordered according to their increasing difficulty. The material is arranged in such a way that it should be possible to use it (or part of it) as a reference for a graduate course.

Part 1 presents several examples and reports on some academic numerical computations. The results presented range from a very basic level (such as the approximation of the one-dimensional Laplace operator), suited to those just starting work in this subject, to more involved examples. In particular, we give a comprehensive review of the Galerkin approximation of the Laplace eigenvalue problem (also in the presence of a singular domain and of non-conforming schemes), of the mixed approximation of the Laplace eigenvalue problem (with stable or unstable schemes), and of the Maxwell eigenvalue problem. Some of the presented material is new, in particular, the numerical results for the one-dimensional mixed Laplacian with the $\mathcal{P}_1 - \mathcal{P}_1$ and the $\mathcal{P}_2 - \mathcal{P}_0$ scheme.

Part 2 contains the main core of the theory concerning the Galerkin approximation of variationally posed eigenvalue problems. With a didactic purpose, we included a direct proof of convergence for the eigenvalues and eigenfunctions of the Laplace equation approximated with piecewise linear elements. By direct proof, we mean a proof which does not make use of the abstract spectral approximation theory, but is based on basic properties of the Rayleigh quotient. This proof is not new, but particular care has been paid to the analysis of the case of multiple eigenfunctions. In Section 9 we describe the so-called Babuška–Osborn theory. As an example of application we analyse the approximation of the eigensolutions of an elliptic operator. Then, we provide another application which involves the non-conforming Crouzeix–Raviart element for the approximation of the Laplace eigenvalue problem. The results of this section are probably not new, but we could not find a reference providing a complete analysis of this form.

Part 3 is devoted to the approximation theory of eigenvalue problems in mixed form. We recall that the natural conditions for the well-posedness and stability of source mixed problems (the classical inf-sup conditions) are not good hypotheses for convergence of the eigensolutions. It is standard to consider two different mixed formulations: problems of the first type (also known as $(f, 0)$ problems) and of the second type $(0, g)$. The first family is used, for instance, when the Stokes system is considered, and an example of an application for the second one is the mixed Laplace eigenvalue problem. The sufficient and necessary conditions for the convergence of eigenvalues and eigenfunctions of either type of mixed problem are discussed.

Finally, Part 4 deals with the homological techniques which lead to the finite element exterior calculus. We recall the Hodge–Laplace eigenvalue problem and show the links between this problem in the language of differential forms and standard eigenvalue problems for differential operators. In particular, we study the Maxwell eigenvalue problem and discuss the main tools for its analysis.

In a project like this one, it is responsibility of the author to make some choices about the material to be included. We acknowledge that we would have added some more subjects, but finally we had to trim our original plan. In particular, we completely ignored the topic of *a posteriori* and adaptive error analysis for eigenvalue problems. For this active and fundamental research field the reader is referred to the following papers and to the references therein: Hackbusch (1979), Larson (2000), Morin, Nochetto and Siebert (2000), Heuveline and Rannacher (2001), Neymeyr (2002), Durán, Padra and Rodríguez (2003), Gardini (2004), Carstensen (2008), Giani and Graham (2009), Grubišić and Ovall (2009) and Garau, Morin and Zuppa (2009). The p and hp version of finite elements is pretty much related to this topic: we give some references on this issue in Section 20 for the approximation of Maxwell's eigenvalue problem. Another area that deserves attention is the discontinuous Galerkin approximation of eigenvalue problems. We refer to the following papers and to the references therein: Hesthaven and Warburton (2004), Antonietti, Buffa and Perugia (2006), Buffa and Perugia (2006), Warburton and Embree (2006), Creusé and Nicaise (2006), Buffa, Houston and Perugia (2007) and Brenner, Li and Sung (2008). Nonstandard approximations, including mimetic schemes (Cangiani, Gardini and Manzini 2010), have not been discussed. Another important result we did not include deals with the approximation of non-compact operators (Descloux, Nassif and Rappaz 1978*a*, 1978*b*). It is interesting to note that such results have often been used for the analysis of the non-conforming approximation of compact operators and, in particular, of the approximation of Maxwell's eigenvalue problem.

Throughout this paper we quote in each section the references we need. We tried to include all significant references we were aware of, but obviously many others have not been included. We apologize for that in advance and encourage all readers to inform the author about results that would have deserved more discussion.

PART ONE
Some preliminary examples

In this section we discuss some numerical results concerning the finite element approximation of eigenvalue problems arising from partial differential equations. The presented examples provide motivation for the rest of this survey and will be used for the applications of the developed theory. We only consider *symmetric* eigenvalue problems, so we are looking for *real* eigenvalues.

2. The one-dimensional Laplace eigenvalue problem

We start with a very basic and well-known one-dimensional example. Let Ω be the open interval $]0,\pi[$ and consider the problem of finding eigenvalues λ and eigenfunctions u with $u \neq 0$ such that

$$-u''(x) = \lambda u(x) \qquad \text{in } \Omega, \tag{2.1a}$$
$$u(0) = u(\pi) = 0. \tag{2.1b}$$

It is well known that the eigenvalues are given by the squares of the integer numbers $\lambda = 1, 4, 9, 16, \ldots$ and that the corresponding eigenspaces are spanned by the eigenfunctions $\sin(kx)$ for $k = 1, 2, 3, 4 \ldots$. A standard finite element approximation of problem (2.1) is obtained by considering a suitable variational formulation. Given $V = H_0^1(\Omega)$, multiplying our equation by $v \in V$, and integrating by parts, yields the following: find $\lambda \in \mathbb{R}$ and a non-vanishing $u \in V$ such that

$$\int_0^\pi u'(x)v'(x)\,\mathrm{d}x = \lambda \int_0^\pi u(x)v(x)\,\mathrm{d}x \quad \forall v \in V. \tag{2.2}$$

A Galerkin approximation of this variational formulation is based on a finite-dimensional subspace $V_h = \mathrm{span}\{\varphi_1, \ldots, \varphi_N\} \subset V$, and consists in looking for discrete eigenvalues $\lambda_h \in \mathbb{R}$ and non-vanishing eigenfunctions $u_h \in V_h$ such that

$$\int_0^\pi u_h'(x)v'(x)\,\mathrm{d}x = \lambda_h \int_0^\pi u_h(x)v(x)\,\mathrm{d}x \quad \forall v \in V_h.$$

It is well known that this gives an algebraic problem of the form

$$A\mathsf{x} = \lambda M\mathsf{x},$$

where the stiffness matrix $A = \{a_{ij}\}_{i,j=1}^N$ is defined as

$$a_{ij} = \int_0^\pi \varphi_j'(x)\varphi_i'(x)\,\mathrm{d}x$$

and the mass matrix $M = \{m_{ij}\}_{i,j=1}^N$ is

$$m_{ij} = \int_0^\pi \varphi_j(x)\varphi_i(x)\,\mathrm{d}x.$$

Given a uniform partition of $[0,\pi]$ of size h, let V_h be the space of continuous piecewise linear polynomials vanishing at the end-points (standard conforming $\mathcal{P}_1$ finite elements); then the associated matrices read

$$a_{ij} = \frac{1}{h}\cdot\begin{cases} 2 & \text{for } i=j,\\ -1 & \text{for } |i-j|=1,\\ 0 & \text{otherwise,}\end{cases} \qquad m_{ij} = h\cdot\begin{cases} 4/6 & \text{for } i=j,\\ 1/6 & \text{for } |i-j|=1,\\ 0 & \text{otherwise,}\end{cases}$$

with $i,j = 1,\dots,N$, where the dimension N is the number of internal nodes in the interval $[0,\pi]$. It is well known that in this case it is possible to compute the explicit eigenmodes: given $k\in\mathbb{N}$, the kth eigenspace is generated by the interpolant of the continuous solution

$$u_h^{(k)}(ih) = \sin(kih), \quad i = 1,\dots,N, \tag{2.3}$$

and the corresponding eigenvalue is

$$\lambda_h^{(k)} = (6/h^2)\frac{1-\cos kh}{2+\cos kh}. \tag{2.4}$$

It is then immediate to deduce the optimal estimates (as $h\to 0$)

$$\|u^{(k)} - u_h^{(k)}\|_V = O(h) \qquad |\lambda^{(k)} - \lambda_h^{(k)}| = O(h^2) \tag{2.5}$$

with $u^{(k)}(x) = \sin(kx)$ and $\lambda^{(k)} = k^2$.

We would like to make some comments about this first example. Although here the picture is very simple and widely known, some of the observations generalize to more complicated situations and will follow from the abstract theory, which is the main object of this survey.

First of all, it is worth noticing that, even if not explicitly stated, estimates (2.5) depend on k. In particular, the estimate on the eigenvalues can be made more precise by observing that

$$\lambda_h^{(k)} = k^2 + (k^4/12)h^2 + O(k^6h^4), \quad \text{as } h\to 0. \tag{2.6}$$

This property has a clear physical meaning: since the eigenfunctions present more and more oscillations when the frequency increases, an increasingly fine mesh is required to keep the approximation error within the same accuracy.

The second important consequence of (2.4) is that all eigenvalues are approximated from above. This behaviour, which is related to the so-called min-max property (see Proposition 7.2), can be stated as follows:

$$\lambda^{(k)} \le \lambda_h^{(k)} \le \lambda^{(k)} + C(k)h^2.$$

The first estimate in (2.5) on the convergence of the eigenfunctions requires some additional comments. It is clear that the solution of the algebraic system arising from (2.2) does not give, in general, the eigenfunctions described in (2.3). Since in this simple example all eigenspaces are one-dimensional, we might expect that the numerical solver will provide us with multiples of the functions in (2.3). It is evident that if we want to perform an automatic error estimation, a first step will be to normalize the computed eigenfunctions so that they have the same norm as the continuous ones. This, however, is not enough, since there can be a difference in sign, so we have to multiply them by ± 1 in order for the scalar product with the continuous eigenfunctions to be positive.

Remark 2.1. If the same eigenvalue computation is performed with V_h equal to the space of continuous piecewise polynomials of degree at most p and vanishing at the end-points (standard conforming $\mathcal{P}_p$ finite elements), then estimates (2.5) become

$$\|u^{(k)} - u_h^{(k)}\|_V = O(h^p) \qquad |\lambda^{(k)} - \lambda_h^{(k)}| = O(h^{2p}).$$

In any case, the order of approximation for the eigenvalues is double with respect to the approximation rate of the corresponding eigenfunctions. This is the typical behaviour of *symmetric* eigenvalue problems.

3. Some numerical results for the two-dimensional Laplace eigenvalue problem

In this section we present some numerical results for the Laplace eigenvalue problem in two dimensions. We use different domains and finite elements.

Given an open Lipschitz domain $\Omega \subset \mathbb{R}^2$, we are interested in the following problem: find eigenvalues λ and eigenfunctions u with $u \neq 0$ such that

$$-\Delta u(x,y) = \lambda u(x,y) \quad \text{in } \Omega, \tag{3.1a}$$

$$u = 0 \quad \text{on } \partial\Omega. \tag{3.1b}$$

Given $V = H_0^1(\Omega)$, a variational formulation of (3.1) can be obtained as follows: find $\lambda \in \mathbb{R}$ and $u \in V$, with $u \neq 0$, such that

$$\int_\Omega \mathbf{grad}\, u(x,y) \cdot \mathbf{grad}\, v(x,y)\, \mathrm{d}x\, \mathrm{d}y = \lambda \int_\Omega u(x,y) v(x,y)\, \mathrm{d}x\, \mathrm{d}y \quad \forall v \in V.$$

A Galerkin approximation based on a finite-dimensional subspace $V_h \subset V$ then reads: find $\lambda_h \in \mathbb{R}$ and $u_h \in V_h$, with $u_h \neq 0$, such that

$$\int_\Omega \mathbf{grad}\, u_h(x,y) \cdot \mathbf{grad}\, v(x,y)\, \mathrm{d}x\, \mathrm{d}y = \lambda_h \int_\Omega u_h(x,y) v(x,y)\, \mathrm{d}x\, \mathrm{d}y \quad \forall v \in V_h.$$

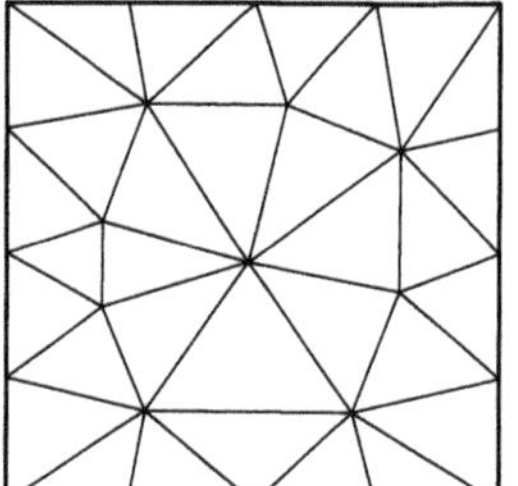
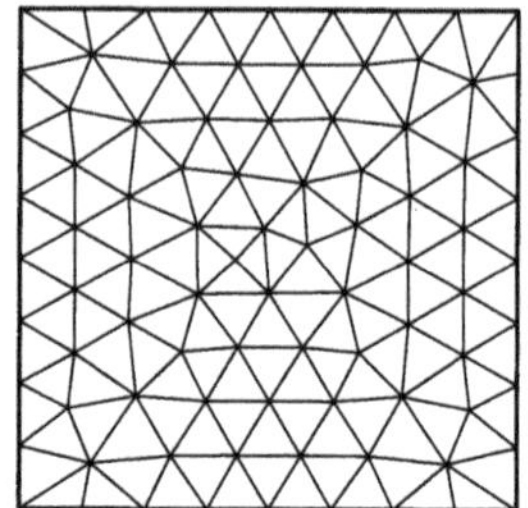
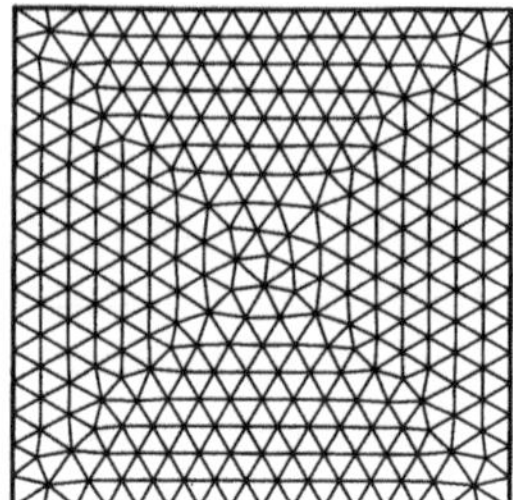

Figure 3.1. Sequence of unstructured meshes ($N = 4, 8, 16$).

3.1. The Laplace eigenvalue problem on the square: continuous piecewise linears

Let Ω be the square $]0,\pi[\times]0,\pi[$. It is well known that the eigenvalues of (3.1) are given by $\lambda_{m,n} = m^2 + n^2$ (with m and n strictly positive integers) and the corresponding eigenfunctions are $u_{m,n} = \sin(mx)\sin(ny)$. Throughout this subsection we are going to use continuous piecewise linear finite elements on triangles.

Our first computation involves a standard sequence of regular unstructured meshes, which is shown in Figure 3.1. Table 3.1 lists the first ten computed eigenvalues and their rate of convergence towards the exact values. It is evident that the scheme is convergent and that the convergence is quadratic. The abstract theory we are going to present will show that the eigenfunctions are first-order convergent in V.

Moreover, from Table 3.1 we can see behaviour similar to that observed in the one-dimensional example: all eigenvalues are approximated from above and the relative error increases with the rank of the eigenvalues in the spectrum (for instance, on the finest mesh, the relative error for the 10th eigenvalue is more than eight times the error for the first one).

This two-dimensional example allows us to make some important comments on multiple eigenvalues. If we look, for instance, at the double eigenvalue $\lambda = 5$, we see that there are two *distinct* discrete eigenvalues $\lambda_h^{(2)} < \lambda_h^{(3)}$ approximating it. Both eigenvalues are good approximations of the exact solution, and on the finest mesh their difference is smaller than 10^{-4}. A natural question concerns the behaviour of the corresponding eigenfunctions. The answer to this question is not trivial: indeed, the exact eigenspace has dimension equal to 2 and it is spanned by the functions $u_{1,2} = \sin x \sin(2y)$ and $u_{2,1} = \sin(2x)\sin y$. On the other hand, since the discrete eigenvalues are distinct, the approximating eigenspace consists of two separate one-dimensional eigenspaces. In particular, we cannot expect an estimate similar to the first one of (2.5) (even after normalization and choice of the sign for each discrete eigenfunction), since there is no rea-

Table 3.1. Eigenvalues computed on the unstructured mesh sequence.

Exact	Computed (rate)				
	$N = 4$	$N = 8$	$N = 16$	$N = 32$	$N = 64$
2	2.2468	2.0463 (2.4)	2.0106 (2.1)	2.0025 (2.1)	2.0006 (2.0)
5	6.5866	5.2732 (2.5)	5.0638 (2.1)	5.0154 (2.0)	5.0038 (2.0)
5	6.6230	5.2859 (2.5)	5.0643 (2.2)	5.0156 (2.0)	5.0038 (2.0)
8	10.2738	8.7064 (1.7)	8.1686 (2.1)	8.0402 (2.1)	8.0099 (2.0)
10	12.7165	11.0903 (1.3)	10.2550 (2.1)	10.0610 (2.1)	10.0152 (2.0)
10	14.3630	11.1308 (1.9)	10.2595 (2.1)	10.0622 (2.1)	10.0153 (2.0)
13	19.7789	14.8941 (1.8)	13.4370 (2.1)	13.1046 (2.1)	13.0258 (2.0)
13	24.2262	14.9689 (2.5)	13.4435 (2.2)	13.1053 (2.1)	13.0258 (2.0)
17	34.0569	20.1284 (2.4)	17.7468 (2.1)	17.1771 (2.1)	17.0440 (2.0)
17		20.2113	17.7528 (2.1)	17.1798 (2.1)	17.0443 (2.0)
DOF	9	56	257	1106	4573

son why, for instance, the eigenspace associated to $\lambda_h^{(2)}$ should provide a good approximation of $u_{1,2}$. The right approach to this problem makes use of the direct sum of the eigenspaces corresponding to $\lambda_h^{(2)}$ and $\lambda_h^{(3)}$, that is, $\text{span}\{u_h^{(2)}, u_h^{(3)}\}$, which does in fact provide a good approximation to the two-dimensional eigenspace associated with $\lambda = 5$. The definition of such an approximation, which relies on the notion of a *gap* between Hilbert spaces, will be made more precise later on. For the moment, we make explicit the concept of convergence in this particular situation which can be stated as follows: there exist constants $\alpha_{1,2}(h)$, $\alpha_{2,1}(h)$, $\beta_{1,2}(h)$ and $\beta_{2,1}(h)$ such that

$$\begin{aligned} \|u_{1,2} - \alpha_{1,2}(h)u_h^{(2)} - \beta_{1,2}(h)u_h^{(3)}\|_V = O(h), \\ \|u_{2,1} - \alpha_{2,1}(h)u_h^{(2)} - \beta_{2,1}(h)u_h^{(3)}\|_V = O(h). \end{aligned} \tag{3.2}$$

It should be clear that the way $u_{1,2}$ and $u_{2,1}$ influence the behaviour of $u_h^{(2)}$ and $u_h^{(3)}$ is mesh-dependent: on the unstructured mesh sequences used for our computations, we cannot expect the α's and the β's to stabilize towards fixed numbers. In order to demonstrate this phenomenon, we display in Figure 3.2 the computed eigenfunctions associated with $\lambda_h^{(2)}$ for $N = 8$, 16, and 32. The corresponding plot for the computed eigenfunctions associated with $\lambda_h^{(3)}$ is shown in Figure 3.3. For the sake of comparison, the exact eigenfunctions $u_{1,2}$ and $u_{2,1}$ are plotted in Figure 3.4.

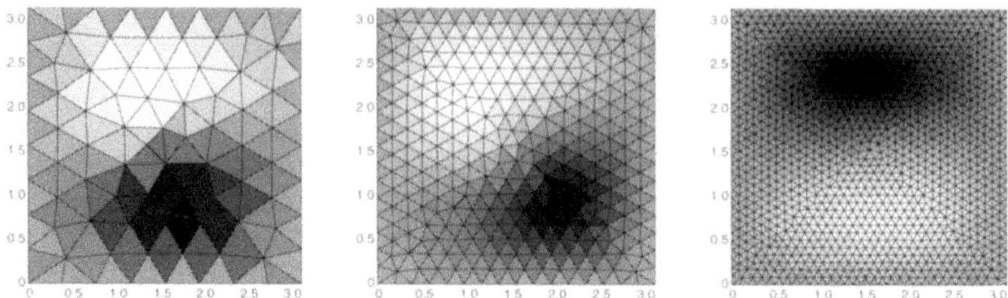

Figure 3.2. Eigenfunction associated with $\lambda_h^{(2)}$ on the unstructured mesh sequence.

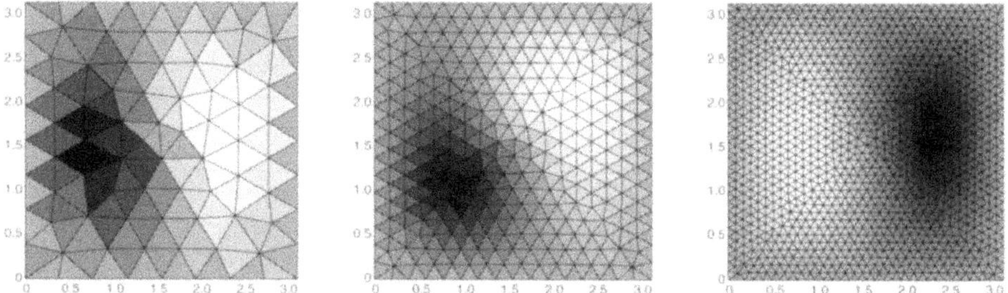

Figure 3.3. Eigenfunction associated with $\lambda_h^{(3)}$ on the unstructured mesh sequence.

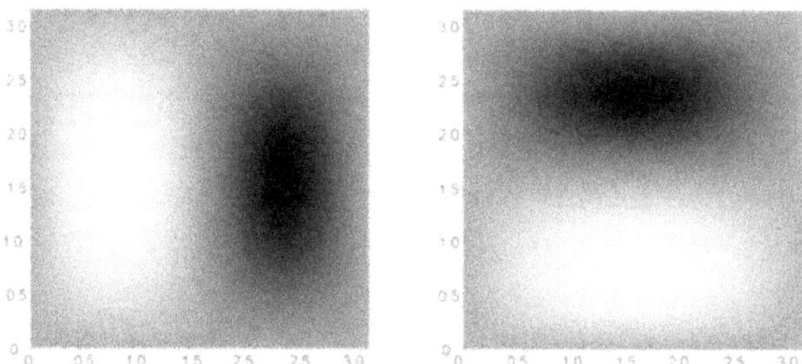

Figure 3.4. Eigenfunctions $u_{1,2}$ and $u_{2,1}$.

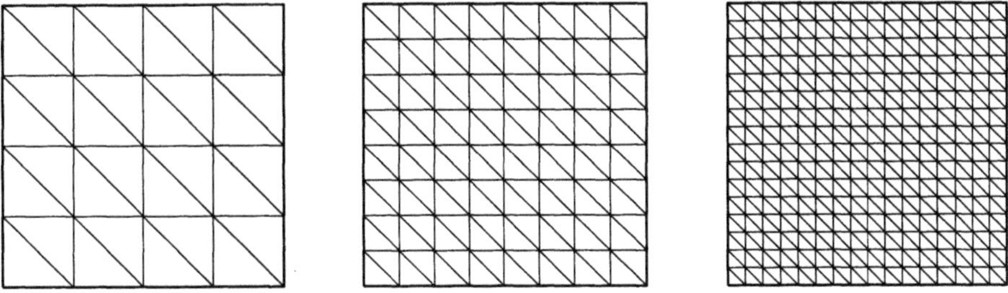

Figure 3.5. Sequence of uniform meshes ($N = 4, 8, 16$).

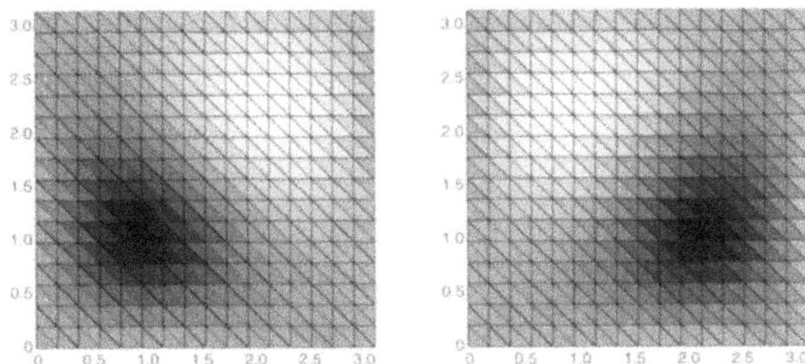

Figure 3.6. Eigenfunctions associated with $\lambda_h^{(2)}$ and $\lambda_h^{(3)}$ on the uniform mesh sequence.

The situation is, however, simpler on a uniform mesh sequence. We consider a mesh sequence of right-angled triangles obtained by bisecting a uniform mesh of squares (see Figure 3.5). Table 3.2 (overleaf) lists the first ten computed eigenvalues and their rate of convergence towards the exact values. This computation does not differ too much from the previous one (besides the fact that the convergence order results are cleaner, since the meshes are now uniform). In particular, the multiple eigenvalues are approximated again by distinct discrete values. The corresponding eigenfunctions are plotted in Figure 3.6 for $N = 16$, where the alignment with the mesh is clearly understood. In order to emphasize the mesh dependence, we performed the same computation on the mesh sequence of Figure 3.7, where the triangles have the opposite orientation from before. The computed eigenvalues are exactly the same as in Table 3.2 (in particular, two distinct eigenvalues approximate $\lambda = 5$) and the eigenfunctions corresponding to the multiple eigenvalue are plotted in Figure 3.8. It is evident that the behaviour has changed due to the change in the orientation of the mesh. This result is not surprising since the problem is invariant under the change of variable induced by the symmetry about the line $y = \pi - x$.

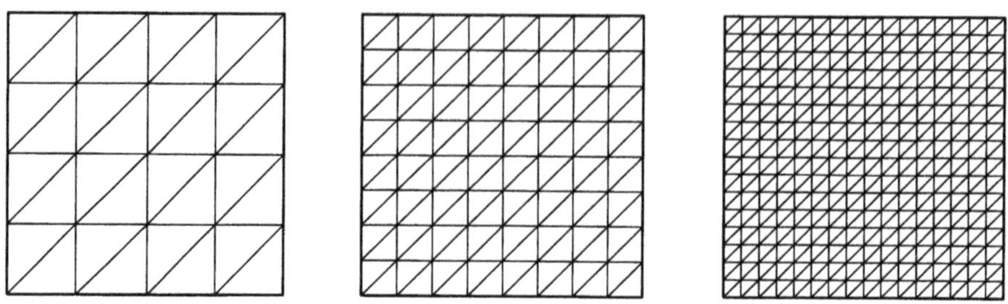

Figure 3.7. Sequence of uniform meshes with reverse orientation ($N = 4, 8, 16$).

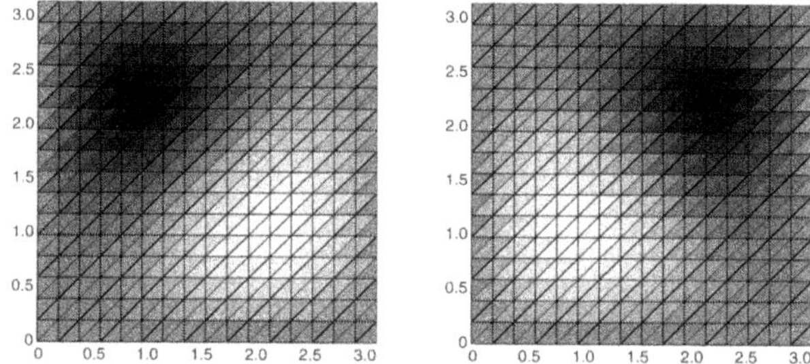

Figure 3.8. Eigenfunctions associated with $\lambda_h^{(2)}$ and $\lambda_h^{(3)}$ on the reversed uniform mesh sequence.

Table 3.2. Eigenvalues computed on the uniform mesh sequence.

Exact	Computed (rate)				
	$N = 4$	$N = 8$	$N = 16$	$N = 32$	$N = 64$
2	2.3168	2.0776 (2.0)	2.0193 (2.0)	2.0048 (2.0)	2.0012 (2.0)
5	6.3387	5.3325 (2.0)	5.0829 (2.0)	5.0207 (2.0)	5.0052 (2.0)
5	7.2502	5.5325 (2.1)	5.1302 (2.0)	5.0324 (2.0)	5.0081 (2.0)
8	12.2145	9.1826 (1.8)	8.3054 (2.0)	8.0769 (2.0)	8.0193 (2.0)
10	15.5629	11.5492 (1.8)	10.3814 (2.0)	10.0949 (2.0)	10.0237 (2.0)
10	16.7643	11.6879 (2.0)	10.3900 (2.1)	10.0955 (2.0)	10.0237 (2.0)
13	20.8965	15.2270 (1.8)	13.5716 (2.0)	13.1443 (2.0)	13.0362 (2.0)
13	26.0989	17.0125 (1.7)	13.9825 (2.0)	13.2432 (2.0)	13.0606 (2.0)
17	32.4184	21.3374 (1.8)	18.0416 (2.1)	17.2562 (2.0)	17.0638 (2.0)
17		21.5751	18.0705 (2.1)	17.2626 (2.0)	17.0653 (2.0)
DOF	9	49	225	961	3969

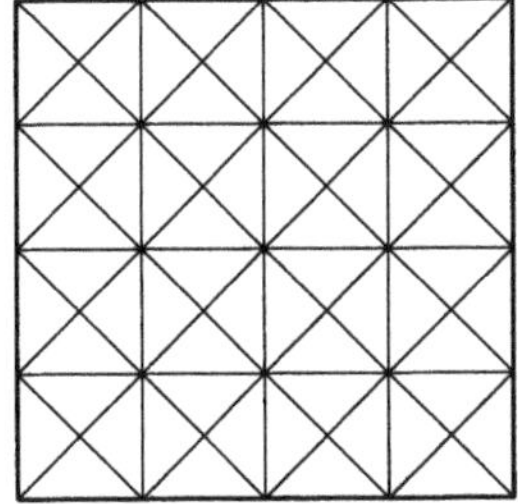
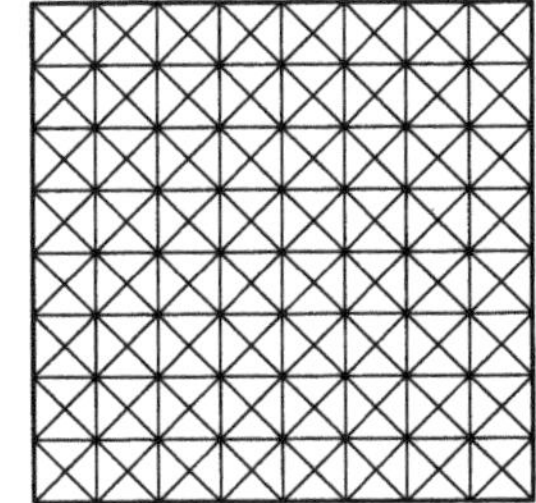
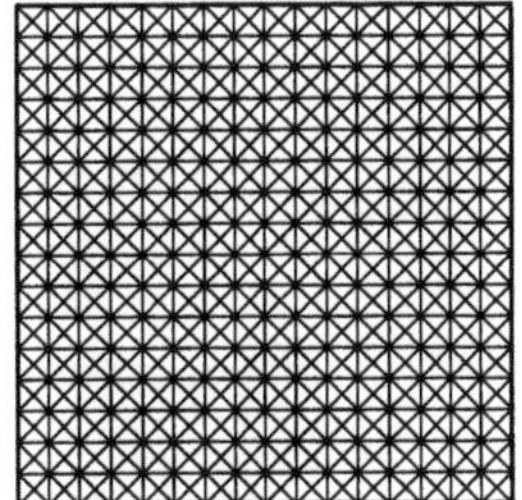

Figure 3.9. Sequence of uniform and symmetric meshes ($N = 4, 8, 16$).

Table 3.3. Eigenvalues computed on the criss-cross mesh sequence.

Exact	Computed (rate)				
	$N = 4$	$N = 8$	$N = 16$	$N = 32$	$N = 64$
2	2.0880	2.0216 (2.0)	2.0054 (2.0)	2.0013 (2.0)	2.0003 (2.0)
5	5.6811	5.1651 (2.0)	5.0408 (2.0)	5.0102 (2.0)	5.0025 (2.0)
5	5.6811	5.1651 (2.0)	5.0408 (2.0)	5.0102 (2.0)	5.0025 (2.0)
8	9.4962	8.3521 (2.1)	8.0863 (2.0)	8.0215 (2.0)	8.0054 (2.0)
10	12.9691	10.7578 (2.0)	10.1865 (2.0)	10.0464 (2.0)	10.0116 (2.0)
10	12.9691	10.7578 (2.0)	10.1865 (2.0)	10.0464 (2.0)	10.0116 (2.0)
13	17.1879	14.0237 (2.0)	13.2489 (2.0)	13.0617 (2.0)	13.0154 (2.0)
13	17.1879	14.0237 (2.0)	13.2489 (2.0)	13.0617 (2.0)	13.0154 (2.0)
17	25.1471	19.3348 (1.8)	17.5733 (2.0)	17.1423 (2.0)	17.0355 (2.0)
17	38.9073	19.3348 (3.2)	17.5733 (2.0)	17.1423 (2.0)	17.0355 (2.0)
18	38.9073	19.8363 (3.5)	18.4405 (2.1)	18.1089 (2.0)	18.0271 (2.0)
20	38.9073	22.7243 (2.8)	20.6603 (2.0)	20.1634 (2.0)	20.0407 (2.0)
20	38.9073	22.7243 (2.8)	20.6603 (2.0)	20.1634 (2.0)	20.0407 (2.0)
25	38.9073	28.7526 (1.9)	25.8940 (2.1)	25.2201 (2.0)	25.0548 (2.0)
25	38.9073	28.7526 (1.9)	25.8940 (2.1)	25.2201 (2.0)	25.0548 (2.0)
DOF	25	113	481	1985	8065

Our last computation is performed on a uniform and symmetric mesh sequence: the criss-cross mesh sequence of Figure 3.9. The results of this computation are shown in Table 3.3. In this case the multiple eigenvalue $\lambda = 5$ is approximated by pairs of coinciding values. The same happens for the double eigenvalues $\lambda = 10$ (modes $(1,3)$ and $(3,1)$) and $\lambda = 13$ (modes $(2,3)$ and $(3,2)$), while the situation seems different for $\lambda = 17$ (modes $(1,4)$ and $(4,1)$) in the case of the coarsest mesh $N = 4$. This

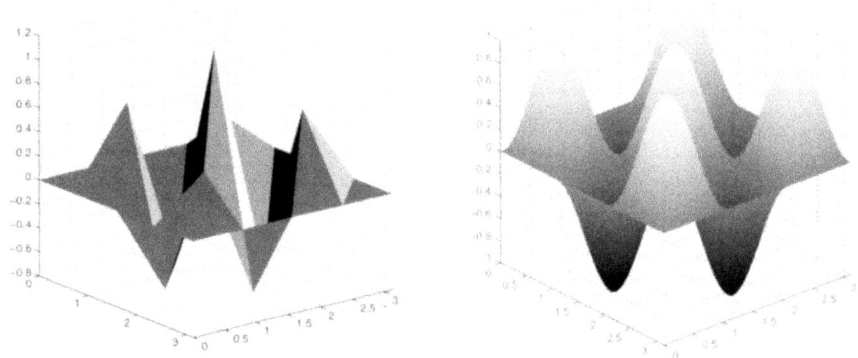

Figure 3.10. Discrete eigenfunction associated to $\lambda_h^{(9)}$ and the exact eigenfunction associated to $\lambda^{(11)}$.

behaviour can be explained as follows: the discrete value $\lambda_h^{(9)} = 25.1471$ is indeed a (bad) approximation of the higher frequency $\lambda^{(11)} = 18$ (mode $(3,3)$). A demonstration of this fact is given by Figure 3.10, which shows the discrete eigenfunction associated to $\lambda_h^{(9)}$ and the exact eigenfunction associated to $\lambda^{(11)}$.

When h is not small enough, we cannot actually expect the order of the discrete eigenvalues to be in a one-to-one correspondence with the continuous ones. For this reason, we include in Table 3.3 five more eigenvalues, which should make the picture clearer.

3.2. The Laplace eigenvalue problem on an L-shaped domain

In all the examples presented so far, the eigenfunctions have been C^∞-functions (they were indeed analytic). We recall here a fundamental example which shows the behaviour of eigenvalue problem approximation when the solution is not smooth.

We consider a domain with a re-entrant corner and the sequence of unstructured triangular meshes shown in Figure 3.11. The shape of our domain is actually a flipped L (the coordinates of the vertices are $(0,0)$, $(1,0)$, $(1,1)$, $(-1,1)$, $(-1,-1)$, and $(0,-1)$), since we use as reference solutions the values proposed in Dauge (2003) where this domain has been considered. In order to compare with the results in the literature, we compute the eigenvalues for the Neumann problem,

$$\begin{aligned} -\Delta u(x,y) &= \lambda u(x,y) && \text{in } \Omega, \\ \frac{\partial u}{\partial n} &= 0 && \text{on } \partial\Omega, \end{aligned}$$

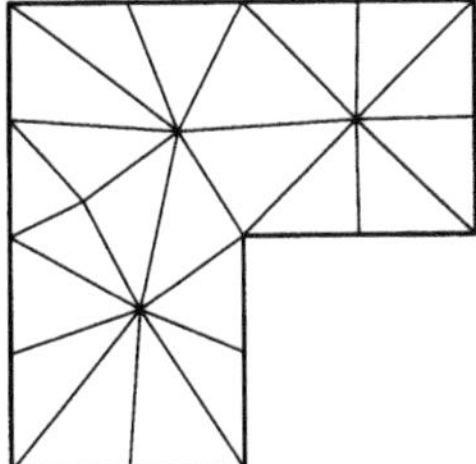
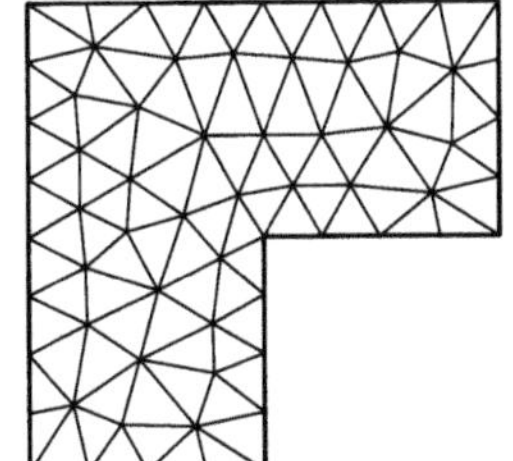
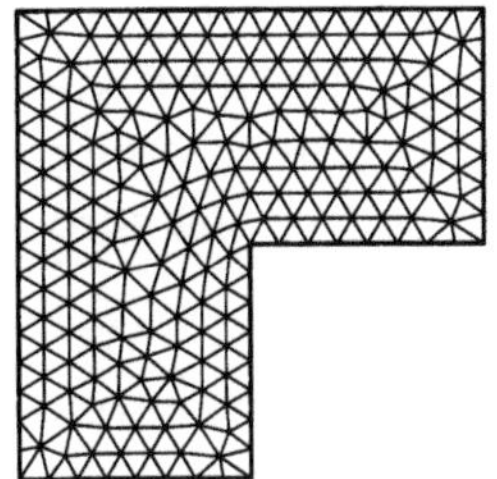

Figure 3.11. Sequence of unstructured mesh for the L-shaped domain ($N = 4, 8, 16$).

Table 3.4. Eigenvalues computed on the L-shaped domain (unstructured mesh sequence).

Exact	Computed (rate)				
	$N = 4$	$N = 8$	$N = 16$	$N = 32$	$N = 64$
0	−0.0000	0.0000	−0.0000	−0.0000	−0.0000
1.48	1.6786	1.5311 (1.9)	1.4946 (1.5)	1.4827 (1.4)	1.4783 (1.4)
3.53	3.8050	3.5904 (2.3)	3.5472 (2.1)	3.5373 (2.0)	3.5348 (2.0)
9.87	12.2108	10.2773 (2.5)	9.9692 (2.0)	9.8935 (2.1)	9.8755 (2.0)
9.87	12.5089	10.3264 (2.5)	9.9823 (2.0)	9.8979 (2.0)	9.8767 (2.0)
11.39	13.9526	12.0175 (2.0)	11.5303 (2.2)	11.4233 (2.1)	11.3976 (2.1)
DOF	20	65	245	922	3626

using the following variational formulation: find $\lambda \in \mathbb{R}$ and $u \in V$, with $u \neq 0$, such that

$$\int_\Omega \mathbf{grad}\, u(x,y) \cdot \mathbf{grad}\, v(x,)\, \mathrm{d}x\, \mathrm{d}y = \lambda \int_\Omega u(x,y)v(x,y)\, \mathrm{d}x\, \mathrm{d}y \quad \forall v \in V,$$

with $V = H^1(\Omega)$.

The results of the numerical computations are shown in Table 3.4, where we can observe the typical lower approximation rate in the presence of singularities: the first eigenvalue is associated to an eigenspace of singular eigenfunctions, so that the convergence rate deteriorates; on the other hand, the other presented eigenvalues are associated to eigenspaces of smooth functions (since the domain is symmetric), and their convergence is quadratic.

As in the previous examples, we observe that all discrete eigenvalues approximate the continuous ones from above, *i.e.*, we have immediate upper bounds for the exact frequencies.

Since we are considering the Neumann problem, there is a vanishing frequency. Its approximation is zero up to machine precision. In Table 3.4 we display the computed values, rounded to four decimal places, and in some occurrences the zero frequencies turn out to be negative.

Remark 3.1. We have chosen not to refine the mesh in the vicinity of the re-entrant corner, since we wanted to emphasize that the convergence rate of the eigenvalues is related to the smoothness of the corresponding eigenfunction. The convergence in the case of singular solutions can be improved by adding more degrees of freedom where they are needed, but this issue is outside the aim of this work.

3.3. The Laplace eigenvalue problem on the square: non-conforming elements

The last scheme we consider for the approximation of the problem discussed in this section is the linear non-conforming triangular element, also known as the Crouzeix–Raviart method. It is clear that there is an intrinsic interest in studying non-conforming elements; moreover, the approximation of mixed problems (which will be the object of the next examples and will constitute an important part of this survey) can be considered as a sort of non-conforming approximation.

We consider the square domain $\Omega =]0,\pi[\times]0,\pi[$ and compute the eigenvalues on the sequence of unstructured meshes presented in Figure 3.1. The computed frequencies are shown in Table 3.5. As expected, we observe an optimal quadratic convergence.

An important difference with respect to the previous computations is that now all discrete frequencies are lower bounds for the exact solutions. In this particular example all eigenvalues are approximated from below. This is typical behaviour for non-conforming approximation and has been reported by several authors. There is an active literature (see Rannacher (1979) and Armentano and Durán (2004), for instance) on predicting whether non-standard finite element schemes provide upper or lower bounds for eigenvalues, but to our knowledge the question has not yet been answered definitively. Numerical results tend to show that the Crouzeix–Raviart method gives values that are below the exact solutions, but so far only partial results are available.

The general theory we are going to present says that *conforming* approximations of eigenvalues are always *above* the exact solutions, while non-conforming ones may be below. In the mixed approximations shown in the next section there are situations where the same computation provides upper bounds for some eigenvalues and lower bounds for others.

Table 3.5. Eigenvalues computed with the Crouzeix–Raviart method (unstructured mesh sequence).

Exact	Computed (rate)				
	$N=4$	$N=8$	$N=16$	$N=32$	$N=64$
2	1.9674	1.9850 (1.1)	1.9966 (2.1)	1.9992 (2.0)	1.9998 (2.0)
5	4.4508	4.9127 (2.7)	4.9787 (2.0)	4.9949 (2.1)	4.9987 (2.0)
5	4.7270	4.9159 (1.7)	4.9790 (2.0)	4.9949 (2.0)	4.9987 (2.0)
8	7.2367	7.7958 (1.9)	7.9434 (1.9)	7.9870 (2.1)	7.9967 (2.0)
10	8.5792	9.6553 (2.0)	9.9125 (2.0)	9.9792 (2.1)	9.9949 (2.0)
10	9.0237	9.6663 (1.5)	9.9197 (2.1)	9.9796 (2.0)	9.9950 (2.0)
13	9.8284	12.4011 (2.4)	12.8534 (2.0)	12.9654 (2.1)	12.9914 (2.0)
13	9.9107	12.4637 (2.5)	12.8561 (1.9)	12.9655 (2.1)	12.9914 (2.0)
17	10.4013	15.9559 (2.7)	16.7485 (2.1)	16.9407 (2.1)	16.9853 (2.0)
17	11.2153	16.0012 (2.5)	16.7618 (2.1)	16.9409 (2.0)	16.9854 (2.0)
DOF	40	197	832	3443	13972

4. The Laplace eigenvalue problem in mixed form

In this section we present examples which, although classical, are probably not widely known, and which sometimes show a substantially different behaviour from the previous examples.

4.1. The mixed Laplace eigenvalue problem in one dimension

It is classical to rewrite the Laplace problem (2.1) as a first-order system: given $\Omega =]0,\pi[$, find eigenvalues λ and eigenfunctions u with $u \neq 0$, such that, for some s,

$$s(x) - u'(x) = 0 \qquad \text{in } \Omega, \tag{4.1a}$$

$$s'(x) = -\lambda u(x) \qquad \text{in } \Omega, \tag{4.1b}$$

$$u(0) = u(\pi) = 0. \tag{4.1c}$$

Remark 4.1. There are two functions involved with problem (4.1): s and u. In the formulation of the problem, we made explicit that the eigenfunctions we are interested in are the ones represented by u. This might seem a useless remark, since of course in problem (4.1), given u, it turns out that s is uniquely determined as its derivative, and analogously u can be uniquely determined from s and the boundary conditions. On the other hand, this might no longer be true for the discrete case (where the counterpart of our problem will be a degenerate algebraic generalized eigenvalue problem). In particular, we want to define the multiplicity of λ as the dimension of the space associated to the solution u; in general it might turn out that there is more than one s associated with u, and we do not want to consider the multiplicity of s when evaluating the multiplicity of λ.

Given $\Sigma = H^1(\Omega)$ and $U = L^2(\Omega)$, a variational formulation of the mixed problem (4.1) reads as follows: find $\lambda \in \mathbb{R}$ and $u \in U$ with $u \neq 0$, such that, for some $s \in \Sigma$,

$$\int_0^\pi s(x)t(x)\,\mathrm{d}x + \int_0^\pi u(x)t'(x)\,\mathrm{d}x = 0 \quad \forall t \in \Sigma,$$
$$\int_0^\pi s'(x)v(x) = -\lambda \int_0^\pi u(x)v(x)\,\mathrm{d}x \quad \forall v \in U.$$

Its Galerkin discretization is based on discrete subspaces $\Sigma_h \subset \Sigma$ and $U_h \subset U$ and reads as follows: find $\lambda_h \in \mathbb{R}$ and $u_h \in U_h$ with $u_h \neq 0$, such that, for some $s_h \in \Sigma_h$,

$$\int_0^\pi s_h(x)t(x)\,\mathrm{d}x + \int_0^\pi u_h(x)t'(x)\,\mathrm{d}x = 0 \quad \forall t \in \Sigma_h, \tag{4.2a}$$
$$\int_0^\pi s_h'(x)v(x) = -\lambda_h \int_0^\pi u_h(x)v(x)\,\mathrm{d}x \quad \forall v \in U_h. \tag{4.2b}$$

If $\Sigma_h = \mathrm{span}\{\varphi_1, \dots, \varphi_{N_s}\}$ and $U_h = \mathrm{span}\{\psi_1, \dots \psi_{N_u}\}$, then we can introduce the matrices $A = \{a_{kl}\}_{k,l=1}^{N_s}$, $M_U = \{m_{ij}\}_{i,j=1}^{N_u}$ and $B = \{b_{jk}\}$ $(j = 1, \dots, N_u,\ k = 1, \dots, N_s)$ as

$$a_{kl} = \int_0^\pi \varphi_l(x)\varphi_k(x)\,\mathrm{d}x,$$
$$m_{ij} = \int_0^\pi \psi_j(x)\psi_i(x)\,\mathrm{d}x,$$
$$b_{jk} = \int_0^\pi \varphi_k'(x)\psi_j(x)\,\mathrm{d}x,$$

so that the algebraic system corresponding to (4.2) has the form

$$\begin{pmatrix} A & B^T \\ B & 0 \end{pmatrix} \begin{pmatrix} \mathsf{x} \\ \mathsf{y} \end{pmatrix} = -\lambda \begin{pmatrix} 0 & 0 \\ 0 & M \end{pmatrix} \begin{pmatrix} \mathsf{x} \\ \mathsf{y} \end{pmatrix}.$$

4.2. The $\mathcal{P}_1 - \mathcal{P}_0$ element

Given a uniform partition of $[0, \pi]$ of size h, we introduce the most natural lowest-order scheme for the resolution of our problem. Observing that Σ_h and U_h need to approximate $H^1(\Omega)$ and L^2, respectively, and taking advantage of the experience coming from the study of the corresponding source problem (see, for instance, Brezzi and Fortin (1991), Boffi and Lovadina (1997) and Arnold, Falk and Winther (2006*a*)), we use continuous piecewise linear finite elements for Σ_h (that is, conforming $\mathcal{P}_1$ elements) and piecewise constants for U_h (that is, standard $\mathcal{P}_0$). The presented element is actually the one-dimensional counterpart of the well-known lowest-order Raviart–Thomas scheme (see the next section for more details). If N is the number of intervals in our decomposition of Ω, then the involved dimensions are $N_s = N+1$ and $N_u = N$. In this case it is possible to compute the eigensolutions explicitly. Given that the exact solutions are $\lambda^{(k)} = k^2$ and $u^{(k)}(x) = \sin(kx)$ $(k = 1, 2, \dots)$, we observe that we have $s^{(k)}(x) = k\cos(kx)$. It turns out that the approximate solution for s is its nodal interpolant, that is, $s_h^{(k)}(ih) = k\cos(kih)$, and that the discrete eigenmodes are given by

$$\lambda_h^{(k)} = (6/h^2)\frac{1-\cos kh}{2+\cos kh}, \quad u_h^{(k)}|_{]ih,(i+1)h[} = \frac{s_h^{(k)}(ih) - s_h^{(k)}((i+1)h)}{h\lambda_h^{(k)}},$$

with $k = 1, \dots, N$.

It is quite surprising that the discrete frequencies are exactly the same as in the first example presented in Section 2. There is actually a slight difference in the number of degrees of freedom: here N is the number of intervals, while in Section 2 N was the number of internal nodes, that is, we compute one value more with the mixed scheme on the same mesh. On the other hand, the eigenfunctions are different, as it must be, since here they are piecewise constants while there they were continuous piecewise linears. More precisely, it can be shown that if we consider the exact solution $u^{(k)}(x) = \sin(kx)$, then we have

$$\int_{ih}^{(i+1)h} (u^{(k)}(x) - u_h^{(k)}(x))\,\mathrm{d}x = \frac{\lambda_h - \lambda}{\lambda_h}\int_{ih}^{(i+1)h} u^{(k)}(x)\,\mathrm{d}x.$$

In particular, it turns out that $u_h^{(k)}$ is not the L^2-projection of $u^{(k)}$ onto the piecewise constants space.

4.3. The $\mathcal{P}_1 - \mathcal{P}_1$ element

It is well known that the $\mathcal{P}_1 - \mathcal{P}_1$ element is not stable for the approximation of the one-dimensional Laplace source problem (Babuška and Narasimhan 1997). In particular, it has been shown that it produces acceptable results for smooth solutions, although it is not convergent in the case of singular data. Even though the eigenfunctions of the problem we consider are regular (indeed, they are analytic), the $\mathcal{P}_1 - \mathcal{P}_1$ does not give good results, as we are going to show in this section.

Let us consider again a uniform partition of the interval $[0, \pi]$ into N subintervals and define both Σ_h and U_h as the space of continuous piecewise linear functions (without any boundary conditions). We then have $N_s = N_u = N + 1$.

Table 4.1. Eigenvalues for the one-dimensional mixed Laplacian computed with the $\mathcal{P}_1 - \mathcal{P}_1$ scheme.

Exact	Computed (rate)				
	$N = 8$	$N = 16$	$N = 32$	$N = 64$	$N = 128$
	0.0000	−0.0000	−0.0000	−0.0000	−0.0000
1	1.0001	1.0000 (4.1)	1.0000 (4.0)	1.0000 (4.0)	1.0000 (4.0)
4	3.9660	3.9981 (4.2)	3.9999 (4.0)	4.0000 (4.0)	4.0000 (4.0)
	7.4257	8.5541	8.8854	8.9711	8.9928
9	8.7603	8.9873 (4.2)	8.9992 (4.1)	9.0000 (4.0)	9.0000 (4.0)
16	14.8408	15.9501 (4.5)	15.9971 (4.1)	15.9998 (4.0)	16.0000 (4.0)
25	16.7900	24.5524 (4.2)	24.9780 (4.3)	24.9987 (4.1)	24.9999 (4.0)
	38.7154	29.7390	34.2165	35.5415	35.8846
36	39.0906	35.0393 (1.7)	35.9492 (4.2)	35.9970 (4.1)	35.9998 (4.0)
49		46.7793	48.8925 (4.4)	48.9937 (4.1)	48.9996 (4.0)

Table 4.2. Eigenvalues for the one-dimensional mixed Laplacian computed with the $\mathcal{P}_1 - \mathcal{P}_1$ scheme (the computed values are truncated to ten decimal places).

Exact	Computed (rate)
	$N = 1000$
	−0.0000000000
1	1.0000000000
4	3.9999999999
	8.9998815658
9	8.9999999992
16	15.9999999971
25	24.9999999784
	35.9981051039
36	35.9999999495
49	48.9999998977

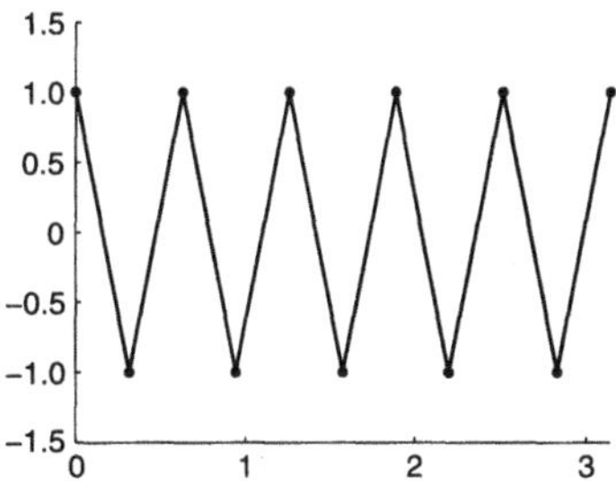

Figure 4.1. Eigenfunction u_h associated to $\lambda_h = 0$.

The results of the numerical computation for increasing N are listed in Table 4.1. The obtained results need some comments. First of all, it is clear that the correct values are well approximated: the rate of convergence is four, meaning that the scheme is of second order (since the rate convergence of the eigenvalues for symmetric eigenproblems, as seen in the previous examples, is doubled). On the other hand, there are some *spurious* solutions which we now describe in more detail.

The zero discrete frequency is related to the fact that the scheme does not satisfy the inf-sup condition. The corresponding eigenfunctions are $s_h(x) \equiv 0$ and $u_h(x)$, as represented in Figure 4.1 in the case $N = 10$. The function u_h is orthogonal in $L^2(0, \pi)$ to all derivatives of functions in Σ_h, and the existence of u_h in this case shows, in particular, that this scheme does not satisfy the classical inf-sup condition. We remark that $\lambda_h = 0$ is a true eigenvalue of our discrete problem even if the corresponding function s_h is vanishing, since the eigenfunction that interests us is u_h (see Remark 4.1).

Besides the zero frequency, there are other spurious solutions: the first one ranges between 7.4257 and 8.9928 in the computations shown in Table 4.1, and is increasing as N increases. Unfortunately, this spurious frequency remains bounded and seems to converge to 9 (which implies the wrong discrete multiplicity for the exact eigenvalue $\lambda = 9$), as is shown in Table 4.2, where we display the results of the computation for $N = 1000$. The same situation occurs for the other spurious value of Tables 4.1 and 4.2, which seems to converge to a value close to 36. The situation is actually more complicated and intriguing: the eigenvalues in the discrete spectrum with rank multiple of four seem spurious, and apparently converge to the value of the next one, that is, $\lambda_h^{(4k)} \to \lambda^{(3k)} = (3k)^2$ for $k = 0, 1, \ldots$. The numerically evaluated order of convergence of the spurious frequencies towards $(3k)^2$ is 2. The eigenfunctions corresponding to $\lambda_h^{(4)}$ and $\lambda_h^{(8)}$ are shown in Figure 4.2.

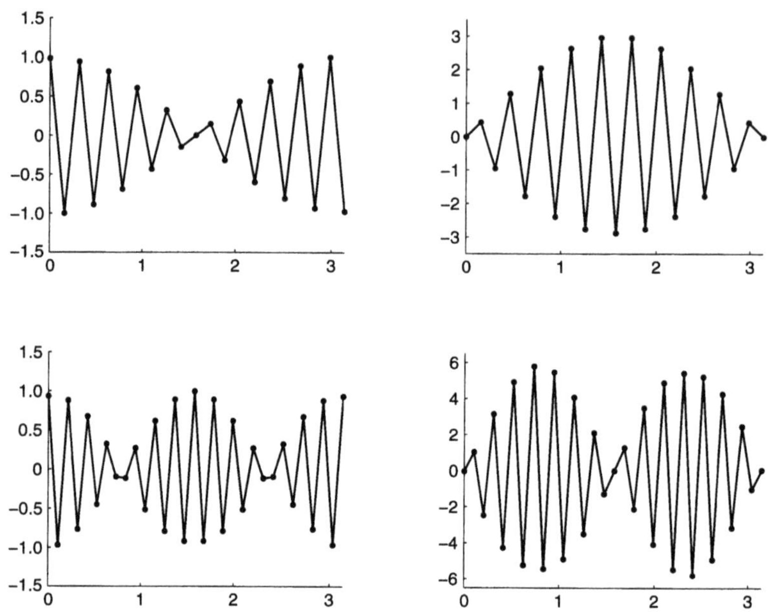

Figure 4.2. $\mathcal{P}_1 - \mathcal{P}_1$ spurious eigenfunctions corresponding to $\lambda_h^{(4)}$ (*above*, $N = 20$) and $\lambda_h^{(8)}$ (*below*, $N = 30$); u_h (*left*) and s_h (*right*).

4.4. The $\mathcal{P}_2 - \mathcal{P}_0$ element

We now discuss briefly the $\mathcal{P}_2 - \mathcal{P}_0$ element, which is known to be unstable for the corresponding source problem (Boffi and Lovadina 1997). The results of the numerical computations on a sequence of successively refined meshes are listed in Table 4.3. In this case there are no spurious solutions, but the computed eigenvalues are wrong by a factor of 6. More precisely, they converge nicely towards six times the exact solutions. The eigenfunctions corresponding to the first two eigenvalues are shown in Figure 4.3: they exhibit behaviour analogous to that observed in the literature for the source problem.

In particular, it turns out that the eigenfunctions u_h are correct approximations of u, while the functions s_h contain spurious components which are clearly associated with a bubble in each element. This behaviour is related to the fact that the ellipticity in the discrete kernel is not satisfied for the presence of the bubble functions in the space $\mathcal{P}_2$. In the case of the source problem, we observed a similar behaviour for s_h, while u_h was a correct approximation of a multiple of u. Here we do not have this phenomenon for u_h since the eigenfunctions are normalized.

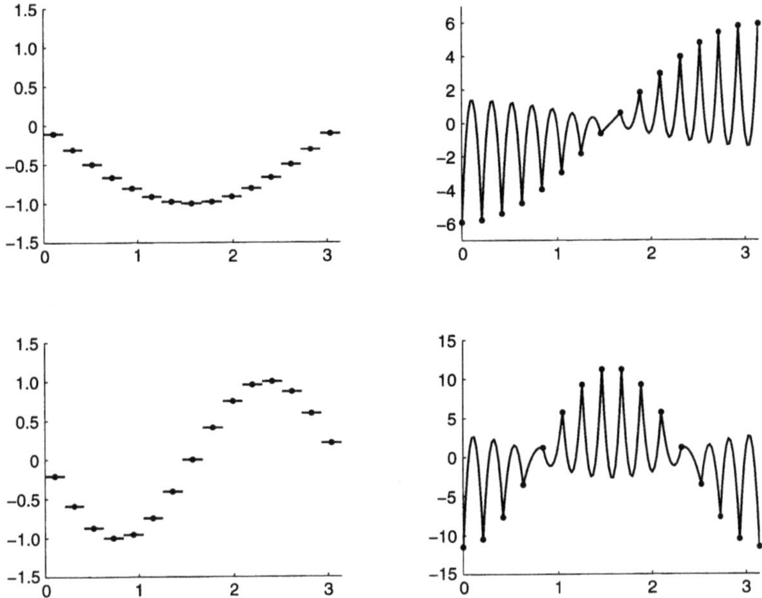

Figure 4.3. $\mathcal{P}_2 - \mathcal{P}_0$ eigenfunctions corresponding to the first (*above*) and the second (*below*) discrete value; u_h (*left*) and s_h (*right*).

Table 4.3. Eigenvalues for the one-dimensional mixed Laplacian computed with the $\mathcal{P}_2 - \mathcal{P}_0$ scheme.

Exact	Computed (rate with respect to 6λ)				
	$N = 8$	$N = 16$	$N = 32$	$N = 64$	$N = 128$
1	5.7061	5.9238 (1.9)	5.9808 (2.0)	5.9952 (2.0)	5.9988 (2.0)
4	19.8800	22.8245 (1.8)	23.6953 (1.9)	23.9231 (2.0)	23.9807 (2.0)
9	36.7065	48.3798 (1.6)	52.4809 (1.9)	53.6123 (2.0)	53.9026 (2.0)
16	51.8764	79.5201 (1.4)	91.2978 (1.8)	94.7814 (1.9)	95.6925 (2.0)
25	63.6140	113.1819 (1.2)	138.8165 (1.7)	147.0451 (1.9)	149.2506 (2.0)
36	71.6666	146.8261 (1.1)	193.5192 (1.6)	209.9235 (1.9)	214.4494 (2.0)
49	76.3051	178.6404 (0.9)	253.8044 (1.5)	282.8515 (1.9)	291.1344 (2.0)
64	77.8147	207.5058 (0.8)	318.0804 (1.4)	365.1912 (1.8)	379.1255 (1.9)
81		232.8461	384.8425 (1.3)	456.2445 (1.8)	478.2172 (1.9)
100		254.4561	452.7277 (1.2)	555.2659 (1.7)	588.1806 (1.9)

4.5. *The mixed Laplace eigenvalue problem in two and three space dimensions*

Given a domain $\Omega \in \mathbb{R}^n$ $(n = 2, 3)$, the Laplace eigenproblem can be formulated as a first-order system in the following way:

$$\begin{aligned} \boldsymbol{\sigma} - \mathbf{grad}\, u &= 0 && \text{in } \Omega,\\ \operatorname{div} \boldsymbol{\sigma} &= -\lambda u && \text{in } \Omega,\\ u &= 0 && \text{on } \partial\Omega, \end{aligned}$$

where we introduced the additional variable $\boldsymbol{\sigma} = \mathbf{grad}\, u$. A variational formulation considers the spaces $\Sigma = \mathbf{H}(\mathrm{div};\Omega)$ and $U = L^2(\Omega)$ and reads as follows: find $\lambda \in \mathbb{R}$ and $u \in U$, with $u \neq 0$, such that, for some $\boldsymbol{\sigma} \in \Sigma$,

$$\int_\Omega \boldsymbol{\sigma} \cdot \boldsymbol{\tau} \,\mathrm{d}\mathbf{x} + \int_\Omega u \operatorname{div} \boldsymbol{\tau} \,\mathrm{d}\mathbf{x} = 0 \quad \forall \boldsymbol{\tau} \in \Sigma, \tag{4.3a}$$

$$\int_\Omega \operatorname{div} \boldsymbol{\sigma} v \,\mathrm{d}\mathbf{x} = -\lambda \int_\Omega u v \,\mathrm{d}\mathbf{x} \quad \forall v \in U. \tag{4.3b}$$

The Galerkin approximation of our problem consists in choosing finite dimensional subspaces $\Sigma_h \subset \Sigma$ and $U_h \subset U$ and in solving the following discrete problem: find $\lambda_h \in \mathbb{R}$ and $u_h \in U_h$, with $u_h \neq 0$ such that, for some $\boldsymbol{\sigma}_h \in \Sigma_h$,

$$\begin{aligned} \int_\Omega \boldsymbol{\sigma}_h \cdot \boldsymbol{\tau} \,\mathrm{d}\mathbf{x} + \int_\Omega u_h \operatorname{div} \boldsymbol{\tau} \,\mathrm{d}\mathbf{x} &= 0 \quad \forall \boldsymbol{\tau} \in \Sigma_h,\\ \int_\Omega \operatorname{div} \boldsymbol{\sigma}_h v \,\mathrm{d}\mathbf{x} &= -\lambda_h \int_\Omega u_h v \,\mathrm{d}\mathbf{x} \quad \forall v \in U_h. \end{aligned}$$

The algebraic structure of the discrete system is the same as that presented in the one-dimensional case:

$$\begin{pmatrix} A & B^T \\ B & 0 \end{pmatrix} \begin{pmatrix} \mathsf{x} \\ \mathsf{y} \end{pmatrix} = -\lambda \begin{pmatrix} 0 & 0 \\ 0 & M \end{pmatrix} \begin{pmatrix} \mathsf{x} \\ \mathsf{y} \end{pmatrix},$$

where M is a symmetric positive definite matrix.

4.6. *Raviart–Thomas elements*

We shall use the Raviart–Thomas (RT) elements, which provide the most natural scheme for the approximation of our problem. Similar comments apply to other well-known mixed finite elements, such as Brezzi–Douglas–Marini (BDM) or Brezzi–Douglas–Fortin–Marini (BDFM). We refer the interested reader to Brezzi and Fortin (1991) for a thorough introduction to this subject, and to Raviart and Thomas (1977), Brezzi, Douglas and Marini (1985), and Brezzi, Douglas, Fortin and Marini (1987*b*) for the original definitions.

Table 4.4. Eigenvalues computed with lowest-order RT elements on the uniform mesh sequence of squares.

Exact	Computed (rate)				
	$N=4$	$N=8$	$N=16$	$N=32$	$N=64$
2	2.1048	2.0258 (2.0)	2.0064 (2.0)	2.0016 (2.0)	2.0004 (2.0)
5	5.9158	5.2225 (2.0)	5.0549 (2.0)	5.0137 (2.0)	5.0034 (2.0)
5	5.9158	5.2225 (2.0)	5.0549 (2.0)	5.0137 (2.0)	5.0034 (2.0)
8	9.7268	8.4191 (2.0)	8.1033 (2.0)	8.0257 (2.0)	8.0064 (2.0)
10	13.8955	11.0932 (1.8)	10.2663 (2.0)	10.0660 (2.0)	10.0165 (2.0)
10	13.8955	11.0932 (1.8)	10.2663 (2.0)	10.0660 (2.0)	10.0165 (2.0)
13	17.7065	14.2898 (1.9)	13.3148 (2.0)	13.0781 (2.0)	13.0195 (2.0)
13	17.7065	14.2898 (1.9)	13.3148 (2.0)	13.0781 (2.0)	13.0195 (2.0)
17	20.5061	20.1606 (0.1)	17.8414 (1.9)	17.2075 (2.0)	17.0517 (2.0)
17	20.5061	20.4666 (0.0)	17.8414 (2.0)	17.2075 (2.0)	17.0517 (2.0)
DOF	16	64	256	1024	4096

The RT space is used for the approximation of Σ. One of the main properties is that the finite element space consists of vector fields that are not globally continuous, but only conforming in $\mathbf{H}(\operatorname{div};\Omega)$. This is achieved by requiring the normal component of the vector to be continuous across the elements, and the main tool for achieving this property is the so-called Piola transform, from the reference to the physical element. The space U is approximated by $\operatorname{div}(\Sigma_h)$. In the case of lowest-order elements, in particular, the space U_h is $\mathcal{P}_0$. We refer to Brezzi and Fortin (1991) for more details. We performed the computation on a square domain $\Omega =]0,\pi[^2$ using a sequence of uniform meshes of squares (the parameter N refers to the number of subdivisions of each side). The results of the computations by means of lowest-order RT elements are displayed in Table 4.4. The number of degrees of freedom is evaluated in terms of the variable u_h, since this is the dimension of the algebraic eigenvalue problem to be solved (in this case equal to the number of elements, since u_h is approximated by piecewise constants). From the computed values we can observe that the convergence is quadratic and that all eigenvalues are approximated from above.

The same computation is then performed on a sequence of unstructured triangular meshes such as that presented in Figure 3.1. The results are shown in Table 4.5. In this case the situation is less clear. The theoretical estimates we present again show second order of convergence in h; the reported values, however, even if they are clearly and rapidly converging, are not exactly consistent with the theoretical bound. The reason is that

Table 4.5. Eigenvalues computed with lowest-order RT elements on the unstructured mesh sequence of triangles.

Exact	Computed (rate)				
	$N=4$	$N=8$	$N=16$	$N=32$	$N=64$
2	2.0138	1.9989 (3.6)	1.9997 (1.7)	1.9999 (2.7)	2.0000 (2.8)
5	4.8696	4.9920 (4.0)	5.0000 (8.0)	4.9999 (-2.1)	5.0000 (3.7)
5	4.8868	4.9952 (4.5)	5.0006 (3.0)	5.0000 (5.8)	5.0000 (2.6)
8	8.6905	7.9962 (7.5)	7.9974 (0.6)	7.9995 (2.5)	7.9999 (2.2)
10	9.7590	9.9725 (3.1)	9.9980 (3.8)	9.9992 (1.3)	9.9999 (3.2)
10	11.4906	9.9911 (7.4)	10.0007 (3.7)	10.0005 (0.4)	10.0001 (2.4)
13	11.9051	12.9250 (3.9)	12.9917 (3.2)	12.9998 (5.4)	12.9999 (1.8)
13	12.7210	12.9631 (2.9)	12.9950 (2.9)	13.0000 (7.5)	13.0000 (1.1)
17	13.5604	16.8450 (4.5)	16.9848 (3.4)	16.9992 (4.3)	16.9999 (2.5)
17	14.1813	16.9659 (6.4)	16.9946 (2.7)	17.0009 (2.6)	17.0000 (5.5)
DOF	32	142	576	2338	9400

Table 4.6. Eigenvalues computed with lowest-order RT elements on the uniform mesh sequence of triangles of Figure 3.5.

Exact	Computed (rate)				
	$N=4$	$N=8$	$N=16$	$N=32$	$N=64$
2	2.0324	2.0084 (1.9)	2.0021 (2.0)	2.0005 (2.0)	2.0001 (2.0)
5	4.8340	4.9640 (2.2)	4.9912 (2.0)	4.9978 (2.0)	4.9995 (2.0)
5	5.0962	5.0259 (1.9)	5.0066 (2.0)	5.0017 (2.0)	5.0004 (2.0)
8	8.0766	8.1185 (-0.6)	8.0332 (1.8)	8.0085 (2.0)	8.0021 (2.0)
10	8.9573	9.7979 (2.4)	9.9506 (2.0)	9.9877 (2.0)	9.9969 (2.0)
10	9.4143	9.8148 (1.7)	9.9515 (1.9)	9.9877 (2.0)	9.9969 (2.0)
13	11.1065	12.8960 (4.2)	12.9828 (2.6)	12.9962 (2.2)	12.9991 (2.0)
13	11.3771	13.4216 (1.9)	13.1133 (1.9)	13.0287 (2.0)	13.0072 (2.0)
17	12.2424	16.1534 (2.5)	16.7907 (2.0)	16.9474 (2.0)	16.9868 (2.0)
17	14.7292	16.1963 (1.5)	16.7992 (2.0)	16.9495 (2.0)	16.9874 (2.0)
DOF	32	128	512	2048	8192

RT elements are quite sensitive to the orientation of the mesh. A clean convergence order can be obtained by using a uniform refinement strategy, as in the mesh sequence of Figure 3.5. The results of this computation are listed in Table 4.6

It is interesting to note that in this case the eigenvalues may be approximated from above or below. Even the same eigenvalue can present numerical lower or upper bounds depending on the chosen mesh.

5. The Maxwell eigenvalue problem

Maxwell's eigenvalue problem can be written as follows by means of Ampère and Faraday's laws: given a domain $\Omega \in \mathbb{R}^3$, find the resonance frequencies $\omega \in \mathbb{R}^3$ (with $\omega \neq 0$) and the electromagnetic fields $(\mathbf{E}, \mathbf{H}) \neq (0, 0)$ such that

$$\begin{aligned} \mathbf{curl}\,\mathbf{E} &= \mathrm{i}\omega\mu\mathbf{H} && \text{in } \Omega, \\ \mathbf{curl}\,\mathbf{H} &= -\mathrm{i}\omega\epsilon\mathbf{E} && \text{in } \Omega, \\ \mathbf{E} \times \mathbf{n} &= 0 && \text{on } \partial\Omega, \\ \mathbf{H} \cdot \mathbf{n} &= 0 && \text{on } \partial\Omega, \end{aligned}$$

where we assumed perfectly conducting boundary conditions, and ϵ and μ denote the dielectric permittivity and magnetic permeability, respectively.

From the assumption $\omega \neq 0$ it is well known that we get the usual divergence equations,

$$\begin{aligned} \operatorname{div} \varepsilon\mathbf{E} &= 0 && \text{in } \Omega, \\ \operatorname{div} \mu\mathbf{H} &= 0 && \text{in } \Omega. \end{aligned}$$

For the sake of simplicity, we consider the material properties ϵ and μ constant and equal to the identity matrix. It is outside the scope of this work to consider more general cases; it is remarkable, however, that major mathematical challenges arise even in this simpler situation.

The classical formulation of the eigenvalue problem is obtained from the Maxwell system by eliminating $\mathbf{H}$ (we let $\mathbf{u}$ denote the unknown eigenfunction $\mathbf{E}$): find $\omega \in \mathbb{R}$ and $\mathbf{u} \neq \mathbf{0}$ such that

$$\mathbf{curl}\,\mathbf{curl}\,\mathbf{u} = \omega^2\mathbf{u} \quad \text{in } \Omega, \tag{5.1a}$$

$$\operatorname{div} \mathbf{u} = 0 \quad \text{in } \Omega, \tag{5.1b}$$

$$\mathbf{u} \times \mathbf{n} = 0 \quad \text{on } \partial\Omega. \tag{5.1c}$$

A standard variational formulation of problem (5.1) reads as follows: find $\omega \in \mathbb{R}$ and $\mathbf{u} \in \mathbf{H}_0(\mathbf{curl}; \Omega)$ with $\mathbf{u} \neq \mathbf{0}$ such that

$$(\mathbf{curl}\,\mathbf{u}, \mathbf{curl}\,\mathbf{v}) = \omega^2(\mathbf{u}, \mathbf{v}) \quad \forall \mathbf{v} \in \mathbf{H}_0(\mathbf{curl}; \Omega), \tag{5.2a}$$

$$(\mathbf{u}, \mathbf{grad}\,\varphi) = 0 \quad \forall \varphi \in H_0^1(\Omega), \tag{5.2b}$$

where, as usual, the space $\mathbf{H}_0(\mathbf{curl};\Omega)$ consists of vector fields in $L^2(\Omega)^3$ with **curl** in $L^2(\Omega)^3$, and with vanishing tangential trace on the boundary. Here $H_0^1(\Omega)$ is the standard Sobolev space of functions in $L^2(\Omega)$ with **grad** in $L^2(\Omega)^3$, and vanishing trace on the boundary.

It is a common practice to consider the following variational formulation for the approximation of problem (5.1): find $\omega \in \mathbb{R}$ and $\mathbf{u} \in \mathbf{H}_0(\mathbf{curl};\Omega)$ with $\mathbf{u} \neq \mathbf{0}$ such that

$$(\mathbf{curl}\,\mathbf{u}, \mathbf{curl}\,\mathbf{v}) = \omega^2(\mathbf{u}, \mathbf{v}) \quad \forall \mathbf{v} \in \mathbf{H}_0(\mathbf{curl};\Omega). \tag{5.3}$$

It is easy to observe that the eigenmodes of (5.3) corresponding to non-vanishing frequencies $\omega \neq 0$ are also solutions to problem (5.2): it is sufficient to choose $\mathbf{v} = \mathbf{grad}\,\varphi$ in (5.3) in order to obtain the second equation of (5.2). When the domain is simply connected, these are the only solutions to problem (5.2): $\omega = 0$ in (5.2) implies $\mathbf{curl}\,\mathbf{u} = \mathbf{0}$ which, together with $\operatorname{div}\mathbf{u} = 0$ and the boundary conditions, means $\mathbf{u} = \mathbf{0}$ if the cohomology is trivial. On the other hand, if there exist non-vanishing vector fields $\mathbf{u}$ with $\mathbf{curl}\,\mathbf{u} = \mathbf{0}$, $\operatorname{div}\mathbf{u} = 0$ in Ω, and $\mathbf{u} \times \mathbf{n}$ on $\partial\Omega$ (harmonic vector fields), then problem (5.2) has solutions with zero frequency $\omega = 0$. These solutions are obviously also present in problem (5.3): in this case the eigenspace corresponding to the zero frequency is made of the harmonic vector fields plus the infinite-dimensional space $\mathbf{grad}(H_0^1(\Omega))$. It is well known that the space of harmonic vector fields is finite-dimensional, its dimension being the first Betti number of Ω.

From now on we assume that Ω is simply connected, and discuss some numerical approximations of the two-dimensional counterpart of problem (5.3).

Following Boffi, Fernandes, Gastaldi and Perugia (1999*b*), it is not difficult to check that problem (5.2) is equivalent to the following: find $\lambda \in \mathbb{R}$ and $\mathbf{p} \in \mathbf{H}_0(\operatorname{div}^0;\Omega)$ with $\mathbf{p} \not\equiv \mathbf{0}$ such that, for some $\boldsymbol{\sigma} \in \mathbf{H}_0(\mathbf{curl};\Omega)$,

$$(\boldsymbol{\sigma}, \boldsymbol{\tau}) + (\mathbf{p}, \mathbf{curl}\,\boldsymbol{\tau}) = 0 \quad \forall \boldsymbol{\tau} \in \mathbf{H}_0(\mathbf{curl};\Omega), \tag{5.4a}$$

$$(\mathbf{curl}\,\boldsymbol{\sigma}, \mathbf{q}) = -\lambda(\mathbf{p}, \mathbf{q}) \quad \forall \mathbf{q} \in \mathbf{H}_0(\operatorname{div}^0;\Omega), \tag{5.4b}$$

where $\mathbf{H}_0(\operatorname{div}^0;\Omega)$ denotes the subspace of $\mathbf{H}_0(\operatorname{div};\Omega)$ consisting of divergence-free vector fields and where the equivalence is given by $\lambda = \omega^2$, $\boldsymbol{\sigma} = \mathbf{u}$, and $\mathbf{p} = -\,\mathbf{curl}\,\boldsymbol{\sigma}/\lambda$. The main property used for the proof of equivalence is that $\mathbf{H}_0(\operatorname{div}^0;\Omega)$ coincides with $\mathbf{curl}(\mathbf{H}_0(\mathbf{curl};\Omega))$.

A Galerkin discretization of Maxwell's eigenproblem usually involves a sequence of finite-dimensional subspaces $\Sigma_h \subset \mathbf{H}_0(\mathbf{curl};\Omega)$ so that the approximate formulation reads as follows: find $\omega_h \in \mathbb{R}$ and $\mathbf{u}_h \in \Sigma_h$ with $\mathbf{u}_h \neq \mathbf{0}$ such that

$$(\mathbf{curl}\,\mathbf{u}_h, \mathbf{curl}\,\mathbf{v}) = \omega_h^2(\mathbf{u}_h, \mathbf{v}) \quad \forall \mathbf{v} \in \Sigma_h. \tag{5.5}$$

The discretization of (5.4) requires two sequences of finite element spaces $\Sigma_h \subset \mathbf{H}_0(\mathbf{curl};\Omega)$ and $U_h \subset \mathbf{H}_0(\mathrm{div}^0;\Omega)$, so that the discrete problem reads as follows: find $\lambda_h \in \mathbb{R}$ and $\mathbf{p}_h \in U_h$ with $\mathbf{p}_h \not\equiv \mathbf{0}$ such that, for some $\boldsymbol{\sigma}_h \in \Sigma_h$,

$$(\boldsymbol{\sigma}_h, \boldsymbol{\tau}) + (\mathbf{p}_h, \mathbf{curl}\,\boldsymbol{\tau}) = 0 \quad \forall \boldsymbol{\tau} \in \Sigma_h, \tag{5.6a}$$

$$(\mathbf{curl}\,\boldsymbol{\sigma}_h, \mathbf{q}) = -\lambda_h(\mathbf{p}_h, \mathbf{q}) \quad \forall \mathbf{q} \in U_h. \tag{5.6b}$$

Boffi *et al.* (1999*b*) showed that, under the assumption

$$\mathbf{curl}(\Sigma_h) = U_h,$$

the same equivalence holds at the discrete level as well: more precisely, all positive frequencies of (5.5) correspond to solutions of (5.6) with the identifications $\lambda_h = \omega_h^2$, $\boldsymbol{\sigma}_h = \mathbf{u}_h$ and $\mathbf{p}_h = -\,\mathbf{curl}\,\boldsymbol{\sigma}_h/\lambda_h$.

Another mixed formulation associated with Maxwell's eigenproblem was introduced in Kikuchi (1987): find $\lambda \in \mathbb{R}$ and $\mathbf{u} \in \mathbf{H}_0(\mathbf{curl};\Omega)$ with $\mathbf{u} \neq \mathbf{0}$ such that, for some $\psi \in H_0^1(\Omega)$,

$$(\mathbf{curl}\,\mathbf{u}, \mathbf{curl}\,\mathbf{v}) + (\mathbf{grad}\,\psi, \mathbf{v}) = \lambda(\mathbf{u}, \mathbf{v}) \quad \forall \mathbf{v} \in \mathbf{H}_0(\mathbf{curl};\Omega), \tag{5.7a}$$

$$(\mathbf{u}, \mathbf{grad}\,\varphi) = 0 \quad \forall \varphi \in H_0^1(\Omega). \tag{5.7b}$$

We shall discuss in Section 17 the analogies between the two proposed mixed formulations.

We conclude this preliminary discussion of Maxwell's eigenvalues with a series of two-dimensional numerical results.

5.1. Approximation of Maxwell's eigenvalues on triangular meshes

The two-dimensional counterpart of (5.3) reads as follows: find $\omega \in \mathbb{R}$ and $\mathbf{u} \in \mathbf{H}_0(\mathrm{rot};\Omega)$ with $\mathbf{u} \neq \mathbf{0}$ such that

$$(\mathrm{rot}\,\mathbf{u}, \mathrm{rot}\,\mathbf{v}) = \omega^2(\mathbf{u}, \mathbf{v}) \quad \forall \mathbf{v} \in \mathbf{H}_0(\mathrm{rot};\Omega), \tag{5.8}$$

where we used the operator

$$\mathrm{rot}\,\mathbf{v} = \frac{\partial v_2}{\partial x_1} - \frac{\partial v_1}{\partial x_2} = -\,\mathrm{div}(\mathbf{v}^\perp).$$

Its discretization involves a finite-dimensional subspace $\Sigma_h \subset \mathbf{H}_0(\mathrm{rot};\Omega)$ and reads as follows: find $\omega_h \in \mathbb{R}$ and $\mathbf{u}_h \in \Sigma_h$ with $\mathbf{u}_h \neq \mathbf{0}$ such that

$$(\mathrm{rot}\,\mathbf{u}_h, \mathrm{rot}\,\mathbf{v}) = \omega_h^2(\mathbf{u}_h, \mathbf{v}) \quad \forall \mathbf{v} \in \Sigma_h. \tag{5.9}$$

The analogous formulation of (5.4) is as follows: find $\lambda \in \mathbb{R}$ and $p \in \mathrm{rot}(\mathbf{H}_0(\mathrm{rot};\Omega)) = L_0^2(\Omega)$ with $p \not\equiv 0$ such that, for some $\boldsymbol{\sigma} \in \mathbf{H}_0(\mathrm{rot};\Omega)$,

$$(\boldsymbol{\sigma}, \boldsymbol{\tau}) + (p, \mathrm{rot}\,\boldsymbol{\tau}) = 0 \quad \forall \boldsymbol{\tau} \in \mathbf{H}_0(\mathrm{rot};\Omega), \tag{5.10a}$$

$$(\mathrm{rot}\,\boldsymbol{\sigma}, q) = -\lambda(p, q) \quad \forall q \in L_0^2(\Omega), \tag{5.10b}$$

where $L_0^2(\Omega)$ is the subspace of $L^2(\Omega)$ of zero mean-valued functions.

Table 5.1. Eigenvalues computed with lowest-order edge elements on the uniform mesh sequence of triangles of Figure 3.5.

Exact	Computed (rate)				
	$N = 4$	$N = 8$	$N = 16$	$N = 32$	$N = 64$
1	0.9702	0.9923 (2.0)	0.9981 (2.0)	0.9995 (2.0)	0.9999 (2.0)
1	0.9960	0.9991 (2.2)	0.9998 (2.1)	0.9999 (2.0)	1.0000 (2.0)
2	2.0288	2.0082 (1.8)	2.0021 (2.0)	2.0005 (2.0)	2.0001 (2.0)
4	3.7227	3.9316 (2.0)	3.9829 (2.0)	3.9957 (2.0)	3.9989 (2.0)
4	3.7339	3.9325 (2.0)	3.9829 (2.0)	3.9957 (2.0)	3.9989 (2.0)
5	4.7339	4.9312 (2.0)	4.9826 (2.0)	4.9956 (2.0)	4.9989 (2.0)
5	5.1702	5.0576 (1.6)	5.0151 (1.9)	5.0038 (2.0)	5.0010 (2.0)
8	7.4306	8.1016 (2.5)	8.0322 (1.7)	8.0084 (1.9)	8.0021 (2.0)
9	7.5231	8.6292 (2.0)	8.9061 (2.0)	8.9764 (2.0)	8.9941 (2.0)
9	7.9586	8.6824 (1.7)	8.9211 (2.0)	8.9803 (2.0)	8.9951 (2.0)
zeros	9	49	225	961	3969
DOF	40	176	736	3008	12160

Since the operators rot and div are isomorphic, formulation (5.10) is indeed equivalent to a Neumann problem for the Laplace operator: find $\lambda \in \mathbb{R}$ and $p \in L^2_0(\Omega))$ with $p \not\equiv 0$ such that, for some $\boldsymbol{\sigma} \in \mathbf{H}_0(\operatorname{div};\Omega)$,

$$
\begin{aligned}
(\boldsymbol{\sigma},\boldsymbol{\tau}) + (p, \operatorname{div}\boldsymbol{\tau}) &= 0 \quad \forall \boldsymbol{\tau} \in \mathbf{H}_0(\operatorname{div};\Omega),\\
(\operatorname{div}\boldsymbol{\sigma}, q) &= -\lambda(p,q) \quad \forall q \in L^2_0(\Omega)),
\end{aligned}
$$

where the difference with respect to formulation (4.3) is in the boundary conditions, *i.e.*, $\mathbf{H}(\operatorname{div};\Omega)$ is replaced by $\mathbf{H}_0(\operatorname{div};\Omega)$ and, consistently in order to have $\operatorname{div}(\Sigma) = U$, $L^2(\Omega)$ is replaced by $L^2_0(\Omega)$.

For theoretical results on problems analogous to (5.3) and (5.8) involving the divergence operator, we refer the interested reader to Bermúdez and Rodríguez (1994) and Bermúdez *et al.* (1995).

The most natural discretization of problem (5.8) makes use of the so-called *edge* finite elements (Nédélec 1980, 1986). In two space dimensions, edge finite elements are simply standard finite elements used in mixed formulations for the approximation of $\mathbf{H}(\operatorname{div};\Omega)$ (such as RT elements, already seen in the approximation of mixed Laplace eigenvalue problem), rotated by the angle $\pi/2$. The name 'edge finite elements' comes from the nature of the degrees of freedom which, for lowest-order approximation, are associated to moments along the edges of the triangulation. Table 5.1 displays the

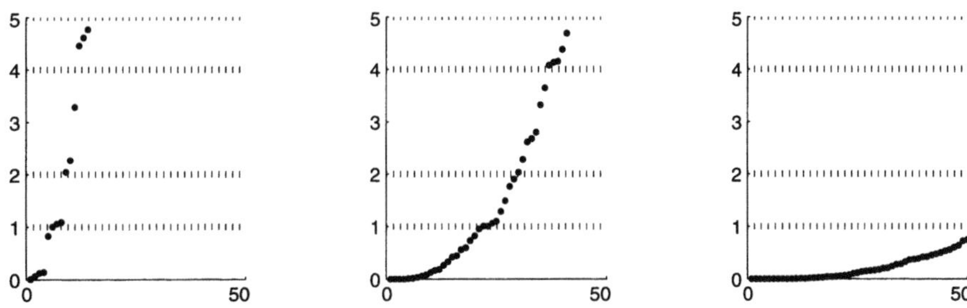

Figure 5.1. First 50 discrete eigenvalues computed with piecewise linears on the unstructured mesh ($N = 4, 8, 16$).

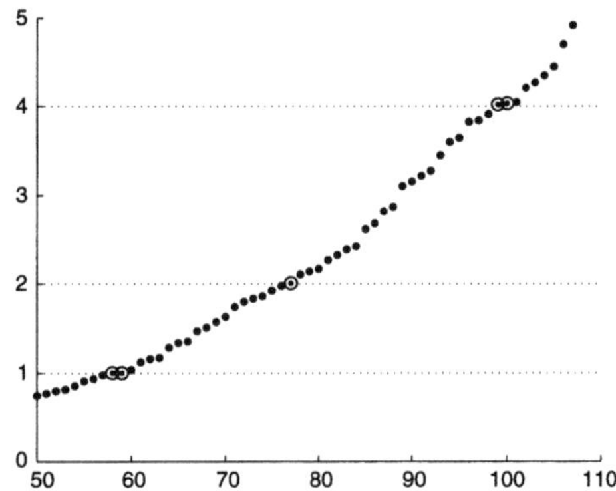

Figure 5.2. Some eigenvalues computed with piecewise linears on the unstructured mesh for $N = 16$.

results of the computation on the square $]0,\pi[^2$ on a sequence of unstructured triangular meshes as in Figure 3.1 with lowest-order edge elements.

Remark 5.1. An important feature of edge element approximation of problem (5.3) is that the zero frequency is approximated by discrete values that are exactly equal to zero (up to machine precision). In the case of lowest-order edge elements, the number of zero frequencies (shown in Table 5.1) is equal to the number of internal vertices of the mesh. This is due to the fact that the elements of Σ_h with vanishing rot coincide with gradients of piecewise linear functions in $H_0^1(\Omega)$.

There have been several attempts to solve problem (5.5) with *nodal* finite elements, that is, standard finite elements in each component with degrees of freedom associated to nodal values. It was soon realized that simulations performed with standard piecewise linears are very sensitive to the used mesh. Figure 5.1 shows the results obtained on the sequence of unstructured triangular meshes of Figure 3.1 with continuous piecewise linear elements in each component. The obtained results can by no means give

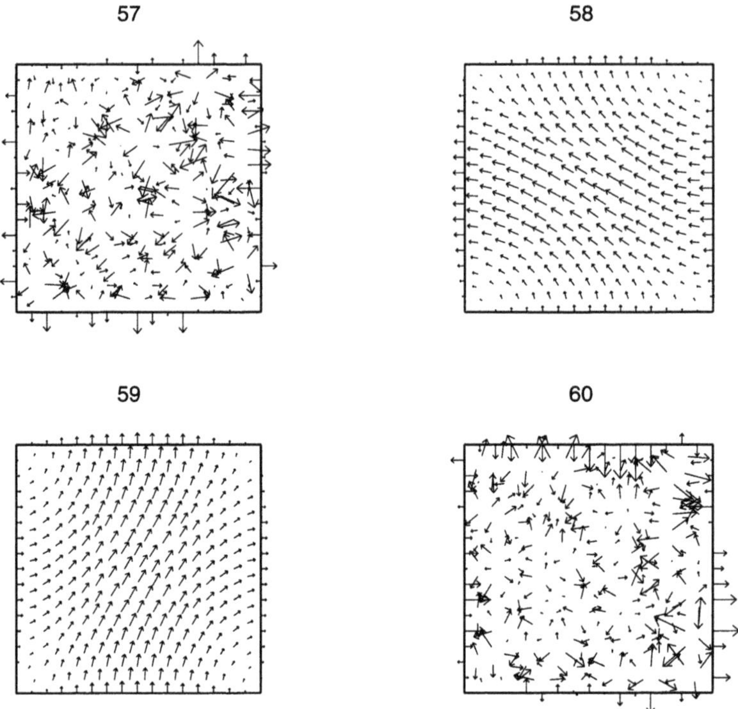

Figure 5.3. Eigenfunctions computed with piecewise linears on the unstructured mesh for $N = 16$.

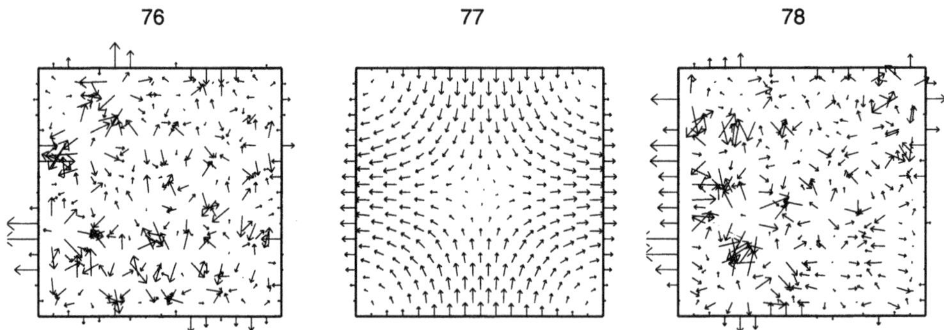

Figure 5.4. More eigenfunctions computed with piecewise linears on the unstructured mesh for $N = 16$.

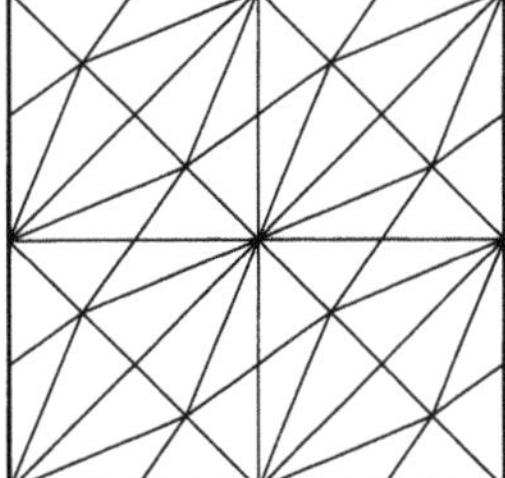
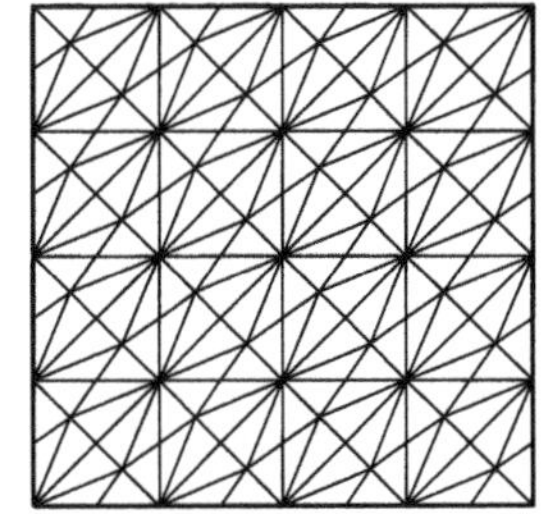
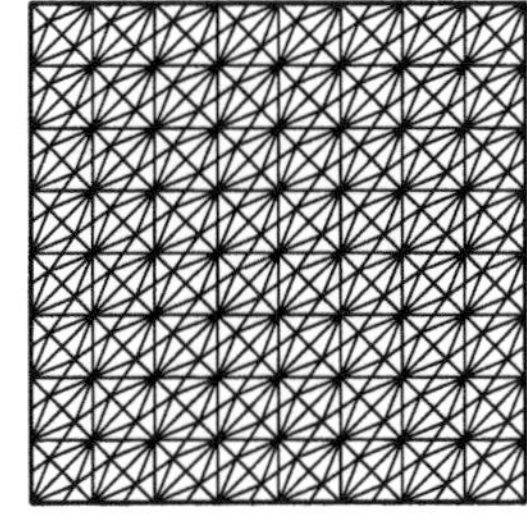

Figure 5.5. Sequence of compatible meshes where gradients are well represented ($N = 2, 4, 8$).

an indication of the exact values. In particular, it is clear that the zero frequency is not well approximated, and it seems that bad approximations of the zero frequency pollute the whole spectrum. Indeed, it can be observed (Boffi *et al.* 1999*b*) that the correct values are well approximated together with their eigenfunctions, but their approximations are hardly distinguishable from the spurious solutions. Figure 5.2 shows the eigenvalues in the range $[50, 110]$ of the spectrum computed with the mesh for $N = 16$. The eigenvalues plotted with different markers are good approximations to the exact solutions. We display some eigenfunctions in Figures 5.3 and 5.4. The different behaviour of the eigenfunctions corresponding to good approximations of the exact solutions and other eigenfunctions can be easily observed.

In view of Remark 5.1, it is clear that a crucial property is that enough gradients are given in the finite element space: this will ensure that the zero frequency is exactly approximated by vanishing discrete eigenvalues. A strategy for designing meshes for which such conditions are satisfied when using piecewise linear elements has been proposed by Wong and Cendes (1988). A sequence of such meshes is plotted in Figure 5.5 and the computed eigenvalues are listed in Table 5.2. It turns out that now several vanishing discrete values correspond to the zero frequency, and that the positive frequencies are optimally approximated.

A rigorous proof of this last statement is not yet available, and for a while there have been researchers who believed that good approximation of the infinite-dimensional kernel was a sufficient condition for the convergence of the eigenmodes. On the other hand, the use of edge elements has to be preferred with respect to nodal elements whenever possible. In order to convince the reader that apparently good results do not necessarily turn out to be correct results, we recall the counter-example presented in Boffi *et al.* (1999*b*). It is actually well known that gradients are well represented by piecewise linears on the criss-cross mesh sequence of Figure 3.9. This is a consequence of results on contour plotting (Powell 1974). The eigenvalues computed with formulation (5.5) using piecewise linears

Table 5.2. Eigenvalues computed with nodal elements on the compatible mesh sequence of triangles of Figure 5.5.

Exact	Computed (rate)				
	$N = 2$	$N = 4$	$N = 8$	$N = 16$	$N = 32$
1	1.0163	1.0045 (1.9)	1.0011 (2.0)	1.0003 (2.0)	1.0001 (2.0)
1	1.0445	1.0113 (2.0)	1.0028 (2.0)	1.0007 (2.0)	1.0002 (2.0)
2	2.0830	2.0300 (1.5)	2.0079 (1.9)	2.0020 (2.0)	2.0005 (2.0)
4	4.2664	4.1212 (1.1)	4.0315 (1.9)	4.0079 (2.0)	4.0020 (2.0)
4	4.2752	4.1224 (1.2)	4.0316 (2.0)	4.0079 (2.0)	4.0020 (2.0)
5	5.2244	5.1094 (1.0)	5.0326 (1.7)	5.0084 (2.0)	5.0021 (2.0)
5	5.5224	5.2373 (1.1)	5.0647 (1.9)	5.0164 (2.0)	5.0041 (2.0)
8	5.8945	8.3376 (2.6)	8.1198 (1.5)	8.0314 (1.9)	8.0079 (2.0)
9	6.3737	9.5272 (2.3)	9.1498 (1.8)	9.0382 (2.0)	9.0096 (2.0)
9	6.8812	9.5911 (1.8)	9.1654 (1.8)	9.0420 (2.0)	9.0105 (2.0)
zeros	7	39	175	735	3007
DOF	46	190	766	3070	12286

(in each component) on the criss-cross mesh sequence are listed in Table 5.3 (page 36). At first glance, the results of the computation might lead to the conclusion that the eigenvalues have been well approximated: the zero frequency is approximated by an increasing number of zero discrete eigenvalues (up to machine precision) and the remaining discrete values are well separated from zero and quadratically converging towards integer numbers. Unfortunately, some limit values do not correspond to exact solutions: *spurious* eigenvalues are computed with this scheme and are indicated with an exclamation mark in Table 5.3. Figure 5.6 shows the eigenfunctions corresponding to the eigenvalues ranging from position 70 to 72 in the spectrum computed with the mesh at level $N = 8$. The checkerboard pattern of the eigenfunction corresponding to eigenvalue number 71, which is the value converging to the spurious solution equal to 6, is evident.

Two more spurious solutions, corresponding to eigenvalues number 79 and 80 (which converge to 15), are displayed in Figure 5.7.

Remark 5.2. All examples presented so far for the approximation of the eigenvalues of Maxwell's equations correspond to standard schemes for the discretization of problem (5.10): find $\lambda_h \in \mathbb{R}$ and $p_h \in \operatorname{rot}(\Sigma_h) = U_h$ with $p_h \not\equiv 0$ such that, for some $\boldsymbol{\sigma}_h \in \Sigma_h$,

$$(\boldsymbol{\sigma}_h, \boldsymbol{\tau}) + (p_h, \operatorname{rot} \boldsymbol{\tau}) = 0 \quad \forall \boldsymbol{\tau} \in \Sigma_h, \tag{5.11a}$$

$$(\operatorname{rot} \boldsymbol{\sigma}_h, q) = -\lambda_h (p_h, q) \quad \forall q \in U_h. \tag{5.11b}$$

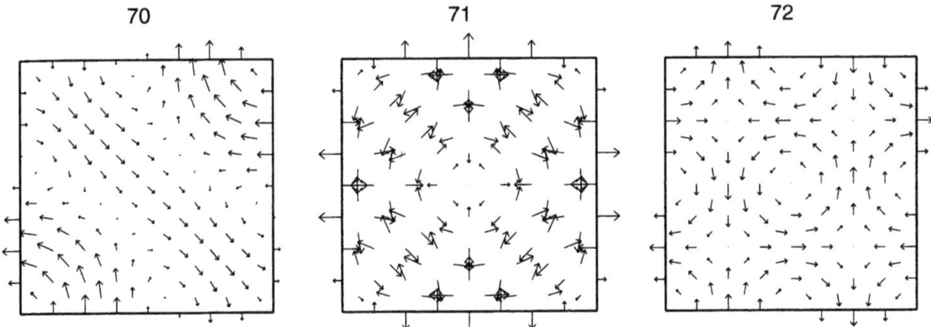

Figure 5.6. The first spurious eigenfunction (*centre*) on the criss-cross mesh for $N = 8$.

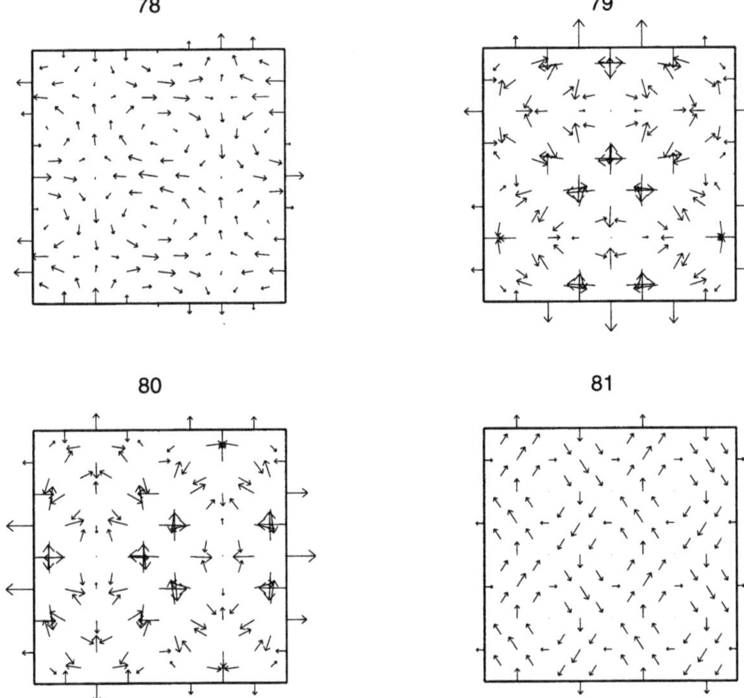

Figure 5.7. The second and third spurious eigenfunctions (numbers 80 and 81) on the criss-cross mesh for $N = 8$.

Table 5.3. Eigenvalues computed with nodal elements on the criss-cross mesh sequence of triangles of Figure 3.9.

Exact	Computed (rate)				
	$N = 2$	$N = 4$	$N = 8$	$N = 16$	$N = 32$
1	1.0662	1.0170 (2.0)	1.0043 (2.0)	1.0011 (2.0)	1.0003 (2.0)
1	1.0662	1.0170 (2.0)	1.0043 (2.0)	1.0011 (2.0)	1.0003 (2.0)
2	2.2035	2.0678 (1.6)	2.0171 (2.0)	2.0043 (2.0)	2.0011 (2.0)
4	4.8634	4.2647 (1.7)	4.0680 (2.0)	4.0171 (2.0)	4.0043 (2.0)
4	4.8634	4.2647 (1.7)	4.0680 (2.0)	4.0171 (2.0)	4.0043 (2.0)
5	6.1338	5.3971 (1.5)	5.1063 (1.9)	5.0267 (2.0)	5.0067 (2.0)
5	6.4846	5.3971 (1.9)	5.1063 (1.9)	5.0267 (2.0)	5.0067 (2.0)
!→ 6	6.4846	5.6712 (0.6)	5.9229 (2.1)	5.9807 (2.0)	5.9952 (2.0)
8	11.0924	8.8141 (1.9)	8.2713 (1.6)	8.0685 (2.0)	8.0171 (2.0)
9	11.0924	10.2540 (0.7)	9.3408 (1.9)	9.0864 (2.0)	9.0217 (2.0)
9	11.1164	10.2540 (0.8)	9.3408 (1.9)	9.0864 (2.0)	9.0217 (2.0)
10		10.9539	10.4193 (1.2)	10.1067 (2.0)	10.0268 (2.0)
10		10.9539	10.4193 (1.2)	10.1067 (2.0)	10.0268 (2.0)
13		11.1347	13.7027 (1.4)	13.1804 (2.0)	13.0452 (2.0)
13		11.1347	13.7027 (1.4)	13.1804 (2.0)	13.0452 (2.0)
!→15		19.4537	13.9639 (2.1)	14.7166 (1.9)	14.9272 (2.0)
!→15		19.4537	13.9639 (2.1)	14.7166 (1.9)	14.9272 (2.0)
16		19.7860	17.0588 (1.8)	16.2722 (2.0)	16.0684 (2.0)
16		19.7860	17.0588 (1.8)	16.2722 (2.0)	16.0684 (2.0)
17		20.9907	18.1813 (1.8)	17.3073 (1.9)	17.0773 (2.0)
zeros	3	15	63	255	1023
DOF	14	62	254	1022	4094

In particular, when Σ_h consists of edge or nodal elements (of lowest order), U_h is the space of piecewise constant functions with zero mean value. All comments made so far then apply to the approximation of the Laplace eigenproblem in mixed form, with the identification discussed above (approximations of the mixed Laplace eigenproblem do not present vanishing discrete values and the eigenfunctions for the formulation in $\mathbf{H}_0(\mathrm{div};\Omega)$ can be obtained from those presented here by rotation of the angle $\pi/2$).

Table 5.4. Eigenvalues computed with edge elements on a sequence of uniform meshes of squares.

Exact	Computed (rate)				
	$N = 4$	$N = 8$	$N = 16$	$N = 32$	$N = 64$
1	1.0524	1.0129 (2.0)	1.0032 (2.0)	1.0008 (2.0)	1.0002 (2.0)
1	1.0524	1.0129 (2.0)	1.0032 (2.0)	1.0008 (2.0)	1.0002 (2.0)
2	2.1048	2.0258 (2.0)	2.0064 (2.0)	2.0016 (2.0)	2.0004 (2.0)
4	4.8634	4.2095 (2.0)	4.0517 (2.0)	4.0129 (2.0)	4.0032 (2.0)
4	4.8634	4.2095 (2.0)	4.0517 (2.0)	4.0129 (2.0)	4.0032 (2.0)
5	5.9158	5.2225 (2.0)	5.0549 (2.0)	5.0137 (2.0)	5.0034 (2.0)
5	5.9158	5.2225 (2.0)	5.0549 (2.0)	5.0137 (2.0)	5.0034 (2.0)
8	9.7268	8.4191 (2.0)	8.1033 (2.0)	8.0257 (2.0)	8.0064 (2.0)
9	12.8431	10.0803 (1.8)	9.2631 (2.0)	9.0652 (2.0)	9.0163 (2.0)
9	12.8431	10.0803 (1.8)	9.2631 (2.0)	9.0652 (2.0)	9.0163 (2.0)
zeros	9	49	225	961	3969
DOF	24	112	480	1984	8064

5.2. Approximation of Maxwell's eigenvalues on quadrilateral meshes

We conclude the discussion of the approximation of Maxwell's eigenvalues with the result of some numerical computations involving quadrilateral meshes.

The first computation, given in Table 5.4, involves edge elements and a sequence of uniform meshes of squares. The discrete eigenvalues converge towards the exact solutions quadratically, as expected, and from above.

In order to warn the reader about possible troubles arising from distorted quadrilateral meshes (in the spirit of the results presented in Arnold, Boffi and Falk (2002, 2005)), in Table 5.5 we present the results of a computation on the sequence of distorted meshes shown in Figure 5.8. In this case the eigenvalues do not converge to the right solution. Indeed, it can be shown that the discrete eigenvalues converge quadratically to incorrect values, which depend on the distortion of the particular mesh used (Gamallo 2002, Bermúdez, Gamallo, Nogueiras and Rodríguez 2006). When using higher-order edge elements, the eigenmodes converge, but with suboptimal rate. Some results on second-order edge elements are reported in Boffi, Kikuchi and Schöberl (2006*c*).

There are several possible cures for this bad behaviour. The first, introduced in Arnold, Boffi and Falk (2005), consists in adding internal degrees of freedom in each element so that the optimal approximation properties

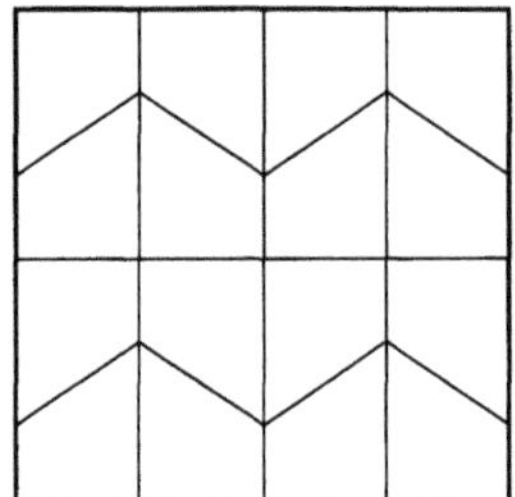 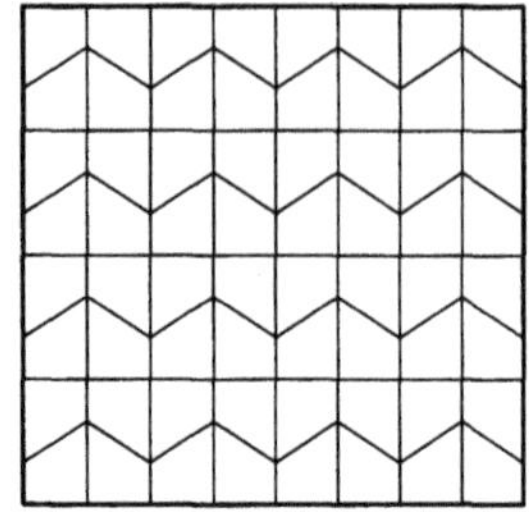 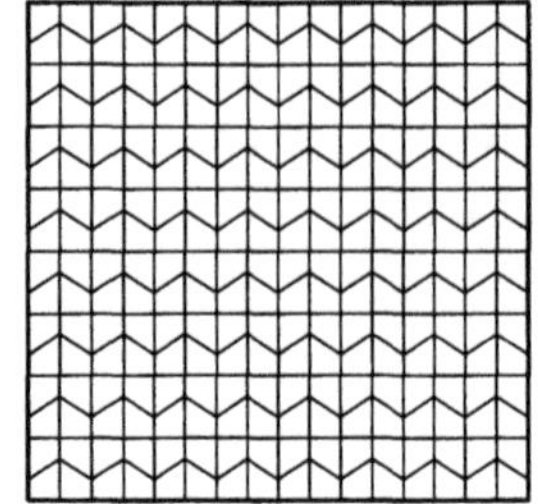

Figure 5.8. Sequence of distorted quadrilateral meshes ($N = 4, 8, 16$).

Table 5.5. Eigenvalues computed with edge elements on the sequence of distorted quadrilaterals of Figure 5.8.

Exact	Computed (rate)				
	$N = 4$	$N = 8$	$N = 16$	$N = 32$	$N = 64$
1	1.0750	1.0484 (0.6)	1.0418 (0.2)	1.0402 (0.1)	1.0398 (0.0)
1	1.0941	1.0531 (0.8)	1.0430 (0.3)	1.0405 (0.1)	1.0399 (0.0)
2	2.1629	2.1010 (0.7)	2.0847 (0.3)	2.0807 (0.1)	2.0797 (0.0)
4	4.6564	4.3013 (1.1)	4.1936 (0.6)	4.1674 (0.2)	4.1609 (0.1)
4	5.0564	4.3766 (1.5)	4.2124 (0.8)	4.1721 (0.3)	4.1621 (0.1)
5	5.8585	5.3515 (1.3)	5.2362 (0.6)	5.2078 (0.2)	5.2007 (0.0)
5	5.9664	5.4232 (1.2)	5.2539 (0.7)	5.2122 (0.3)	5.2019 (0.1)
8	9.5155	8.6688 (1.2)	8.4046 (0.7)	8.3390 (0.3)	8.3228 (0.1)
9	11.5509	10.0919 (1.2)	9.5358 (1.0)	9.4011 (0.4)	9.3681 (0.1)
9	12.9986	10.4803 (1.4)	9.6307 (1.2)	9.4250 (0.6)	9.3741 (0.2)
zeros	9	49	225	961	3969
DOF	24	112	480	1984	8064

are restored. The convergence analysis for the eigenvalues computed with this new element (sometimes referred to as the ABF element) can be found in Gardini (2005).

Another, cheaper, cure consists in using a projection technique, which can also be interpreted as a reduced integration strategy (Boffi *et al.* 2006*c*). In the lowest-order case it reduces to projecting $\operatorname{rot} \mathbf{u}_h$ onto piecewise constants in formulation (5.9), or, equivalently, to using the midpoint rule in order to evaluate the integral $(\operatorname{rot} \mathbf{u}_h, \operatorname{rot} \mathbf{v})$.

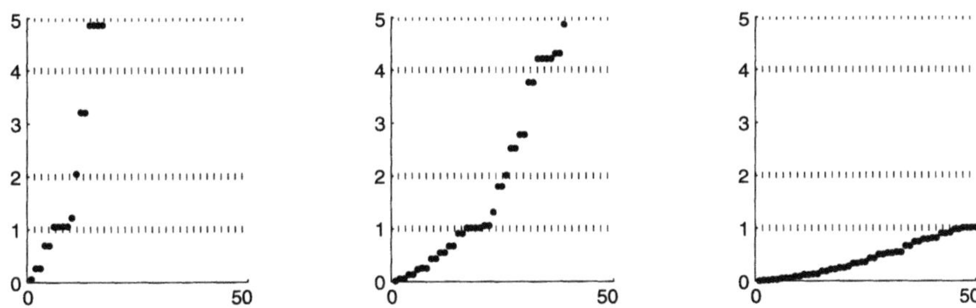

Figure 5.9. First 50 discrete eigenvalues computed with piecewise bilinear elements on the uniform mesh of squares ($N = 4, 8, 16$).

Remark 5.3. On distorted quadrilateral meshes, the equivalence between our original formulation and the mixed formulation (5.11) is no longer true. More precisely, the equivalence is true by choosing $U_h = \mathrm{rot}(\Sigma_h)$, and U_h in this case is not a standard finite element space. For instance, in the case of lowest-order edge elements, U_h in each element K is made of functions $C/|J|$, where $|J|$ is the determinant of the Jacobian of the mapping from the reference cube $\hat{K}$ to K, and C is a generic constant. The projection procedure just described has the effect of changing the discrete formulation, so that in the lowest-order case it turns out to be equivalent to the mixed problem (5.11), with U_h equal to the space of piecewise constant functions. A similar procedure also holds for higher-order edge elements; we refer the interested reader to Boffi *et al.* (2006*c*) for more details.

We conclude this section with some comments on nodal element approximation of Maxwell's eigenvalues on rectangular meshes (Boffi, Durán and Gastaldi 1999*a*). The presented results, in particular, will give some explanations for the spurious eigenvalues shown in Table 5.3.

We start by using classical bilinear elements $\mathcal{Q}_1$ in each component, on a sequence of meshes obtained by dividing the square $\Omega =]0, \pi[$ into N^2 sub-squares. The results, which are similar to those obtained by linear elements on unstructured triangular meshes, are given in Figure 5.9. It is clear that these results cannot provide any reasonable approximation to the problem we are interested in.

Another possible scheme consists in projecting $\mathrm{rot}\,\mathbf{u}_h$ onto piecewise constants in formulation (5.9). From the comments made in Remark 5.3, it turns out that this is indeed analogous to considering the $\mathcal{Q}_1 - \mathcal{P}_0$ scheme for mixed Laplacian. The eigenvalues computed with this scheme are given in Table 5.6 and we would like to point out the analogies with the criss-cross computation shown in Table 5.3. It is clear that there is a spurious discrete eigenvalue which converges to 18 (here the term 'spurious' is meant with respect to the multiplicity, since there is in fact an exact solution with value 18 and multiplicity 1). As in the criss-cross example of Table 5.3, there are

Table 5.6. Eigenvalues computed with the projected $\mathcal{Q}_1$ scheme ($\mathcal{Q}_1 - \mathcal{P}_0$) on the sequence of uniform meshes of squares.

Exact	Computed (rate)				
	$N = 4$	$N = 8$	$N = 16$	$N = 32$	$N = 64$
1	1.0524	1.0129 (2.0)	1.0032 (2.0)	1.0008 (2.0)	1.0002 (2.0)
1	1.0524	1.0129 (2.0)	1.0032 (2.0)	1.0008 (2.0)	1.0002 (2.0)
2	1.9909	1.9995 (4.1)	2.0000 (4.0)	2.0000 (4.0)	2.0000 (4.0)
4	4.8634	4.2095 (2.0)	4.0517 (2.0)	4.0129 (2.0)	4.0032 (2.0)
4	4.8634	4.2095 (2.0)	4.0517 (2.0)	4.0129 (2.0)	4.0032 (2.0)
5	5.3896	5.1129 (1.8)	5.0288 (2.0)	5.0072 (2.0)	5.0018 (2.0)
5	5.3896	5.1129 (1.8)	5.0288 (2.0)	5.0072 (2.0)	5.0018 (2.0)
8	7.2951	7.9636 (4.3)	7.9978 (4.1)	7.9999 (4.0)	8.0000 (4.0)
9	8.7285	10.0803 (-2.0)	9.2631 (2.0)	9.0652 (2.0)	9.0163 (2.0)
9	11.2850	10.0803 (1.1)	9.2631 (2.0)	9.0652 (2.0)	9.0163 (2.0)
10	11.2850	10.8308 (0.6)	10.2066 (2.0)	10.0515 (2.0)	10.0129 (2.0)
10	12.5059	10.8308 (1.6)	10.2066 (2.0)	10.0515 (2.0)	10.0129 (2.0)
13	12.5059	13.1992 (1.3)	13.0736 (1.4)	13.0197 (1.9)	13.0050 (2.0)
13	12.8431	13.1992 (−0.3)	13.0736 (1.4)	13.0197 (1.9)	13.0050 (2.0)
16	12.8431	14.7608 (1.3)	16.8382 (0.6)	16.2067 (2.0)	16.0515 (2.0)
16		17.5489	16.8382 (0.9)	16.2067 (2.0)	16.0515 (2.0)
17		19.4537	17.1062 (4.5)	17.1814 (−0.8)	17.0452 (2.0)
17		19.4537	17.7329 (1.7)	17.1814 (2.0)	17.0452 (2.0)
!→ 18		19.9601	17.7329 (2.9)	17.7707 (0.2)	17.9423 (2.0)
18		19.9601	17.9749 (6.3)	17.9985 (4.0)	17.9999 (4.0)
20		21.5584	20.4515 (1.8)	20.1151 (2.0)	20.0289 (2.0)
20		21.5584	20.4515 (1.8)	20.1151 (2.0)	20.0289 (2.0)
zeros	15	63	255	1023	4095
DOF	30	126	510	2046	8190

other spurious solutions with higher frequencies. In this particular case, the closed form of the computed solutions was computed in Boffi *et al.* (1999*a*). It has been shown that the $2(N^2-1)$ degrees of freedom are split into two equal parts: N^2-1 of them correspond to the zero frequency, while the remaining N^2-1 can be ordered in the following way, by means of two indices m, n ranging from 0 to $N-1$ with $m+n \neq 0$:

$$\lambda_h^{(m,n)} = (4/h^2)\frac{\sin^2(\frac{mh}{2}) + \sin^2(\frac{nh}{2}) - 2\sin^2(\frac{mh}{2})\sin^2(\frac{nh}{2})}{1-(2/3)(\sin^2(\frac{mh}{2})+\sin^2(\frac{nh}{2})) + (4/9)\sin^2(\frac{mh}{2})\sin^2(\frac{nh}{2})}.$$

The corresponding eigenfunctions $\mathbf{u}_h^{(m,n)} = (u^{(m,n)}, v^{(m,n)})$ are given by

$$u^{(m,n)}(x_i, y_j) = -\frac{2}{h}\sin\left(\frac{mh}{2}\right)\cos\left(\frac{nh}{2}\right)\sin(mx_i)\cos(ny_j), \qquad (5.12a)$$

$$v^{(m,n)}(x_i, y_j) = -\frac{2}{h}\cos\left(\frac{mh}{2}\right)\sin\left(\frac{nh}{2}\right)\cos(mx_i)\sin(ny_j). \qquad (5.12b)$$

Looking at the formulae (5.12), it seems at first glance that the eigenmodes converge to the exact solution with the correct multiplicity and that there are no spurious solutions. Indeed, given a fixed pair (m, n), it is easy

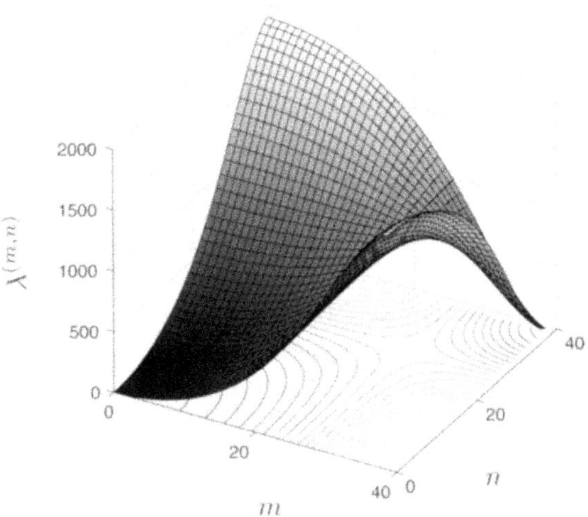

Figure 5.10. Discrete eigenvalues of the $\mathcal{Q}_1 - \mathcal{P}_0$ scheme as a function of (m, n).

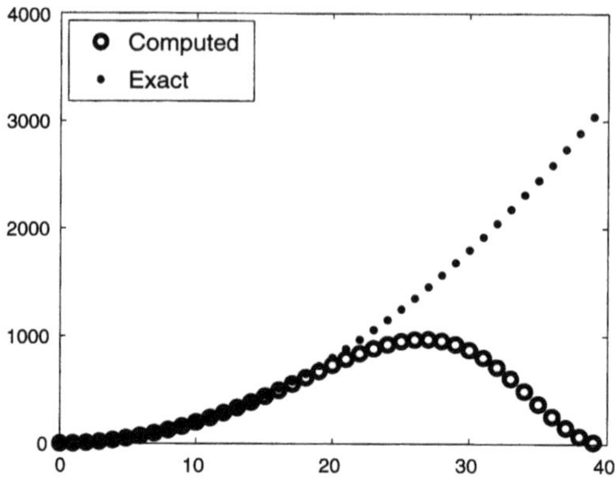

Figure 5.11. Discrete eigenvalues of the $\mathcal{Q}_1 - \mathcal{P}_0$ scheme for $m = n$.

to see that $\lambda_h^{(m,n)}$ tends to $m^2 + n^2$ and $\mathbf{u}_h^{(m.n)}$ to $\mathbf{grad}(\cos(mx)\cos(nx)) = -(m\sin(mx)\cos(nx), n\cos(mx)\sin(nx))$. On the other hand, it can also be easily observed that

$$\lim_{N\to\infty} \lambda_h^{(N-1,N-1)} = 18,$$

where $N = \pi/h$. A clear picture of this phenomenon can be seen in Figure 5.10, where the surface defined by $\lambda_h^{(m,n)}$ is plotted as a function of (m, n). The surface is not convex; in particular, it is not monotone in m and n and, moreover, the value at the corner opposite to the origin tends to 18 as h goes to zero. In Figure 5.11 we also show the section of the surface along the diagonal $m = n$.

With the help of these analytical results, it is possible to sort the eigenvalues of Table 5.6 in a different way, so that the rate of convergence of the spurious eigenvalue can be better evaluated (see Table 5.7).

The behaviour of the presented $\mathcal{Q}_1 - \mathcal{P}_0$ scheme is very similar to that already seen in Table 5.3 for the triangular criss-cross mesh. In that case a closed form of the discrete solution is not available, but can be found for a slight modification of the method (Boffi and Gastaldi 2004).

Remark 5.4. A possible cure for the pathology of the $\mathcal{Q}_1 - \mathcal{P}_0$ scheme was proposed in Chen and Taylor (1990) and analysed for a square domain in Boffi *et al.* (1999*a*). Unfortunately, this method does not seem to provide good results in the case of singular solutions (such as those obtained in an L-shaped domain).

Table 5.7. Eigenvalues computed with the projected $\mathcal{Q}_1$ scheme ($\mathcal{Q}_1 - \mathcal{P}_0$) on the sequence of uniform meshes of squares with the spurious eigenvalue sorted in a different way.

Exact	Computed (rate)				
	$N = 4$	$N = 8$	$N = 16$	$N = 32$	$N = 64$
1	1.0524	1.0129 (2.0)	1.0032 (2.0)	1.0008 (2.0)	1.0002 (2.0)
1	1.0524	1.0129 (2.0)	1.0032 (2.0)	1.0008 (2.0)	1.0002 (2.0)
2	1.9909	1.9995 (4.1)	2.0000 (4.0)	2.0000 (4.0)	2.0000 (4.0)
4	4.8634	4.2095 (2.0)	4.0517 (2.0)	4.0129 (2.0)	4.0032 (2.0)
4	4.8634	4.2095 (2.0)	4.0517 (2.0)	4.0129 (2.0)	4.0032 (2.0)
5	5.3896	5.1129 (1.8)	5.0288 (2.0)	5.0072 (2.0)	5.0018 (2.0)
5	5.3896	5.1129 (1.8)	5.0288 (2.0)	5.0072 (2.0)	5.0018 (2.0)
8	7.2951	7.9636 (4.3)	7.9978 (4.1)	7.9999 (4.0)	8.0000 (4.0)
9	11.2850	10.0803 (1.1)	9.2631 (2.0)	9.0652 (2.0)	9.0163 (2.0)
9	11.2850	10.0803 (1.1)	9.2631 (2.0)	9.0652 (2.0)	9.0163 (2.0)
10	12.5059	10.8308 (1.6)	10.2066 (2.0)	10.0515 (2.0)	10.0129 (2.0)
10	12.5059	10.8308 (1.6)	10.2066 (2.0)	10.0515 (2.0)	10.0129 (2.0)
13	12.8431	13.1992 (-0.3)	13.0736 (1.4)	13.0197 (1.9)	13.0050 (2.0)
13	12.8431	13.1992 (-0.3)	13.0736 (1.4)	13.0197 (1.9)	13.0050 (2.0)
16		17.5489	16.8382 (0.9)	16.2067 (2.0)	16.0515 (2.0)
16		19.4537	16.8382 (2.0)	16.2067 (2.0)	16.0515 (2.0)
17		19.4537	17.7329 (1.7)	17.1814 (2.0)	17.0452 (2.0)
17		19.9601	17.7329 (2.0)	17.1814 (2.0)	17.0452 (2.0)
$! \rightarrow$ 18	8.7285	14.7608 (1.5)	17.1062 (1.9)	17.7707 (2.0)	17.9423 (2.0)
18		19.9601	17.9749 (6.3)	17.9985 (4.0)	17.9999 (4.0)
20		21.5584	20.4515 (1.8)	20.1151 (2.0)	20.0289 (2.0)
20		21.5584	20.4515 (1.8)	20.1151 (2.0)	20.0289 (2.0)
zeros	15	63	255	1023	4095
DOF	30	126	510	2046	8190

PART TWO
Galerkin approximation of compact eigenvalue problems

This part of our survey contains the classical spectral approximation theory for compact operators. It is the core of our work, since all eigenvalue problems we are going to consider are related to compact operators, and we will constantly rely on the fundamental tools described in Section 9. The approximation theory for compact eigenvalue problems has been the object of a wide investigation. A necessarily incomplete list of the most relevant references is Vaĭnikko (1964, 1966), Kato (1966), Anselone and Palmer (1968), Stummel (1970, 1971, 1972), Anselone (1971), Bramble and Osborn (1973), Chatelin (1973), Osborn (1975), Grigorieff (1975*a*, 1975*b*, 1975*c*), Chatelin and Lemordant (1975), Chatelin (1983), Babuška and Osborn (1989, 1991), Kato (1995) and Knyazev and Osborn (2006).

The Galerkin approximation of the Laplace eigenvalue problem, for which we present in Section 8 a rigorous analysis that makes use of standard tools, fits within the framework of the numerical approximation of variationally posed eigenvalue problems, discussed in Sections 7 and 9.

An example of the non-conforming approximation of eigenvalue problems is analysed in Section 11, where a new convergence analysis for Crouzeix–Raviart approximation of the Laplace eigenvalue problem is provided.

6. Spectral theory for compact operators

In this section we present the main definitions we shall need.

Let X be a complex Hilbert space and let $T : X \to X$ be a compact linear operator. The resolvent set $\rho(T)$ is given by the complex numbers $z \in \mathbb{C}$ such that $(zI - T)$ is bijective. We shall use the standard notation $z - T = zI - T$ and the resolvent operator is given by $(z - T)^{-1}$. The spectrum of T is $\sigma(T) = \mathbb{C} \setminus \rho(T)$, which is well known to be a countable set with no limit points different from zero. All non-zero values in $\sigma(T)$ are eigenvalues (that is, $z - T$ is one-to-one); zero may or may not be an eigenvalue.

If λ is a non-vanishing eigenvalue of T, then the ascent multiplicity α of $\lambda - T$ is the smallest integer such that $\ker(\lambda - T)^\alpha = \ker(\lambda - T)^{\alpha+1}$. The terminology comes from the fact that there also exists a similar definition for the descent multiplicity, which makes use of the range instead of the kernel; for compact operators ascent and descent multiplicities coincide. The dimension of $\ker(\lambda - T)^\alpha$ is called the algebraic multiplicity of λ, and the elements of $\ker(\lambda - T)^\alpha$ are the generalized eigenvectors of T associated with λ. A generalized eigenvector is of order k if it is in $\ker(\lambda - T)^k$, but not in $\ker(\lambda - T)^{k-1}$. The generalized eigenvectors of order 1 are called eigenvectors of T associated with λ, and are the elements of $\ker(\lambda - T)$.

The dimension of $\ker(\lambda - T)$ is called the geometric multiplicity of λ, which is always less than or equal to the algebraic multiplicity. If T is self-adjoint, which will be the case for all examples discussed in this work, then the ascent multiplicity of each eigenvalue is equal to one. This implies that all generalized eigenvectors are eigenvectors, and that the geometric and the algebraic multiplicities coincide.

Given a closed smooth curve $\Gamma \subset \rho(T)$ which encloses $\lambda \in \sigma(T)$, and no other elements of $\sigma(T)$, the Riesz spectral projection associated with λ is defined by

$$E(\lambda) = \frac{1}{2\pi \mathrm{i}} \int_\Gamma (z - T)^{-1}\, \mathrm{d}z. \tag{6.1}$$

The definition clearly does not depend on the chosen curve, and it can be checked that $E(\lambda) : X \to X$, that $E(\lambda) \circ E(\lambda) = E(\lambda)$ (which means it is actually a projection), that $E(\lambda) \circ E(\mu) = 0$ if $\lambda \neq \mu$, that $T \circ E(\lambda) = E(\lambda) \circ T$, and that the range of $E(\lambda)$ is equal to $\ker(\lambda - T)^\alpha$, the space of generalized eigenvectors (which is an invariant subspace for T). This last property will be of fundamental importance for the study of eigenvector approximation, and we emphasize it in the following formula:

$$E(\lambda)X = \ker(\lambda - T)^\alpha.$$

In general, if $\Gamma \subset \rho(T)$ encloses more eigenvalues $\lambda_1, \lambda_2, \ldots, \lambda_n$, then we have that

$$E(\lambda_1, \lambda_2, \ldots, \lambda_n)X = \bigoplus_{i=1}^{n} \ker(\lambda_i - T)^{\alpha_i},$$

where α_i denotes the ascent multiplicity of $\lambda_i - T$, so that the dimension of the range of the spectral projection is in general the sum of the algebraic multiplicities of the eigenvalues that lie inside Γ.

Let $T^* : X \to X$ denote the adjoint of T. Then $\lambda \in \sigma(T^*)$ if and only if $\overline{\lambda} \in \sigma(T)$, where $\overline{\lambda}$ denotes the conjugate of λ. In particular, the eigenvalues of self-adjoint operators are real. The algebraic multiplicity of $\lambda \in \sigma(T^*)$ is equal to the algebraic multiplicity of $\overline{\lambda} \in \sigma(T)$ and the ascent multiplicity of $\lambda - T^*$ is equal to that of $\overline{\lambda} - T$.

7. Variationally posed eigenvalue problems

In this section we introduce some preliminary results on variationally posed eigenvalue problems. The main theoretical results are presented in Section 9.

The main focus of this survey is on *symmetric* eigenvalue problems, and for this reason we start with symmetric variationally posed eigenvalue problems. Hence, we are dealing with real Hilbert spaces and real-valued bilinear forms. Some discussion of the non-symmetric case can be found in Section 9.

Let V and H be real Hilbert spaces. We suppose that $V \subset H$ with dense and continuous embedding. Let $a : V \times V \to \mathbb{R}$ and $b : H \times H \to \mathbb{R}$

be symmetric and continuous bilinear forms, and consider the following problem: find $\lambda \in \mathbb{R}$ and $u \in V$, with $u \neq 0$, such that

$$a(u, v) = \lambda b(u, v) \quad \forall v \in V. \tag{7.1}$$

We suppose that a is V-elliptic, that is, there exists $\alpha > 0$ such that

$$a(v, v) \geq \alpha \|v\|_V^2 \quad \forall v \in V.$$

The ellipticity condition could be weakened by assuming a Gårding-like inequality, that is, $a(\cdot, \cdot) + \mu b(\cdot, \cdot)$ is elliptic for a suitable positive μ. We will not detail this situation, which can be reduced to the elliptic case by a standard shift procedure.

For the sake of simplicity we assume that b defines a scalar product in H. In several applications, H will be $L^2(\Omega)$ and b its standard inner product.

An important tool for the analysis of (7.1) is the solution operator $T : H \to H$: given $f \in H$, our hypotheses guarantee the existence of a unique $Tf \in V$ such that

$$a(Tf, v) = b(f, v) \quad \forall v \in V.$$

Since we are interested in *compact* eigenvalue problems, we make the assumption that

$$T : H \to H \quad \text{is a compact operator,}$$

which is often a consequence of a compact embedding of V into H. We have already observed that we consider T to be self-adjoint.

We assume that the reader is familiar with the spectral theory of compact operators; we recall in particular that the spectrum $\sigma(T)$ of T is a countable or finite set of real numbers with a cluster point possible only at zero. All positive elements of $\sigma(T)$ are eigenvalues with finite multiplicity, and their reciprocals are exactly the eigenvalues of (7.1); moreover, the eigensolutions of (7.1) have the same eigenspaces as those of T.

We let $\lambda^{(k)}$, $k \in \mathbb{N}$, denote the eigenvalues of (7.1) with the natural numbering

$$\lambda^{(1)} \leq \lambda^{(2)} \leq \cdots \leq \lambda^{(k)} \leq \cdots,$$

where the same eigenvalue can be repeated several times according to its multiplicity. We let $u^{(k)}$ denote the corresponding eigenfunctions, with the standard normalization $b(u^{(k)}, u^{(k)}) = 1$, and let $E^{(k)} = \mathrm{span}\{u^{(k)}\}$ denote the associated eigenspaces (see below for multiple eigenvalues). We explicitly observe that even for simple eigenvalues the normalization procedure does not identify $u^{(k)}$ uniquely, but only up to its sign.

It is well known that the eigenfunctions enjoy the orthogonalities

$$a(u^{(m)}, u^{(n)}) = b(u^{(m)}, u^{(n)}) = 0 \quad \text{if } m \neq n, \tag{7.2}$$

which can be deduced easily from (7.1) if $\lambda^{(m)} \neq \lambda^{(n)}$; otherwise, for multiple eigenvalues, when $\lambda^{(m)} = \lambda^{(n)}$, the eigenfunctions $u^{(m)}$ and $u^{(n)}$ can be chosen such that the orthogonalities (7.2) hold.

The Rayleigh quotient is an important tool for the study of eigenvalues: it turns out that

$$\lambda^{(1)} = \min_{v\in V} \frac{a(v,v)}{b(v,v)}, \qquad u^{(1)} = \arg\min_{v\in V} \frac{a(v,v)}{b(v,v)},$$

$$\lambda^{(k)} = \min_{v\in\left(\bigoplus_{i=1}^{k-1} E^{(i)}\right)^{\perp}} \frac{a(v,v)}{b(v,v)}, \qquad u^{(k)} = \arg\min_{v\in\left(\bigoplus_{i=1}^{k-1} E^{(i)}\right)^{\perp}} \frac{a(v,v)}{b(v,v)}, \tag{7.3}$$

where it has been implicitly understood (here and in the rest of the paper) that the minima are taken for $v \neq 0$, so that quantities in the denominators do not vanish. The symbol '$\perp$' denotes the orthogonal complement in V with respect to the scalar product induced by the bilinear form b. Due to the orthogonalities (7.2), it turns out that the orthogonal complement could also be taken with respect to the scalar product induced by a.

The Galerkin discretization of problem (7.1) is based on a finite-dimensional space $V_h \subset V$ and reads as follows: find λ_h and $u_h \in V_h$, with $u_h \neq 0$, such that

$$a(u_h, v) = \lambda_h b(u_h, v) \quad \forall v \in V_h. \tag{7.4}$$

Remark 7.1. For historical reasons, we adopt the notation of the h-version of finite elements, and we understand that h is a parameter which tends to zero. Nevertheless, if not explicitly stated, the theory we describe applies to a general Galerkin approximation.

Since V_h is a Hilbert subspace of V, we can repeat the same comments we made for problem (7.1), starting from the definition of the discrete solution operator $T_h : H \to H$: given $f \in H$, $T_h f \in V_h$ is uniquely defined by

$$a(T_h f, v) = b(f, v) \quad \forall v \in V_h.$$

Since V_h is finite-dimensional, T_h is compact; the eigenmodes of T_h are in one-to-one correspondence with those of (7.4) (the equivalence being that the eigenvalues are inverse to each other and the eigenspaces are the same), and we can order the discrete eigenvalues of (7.4) as follows:

$$\lambda_h^{(1)} \leq \lambda_h^{(2)} \leq \cdots \leq \lambda_h^{(k)} \leq \cdots,$$

where eigenvalues can be repeated according to their multiplicity. We use $u_h^{(k)}$, with the normalization $b(u_h^{(k)}, u_h^{(k)}) = 1$, to denote the discrete eigenfunctions, and $E_h^{(k)} = \operatorname{span}\{u_h^{(k)}\}$ for the associated eigenspaces. Discrete eigenfunctions satisfy the same orthogonalities as the continuous ones,

$$a(u_h^{(m)}, u_h^{(n)}) = b(u_h^{(m)}, u_h^{(n)}) = 0 \quad \text{if } m \neq n,$$

where again this property is a theorem when $\lambda_h^{(m)} \neq \lambda_h^{(n)}$ or a definition when $\lambda_h^{(m)} = \lambda_h^{(n)}$.

Moreover, the properties of the Rayleigh quotient can be applied to discrete eigenmodes as follows:

$$\begin{aligned} \lambda_h^{(1)} &= \min_{v \in V_h} \frac{a(v,v)}{b(v,v)}, & u_h^{(1)} &= \arg\min_{v \in V_h} \frac{a(v,v)}{b(v,v)}, \\ \lambda_h^{(k)} &= \min_{v \in \left(\bigoplus_{i=1}^{k-1} E_h^{(i)}\right)^{\perp}} \frac{a(v,v)}{b(v,v)}, & u_h^{(k)} &= \arg\min_{v \in \left(\bigoplus_{i=1}^{k-1} E_h^{(i)}\right)^{\perp}} \frac{a(v,v)}{b(v,v)}, \end{aligned} \tag{7.5}$$

where the symbol '$\perp$' now denotes the orthogonal complement in V_h. An easy consequence of the inclusion $V_h \subset V$ and of the Rayleigh quotient properties is

$$\lambda^{(1)} \leq \lambda_h^{(1)},$$

that is, the first discrete eigenvalue is always an upper bound of the first continuous eigenvalue. Unfortunately, equations (7.3) and (7.5) do not allow us to infer any bound between the other eigenvalues, since it is not true in general that $\left(\oplus_{i=1}^{k-1} E_h^{(i)}\right)^{\perp}$ is a subset of $\left(\oplus_{i=1}^{k-1} E^{(i)}\right)^{\perp}$. For this reason, we recall the important *min-max* characterization of the eigenvalues.

Proposition 7.2. The kth eigenvalue $\lambda^{(k)}$ of problem (7.1) satisfies

$$\lambda^{(k)} = \min_{E \in V^{(k)}} \max_{v \in E} \frac{a(v,v)}{b(v,v)},$$

where $V^{(k)}$ denotes the set of all subspaces of V with dimension equal to k.

Proof. In order to show that $\lambda^{(k)}$ is greater than or equal to the min-max, take $E = \oplus_{i=1}^{k} E^{(i)}$, so that $v = \sum_{i=1}^{k} \alpha_i u^{(i)}$. From the orthogonalities and the normalization of the eigenfunctions, it is easy to obtain the inequality $a(v,v)/b(v,v) \leq \lambda^{(k)}$.

The proof of the opposite inequality also gives the additional information that the minimum is attained for $E = \oplus_{i=1}^{k} E^{(i)}$ and the choice $v = u^{(k)}$. It is clear that if $E = \oplus_{i=1}^{k} E^{(i)}$ then the optimal choice for v is $u^{(k)}$. On the other hand, if $E \neq \oplus_{i=1}^{k} E^{(i)}$ then there exists $v \in E$ with v orthogonal to $u^{(i)}$ for all $i \leq k$, and hence $a(v,v)/b(v,v) \geq \lambda^{(k)}$, which shows that $E = \oplus_{i=1}^{k} E^{(i)}$ is the optimal choice for E. □

The analogous min-max condition for the discrete problem (7.4) states

$$\lambda_h^{(k)} = \min_{E_h \in V_h^{(k)}} \max_{v \in E_h} \frac{a(v,v)}{b(v,v)}, \tag{7.6}$$

where $V_h^{(k)}$ denotes the set of all subspaces of V_h with dimension equal to k.

It is then an easy consequence that a conforming approximation $V_h \subset V$ implies that all eigenvalues are approximated from above,

$$\lambda^{(k)} \leq \lambda_h^{(k)} \quad \forall k,$$

since all sets $E_h \in V_h^{(k)}$ in the discrete min-max property are also in $V^{(k)}$, and hence the discrete minimum is evaluated over a smaller set than the continuous one.

Monotonicity is an interesting property (and this behaviour was observed in the numerical examples of Part 1), but it is not enough to show the convergence. The definition of convergence for eigenvalues/eigenfunctions is an intuitive concept, which requires a careful formalism. First of all, we would like the kth discrete eigenvalue to converge towards the kth continuous one. This implies two important facts: all solutions are well approximated and no spurious eigenvalues pollute the spectrum. The numbering we have chosen for the eigenvalues, moreover, implies that the eigenvalues are approximated correctly with their multiplicity. The convergence of eigenfunctions is a little more involved, since we cannot simply require $u_h^{(k)}$ to converge to $u^{(k)}$ in a suitable norm. This type of convergence cannot be expected for at least two good reasons. First of all, the eigenspace associated with multiple eigenvalues can be approximated by the eigenspaces of distinct discrete eigenvalues (see, for example, (3.2)). Then, even in the case of simple eigenvalues, the normalization of the eigenfunctions is not enough to ensure convergence, since they might have the wrong sign. The natural definition of convergence makes use of the notion of the gap between Hilbert spaces, defined by

$$\begin{aligned} \delta(E,F) &= \sup_{\substack{u \in E \\ \|u\|_H = 1}} \inf_{v \in F} \|u - v\|_H, \\ \hat{\delta}(E,F) &= \max(\delta(E,F), \delta(F,E)). \end{aligned}$$

A possible definition of the convergence of eigensolutions was introduced in Boffi, Brezzi and Gastaldi (2000*a*). For every positive integer k, let $m(k)$ denote the dimension of the space spanned by the first distinct k eigenspaces. Then we say that the discrete eigenvalue problem (7.4) converges to the continuous one (7.1) if, for any $\varepsilon > 0$ and $k > 0$, there exists $h_0 > 0$ such that, for all $h < h_0$, we have

$$\begin{aligned} &\max_{1 \leq i \leq m(k)} |\lambda^{(i)} - \lambda_h^{(i)}| \leq \varepsilon, \\ &\hat{\delta}\left(\bigoplus_{i=1}^{m(k)} E^{(i)}, \bigoplus_{i=1}^{m(k)} E_h^{(i)} \right) \leq \varepsilon. \end{aligned} \tag{7.7}$$

It is remarkable that this definition includes all properties that we need: convergence of eigenvalues and eigenfunctions with correct multiplicity, and absence of spurious solutions.

Remark 7.3. The definition of convergence (7.7) does not give any indication of the approximation rate. It is indeed quite common to separate the convergence analysis for eigenvalue problems into two steps: firstly, the convergence and the absence of spurious modes is investigated in the spirit of (7.7), then suitable approximation rates are proved.

Proposition 7.4. Problem (7.4) converges to (7.1) in the spirit of (7.7) if and only if the following norm convergence holds true:

$$\|T - T_h\|_{\mathcal{L}(H)} \to 0 \quad \text{when } h \to 0. \tag{7.8}$$

Proof. The sufficient part of the proposition is a well-known result in the spectral approximation theory of linear operators (Kato 1995, Chapter IV). The necessity of norm convergence for good spectral approximation of symmetric compact operators was shown in Boffi *et al.* (2000*a*, Theorem 5.1). □

Remark 7.5. Our compactness assumption can be modified by assuming that $T : V \to V$ is compact. In this case norm convergence similar to (7.8) in $\mathcal{L}(V)$ would ensure an analogous eigenmodes convergence of (7.7) with the natural modifications.

In order to show the convergence in norm (7.8), it is useful to recall that the discrete operator T_h can be seen as $T_h = P_h T$, where $P_h : V \to V_h$ is the elliptic projection associated to the bilinear form a. This fact is a standard consequence of Galerkin orthogonality and implies that $T - T_h$ can be written as $(I - P_h)T$, where I denotes the identity operator.

The next proposition can be used to prove convergence in norm.

Proposition 7.6. If T is compact from H to V and P_h converges strongly to the identity operator from V to H, then the norm convergence (7.8) from H to H holds true.

Proof. First we show that the sequence $\{\|I - P_h\|_{\mathcal{L}(V,H)}\}$ is bounded. Define $c(h,u)$ by $\|(I - P_h)u\|_H = c(h,u)\|u\|_V$. Strong (pointwise) convergence means that for each u we have $c(h,u) \to 0$. Thus $M(u) = \max_h c(h,u)$ is finite. By the uniform boundedness principle (or Banach–Steinhaus theorem), there exists C such that, for all h, $\|I - P_h\|_{\mathcal{L}(V,H)} \leq C$.

Consider a sequence $\{f_h\}$ such that, for each h, $\|f_h\|_H = 1$ and $\|T - T_h\|_{\mathcal{L}(H)} = \|(T - T_h)f_h\|_H$. Since $\{f_h\}$ is bounded in H and T is compact from H to V, there exists a subsequence, which we again denote by $\{f_h\}$, such that $Tf_h \to w$ in V. We claim that $\|(I - P_h)Tf_h\|_H \to 0$ for the subsequence, and hence for the sequence itself. T is a closed operator: there exists $v \in H$ such that $Tv = w$. By hypothesis $T_h v \to w$; furthermore, $T_h v = P_h T v = P_h w$. The strong convergence of $Tf_h \to w$ in V and

$P_h w \to w$ in H, the triangle inequality and the boundedness of $\{\|I - P_h\|_{\mathcal{L}(V,H)}\}$ imply that, for any $\varepsilon > 0$, there exists h small enough such that

$$\begin{aligned}\|(I - P_h)Tf_h\|_H &\le \|(I - P_h)(Tf_h - w)\|_H + \|(I - P_h)w\|_H \\ &\le C\|Tf_h - w\|_V + \|(I - P_h)w\|_H \le \varepsilon. \qquad \square\end{aligned}$$

Remark 7.7. Results such as that presented in Proposition 7.6 have been used very often in the literature dealing with the approximation theory for linear operators. There are many variants of this, and it is often said that compact operators transform strong (pointwise) into norm (uniform) convergence. It is worth noticing that this result is only true when the compact operator is applied to the right of the converging sequence. For abstract results in this context, we refer, for instance, to Anselone (1971); the same statement as in Proposition 7.6 in the framework of variationally posed eigenvalue problems can be found in Kolata (1978).

Remark 7.8. The same proof as in Proposition 7.6 can be used to show that, if T is compact from V into V, then a stronger pointwise convergence of P_h to the identity, from V into V, is sufficient to ensure the norm convergence

$$\|T - T_h\|_{\mathcal{L}(V)} \to 0 \quad \text{when } h \to 0.$$

Such convergence is equivalent to a type of convergence of eigenvalues and eigenfunctions analogous to (7.7).

8. A direct proof of convergence for Laplace eigenvalues

A fundamental example of elliptic partial differential equation is given by the Laplace operator. Although the convergence theory of the finite element approximation of Laplace eigenmodes is a particular case of the analysis presented in Sections 7 and 9, we now study this basic example. The analysis will be performed with standard tools in the case of Dirichlet boundary conditions and piecewise linear finite elements, but can be applied with minor modifications to Neumann or mixed boundary conditions and to higher-order finite elements. The same technique extends to more general linear eigenvalue problems associated with elliptic operators. Similar arguments can be found in Strang and Fix (1973) and are based on the pioneering work of Birkhoff, de Boor, Swartz and Wendroff (1966) (see also Raviart and Thomas (1983) or Larsson and Thomée (2003)).

Given a polyhedral domain in $\mathbb{R}^3$ (respectively, a polygonal domain in $\mathbb{R}^2$), we are interested in the solution of the following problem: find eigenvalues λ and eigenfunctions u with $u \ne 0$ such that

$$\begin{aligned}-\Delta u &= \lambda u \quad \text{in } \Omega,\\ u &= 0 \qquad \text{on } \partial\Omega.\end{aligned}$$

A variational formulation of our problem can be obtained by introducing the space $V = H^1_0(\Omega)$ and in looking for $\lambda \in \mathbb{R}$ and $u \in V$, with $u \neq 0$, such that

$$(\mathbf{grad}\, u, \mathbf{grad}\, v) = \lambda(u, v) \quad \forall v \in V. \tag{8.1}$$

Let $V_h \subset V$ be the space of piecewise linear finite elements with vanishing boundary conditions. Then we consider the following approximating problem: find $\lambda_h \in \mathbb{R}$ and $u_h \in V_h$, with $u_h \neq 0$, such that

$$(\mathbf{grad}\, u_h, \mathbf{grad}\, v) = \lambda_h(u_h, v) \quad \forall v \in V_h. \tag{8.2}$$

We use the notation of the previous section for the eigensolutions of our continuous and discrete problems. In particular, we adopt the enumeration convention that eigenvalues are repeated according to their multiplicity. We already know from the min-max principle stated in Proposition 7.2 that all eigenvalues are approximated from above, that is,

$$\lambda^{(k)} \leq \lambda_h^{(k)} \quad \forall k,$$

so that, in order to show the convergence of the eigenvalues, we need the upper bound

$$\lambda_h^{(k)} \leq \lambda^{(k)} + \varepsilon(h)$$

with $\varepsilon(h)$ tending to zero as h tends to zero.

We shall use

$$E_h = \Pi_h V^{(k)}$$

in the min-max characterization of the discrete eigenvalues (7.6), where

$$V^{(k)} = \bigoplus_{i=1}^{k} E^{(i)}$$

and $\Pi_h : V \to V_h$ denotes the elliptic projection

$$(\mathbf{grad}(u - \Pi_h u), \mathbf{grad}\, v_h) = 0 \quad \forall v_h \in V_h.$$

In order to do so, we need to check whether the dimension of E_h is equal to k. This can be false in general (for instance, the entire dimension of V_h could be smaller than k), but it is true if h is small enough, as a consequence of the bound

$$\|\Pi_h v\|_{L^2(\Omega)} \geq \|v\|_{L^2(\Omega)} - \|v - \Pi_h v\|_{L^2(\Omega)} \quad \forall v \in V. \tag{8.3}$$

Indeed, if we take h such that

$$\|v - \Pi_h v\|_{L^2(\Omega)} \leq \frac{1}{2}\|v\|_{L^2(\Omega)} \quad \forall v \in V^{(k)}, \tag{8.4}$$

then (8.3) implies that Π_h is injective from $V^{(k)}$ to E_h. It is clear that (8.4) is satisfied for sufficiently small h (how small depending on k).

Taking E_h in the discrete min-max equation (7.6) gives

$$\begin{aligned}
\lambda_h^{(k)} &\le \max_{w\in E_h} \frac{\|\mathbf{grad}\, w\|^2_{L^2(\Omega)}}{\|w\|^2_{L^2(\Omega)}} = \max_{v\in V^{(k)}} \frac{\|\mathbf{grad}(\Pi_h v)\|^2_{L^2(\Omega)}}{\|\Pi_h v\|^2_{L^2(\Omega)}} \\
&\le \max_{v\in V^{(k)}} \frac{\|\mathbf{grad}\, v\|^2_{L^2(\Omega)}}{\|\Pi_h v\|^2_{L^2(\Omega)}} = \max_{v\in V^{(k)}} \frac{\|\mathbf{grad}\, v\|^2_{L^2(\Omega)}}{\|v\|^2_{L^2(\Omega)}} \frac{\|v\|^2_{L^2(\Omega)}}{\|\Pi_h v\|^2_{L^2(\Omega)}} \\
&\le \lambda^{(k)} \max_{v\in V^{(k)}} \frac{\|v\|^2_{L^2(\Omega)}}{\|\Pi_h v\|^2_{L^2(\Omega)}}.
\end{aligned}$$

In order to estimate the last term, let us suppose that Ω is convex. Then, it is well known that $V^{(k)}$ is contained in $H^2(\Omega)$ and that

$$\|v - \Pi_h v\|_{L^2(\Omega)} \le Ch^2\|\Delta v\|_{L^2(\Omega)} \le C\lambda^{(k)}h^2\|v\|_{L^2(\Omega)} = C(k)h^2\|v\|_{L^2(\Omega)}.$$

Hence, from (8.3), we obtain

$$\|\Pi_h v\|_{L^2(\Omega)} \ge \|v\|_{L^2(\Omega)}(1 - C(k)h^2),$$

which gives the final estimate,

$$\lambda_h^{(k)} \le \lambda^{(k)}\left(\frac{1}{1 - C(k)h^2}\right)^2 \simeq \lambda^{(k)}(1 + C(k)h^2)^2 \simeq \lambda^{(k)}(1 + 2C(k)h^2).$$

In the case of a general domain Ω, it is possible to obtain the following more general result (Raviart and Thomas 1983):

$$\lambda_h^{(k)} \le \lambda^{(k)}\left(1 + C(k) \sup_{\substack{v\in V^{(k)}\\ \|v\|=1}} \|v - \Pi_h v\|^2_{H^1(\Omega)}\right). \tag{8.5}$$

We conclude this section with an estimate for the eigenfunctions. It should be clear from the discussion related to estimate (3.2) that the study of the case of multiple eigenvalues is not so simple. For this reason, we start with the situation when $\lambda^{(i)} \ne \lambda^{(k)}$ for all $i \ne k$ (that is, $\lambda^{(k)}$ is a simple eigenvalue).

We introduce the following quantity (Raviart and Thomas 1983):

$$\rho_h^{(k)} = \max_{i\ne k} \frac{\lambda^{(k)}}{|\lambda^{(k)} - \lambda_h^{(i)}|},$$

which makes sense for sufficiently small h since $\lambda^{(k)}$ is a simple eigenvalue and we already know that $\lambda_h^{(i)}$ tends to $\lambda^{(i)} \ne \lambda^{(k)}$.

We also consider the $L^2(\Omega)$-projection of $\Pi_h u^{(k)}$ onto the space spanned by $u_h^{(k)}$,

$$w_h^{(k)} = (\Pi_h u^{(k)}, u_h^{(k)})u_h^{(k)},$$

in order to estimate the difference $(u^{(k)} - u_h^{(k)})$ as follows:

$$\|u^{(k)} - u_h^{(k)}\|_{L^2(\Omega)} \leq \|u^{(k)} - \Pi_h u^{(k)}\| + \|\Pi_h u^{(k)} - w_h^{(k)}\| + \|w_h^{(k)} - u_h^{(k)}\|. \quad (8.6)$$

The first term in (8.6) can be easily estimated in terms of powers of h using the properties of Π_h; let us start with the analysis of the second term. From the definition of $w_h^{(k)}$, we have

$$\Pi_h u^{(k)} - w_h^{(k)} = \sum_{i \neq k} (\Pi_h u^{(k)}, u_h^{(i)}) u_h^{(i)},$$

which gives

$$\|\Pi_h u^{(k)} - w_h^{(k)}\|^2 = \sum_{i \neq k} (\Pi_h u^{(k)}, u_h^{(i)})^2. \quad (8.7)$$

We have

$$\begin{aligned}(\Pi_h u^{(k)}, u_h^{(i)}) &= \frac{1}{\lambda_h^{(i)}} (\mathbf{grad}(\Pi_h u^{(k)}), \mathbf{grad}\, u_h^{(i)}) \\ &= \frac{1}{\lambda_h^{(i)}} (\mathbf{grad}\, u^{(k)}, \mathbf{grad}\, u_h^{(i)}) = \frac{\lambda^{(k)}}{\lambda_h^{(i)}} (u^{(k)}, u_h^{(i)}),\end{aligned}$$

that is,

$$\lambda_h^{(i)} (\Pi_h u^{(k)}, u_h^{(i)}) = \lambda^{(k)} (u^{(k)}, u_h^{(i)}).$$

Subtracting $\lambda^{(k)}(\Pi_h u^{(k)}, u_h^{(i)})$ from both sides of the equality, we obtain

$$(\lambda_h^{(i)} - \lambda^{(k)})(\Pi_h u^{(k)}, u_h^{(i)}) = \lambda^{(k)} (u^{(k)} - \Pi_h u^{(k)}, u_h^{(i)}),$$

which gives

$$|(\Pi_h u^{(k)}, u_h^{(i)})| \leq \rho_h^{(k)} |(u^{(k)} - \Pi_h u^{(k)}, u_h^{(i)})|.$$

From (8.7) we finally get

$$\begin{aligned}\|\Pi_h u^{(k)} - w_h^{(k)}\|^2 &\leq (\rho_h^{(k)})^2 \sum_{i \neq k} (u^{(k)} - \Pi_h u^{(k)}), u_h^{(i)})^2 \\ &\leq (\rho_h^{(k)})^2 \|u^{(k)} - \Pi_h u^{(k)}\|^2.\end{aligned} \quad (8.8)$$

In order to bound the final term in (8.6), we observe that if we show that

$$\|u_h^{(k)} - w_h^{(k)}\| \leq \|u^{(k)} - w_h^{(k)}\|, \quad (8.9)$$

then we can conclude that

$$\|u_h^{(k)} - w_h^{(k)}\| \leq \|u^{(k)} - \Pi_h u^{(k)}\| + \|\Pi_h u^{(k)} - w_h^{(k)}\|, \quad (8.10)$$

and we have already estimated the last two terms. From the definition of $w_h^{(k)}$, we have

$$u_h^{(k)} - w_h^{(k)} = u_h^{(k)}(1 - ((\Pi_h u^{(k)}, u_h^{(k)})).$$

Moreover,

$$\|u^{(k)}\| - \|u^{(k)} - w_h^{(k)}\| \le \|w_h^{(k)}\| \le \|u^{(k)}\| + \|u^{(k)} - w_h^{(k)}\|,$$

and the normalization of $u^{(k)}$ and $u_h^{(k)}$ gives

$$1 - \|u^{(k)} - w_h^{(k)}\| \le |(\Pi_h u^{(k)}, u_h^{(k)})| \le 1 + \|u^{(k)} - w_h^{(k)}\|,$$

that is,

$$\big||(\Pi_h u^{(k)}, u_h^{(k)})| - 1\big| \le \|u^{(k)} - w_h^{(k)}\|. \tag{8.11}$$

Now comes a crucial point concerning the uniqueness of the normalized eigenfunctions. We have already observed that the normalization of the eigenfunctions does not identify them in a unique way (even in the case of simple eigenvalues), but only up to their sign. Here we have to choose the appropriate sign of $u_h^{(k)}$ in order to have a good approximation of $u^{(k)}$. The correct choice in this case is the one that provides

$$(\Pi_h u^{(k)}, u_h^{(k)}) \ge 0,$$

so that we can conclude that the left-hand side of (8.11) is equal to $\|w_h^{(k)} - u_h^{(k)}\|$ and (8.9) is satisfied.

Putting together all the previous considerations, that is, (8.6), (8.8) and (8.10), we can conclude that, in the case of a simple eigenfunction $u^{(k)}$, there exists an appropriate choice of the sign of $u_h^{(k)}$ such that

$$\|u^{(k)} - u_h^{(k)}\|_{L^2(\Omega)} \le 2(1 + \rho_h^{(k)})\|u^{(k)} - \Pi_h u^{(k)}\|_{L^2(\Omega)}.$$

In particular, in the case of a convex domain this gives the optimal bound

$$\|u^{(k)} - u_h^{(k)}\|_{L^2(\Omega)} \le Ch^2.$$

The error in the energy norm can be estimated in a standard way as follows:

$$\begin{aligned}
C\|u^{(k)} - u_h^{(k)}\|^2_{H^1(\Omega)} &\le \|\mathbf{grad}(u^{(k)} - u_h^{(k)})\|^2_{L^2(\Omega)} \\
&= \|\mathbf{grad}\, u^{(k)}\|^2 - 2(\mathbf{grad}\, u^{(k)}, \mathbf{grad}\, u_h^{(k)}) + \|\mathbf{grad}\, u_h^{(k)}\|^2 \\
&= \lambda^{(k)} - 2\lambda^{(k)}(u^{(k)}, u_h^{(k)}) + \lambda_h^{(k)} \\
&= \lambda^{(k)} - 2\lambda^{(k)}(u^{(k)}, u_h^{(k)}) + \lambda^{(k)} - (\lambda^{(k)} - \lambda_h^{(k)}) \\
&= \lambda^{(k)}\|u^{(k)} - u_h^{(k)}\|^2_{L^2(\Omega)} - (\lambda^{(k)} - \lambda_h^{(k)}).
\end{aligned}$$

The leading term in the last estimate is the second one, which gives the following optimal bound (see (8.5)):

$$\|u^{(k)} - u_h^{(k)}\|_{H^1(\Omega)} \le C(k) \sup_{\substack{v\in V^{(k)}\\ \|v\|=1}} \|v - \Pi_h v\|_{H^1(\Omega)}.$$

In order to conclude the convergence analysis of problem (8.2), it remains to discuss the convergence of eigenfunctions in the case of multiple eigensolutions. As we have already remarked several times, one of the most important issues consists in the appropriate definition of convergence. For the sake of simplicity, we shall discuss the case of a double eigenvalue, but our analysis generalizes easily to any multiplicity. Some of the technical details are identical to the arguments used in the case of an eigenfunction of multiplicity 1, but we repeat them here for the sake of completeness.

Let $\lambda^{(k)}$ be an eigenvalue of multiplicity 2, that is, $\lambda^{(k)} = \lambda^{(k+1)}$ and $\lambda^{(i)} \ne \lambda^{(k)}$ for $i \ne k, k+1$. We would like to find a good approximation for $u^{(k)}$ trying to mimic what has been done in the case of multiplicity 1. Analogous considerations hold for the approximation of $u^{(k+1)}$. It is clear that we cannot expect $u_h^{(k)}$ to converge to $u^{(k)}$, as was observed in the discussion related to (3.2); hence we look for an appropriate linear combination of two discrete eigenfunctions:

$$w_h^{(k)} = \alpha_h u_h^{(k)} + \beta_h u_h^{(k+1)}.$$

From the above study, it seems reasonable to make the following choice:

$$\alpha_h = (\Pi_h u^{(k)}, u_h^{(k)}), \quad \beta_h = (\Pi_h u^{(k)}, u_h^{(k+1)}),$$

so that $w_h^{(k)}$ will be the $L^2(\Omega)$-projection of $\Pi_h u^{(k)}$ onto the space spanned by $u_h^{(k)}$ and $u_h^{(k+1)}$. The analogue of (8.6) then contains two terms,

$$\|u^{(k)} - w_h^{(k)}\|_{L^2(\Omega)} \le \|u^{(k)} - \Pi_h u^{(k)}\| + \|\Pi_h u^{(k)} - w_h^{(k)}\|,$$

and only the last one needs to be estimated. We have

$$\Pi_h u^{(k)} - w_h^{(k)} = \sum_{i\ne k,k+1} (\Pi_h u^{(k)}, u_h^{(i)}) u_h^{(i)},$$

which gives

$$\|\Pi_h u^{(k)} - w_h^{(k)}\|^2 = \sum_{i\ne k,k+1} (\Pi_h u^{(k)}, u_h^{(i)})^2. \tag{8.12}$$

It follows that

$$\begin{aligned}(\Pi_h u^{(k)}, u_h^{(i)}) &= \frac{1}{\lambda_h^{(i)}} (\mathbf{grad}(\Pi_h u^{(k)}), \mathbf{grad}\, u_h^{(i)})\\ &= \frac{1}{\lambda_h^{(i)}} (\mathbf{grad}\, u^{(k)}, \mathbf{grad}\, u_h^{(i)}) = \frac{\lambda^{(k)}}{\lambda_h^{(i)}} (u^{(k)}, u_h^{(i)}),\end{aligned}$$

that is,

$$\lambda_h^{(i)}(\Pi_h u^{(k)}, u_h^{(i)}) = \lambda^{(k)}(u^{(k)}, u_h^{(i)}).$$

Subtracting $\lambda^{(k)}(\Pi_h u^{(k)}, u_h^{(i)})$ from both sides of the equality, we obtain

$$(\lambda_h^{(i)} - \lambda^{(k)})(\Pi_h u^{(k)}, u_h^{(i)}) = \lambda^{(k)}(u^{(k)} - \Pi_h u^{(k)}, u_h^{(i)}),$$

which gives

$$|(\Pi_h u^{(k)}, u_h^{(i)})| \le \rho_h^{(k)} |(u^{(k)} - \Pi_h u^{(k)}, u_h^{(i)})|$$

with the appropriate definition of $\rho_h^{(k)}$,

$$\rho_h^{(k)} = \max_{i \neq k,k+1} \frac{\lambda^{(k)}}{|\lambda^{(k)} - \lambda_h^{(i)}|},$$

which makes sense for sufficiently small h, since we know that $\lambda_h^{(i)}$ tends to $\lambda^{(i)} \neq \lambda^{(k)}$ for $i \neq k, k+1$. From (8.12) we finally get

$$\begin{aligned} \|\Pi_h u^{(k)} - w_h^{(k)}\|^2 &\le (\rho_h^{(k)})^2 \sum_{i \neq k,k+1} (u^{(k)} - \Pi_h u^{(k)}), u_h^{(i)})^2 \\ &\le (\rho_h^{(k)})^2 \|u^{(k)} - \Pi_h u^{(k)}\|^2, \end{aligned}$$

which gives the optimal bound

$$\|u^{(k)} - w_h^{(k)}\|_{L^2(\Omega)} \le (1 + \rho_h^{(k)}) \|u^{(k)} - \Pi_h u^{(k)}\|_{L^2(\Omega)}.$$

The derivation of convergence estimates in the energy norm is less immediate, since we cannot repeat the argument used for the case of eigensolutions of multiplicity 1. The main difference is that the approximating eigenfunction $w_h^{(k)}$ is not normalized. However, the proof can be modified as follows:

$$\begin{aligned} C\|u^{(k)} - w_h^{(k)}\|_{H^1(\Omega)}^2 &\le \|\mathbf{grad}(u^{(k)} - w_h^{(k)})\|_{L^2(\Omega)}^2 \\ &= \|\mathbf{grad}\, u^{(k)}\|^2 - 2(\mathbf{grad}\, u^{(k)}, \mathbf{grad}\, w_h^{(k)}) + \|\mathbf{grad}\, w_h^{(k)}\|^2 \\ &= \lambda^{(k)} - 2\lambda^{(k)}(u^{(k)}, w_h^{(k)}) + \alpha_h^2 \lambda_h^{(k)} + \beta_h^2 \lambda_h^{(k+1)} \\ &= \lambda^{(k)} - 2\lambda^{(k)}(u^{(k)}, w_h^{(k)}) + (\alpha_h^2 + \beta_h^2)\lambda^{(k)} \\ &\qquad - ((\alpha_h^2 + \beta_h^2)\lambda^{(k)} - \alpha_h^2 \lambda_h^{(k)} - \beta_h^2 \lambda_h^{(k+1)}) \\ &= \lambda^{(k)} \|u^{(k)} - w_h^{(k)}\|_{L^2(\Omega)}^2 - \alpha_h^2(\lambda^{(k)} - \lambda_h^{(k)}) - \beta_h^2(\lambda^{(k)} - \lambda_h^{(k+1)}) \end{aligned}$$

and we get the optimal estimate

$$\|u^{(k)} - w_h^{(k)}\|_{H^1(\Omega)} \le C(k) \sup_{\substack{v \in V^{(k+1)} \\ \|v\|=1}} \|v - \Pi_h v\|_{H^1(\Omega)}.$$

The following theorem summarizes the results obtained so far.

Theorem 8.1. Let $(\lambda^{(i)}, u^{(i)})$ be the solutions of problem (8.1) with the notation of Section 7, and let $(\lambda_h^{(i)}, u_h^{(i)})$ be the corresponding discrete solutions of problem (8.2). Let $\Pi_h : V \to V_h$ denote the elliptic projection. Then, for any k not larger than the dimension of V_h, for sufficiently small h, we have

$$\lambda^{(k)} \leq \lambda_h^{(k)} \leq \lambda^{(k)} + C(k) \sup_{\substack{v \in V^{(k)} \\ \|v\|=1}} \|v - \Pi_h v\|^2_{H^1(\Omega)},$$

with $V^{(k)} = \oplus_{i \leq k} E^{(k)}$.

Moreover, let $\lambda^{(k)}$ be an eigenvalue of multiplicity $m \geq 1$, so that

$$\lambda^{(k)} = \cdots = \lambda^{(k+m-1)} \quad \text{and} \quad \lambda^{(i)} \neq \lambda^{(k)}$$

for $i \neq k, \ldots, k+m-1$. Then there exists

$$\{w_h^{(k)}\} \subset E_h^{(k)} \oplus \cdots \oplus E_h^{(k+m-1)}$$

such that

$$\|u^{(k)} - w_h^{(k)}\|_{H^1(\Omega)} \leq C(k) \sup_{\substack{v \in V^{(k+m-1)} \\ \|v\|=1}} \|v - \Pi_h v\|_{H^1(\Omega)}$$

and

$$\|u^{(k)} - w_h^{(k)}\|_{L^2(\Omega)} \leq C(k) \|u^{(k)} - \Pi_h u^{(k)}\|_{L^2(\Omega)}.$$

The results presented so far are optimal when all the eigenfunctions are smooth. For instance, if the domain is convex, it is well known that $\|v - \Pi_h v\|_{H^1(\Omega)} = O(h)$ for all eigenfunctions v, so that Theorem 8.1 gives the optimal second order of convergence for the eigenvalues, first order in $H^1(\Omega)$ for the eigenfunctions, and second order in $L^2(\Omega)$ for the eigenfunctions. On the other hand, if the domain is not regular (see, for instance, the computations presented in Section 3.2) it usually turns out that some eigenspaces contain smooth eigenfunctions, while others may contain singular eigenfunctions. In such cases, we obtain from Theorem 8.1 a sub-optimal estimate, since some bounds are given in terms of the approximability of $V^{(k)}$. Hence, if we are interested in the kth eigenvalue, we have to consider the regularity properties of all the eigenspaces up to the kth one. This sub-optimal behaviour is not observed in practice (see, for instance, Table 3.4); the theoretical investigations presented in the subsequent sections will confirm that the rate of convergence of the kth eigenvalue/eigenfunction is indeed related to the approximability of the eigenfunctions associated with the kth eigenvalue, alone. For sharper results concerning multiple eigenvalues, the reader is referred to Knyazev and Osborn (2006).

9. The Babuška–Osborn theory

It is generally understood that the basic reference for the finite element approximation of compact eigenvalue problems is the so-called Babuška–Osborn theory (Babuška and Osborn 1991). In this section, we recall the main results of the theory, and refer the reader to the original reference for more details.

While the main focus of this survey is on symmetric eigenvalue problems, it is more convenient to embed the discussion of the present section in the complex field $\mathbb{C}$ and to study a generic non-symmetric problem. Those interested in wider generality may observe that the original theory can be developed in Banach spaces; we shall, however, limit ourselves to the interesting case of Hilbert spaces.

We follow the notation introduced in Section 6.

Let X be a Hilbert space and let $T : X \to X$ be a compact linear operator. We consider a family of compact operators $T_h : X \to X$ such that

$$\|T - T_h\|_{\mathcal{L}(X)} \to 0 \quad \text{as } h \to 0. \tag{9.1}$$

In our applications, T_h will be a finite rank operator. We have already seen examples of this situation in Section 7.

As a consequence of (9.1), if $\lambda \in \sigma(T)$ is a non-zero eigenvalue with algebraic multiplicity m, then exactly m discrete eigenvalues of T_h (counted with their algebraic multiplicities), converge to λ as h tends to zero. This follows from the well-known fact that, given an arbitrary closed curve $\Gamma \subset \rho(T)$ as in the definition (6.1) of the projection $E(\lambda)$, for sufficiently small h we have $\Gamma \subset \rho(T_h)$, and Γ encloses exactly m eigenvalues of T_h, counted with their algebraic multiplicities. More precisely, for sufficiently small h it makes sense to consider the discrete spectral projection

$$E_h(\lambda) = \frac{1}{2\pi \mathrm{i}} \int_\Gamma (z - T_h)^{-1} \, \mathrm{d}z,$$

and it turns out that the dimension of $E_h(\lambda)X$ is equal to m. Moreover,

$$\|E(\lambda) - E_h(\lambda)\|_{\mathcal{L}(X)} \to 0 \quad \text{as } h \to 0,$$

which implies the convergence of the generalized eigenvectors.

It is common practice, when studying the approximation of eigenmodes, to split the convergence analysis into two parts: the first step consists in showing that the eigenmodes converge and that there are no spurious solutions, the second one deals with the order of convergence.

The above considerations give an answer to the first question and we summarize these results in the following statement.

Theorem 9.1. Let us assume that the convergence in norm (9.1) is satisfied. For any compact set $K \subset \rho(T)$, there exists $h_0 > 0$ such that, for all $h < h_0$, we have $K \subset \rho(T_h)$ (absence of spurious modes). If λ is a nonzero eigenvalue of T with algebraic multiplicity equal to m, then there are m eigenvalues $\lambda_{1,h}, \lambda_{2,h}, \dots, \lambda_{m,h}$ of T_h, repeated according to their algebraic multiplicities, such that each $\lambda_{i,h}$ converges to λ as h tends to 0.

Moreover, the gap between the direct sum of the generalized eigenspaces associated with $\lambda_{1,h}, \lambda_{2,h}, \dots, \lambda_{m,h}$ and the generalized eigenspace associated to λ tends to zero as h tends to 0.

We now report the main results of the Babuška–Osborn theory (Babuška and Osborn 1991, Theorems 7.1–7.4) which deal with the convergence order of eigenvalues and eigenvectors. One of the main applications of the theory consists in the convergence analysis for variationally posed eigenvalue problems (Babuška and Osborn 1991, Theorems 8.1–8.4); this is the correct setting for the general analysis of the problems discussed in Sections 7 and 8. We start with the generalization of the framework of Section 7 to non-symmetric variationally posed eigenvalue problems.

Let V_1 and V_2 be complex Hilbert spaces. We are interested in the following eigenvalue problem: find $\lambda \in \mathbb{C}$ and $u \in V_1$, with $u \neq 0$, such that

$$a(u, v) = \lambda b(u, v) \quad \forall v \in V_2, \tag{9.2}$$

where $a : V_1 \times V_2 \to \mathbb{C}$ and $b : V_1 \times V_2 \to \mathbb{C}$ are sesquilinear forms. The form a is assumed to be continuous,

$$|a(v_1, v_2)| \leq C\|v_1\|_{V_1}\|v_2\|_{V_2} \quad \forall v_1 \in V_1\ \forall v_2 \in V_2,$$

and the form b is continuous with respect to a *compact norm*: there exists a norm $\|\cdot\|_{H_1}$ in V_1 such that any bounded sequence in V_1 has a Cauchy subsequence with respect to $\|\cdot\|_{H_1}$ and

$$|b(v_1, v_2)| \leq C\|v_1\|_{H_1}\|v_2\|_{V_2} \quad \forall v_1 \in V_1\ \forall v_2 \in V_2.$$

The Laplace eigenvalue problem considered in Sections 8 and 10 fits within this setting with the choices $V_1 = V_2 = H^1_0(\Omega)$ and $H_1 = L^2(\Omega)$.

In order to define the solution operators, we assume the inf-sup condition

$$\inf_{v_1 \in V_1} \sup_{v_2 \in V_2} \frac{|a(v_1, v_2)|}{\|v_1\|_{V_1}\|v_2\|_{V_2}} \geq \gamma > 0,$$
$$\sup_{v_1 \in V_1} |a(v_1, v_2)| > 0 \quad \forall v_2 \in V_2 \setminus \{0\},$$

so that we can introduce $T : V_1 \to V_1$ and $T_* : V_2 \to V_2$ by

$$a(Tf, v) = b(f, v) \quad \forall f \in V_1\ \forall v \in V_2,$$
$$a(v, T_*g) = b(v, g) \quad \forall g \in V_2\ \forall v \in V_1.$$

From our assumptions it follows that T and T_* are compact operators (Babuška and Osborn 1991); moreover, the adjoint of T on V_1 is given by $T^* = A^* \circ T_* \circ A^{*-1}$, where $A : V_1 \to V_2$ is the standard linear operator associated to the bilinear form a.

Remark 9.2. In some applications (for instance those involving spaces like $\mathbf{H}_0(\mathrm{div};\Omega)$ or $\mathbf{H}_0(\mathbf{curl};\Omega)$) it might be difficult to satisfy the compactness assumption on the bilinear form b. The theory can, however, be applied without modifications, by directly assuming the compactness of T and T_*.

A pair (λ, u) is an eigenmode of problem (9.2) if and only if it satisfies $\lambda T u = u$, that is, (μ, u) is an eigenpair of the operator T with $\mu = \lambda^{-1}$. The concepts of ascent multiplicity, algebraic multiplicity and generalized eigenfunctions of problem (9.2) are then defined in terms of the analogous properties for the operator T.

We shall also make use of the following adjoint eigenvalue problem: find $\lambda \in \mathbb{C}$ and $u \in V_2$, with $u \neq 0$, such that

$$a(v, u) = \lambda b(v, u) \quad \forall v \in V_1. \tag{9.3}$$

The discretization of problem (9.2) consists in selecting finite-dimensional subspaces $V_{1,h}$ and $V_{2,h}$, and in considering the following problem: find $\lambda_h \in \mathbb{C}$ and $v_{1,h} \in V_{1,h}$ with $v_{1,h} \neq 0$ such that

$$a(v_{1,h}, v_2) = \lambda_h b(v_{1,h}, v_2) \quad \forall v_2 \in V_{2,h}. \tag{9.4}$$

We suppose that

$$\dim(V_{1,h}) = \dim(V_{2,h}),$$

so that (9.4) is actually a generalized (square) eigenvalue problem.

We assume that the discrete uniform inf-sup conditions are satisfied,

$$\inf_{v_1 \in V_{1,h}} \sup_{v_2 \in V_{2,h}} \frac{|a(v_1, v_2)|}{\|v_1\|_{V_1} \|v_2\|_{V_2}} \geq \gamma > 0,$$

$$\sup_{v_1 \in V_{1,h}} |a(v_1, v_2)| > 0 \quad \forall v_2 \in V_{2,h} \setminus \{0\},$$

so that the discrete solution operators T_h and $T_{*,h}$ can be defined in analogy to T and T_*. It is clear that the convergence of the eigensolutions of (9.4) towards those of (9.2) can be analysed by means of the convergence of T_h and $T_{*,h}$ to T and T_*.

We are now ready to report the four main results of the theory. For each result, we state a theorem concerning the approximation of the eigenpairs of T followed by a corollary containing the consequences for the approximation of the eigensolutions to (9.2).

We consider an eigenvalue λ of (9.2) ($\mu = \lambda^{-1}$ in the case of the operator T) of algebraic multiplicity m and with ascent of $\mu - T$ equal to α. For the sake of generality, we let X denote the domain V_1 of the operator T, so we shall revert to the notation of Section 6.

The first theorem concerns the approximation of eigenvectors.

Theorem 9.3. Let μ be a non-zero eigenvalue of T, let $E = E(\mu)X$ be its generalized eigenspace, and let $E_h = E_h(\mu)X$. Then

$$\hat{\delta}(E, E_h) \leq C\|(T - T_h)_{|E}\|_{\mathcal{L}(X)}.$$

Corollary 9.4. Let λ be an eigenvalue of (9.2), let $E = E(\lambda^{-1})V_1$ be its generalized eigenspace and let $E_h = E_h(\lambda^{-1})V_1$. Then

$$\hat{\delta}(E, E_h) \leq C \sup_{\substack{u\in E\\ \|u\|=1}} \inf_{v\in V_{1,h}} \|u - v\|_{V_1}.$$

In the case of multiple eigenvalues it has been observed that it is convenient to introduce the arithmetic mean of the approximating eigenvalues.

Theorem 9.5. Let μ be a non-zero eigenvalue of T with algebraic multiplicity equal to m and let $\widehat{\mu}_h$ denote the arithmetic mean of the m discrete eigenvalues of T_h converging towards μ. Let $\phi_1, \dots, \phi_m$ be a basis of generalized eigenvectors in $E = E(\mu)X$ and let $\phi_1^*, \dots, \phi_m^*$ be a dual basis of generalized eigenvectors in $E^* = E^*(\overline{\mu})X$. Then

$$|\mu - \widehat{\mu}_h| \leq \frac{1}{m}\sum_{i=1}^{m} |((T - T_h)\phi_i, \phi_i^*)| + C\|(T - T_h)_{|E}\|_{\mathcal{L}(X)}\|(T^* - T_h^*)_{|E^*}\|_{\mathcal{L}(X)}.$$

Corollary 9.6. Let λ be an eigenvalue of (9.2) and let $\widehat{\lambda}_h$ denote the arithmetic mean of the m discrete eigenvalues of (9.4) converging towards λ. Then

$$|\lambda - \widehat{\lambda}_h| \leq C \sup_{\substack{u\in E\\ \|u\|=1}} \inf_{v\in V_{1,h}} \|u - v\|_{V_1} \sup_{\substack{u\in E^*\\ \|u\|=1}} \inf_{v\in V_{2,h}} \|u - v\|_{V_2},$$

where E is the space of generalized eigenfunctions associated with λ and E^* is the space of generalized adjoint eigenfunctions associated with λ (see the adjoint problem (9.3)).

The estimate of the error in the eigenvalues involves the ascent multiplicity α.

Theorem 9.7. Let $\phi_1, \dots, \phi_m$ be a basis of the generalized eigenspace $E = E(\mu)X$ of T and let $\phi_1^*, \dots, \phi_m^*$ be a dual basis. Then, for $i = 1, \dots m$,

$$|\mu - \mu_{i,h}|^\alpha \leq C\left\{\sum_{j,k=1}^{m} |((T - T_h)\phi_j, \phi_k^*)| + \|(T - T_h)_{|E}\|_{\mathcal{L}(X)}\|(T^* - T_h^*)_{|E^*}\|_{\mathcal{L}(X)}\right\},$$

where $\mu_{1,h}, \dots, \mu_{m,h}$ are the m discrete eigenvalues (repeated according to their algebraic multiplicity) converging to μ, and E^* is the space of generalized eigenvectors of T^* associated with $\overline{\mu}$.

Corollary 9.8. With notation analogous to that of the previous theorem, for $i = 1, \dots, m$ we have

$$|\lambda - \lambda_{i,h}|^\alpha \leq C \sup_{\substack{u \in E \\ \|u\|=1}} \inf_{v \in V_{1,h}} \|u - v\|_{V_1} \sup_{\substack{u \in E^* \\ \|u\|=1}} \inf_{v \in V_{2,h}} \|u - v\|_{V_2}, \tag{9.5}$$

where E is the space of generalized eigenfunctions associated with λ and E^* is the space of generalized adjoint eigenfunctions associated with λ (see the adjoint problem (9.3)).

Remark 9.9. Apparently, the estimates of Corollaries 9.6 and 9.8 are not immediate consequences of Theorems 9.5 and 9.7. In Section 10 we give a proof of these results in the particular case of the Laplace eigenvalue problem. The interested reader is referred to Babuška and Osborn (1991) for the general case.

The last result is more technical than the previous ones and complements Theorem 9.3 on the description of the approximation of the generalized eigenvectors. In particular, for $k = \ell = 1$, the theorem applies to the eigenvectors.

Theorem 9.10. Let $\{\mu_h\}$ be a sequence of discrete eigenvalues of T_h converging to a non-zero eigenvalue μ of T. Consider a sequence $\{u_h\}$ of unit vectors in $\ker(\mu_h - T_h)^k$ for some $k \leq \alpha$ (discrete generalized eigenvectors of order k). Then, for any integer ℓ with $k \leq \ell \leq \alpha$, there exists a generalized eigenvector $u(h)$ of T of order ℓ such that

$$\|u(h) - u_h\|_X^{\alpha/(\ell-k+1)} \leq C\|(T - T_h)_{|E}\|_{\mathcal{L}(X)}.$$

Corollary 9.11. Let $\{\lambda_h\}$ be a sequence of discrete eigenvalues of (9.4) converging to an eigenvalue λ of (9.2). Consider a sequence $\{u_h\}$ of unit eigenfunctions in $\ker(\lambda_h^{-1} - T_h)^k$ for some $k \leq \alpha$ (discrete generalized eigenfunctions of order k). Then, for any integer ℓ with $k \leq \ell \leq \alpha$, there exists a generalized eigenvector $u(h)$ of (9.2) of order ℓ such that

$$\|u(h) - u_h\|_{V_1}^{\alpha/(\ell-k+1)} \leq C \sup_{\substack{u \in E \\ \|u\|=1}} \inf_{v \in V_{1,h}} \|u - v\|_{V_1}.$$

We conclude this section with the application of the present theory to the case of symmetric variationally posed eigenvalue problems. In particular, the presented results will give a more comprehensive and precise treatment of the problems discussed in Sections 7 and 8.

We suppose that $V_1 = V_2$ are identical real Hilbert spaces, denoted by V, and that a and b are real and symmetric. We assume that a is V-elliptic,

$$a(v,v) \geq \gamma > 0 \quad \forall v \in V,$$

and that b can be extended to a continuous bilinear form on $H \times H$ with a Hilbert space H such that $V \subset H$ is a compact inclusion. We assume, moreover, that b is positive on $V \times V$:

$$b(v,v) > 0 \quad \forall v \in V \setminus \{0\}.$$

In this case, as has already been observed in Section 7, the eigenvalues of (9.2) are positive and can be ordered in a sequence tending to infinity,

$$0 < \lambda^{(1)} \leq \lambda^{(2)} \leq \cdots \leq \lambda^{(k)} \leq \cdots,$$

where we repeat the eigenvalues according to their multiplicities (we recall that geometric and algebraic multiplicities are now the same and that the ascent multiplicity of $1/\lambda^{(k)} - T$ is 1 for all k).

Let $V_h \subset V$ be the finite-dimensional space used for the eigenmodes approximation and denote by $\lambda_h^{(k)}$ $(k = 1, \ldots, \dim(V_h))$ the discrete eigenvalues. The min-max principle (see Section 7, in particular Proposition 7.2 and the discussion thereafter) and Corollary 9.8 give the following result.

Theorem 9.12. For each k, we have

$$\lambda^{(k)} \leq \lambda_h^{(k)} \leq \lambda^{(k)} + C \sup_{\substack{u \in E \\ \|u\|=1}} \inf_{v \in V_h} \|u - v\|_V^2,$$

where E denotes the eigenspace associated with $\lambda^{(k)}$.

Corollary 9.4 reads as follows in the symmetric case.

Theorem 9.13. Let $u^{(k)}$ be a unit eigenfunction associated with an eigenvalue $\lambda^{(k)}$ of multiplicity m, such that $\lambda^{(k)} = \cdots = \lambda^{(k+m-1)}$, and denote by $u_h^{(k)}, \ldots, u_h^{(k+m-1)}$ the eigenfunctions associated with the m discrete eigenvalues converging to $\lambda^{(k)}$. Then, there exists

$$w_h^{(k)} \in \operatorname{span}\{u_h^{(k)}, \ldots, u_h^{(k+m-1)}\}$$

such that

$$\|u^{(k)} - w_h^{(k)}\|_V \leq C \sup_{\substack{u \in E \\ \|u\|=1}} \inf_{v \in V_h} \|u - v\|_V,$$

where E denotes the eigenspace associated with $\lambda^{(k)}$.

The results of the present section contain the basic estimates for eigenvalues and eigenfunctions of compact variationally posed eigenvalue problems. Several other refined results are available.

For instance, it is possible to obtain sharper estimates in the case of multiple eigenvalues (see Knyazev and Osborn (2006) and the references therein): in particular, these can be useful when a multiple eigenvalue is associated with eigenfunctions with different regularities. Estimate (9.5) would predict that in such a case the eigenvalue is approximated with the order of convergence dictated by the lowest regularity of the eigenfunctions; on the other hand it is possible that the approximating eigenvalues have different speeds according to the different regularities of the associated eigenfunctions.

10. The Laplace eigenvalue problem

In this section we apply the Babuška–Osborn theory to the convergence analysis of conforming finite element approximation to Laplace eigenvalue problem.

The Laplace eigenvalue problem has already been analysed in several parts of this paper, but we recall here the related variational formulations for completeness. Given $\Omega \subset \mathbb{R}^n$ and the real Sobolev space $H_0^1(\Omega)$, we look for eigenvalues $\lambda \in \mathbb{R}$ and eigenfunctions $u \in H_0^1(\Omega)$, with $u \neq 0$, such that

$$(\mathbf{grad}\, u, \mathbf{grad}\, v) = \lambda(u, v) \quad \forall v \in H_0^1(\Omega).$$

The Riesz–Galerkin discretization makes use of a finite-dimensional space $V_h \subset H_0^1(\Omega)$, and consists in looking for eigenvalues $\lambda_h \in \mathbb{R}$ and eigenfunctions $u_h \in V_h$, with $u_h \neq 0$, such that

$$(\mathbf{grad}\, u_h, \mathbf{grad}\, v) = \lambda_h(u_h, v) \quad \forall v \in V_h.$$

We denote by $a(\cdot,\cdot)$ the bilinear form $(\mathbf{grad}\,\cdot, \mathbf{grad}\,\cdot)$ and remark that the considerations of this section can be easily generalized to any bilinear form a which is equivalent to the scalar product of $H_0^1(\Omega)$. Moreover, other types of homogeneous boundary conditions might be considered as well.

Using the notation of the previous section, the starting point for the analysis consists in a suitable definition of the solution operator $T : X \to X$ and, in particular, of the functional space X. Let V and H denote the spaces $H_0^1(\Omega)$ and $L^2(\Omega)$, respectively. The first, natural, definition consists in taking $X = V$ and in defining $T : V \to V$ by

$$a(Tf, v) = (f, v) \quad \forall v \in V. \tag{10.1}$$

Of course, the above definition can be easily extended to $X = H$, since it makes perfect sense to consider the solution to the source Laplace problem with f in $L^2(\Omega)$. We then have at least two admissible definitions: $T_V : V \to V$ and $T_H : H \to H$. Clearly, T_H can be defined analogously as for T_V by $T_H f \in V \subset H$ and (10.1): the only difference between T_V and T_H is the underlying spaces.

It is clear that T_V and T_H are self-adjoint. Since we are dealing with a basic example, we stress the details of this result.

Lemma 10.1. If V is endowed with the norm induced by the scalar product given by the bilinear form a, then the operator T_V is self-adjoint.

Proof. The result follows from the identities

$$a(T_V x, y) = (x, y) = (y, x) = a(T_V y, x) = a(x, T_V y) \quad \forall x, y \in V. \qquad \square$$

Lemma 10.2. The operator T_H is self-adjoint.

Proof. The result follows from the identities

$$(T_H x, y) = (y, T_H x) = a(T_H y, T_H x) = a(T_H x, T_H y) = (x, T_H y) \quad \forall x, y \in H.$$

$\square$

Moreover, it is clear that the eigenvalues/eigenfunctions of the operators T_V and T_H coincide, so that either operator can be used for the analysis.

The discussion of Section 9 shows that two main steps are involved. First of all we have to define a suitable discrete solution operator T_h satisfying a convergence in norm results like (9.1). As a second step, only after we know that the eigenvalues/eigenfunctions are well approximated can we estimate the order of convergence.

For the convergence analysis performed in the next sections, we assume that V_h is such that the following best approximation holds:

$$\inf_{v \in V_h} \|u - v\|_{L^2(\Omega)} \leq C h^{\min\{k+1, r\}} \|u\|_{H^r(\Omega)},$$

$$\inf_{v \in V_h} \|u - v\|_{H^1(\Omega)} \leq C h^{\min\{k, r-1\}} \|u\|_{H^r(\Omega)}.$$

Such estimates are standard when V_h contains piecewise polynomials of degree k.

10.1. Analysis for the choice $T = T_V$

We now show how to use the results of Section 9 with the choice $T = T_V$. The discrete solution operator can be defined in a coherent way as $T_h : V \to V$ by $T_h f \in V_h \subset V$, and

$$a(T_h f, v) = (f, v) \quad \forall v \in V_h.$$

The standard error estimate for the solution of the *source* Laplace equation implies that the norm convergence (9.1) is satisfied for all reasonable domains. If Ω is Lipschitz-continuous, for instance, the following estimate is well known: there exists $\varepsilon > 0$ such that

$$\|Tf - T_h f\|_{H^1_0(\Omega)} \leq C h^{\varepsilon} \|f\|_{H^1_0(\Omega)}.$$

We can then conclude that the consequences of Theorem 9.1 are valid: all continuous eigenvalues/eigenfunctions are correctly approximated and all discrete eigenvalues/eigenfunctions approximate some continuous eigenvalues/eigenfunctions with the correct multiplicity.

We now come to the task of estimating the rate of convergence. Let us suppose that we are interested in the convergence rate for the approximation of the eigenvalue λ, and that the regularity of the eigenspace E associated with λ is r, that is, $E \subset H^r(\Omega)$, which implies

$$\|(T - T_h)_{|E}\|_{\mathcal{L}(V)} = O(h^{\min\{k,r-1\}}). \tag{10.2}$$

Let us denote by τ the quantity $\min\{k, r-1\}$.

The estimate for the eigenfunctions is a more-or-less immediate consequence of Corollary 9.4: we easily deduce the result of Theorem 9.13, which can be summarized in this case by the next theorem.

Theorem 10.3. Let u be a unit eigenfunction associated with the eigenvalue λ of multiplicity m, and let $w_h^{(1)}, \dots, w_h^{(m)}$ denote eigenfunctions associated with the m discrete eigenvalues converging to λ. Then there exists

$$u_h \in \operatorname{span}\{w_h^{(1)}, \dots, w_h^{(m)}\}$$

such that

$$\|u - u_h\|_V \le C h^\tau \|u\|_{H^{1+\tau}(\Omega)}.$$

We now see how the result of Theorem 9.12 can be obtained from Theorem 9.7. The proposed arguments will also provide a proof of Corollary 9.8 in this particular case.

Theorem 10.4. Let λ_h be an eigenvalue converging to λ. Then the following optimal double order of convergence holds:

$$\lambda \le \lambda_h \le \lambda + C h^{2\tau}.$$

Proof. From (10.2) and the conclusion of Theorem 9.7 it is clear that, since T is self-adjoint, we only need to bound the term

$$\sum_{j,k=1}^{m} |((T - T_h)\phi_j, \phi_k)_V|,$$

where $\{\phi_1, \dots, \phi_m\}$ is a basis for the eigenspace E.

We have

$$\begin{aligned}
|((T - T_h)u, v)_V| &\le C|a((T - T_h)u, v)| = C \inf_{v_h \in V_h} |a((T - T_h)u, v - v_h)| \\
&\le \|(T - T_h)u\|_V \inf_{v_h \in V_h} \|v - v_h\|_V \\
&\le C h^\tau \|u\|_V h^\tau \|v\|_{H^{1+\tau}(\Omega)} \le C h^{2\tau} \|u\|_V \|v\|_V,
\end{aligned}$$

which is valid for any $u, v \in E$ since $v = \lambda T v$ implies

$$\|v\|_{H^{1+\tau}(\Omega)} \le C \|v\|_V. \qquad \square$$

10.2. Analysis for the choice $T = T_H$

If we choose to perform our analysis in the space $H = L^2(\Omega)$, then we have to define the discrete operator $T_h : H \to H$, which can be done by taking $T_h f \in V_h \subset V \subset H$ as

$$a(T_h f, v) = (f, v) \quad \forall v \in V_h.$$

As in the previous case, it is immediate to obtain that the convergence in norm (9.1) is satisfied for any reasonable domain. Namely, for Ω Lipschitz-continuous, we have that there exists $\varepsilon > 0$ with

$$\|Tf - T_h f\|_{L^2(\Omega)} \le Ch^{1+\varepsilon}\|f\|_{L^2(\Omega)}.$$

Using the same definition of τ as in the previous case, we can easily deduce from Theorem 9.13 the optimal convergence estimate for the eigenfunctions.

Theorem 10.5. Let u be a unit eigenfunction associated with the eigenvalue λ of multiplicity m and let $w_h^{(1)}, \dots, w_h^{(m)}$ denote eigenfunctions associated with the m discrete eigenvalues converging to λ. Then there exists $u_h \in \operatorname{span}\{w_h^{(1)}, \dots, w_h^{(m)}\}$ such that

$$\|u - u_h\|_H \le Ch^{1+\tau}\|u\|_{H^{1+\tau}(\Omega)}.$$

We conclude this section by showing that the estimates already proved for the eigenvalues (optimal double order of convergence) and the eigenfunctions (optimal order of convergence in $H^1(\Omega)$) can also be obtained in this setting. In order to get a rate of convergence for the eigenvalues, we need, as in the proof of Theorem 10.4, an estimate for the term

$$\sum_{j,k=1}^{m} |((T - T_h)\phi_j, \phi_k)_H|,$$

where $\{\phi_1, \dots, \phi_m\}$ is a basis for the eigenspace E.

The conclusion is a consequence of the following estimate:

$$\begin{aligned}
|((T - T_h)u, v)| &= |(v, (T - T_h)u)| = |a(Tv, (T - T_h)u)| \\
&= |a((T - T_h)u, Tv)| = |a((T - T_h)u, Tv - T_h v)| \\
&\le \|(T - T_h)u\|_V \|(T - T_h)v\|_V \\
&\le Ch^{2\tau},
\end{aligned}$$

which is valid for any $u, v \in E$ with $\|u\|_H = \|v\|_H = 1$.

Finally, using the same notation as in Theorem 10.5, the estimate in V for the eigenfunctions associated with *simple* eigenvalues follows from the identity

$$a(u - u_h, u - u_h) = \lambda(u - u_h, u - u_h) - (\lambda - \lambda_h)(u_h, u_h),$$

which can be obtained directly from the definitions of u, u_h, λ, and λ_h. The case of eigenfunctions with higher multiplicity is less immediate, but can be handled with similar tools, using for λ_h a suitable linear combination of the discrete eigenvalues converging to λ.

11. Non-conforming approximation of eigenvalue problems

The aim of this section is to see how the theory developed so far changes when it is applied to non-conforming approximations, in particular when V_h is not contained in V. Our interest in this topic lies in the fact that mixed discretizations of partial differential equations can be seen as particular situations of non-conforming approximations. We shall devote Part 3 of this paper to the analysis of mixed finite elements for eigenvalue problems.

The question of the non-conforming approximation of compact eigenvalue problems has been raised often in the literature, and several possible answers are available. Without attempting to be complete, we refer the interested reader to Rannacher (1979), Stummel (1980), Werner (1981), Armentano and Durán (2004) and Alonso and Dello Russo (2009). Non-conforming approximations can also be analysed in the nice setting introduced in Descloux, Nassif and Rappaz (1978*a*, 1978*b*) for the approximation of non-compact operators.

We start with a basic example: triangular Crouzeix–Raviart elements for the Laplace eigenvalue. This example has already been discussed from the numerical point of view in Table 3.5. For this example, probably the most complete reference can be found in Durán, Gastaldi and Padra (1999), where this element is studied for the solution of an auxiliary problem. The continuous problem is the same as in the previous section: find $\lambda \in \mathbb{R}$ and $u \in V$ with $u \neq 0$ such that

$$a(u, v) = \lambda(u, v) \quad \forall v \in V,$$

with $V = H_0^1(\Omega)$ and $a(u, v) = (\mathbf{grad}\, u, \mathbf{grad}\, v)$. More general situations might be considered with the same arguments; in particular, the analysis is not greatly affected by the presence of a generic elliptic bilinear form a or the case of a right-hand side of the equation where the the scalar product in $L^2(\Omega)$ is replaced by a bilinear form b which is equivalent to the scalar product (this apparently small change may, however, introduce an additional source of non-conformity). Other non-conforming finite elements might be considered as well, as long as suitable estimates for the consistency terms we are going to introduce are available.

Let V_h be the space of lowest-order Crouzeix–Raviart elements: given a triangular mesh $\mathcal{T}_h$ of the domain Ω with mesh size h, the space V_h consists of piecewise linear elements which are continuous at the midpoint of the inter-elements. The discrete problem is as follows: find $\lambda_h \in \mathbb{R}$ and $u_h \in V_h$

with $u_h \neq 0$ such that

$$a_h(u_h, v) = \lambda_h(u_h, v) \quad \forall v \in V_h,$$

where the *discrete* bilinear form a_h is defined as

$$a_h(u, v) = \sum_{K \in \mathcal{T}_h} \int_K \mathbf{grad}\, u \cdot \mathbf{grad}\, v \, \mathrm{d}\mathbf{x} \quad \forall u, v \in V + V_h.$$

It is clear that $a_h(u, v) = a(u, v)$ if u and v belong to V. It is natural to introduce a discrete energy norm on the space $V + V_h$:

$$\|v\|_h^2 = \|v\|_{L^2(\Omega)}^2 + a_h(v, v) \quad \forall v \in V + V_h.$$

In the previous section we saw that the Babuška–Osborn theory can be applied to the analysis of the approximation of the Laplace eigenvalue problem with two different choices of the solution operator T. The first (and more standard) approach consists in choosing $T : V \to V$, while the second one makes use of $H = L^2(\Omega)$ and defines $T : H \to H$. It is clear that for the non-conforming approximation the first approach cannot produce any useful result, since it is impossible to construct an operator valued in V which represents the solution to our discrete problem which is defined in $V_h \not\subset V$. We shall then use the latter approach and define $T : H \to H$ as $Tf \in V$ given by

$$a(Tf, v) = (f, v) \quad \forall f \in H \text{ and } \forall v \in V.$$

The corresponding choice for the discrete operator $T_h : H \to H$ is $T_h f \in V_h \subset H$ given by

$$a_h(T_h f, v) = (f, v) \quad \forall f \in H \text{ and } \forall v \in V_h.$$

It is well known that T_h is well-defined, and that the following optimal estimate holds:

$$\|(T - T_h)f\|_h \leq Ch\|f\|_{L^2(\Omega)}. \tag{11.1}$$

The optimal estimate of $(T - T_h)f$ in $L^2(\Omega)$ requires more regularity than simply $f \in L^2(\Omega)$ (Durán *et al.* 1999, Lemmas 1 and 2). In general we have

$$\|(T - T_h)f\|_{L^2(\Omega)} \leq Ch^2\|f\|_{H^1(\Omega)},$$

which gives the optimal estimate if f is an eigenfunction, since in that case $\|f\|_{H^1(\Omega)}$ can be bounded by $\|f\|_{L^2(\Omega)}$:

$$\|(T - T_h)f\|_{L^2(\Omega)} \leq Ch^2\|f\|_{L^2(\Omega)} \quad \forall f \text{ eigenfunction of } T. \tag{11.2}$$

Remark 11.1. For the sake of simplicity, in the last estimates and in the following analysis we are assuming that the domain is convex, so we do not have to worry about the regularity of the solutions. Our arguments, however, cover the general case of lower regularity as well.

Estimate (11.1), in particular, shows that the convergence in norm (9.1) is satisfied with $X = H$, so we can conclude that the discrete eigenmodes converge to the continuous ones in the spirit of Theorem 9.1.

Let us now study the rate of convergence of the eigenvalues and eigenfunctions. Following what we have done in the previous section, we start with the analysis involving the eigenfunctions.

Theorem 11.2. Let u be a unit eigenfunction associated with the eigenvalue λ of multiplicity m and let $w_h^{(1)}, \dots, w_h^{(m)}$ denote linearly independent eigenfunctions associated with the m discrete eigenvalues converging to λ. Then there exists $u_h \in \text{span}\{w_h^{(1)}, \dots, w_h^{(m)}\}$ such that

$$\|u - u_h\|_{L^2(\Omega)} \le Ch^2 \|u\|_{L^2(\Omega)}.$$

Proof. The proof is an immediate consequence of Theorem 9.3 and estimate (11.2). □

Theorem 11.3. Let λ_h be an eigenvalue converging to λ. Then the following optimal double order of convergence holds:

$$|\lambda - \lambda_h| \le Ch^2.$$

Proof. We are going to use Theorem 9.7 with $X = H$. It is clear that in our case T and T_h are self-adjoint, so we have $T^* = T$, $T_h^* = T_h$, and $\alpha = 1$. The second term in the estimate of Theorem 9.7 is of order h^4 due to (11.2), so we analyse in detail the term $((T - T_h)\phi_j, \phi_k)$. Let u and v be eigenfunctions associated with λ; we have to estimate $((T - T_h)u, v)$. It is clear that we can use the direct estimate

$$\|((T - T_h)u, v)\|_H \le \|(T - T_h)u\|_H \|v\|_H,$$

and the result follows from (11.2).

We now present an alternative estimate of the term $((T - T_h)u, v)$ which emphasizes the role of the consistency error, and offers more flexibility for generalization to other types of non-conforming approximations,

$$\begin{aligned} ((T - T_h)u, v) &= a_h((T - T_h)u, Tv) + a_h(T_h u, (T - T_h)v) \\ &= a_h((T - T_h)u, (T - T_h)v) \\ &\quad + a_h((T - T_h)u, T_h v) + a_h(T_h u, (T - T_h)v). \end{aligned} \tag{11.3}$$

The first term on the right-hand side of (11.3) is of order h^2 from (11.1), so we need to bound the second term, which is analogous to the third one from the symmetry of a_h. We have

$$\begin{aligned} a_h((T - T_h)u, T_h v) &= a_h(Tu, T_h v) - (u, T_h v) \\ &= \sum_{K \in \mathcal{T}_h} \int_{\partial K} (\mathbf{grad}\, Tu \cdot \mathbf{n}) T_h v, \end{aligned} \tag{11.4}$$

which by standard arguments is equal to

$$\sum_{e\in\mathcal{E}_h}\int_e \big((\mathbf{grad}\,Tu\cdot\mathbf{n}_e)-P_e(\mathbf{grad}\,Tu\cdot\mathbf{n}_e)\big)(T_hv-P_eT_hv)$$
$$=\sum_{e\in\mathcal{E}_h}\int_e \big((\mathbf{grad}\,Tu\cdot\mathbf{n}_e)-P_e(\mathbf{grad}\,Tu\cdot\mathbf{n}_e)\big)$$
$$\times\big((T_hv-P_eT_hv)-(Tv-P_eTv)\big),$$

where the set $\mathcal{E}_h$ contains all edges of the triangulation $\mathcal{T}_h$: the internal edges are repeated twice with opposite orientation of the normal and with appropriate definition of $T_hv_{|e}$ (which jumps from one triangle to the other); P_e denotes the $L^2(e)$-projection onto constant functions on e. We deduce

$$|a_h((T-T_h)u,T_hv)|$$
$$\leq\sum_{e\in\mathcal{E}_h}\|(I-P_e)(\mathbf{grad}\,Tu\cdot\mathbf{n}_e)\|_{L^2(e)}\|(I-P_e)(Tv-T_hv)\|_{L^2(e)}.$$

Putting all the pieces together gives the desired results from

$$\sum_{e\in\mathcal{E}_h}\|(I-P_e)(\mathbf{grad}\,Tu\cdot\mathbf{n}_e)\|_{L^2(e)}\leq Ch^{1/2}\sum_{K\in\mathcal{T}_h}\|Tu\|_{H^2(K)}$$

and

$$\sum_{e\in\mathcal{E}_h}\|(I-P_e)(Tv-T_hv)\|_{L^2(e)}\leq Ch^{1/2}\|(T-T_h)v\|_h$$
$$\leq Ch^{3/2}\|v\|_{L^2(\Omega)}.$$ □

We now deduce an optimal error estimate for the eigenfunctions in the discrete energy norm.

Theorem 11.4. With the same notation as in Theorem 11.2, we have

$$\|u-u_h\|_h\leq Ch\|u\|_{L^2(\Omega)}.$$

Proof. We consider the case of a simple eigenvalue λ. The generalization to multiple eigenvalues is technical and needs no significant new arguments. We have the identity

$$u-u_h=\lambda Tu-\lambda_hT_hu_h$$
$$=(\lambda-\lambda_h)Tu+\lambda_h(T-T_h)u+\lambda_hT_h(u-u_h),$$

and hence

$$\|u-u_h\|_h\leq|\lambda-\lambda_h|\|Tu\|_{H^1(\Omega)}+\lambda_h\|(T-T_h)u\|_h+\lambda_h\|T_h(u-u_h)\|_h.$$

The first two terms are easily bounded by Theorem 11.3 and (11.1), respectively. The last term can be estimated by observing that we have

$$C\|T_h(u-u_h)\|_h^2\leq a_h(T_h(u-u_h),T_h(u-u_h))=(u-u_h,T_h(u-u_h))$$

and using the estimate for $\|u-u_h\|_{L^2(\Omega)}$ in Theorem 11.2. □

PART THREE
Approximation of eigenvalue problems in mixed form

In this part we study the approximation of eigenvalue problems which have a particular structure, and which are often referred to as eigenvalue problems in mixed form.

12. Preliminaries

Given two Hilbert spaces Φ and Ξ, and two bilinear forms $a : \Phi \times \Phi \to \mathbb{R}$ and $b : \Phi \times \Xi \to \mathbb{R}$, the standard form of a *source* mixed problem is as follows: given $f \in \Phi'$ and $g \in \Xi'$, find $\psi \in \Phi$ and $\chi \in \Xi$ such that

$$a(\psi, \varphi) + b(\varphi, \chi) = \langle f, \varphi \rangle \quad \forall \varphi \in \Phi, \tag{12.1a}$$

$$b(\psi, \xi) = \langle g, \xi \rangle \quad \forall \xi \in \Xi. \tag{12.1b}$$

It is widely known that the natural conditions for the well-posedness of problem (12.1) are suitable inf-sup conditions, and that the discrete versions of the inf-sup conditions guarantee the stability of its approximation (Brezzi 1974, Babuška 1973, Brezzi and Fortin 1991).

If we suppose that there exist Hilbert spaces H_Φ and H_Ξ such that the following dense and continuous embeddings hold in a compatible way,

$$\Phi \subset H_\Phi \simeq H_\Phi' \subset \Phi',$$
$$\Xi \subset H_\Xi \simeq H_\Xi' \subset \Xi'$$

and we assume that the solution operator $T \in \mathcal{L}(H_\Phi \times H_\Xi)$ defined by

$$T(f, g) = (\psi, \chi) \tag{12.2}$$

(see (12.1)) is a compact operator, then a straightforward application of the theory presented in Part 2 might provide a straightforward convergence analysis for the approximation of the following eigenvalue problem: find $\lambda \in \mathbb{R}$ and $(\phi, \xi) \in \Phi \times \Xi$ with $(\phi, \xi) \neq (0, 0)$ such that

$$a(\psi, \varphi) + b(\varphi, \chi) = \lambda(\psi, \varphi)_{H_\Phi} \quad \forall \varphi \in \Phi,$$
$$b(\psi, \xi) = \lambda(\chi, \xi)_{H_\Xi} \quad \forall \xi \in \Xi.$$

On the other hand, the most common eigenvalue problems in mixed form correspond to a mixed formulation (12.1) where either f or g vanishes. Moreover, it is not obvious (the Laplace eigenvalue problem being the most famous counter-example) that the operator T defined on $H_\Phi \times H_\Xi$ is compact, so that, in general, eigenvalue problems in mixed form have to be studied with particular care.

Following Boffi, Brezzi and Gastaldi (1997), we consider two types of mixed eigenvalue problems. The first one (also known as the $(f, 0)$-type) consists of eigenvalue problems associated with the system (12.1) when $g = 0$. A fundamental example for this class is the Stokes eigenvalue problem. The second family (also known as the $(0, g)$-type) corresponds to the situation when in (12.1a) the right-hand side f vanishes. The Laplace eigenvalue problem in mixed form, for instance, belongs to this class.

The approximation of eigenvalue problems in mixed form has been the object of several papers; among them we refer to Canuto (1978), to Mercier, Osborn, Rappaz and Raviart (1981), which provides useful results for the estimate of the order of convergence, and to Babuška and Osborn (1991). The approach of this survey is taken from Boffi *et al.* (1997), where a comprehensive theory has been developed under the influence of the application presented in Boffi *et al.* (1999*b*) and the counter-example of Boffi *et al.* (2000*a*).

Since the notation used so far might appear cumbersome, in the next two sections we use notation which should resemble the Stokes problem in the case of problems of the first type and the mixed Laplace problem in the case of problems of the second type.

13. Problems of the first type

Let V, Q, and H be Hilbert spaces, suppose that the standard inclusions

$$V \subset H \simeq H' \subset V'$$

hold with continuous and dense embeddings, and let us consider two bilinear forms, $a : V \times V \to \mathbb{R}$ and $b : V \times Q \to \mathbb{R}$. We are interested in the following symmetric eigenvalue problem: find $\lambda \in \mathbb{R}$ and $u \in V$ with $u \neq 0$ such that, for some $p \in Q$,

$$a(u, v) + b(v, p) = \lambda(u, v) \quad \forall v \in V, \tag{13.1a}$$
$$b(u, q) = 0 \quad \forall q \in Q, \tag{13.1b}$$

where $(\cdot, \cdot)$ denotes the scalar product of H. We assume that a and b are continuous and that a is symmetric and positive semidefinite.

Given finite-dimensional subspaces $V_h \subset V$ and $Q_h \subset Q$, the discretization of (13.1) reads as follows: find $\lambda_h \in \mathbb{R}$ and $u_h \in V_h$ such that, for some $p_h \in Q$,

$$a(u_h, v) + b(v, p_h) = \lambda_h(u_h, v) \quad \forall v \in V_h, \tag{13.2a}$$
$$b(u_h, q) = 0 \quad \forall q \in Q_h. \tag{13.2b}$$

We start by studying the convergence of the eigensolution to (13.2) towards those of (13.1) and we shall discuss the rate of approximation afterwards. We aim to apply the spectral theory recalled in Section 6: in particular, we need to define a suitable solution operator. Let us consider the source problem associated with (13.1) (which corresponds to problem (12.1) with $g = 0$). Given $f \in H$, find $u \in V$ and $p \in Q$ such that

$$a(u, v) + b(v, p) = (f, v) \quad \forall v \in V, \tag{13.3a}$$
$$b(u, q) = 0 \quad \forall q \in Q. \tag{13.3b}$$

Under the assumption that (13.3) is solvable for any $f \in H$ and that the component u of the solution is unique, we define $T : H \to V$ as $Tf = u$. We assume that

$$T \text{ is compact from } H \text{ to } V.$$

The discrete counterpart of (13.3) is as follows: find $u_h \in V_h$ and $p_h \in Q_h$ such that

$$a(u_h, v) + b(v, p_h) = (f, v) \quad \forall v \in V_h, \tag{13.4a}$$
$$b(u_h, q) = 0 \quad \forall q \in Q_h. \tag{13.4b}$$

Assuming that the component u_h of the discrete solution of (13.4) exists and is unique, we can define the discrete operator $T_h : H \to V$ as $Tf = u_h \in V_h \subset V$.

It is clear that the eigenvalue problems (13.1) and (13.2), respectively, can be written in the equivalent form

$$\lambda T u = u, \quad \lambda_h T_h u_h = u_h.$$

We now introduce some abstract conditions that will guarantee the convergence of T_h to T in $\mathcal{L}(H, V)$. It follows from the discussion of Section 6 that this is a sufficient condition for the eigenmodes convergence. We denote by V_0 and Q_0 the subspaces of V and Q, respectively, containing all the solutions $u \in V$ and $p \in Q$ of (13.3) when f varies in H. In particular, we have $V_0 = T(H)$, and the inclusion $V_0 \subset \mathbb{K}$ holds true, where the kernel $\mathbb{K}$ of the operator associated with the bilinear form b is defined as usual by

$$\mathbb{K} = \{v \in V : b(v, q) = 0 \ \forall q \in Q\}.$$

We need also to introduce the discrete kernel as

$$\mathbb{K}_h = \{v_h \in V_h : b(v_h, q_h) = 0 \ \forall q_h \in Q_h\}.$$

It is well known that in general $\mathbb{K}_h \not\subset \mathbb{K}$. We shall also make use of suitable norms in V_0 and Q_0, which can be defined in a canonical way as

$$\|v\|_{V_0} = \inf\{\|f\|_H : Tf = v\},$$
$$\|q\|_{Q_0} = \inf\{\|f\|_H : q \text{ is the second component of the solution of (13.3)}\}.$$

Definition 13.1. The *ellipticity in the kernel* of the bilinear form a is satisfied if there exists $\alpha > 0$ such that

$$a(v, v) \geq \alpha \|v\|_V^2 \quad \forall v \in \mathbb{K}_h.$$

It can be shown that the ellipticity in the kernel property is sufficient for the well-posedness of the operator T_h (Boffi *et al.* 1997, Proposition 2).

Definition 13.2. We say that the *weak approximability* of Q_0 is satisfied if there exists $\rho_W(h)$, tending to zero as h tends to zero, such that

$$\sup_{v_h\in\mathbb{K}_h}\frac{b(v_h,q)}{\|v_h\|_V}\le\rho_W(h)\|q\|_{Q_0}\quad\forall q\in Q_0.$$

Definition 13.3. We say that the *strong approximability* of V_0 is satisfied if there exists $\rho_S(h)$, tending to zero as h tends to zero, such that, for any $v\in V_0$, there exists $v^I\in\mathbb{K}_h$ with

$$\|v-v^I\|_V\le\rho_S(h)\|v\|_{V_0}.$$

The next theorem says that the conditions introduced with the above definitions are sufficient for the eigenmode convergence of eigenvalue problems of the first kind.

Theorem 13.4. If the ellipticity in the kernel of the bilinear form a, the weak approximability of Q_0, and the strong approximability of V_0 are satisfied (see Definitions 13.1, 13.2 and 13.3), then there exists $\rho(h)$, tending to zero as h tends to zero, such that

$$\|(T-T_h)f\|_V\le\rho(h)\|f\|_H\quad\forall f\in H. \tag{13.5}$$

Proof. Take $f\in H$ and consider the solutions $(u,p)\in V_0\times Q_0$ of (13.3) and $(u_h,p_h)\in\mathbb{K}_h\times Q_h$ of (13.4) (p and p_h might not be unique). We need to estimate the difference $\|(T-T_h)f\|_V=\|u-u_h\|_V$. From the strong approximability of V_0, this can be performed by bounding the difference $\|u^I-u_h\|_V$. From the ellipticity in the kernel of a and the error equations, we have

$$\begin{aligned}\alpha\|u^I-u_h\|_V^2&\le a(u^I-u_h,u^I-u_h)\\&=a(u^I-u,u^I-u_h)+a(u-u_h,u^I-u_h)\\&=a(u^I-u,u^I-u_h)-b(u^I-u_h,p-p_h)\\&\le C\|u^I-u\|_V\|u^I-u_h\|_V-|b(u^I-u_h,p-p_h)|\\&\le\left(C\|u^I-u\|_V+\sup_{v_h\in\mathbb{K}_h}\frac{b(v_h,p-p_h)}{\|v_h\|_V}\right)\|u^I-u_h\|_V\\&\le\left(C\|u^I-u\|_V+\sup_{v_h\in\mathbb{K}_h}\frac{b(v_h,p)}{\|v_h\|_V}\right)\|u^I-u_h\|_V,\end{aligned}$$

which gives the required estimate thanks to the strong approximability of V_0 and the weak approximability of Q_0. □

Remark 13.5. The result of Theorem 13.4 can be essentially inverted by showing that the assumptions are necessary for the convergence in norm (13.5). This analysis is performed in Theorem 2 of Boffi *et al.* (1997), under the additional assumption that the operator T can be extended to a bounded operator in $\mathcal{L}(V', V)$.

Remark 13.6. So far, we have not assumed the well-posedness of the source problems (13.3) and (13.4). Indeed, we assumed existence and uniqueness of the first component of the solutions (u and u_h, respectively) in order to be able to define the solution operators T and T_h. On the other hand, the second component of solutions p and p_h might be non-unique. Examples of this situations (where p is unique, but p_h is not) are presented at the end of this section.

Remark 13.7. The presented result might look too strong, since we are considering the convergence in norm from H to V. Indeed, although our result immediately implies the convergence in norm from H into itself, it might be interesting to investigate directly the behaviour of $\|T - T_h\|$ in $\mathcal{L}(H)$ or $\mathcal{L}(V)$. On the other hand, the analysis presented in Theorem 13.4 is quite natural, and we are not aware of applications where a sharper result is needed.

Theorem 13.4 concerns good approximation of eigenvalues and eigenfunctions, but does not answer the question of estimating the rate of convergence. In most practical situations, this issue can be solved with the help of Babuška–Osborn theory, as developed in Section 9. This task was performed in Mercier *et al.* (1981) in the general situation of *non-symmetric* eigenvalue problems in mixed form; we report here the main results of this theory in the symmetric case (see Mercier *et al.* (1981, Section 5)).

Let λ be an eigenvalue of (13.1) of multiplicity m and let $E \subset V$ be the corresponding eigenspace. We denote by $\lambda_{1,h}, \lambda_{2,h}, \ldots, \lambda_{m,h}$ the discrete eigenvalues converging to λ and by E_h the direct sum of the corresponding eigenspaces.

The convergence of eigenfunctions is a direct application of the results of the abstract theory and is summarized in the next theorem.

Theorem 13.8. Under the hypotheses of Theorem 13.4, there exists constant C such that

$$\hat{\delta}(E, E_h) \leq C\|(T - T_h)_{|E}\|_{\mathcal{L}(H,H)},$$

where the gap is evaluated in the H-norm.

Remark 13.9. The analogous estimate for the error in the norm of V can be obtained directly from Theorem 9.3 by assuming a uniform convergence of T_h to T in $\mathcal{L}(V)$. Such a convergence is usually a consequence of standard estimates for the source problem.

In order to estimate the convergence order for the eigenvalues, we consider the case when problems (13.3) and (13.4) are well-posed (see Remark 13.6). This is typically true if the standard inf-sup conditions are satisfied (Brezzi and Fortin 1991). According to Mercier *et al.* (1981), we can define the operator $B : H \to Q$ as $Bf = p$, where p is the second component of the solution of (13.3). Analogously, we define the discrete operator $B_h : H \to Q_h$ by $B_h f = p_h$, with p_h coming from (13.4). The results of Mercier *et al.* (1981, Theorem 5.1) are rewritten in the next theorem in this particular case.

Theorem 13.10. Under the hypotheses of Theorem 13.4 and the additional assumption that the operators B and B_h are well-defined, there exists C such that, for sufficiently small h,

$$|\lambda - \lambda_{i,h}| \le C\big(\|(T - T_h)_{|E}\|^2_{\mathcal{L}(H,V)} + \|(T - T_h)_{|E}\|_{\mathcal{L}(H,V)}\|(B - B_h)_{|E}\|_{\mathcal{L}(H,Q)}\big), \tag{13.6}$$

for $i = 1, \dots, m$.

We conclude this section with some applications of the developed theory.

13.1. The Stokes problem

Let Ω be an open bounded domain in $\mathbb{R}^n$. The eigenvalue problem associated with the Stokes equations fits within the developed theory with the following definitions:

$$\begin{aligned} H &= L^2(\Omega), \\ V &= (H^1_0(\Omega))^n, \\ Q &= L^2_0(\Omega), \\ a(\mathbf{u}, \mathbf{v}) &= \int_\Omega \underline{\varepsilon}(\mathbf{u}) : \underline{\varepsilon}(\mathbf{v})\,\mathrm{d}\mathbf{x}, \\ b(\mathbf{v}, q) &= \int_\Omega q \operatorname{div} \mathbf{v}\,\mathrm{d}\mathbf{x}. \end{aligned}$$

It is well known that the bilinear form a is coercive in V, so the ellipticity in the kernel property (see Definition 13.1) holds true for any finite element choice. The weak approximability of Q_0 (see Definition 13.2) is satisfied by any reasonable approximating scheme as well. Namely, it can be easily seen that there exists $\varepsilon > 0$ such that the solution space Q_0 is contained in $H^\varepsilon(\Omega)$ for a wide class of domains (in particular, ε can be taken equal to one if Ω is a convex polygon or polyhedron). The weak approximability property then follows from standard approximation properties as follows:

$$|b(\mathbf{v}_h, q)| = |b(\mathbf{v}_h, q - q^I)| \le C\|\mathbf{v}_h\|_{H^1_0(\Omega)}\|q - q^I\|_{L^2(\Omega)},$$

where $q^I \in Q_h$ is any discrete function (recall that $\mathbf{v}_h \in \mathbb{K}_h$).

In this case the strong approximability of V_0 is the crucial condition for the convergence. In general V_0 consists of divergence-free functions which are in $(H^{1+\varepsilon}(\Omega))^n \cap V$ if the domain is smooth enough (for a suitable $\varepsilon > 0$, in particular $\varepsilon = 1$ if Ω is a convex polygon or polyhedron). The strong approximability means that, for any $\mathbf{v} \in V_0$, there exists $\mathbf{v}^I \in \mathbb{K}_h$ such that

$$\|\mathbf{v} - \mathbf{v}^I\|_{H_0^1(\Omega)} \leq \rho(h)\|\mathbf{v}\|_{V_0},$$

with $\rho(h)$ tending to zero as h tends to zero. This property is well known to be valid if, for instance, the discrete space V_h discretizes V with standard approximation properties and the classical inf-sup condition links V_h and Q_h: there exists $\beta > 0$ such that

$$\inf_{q_h \in Q_h} \sup_{\mathbf{v}_h \in V_h} \frac{b(\mathbf{v}_h, q_h)}{\|\mathbf{v}_h\|_{H_0^1(\Omega)}\|q_h\|_{L^2(\Omega)}} > \beta.$$

This implies that all inf-sup stable approximations of the Stokes problem provide convergent discretization of the corresponding eigenvalue problem. As for the estimation of the rate of convergence (which was not considered in Boffi *et al.* (1997)), we can use Theorems 13.8 and 13.10. For instance, using a popular stable scheme of order k such as the generalized Hood–Taylor method (Boffi 1997), that is, continuous piecewise polynomials of degree k and $k-1$ for velocities and pressures, respectively, we get that the order of convergence of the eigenvalues is h^{2k}, as expected, if the domain is smooth enough in both two and three space dimensions. Indeed, looking at the terms appearing in formula (13.6), we essentially have to estimate two items: $\|(T - T_h)_{|E}\|_{\mathcal{L}(H,V)}$ and $\|(B - B_h)_{|E}\|_{\mathcal{L}(H,Q)}$. The classical error analysis for the source problem gives

$$\|\mathbf{u} - \mathbf{u}_h\|_V + \|p - p_h\|_Q \leq Ch^k\|f\|_H$$

if we assume that we have no limitation in the regularity of the solution $(\mathbf{u}, p)$. This implies

$$\|(T - T_h)_{|E}\|_{\mathcal{L}(H,V)} = O(h^k),$$
$$\|(B - B_h)_{|E}\|_{\mathcal{L}(H,Q)} = O(h^k),$$

which gives the result.

For the eigenfunctions, the result is similar. In particular, Theorem 13.8 and Remark 13.9 give that the gap between discrete and continuous eigenfunctions is of optimal order h^{k+1} in $L^2(\Omega)$ and h^k in $H^1(\Omega)$.

On the other hand, Theorem 13.4 allows us to conclude that we have the convergence of the eigenvalues even in cases when the source problem might not be solvable. This is the case, for instance, of the widely studied two-dimensional $\mathcal{Q}_1$–$\mathcal{P}_0$ element: that is, the velocities are continuous piecewise bilinear functions and the pressures are piecewise constants. It is well known that this element does not satisfy the inf-sup condition: a checkerboard spurious mode is present and the filtered inf-sup constant tends to zero as $O(h)$ (Johnson and Pitkäranta 1982). On the other hand, it can be proved that the hypotheses of Theorem 13.4 are satisfied (Boffi *et al.* 1997), so the eigenvalues are well approximated in spite of the fact that the approximation of the source problem is affected by spurious pressure modes.

13.2. Dirichlet problem with Lagrange multipliers

Following Babuška (1973), the Dirichlet problem with Lagrange multipliers can be studied with the theory developed in this section and the following identifications. Let Ω be a two-dimensional polygonal domain and define

$$\begin{aligned} H &= L^2(\Omega), \\ V &= H^1(\Omega), \\ Q &= H^{-1/2}(\partial\Omega), \\ a(u,v) &= \int_\Omega \mathbf{grad}\, u \cdot \mathbf{grad}\, v \,\mathrm{d}\mathbf{x}, \\ b(v,\mu) &= \langle \mu, v_{|\partial\Omega} \rangle, \end{aligned}$$

where $\langle \cdot,\cdot \rangle$ denotes the duality between $H^{1/2}(\partial\Omega)$ and $H^{-1/2}(\partial\Omega)$.

Given a regular decomposition of Ω and a regular decomposition of $\partial\Omega$, we let V_h be the space of continuous piecewise polynomials of degree k_1 defined in Ω, and let Q_h be the space of continuous piecewise polynomials of degree k_2 defined in $\partial\Omega$, with $k_1 \geq 0$ and $k_2 \geq 0$. It turns out that the weak approximability condition (see Definition 13.2) and the strong approximability condition (see Definition 13.3) are satisfied for any choice of triangulation and degrees k_1 and k_2. On the other hand, the ellipticity in the kernel property (see Definition 13.1) is satisfied under the weak assumption that Q_h contains μ_h such that $\langle \mu_h, 1 \rangle \neq 0$.

It is interesting to notice that in this example we have good convergence of the discretized eigenvalue problem under very general assumptions, while the corresponding source problem requires rather strict compatibility conditions between the meshes of Ω and $\partial\Omega$, and k_1 and k_2.

14. Problems of the second type

Let Σ, U, and H be Hilbert spaces, suppose that the standard inclusions

$$U \subset H \simeq H' \subset U'$$

hold with continuous and dense embeddings, and let us consider two bilinear forms $a : \Sigma \times \Sigma \to \mathbb{R}$ and $b : \Sigma \times U \to \mathbb{R}$. We are interested in the following symmetric eigenvalue problem: find $\lambda \in \mathbb{R}$ and $u \in U$ with $u \neq 0$ such that, for some $\sigma \in \Sigma$,

$$a(\sigma, \tau) + b(\tau, u) = 0 \quad \forall \tau \in \Sigma, \tag{14.1a}$$

$$b(\sigma, v) = -\lambda(u, v) \quad \forall v \in U, \tag{14.1b}$$

where $(\cdot, \cdot)$ denotes the scalar product in H. We assume that a and b are continuous and that a is symmetric and positive semidefinite. The hypotheses on a imply that the seminorm

$$|v|_a = (a(v, v))^{1/2}$$

is well-defined, so that we have

$$a(u, v) \leq |u|_a |v|_a \quad \forall u, v \in U.$$

Moreover, we assume that the following source problem associated with (14.1) has a unique solution $(\sigma, u) \in \Sigma \times U$ to

$$a(\sigma, \tau) + b(\tau, u) = 0 \quad \forall \tau \in \Sigma, \tag{14.2a}$$

$$b(\sigma, v) = -\langle g, v \rangle \quad \forall v \in U \tag{14.2b}$$

satisfying the *a priori* bound

$$\|\sigma\|_\Sigma + \|u\|_U \leq C \|g\|_{U'},$$

where the symbol $\langle \cdot, \cdot \rangle$ in (14.2) denotes the duality pairing between U' and U.

Given finite-dimensional subspaces $\Sigma_h \subset \Sigma$ and $U_h \subset U$, the Galerkin discretization of (14.1) reads as follows: find $\lambda_h \in \mathbb{R}$ and $u_h \in U_h$ with $u_h \neq 0$ such that, for some $\sigma_h \in \Sigma_h$,

$$a(\sigma_h, \tau) + b(\tau, u_h) = 0 \quad \forall \tau \in \Sigma_h, \tag{14.3a}$$

$$b(\sigma_h, v) = -\lambda_h(u_h, v) \quad \forall v \in U_h. \tag{14.3b}$$

Following the same lines as the previous section, we start by analysing how the solutions of (14.3) converge towards those of (14.1), and postpone the question of the rate of convergence to the end of this section.

We define the solution operator $T : H \to H$ by $Tg = u \in U \subset H$, where $u \in U$ is the second component of the solution to (14.2). It is clear that when g belongs to H the duality pairing in the right-hand side of (14.2) is equivalent to the scalar product (g, v).

The discrete counterpart of (14.2) when g belongs to H is as follows: find $(\sigma_h, u_h) \in \Sigma_h \times U_h$ such that

$$a(\sigma_h, \tau) + b(\tau, u_h) = 0 \quad \forall \tau \in \Sigma_h, \tag{14.4a}$$

$$b(\sigma_h, v) = -(g, v) \quad \forall v \in U_h. \tag{14.4b}$$

We suppose that the second component u_h of the solution of (14.4) exists and is unique, so we can define the discrete solution operator $T_h : H \to H$ as $T_h g = u_h \in U_h \subset U \subset H$.

We now introduce some abstract conditions that will be used in an initial theorem in order to show the convergence of T_h to T in $\mathcal{L}(H,H)$ and, in a second theorem, in $\mathcal{L}(H,U)$. According to the results of Section 9, this is enough to show the convergence of the eigensolutions of (14.3) towards those of (14.1). We let Σ_0 and U_0 denote the subspaces of Σ and U, respectively, containing all solutions $\sigma \in \Sigma$ and $u \in U$ of (14.2) when g varies in H. In particular, we have $U_0 = T(H)$. We shall also make use of the space $\Sigma_{0'} \subset \Sigma$ containing the second components of the solution $\sigma \in \Sigma$ of (14.2) when g varies in U'. The spaces Σ_0, U_0, and $\Sigma_{0'}$ will be endowed with their natural norms:

$$\begin{aligned}
\|\tau\|_{\Sigma_0} &= \inf\{\|g\|_H : \tau \text{ is solution of (14.2) with datum } g\},\\
\|v\|_{U_0} &= \inf\{\|g\|_H : Tg = v\},\\
\|\tau\|_{\Sigma_{0'}} &= \inf\{\|g\|_{U'} : \tau \text{ is solution of (14.2) with datum } g\}.
\end{aligned}$$

Finally, the discrete kernel is given by

$$\mathbb{K}_h = \{\tau_h \in \Sigma_h : b(\tau_h, v) = 0 \ \forall v \in U_h\}.$$

Definition 14.1. We say that the *weak approximability* of U_0 with respect to a is satisfied if there exists $\rho_W(h)$, tending to zero as h tends to zero, such that

$$b(\tau_h, v) \le \rho_W(h)|\tau_h|_a\|v\|_{U_0} \quad \forall v \in U_0 \ \forall \tau_h \in \mathbb{K}_h.$$

Definition 14.2. We say that the *strong approximability* of U_0 is satisfied if there exists $\rho_S(h)$, tending to zero as h tends to zero, such that, for every $v \in U_0$, there exists $v^I \in U_h$ such that

$$\|v - v^I\|_U \le \rho_S(h)\|v\|_{U_0}.$$

Remark 14.3. The same terms, *weak* and *strong approximability*, were used in the framework of problems of the first kind (see Definitions 13.2 and 13.3) and of the second kind (see Definitions 14.1 and 14.2). In the applications, it should be clear from the context which definition the terms refer to.

A powerful and commonly used tool for the analysis of mixed approximation is the Fortin operator (Fortin 1977), that is, a linear operator $\Pi_h : \Sigma_0 \to \Sigma_h$ that satisfies

$$b(\tau - \Pi_h\tau, v) = 0 \quad \forall \tau \in \Sigma_0 \ \forall v \in U_h. \tag{14.5}$$

Definition 14.4. A *bounded Fortin operator* is a Fortin operator that can be extended from Σ_0 to $\Sigma_{0'}$ and that is uniformly bounded in $\mathcal{L}(\Sigma_{0'}, \Sigma)$.

Definition 14.5. The *Fortid condition* is satisfied if there exists a Fortin operator which converges to the identity in the following sense. There exists $\rho_F(h)$, tending to zero as h tends to zero, such that

$$|\sigma - \Pi_h\sigma|_a \leq \rho_F(h)\|\sigma\|_{\Sigma_0} \quad \forall\sigma \in \Sigma_0.$$

The next theorem presents sufficient conditions for good approximation of the eigensolutions of (14.3) towards those of (14.1).

Theorem 14.6. If the Fortid condition, the weak approximability of U_0 with respect to a, and the strong approximability of U_0 are satisfied (see Definitions 14.5, 14.1 and 14.2), then there exists $\rho(h)$, tending to zero as h tends to zero, such that

$$\|(T - T_h)g\|_H \leq \rho(h)\|g\|_H \quad \forall g \in H.$$

Proof. Let g be in H and consider the solution (σ, u) of (14.2). In particular, we have $u = Tg$. Let us define $u_h = T_hg$ and let $\sigma_h \in \Sigma_h$ be such that (σ_h, u_h) solves (14.4) (such a σ_h might not be unique). Using a duality argument, let $(\sigma(h), u(h)) \in \Sigma \times U$ be the solution of (14.2) with $g = u - u_h \in U \subset H$. By the definition of the norm in Σ_0, we have $\|\sigma(h)\|_{\Sigma_0} \leq \|u - u_h\|_H$. Moreover,

$$\begin{aligned}
\|u - u_h\|_H^2 &= (u - u_h, u - u_h) = -b(\sigma(h), u - u_h)\\
&= b(\sigma(h) - \Pi_h\sigma(h), u) + b(\Pi_h\sigma(h), u - u_h)\\
&= a(\sigma, \sigma(h) - \Pi_h\sigma(h)) + a(\sigma - \sigma_h, \Pi_h\sigma(h))\\
&\leq |\sigma|_a|\sigma(h) - \Pi_h\sigma(h)|_a + |\sigma - \sigma_h|_a|\Pi_h\sigma(h)|_a\\
&\leq |\sigma|_a|\sigma(h) - \Pi_h\sigma(h)|_a + C|\sigma - \sigma_h|_a\|\sigma(h)\|_{\Sigma_0}\\
&\leq |\sigma|_a\rho_F(h)\|\sigma(h)\|_{\Sigma_0} + C|\sigma - \sigma_h|_a\|\sigma(h)\|_{\Sigma_0}\\
&\leq (\rho_F(h)|\sigma|_a + C|\sigma - \sigma_h|_a)\|u - u_h\|_H,
\end{aligned}$$

where $\rho_F(h)$ was introduced in Definition 14.5. This implies

$$\|u - u_h\|_H \leq \rho_F(h)|\sigma|_a + C|\sigma - \sigma_h|_a.$$

It remains to estimate the term $|\sigma - \sigma_h|_a$, which we do by triangular inequality after summing and subtracting $\Pi_h\sigma$. Since we assumed the Fortid condition, it is enough to bound $|\Pi_h\sigma - \sigma_h|_a$. We notice that $|\Pi_h\sigma - \sigma_h|_a$ belongs to $\mathbb{K}_h$; we have

$$\begin{aligned}
|\Pi_h\sigma - \sigma_h|_a^2 &= a(\Pi_h\sigma - \sigma, \Pi_h\sigma - \sigma_h) + a(\sigma - \sigma_h, \Pi_h\sigma - \sigma_h)\\
&\leq |\Pi_h\sigma - \sigma|_a|\Pi_h\sigma - \sigma_h|_a - b(\Pi_h\sigma - \sigma_h, u - u_h)\\
&= |\Pi_h\sigma - \sigma|_a|\Pi_h\sigma - \sigma_h|_a - b(\Pi_h\sigma - \sigma_h, u)\\
&\leq |\Pi_h\sigma - \sigma_h|_a(|\Pi_h\sigma - \sigma|_a + \rho_W(h)\|u\|_{U_0}),
\end{aligned}$$

where $\rho_W(h)$ was introduced in Definition 14.1. This gives

$$|\sigma - \sigma_h|_a \le 2|\Pi_h\sigma - \sigma|_a + \rho_W(h)\|u\|_{U_0}$$
$$\le 2\rho_S(h)\|\sigma\|_{\Sigma_0} + \rho_W(h)\|u\|_{U_0},$$

where $\rho_S(h)$ was introduced in Definition 14.2. The theorem then follows from the definition of the norms of Σ_0 and U_0. □

Before moving to the estimate of the rate of convergence for eigenvalues and eigenfunctions, we present a slight modification of the previous theorem which implies the convergence of T_h to T in $\mathcal{L}(H, U)$.

Theorem 14.7. If there exists a bounded Fortin operator (see Definition 14.4) and if the Fortid condition, the weak approximability of U_0 and the strong approximability of U_0 are satisfied (see Definitions 14.5, 14.1 and 14.2), then there exists $\rho(h)$, tending to zero as h goes to zero, such that

$$\|(T - T_h)g\|_U \le \rho(h)\|g\|_H \quad \forall g \in H.$$

Proof. Let g be in H and consider the solution (σ, u) of (14.2). In particular, we have $u = Tg$. Let us define $u_h = T_h g$ and let σ_h be such that (σ_h, u_h) solves (14.4) (such a σ_h might not be unique). Let $g(h) \in U'$ be such that $\langle g(h), u - u_h\rangle = \|u - u_h\|_U$ and $\|g(h)\|_{U'} = 1$. Let $\sigma(h)$ be the first component of the solution to (14.2) with datum $g(h)$, so that $\sigma(h) \in \Sigma_{0'}$ and $\|\sigma(h)\|_{\Sigma_{0'}} \le \|g(h)\|_{U'} = 1$. We have

$$\|u - u_h\|_U = \langle g(h), u - u_h\rangle = -b(\sigma(h), u - u_h)$$
$$= -b(\sigma(h) - \Pi_h\sigma(h), u - u_h) - b(\Pi_h\sigma(h), u - u_h)$$
$$= -b(\sigma(h) - \Pi_h\sigma(h), u - u^I) + a(\sigma - \sigma_h, \Pi_h\sigma(h)).$$

We estimate the two terms on the right-hand side separately:

$$b(\sigma(h) - \Pi_h\sigma(h), u - u^I) \le C\|\sigma(h) - \Pi_h\sigma(h)\|_\Sigma\|u - u^I\|_U$$
$$\le C(\|\sigma(h)\|_\Sigma + \|\Pi_h\sigma(h)\|_\Sigma)\|u - u^I\|_U$$

and

$$a(\sigma - \sigma_h, \Pi_h\sigma(h)) \le |\Pi_h\sigma(h)|_a|\sigma - \sigma_h|_a.$$

The proof is then easily concluded from the definition of the norms in Σ_0, U_0, $\Sigma_{0'}$, and by using the strong approximability to bound $\|u - u^I\|_U$, the boundedness of the Fortin operator and the definition of $\sigma(h)$ to bound $\|\Pi_h\sigma(h)\|_\Sigma$ and $|\Pi_h\sigma(h)|_a$, and the same argument as in the proof of Theorem 14.6 to estimate $|\sigma - \sigma_h|_a$. □

Remark 14.8. The results of Theorems 14.6 and 14.7 can be essentially inverted by stating that suitable norm convergences of T_h to T imply the three main conditions: weak approximability of Definition 14.1, strong approximability of Definition 14.2, and Fortid condition of Definition 14.5. We refer the interested reader to Boffi *et al.* (1997, Theorems 5–7) for the technical details.

We now give basic estimates for the rate of convergence of eigenvalues and eigenfunctions in the spirit of Mercier *et al.* (1981). Let λ be an eigenvalue of (14.1) of multiplicity m and let $E \subset U$ be the corresponding eigenspace. We denote by $\lambda_{1,h}, \lambda_{2,h}, \ldots, \lambda_{m,h}$ the discrete eigenvalues converging to λ and by E_h the direct sum of the corresponding eigenspaces.

The eigenfunction convergence follows directly from the results of the abstract theory presented in Section 6.

Theorem 14.9. Under the hypotheses of Theorem 14.6 or 14.7, there is a constant C such that

$$\hat{\delta}(E, E_h) \leq C\|T - T_h\|_{\mathcal{L}(H,H)},$$

where the gap is evaluated in the H-norm.

Remark 14.10. The same comment as in Remark 13.9 applies to this situation as well. In particular, the convergence of the eigenfunctions in the norm of U would follow from a uniform convergence of T_h to T in $\mathcal{L}(V)$.

In order to estimate the rate of convergence for the eigenvalues, we invoke Mercier *et al.* (1981, Theorem 6.1), where the more general situation of non-symmetric problems is discussed. We assume that the source problems (14.2) and (14.4) are well-posed (with $g \in H$), so that we can define the operator $A : H \to \Sigma$ associated with the first component of the solution as $Ag = \sigma$. Analogously, $A_h : H \to \Sigma_h$ denotes the discrete operator associated to the first component of the solution to problem (14.4): $A_h g = \sigma_h$.

Theorem 14.11. Under the hypotheses of Theorem 14.6 or 14.7 and the additional assumption that the operators A and A_h are well-defined, there exists C such that, for sufficiently small h,

$$\begin{aligned}|\lambda - \lambda_{i,h}| \leq C\big(&\|(T - T_h)_{|E}\|^2_{\mathcal{L}(H,U)} \\ &+ \|(T - T_h)_{|E}\|_{\mathcal{L}(H,U)}\|(A - A_h)_{|E}\|_{\mathcal{L}(H,\Sigma)} \\ &+ \|(A - A_h)_{|E}\|^2_{\mathcal{L}(H,H)}\big),\end{aligned}$$

for $i = 1, \ldots, m$.

We conclude this section with the application of the presented theory to two fundamental examples: the Laplace eigenvalue problem and the biharmonic problem.

14.1. The Laplace problem

Let Ω be an open bounded domain in $\mathbb{R}^n$. The eigenvalue problem associated with the Laplace operator fits within the developed theory with the following definitions:

$$\begin{aligned} H &= L^2(\Omega), \\ \Sigma &= \mathbf{H}(\mathrm{div};\Omega), \\ U &= L^2(\Omega), \\ a(\boldsymbol{\sigma},\boldsymbol{\tau}) &= \int_\Omega \boldsymbol{\sigma}\cdot\boldsymbol{\tau}\,\mathrm{d}\mathbf{x}, \\ b(\boldsymbol{\tau},v) &= \int_\Omega v\,\mathrm{div}\,\boldsymbol{\tau}\,\mathrm{d}\mathbf{x}. \end{aligned}$$

It follows, in particular, that the seminorm induced by the form a is indeed the $L^2(\Omega)$-norm,

$$|\cdot|_a = \|\cdot\|_{L^2(\Omega)}.$$

The solution spaces Σ_0 and U_0 contain functions with higher regularity than Σ and U: for a wide class of domains Ω there exists $\varepsilon > 0$ such that $\Sigma_0 \subset H^\varepsilon(\Omega)^n$ and $U_0 \subset H^{1+\varepsilon}(\Omega)$ (if Ω is a convex polygon or polyhedron, the inclusions hold with $\varepsilon = 1$). We should, however, pay attention to the fact that functions $\boldsymbol{\sigma}$ in Σ_0 do not have a more regular divergence than $\mathrm{div}\,\boldsymbol{\sigma} \in L^2(\Omega)$, since from (14.2) we have $\mathrm{div}\,\boldsymbol{\sigma} = -g$, and g varies in $L^2(\Omega)$.

Remark 14.12. The presented setting applies to the Dirichlet problem for Laplace operator. With natural modifications the Neumann problem could be studied as well: $\Sigma = \mathbf{H}_0(\mathrm{div};\Omega)$, $U = L^2_0(\Omega)$.

Several choices of discrete spaces have been proposed for the approximation of Σ and U in two and three space dimensions. In general, U_h is defined as $\mathrm{div}(\Sigma_h)$, where Σ_h is a suitable discretization of $\mathbf{H}(\mathrm{div};\Omega)$. RT (Raviart and Thomas 1977, Nédélec 1980), BDM (Brezzi *et al.* 1985, Brezzi, Douglas and Marini 1986), BDFM (Brezzi *et al.* 1987*b*) elements are possible choices. On quadrilateral meshes, ABF elements (Arnold *et al.* 2005) are a possible solution in order to avoid the lack of convergence arising from the distortion of the elements. In this case the identity $\mathrm{div}(\Sigma_h) = U_h$ is no longer true and the analysis of convergence requires particular care (Gardini 2005).

From the equality $\mathrm{div}(\Sigma_h) = U_h$, it follows that the discrete kernel $\mathbb{K}_h$ contains divergence-free functions, so that the weak approximability (see Definition 14.1) is satisfied:

$$\begin{aligned} b(\boldsymbol{\tau}_h,v) = b(\boldsymbol{\tau}_h,v-v^I) &\le C\|\boldsymbol{\tau}_h\|_{\mathbf{H}(\mathrm{div};\Omega)}\|v-v^I\|_{L^2(\Omega)} \\ &= C\|\boldsymbol{\tau}_h\|_{L^2(\Omega)}\|v-v^I\|_{L^2(\Omega)}, \end{aligned}$$

for $\boldsymbol{\tau}_h \in \mathbb{K}_h$, $v \in U_0$, and $v^I \in U_h$ suitable approximation of v.

Strong approximability (see Definition 14.2) is a consequence of standard approximation properties in U_0.

It turns out that the main condition for good approximation of the Laplace eigenvalue problem is the Fortid condition (see Definition 14.5). For the schemes mentioned so far (RT, BDM, BDFM), it is not difficult to see that the standard interpolation operator (Brezzi and Fortin 1991) is indeed a Fortin operator (see Definition 14.5): in general, if we denote by $\boldsymbol{\tau}^I$ the interpolant of $\boldsymbol{\tau}$, we have

$$\begin{aligned} b(\boldsymbol{\tau}-\boldsymbol{\tau}^I, v) &= \int_\Omega v \operatorname{div}(\boldsymbol{\tau}-\boldsymbol{\tau}^I)\,\mathrm{d}\mathbf{x} = \sum_K \int_K v \operatorname{div}(\boldsymbol{\tau}-\boldsymbol{\tau}^I)\,\mathrm{d}\mathbf{x} \\ &= -\int_K \mathbf{grad}\, v \cdot (\boldsymbol{\tau}-\boldsymbol{\tau}^I)\,\mathrm{d}\mathbf{x} + \int_{\partial K} v(\boldsymbol{\tau}-\boldsymbol{\tau}^I)\cdot \mathbf{n}\,\mathrm{d}s, \end{aligned}$$

and the degrees of freedom for $\boldsymbol{\tau}^I$ are usually chosen so that the last two terms vanish for all $v \in U_h$. The Fortid condition (see Definition 14.5) is then also a consequence of standard approximation properties. We explicitly notice that a Fortid condition where the term $|\sigma - \Pi_h\sigma|_a$ is replaced by $\|\sigma - \Pi_h\sigma\|_\Sigma$ does not hold in this situation, since this would imply an estimate for $\|\boldsymbol{\tau} - \boldsymbol{\tau}^I\|_{\mathbf{H}(\mathrm{div};\Omega)}$, with $\boldsymbol{\tau} \in \Sigma_0$, but this cannot provide any uniform bound, since $\operatorname{div}\boldsymbol{\tau}$ is a generic element of $L^2(\Omega)$.

We shall return to this example in Section 15.

14.2. The biharmonic problem

For the sake of simplicity, let Ω be a convex polygon in $\mathbb{R}^2$. The biharmonic problem

$$\begin{aligned} \Delta^2 u &= -g \quad \text{in } \Omega, \\ u = \frac{\partial u}{\partial \mathbf{n}} &= 0 \qquad \text{on } \partial\Omega \end{aligned}$$

has been widely studied in the framework of mixed approximations. In particular, it fits within our setting with the following choices:

$$\begin{aligned} H &= L^2(\Omega), \\ \Sigma &= H^1(\Omega), \\ U &= H^1_0(\Omega), \\ a(\sigma,\tau) &= \int_\Omega \sigma\tau\,\mathrm{d}\mathbf{x}, \\ b(\tau,v) &= -\int_\Omega \mathbf{grad}\,\tau\cdot\mathbf{grad}\,v\,\mathrm{d}\mathbf{x}, \end{aligned}$$

where the auxiliary variable $\sigma = -\Delta u$ has been introduced. It follows that the seminorm associated with the bilinear form a is the $L^2(\Omega)$-norm

$$|\cdot|_a = \|\cdot\|_{L^2(\Omega)}.$$

A possible discretization of the biharmonic problem consists in an equal-order approximation, where Σ_h and U_h are made of continuous piecewise polynomials of degree k (Glowinski 1973, Mercier 1974, Ciarlet and Raviart 1974).

In this case the most delicate condition to be checked is weak approximability (see Definition 14.1). Indeed, strong approximability (see Definition 14.2) is a simple consequence of standard approximation properties. Given $\sigma \in \Sigma_0$, a Fortin operator (14.5) can be defined by $\Pi_h\sigma \in \Sigma_h$ and

$$(\mathbf{grad}\,\Pi_h\sigma, \mathbf{grad}\,\tau) = (\mathbf{grad}\,\sigma, \mathbf{grad}\,\tau) \quad \forall\tau \in \Sigma_h.$$

It is clear that the Fortid condition (see Definition 14.5) holds true.

A direct proof of the weak approximability property requires an inverse estimate (which is valid, for instance, if the mesh is quasi-uniform) and $k \geq 2$:

$$\begin{aligned} b(\tau, v) &= (\mathbf{grad}\,\tau, \mathbf{grad}\,v) = (\mathbf{grad}\,\tau, \mathbf{grad}(v - v^I)) \\ &\leq Ch^{-1}\|\tau\|_{L^2(\Omega)}h^2\|v\|_{H^3(\Omega)}. \end{aligned}$$

A more refined analysis (Scholz 1978) shows that the weak approximability property is valid for $k = 1$ as well.

15. Inf-sup condition and eigenvalue approximation

In the last section of this part we review the connections between the conditions for good approximation of a source mixed problem (inf-sup conditions) and the conditions for good approximation of the corresponding eigenvalue problem in mixed form (see Sections 13 and 14).

Going back to the notation used at the beginning of Part 3 (the discrete spaces will be denoted by $\Phi_h \subset \Phi$ and $\Xi_h \subset \Xi$), we consider as the two keystones for the approximation of the source problem (12.1) the *ellipticity in the kernel property*: there exists $\alpha > 0$ such that

$$a(\varphi, \varphi) \geq \alpha\|\varphi\|_\Phi^2 \quad \forall\varphi \in \mathbb{K}_h, \tag{15.1}$$

where

$$\mathbb{K}_h = \{\varphi \in \Phi_h : b(\varphi, \xi) = 0\ \forall\xi \in \Xi_h\}$$

and the *inf-sup condition*: there exists $\beta > 0$ such that

$$\inf_{\xi\in\Xi_h} \sup_{\varphi_h\in\Phi_h} \frac{b(\varphi, \xi)}{\|\varphi\|_\Phi\|\xi\|_\Xi} \geq \beta \tag{15.2}$$

(Brezzi and Fortin 1991). The ellipticity in the kernel might actually be weakened and written in the form of an inf-sup condition as well, but for our purposes, in the case of symmetric problems, ellipticity in the kernel is a quite general condition.

The conditions presented in Sections 13 and 14 involve quantities other than simply ellipticity in the kernel and the inf-sup condition. Loosely speaking, it turns out that the conditions for good approximation of eigenvalue problems of the first kind require more than ellipticity in the kernel but a weaker inf-sup condition, while the conditions for good approximation of eigenvalue problems of the second kind require the opposite: a stronger inf-sup condition and less than ellipticity in the kernel.

The aim of this section is to review how the conditions introduced in Sections 13 and 14 are actually not equivalent to the ellipticity in the kernel and the inf-sup condition. This is not surprising, for several reasons. First of all, the conditions for the source problem refer to equation (12.1), where the right-hand side consists of the generic functions f and g, which are present in both equations of the mixed problem; in contrast, the eigenvalue problems (13.1) and (14.1) are of different types, since in one equation the right-hand side vanishes. Moreover, there is an intrinsic difference between source problem and eigenvalue problem: the convergence of a source problem is usually reduced to a *pointwise* estimate (*i.e.*, for a generic right-hand side we look for a discrete solution which converges to the continuous one), while the convergence of the eigenvalue problem is related to a *uniform* estimate (see equation (7.8), for instance).

The uniform convergence is usually implied by suitable compactness assumptions and standard pointwise convergence. In the case of a mixed problem, it may be that compactness is not enough to turn pointwise convergence into uniform convergence (see the discussion at the beginning of Part 3 and, in particular, the definition of the solution operator T in (12.2)). A typical example of this situation is the Laplace eigenvalue problem in mixed form, where the operator (12.2) is not compact. It may occur in this situation that ellipticity in the kernel and the inf-sup condition hold true, while the eigenvalues are not correctly approximated. We recall an important counter-example (Boffi *et al.* 2000*a*), already mentioned in Section 5 in the framework of Maxwell's eigenvalue problem (see Table 5.3).

We consider the Laplace eigenvalue problem in mixed form: find $\lambda \in \mathbb{R}$ and $\boldsymbol{\sigma} \in \mathbf{H}(\mathrm{div};\Omega)$ such that, for $u \in L^2(\Omega)$,

$$(\boldsymbol{\sigma},\boldsymbol{\tau}) + (\operatorname{div}\boldsymbol{\tau}, u) = 0 \quad \forall \boldsymbol{\tau} \in \mathbf{H}(\mathrm{div};\Omega), \tag{15.3a}$$

$$(\operatorname{div}\boldsymbol{\sigma}, v) = -\lambda(u,v) \quad \forall v \in L^2(\Omega). \tag{15.3b}$$

Its approximation ($\Sigma_h \subset \mathbf{H}(\mathrm{div};\Omega)$, $U_h \subset L^2(\Omega)$) is then as follows: find $\lambda_h \in \mathbb{R}$ and $\boldsymbol{\sigma}_h \in \Sigma_h$ such that, for $u_h \in U_h$,

$$(\boldsymbol{\sigma}_h,\boldsymbol{\tau}) + (\operatorname{div}\boldsymbol{\tau}, u_h) = 0 \quad \forall \boldsymbol{\tau} \in \Sigma_h, \tag{15.4a}$$

$$(\operatorname{div}\boldsymbol{\sigma}_h, v) = -\lambda_h(u_h,v) \quad \forall v \in U_h. \tag{15.4b}$$

Let us consider a square domain $\Omega =]0,\pi[\times]0,\pi[$ and a criss-cross mesh sequence such as that presented in Figure 3.9. Let Σ_h be the space of continuous piecewise linear finite elements in each component and let U_h be the space containing all the divergences of the elements of Σ_h. It turns out that the equality $U_h = \operatorname{div}(\Sigma_h)$ easily implies that the ellipticity in the kernel property (15.1) is satisfied. Moreover, it can be proved that the proposed scheme satisfies the inf-sup condition (15.2) as well (Boffi *et al.* 2000*a*, Fix, Gunzburger and Nicolaides 1981).

Remark 15.1. Those familiar with the Stokes problem might recognize the inf-sup condition (15.2). It is well known that the $\mathcal{P}_1 - \mathcal{P}_0$ element does not provide a stable Stokes scheme. On the other hand, we are using a very particular mesh sequence (the inf-sup constant would tend to zero on a general mesh sequence) and a norm different to that in the case of the Stokes problem: here we consider the $\mathbf{H}(\operatorname{div};\Omega)$-norm, and not the full $H^1(\Omega)$-norm (even on the criss-cross mesh sequence, the inf-sup constant for the Stokes problem tends to zero when the $H^1(\Omega)$-norm is considered).

The classical theory implies that a quasi-optimal error estimate holds true for the approximation of the source problem associated with problem (15.3). More precisely, we consider the following source problem and its approximation: given $g \in L^2(\Omega)$, find $(\boldsymbol{\sigma}, u) \in \mathbf{H}(\operatorname{div};\Omega) \times L^2(\Omega)$ such that

$$(\boldsymbol{\sigma}, \boldsymbol{\tau}) + (\operatorname{div} \boldsymbol{\tau}, u) = 0 \quad \forall \boldsymbol{\tau} \in \mathbf{H}(\operatorname{div};\Omega), \tag{15.5a}$$

$$(\operatorname{div} \boldsymbol{\sigma}, v) = -(g, v) \quad \forall v \in L^2(\Omega), \tag{15.5b}$$

and find $(\boldsymbol{\sigma}_h, u_h) \in \Sigma_h \times U_h$ such that

$$(\boldsymbol{\sigma}_h, \boldsymbol{\tau}) + (\operatorname{div} \boldsymbol{\tau}, u_h) = 0 \quad \forall \boldsymbol{\tau} \in \Sigma_h, \tag{15.6a}$$

$$(\operatorname{div} \boldsymbol{\sigma}_h, v) = -(g, v) \quad \forall v \in U_h. \tag{15.6b}$$

Then we have the following error estimate for the solution of problem (15.5):

$$\begin{aligned} &\|\boldsymbol{\sigma} - \boldsymbol{\sigma}_h\|_{\mathbf{H}(\operatorname{div};\Omega)} + \|u - u_h\|_{L^2(\Omega)} \\ &\qquad \le C \inf_{\substack{\boldsymbol{\tau}_h \in \Sigma_h \\ v_h \in U_h}} \big(\|\boldsymbol{\sigma} - \boldsymbol{\tau}_h\|_{\mathbf{H}(\operatorname{div};\Omega)} + \|u - v_h\|_{L^2(\Omega)}\big). \end{aligned} \tag{15.7}$$

On the other hand, with our choice for the discrete spaces, problem (15.4) does not provide a good approximation of problem (15.3). The results of the computations are shown in Table 15.1. After a transient situation for the smallest meshes, a clear second order of convergence is detected towards the eigenvalues tagged as 'Exact'. Unfortunately, some of them (emphasized by the exclamation mark) do not correspond to eigenvalues of the continuous problem (15.3). This situation is very close to that presented in Table 5.3; it can be observed that the only difference between this computation and that of Section 5 consists in the boundary conditions.

Table 15.1. Eigenvalues computed with nodal elements on the criss-cross mesh sequence of triangles of Figure 3.9.

Exact	Computed (rate)				
	$N = 2$	$N = 4$	$N = 8$	$N = 16$	$N = 32$
2	2.2606	2.0679 (1.9)	2.0171 (2.0)	2.0043 (2.0)	2.0011 (2.0)
5	4.8634	5.4030 (−1.6)	5.1064 (1.9)	5.0267 (2.0)	5.0067 (2.0)
5	5.6530	5.4030 (0.7)	5.1064 (1.9)	5.0267 (2.0)	5.0067 (2.0)
!→ 6	5.6530	5.6798 (0.1)	5.9230 (2.1)	5.9807 (2.0)	5.9952 (2.0)
8	11.3480	9.0035 (1.7)	8.2715 (1.9)	8.0685 (2.0)	8.0171 (2.0)
10	11.3480	11.3921 (−0.0)	10.4196 (1.7)	10.1067 (2.0)	10.0268 (2.0)
10	12.2376	11.4495 (0.6)	10.4197 (1.8)	10.1067 (2.0)	10.0268 (2.0)
13	12.2376	11.6980 (−0.8)	13.7043 (0.9)	13.1804 (2.0)	13.0452 (2.0)
13	12.9691	11.6980 (−5.4)	13.7043 (0.9)	13.1804 (2.0)	13.0452 (2.0)
!→15	13.9508	15.4308 (1.3)	13.9669 (−1.3)	14.7166 (1.9)	14.9272 (2.0)
!→15	16.1534	15.4308 (1.4)	13.9669 (−1.3)	14.7166 (1.9)	14.9272 (2.0)
17	16.1534	17.0972 (3.1)	18.1841 (−3.6)	17.3073 (1.9)	17.0773 (2.0)
17		18.2042	18.1841 (0.0)	17.3073 (1.9)	17.0773 (2.0)
18		18.3067	19.3208 (−2.1)	18.3456 (1.9)	18.0867 (2.0)
20		20.1735	21.5985 (−3.2)	20.4254 (1.9)	20.1070 (2.0)
20		20.1735	21.6015 (−3.2)	20.4254 (1.9)	20.1070 (2.0)
!→24		27.5131	22.7084 (1.4)	23.6919 (2.1)	23.9230 (2.0)
25		27.6926	24.8559 (4.2)	25.6644 (−2.2)	25.1672 (2.0)
25		28.0122	24.8586 (4.4)	25.6644 (−2.2)	25.1672 (2.0)
26		30.4768	27.3758 (1.7)	26.7152 (0.9)	26.1805 (2.0)

From the computed eigenvalues, it is clear that the conclusions of neither Theorem 14.6 nor 14.7 hold. Let us try to better understand this phenomenon.

Let us forget for a moment the theory developed in Section 14 and try to use the argument of Proposition 7.6, to see whether the pointwise convergence arising from the stability of the source problem approximation implies the required uniform convergence. One of the main issues concerns the appropriate definition of the solution operator.

A first possible (but not useful) definition for the solution operator is given by $T_{\Sigma U} : L^2(\Omega) \times L^2(\Omega) \to L^2(\Omega) \times L^2(\Omega)$, $T(f,g) = (\sigma, u)$, where (σ, u) is the solution of (15.5) with datum g. We can define the discrete solution operator $T_{\Sigma U,h}$ in a natural way and try to show that the hypotheses of Proposition 7.6 are satisfied. With the notation of Section 7, we have $H = L^2(\Omega) \times L^2(\Omega)$ and $V = \mathbf{H}(\mathrm{div};\Omega) \times L^2(\Omega)$. Unfortunately, it turns out that the operator $T_{\Sigma U}$ is not compact in $\mathcal{L}(L^2(\Omega) \times L^2(\Omega), \mathbf{H}(\mathrm{div};\Omega) \times L^2(\Omega))$ as required; indeed, the component σ of the solution of (15.5) cannot be in a compact subset of $\mathbf{H}(\mathrm{div};\Omega)$ since the divergence of $\boldsymbol{\sigma}$ is equal to g which is only in $L^2(\Omega)$. We can try to use the modified version of Proposition 7.6 introduced in Remark 7.5; however, this does not help since the operator $T_{\Sigma U}$ is not compact in $\mathcal{L}(\mathbf{H}(\mathrm{div};\Omega)\times L^2(\Omega),\mathbf{H}(\mathrm{div};\Omega)\times L^2(\Omega))$ either, for the same reason as before ($\mathrm{div}\,\boldsymbol{\sigma}=-g\in L^2(\Omega)$). On the other hand, the error estimate (15.7) does not give any significant improvement: we have

$$\|(T_{\Sigma U}-T_{\Sigma U,h})(f,g)\|_{\mathbf{H}(\mathrm{div};\Omega)\times L^2(\Omega)} \le C \inf_{\substack{\boldsymbol{\tau}_h\in\Sigma_h\\ v_h\in U_h}} \big(\|\boldsymbol{\sigma}-\boldsymbol{\tau}_h\|_{\mathbf{H}(\mathrm{div};\Omega)}+\|u-v_h\|_{L^2(\Omega)}\big).$$

For the same reason as in the previous comments ($\mathrm{div}\,\sigma=-g\in L^2(\Omega)$), we cannot hope to get a uniform bound of the term $\|\boldsymbol{\sigma}-\boldsymbol{\tau}_h\|_{\mathbf{H}(\mathrm{div};\Omega)}$, for which a higher regularity of $\mathrm{div}\,\boldsymbol{\sigma}$ would be required.

According to Section 14, now let $T: L^2(\Omega)\to L^2(\Omega)$ be the operator associated with the second component of the solution to problem (15.5): $Tg=u$; let $T_h: L^2(\Omega)\to L^2(\Omega)$ be its discrete counterpart, $T_hg=u_h$. A direct application of estimate (15.7) gives, as before,

$$\|(T-T_h)g\|_{L^2(\Omega)}\le C \inf_{\substack{\boldsymbol{\tau}_h\in\Sigma_h\\ v_h\in U_h}} \big(\|\boldsymbol{\sigma}-\boldsymbol{\tau}_h\|_{\mathbf{H}(\mathrm{div};\Omega)}+\|u-v_h\|_{L^2(\Omega)}\big).$$

Again, we cannot hope to obtain a uniform convergence since $\mathrm{div}\,\boldsymbol{\sigma}$ is only in $L^2(\Omega)$. Indeed, the profound meaning of Theorems 14.6 and 14.7 in this particular situation is that we try to obtain a uniform estimate by bounding the term $\boldsymbol{\sigma}-\boldsymbol{\sigma}_h$ in $L^2(\Omega)$ and not in $\mathbf{H}(\mathrm{div};\Omega)$; this task is performed by the Fortin operator through the Fortid condition (see Definition 14.5).

A more careful analysis (Boffi *et al.* 2000*a*) of the particular scheme we discuss in this section ($\mathcal{P}_1-\mathcal{P}_0$ element on the criss-cross mesh), shows that, indeed, T_h does not converge uniformly to T.

Proposition 15.2. There exists a sequence $\{v_h^*\}$ with $v_h^*\in U_h$ such that $\|v_h^*\|_{L^2(\Omega)}=1$ for any h and $\|(T-T_h)v_h^*\|_{L^2(\Omega)}$ does not tend to zero as h tends to zero.

Proof. We follow the proof of Boffi *et al.* (2000*a*, Theorem 5.2). Qin (1994), following an idea of Boland and Nicolaides (1985), showed that there exists a sequence $\{\tilde v_h\}$ such that

$$(\mathrm{div}\,\boldsymbol{\tau}_h,\tilde v_h)\le C\|\boldsymbol{\tau}_h\|_{L^2(\Omega)}\|\tilde v_h\|_{L^2(\Omega)}\quad \forall\boldsymbol{\tau}_h\in\Sigma_h.$$

Defining $v_h^*=\tilde v_h/\|\tilde v_h\|_{L^2(\Omega)}$ gives

$$(\mathrm{div}\,\boldsymbol{\tau}_h,v_h^*)\le C\|\boldsymbol{\tau}_h\|_{L^2(\Omega)}\quad \forall\boldsymbol{\tau}_h\in\Sigma_h. \tag{15.8}$$

It turns out that v_h^* has zero mean value in each square macro-element composed of four triangles (it is indeed constructed by using a suitable checkerboard structure) and hence its weak limit is zero. From the compactness of T it follows that Tv_h^* tends to zero strongly in $L^2(\Omega)$. By the definition of T_h, we have that $T_h v_h^*$ is the second component u_h of the solution to the following mixed problem:

$$(\boldsymbol{\sigma}_h, \boldsymbol{\tau}) + (\operatorname{div} \boldsymbol{\tau}, u_h) = 0 \qquad \forall \boldsymbol{\tau} \in \Sigma_h,$$
$$(\operatorname{div} \boldsymbol{\sigma}_h, v) = -(v_h^*, v) \qquad \forall v \in U_h.$$

We have

$$|(\operatorname{div} \boldsymbol{\sigma}_h, u_h)| = |(v_h^*, u_h)| \leq \|u_h\|_{L^2(\Omega)},$$
$$|(\operatorname{div} \boldsymbol{\sigma}_h, u_h)| = \|\boldsymbol{\sigma}_h\|_{L^2(\Omega)}^2.$$

Moreover, from (15.8) we have

$$\|\boldsymbol{\sigma}_h\|_{L^2(\Omega)} \geq \frac{1}{C} |(\operatorname{div} \boldsymbol{\sigma}_h, v_h^*)| = \frac{1}{C}.$$

Putting together the last equations gives the final result,

$$\|u_h\|_{L^2(\Omega)} \geq \frac{1}{C^2}. \qquad \square$$

PART FOUR
The language of differential forms

The use of differential forms and homological techniques for the analysis of the finite element approximation of partial differential equations has become a popular and effective tool in the recent literature.

The aim of this part is to provide the reader with some basic notions about the de Rham complex and its role in the analysis of eigenvalue problems arising from partial differential equations. The experience of the author in this field comes from the approximation of Maxwell's eigenvalue problem; we shall, however, see how the abstract setting can be used for the analysis of a wider class of applications. For a thorough introduction to this subject in the framework of the numerical analysis of partial differential equations, we refer the interested reader to Arnold, Falk and Winther (2006*b*).

16. Preliminaries

In this section we recall the definitions of the main entities we are going to use in our analysis.

For any integer k with $0 \leq k \leq n$, let $\Lambda^k = \Lambda^k(\mathbb{R}^n)$ be the vector space of alternating k-linear forms on $\mathbb{R}^n$, so $\dim(\Lambda^k) = \binom{n}{k}$. The Euclidean norm of $\mathbb{R}^n$ induces a norm on Λ^k for any k and the wedge product '$\wedge$' acts from $\Lambda^i \times \Lambda^j$ to Λ^{i+j}. Given a domain $\Omega \subset \mathbb{R}^n$, we denote by $C^\infty(\Omega, \Lambda^k)$ the space of smooth differential forms of order k in Ω and by $\mathbf{\Lambda}(\Omega)$ the corresponding anti-commutative graded algebra

$$\mathbf{\Lambda}(\Omega) = \bigoplus_k C^\infty(\Omega, \Lambda^k).$$

An exterior derivative $\mathbf{d} : \mathbf{\Lambda}(\Omega) \to \mathbf{\Lambda}(\Omega)$ is a linear graded operator of degree one, that is, for any k, we are given a map

$$d_k : C^\infty(\Omega, \Lambda^k) \to C^\infty(\Omega, \Lambda^{k+1}).$$

We assume that $d_{k+1} \circ d_k = 0$, that is, $\mathbf{d}$ is a differential.[1]

The set $\mathbf{L}^2(\Omega, \Lambda^k)$ is the space of differential k-forms on Ω with square integrable coefficients in their canonical basis representation; its inner product is given by

$$(\mathbf{u}, \mathbf{v}) = \int_\Omega \mathbf{u} \wedge \star\mathbf{v} \quad \forall \mathbf{u}, \mathbf{v} \in \mathbf{L}^2(\Omega, \Lambda^k),$$

[1] The external derivative we are going to use can be naturally defined as follows: given $\omega \in C^\infty(\Omega, \Lambda^k)$, $d_k\omega_x(v_1, \ldots, v_{k+1}) = \sum_{j=1}^{k+1}(-1)^{j+1}\partial_{v_j}\omega_x(v_1, \ldots, \hat{v}_j, \ldots, v_{k+1})$ (Arnold *et al.* 2006*b*, p. 15)

where $\star$ denotes the Hodge star operator mapping k-forms to $(n-k)$-forms. Differential k-forms with different regularity can be considered by using standard Sobolev spaces: the corresponding spaces will be denoted by $\mathbf{H}^s(\Omega, \Lambda^k)$. The main spaces of our construction are based on the exterior derivative

$$\mathbf{H}(d_k;\Omega) = \{\mathbf{v} \in \mathbf{L}^2(\Omega, \Lambda^k) : d_k\mathbf{v} \in \mathbf{L}^2(\Omega, \Lambda^{k+1})\},$$

and are endowed with their natural scalar product

$$(\mathbf{u}, \mathbf{v})^2_{\mathbf{H}(d_k;\Omega)} = (\mathbf{u}, \mathbf{v})^2_{\mathbf{L}^2(\Omega,\Lambda^k)} + (d_k\mathbf{u}, d_k\mathbf{v})^2_{\mathbf{L}^2(\Omega,\Lambda^{k+1})} \quad \forall \mathbf{u}, \mathbf{v} \in \mathbf{H}(d_k;\Omega).$$

It can be shown that a trace operator $\mathbf{tr}_{\partial\Omega}$ is well defined on $\mathbf{H}(d_k;\Omega)$, so that it makes sense to introduce the subspace of $\mathbf{H}(d_k;\Omega)$ consisting of differential forms with vanishing boundary conditions:

$$\mathbf{H}_0(d_k;\Omega) = \{\mathbf{v} \in \mathbf{H}(d_k;\Omega) : \mathbf{tr}_{\partial\Omega}\,\mathbf{v} = 0\}.$$

The coderivative operator $\delta_k : C^\infty(\Omega, \Lambda^k) \to C^\infty(\Omega, \Lambda^{k-1})$ is defined by

$$\delta_k = \star d_{n-k}\star,$$

and is the formal adjoint of d_{k-1}. Indeed, we have the following generalization of the integration by parts:

$$(d_{k-1}\mathbf{p}, \mathbf{u}) = (\mathbf{p}, \delta_k\mathbf{u}) + \langle \mathbf{tr}_{\partial\Omega}\,\mathbf{p}, \mathbf{tr}_{\partial\Omega} \star\mathbf{u}\rangle.$$

We can define Hilbert spaces associated with the coderivative

$$\mathbf{H}(\delta_k;\Omega) = \{\mathbf{v} \in \mathbf{L}^2(\Omega, \Lambda^k) : \delta_k\mathbf{v} \in \mathbf{L}^2(\Omega, \Lambda^{k-1})\}.$$

For $\mathbf{u} \in \mathbf{H}(\delta_k;\Omega)$ we have $\star\mathbf{u} \in \mathbf{H}(d_{n-k};\Omega)$, so it makes sense to consider $\mathbf{tr}_{\partial\Omega}(\star\mathbf{u})$ and to define $\mathbf{H}_0(\delta_k;\Omega)$ as $\star\mathbf{H}_0(d_{n-k};\Omega)$, that is,

$$\mathbf{H}_0(\delta_k;\Omega) = \{\mathbf{v} \in \mathbf{H}(\delta_k;\Omega) : \mathbf{tr}_{\partial\Omega}(\star\mathbf{v}) = 0\}.$$

Before introducing additional definitions and the fundamental notion of the de Rham complex, it might be useful to recall how functions of differential forms can be identified with standard functional spaces in two and three space dimensions. Following Arnold *et al.* (2006*b*), the identification is performed in a standard way by means of Euclidean vector proxies and is reported in Table 16.1 (Boffi *et al.* 2009).

The de Rham complex is given by the chain

$$0 \longrightarrow \mathbf{H}(d_0;\Omega) \xrightarrow{d_0} \mathbf{H}(d_1;\Omega) \xrightarrow{d_1} \cdots \xrightarrow{d_{n-1}} \mathbf{H}(d_n;\Omega) \longrightarrow 0.$$

The analogous complex when boundary conditions are considered is

$$0 \longrightarrow \mathbf{H}_0(d_0;\Omega) \xrightarrow{d_0} \mathbf{H}_0(d_1;\Omega) \xrightarrow{d_1} \cdots \xrightarrow{d_{n-1}} \mathbf{H}_0(d_n;\Omega) \longrightarrow 0.$$

Table 16.1. Identification between differential forms and vector proxies in $\mathbb{R}^2$ and $\mathbb{R}^3$.

	Differential form	Proxy representation $n=2$	Proxy representation $n=3$
$k=0$	d_0	$\mathbf{grad}$	$\mathbf{grad}$
	$\mathbf{tr}_{\partial\Omega}\,\phi$	$\phi_{\vert\partial\Omega}$	$\phi_{\vert\partial\Omega}$
	$H_0(d_0,\Omega)$	$H_0^1(\Omega)$	$H_0^1(\Omega)$
	δ_1	$-\operatorname{div}$	$-\operatorname{div}$
$k=1$	d_1	rot	$\mathbf{curl}$
	$\mathbf{tr}_{\partial\Omega}\,\mathbf{u}$	$(\mathbf{u}\cdot\mathbf{t})_{\vert\partial\Omega}$	$\mathbf{n}\times(\mathbf{u}\times\mathbf{n})_{\vert\partial\Omega}$
	$H_0(d_1,\Omega)$	$\mathbf{H}_0(\mathrm{rot})$	$\mathbf{H}_0(\mathbf{curl})$
	δ_2	$\overrightarrow{\mathrm{rot}}$	$\mathbf{curl}$
$k=2$	d_2	0	div
	$\mathbf{tr}_{\partial\Omega}\,\mathbf{q}$	0	$(\mathbf{q}\cdot\mathbf{n})_{\vert\partial\Omega}$
	$H_0(d_2,\Omega)$	$L_0^2(\Omega)$	$\mathbf{H}_0(\mathrm{div})$
	δ_3		$-\,\mathbf{grad}$

Since $d_{k+1}\circ d_k = 0$, the de Rham complex is a cochain complex, that is, the kernel of d_{k+1} contains the range of d_k. The quotient spaces between the kernels and the ranges of the exterior derivatives are called cohomology spaces and have finite dimension, which is related to the topology of Ω (the dimension of the kth cohomology space is called kth Betti number). The kth cohomology space is given by the set of harmonic differential forms:

$$\mathcal{H}^k = \{\mathbf{v}\in\mathbf{H}(d_k;\Omega)\cap\mathbf{H}_0(\delta_k;\Omega) : d_k\mathbf{v}=0,\ \delta_k\mathbf{v}=0\}$$

and

$$\mathcal{H}_0^k = \{\mathbf{v}\in\mathbf{H}_0(d_k;\Omega)\cap\mathbf{H}(\delta_k;\Omega) : d_k\mathbf{v}=0,\ \delta_k\mathbf{v}=0\},$$

respectively.

In this survey, we are going to consider the case when the Betti numbers corresponding to k different from 0 and n vanish. This essentially means that the de Rham complex is exact[2] and corresponds to the case when the domain Ω is contractible. On the other hand, even if this assumption

[2] The de Rham complex is exact in the case of trivial cohomology if, in the definition of the first (last, respectively, when boundary conditions are considered) space, 0 is replaced by $\mathbb{R}$. We consider this modification in the rest of our survey.

considerably simplifies the exposition, the results we are going to present generalize to the case when the topology may not be trivial; the techniques for dealing with the more general case correspond to those used, for instance, in Arnold *et al.* (2006*b*).

For the discretization of the spaces of differential forms, we introduce spaces of discrete differential forms $\mathcal{V}_h^k \subset \mathbf{H}(d_k;\Omega)$ $(k=0,\dots,n)$. A typical setting involves appropriate projection operators $\pi_h^k : \mathbf{H}(d_k;\Omega) \to \mathcal{V}_h^k$ such that the following diagram commutes:

$$\begin{array}{ccccccccccc}
\mathbb{R} & \longrightarrow & \mathbf{H}(d_0;\Omega) & \xrightarrow{d_0} & \mathbf{H}(d_1;\Omega) & \xrightarrow{d_1} & \cdots & \xrightarrow{d_{n-1}} & \mathbf{H}(d_n;\Omega) & \longrightarrow & 0 \\
 & & \pi_h^0 \big\downarrow & & \pi_h^1 \big\downarrow & & & & \pi_h^n \big\downarrow & & \\
\mathbb{R} & \longrightarrow & \mathcal{V}_h^0 & \xrightarrow{d_0} & \mathcal{V}_h^1 & \xrightarrow{d_1} & \cdots & \xrightarrow{d_{n-1}} & \mathcal{V}_h^n & \longrightarrow & 0.
\end{array}$$

We have an analogous diagram when boundary conditions are considered. If $\mathcal{V}_h^k \subset \mathbf{H}_0(d_k;\Omega)$ $(k=0,\dots,n)$ and suitable projection operators π_h^k : $\mathbf{H}_0(d_k;\Omega) \to \mathcal{V}_h^k$ are considered, then we use

$$\begin{array}{ccccccccccc}
0 & \longrightarrow & \mathbf{H}_0(d_0;\Omega) & \xrightarrow{d_0} & \mathbf{H}_0(d_1;\Omega) & \xrightarrow{d_1} & \cdots & \xrightarrow{d_{n-1}} & \mathbf{H}_0(d_n;\Omega) & \longrightarrow & \mathbb{R} \\
 & & \pi_h^0 \big\downarrow & & \pi_h^1 \big\downarrow & & & & \pi_h^n \big\downarrow & & \\
0 & \longrightarrow & \mathcal{V}_h^0 & \xrightarrow{d_0} & \mathcal{V}_h^1 & \xrightarrow{d_1} & \cdots & \xrightarrow{d_{n-1}} & \mathcal{V}_h^n & \longrightarrow & \mathbb{R}.
\end{array} \tag{16.1}$$

Another important tool we use is the Hodge decomposition, which can be easily expressed by means of the cycles and the boundaries coming from the de Rham complex (Arnold *et al.* 2006*b*, equation 2.18). Roughly speaking, the Hodge decomposition states that every k-form $\mathbf{u}$ can be split as the sum of three components:

$$\mathbf{u} = d_{k-1}\boldsymbol{\alpha} + \delta_{k+1}\boldsymbol{\beta} + \boldsymbol{\gamma},$$

where $\boldsymbol{\alpha}$ and $\boldsymbol{\beta}$ are $(k-1)$- and $(k+1)$-forms, respectively, and $\boldsymbol{\gamma} \in \mathcal{H}^k$ is a harmonic k-form. More precisely, it turns out that exact (*i.e.*, in the range of d_{k-1}) and co-exact (*i.e.*, in the range of δ_{k+1}) k-forms are orthogonal in $\mathbf{L}^2(\Omega,\Lambda^k)$: it follows that the orthogonal complement of exact and co-exact k-forms consists of forms that are simultaneously closed (*i.e.*, in the kernel of d_k) and co-closed (*i.e.*, in the kernel of δ_k), that is, of harmonic k-forms. In the particular case we consider, there are no harmonic forms and the Hodge decomposition says that $\mathbf{L}^2(\Omega,\Lambda^k)$ can be presented as the direct sum of $d_{k-1}(\mathbf{H}(d_{k-1};\Omega))$ and $\delta_{k+1}(\mathbf{H}_0(\delta_{k+1};\Omega))$. A second Hodge decomposition with different boundary conditions says

$$\mathbf{L}^2(\Omega,\Lambda^k) = d_{k-1}(\mathbf{H}_0(d_{k-1};\Omega)) \oplus \delta_{k+1}(\mathbf{H}(\delta_{k+1};\Omega)). \tag{16.2}$$

17. The Hodge–Laplace eigenvalue problem

The main object of our analysis is the following symmetric variational eigenvalue problem: find $\lambda \in \mathbb{R}$ and $\mathbf{u} \in \mathbf{H}_0(d_k;\Omega)$ with $\mathbf{u} \neq 0$ such that

$$(d_k\mathbf{u}, d_k\mathbf{v}) = \lambda(\mathbf{u},\mathbf{v}) \quad \forall \mathbf{v} \in \mathbf{H}_0(d_k;\Omega). \tag{17.1}$$

Taking $\mathbf{v} = \mathbf{u}$, it follows that λ cannot be negative. For $k = 0$, problem (17.1) reduces to the standard Laplace eigenvalue problem. For $k \geq 1$, which is the most interesting case, $\lambda = 0$ implies $d_k\mathbf{u} = 0$, that is, $\mathbf{u} = d_{k-1}\boldsymbol{\alpha}$ for some $\boldsymbol{\alpha} \in \mathbf{H}_0(d_{k-1};\Omega)$. On the other hand, $\lambda > 0$ implies $\delta_k\mathbf{u} = 0$, as can be seen by taking $\mathbf{v} = d_{k-1}\boldsymbol{\alpha}$ in (17.1) for an arbitrary $\boldsymbol{\alpha} \in \mathbf{H}_0(d_{k-1};\Omega)$ and using the orthogonalities discussed when introducing the Hodge decomposition (16.2). This means that intrinsic constraints are hidden in the formulation of problem (17.1): either λ vanishes and is associated to the infinite-dimensional eigenspace $d_{k-1}(\mathbf{H}_0(d_{k-1};\Omega))$ (that is, $\mathbf{u}$ is a closed form), or the eigenfunctions $\mathbf{u}$ associated with positive values of λ are coclosed forms (that is, $\delta_k\mathbf{u} = 0$).

One motivation for the study of problem (17.1) comes from the fact that, for $k = 1$ and $n = 3$, it corresponds to the Maxwell eigenvalue problem (5.3) (see the identifications in Table 16.1). Moreover, for $k = 0$, problem (17.1) reduces to the well-known eigenvalue problem for the Laplace operator. Another interesting application is given for $k = 2$ and $n = 3$, where the eigenvalue problem associated with the **grad** div operator is obtained: this operator plays an important role in the approximation of fluid–structure interaction and acoustic problems (Bermúdez and Rodríguez 1994, Bermúdez *et al.* 1995, Bathe, Nitikitpaiboon and Wang 1995, Gastaldi 1996, Boffi, Chinosi and Gastaldi 2000*b*).

Remark 17.1. We have introduced the eigenvalue problem (17.1) in the space $\mathbf{H}_0(d_k;\Omega)$, which involves essential Dirichlet boundary conditions in the sense that the definition of our functional space implies $\mathbf{tr}_{\partial\Omega}\,\mathbf{u} = 0$. The same problem can be considered in the space $\mathbf{H}(d_k;\Omega)$ and would correspond to natural Neumann boundary conditions. Since the modifications involved with the analysis of the Neumann problem are standard, we limit our presentation to the Dirichlet problem.

Remark 17.2. The term 'Dirichlet' used in the previous remark needs a more precise explanation. From the technical point of view, problem (17.1) is a Dirichlet problem, since essential boundary conditions are imposed in the space $\mathbf{H}_0(d_k;\Omega)$ and we are sticking with this terminology.

On the other hand, a Dirichlet problem in the framework of differential forms might correspond to a different type of boundary conditions when translated to a more conventional language. This is sometimes the case when using the proxy identification of Table 16.1 to reduce problem (17.1) to standard mixed formulations. For instance, the case $k = 2$ and $n = 3$

corresponds to an eigenvalue problem associated with the **grad** div operator, which turns out to be equivalent to the Neumann problem for the Laplace eigenvalue problem. A similar situation occurs in the case $k = 1$ and $n = 2$, which corresponds to the Maxwell eigenvalue problem in two space dimensions with perfectly conducting boundary conditions $\mathbf{u} \cdot \mathbf{t} = 0$ on the boundary. This problem is equivalent to the mixed formulation of the Neumann–Laplace eigenvalue problem ($\mathbf{u}$ corresponds to the gradient of the solution u rotated by the angle $\pi/2$, so that $\mathbf{u} \cdot \mathbf{t}$ means $\partial u/\partial n$).

Problem (17.1) is strictly related to the Hodge–Laplace elliptic eigenvalue problem (Arnold *et al.* 2006*b*, Arnold, Falk and Winther 2010): find $\omega \in \mathbb{R}$ and $\boldsymbol{\sigma} \in \mathbf{H}_0(d_k;\Omega) \cap \mathbf{H}(\delta_k;\Omega)$ with $\boldsymbol{\sigma} \neq 0$ such that

$$(d_k\boldsymbol{\sigma}, d_k\boldsymbol{\tau}) + (\delta_k\boldsymbol{\sigma}, \delta_k\boldsymbol{\tau}) = \omega(\boldsymbol{\sigma}, \boldsymbol{\tau}) \quad \forall \boldsymbol{\tau} \in \mathbf{H}_0(d_k;\Omega) \cap \mathbf{H}(\delta_k;\Omega). \qquad (17.2)$$

It is clear that all solutions to (17.2) have positive frequency $\omega > 0$ (since Ω is assumed to have trivial topology). Moreover, problem (17.2) is associated with a *compact* solution operator; this important property is a consequence of the compact embedding of $\mathbf{H}_0(d_k;\Omega) \cap \mathbf{H}(\delta_k;\Omega)$ into $\mathbf{L}^2(\Omega, \Lambda^k)$ (Picard 1984).

It is possible to classify the solutions to (17.2) into two distinct families. The first family corresponds to the solutions $(\lambda, \mathbf{u})$ to (17.1) with positive frequency (we have already observed that in this case $\delta_k\mathbf{u} = 0$, so that it is clear that $(\omega, \boldsymbol{\sigma}) = (\lambda, \mathbf{u})$ is a solution to (17.2)). The second family is given by forms $\boldsymbol{\sigma} \in \mathbf{H}(\delta_k;\Omega)$ satisfying

$$(\delta_k\boldsymbol{\sigma}, \delta_k\boldsymbol{\tau}) = \omega(\boldsymbol{\sigma}, \boldsymbol{\tau}) \quad \forall \boldsymbol{\tau} \in \mathbf{H}(\delta_k;\Omega)$$

with $\omega > 0$, which implies, in particular, $d_k\boldsymbol{\sigma} = 0$.

We are interested in the first family of solutions to the Hodge–Laplace eigenvalue problem (17.2) in the case $k \geq 1$. From the above discussion, the problem can be written in the following way: find $\lambda \in \mathbb{R}$ and $\mathbf{u} \in \mathbf{H}_0(d_k;\Omega)$ with $\mathbf{u} \neq 0$ such that

$$(d_k\mathbf{u}, d_k\mathbf{v}) = \lambda(\mathbf{u}, \mathbf{v}) \quad \forall \mathbf{v} \in \mathbf{H}_0(d_k;\Omega), \qquad (17.3a)$$

$$(\mathbf{u}, d_{k-1}\mathbf{q}) = 0 \quad \forall \mathbf{q} \in \mathbf{H}_0(d_{k-1};\Omega). \qquad (17.3b)$$

A natural mixed formulation associated with problem (17.3) can be constructed as follows: find $\lambda \in \mathbb{R}$ and $\mathbf{u} \in \mathbf{H}_0(d_k;\Omega)$ with $\mathbf{u} \neq 0$ such that, for $\mathbf{p} \in \mathbf{H}_0(d_{k-1};\Omega)$,

$$(d_k\mathbf{u}, d_k\mathbf{v}) + (\mathbf{v}, d_{k-1}\mathbf{p}) = \lambda(\mathbf{u}, \mathbf{v}) \quad \forall \mathbf{v} \in \mathbf{H}_0(d_k;\Omega), \qquad (17.4a)$$

$$(\mathbf{u}, d_{k-1}\mathbf{q}) = 0 \quad \forall \mathbf{q} \in \mathbf{H}_0(d_{k-1};\Omega). \qquad (17.4b)$$

Taking $\mathbf{v} = d_{k-1}\mathbf{p}$ in (17.4a) and using (17.4b) easily give $d_{k-1}\mathbf{p} = 0$, which shows that all solutions to (17.4) solve (17.3) as well. *Vice versa*, it is clear that a solution to (17.3) solves (17.4) with $\mathbf{p} = 0$.

In the case of Maxwell's eigenvalue problem, formulation (17.4) is often referred to as Kikuchi's formulation (Kikuchi 1987).

It is clear that the value of $\mathbf{p}$ cannot be uniquely determined when $k \geq 2$, since if $(\lambda, \mathbf{u}, \mathbf{p})$ satisfies (17.4), then $(\lambda, \mathbf{u}, \mathbf{p} + d_{k-2}\boldsymbol{\alpha})$ satisfies (17.4) as well for any $\boldsymbol{\alpha} \in \mathbf{H}_0(d_{k-2}; \Omega)$. It might be then interesting to consider the following modified problem, which avoids the indeterminacy of $\mathbf{p}$. Given the space

$$\mathbf{K}^{\delta}_{k-1} = \{\mathbf{v} \in \mathbf{H}_0(d_{k-1}; \Omega) \cap \mathbf{H}(\delta_{k-1}; \Omega) : \delta_{k-1}\mathbf{v} = 0\},$$

find $\lambda \in \mathbb{R}$ and $\mathbf{u} \in \mathbf{H}_0(d_k; \Omega)$ with $\mathbf{u} \neq 0$ such that, for $\mathbf{p} \in \mathbf{K}^{\delta}_{k-1}$,

$$(d_k\mathbf{u}, d_k\mathbf{v}) + (\mathbf{v}, d_{k-1}\mathbf{p}) = \lambda(\mathbf{u}, \mathbf{v}) \quad \forall \mathbf{v} \in \mathbf{H}_0(d_k; \Omega), \tag{17.5a}$$

$$(\mathbf{u}, d_{k-1}, \mathbf{q}) = 0 \qquad \forall \mathbf{q} \in \mathbf{K}^{\delta}_{k-1}. \tag{17.5b}$$

Formulation (17.5) is, however, not suited to the numerical approximation, since it is not obvious how to introduce a conforming approximation of the space $\mathbf{K}^{\delta}_{k-1}$. For this reason, we are going to use formulation (17.4); the fact that $\mathbf{p}$ might not be uniquely determined by $\mathbf{u}$ is not a problem, since we are interested in the eigenfunction $\mathbf{u}$.

Following Boffi *et al.* (1999*b*), a second mixed formulation can be obtained as follows. Given the space

$$\mathbf{K}^{d}_{k+1} = d_k(\mathbf{H}_0(d_k; \Omega)) \subset \mathbf{H}_0(d_{k+1}; \Omega),$$

find $\lambda \in \mathbb{R}$ and $\mathbf{s} \in \mathbf{K}^{d}_{k+1}$ such that, for $\mathbf{u} \in \mathbf{H}_0(d_k; \Omega)$,

$$(\mathbf{u}, \mathbf{v}) + (d_k\mathbf{v}, \mathbf{s}) = 0 \qquad \forall \mathbf{v} \in \mathbf{H}_0(d_k; \Omega), \tag{17.6a}$$

$$(d_k\mathbf{u}, \mathbf{t}) = -\lambda(\mathbf{s}, \mathbf{t}) \quad \forall \mathbf{t} \in \mathbf{K}^{d}_{k+1}. \tag{17.6b}$$

It is clear that any solution to problem (17.6) is associated with a positive frequency $\lambda > 0$. Indeed, if $\lambda = 0$ then (17.6b) implies $d_k\mathbf{u} = 0$, and taking $\mathbf{v} = \mathbf{u}$ in (17.6a) gives $\mathbf{u} = 0$, which contradicts the existence of solutions with vanishing frequency. Hence, (17.6b) gives the fundamental relation $\mathbf{s} = -d_k\mathbf{u}/\lambda$, which yields the equivalence between (17.6a) and (17.3a).

Remark 17.3. The similar notations used for the spaces $\mathbf{K}^{\delta}_{k-1}$ and $\mathbf{K}^{d}_{k+1}$ are compatible, in the sense that the space $\mathbf{K}^{d}_{k+1}$ is made of functions in the kernel of the operator d_{k+1}, in analogy to the space $\mathbf{K}^{\delta}_{k-1}$, which contains functions in the kernel of the operator δ_{k-1}.

The equivalence between the mixed formulations and problem (17.1) is stated in the following proposition.

Proposition 17.4. If $(\lambda, \mathbf{u}) \in \mathbb{R} \times \mathbf{H}_0(d_k; \Omega)$ is a solution of (17.1) with $\lambda > 0$, that is, if $(\lambda, \mathbf{u})$ is a solution of (17.3), then $(\lambda, \mathbf{u})$ is a solution of (17.4), and there exists $\mathbf{s} \in \mathbf{K}^{d}_{k+1}$ such that $(\lambda, \mathbf{s})$ is a solution of (17.6).

Conversely, if $(\lambda, \mathbf{u}) \in \mathbb{R} \times \mathbf{H}_0(d_k; \Omega)$ is a solution of (17.4), then $\lambda > 0$ and $(\lambda, \mathbf{u})$ solves (17.3) and (17.1); if $(\lambda, \mathbf{s}) \in \mathbb{R} \times \mathbf{K}^d_{k+1}$ is a solution of (17.6) for some $\mathbf{u} \in \mathbf{H}_0(d_k; \Omega)$, then $(\lambda, \mathbf{u})$ solves (17.3) and (17.1).

18. Approximation of the mixed formulations

The aim of this section is to translate the abstract theory presented in Part 3 into the language of differential forms and to apply it to the approximation of problems (17.4) and (17.6).

18.1. Approximation of problem (17.4)

With the notation introduced at the beginning of this part, the discretization of problem (17.4) involves the spaces $\mathcal{V}_h^k \subset \mathbf{H}_0(d_k; \Omega)$ and $\mathcal{V}_h^{k-1} \subset \mathbf{H}_0(d_{k-1}; \Omega)$. The discrete formulation is as follows: find $\lambda_h \in \mathbb{R}$ and $\mathbf{u}_h \in \mathcal{V}_h^k$ with $\mathbf{u}_h \neq 0$ such that, for $\mathbf{p}_h \in \mathcal{V}_h^{k-1}$,

$$(d_k \mathbf{u}_h, d_k \mathbf{v}) + (\mathbf{v}, d_{k-1} \mathbf{p}_h) = \lambda_h (\mathbf{u}_h, \mathbf{v}) \quad \forall \mathbf{v} \in \mathcal{V}_h^k, \tag{18.1a}$$

$$(\mathbf{u}_h, d_{k-1} \mathbf{q}) = 0 \quad \forall \mathbf{q} \in \mathcal{V}_h^{k-1}. \tag{18.1b}$$

This is a mixed problem of the first kind, so we can analyse it with the tools introduced in Section 13. According to the discussion of Section 13, we can define a continuous operator $T^{(1)} : \mathbf{L}^2(\Omega, \Lambda^k) \to \mathbf{H}_0(d_k; \Omega)$ and a discrete operator $T_h^{(1)} : \mathbf{L}^2(\Omega, \Lambda^k) \to \mathcal{V}_h^k$ related to the first component of the solution of the corresponding (continuous and discrete) source problems, which we now write explicitly for the reader's convenience. The continuous source problem is as follows: given $\mathbf{f} \in \mathbf{L}^2(\Omega, \Lambda^k)$, find $\mathbf{u} \in \mathbf{H}_0(d_k; \Omega)$ and $\mathbf{p} \in \mathbf{H}_0(d_{k-1}; \Omega)$ such that

$$\begin{aligned} (d_k \mathbf{u}, d_k \mathbf{v}) + (\mathbf{v}, d_{k-1} \mathbf{p}) &= (\mathbf{f}, \mathbf{v}) \quad \forall \mathbf{v} \in \mathbf{H}_0(d_k; \Omega), \\ (\mathbf{u}, d_{k-1} \mathbf{q}) &= 0 \qquad\quad \forall \mathbf{q} \in \mathbf{H}_0(d_{k-1}; \Omega), \end{aligned}$$

and its discrete counterpart is to find $\mathbf{u}_h \in \mathcal{V}_h^k$ and $\mathbf{p}_h \in \mathcal{V}_h^{k-1}$ such that

$$\begin{aligned} (d_k \mathbf{u}_h, d_k \mathbf{v}) + (\mathbf{v}, d_{k-1} \mathbf{p}_h) &= (\mathbf{f}, \mathbf{v}) \quad \forall \mathbf{v} \in \mathcal{V}_h^k, \\ (\mathbf{u}_h, d_{k-1} \mathbf{q}) &= 0 \qquad\quad \forall \mathbf{q} \in \mathcal{V}_h^{k-1}. \end{aligned}$$

To apply the theory developed in Section 13, we need to show that

$$T^{(1)} \text{ is compact from } \mathbf{L}^2(\Omega, \Lambda^k) \text{ to } \mathbf{H}_0(d_k; \Omega).$$

From compactness of the embedding of $\mathbf{H}_0(d_k; \Omega) \cap \mathbf{H}(\delta_k; \Omega)$ into $\mathbf{L}^2(\Omega, \Lambda^k)$, it follows that $T^{(1)}$ is a compact operator from $\mathbf{L}^2(\Omega, \Lambda^k)$ into $\mathbf{L}^2(\Omega, \Lambda^k)$. Moreover, it turns out that $d_k(T^{(1)} \mathbf{L}^2(\Omega, \Lambda^k))$ is contained in $\mathbf{H}_0(d_{k+1}; \Omega) \cap \mathbf{H}(\delta_{k+1}; \Omega)$, which is compactly embedded into $\mathbf{L}^2(\Omega, \Lambda^{k+1})$. This implies that $T^{(1)}$ has the required compactness from $\mathbf{L}^2(\Omega, \Lambda^k)$ into $\mathbf{H}_0(d_k; \Omega)$.

In agreement with Section 13, we introduce some spaces. Let $\mathbb{K}$ be the kernel of the operator δ_k, that is,

$$\mathbb{K} = \{\mathbf{v} \in \mathbf{H}_0(d_k;\Omega) : (\mathbf{v}, d_{k-1}\mathbf{q}) = 0 \ \forall \mathbf{q} \in \mathbf{H}_0(d_{k-1};\Omega)\}$$

and its discrete counterpart

$$\mathbb{K}_h = \{\mathbf{v} \in \boldsymbol{\mathcal{V}}_h^k : (\mathbf{v}, d_{k-1}\mathbf{q}) = 0 \ \forall \mathbf{q} \in \boldsymbol{\mathcal{V}}_h^{k-1}\}.$$

Moreover, $\mathcal{V}_0$ and $\mathcal{Q}_0$ denote the spaces containing all solutions $\mathbf{u}$ and $\mathbf{p}$, respectively, of the continuous source problem (remember that the component $\mathbf{p}$ of the solution might not be unique).

The three fundamental hypotheses of Theorem 13.4 are the *ellipticity in the kernel*, the *weak approximability* of $\mathcal{Q}_0$ and the *strong approximability* of $\mathcal{V}_0$ (see Definitions 13.1, 13.2, and 13.3). For the reader's convenience, we recall these properties with the actual notation.

The ellipticity in the kernel states that there exists $\alpha > 0$ such that

$$(d_k\mathbf{v}, d_k\mathbf{v}) \geq \alpha(\mathbf{v}, \mathbf{v}) \quad \forall \mathbf{v} \in \mathbb{K}_h. \tag{18.2}$$

The weak approximability of $\mathcal{Q}_0$ means that there exists $\rho_W(h)$, tending to zero as h tends to zero, such that

$$\sup_{\mathbf{v}\in\mathbb{K}_h} \frac{(\mathbf{v}, d_{k-1}\mathbf{q})}{\|\mathbf{v}\|_{\mathbf{H}(d_k;\Omega)}} \leq \rho_W(h)\|\mathbf{q}\|_{\mathcal{Q}_0}. \tag{18.3}$$

The strong approximability of $\mathcal{V}_0$ means that there exists $\rho_S(h)$, tending to zero as h tends to zero, such that, for any $\mathbf{u} \in \mathcal{V}_0$, there exists $\mathbf{u}^I \in \mathbb{K}_h$ with

$$\|\mathbf{u} - \mathbf{u}^I\|_{\mathbf{H}(d_k;\Omega)} \leq \rho_S(h)\|\mathbf{u}\|_{\mathcal{V}_0}. \tag{18.4}$$

The next proposition is the analogue of Theorem 13.4 in the setting of this section.

Proposition 18.1. If the ellipticity in the kernel (18.2), the weak approximability of $\mathcal{Q}_0$ (18.3), and the strong approximability of $\mathcal{V}_0$ (18.4) hold true, then there exists $\rho(h)$, tending to zero as h tends to zero, such that

$$\|(T^{(1)} - T_h^{(1)})\mathbf{f}\|_{\mathbf{H}(d_k;\Omega)} \leq \rho(h)\|\mathbf{f}\|_{\mathbf{L}^2(\Omega,\Lambda^k)} \quad \forall \mathbf{f} \in \mathbf{L}^2(\Omega, \Lambda^k).$$

18.2. Approximation of problem (17.6)

The approximation of the second mixed formulation (17.6) reads as follows: find $\lambda_h \in \mathbb{R}$ and $\mathbf{s}_h \in d_k(\boldsymbol{\mathcal{V}}_h^k)$ such that, for $\mathbf{u}_h \in \boldsymbol{\mathcal{V}}_h^k$,

$$(\mathbf{u}_h, \mathbf{v}) + (d_k\mathbf{v}, \mathbf{s}_h) = 0 \quad \forall \mathbf{v} \in \boldsymbol{\mathcal{V}}_h^k, \tag{18.5a}$$

$$(d_k\mathbf{u}_h, \mathbf{t}) = -\lambda_h(\mathbf{s}_h, \mathbf{t}) \quad \forall \mathbf{t} \in d_k(\boldsymbol{\mathcal{V}}_h^k). \tag{18.5b}$$

This is a problem of the second kind according to the classification of Part 3; thus it can be analysed using the tools of Section 14.

The first step consists in the introduction of suitable operators $T^{(2)}$: $\mathbf{L}^2(\Omega, \Lambda^{k+1}) \to \mathbf{L}^2(\Omega, \Lambda^{k+1})$ and $T_h^{(2)} : \mathbf{L}^2(\Omega, \Lambda^{k+1}) \to d_k(\boldsymbol{\mathcal{V}}_h^k)$, by using the second components of the solutions to the source problems corresponding to (17.6) and (18.5), respectively. For the reader's convenience, these source problems are as follows: given $\mathbf{g} \in \mathbf{L}^2(\Omega, \Lambda^{k+1})$, find $\mathbf{u} \in \mathbf{H}_0(d_k; \Omega)$ and $\mathbf{s} \in \mathbf{K}_{k+1}^d$ such that

$$
\begin{aligned}
(\mathbf{u}, \mathbf{v}) + (d_k\mathbf{v}, \mathbf{s}) = 0 \quad & \forall \mathbf{v} \in \mathbf{H}_0(d_k; \Omega), \\
(d_k\mathbf{u}, \mathbf{t}) = -(\mathbf{g}, \mathbf{t}) \quad & \forall \mathbf{t} \in \mathbf{K}_{k+1}^d,
\end{aligned}
$$

and find $\mathbf{u}_h \in \boldsymbol{\mathcal{V}}_h^k$ and $\mathbf{s}_h \in d_k(\boldsymbol{\mathcal{V}}_h^k)$ such that

$$
\begin{aligned}
(\mathbf{u}_h, \mathbf{v}) + (d_k\mathbf{v}, \mathbf{s}_h) = 0 \quad & \forall \mathbf{v} \in \boldsymbol{\mathcal{V}}_h^k, \\
(d_k\mathbf{u}_h, \mathbf{t}) = -(\mathbf{g}, \mathbf{t}) \quad & \forall \mathbf{t} \in d_k(\boldsymbol{\mathcal{V}}_h^k),
\end{aligned}
$$

respectively.

The theory of Section 14 uses the following spaces: $\mathcal{U}_0 \subset \mathbf{H}_0(d_k; \Omega)$ and $\mathcal{S}_0 \subset \mathbf{K}_{k+1}^d$ denote the spaces containing all solutions $\mathbf{u}$ and $\mathbf{s}$, respectively, of the continuous source problem when the datum $\mathbf{g}$ varies in $\mathbf{L}^2(\Omega, \Lambda^{k+1})$; the discrete kernel of the operator d_k is given by

$$
\mathbb{K}_h = \{\mathbf{v} \in \boldsymbol{\mathcal{V}}_h^k : (d_k\mathbf{v}, \mathbf{t}) = 0 \ \forall \mathbf{t} \in d_k(\boldsymbol{\mathcal{V}}_h^k)\},
$$

and in this particular case it is clearly included in the continuous kernel, that is, $d_k\mathbf{v} = 0$ for all $\mathbf{v} \in \mathbb{K}_h$. In this setting, the three fundamental conditions for the convergence of the eigensolution of (18.5) towards those of (17.6) are the *weak approximability* of $\mathcal{S}_0$, the *strong approximability* of $\mathcal{S}_0$, and the *Fortid condition* (see Definitions 14.1, 14.2 and 14.5, respectively).

The weak approximability requires the existence of $\rho_W(h)$, tending to zero as h tends to zero, such that

$$
(d_k\mathbf{v}, \mathbf{t}) \le \rho_W(h) \|\mathbf{v}\|_{\mathbf{L}^2(\Omega, \Lambda^k)} \|\mathbf{t}\|_{\mathcal{S}_0}. \tag{18.6}
$$

The strong approximability means that there exists $\rho_S(h)$, tending to zero as h tends to zero, such that, for any $\mathbf{s} \in \mathcal{S}_0$, there exists $\mathbf{s}^I \in d_k(\boldsymbol{\mathcal{V}}_h^k)$ with

$$
\|\mathbf{s} - \mathbf{s}^I\|_{\mathbf{L}^2(\Omega, \Lambda^{k+1})} \le \rho_S(h) \|\mathbf{s}\|_{\mathcal{S}_0}. \tag{18.7}
$$

The last property is related to the Fortin operator, that is, an operator $\Pi_h : \mathcal{U}_0 \to \boldsymbol{\mathcal{V}}_h^k$ such that

$$
\begin{aligned}
(d_k(\mathbf{u} - \Pi_h\mathbf{u}), \mathbf{t}) = 0 \quad & \forall \mathbf{u} \in \mathcal{U}_0 \ \forall \mathbf{t} \in d_k(\boldsymbol{\mathcal{V}}_h^k), \\
\|\Pi_h\mathbf{u}\|_{\mathbf{L}^2(\Omega, \Lambda^k)} \le C\|u\|_{\mathcal{U}_0} \quad & \forall \mathbf{u} \in \mathcal{U}_0.
\end{aligned}
$$

The Fortid property expresses the existence of $\rho_F(h)$, tending to zero as h tends to zero, such that

$$\|\mathbf{u} - \Pi_h \mathbf{u}\|_{\mathbf{L}^2(\Omega,\Lambda^k)} \le \rho_F(h)\|\mathbf{u}\|_{\mathcal{U}_0}. \tag{18.8}$$

The next proposition is the analogue of Theorem 14.6 in the setting of this section.

Proposition 18.2. If the weak approximability of $\mathcal{S}_0$ (18.6), the strong approximability of $\mathcal{S}_0$ (18.7), and the Fortid property (18.8) hold true, then there exists $\rho(h)$, tending to zero as h tends to zero, such that

$$\|(T^{(2)} - T_h^{(2)})\mathbf{g}\|_{\mathbf{L}^2(\Omega,\Lambda^{k+1})} \le \rho(h)\|\mathbf{g}\|_{\mathbf{L}^2(\Omega,\Lambda^{k+1})}.$$

19. Discrete compactness property

We introduce the approximation of problem (17.1) as follows: find $\lambda_h \in \mathbb{R}$ and $\mathbf{u}_h \in \mathcal{V}_h^k \subset \mathbf{H}_0(d_k;\Omega)$ with $\mathbf{u} \neq 0$ such that

$$(d_k\mathbf{u}_h, d_k\mathbf{v}) = \lambda_h(\mathbf{u}_h, \mathbf{v}) \quad \forall \mathbf{v} \in \mathcal{V}_h^k. \tag{19.1}$$

The results of Section 18 can be used for the analysis of (19.1) with the following strategy. Proposition 17.4 states that all solutions to problem (17.1) with positive frequency are in one-to-one correspondence with the solutions of problems (17.4) and (17.6). If a similar result is true for the discrete solutions of (19.1), (18.1) and (18.5), then we can use the theory of the approximation of mixed formulations.

The equivalence of the discrete problems is true in the setting of the de Rham diagram (16.1). Indeed, the much weaker assumption

$$d_{k-1}(\mathcal{V}_h^{k-1}) \subset \mathcal{V}_h^k$$

is sufficient for proving the following result, which is the discrete version of Proposition 17.4.

Proposition 19.1. Let $(\lambda_h, \mathbf{u}_h) \in \mathbb{R} \times \mathcal{V}_h^k$ be a solution of (19.1) with $\lambda_h > 0$; then $(\lambda_h, \mathbf{u}_h)$ is the solution of (18.1) and there exists $\mathbf{s}_h \in d_k(\mathcal{V}_h^k)$ such that $(\lambda_h, \mathbf{s}_h)$ is the solution of (18.5). Conversely, if $(\lambda_h, \mathbf{u}_h) \in \mathbb{R}\times\mathcal{V}_h^k$ is a solution of (18.1), then $\lambda_h > 0$ and $(\lambda_h, \mathbf{u}_h)$ solves (19.1); if $(\lambda_h, \mathbf{s}_h) \in \mathbb{R} \times d_k(\mathcal{V}_h^k)$ is a solution of (18.5) for some $\mathbf{u}_h \in \mathcal{V}_h^k$, then $(\lambda_h, \mathbf{u}_h)$ solves (19.1).

In this section we show how it is possible to introduce suitable conditions that ensure the convergence of the solutions of (19.1) towards those of (17.1). The main condition is the so-called discrete compactness property. Given a finite-dimensional subspace $\mathcal{V}_h^k$ of $\mathbf{H}_0(d_k;\Omega)$, we introduce the subspace of *discretely co-closed* k-forms,

$$\mathcal{Z}_h^k = \{\mathbf{v} \in \mathcal{V}_h^k : (\mathbf{v}, \mathbf{w}) = 0 \ \forall \mathbf{w} \in \mathcal{V}_h^k \text{ with } d_k\mathbf{w} = 0\}.$$

In the case of the de Rham diagram (16.1) and trivial cohomologies, the space $\mathcal{Z}_h^k$ can also be expressed in terms of orthogonalities with respect to $d_{k-1}(\mathcal{V}_h^{k-1})$; more precisely,

$$\mathcal{Z}_h^k = \{\mathbf{v} \in \mathcal{V}_h^k : (\mathbf{v}, d_{k-1}\mathbf{q}) = 0 \ \forall \mathbf{q} \in \mathcal{V}_h^{k-1}\}.$$

If the cohomologies are not trivial, then the two descriptions of $\mathcal{Z}_h^k$ differ by a finite-dimensional space; in particular, the following definition can easily be proved invariant from this choice.

Definition 19.2. We say that the *discrete compactness property* holds for a family $\{\mathcal{V}_h^k\}_h$ of finite-dimensional subspaces of $\mathbf{H}_0(d_k;\Omega)$ if any sequence $\{\mathbf{u}_n\} \subset \mathbf{H}_0(d_k;\Omega)$, with $\mathbf{u}_n \in \mathcal{Z}_{h_n}^k$, which is bounded in $\mathbf{H}_0(d_k;\Omega)$ contains a subsequence which converges in $\mathbf{L}^2(\Omega,\Lambda^k)$.

Remark 19.3. The definition of discrete compactness is often found in the literature with the following formulation: ... *the discrete compactness property holds* ... *if any sequence* $\{\mathbf{u}_h\}$, *with* $\mathbf{u}_h \in \mathcal{Z}_h^k$, *which is bounded in* $\mathbf{H}_0(d_k;\Omega)$ *contains a subsequence which converges in* $\mathbf{L}^2(\Omega,\Lambda^k)$. Here we make it explicit that the sequence $\mathbf{u}_n$ refers to an arbitrary index choice h_n. This is needed to avoid abstract situations occurring in cases such as when the family $\{\mathcal{V}_h^k\}_h$ comprises *good* spaces interspersed with an infinite number of *bad* spaces. Without extracting the first arbitrary subsequence associated with h_n, the negative effect of the *bad* spaces might be annihilated by a suitable subsequence choice (Christiansen 2009).

It can easily be shown that the limit $\mathbf{u}$ of the subsequence appearing in Definition 19.2 is in $\mathbf{H}_0(d_k;\Omega)$, and that $\delta_k\mathbf{u} = 0$ whenever $d_{k-1}(\mathcal{V}_h^{k-1})$ provides a good approximation of $d_{k-1}(\mathbf{H}_0(d_{k-1};\Omega))$. This motivates the following definition.

Definition 19.4. We say that the *strong compactness property* holds for a family $\{\mathcal{V}_h^k\}_h$ of finite-dimensional subspaces of $\mathbf{H}_0(d_k;\Omega)$ if it satisfies the discrete compactness property, and the limit $\mathbf{u}$ of the subsequence appearing in Definition 19.2 is a co-closed form, that is, $\delta_k\mathbf{u} = 0$.

Remark 19.5. It is worth noticing that, in general, the strong discrete compactness property is not 'much stronger' than the standard discrete compactness property. Indeed, if the space sequence $\{\mathcal{V}_h^{k-1}\}$ has good approximation properties and the discrete compactness property holds for $\{\mathcal{V}_h^k\}$, then passing to the limit in the orthogonality $(\mathbf{v}, d_{k-1}\mathbf{q}) = 0$ which defines the space $\mathcal{Z}_h^k$ gives the strong discrete compactness.

The main result of this section is stated in the next theorem. We consider k-forms in $\mathbf{H}_0(d_k;\Omega)$ and a sequence of finite element spaces $\mathcal{V}_h^k \subset \mathbf{H}_0(d_k;\Omega)$. Moreover, we suppose that we can write problems (17.4) and (17.6), that is, we have $\mathbf{H}_0(d_{k-1};\Omega)$, $\mathbf{K}_{k+1}^d$, and their approximations $\mathcal{V}_h^{k-1}$ and $d_k(\mathcal{V}_h^k)$.

Theorem 19.6. The following three sets of conditions are equivalent.

(i) The strong discrete compactness property (see Definition 19.4) and the following standard approximation property: for any $\mathbf{v} \in \mathbf{H}_0(d_k;\Omega)$ with $\delta_k \mathbf{v} = 0$, there exists a discrete sequence $\{\mathbf{v}_h\} \subset \mathcal{V}_h^k$ such that

$$\|\mathbf{v} - \mathbf{v}_h\|_{\mathbf{H}(d_k;\Omega)} \to 0 \quad \text{as } h \to 0.$$

(ii) The ellipticity in the kernel (18.2), the weak approximability of $\mathcal{Q}_0$ (18.3), and the strong approximability of $\mathcal{V}_0$ (18.4).

(iii) The weak approximability of $\mathcal{S}_0$ (18.6), the strong approximability of $\mathcal{S}_0$ (18.7), and the existence of a Fortin operator satisfying the Fortid property (18.8).

Proof. The proof is a generalization of the result of Boffi (2007, Theorem 3) where it is split into a series of propositions. We report here the main arguments for the sake of completeness.

Let us start with the implication (i) $\Rightarrow$ (ii).

The strong discrete compactness property implies the ellipticity in the kernel by the generalization of Monk and Demkowicz (2001, Corollary 4.2).

The strong discrete compactness property implies the weak approximability of $\mathcal{Q}_0$ from the following argument. By contradiction, let $\{\mathbf{q}_h\} \subset \mathcal{Q}_0$ be a sequence such that there exists $\{\mathbf{v}_h\} \subset \mathbb{K}_h$ with

$$\begin{aligned} \|\mathbf{q}_h\|_{\mathcal{Q}_0} &= 1, \\ \|\mathbf{v}_h\|_{\mathbf{H}(d_k;\Omega)} &= 1, \\ (\mathbf{v}_h, d_{k-1}\mathbf{q}_h) &\geq \varepsilon_0 > 0. \end{aligned}$$

From the boundedness of $\{\mathbf{q}_h\}$ and the strong discrete compactness, we can extract subsequences (denoted with the same notation) $\{\mathbf{q}_h\}$ and $\{\mathbf{v}_h\}$, and there exist $\mathbf{q} \in \mathbf{H}_0(d_{k-1};\Omega)$ and $\mathbf{v} \in \mathbf{L}^2(\Omega, \Lambda^k)$ such that $\mathbf{q}_h \rightharpoonup \mathbf{q}$ weakly in $\mathbf{H}_0(d_{k-1};\Omega)$ and $\mathbf{v}_h \to \mathbf{v}$ strongly in $\mathbf{L}^2(\Omega, \Lambda^k)$. Moreover, $\delta_k \mathbf{v} = 0$. Passing to the limit gives

$$(\mathbf{v}, d_{k-1}\mathbf{q}) \geq \varepsilon_0,$$

which contradicts $\delta_k \mathbf{v} = 0$.

The strong approximability of $\mathcal{V}_0$ is a consequence of (i) by the following argument. By contradiction we assume that the strong approximability of $\mathcal{V}_0$ is not satisfied. Let $\{\mathbf{u}_n\} \subset \mathcal{V}_0$ be a sequence such that

$$\begin{aligned} \|\mathbf{u}_n\|_{\mathcal{V}_0} &= 1, \\ \inf_{\mathbf{v}_{h_n} \in \mathbb{K}_{h_n}} \|\mathbf{u}_n - \mathbf{v}_{h_n}\|_{\mathbf{H}(d_k;\Omega)} &\geq \varepsilon_0 > 0 \quad \forall n, \end{aligned}$$

where h_n is a sequence of mesh sizes tending to zero. From the compact embedding of $\mathcal{V}_0$ in $\mathbf{H}_0(d_k;\Omega)$, it follows that up to a subsequence there exists $\mathbf{u} \in \mathbf{H}_0(d_k;\Omega)$ such that $\mathbf{u}_n \to \mathbf{u}$ in $\mathbf{H}(d_k;\Omega)$. Moreover, we have $\delta_k\mathbf{u} = 0$. We reach a contradiction if we are able to approximate $\mathbf{u}$ in $\mathbf{H}(d_k;\Omega)$ with a sequence in $\mathbb{K}_{h_n}$. From the approximation property in (i) there exists $\mathbf{u}_{h_n} \subset \boldsymbol{\mathcal{V}}^k_{h_n}$ such that $\mathbf{u}_{h_n} \to \mathbf{u}$ in $\mathbf{H}(d_k;\Omega)$. We perform a discrete Hodge decomposition,

$$\mathbf{u}_{h_n} = d_{k-1}\mathbf{p}_{h_n} + \mathbf{u}^I_{h_n},$$

as follows. We take $\mathbf{p}_{h_n} \in \boldsymbol{\mathcal{V}}^{k-1}_{h_n}$ such that

$$(d_{k-1}\mathbf{p}_{h_n}, d_{k-1}\mathbf{q}) = (\mathbf{u}_{h_n}, d_{k-1}\mathbf{q}) \quad \forall \mathbf{q} \in \boldsymbol{\mathcal{V}}^{k-1}_{h_n}$$

(this $\mathbf{p}_{h_n}$ is not unique if $k > 1$), and define $\mathbf{u}^I_{h_n}$ by difference. By definition $\{\mathbf{u}^I_{h_n}\}$ is bounded in $\mathbf{H}(d_k;\Omega)$ and belongs to $\boldsymbol{\mathcal{Z}}^k_{h_n}$, so that up to a subsequence it converges in $\mathbf{L}^2(\Omega,\Lambda^k)$ to a limit $\mathbf{u}^*$ satisfying $\delta_k\mathbf{u}^* = 0$. Since $\{\mathbf{u}^I_{h_n}\}$ belongs to $\mathbb{K}_{h_n}$, it is enough to prove that $\mathbf{u} = \mathbf{u}^*$. We have that $\mathbf{u} - \mathbf{u}^* = d_{k-1}\mathbf{p}$, with $\mathbf{p} \in \mathbf{H}_0(d_{k-1};\Omega)$ and $d_{k-1}\mathbf{p}_{h_n} \to d_{k-1}\mathbf{p}$ in $\mathbf{L}^2(\Omega,\Lambda^k)$. On the other hand, $\delta_k(\mathbf{u} - \mathbf{u}^*) = 0$ and $\mathbf{u} - \mathbf{u}^* = d_{k-1}\mathbf{p}$ imply $\mathbf{u} - \mathbf{u}^* = 0$.

We now consider the implication (ii) $\Rightarrow$ (iii).

The hypotheses in (ii), according to Proposition 18.1, imply that the eigensolutions of (18.1) converge towards those of (17.4). From the equivalences stated in Propositions 17.4 and 19.1, it follows that the eigensolutions of (18.5) converge towards those of (17.6) as well. Hence, the conditions in (iii) are satisfied since they are necessary for the norm convergence of $T^{(2)}_h$ to $T^{(2)}$ in $\mathcal{L}(\mathbf{L}^2(\Omega,\Lambda^{k+1}), \mathbf{K}^d_{k+1})$ (Boffi *et al.* 1997, Theorem 7).

Finally, let us show that (iii) $\Rightarrow$ (i).

First we prove that (iii) implies the strong discrete compactness property. Let $\{\mathbf{u}_n\}$ be a sequence as in Definition 19.2. It follows that $\mathbf{u}_n$ satisfies

$$(\mathbf{u}_n, \mathbf{v}) + (d_k\mathbf{v}, \mathbf{s}_n) = 0 \quad \forall \mathbf{v} \in \boldsymbol{\mathcal{V}}^k_{h_n} \tag{19.2}$$

for a suitable $\mathbf{s}_n \in d_k(\boldsymbol{\mathcal{V}}^k_{h_n})$. We define $\mathbf{u}(n) \in \mathbf{H}_0(d_k;\Omega)$ and $\mathbf{s}(n) \in \mathbf{K}^d_{k+1}$ by

$$(\mathbf{u}(n), \mathbf{v}) + (d_k\mathbf{v}, \mathbf{s}(n)) = 0 \quad \forall \mathbf{v} \in \mathbf{H}_0(d_k;\Omega), \tag{19.3a}$$

$$(d_k\mathbf{u}(n), \mathbf{t}) = (d_k\mathbf{u}_n, \mathbf{t}) \quad \forall \mathbf{t} \in \mathbf{K}^d_{k+1}. \tag{19.3b}$$

In particular, $\{\mathbf{u}(n)\}$ is bounded in $\mathbf{H}_0(d_k;\Omega) \cap \mathbf{H}(\delta_k;\Omega)$, which is compact in $\mathbf{L}^2(\Omega,\Lambda^k)$, so that there exists a limit $\mathbf{u}$ with $\delta_k\mathbf{u} = 0$ such that $\mathbf{u}(n) \to \mathbf{u}$ in $\mathbf{L}^2(\Omega,\Lambda^k)$ (up to a subsequence). We can show the strong discrete compactness property if we can prove that $\mathbf{u}_n$ tends to $\mathbf{u}$ in $\mathbf{L}^2(\Omega,\Lambda^k)$. The conditions in (iii) guarantee the norm convergence of $T^{(2)}_h$ to $T^{(2)}$ in $\mathcal{L}(\mathbf{L}^2(\Omega,\Lambda^{k+1}), \mathbf{L}^2(\Omega,\Lambda^{k+1}))$, which implies

$$\|\mathbf{s}(n) - \mathbf{s}_n\|_{\mathbf{L}^2(\Omega,\Lambda^{k+1})} \le \rho(n)\|d_k\mathbf{u}_n\|_{\mathbf{L}^2(\Omega,\Lambda^{k+1})},$$

with $\rho(n)$ tending to zero as n tends to infinity. Choosing $\mathbf{v} = \mathbf{u}(n) - \mathbf{u}_n$

in (19.3) gives

$$(\mathbf{u}(n), \mathbf{u}(n) - \mathbf{u}_n) = 0,$$

and the difference between the first equation in (19.3) and (19.2) leads to

$$-(\mathbf{u}(n) - \mathbf{u}_n, \mathbf{v}) + (d_k\mathbf{v}, \mathbf{s}(n) - \mathbf{s}_n) = 0 \quad \forall \mathbf{v} \in \mathcal{V}^k_{h_n}.$$

The last two equations and the choice $\mathbf{v} = \mathbf{u}_n$ give

$$\begin{aligned}\|\mathbf{u}(n) - \mathbf{u}_n\|^2_{\mathbf{L}^2(\Omega,\Lambda^k)} &\leq \|\mathbf{s}(n) - \mathbf{s}_n\|_{\mathbf{L}^2(\Omega,\Lambda^{k+1})}\|d_k\mathbf{u}_n\|_{\mathbf{L}^2(\Omega,\Lambda^{k+1})}\\ &\leq \rho(n)\|d_k\mathbf{u}_n\|^2_{\mathbf{L}^2(\Omega,\Lambda^{k+1})},\end{aligned}$$

which implies $\mathbf{u}_n \to \mathbf{u}$ in $\mathbf{L}^2(\Omega, \Lambda^k)$. Finally, the approximation property in (i) follows from the fact that (iii) implies the correct approximation of the eigenfunctions of (17.6), which is equivalent to (17.4) in the spirit of Proposition 17.4. Since the eigenfunctions are a dense space in the set of functions $\mathbf{v}$ in $\mathbf{H}(d_k;\Omega)$ with $\delta_k\mathbf{v} = 0$, the approximation property follows from the approximation of the eigenfunctions. □

The main consequence of Theorem 19.6 is that the discrete compactness property and standard approximabilities are designated as the natural conditions for good convergence of the eigensolutions of (19.1) towards those of (17.1). In the next section we are going to show how this theory can be applied to the approximation of Maxwell's eigenvalue problem.

20. Edge elements for the approximation of Maxwell's eigenvalue problem

We conclude this part with a discussion about the relationships between the results presented so far and the approximation of Maxwell's eigenvalue problem, which has been the main motivation for the author's study of the finite element approximation of eigenvalue problems in the setting of differential forms.

In Section 5 we recalled the definition of Maxwell's eigenvalue problem and presented some numerical examples concerning its approximation. We explained how edge finite elements are the correct choice for the discretization of problem (5.3), which is a particular case of problem (19.1) ($k = 1$). It was also discussed that the direct use of standard (nodal) finite element produces unacceptable results (see, in particular, Figures 5.2, 5.3, 5.4, and Table 5.3 with Figures 5.6 and 5.7). Some modifications of problem (5.3) are available that allow the use of standard finite elements (Costabel and Dauge 2002) or of standard finite elements enriched with bubble functions (Bramble, Kolev and Pasciak 2005).

In this section we review some basic literature about edge finite elements and show how Theorem 19.6 applies to this situation. We also discuss the difference between the discrete compactness property (see Definition 19.2) and the strong discrete compactness property (see Definition 19.4).

Edge finite elements were introduced by Nédélec (1980, 1986). The entire family of mixed finite elements is often referred to as the Nédélec–Raviart–Thomas family, since Raviart–Thomas elements (Raviart and Thomas 1977) also belong to this family. Other families are available: among them we recall Brezzi–Douglas–Marini elements (Brezzi *et al.* 1985, 1986), Brezzi–Douglas–Fortin–Marini elements (Brezzi *et al.* 1987*a*, 1987*b*), and the *hp* adaptive family presented by Demkowicz and co-workers (Vardapetyan and Demkowicz 1999, Demkowicz, Monk, Vardapetyan and Rachowicz 2000*b*). The merit of linking edge elements to the de Rham complex comes from the fundamental work of Bossavit (1988, 1989, 1990, 1998). The idea is intrinsically related to the concept of Whitney forms (Whitney 1957) and it should be acknowledged that lowest-order edge finite elements are often referred to as Whitney elements. The de Rham complex is a more complete viewpoint than the so-called *commuting diagram property* (Douglas and Roberts 1982), which was introduced in the framework of mixed approximations. Many authors have discussed the relationship between finite elements for electromagnetic problems and differential forms. Among others, the following papers and the references therein give an idea of the underlying discussion: Hiptmair (1999*a*, 2002), Gross and Kotiuga (2004), Christiansen (2007) and Boffi (2001). A deep understanding of the subject which leads to the formalism of the finite element exterior calculus is presented in the following works: Arnold (2002) and Arnold *et al.* (2006*a*, 2006*b*, 2010).

Proposition 17.4 in the context of Maxwell's eigenvalue problem states that there are three equivalent formulations.

(1) The standard Maxwell eigenvalue problem: find $\lambda \in \mathbb{R}$ and $\mathbf{u} \in \mathbf{H}_0(\mathbf{curl};\Omega)$ with $\mathbf{u} \neq 0$ such that

$$(\boldsymbol{\mu}^{-1}\,\mathbf{curl}\,\mathbf{u}, \mathbf{curl}\,\mathbf{v}) = \lambda(\boldsymbol{\varepsilon}\mathbf{u}, \mathbf{v}) \quad \forall \mathbf{v} \in \mathbf{H}_0(\mathbf{curl};\Omega).$$

(2) The mixed formulation of the first type (Kikuchi 1987): find $\lambda \in \mathbb{R}$ and $\mathbf{u} \in \mathbf{H}_0(\mathbf{curl};\Omega)$ with $\mathbf{u} \neq 0$ such that, for $p \in H^1_0(\Omega)$,

$$\begin{aligned}(\boldsymbol{\mu}^{-1}\,\mathbf{curl}\,\mathbf{u}, \mathbf{curl}\,\mathbf{v}) + (\boldsymbol{\varepsilon}\mathbf{v}, \mathbf{grad}\,p) &= \lambda(\boldsymbol{\varepsilon}\mathbf{u}, \mathbf{v}) \quad \forall \mathbf{v} \in \mathbf{H}_0(\mathbf{curl};\Omega),\\ (\boldsymbol{\varepsilon}\mathbf{u}, \mathbf{grad}\,q) &= 0 \quad \forall q \in H^1_0(\Omega).\end{aligned}$$

(3) The mixed formulation of the second type (Boffi *et al.* 1999*b*): find $\lambda \in \mathbb{R}$ and $\mathbf{s} \in \Sigma$ with $\mathbf{s} \neq 0$ such that, for $\mathbf{u} \in \mathbf{H}_0(\mathbf{curl};\Omega)$,

$$\begin{aligned}(\boldsymbol{\varepsilon}\mathbf{u}, \mathbf{v}) + (\boldsymbol{\mu}^{-1/2}\,\mathbf{curl}\,\mathbf{v}, \mathbf{s}) &= 0 \quad \forall \mathbf{v} \in \mathbf{H}_0(\mathbf{curl};\Omega),\\ (\boldsymbol{\mu}^{-1/2}\,\mathbf{curl}\,\mathbf{u}, \mathbf{t}) &= -\lambda(\mathbf{s}, \mathbf{t}) \quad \forall \mathbf{t} \in \Sigma,\end{aligned}$$

with $\Sigma = \boldsymbol{\mu}^{-1/2}\,\mathbf{curl}(\mathbf{H}_0(\mathbf{curl};\Omega))$.

From Proposition 19.1 the equivalence holds at the discrete level as well. We would like to point out that, from the computational point of view, the standard formulation is the most convenient and is the one commonly used. The two mixed formulations have essentially been introduced for the theoretical analysis of the finite element approximation. Some comments on the computational issues can be found, for instance, in Simoncini (2003) and Arbenz and Geus (1999). For multigrid solvers, the reader is referred to Hiptmair (1999*b*), Arnold, Falk and Winther (2000), Reitzinger and Schöberl (2002), and to the references therein.

Theorem 19.6 is the main tool for the analysis of the problem we are interested in. In particular, conditions stated in items (i), (ii) and (iii) refer to the standard formulation, the first mixed formulation, and the second mixed formulation, respectively. In the literature, conditions (i) and (iii) have mostly been used. Condition (i) was used in Boffi, Conforti and Gastaldi (2006*a*) for the analysis of a modification of Maxwell's eigenvalue problem for the approximation of band gaps in photonic crystals.

The case of a two-dimensional domain $\Omega \subset \mathbb{R}^2$ can easily be analysed. For instance, using the second mixed formulation and the equivalence of rot and div operators, we are led to the problem of approximating the eigensolutions of the Neumann problem for the Laplace operator in mixed form with Raviart–Thomas elements (see Section 5.1). The analysis of this problem was performed by Falk and Osborn (1980) (see also Demkowicz, Monk, Schwab and Vardapetyan (2000*a*)).

For a three-dimensional domain $\Omega \subset \mathbb{R}^3$, the discrete compactness property was proved by Kikuchi (1989) in the case of lowest-order tetrahedral elements. The convergence of the h method for practically all known families of edge finite elements follows from the arguments of Boffi (2000), where the Fortid condition, which makes it possible to use the second mixed formulation (see Theorem 19.6(iii)), was proved. A direct proof of convergence of the eigensolutions, which makes use of the discrete compactness property and of the abstract results of Anselone (1971), was given by Monk and Demkowicz (2001) under the assumption of quasi-uniformity of the mesh (see also Ciarlet and Zou (1999), Hiptmair (2002), Monk (2003) and Costabel and Dauge (2003)).

In the work of Caorsi, Fernandes and Raffetto (2000) (see also Caorsi, Fernandes and Raffetto (2001)) it was proved that the discrete compactness property is a necessary and sufficient condition for good convergence of the eigensolutions of Maxwell's system. In that paper the strong discrete compactness is not explicitly addressed, but suitable approximation properties are considered (see Remark 19.5)

All the results presented so far fit very well into the theoretical setting of Arnold *et al.* (2010), where a more general theory is developed for the analysis of the approximation of the Hodge–Laplace eigenvalue problem (17.2).

In that very recent paper it has been proved that a sufficient condition for the convergence of the eigensolution is the existence of projection operators that are uniformly bounded in $\mathbf{L}^2(\Omega, \Lambda^k)$. A construction of such operators is presented (for any admissible k and n), which is performed by means of a suitable extension–regularization procedure. It is also shown that the existence of such projections implies the Fortid property and the discrete compactness property.

In practical applications the computational domains and the material properties related to Maxwell's cavities are such that the eigensolutions often correspond to non-smooth eigenfunctions. It has been observed that in such cases it may be convenient to use anisotropic elements (Nicaise 2001, Buffa, Costabel and Dauge 2005), suitable spaces that take care of the singular functions (Assous, Ciarlet and Sonnendrücker 1998), or an adaptive *hp* strategy (Demkowicz 2005, Ainsworth and Coyle 2003, Hiptmair and Ledger 2005).

It would be nice to construct suitable projections which are bounded in $\mathbf{L}^2(\Omega, \Lambda^k)$, uniformly in p, so that the theory of Arnold *et al.* (2010) can be applied to the analysis of the p and perhaps of the *hp* version of edge finite elements. Unfortunately, such projections are not yet available, and it is not clear whether they exist.

The first result concerning the *hp* version of edge finite elements was by Boffi, Demkowicz and Costabel (2003), where the triangular case was analysed and the discrete compactness property was proved modulo a conjectured estimate, which was only demonstrated numerically. The first rigorous proof of the discrete compactness property for the *hp* version of edge elements was by Boffi, Costabel, Dauge and Demkowicz (2006*b*) in the case of rectangular meshes allowing for one-irregular hanging nodes. It is interesting to note that the plain p version of edge finite elements (pure spectral elements) had never been analysed before that paper, even though there was evidence of good performance (Wang, Monk and Szabo 1996). Finally, a fairly general result concerning the p version of edge finite elements can be found in Boffi *et al.* (2009), where the discussion is performed in the framework of differential forms, and where the role of the discrete compactness property has also been studied.

We conclude this presentation with a comment on the role of the discrete compactness property and the strong discrete compactness property. In particular, we want to emphasize the differences between the two conditions (see Boffi (2007, Section 5) and Boffi *et al.* (2009, Section 2.3)). The behaviour we are going to describe is strictly related to the concepts of spectral correctness and spurious-free approximation introduced in Caorsi *et al.* (2000, Section 4). We go back to the notation of the h version of finite elements, but our comments apply to the p and *hp* versions as well. It is clear that the main difficulties in the approximation of problem (17.1) come from the discretization of the infinite-dimensional kernel that occurs for $k \geq 1$. The discrete compactness property is related to the spectrally correct approximation, that is, all continuous eigenvalues (including the zero frequency) are approximated by a correct number of discrete eigenvalues and the corresponding eigenspaces are well approximated. For $k = 0$, in the case of a compact resolvent, this is an optimal notion, and implies that the numerical scheme is capable of providing a good approximation of the eigenmodes. For $k \geq 1$, however, the eigenvalues approximating the infinite-dimensional kernel may pollute the whole spectrum, and the numerical scheme becomes unusable. Moreover, Caorsi *et al.* (2000) showed that, if $\mathbf{H}_0(d_k; \Omega)$ is well approximated by $\boldsymbol{\mathcal{V}}_h^k$, then the eigenvalues approximating the zero frequency are confined to a region close to zero, which can be made arbitrarily small, for sufficiently small h. The big improvement given by strong discrete compactness (or, analogously, by discrete compactness and completeness of the discrete kernel (CDK) of Caorsi *et al.* (2000)), consists in the conclusion that the discrete frequencies approximating zero are exactly at zero, meaning that all non-physical frequencies are well separated from the rest of the spectrum.

Acknowledgements

I would like to thank all friends who encouraged and helped me during the preparation of this paper. Their valuable support has been greatly appreciated. In particular, Lucia Gastaldi gave me her opinion on each chapter of my paper while I was writing it; Rodolfo Rodríguez carefully proofread the first draft of my paper; Leszek Demkowicz, Ricardo Durán and Ilaria Perugia gave me several hints; Doug Arnold helped me refine the final part.

REFERENCES

M. Ainsworth and J. Coyle (2003), Computation of Maxwell eigenvalues on curvilinear domains using *hp*-version Nédélec elements. In *Numerical Mathematics and Advanced Applications*, Springer Italia, Milan, pp. 219–231.

A. Alonso and A. Dello Russo (2009), 'Spectral approximation of variationally-posed eigenvalue problems by nonconforming methods', *J. Comput. Appl. Math.* **223**, 177–197.

P. M. Anselone (1971), *Collectively Compact Operator Approximation Theory and Applications to Integral Equations*, Prentice-Hall Series in Automatic Computation, Prentice-Hall, Englewood Cliffs, NJ.

P. M. Anselone and T. W. Palmer (1968), 'Spectral properties of collectively compact sets of linear operators.', *J. Math. Mech.* **17**, 853–860.

P. F. Antonietti, A. Buffa and I. Perugia (2006), 'Discontinuous Galerkin approximation of the Laplace eigenproblem', *Comput. Methods Appl. Mech. Engrg* **195**, 3483–3503.

P. Arbenz and R. Geus (1999), 'A comparison of solvers for large eigenvalue problems occurring in the design of resonant cavities', *Numer. Linear Algebra Appl.* **6**, 3–16.

M. G. Armentano and R. G. Durán (2004), 'Asymptotic lower bounds for eigenvalues by nonconforming finite element methods', *Electron. Trans. Numer. Anal.* **17**, 93–101 (electronic).

D. N. Arnold (2002), Differential complexes and numerical stability. In *Proc. International Congress of Mathematicians, Vol. I: Beijing 2002*, Higher Education Press, Beijing, pp. 137–157.

D. N. Arnold, D. Boffi and R. S. Falk (2002), 'Approximation by quadrilateral finite elements', *Math. Comp.* **71**, 909–922 (electronic).

D. N. Arnold, D. Boffi and R. S. Falk (2005), 'Quadrilateral $H(\mathrm{div})$ finite elements', *SIAM J. Numer. Anal.* **42**, 2429–2451 (electronic).

D. N. Arnold, R. S. Falk and R. Winther (2000), 'Multigrid in $H(\mathrm{div})$ and $H(\mathrm{curl})$', *Numer. Math.* **85**, 197–217.

D. N. Arnold, R. S. Falk and R. Winther (2006*a*), Differential complexes and stability of finite element methods I: The de Rham complex. In *Compatible Spatial Discretizations* (D. Arnold, P. Bochev, R. Lehoucq, R. Nicolaides and M. Shaskov, eds), Vol. 142 of *The IMA Volumes in Mathematics and its Applications*, Springer, Berlin, pp. 23–46.

D. N. Arnold, R. S. Falk and R. Winther (2006*b*), Finite element exterior calculus, homological techniques, and applications. In *Acta Numerica*, Vol. 15, Cambridge University Press, pp. 1–155.

D. N. Arnold, R. S. Falk and R. Winther (2010), 'Finite element exterior calculus: From Hodge theory to numerical stability', *Bull. Amer. Math. Soc.* NS **47**, 281–353.

F. Assous, P. Ciarlet, Jr. and E. Sonnendrücker (1998), 'Resolution of the Maxwell equations in a domain with reentrant corners', *RAIRO Modél. Math. Anal. Numér.* **32**, 359–389.

I. Babuška (1973), 'The finite element method with Lagrangian multipliers', *Numer. Math.* **20**, 179–192.

I. Babuška and R. Narasimhan (1997), 'The Babuška–Brezzi condition and the patch test: An example', *Comput. Methods Appl. Mech. Engrg* **140**, 183–199.

I. Babuška and J. Osborn (1991), Eigenvalue problems. In *Handbook of Numerical Analysis, Vol. II*, North-Holland, Amsterdam, pp. 641–787.

I. Babuška and J. E. Osborn (1989), 'Finite element-Galerkin approximation of the eigenvalues and eigenvectors of selfadjoint problems', *Math. Comp.* **52**, 275–297.

K.-J. Bathe, C. Nitikitpaiboon and X. Wang (1995), 'A mixed displacement-based finite element formulation for acoustic fluid–structure interaction', *Comput. & Structures* **56**, 225–237.

A. Bermúdez and R. Rodríguez (1994), 'Finite element computation of the vibration modes of a fluid–solid system', *Comput. Methods Appl. Mech. Engrg* **119**, 355–370.

A. Bermúdez, R. Durán, M. A. Muschietti, R. Rodríguez and J. Solomin (1995), 'Finite element vibration analysis of fluid–solid systems without spurious modes', *SIAM J. Numer. Anal.* **32**, 1280–1295.

A. Bermúdez, P. Gamallo, M. R. Nogueiras and R. Rodríguez (2006), 'Approximation of a structural acoustic vibration problem by hexahedral finite elements', *IMA J. Numer. Anal.* **26**, 391–421.

G. Birkhoff, C. de Boor, B. Swartz and B. Wendroff (1966), 'Rayleigh–Ritz approximation by piecewise cubic polynomials', *SIAM J. Numer. Anal.* **3**, 188–203.

D. Boffi (1997), 'Three-dimensional finite element methods for the Stokes problem', *SIAM J. Numer. Anal.* **34**, 664–670.

D. Boffi (2000), 'Fortin operator and discrete compactness for edge elements', *Numer. Math.* **87**, 229–246.

D. Boffi (2001), 'A note on the de Rham complex and a discrete compactness property', *Appl. Math. Lett.* **14**, 33–38.

D. Boffi (2007), 'Approximation of eigenvalues in mixed form, discrete compactness property, and application to *hp* mixed finite elements', *Comput. Methods Appl. Mech. Engrg* **196**, 3672–3681.

D. Boffi and L. Gastaldi (2004), 'Analysis of finite element approximation of evolution problems in mixed form', *SIAM J. Numer. Anal.* **42**, 1502–1526 (electronic).

D. Boffi and C. Lovadina (1997), 'Remarks on augmented Lagrangian formulations for mixed finite element schemes', *Boll. Un. Mat. Ital.* A (7) **11**, 41–55.

D. Boffi, F. Brezzi and L. Gastaldi (1997), 'On the convergence of eigenvalues for mixed formulations', *Ann. Scuola Norm. Sup. Pisa Cl. Sci.* (4) **25**, 131–154.

D. Boffi, F. Brezzi and L. Gastaldi (2000*a*), 'On the problem of spurious eigenvalues in the approximation of linear elliptic problems in mixed form', *Math. Comp.* **69**, 121–140.

D. Boffi, C. Chinosi and L. Gastaldi (2000*b*), 'Approximation of the grad div operator in nonconvex domains', *Comput. Model. Eng. Sci.* **1**, 31–43.

D. Boffi, M. Conforti and L. Gastaldi (2006*a*), 'Modified edge finite elements for photonic crystals', *Numer. Math.* **105**, 249–266.

D. Boffi, M. Costabel, M. Dauge and L. Demkowicz (2006*b*), 'Discrete compactness for the *hp* version of rectangular edge finite elements', *SIAM J. Numer. Anal.* **44**, 979–1004 (electronic).

D. Boffi, M. Costabel, M. Dauge, L. Demkowicz and R. Hiptmair (2009), Discrete compactness for the *p*-version of discrete differential forms.
HAL: hal-00420150, arXiv:0909.5079v2.

D. Boffi, L. Demkowicz and M. Costabel (2003), 'Discrete compactness for *p* and *hp* 2D edge finite elements', *Math. Models Methods Appl. Sci.* **13**, 1673–1687.

D. Boffi, R. G. Durán and L. Gastaldi (1999*a*), 'A remark on spurious eigenvalues in a square', *Appl. Math. Lett.* **12**, 107–114.

D. Boffi, P. Fernandes, L. Gastaldi and I. Perugia (1999*b*), 'Computational models of electromagnetic resonators: Analysis of edge element approximation', *SIAM J. Numer. Anal.* **36**, 1264–1290 (electronic).

D. Boffi, F. Kikuchi and J. Schöberl (2006*c*), 'Edge element computation of Maxwell's eigenvalues on general quadrilateral meshes', *Math. Models Methods Appl. Sci.* **16**, 265–273.

J. M. Boland and R. A. Nicolaides (1985), 'Stable and semistable low order finite elements for viscous flows', *SIAM J. Numer. Anal.* **22**, 474–492.

A. Bossavit (1988), 'Whitney forms: A class of finite elements for three-dimensional computations in electromagnetism', *IEEE Proc.* A **135**, 493–500.

A. Bossavit (1989), 'Un nouveau point de vue sur les éléments mixtes', *Bull. Soc. Math. Appl. Industr.* **20**, 22–35.

A. Bossavit (1990), 'Solving Maxwell equations in a closed cavity and the question of spurious modes', *IEEE Trans. Magnetics* **26**, 702–705.

A. Bossavit (1998), *Computational Electromagnetism: Variational Formulations, Complementarity, Edge Elements*, Academic Press, San Diego, CA.

J. H. Bramble and J. E. Osborn (1973), 'Rate of convergence estimates for non-selfadjoint eigenvalue approximations', *Math. Comp.* **27**, 525–549.

J. H. Bramble, T. V. Kolev and J. E. Pasciak (2005), 'The approximation of the Maxwell eigenvalue problem using a least-squares method', *Math. Comp.* **74**, 1575–1598 (electronic).

S. C. Brenner, F. Li and L. Y. Sung (2008), 'A locally divergence-free interior penalty method for two-dimensional curl-curl problems', *SIAM J. Numer. Anal.* **46**, 1190–1211.

F. Brezzi (1974), 'On the existence, uniqueness and approximation of saddle-point problems arising from Lagrangian multipliers', *Rev. Française Automat. Informat. Recherche Opérationnelle Sér. Rouge* **8**, 129–151.

F. Brezzi and M. Fortin (1991), *Mixed and Hybrid Finite Element Methods*, Vol. 15 of *Springer Series in Computational Mathematics*, Springer, New York.

F. Brezzi, J. Douglas, R. Duran and M. Fortin (1987*a*), 'Mixed finite elements for second order elliptic problems in three variables', *Numer. Math.* **51**, 237–250.

F. Brezzi, J. Douglas, Jr. and L. D. Marini (1985), 'Two families of mixed finite elements for second order elliptic problems', *Numer. Math.* **47**, 217–235.

F. Brezzi, J. Douglas, Jr. and L. D. Marini (1986), Recent results on mixed finite element methods for second order elliptic problems. In *Vistas in Applied Mathematics*, Optimization Software, New York, pp. 25–43.

F. Brezzi, J. Douglas, Jr., M. Fortin and L. D. Marini (1987*b*), 'Efficient rectangular mixed finite elements in two and three space variables', *RAIRO Modél. Math. Anal. Numér.* **21**, 581–604.

A. Buffa and I. Perugia (2006), 'Discontinuous Galerkin approximation of the Maxwell eigenproblem', *SIAM J. Numer. Anal.* **44**, 2198–2226 (electronic).

A. Buffa, M. Costabel and M. Dauge (2005), 'Algebraic convergence for anisotropic edge elements in polyhedral domains', *Numer. Math.* **101**, 29–65.

A. Buffa, P. Houston and I. Perugia (2007), 'Discontinuous Galerkin computation of the Maxwell eigenvalues on simplicial meshes', *J. Comput. Appl. Math.* **204**, 317–333.

A. Buffa, I. Perugia and T. Warburton (2009), 'The mortar-discontinuous Galerkin method for the 2D Maxwell eigenproblem', *J. Sci. Comput.* **40**, 86–114.

A. Cangiani, F. Gardini and G. Manzini (2010), 'Convergence of the mimetic finite difference method for eigenvalue problems in mixed form', *Comput. Methods Appl. Mech. Engrg.* Submitted.

C. Canuto (1978), 'Eigenvalue approximations by mixed methods', *RAIRO Anal. Numér.* **12**, 27–50.

S. Caorsi, P. Fernandes and M. Raffetto (2000), 'On the convergence of Galerkin finite element approximations of electromagnetic eigenproblems', *SIAM J. Numer. Anal.* **38**, 580–607 (electronic).

S. Caorsi, P. Fernandes and M. Raffetto (2001), 'Spurious-free approximations of electromagnetic eigenproblems by means of Nedelec-type elements', *M2AN Math. Model. Numer. Anal.* **35**, 331–354.

C. Carstensen (2008), 'Convergence of an adaptive FEM for a class of degenerate convex minimization problems', *IMA J. Numer. Anal.* **28**, 423–439.

F. Chatelin (1973), 'Convergence of approximation methods to compute eigenelements of linear operations', *SIAM J. Numer. Anal.* **10**, 939–948.

F. Chatelin (1983), *Spectral Approximation of Linear Operators*, Computer Science and Applied Mathematics, Academic Press, New York.

F. Chatelin and M. J. Lemordant (1975), 'La méthode de Rayleigh–Ritz appliquée à des opérateurs différentiels elliptiques: Ordres de convergence des éléments propres', *Numer. Math.* **23**, 215–222.

H. Chen and R. Taylor (1990), 'Vibration analysis of fluid–solid systems using a finite element displacement formulation', *Internat. J. Numer. Methods Engrg* **29**, 683–698.

S. H. Christiansen (2007), 'Stability of Hodge decompositions in finite element spaces of differential forms in arbitrary dimension', *Numer. Math.* **107**, 87–106.

S. H. Christiansen (2009), personal communication.

P. Ciarlet, Jr. and J. Zou (1999), 'Fully discrete finite element approaches for time-dependent Maxwell's equations', *Numer. Math.* **82**, 193–219.

P. G. Ciarlet and P.-A. Raviart (1974), A mixed finite element method for the biharmonic equation. In *Mathematical Aspects of Finite Elements in Partial Differential Equations: Proc. Sympos., Madison 1974*, Academic Press, New York, pp. 125–145.

M. Costabel and M. Dauge (2002), 'Weighted regularization of Maxwell equations in polyhedral domains: A rehabilitation of nodal finite elements', *Numer. Math.* **93**, 239–277.

M. Costabel and M. Dauge (2003), Computation of resonance frequencies for Maxwell equations in non-smooth domains. In *Topics in Computational Wave Propagation*, Vol. 31 of *Lecture Notes in Computational Science and Engineering*, Springer, Berlin, pp. 125–161.

E. Creusé and S. Nicaise (2006), 'Discrete compactness for a discontinuous Galerkin approximation of Maxwell's system', *M2AN Math. Model. Numer. Anal.* **40**, 413–430.

M. Dauge (2003), Benchmark computations for Maxwell equations. http://perso.univ-rennes1.fr/monique.dauge/benchmax.html.

L. Demkowicz (2005), 'Fully automatic *hp*-adaptivity for Maxwell's equations', *Comput. Methods Appl. Mech. Engrg* **194**, 605–624.

L. Demkowicz, P. Monk, C. Schwab and L. Vardapetyan (2000*a*), 'Maxwell eigenvalues and discrete compactness in two dimensions', *Comput. Math. Appl.* **40**, 589–605.

L. Demkowicz, P. Monk, L. Vardapetyan and W. Rachowicz (2000*b*), 'De Rham diagram for *hp* finite element spaces', *Comput. Math. Appl.* **39**, 29–38.

J. Descloux, N. Nassif and J. Rappaz (1978*a*), 'On spectral approximation I: The problem of convergence', *RAIRO Anal. Numér.* **12**, 97–112.

J. Descloux, N. Nassif and J. Rappaz (1978*b*), 'On spectral approximation II: Error estimates for the Galerkin method', *RAIRO Anal. Numér.* **12**, 113–119.

J. Douglas, Jr. and J. E. Roberts (1982), 'Mixed finite element methods for second order elliptic problems', *Mat. Apl. Comput.* **1**, 91–103.

R. G. Durán, L. Gastaldi and C. Padra (1999), '*A posteriori* error estimators for mixed approximations of eigenvalue problems', *Math. Models Methods Appl. Sci.* **9**, 1165–1178.

R. G. Durán, C. Padra and R. Rodríguez (2003), '*A posteriori* error estimates for the finite element approximation of eigenvalue problems', *Math. Models Methods Appl. Sci.* **13**, 1219–1229.

R. S. Falk and J. E. Osborn (1980), 'Error estimates for mixed methods', *RAIRO Anal. Numér.* **14**, 249–277.

G. J. Fix, M. D. Gunzburger and R. A. Nicolaides (1981), 'On mixed finite element methods for first-order elliptic systems', *Numer. Math.* **37**, 29–48.

M. Fortin (1977), 'An analysis of the convergence of mixed finite element methods', *RAIRO Anal. Numér.* **11**, 341–354.

P. Gamallo (2002), Contribución al estudio matemático de problemas de simulación elastoacústica y control activo del ruido. PhD thesis, Universidade de Santiago de Compostela, Spain.

E. M. Garau, P. Morin and C. Zuppa (2009), 'Convergence of adaptive finite element methods for eigenvalue problems', *Math. Models Methods Appl. Sci.* **19**, 721–747.

F. Gardini (2004), '*A posteriori* error estimates for an eigenvalue problem arising from fluid–structure interaction', *Istit. Lombardo Accad. Sci. Lett. Rend.* A **138**, 17–34.

F. Gardini (2005), 'Discrete compactness property for quadrilateral finite element spaces', *Numer. Methods Partial Differential Equations* **21**, 41–56.

L. Gastaldi (1996), 'Mixed finite element methods in fluid structure systems', *Numer. Math.* **74**, 153–176.

S. Giani and I. G. Graham (2009), 'A convergent adaptive method for elliptic eigenvalue problems', *SIAM J. Numer. Anal.* **47**, 1067–1091.

R. Glowinski (1973), Approximations externes, par éléments finis de Lagrange d'ordre un et deux, du problème de Dirichlet pour l'opérateur biharmonique: Méthode itérative de résolution des problèmes approchés. In *Topics in Numerical Analysis: Proc. Roy. Irish Acad. Conf., University Coll., Dublin, 1972*, Academic Press, London, pp. 123–171.

R. D. Grigorieff (1975*a*), 'Diskrete Approximation von Eigenwertproblemen I: Qualitative Konvergenz', *Numer. Math.* **24**, 355–374.

R. D. Grigorieff (1975*b*), 'Diskrete Approximation von Eigenwertproblemen II: Konvergenzordnung', *Numer. Math.* **24**, 415–433.

R. D. Grigorieff (1975*c*), 'Diskrete Approximation von Eigenwertproblemen III: Asymptotische Entwicklungen', *Numer. Math.* **25**, 79–97.

P. W. Gross and P. R. Kotiuga (2004), *Electromagnetic Theory and Computation: A Topological Approach*, Vol. 48 of *Mathematical Sciences Research Institute Publications*, Cambridge University Press, Cambridge.

L. Grubišić and J. S. Ovall (2009), 'On estimators for eigenvalue/eigenvector approximations', *Math. Comp.* **78**, 739–770.

W. Hackbusch (1979), 'On the computation of approximate eigenvalues and eigenfunctions of elliptic operators by means of a multi-grid method', *SIAM J. Numer. Anal.* **16**, 201–215.

J. S. Hesthaven and T. Warburton (2004), 'High-order nodal discontinuous Galerkin methods for the Maxwell eigenvalue problem', *Philos. Trans. Roy. Soc. Lond. Ser. A: Math. Phys. Eng. Sci.* **362**, 493–524.

V. Heuveline and R. Rannacher (2001), '*A posteriori* error control for finite approximations of elliptic eigenvalue problems', *Adv. Comput. Math.* **15**, 107–138.

R. Hiptmair (1999*a*), 'Canonical construction of finite elements', *Math. Comp.* **68**, 1325–1346.

R. Hiptmair (1999*b*), 'Multigrid method for Maxwell's equations', *SIAM J. Numer. Anal.* **36**, 204–225 (electronic).

R. Hiptmair (2002), Finite elements in computational electromagnetism. In *Acta Numerica*, Vol. 11, Cambridge University Press, pp. 237–339.

R. Hiptmair and P. D. Ledger (2005), 'Computation of resonant modes for axisymmetric Maxwell cavities using *hp*-version edge finite elements', *Internat. J. Numer. Methods Engrg* **62**, 1652–1676.

C. Johnson and J. Pitkäranta (1982), 'Analysis of some mixed finite element methods related to reduced integration', *Math. Comp.* **38**, 375–400.

T. Kato (1966), *Perturbation Theory for Linear Operators*, Vol. 132 of *Die Grundlehren der Mathematischen Wissenschaften*, Springer, New York.

T. Kato (1995), *Perturbation Theory for Linear Operators*, Classics in Mathematics, Springer, Berlin. Reprint of the 1980 edition.

F. Kikuchi (1987), 'Mixed and penalty formulations for finite element analysis of an eigenvalue problem in electromagnetism', *Comput. Methods Appl. Mech. Engrg* **64**, 509–521.

F. Kikuchi (1989), 'On a discrete compactness property for the Nédélec finite elements', *J. Fac. Sci. Univ. Tokyo Sect. IA Math.* **36**, 479–490.

A. V. Knyazev and J. E. Osborn (2006), 'New *a priori* FEM error estimates for eigenvalues', *SIAM J. Numer. Anal.* **43**, 2647–2667 (electronic).

W. G. Kolata (1978), 'Approximation in variationally posed eigenvalue problems', *Numer. Math.* **29**, 159–171.

M. G. Larson (2000), '*A posteriori* and *a priori* error analysis for finite element approximations of self-adjoint elliptic eigenvalue problems', *SIAM J. Numer. Anal.* **38**, 608–625 (electronic).

S. Larsson and V. Thomée (2003), *Partial Differential Equations with Numerical Methods*, Vol. 45 of *Texts in Applied Mathematics*, Springer, Berlin.

B. Mercier (1974), 'Numerical solution of the biharmonic problem by mixed finite elements of class C^0', *Boll. Un. Mat. Ital.* (4) **10**, 133–149.

B. Mercier, J. Osborn, J. Rappaz and P.-A. Raviart (1981), 'Eigenvalue approximation by mixed and hybrid methods', *Math. Comp.* **36**, 427–453.

P. Monk (2003), *Finite Element Methods for Maxwell's Equations*, Numerical Mathematics and Scientific Computation, Oxford University Press.

P. Monk and L. Demkowicz (2001), 'Discrete compactness and the approximation of Maxwell's equations in $\mathbb{R}^3$', *Math. Comp.* **70**, 507–523.

P. Morin, R. H. Nochetto and K. G. Siebert (2000), 'Data oscillation and convergence of adaptive FEM', *SIAM J. Numer. Anal.* **38**, 466–488 (electronic).

J.-C. Nédélec (1980), 'Mixed finite elements in $\mathbb{R}^3$', *Numer. Math.* **35**, 315–341.

J.-C. Nédélec (1986), 'A new family of mixed finite elements in $\mathbb{R}^3$', *Numer. Math.* **50**, 57–81.

K. Neymeyr (2002), '*A posteriori* error estimation for elliptic eigenproblems', *Numer. Linear Algebra Appl.* **9**, 263–279.

S. Nicaise (2001), 'Edge elements on anisotropic meshes and approximation of the Maxwell equations', *SIAM J. Numer. Anal.* **39**, 784–816 (electronic).

J. E. Osborn (1975), 'Spectral approximation for compact operators', *Math. Comput.* **29**, 712–725.

R. Picard (1984), 'An elementary proof for a compact imbedding result in generalized electromagnetic theory', *Math. Z.* **187**, 151–161.

M. J. D. Powell (1974), Piecewise quadratic surface fitting for contour plotting. In *Software for Numerical Mathematics: Proc. Conf., Inst. Math. Appl., Loughborough 1973*, Academic Press, London, pp. 253–271.

J. Qin (1994), On the convergence of some low order mixed finite elements for incompressible fluids. PhD thesis, The Pennsylvania State University, Department of Mathematics.

R. Rannacher (1979), 'Nonconforming finite element methods for eigenvalue problems in linear plate theory', *Numer. Math.* **33**, 23–42.

P.-A. Raviart and J. M. Thomas (1977), A mixed finite element method for 2nd order elliptic problems. In *Mathematical Aspects of Finite Element Methods: Proc. Conf., Rome 1975*, Vol. 606 of *Lecture Notes in Mathematics*, Springer, Berlin, pp. 292–315.

P.-A. Raviart and J.-M. Thomas (1983), *Introduction à l'Analyse Numérique des Équations aux Dérivées Partielles*, Collection Mathématiques Appliquées pour la Maîtrise, Masson, Paris.

S. Reitzinger and J. Schöberl (2002), 'An algebraic multigrid method for finite element discretizations with edge elements', *Numer. Linear Algebra Appl.* **9**, 223–238.

R. Scholz (1978), 'A mixed method for 4th order problems using linear finite elements', *RAIRO Anal. Numér.* **12**, 85–90.

V. Simoncini (2003), 'Algebraic formulations for the solution of the nullspace-free eigenvalue problem using the inexact shift-and-invert Lanczos method', *Numer. Linear Algebra Appl.* **10**, 357–375.

G. Strang and G. J. Fix (1973), *An Analysis of the Finite Element method*, Prentice-Hall Series in Automatic Computation, Prentice-Hall, Englewood Cliffs, NJ.

F. Stummel (1970), 'Diskrete Konvergenz linearer Operatoren I', *Math. Ann.* **190**, 45–92.

F. Stummel (1971), 'Diskrete Konvergenz linearer Operatoren II', *Math. Z.* **120**, 231–264.

F. Stummel (1972), Diskrete Konvergenz linearer Operatoren III. In *Linear Operators and Approximation: Proc. Conf., Oberwolfach 1971*, Vol. 20 of *Internat. Ser. Numer. Math.*, Birkhäuser, Basel, pp. 196–216.

F. Stummel (1980), 'Basic compactness properties of nonconforming and hybrid finite element spaces', *RAIRO Anal. Numér.* **14**, 81–115.

L. Vardapetyan and L. Demkowicz (1999), '*hp*-adaptive finite elements in electromagnetics', *Comput. Methods Appl. Mech. Engrg* **169**, 331–344.

G. M. Vaĭnikko (1964), 'Asymptotic error bounds for projection methods in the eigenvalue problem', *Ž. Vyčisl. Mat. i Mat. Fiz.* **4**, 405–425.

G. M. Vaĭnikko (1966), 'On the rate of convergence of certain approximation methods of Galerkin type in eigenvalue problems', *Izv. Vysš. Učebn. Zaved. Matematika* **2**, 37–45.

Y. Wang, P. Monk and B. Szabo (1996), 'Computing cavity modes using the p version of the finite element method', *IEEE Trans. Magnetics* **32**, 1934–1940.

T. Warburton and M. Embree (2006), 'The role of the penalty in the local discontinuous Galerkin method for Maxwell's eigenvalue problem', *Comput. Methods Appl. Mech. Engrg* **195**, 3205–3223.

B. Werner (1981), Complementary variational principles and nonconforming Trefftz elements. In *Numerical Treatment of Differential Equations, Vol. 3: Clausthal 1980*, Vol. 56 of *Internat. Schriftenreihe Numer. Math.*, Birkhäuser, Basel, pp. 180–192.

H. Whitney (1957), *Geometric Integration Theory*, Princeton University Press.

S. H. Wong and Z. J. Cendes (1988), 'Combined finite element-modal solution of three-dimensional eddy current problems', *IEEE Trans. Magnetics* **24**, 2685–2687.

Acta Numerica (2010), pp. 121–158
doi:10.1017/S0962492910000024

Binary separation and training support vector machines*

Roger Fletcher†
Department of Mathematics,
University of Dundee,
Dundee DD1 4HN, UK
E-mail: fletcher@maths.dundee.ac.uk

Gaetano Zanghirati‡
Department of Mathematics and Math4Tech Center,
University of Ferrara,
44100 Ferrara, Italy
E-mail: g.zanghirati@unife.it

We introduce basic ideas of binary separation by a linear hyperplane, which is a technique exploited in the *support vector machine* (SVM) concept. This is a decision-making tool for pattern recognition and related problems. We describe a fundamental *standard problem* (SP) and show how this is used in most existing research to develop a dual-based algorithm for its solution. This algorithm is shown to be deficient in certain aspects, and we develop a new primal-based SQP-like algorithm, which has some interesting features. Most practical SVM problems are not adequately handled by a linear hyperplane. We describe the *nonlinear SVM* technique, which enables a nonlinear separating surface to be computed, and we propose a new primal algorithm based on the use of low-rank Cholesky factors.

It may be, however, that exact separation is not desirable due to the presence of uncertain or mislabelled data. Dealing with this situation is the main challenge in developing suitable algorithms. Existing dual-based algorithms use the idea of L_1 penalties, which has merit. We suggest how penalties can be incorporated into a primal-based algorithm. Another aspect of practical SVM problems is often the huge size of the data set, which poses severe challenges both for software package development and for control of ill-conditioning. We illustrate some of these issues with numerical experiments on a range of problems.

* An early version of this paper was presented at the 22nd Dundee Numerical Analysis Conference NA07, June 2007.

† Partially supported by the University of Ferrara under the Copernicus Visiting Professor Programme 2008.

‡ Partially funded by the HPC-EUROPA initiative (RII3-CT-2003-506079), with the support of the European Community Research Infrastructure Action under the FP6 'Structuring the European Research Area' Programme.

CONTENTS

1. Introduction

In this paper we deal with the problem of separating two given non-empty clusters within a data set of points $\boldsymbol{v}_i \in \mathbb{R}^n$, $i = 1, \ldots, m$. To distinguish between the clusters we assign a label a_i to each point which is either $+1$ or -1. In its most basic form the problem is to find a hyperplane

$$f(\boldsymbol{v}) = \boldsymbol{w}^T \boldsymbol{v} + b = 0 \qquad \|\boldsymbol{w}\| = 1 \tag{1.1}$$

in $\mathbb{R}^n$ which best separates the two clusters, with the 'plus' points on the plus side of the hyperplane (that is, $f(\boldsymbol{v}_i) > 0 \; \forall i : a_i = +1$) and the 'minus' points on the minus side. This is referred to as *linear separation.* If the points cannot be separated we seek a solution which provides the minimum overlap in a certain sense.

This basic notation is developed in the *support vector machine* (SVM) concept (Vapnik 1998), which is a decision-making tool for pattern recognition and related problems. Training an SVM is the activity of using existing data (the $\boldsymbol{v}_i$) to fix the parameters in the SVM ($\boldsymbol{w}$ and b) in an optimal way. An important concept is the existence of so-called *support vectors,* which are the subset of points that are active in defining the separation. Subsequently the SVM would be used to classify a previously unseen instance $\boldsymbol{v}$ by evaluating the sign of the *classification function* $f(\boldsymbol{v})$. In practice, binary separation very rarely occurs in a form which permits satisfactory separation by a hyperplane. Most interest therefore lies in what is referred to as *nonlinear SVMs,* in which the points are separated by a nonlinear hypersurface which is the zero contour of a nonlinear classification function.

In Section 2 of the paper we first develop the basic formulation of linear separation, and observe that it leads to a certain optimization problem which is a linear programming problem plus a single nonlinear constraint, which we refer to as the *standard problem.* The usual approach at this stage is to transform this nonlinear programming (NLP) problem into a convex quadratic programming (QP) problem, and we provide a simple explanation of how this transformation is carried out. Unfortunately this transformation is only valid when the data are linearly separable. The convex QP has a dual which is a QP with non-negative variables and a single linear constraint. This is the approach commonly taken by most existing research into SVMs. The non-separable case is handled by adding penalties, usually of an L_1 type, which is readily achieved by imposing simple upper bounds on the dual variables, and may lead to some data points being misclassified.[1]

However, we point out some unsatisfactory features of the usual approach, and give some attention to solving the standard problem (SP) directly. Based on some aspects of the KT conditions, we are able, in Sections 3 and 4, to propose a new SQP-like algorithm for its solution. Unusually for an NLP solver, our algorithm maintains feasibility throughout, and we prove that it monotonically increases the objective function without any need for line search, trust-region or filter strategies. Also, we observe that the algorithm terminates at a local solution, for which at present we do not have a good explanation.

In Section 5 we provide a simple introduction to the nonlinear SVM concept. We show that an approach to solving nonlinear problems by means of the new SQP-like algorithm is entirely practical, and we give some details of suitable strategies. The concept of *low-rank Cholesky factors* is explained and plays an important part. Some new proposals as to how pivots may be chosen are suggested.

In Section 6 we report preliminary numerical experiments on both randomly generated and well-known benchmark data sets for binary classification, where the focus is on how well the proposed approach works, rather than on running time performance.

The goal of Section 7 is to explore briefly some of the critical situations that can happen due to 'bad' data. We show by examples that in cases where the training data contain points that are mislabelled, obtaining the exact solution of the SP is possible, but provides an undesirable separating contour. This is one of the main reasons for introducing the regularization term in the dual, which has interpretations in the context of Statistical Learning. However, we try to suggest a different way to look at this problem in the context of our primal approach.

In Section 8 we briefly recall some of the most relevant issues related to the probabilistic view of SVMs, which we do not otherwise consider in this paper. Section 9 contains some final discussions and directions of future work.

[1] We distinguish between a data point $\boldsymbol{v}$ being *mislabelled* when the incorrect label has been assigned *a priori*, and *misclassified* when, as a result of an SVM calculation, $\boldsymbol{v}$ falls on the wrong side of the separating surface, so that its label is opposite to the sign of $f(\boldsymbol{v})$ in (1.1).

There is an extensive literature dealing with *binary separation* (or *binary classification*) and SVMs. When the best hyperplane is defined in a max-min sense (as in Section 2 here), we have the *maximal margin* problem, for which a number of methods have been proposed in the past, ranging from regression to neural networks, from principal components analysis (PCA) to Fisher discriminant analysis, *etc.*; see, *e.g.*, Hastie, Tibshirani and Friedman (2001) and Shawe-Taylor and Cristianini (2004). Also, when binary classifiers have to be constructed, additional probabilistic properties are usually considered together with the separating properties (we say more on this in the final discussion). The SVM approach is one of the most successful techniques so far developed; see Burges (1998), Osuna, Freund and Girosi (1997) and Cristianini and Shawe-Taylor (2000) for good introductions. This approach has received increasing attention in recent years from both the theoretical and computational viewpoints. On the theoretical side, SVMs have well-understood foundations both in probability theory and in approximation theory, the former mainly due to Vapnik (see, for instance, Vapnik (1998) and Shawe-Taylor and Cristianini (2004)), with the latter being developed by different researchers with strong connections with regularization theory in certain Hilbert spaces; see, for instance, Evgeniou, Pontil and Poggio (2000), Cucker and Smale (2001), De Vito, Rosasco, Caponnetto, De Giovannini and Odone (2005), De Vito, Rosasco, Caponnetto, Piana and Verri (2004) and Cucker and Zhou (2007), and the many references therein. On the computational side, the SVM methodology has attracted the attention of people from the numerical optimization field, given that the solution of the problem is most often sought by solving certain quadratic programming (QP) problems. Many contributions have been given here by Boser, Guyon and Vapnik (1992), Platt (1999), Joachims (1999), Chang, Hsu and Lin (2000), Mangasarian (2000), Lee and Mangasarian (2001*a*), Lin (2001*a*, 2001*b*), Mangasarian and Musicant (2001), Lin (2002), Keerthi and Gilbert (2002), Hush and Scovel (2003), Caponnetto and Rosasco (2004), Serafini, Zanghirati and Zanni (2005), Hush, Kelly, Scovel and Steinwart (2006), Zanni (2006) and Mangasarian (2006), just to mention a few. Low-rank approximations to the Hessian matrix (such as we use in Section 5) have also been used, for example, by Fine and Scheinberg (2001), Goldfarb and Scheinberg (2004), Keerthi and DeCoste (2005), Keerthi, Chapelle and DeCoste (2006) and Woodsend and Gondzio (2007*a*, 2007*b*), and the interested reader can find an extensive updated account of the state of the art in Bennett and Parrado-Hernández (2006). Moreover, real-world applications are often highly demanding, so that clever strategies have been developed to handle large-scale and huge-scale problems with up to millions of data points; see, for instance, Ferris and Munson (2002), Graf, Cosatto, Bottou, Dourdanovic and Vapnik (2005), Zanni, Serafini and Zanghirati (2006) and Woodsend and Gondzio (2007*a*). These strategies are implemented in a number of effective software packages, such as SMO (Platt 1998, Keerthi, Shevade, Bhattacharyya and Murthy 2001, Chen, Fan and Lin 2006), SVM^{light} (Joachims 1999), LIBSVM (Chang and Lin 2001), GPDT (Serafini and Zanni 2005), SVM-QP (Scheinberg 2006), SVMTorch (Collobert and Bengio 2001), HeroSVM (Dong, Krzyzak and Suen 2003, 2005), Core SVM (Tsang, Kwok and Cheung 2005), SVM-Maj (Groenen, Nalbantov and Bioch 2008), SVM-HOPDM (Woodsend and Gondzio 2007*a*), LIBLINEAR (Fan, Chang, Hsieh, Wang and Lin 2008), SVM^{perf} (Joachims 2006), LASVM (Bordes, Ertekin, Weston and Bottou 2005), LS-SVMlab (Suykens, Van Gestel, De Brabanter, De Moor and Vandewalle 2002), LIBOCAS (Franc and Sonnenburg 2008*a*) and INCAS (Fine and Scheinberg 2002). Furthermore, out-of-core computations are considered by Ferris and Munson (2002). Also, effective parallel implementations exist: PGPDT (Zanni *et al.* 2006; multiprocessor MPI-based version of the GPDT scheme, available at http://dm.unife.it/gpdt), Parallel SVM (Chang *et al.* 2008), Milde (Durdanovic, Cosatto and Graf 2007), Cascade SVM (Graf *et al.* 2005), BMRM (Teo, Le, Smola and Vishwanathan 2009; based on the PETSc and TAO technologies), and SVM-OOPS (Woodsend and Gondzio 2009; a hybrid MPI-OpenMP implementation based on the OOPS solver; see also Ferris and Munson (2002) and Gertz and Griffin (2005) for other interior-point-based parallel implementations). Moreover, some codes exist for emerging massively parallel architectures: PSMO-GPU (Catanzaro, Sundaram and Keutzer 2008) is an implementation of the SMO algorithm on graphics processing units (GPUs), while PGPDT-Cell (Wyganowski 2008) is a version of PGPDT for Cell-processor-based computers. Software repositories for SVM and other Machine Learning methods are mloss.org and kernel-machines.org.

The relevance of this subject is demonstrated by the extremely wide range of applications to which SVMs are successfully applied: from classical fields such as object detection in industry, medicine and surveillance, to text categorization, genomics and proteomics, up to the most recently emerging areas such as brain activity interpretation via the estimation of functional magnetic resonance imaging (fMRI) data (Prato *et al.* 2007). The list is continuously growing. Finally, we should mention that the SVM approach is not only used for binary classifications: a number of variations address other relevant pattern recognition problems, such as regression estimation, novelty detection (or single-class classification), multi-class classification and regression, on-line learning, semisupervised learning, multi-task reinforcement learning, and many others, which are beyond the scope of this paper.

The aim of this paper is to revisit the classical way to answer questions such as whether or not the two classes can be separated by a hyperplane, which is the best hyperplane, or what hyperplane comes closest to separating the classes if the points are not separable.

The key feature of our formulation is that we address the primal problem directly. Solution of a primal formulation of the SVM problem has already been addressed by some authors, first for the linear case (Mangasarian 2002, Lee and Mangasarian 2001*b*, Keerthi and DeCoste 2005, Groenen *et al.* 2008, Franc and Sonnenburg 2008*b*) and then for the nonlinear case (Keerthi *et al.* 2006, Chapelle 2007, Groenen, Nalbantov and Bioch 2007), but our approach differs from the above because it can treat both the separable and non-separable cases in the same way, without introducing the penalization term. Furthermore, the method we propose in this paper always provides primal feasible solutions, even if numerically we are not able to locate an exact solution, whereas this is not the case for the other approaches.

Notation
Vectors are intended as column vectors and are represented by bold symbols; lower-case letters are for scalars, vectors and function names; calligraphic upper-case letters are for sets, and roman upper-case letters are for matrices. When not otherwise stated, $\|\cdot\|$ will be the Euclidean norm.

2. Linear separation

Suppose the set $\mathcal{D}$ of m data points $\boldsymbol{v}_i \in \mathbb{R}^n$, $i = 1, \ldots, m$, is given, where the points fall into two classes labelled by $a_i \in \{-1, +1\}$, $i = 1, \ldots, m$. It is required that $m \geq 2$ with at least one point per class (see De Vito *et al.* (2004)), but interesting cases typically have $m \gg n$.

Assume first that the points can be separated by a hyperplane $\boldsymbol{w}^T\boldsymbol{v}+b = 0$ where $\boldsymbol{w}$ ($\|\boldsymbol{w}\| = 1$) and b are fixed, with $\boldsymbol{w}^T\boldsymbol{v}_i + b \geq 0$ for the 'plus' points and $\boldsymbol{w}^T\boldsymbol{v}_i + b \leq 0$ for the 'minus' points. We can express this as $a_i(\boldsymbol{w}^T\boldsymbol{v}_i + b) \geq 0$, $i = 1, \ldots, m$. Now we shift each point a distance h along the vector $-a_i\boldsymbol{w}$ until one point reaches the hyperplane (Figure 2.1), that is, we maximize h subject to $a_i(\boldsymbol{w}^T(\boldsymbol{v}_i - ha_i\boldsymbol{w}) + b) \geq 0$ (or equivalently $a_i(\boldsymbol{w}^T\boldsymbol{v}_i + b) - h \geq 0$), for $i = 1, \ldots, m$. The solution is clearly $h = \min_{i=1,\ldots,m} a_i(\boldsymbol{w}^T\boldsymbol{v}_i + b)$.

If only $\boldsymbol{w}$ is fixed, the best solution is $h = \max_b \min_i a_i(\boldsymbol{w}^T\boldsymbol{v}_i + b)$, which equates the best distances moved by the plus and minus points separately. Then the best solution over all $\boldsymbol{w}$, $\|\boldsymbol{w}\| = 1$, is

$$h_* = \max_{\boldsymbol{w}\,|\,\boldsymbol{w}^T\boldsymbol{w}=1} \max_{b} \min_{i=1,\ldots,m} a_i(\boldsymbol{w}^T\boldsymbol{v}_i + b), \tag{2.1}$$

which can be determined by solving the problem

$$\begin{aligned} & \underset{\boldsymbol{w},b,h}{\text{maximize}} \ \ h & (2.2\text{a}) \\ \text{SP:} \quad & \text{subject to } a_i(\boldsymbol{w}^T\boldsymbol{v}_i + b) - h \geq 0 \quad i = 1, \ldots, m, & (2.2\text{b}) \\ & \boldsymbol{w}^T\boldsymbol{w} = 1. & (2.2\text{c}) \end{aligned}$$

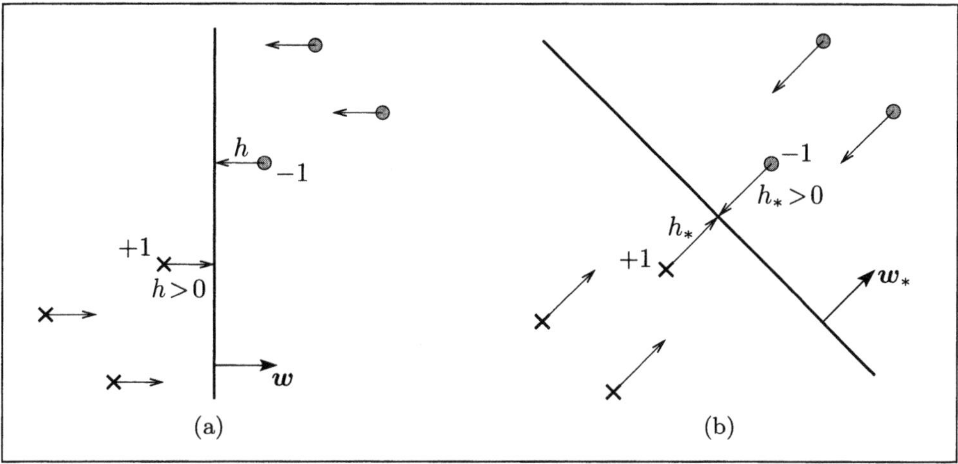

Figure 2.1. In the separable case, shift the points along the vector $-a_i\boldsymbol{w}$ towards the hyperplane. (a) A sub-optimal solution. (b) The optimal separating hyperplane (OSH).

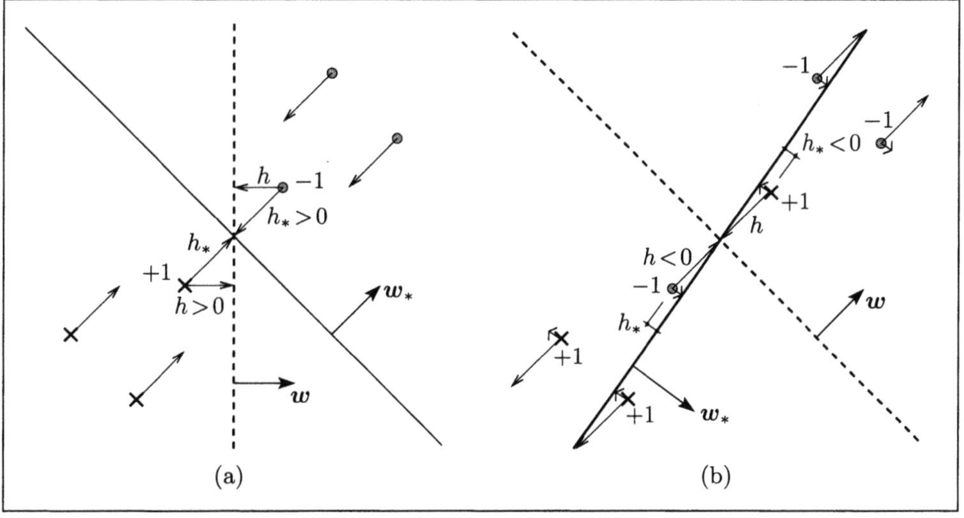

Figure 2.2. (a) In the linearly separable case $h > 0$, so any change to $\boldsymbol{w}_*$ gives a smaller h, that is, a worse non-optimal 'solution'. (b) In the linearly non-separable case the optimal separating hyperplane is still correctly identified, but $h_* < 0$, thus any change to $\boldsymbol{w}_*$ decreases h by increasing its modulus.

The solution $\boldsymbol{w}_*$, b_*, h_* of (2.2) defines the so-called *optimal separating hyperplane* (OSH). In matrix notation the constraints (2.2b) become

$$AV^T\boldsymbol{w} + \boldsymbol{a}b - \boldsymbol{e}h \geq \boldsymbol{0},$$

where

$$\boldsymbol{a} = (a_1, \ldots, a_m)^T, \quad A = \mathrm{diag}(\boldsymbol{a}), \quad V = [\boldsymbol{v}_1\ \boldsymbol{v}_2\ \cdots\ \boldsymbol{v}_m], \quad \boldsymbol{e} = (1, \ldots, 1)^T.$$

We refer to (2.2) as the *standard problem* (SP). Unfortunately $\boldsymbol{w}^T\boldsymbol{w} = 1$ is a nonlinear constraint. Nonetheless, this is the problem we would most like to solve.

If the points are not separable, the best solution is obtained by shifting the points by the least distance in the opposite direction. This leads to the same standard problem, but the solution has $h_* < 0$ (Figure 2.2). Hence, we shall say that if $h_* > 0$ then the points are *strictly separable*, if $h_* < 0$ then the points are *not separable*, and if $h_* = 0$ the points are *weakly separable*.

Note that in the solution of the separable case, the active constraints in (2.2b) identify the points of the two clusters which are nearest to the OSH (that is, at the distance h_* from it): these points are called *support vectors*.

2.1. The separable case

In this case, existing theory has a clever way of reducing the standard problem to a convex QP. We let $\boldsymbol{w} \neq \boldsymbol{0}$ be non-normalized and shift the points along the normalized vector $\boldsymbol{w}/\|\boldsymbol{w}\|$ as before, giving the problem

$$\begin{aligned} &\underset{\boldsymbol{w},b,h}{\text{maximize}} \quad h \\ &\text{subject to} \quad a_i\big(\boldsymbol{w}^T(\boldsymbol{v}_i - ha_i\boldsymbol{w}/\|\boldsymbol{w}\|) + b\big) \geq 0 \quad i = 1, \ldots, m, \end{aligned} \tag{2.3}$$

or equivalently

$$\begin{aligned} &\underset{\boldsymbol{w},b,h}{\text{maximize}} \quad h \\ &\text{subject to} \quad a_i\big(\boldsymbol{w}^T\boldsymbol{v}_i + b\big) \geq h\|\boldsymbol{w}\| \quad i = 1, \ldots, m. \end{aligned} \tag{2.4}$$

Solving this problem, followed by dividing $\boldsymbol{w}$ and b by $\|\boldsymbol{w}\|$, yields the same solution as above. We now fix $\|\boldsymbol{w}\|$ by

$$h\|\boldsymbol{w}\| = 1 \quad \text{or} \quad h = 1/\|\boldsymbol{w}\|.$$

Then the problem becomes

$$\begin{aligned} &\underset{\boldsymbol{w},b}{\text{maximize}} \quad \|\boldsymbol{w}\|^{-1} \\ &\text{subject to} \quad AV^T\boldsymbol{w} + \boldsymbol{a}b \geq \boldsymbol{e}. \end{aligned} \tag{2.5}$$

But maximizing $\|\boldsymbol{w}\|^{-1}$ can be solved by minimizing $\|\boldsymbol{w}\|$ and hence by minimizing $\frac{1}{2}\boldsymbol{w}^T\boldsymbol{w}$. Hence we can equivalently solve the convex QP,

$$\text{CQP:} \quad \begin{array}{ll} \underset{\boldsymbol{w},b}{\text{minimize}} & \frac{1}{2}\boldsymbol{w}^T\boldsymbol{w} \\ \text{subject to} & AV^T\boldsymbol{w} + \boldsymbol{a}b \geq \boldsymbol{e}. \end{array} \tag{2.6}$$

Denoting the multipliers of (2.6) by $\boldsymbol{x}$, this problem has a useful dual,

$$\text{CQD:} \quad \begin{array}{ll} \underset{\boldsymbol{x}}{\text{minimize}} & \frac{1}{2}\boldsymbol{x}^TQ\boldsymbol{x} - \boldsymbol{e}^T\boldsymbol{x} \\ \text{subject to} & \boldsymbol{a}^T\boldsymbol{x} = 0, \quad \boldsymbol{x} \geq \boldsymbol{0}, \end{array} \tag{2.7}$$

where $Q = AV^TVA$. Because the normalization of $\boldsymbol{w}$ requires $h > 0$, this development only applies to the strictly separable case.

In comparing (2.6) and (2.7) with the solution of the SP, we see that if we take a sequence of separable problems in which $h_* \to 0$, then $\|\boldsymbol{w}_*\| \to +\infty$ (since $h_*\|\boldsymbol{w}_*\| = 1$): in fact all non-zero values of b_* and $\boldsymbol{x}_*$ converge to $\pm\infty$. For the limiting weakly separable problem, the convex QP (2.6) is infeasible and the dual (2.7) is unbounded. However, solution of all these problems by the SP (2.2) is well behaved, including the weakly separable case. The solution of the limiting problem could also be obtained by scaling the solution values obtained by (2.6) or (2.7) (see Section 3) and then taking the limit, but it seems more appropriate to solve the SP directly.

2.2. The non-separable case

For a non-separable problem ($h < 0$), one might proceed by normalizing by $\|\boldsymbol{w}\| = -1/h$ in (2.4), giving

$$\begin{array}{ll} \underset{\boldsymbol{w},b}{\text{maximize}} & -\|\boldsymbol{w}\|^{-1} \\ \text{subject to} & AV^T\boldsymbol{w} + \boldsymbol{a}b \geq -\boldsymbol{e}. \end{array} \tag{2.8}$$

As for (2.5), maximizing $-\|\boldsymbol{w}\|^{-1}$ can be replaced by maximizing $\|\boldsymbol{w}\|$ and hence by maximizing $\frac{1}{2}\boldsymbol{w}^T\boldsymbol{w}$. Unfortunately we now have a non-convex QP. This can sometimes be solved by careful choice of initial approximation. However, there is no dual, and ill-conditioning happens as $h \to 0$. It does not solve the $h = 0$ problem. We therefore look for alternatives.

Currently (see Section 1) the preferred approach in the literature is to solve a dual,

$$L_1\text{QD:} \quad \begin{array}{ll} \underset{\boldsymbol{x}}{\text{minimize}} & \frac{1}{2}\boldsymbol{x}^TQ\boldsymbol{x} - \boldsymbol{e}^T\boldsymbol{x} \\ \text{subject to} & \boldsymbol{a}^T\boldsymbol{x} = 0, \quad \boldsymbol{0} \leq \boldsymbol{x} \leq c\boldsymbol{e}, \end{array} \tag{2.9}$$

which is the dual of the L_1-penalized primal

$$L_1\text{QP:} \quad \begin{array}{ll} \underset{\boldsymbol{\xi},\boldsymbol{w},b}{\text{minimize}} & \frac{1}{2}\boldsymbol{w}^T\boldsymbol{w} + c\boldsymbol{e}^T\boldsymbol{\xi} \\ \text{subject to} & AV^T\boldsymbol{w} + \boldsymbol{a}b \geq \boldsymbol{e} - \boldsymbol{\xi}, \quad \boldsymbol{\xi} \geq \boldsymbol{0}. \end{array} \tag{2.10}$$

The advantage is that it is easy to solve. Some major disadvantages are:

- $h > 0$ case: the penalty parameter c must be not smaller than the maximum multiplier if the solution of the SP is to be recovered. This requires $c \to \infty$ in the limiting case.
- $h = 0$ and $h < 0$ cases: the solution of the SP is not recovered. Some experiments indicate that very poor 'solutions' may be obtained.

We are concerned that the solutions obtained by this method may be significantly sub-optimal in comparison to the SP solution. Note that approximate solutions of L_1QD that are feasible in the dual (such as are obtained by most existing software) will not be feasible in the primal, whereas our methods always provide primal feasible solutions, even if numerically we are not able to locate an exact solution. A problem similar to (2.10) is considered directly by Chapelle (2007), but using different (quadratic) penalties.

3. KT conditions for the standard problem

In view of the difficulties inherent in the existing approach based on (2.7), when the problem is not strictly separable, we shall investigate an approach based on solving the SP (2.2). First we need to identify KT conditions for the SP. In doing this we again use $\boldsymbol{x}$ to denote the multipliers of the inequality constraints, and we write the normalization constraint in a particular way which relates its multiplier, π say, to the solution value h. Thus we write the SP as

$$\underset{\boldsymbol{w},b,h}{\text{minimize}} \quad -h \tag{3.1a}$$

$$\text{subject to} \quad AV^T\boldsymbol{w} + \boldsymbol{a}b - \boldsymbol{e}h \geq \boldsymbol{0} \tag{3.1b}$$

$$\frac{1}{2}(1 - \boldsymbol{w}^T\boldsymbol{w}) = 0. \tag{3.1c}$$

The Lagrangian function for this problem is

$$\mathcal{L}(\boldsymbol{w}, b, h, \boldsymbol{x}, \pi) = -h - \boldsymbol{x}^T(AV^T\boldsymbol{w} + \boldsymbol{a}b - \boldsymbol{e}h) - \frac{\pi}{2}(1 - \boldsymbol{w}^T\boldsymbol{w}). \tag{3.2}$$

KT conditions for optimality are feasibility in (3.1b) and (3.1c), and stationarity of $\mathcal{L}$ with respect to $\boldsymbol{w}$, b and h, giving, respectively,

$$VA\boldsymbol{x} = \pi\boldsymbol{w}, \tag{3.3a}$$

$$\boldsymbol{a}^T\boldsymbol{x} = 0, \tag{3.3b}$$

$$\boldsymbol{e}^T\boldsymbol{x} = 1, \tag{3.3c}$$

together with the complementarity condition

$$\boldsymbol{x}^T\left(AV^T\boldsymbol{w} + \boldsymbol{a}b - \boldsymbol{e}h\right) = 0, \tag{3.4}$$

and non-negative multipliers

$$\boldsymbol{x} \geq \mathbf{0}. \tag{3.5}$$

An interesting interpretation of (3.3b) and (3.3a) in terms of forces and torques acting on a rigid body is given by Burges and Schölkopf (1997). We note that (3.4) simplifies to give $\boldsymbol{x}^T AV^T\boldsymbol{w} = h$, and it follows from (3.3a) and (3.1c) that

$$\pi = h. \tag{3.6}$$

This simple relationship is the motivation for expressing the normalization constraint as (3.1c).

These KT conditions are closely related to those for the CQP (2.6) when $h > 0$. For the CQP the Lagrangian function is

$$\mathcal{L}(\boldsymbol{w}, b, h) = \frac{1}{2}\boldsymbol{w}^T\boldsymbol{w} - \boldsymbol{x}^T\left(AV^T\boldsymbol{w} + \boldsymbol{a}b - \boldsymbol{e}\right), \tag{3.7}$$

and stationarity with respect to $\boldsymbol{w}$ and b gives $AV\boldsymbol{x} = \boldsymbol{w}$ and $\boldsymbol{a}^T\boldsymbol{x} = 0$. The complementarity condition $\boldsymbol{x}^T\left(AV^T\boldsymbol{w} + \boldsymbol{a}b - \boldsymbol{e}\right) = 0$ then yields $\boldsymbol{e}^T\boldsymbol{x} = \boldsymbol{w}^T\boldsymbol{w}$. If $\boldsymbol{w}_*$, b_*, h_*, $\boldsymbol{x}_*$ denote the solution and multiplier of the SP, we see that $\boldsymbol{w}_*/h_*$, b_*/h_* and $\boldsymbol{x}_*/h_*^2$ solve the KT conditions of the CQP, and conversely $h_* = 1/\|\boldsymbol{w}\|$, $\boldsymbol{w}_* = \boldsymbol{w}/\|\boldsymbol{w}\|$, $b_* = b/\|\boldsymbol{w}\|$, $\boldsymbol{x}_* = \boldsymbol{x}/\left(\boldsymbol{w}^T\boldsymbol{w}\right)$ determine the solution of the SP from that of the CQP.

If the CQD is solved, the solution of the CQP can be recovered from $\boldsymbol{w} = VA\boldsymbol{x}$, $h = 1/\|\boldsymbol{w}\|$, and b can be obtained by solving $a_i\left(\boldsymbol{w}^T\boldsymbol{v}_i + b\right) = 1$ for any i such that $x_i > 0$.

4. A new SQP-like algorithm

We now develop a new SQP-like algorithm for computing the solution $\boldsymbol{w}_*$, b_*, h_* of the standard problem (2.2). The iterates in our algorithm are $\boldsymbol{w}_k$, b_k, h_k, $k = 0, 1, \ldots,$ and we shall maintain the property that the iterates are feasible in (2.2) for all k. Initially $\boldsymbol{w}_0$ ($\boldsymbol{w}_0^T\boldsymbol{w}_0 = 1$) is arbitrary, and we choose b_0 and h_0 by solving the 2-variable LP

$$\underset{b,h}{\text{minimize}} \quad -h \tag{4.1a}$$

$$\text{subject to } AV^T\boldsymbol{w}_0 + \boldsymbol{a}b - \boldsymbol{e}h \geq \mathbf{0}. \tag{4.1b}$$

At the kth iteration we solve a QP problem formulated in terms of the correction $\boldsymbol{d} = \left(\boldsymbol{d}_{\boldsymbol{w}}^T, d_b, d_h\right)^T$. We shall need the Hessian of the Lagrangian, which is

$$W = \begin{pmatrix} \pi I_{n\times n} & \mathbf{0}_{n\times 2} \\ \mathbf{0}_{n\times 2}^T & \mathbf{0}_{2\times 2} \end{pmatrix}. \tag{4.2}$$

We also need to specify an estimate for the multiplier π_k, which we shall do below. Thus the standard QP subproblem (see, *e.g.*, Fletcher (1987)) becomes

$$\underset{\boldsymbol{d}}{\text{minimize}} \quad \frac{1}{2}\pi_k \boldsymbol{d}_{\boldsymbol{w}}^T \boldsymbol{d}_{\boldsymbol{w}} - d_h \tag{4.3a}$$

$$\text{subject to} \quad AV^T(\boldsymbol{w}_k + \boldsymbol{d}_{\boldsymbol{w}}) + \boldsymbol{a}(b_k + d_b) - \boldsymbol{e}(h_k + d_h) \geq \boldsymbol{0} \tag{4.3b}$$

$$\frac{1}{2}(1 - \boldsymbol{w}_k^T \boldsymbol{w}_k) - \boldsymbol{w}_k^T \boldsymbol{d}_{\boldsymbol{w}} = 0. \tag{4.3c}$$

Because we maintain feasibility, so $\boldsymbol{w}_k^T \boldsymbol{w}_k = 1$, and the equality constraint simplifies to give $-\boldsymbol{w}_k^T \boldsymbol{d}_{\boldsymbol{w}} = 0$.

It is also the case, in our algorithm, that the possibility of (4.3) being *unbounded* can occur. Thus we also restrict the size of $\boldsymbol{d}_{\boldsymbol{w}}$ by a trust-region-like constraint $\|\boldsymbol{d}_{\boldsymbol{w}}\| \leq \Delta$ where $\Delta > 0$ is fixed. The aim is not to use this to force convergence, but merely to prevent any difficulties caused by unboundedness. Since we shall subsequently normalize $\boldsymbol{w}_k + \boldsymbol{d}_{\boldsymbol{w}}$, the actual value of Δ is not critical, but we have chosen $\Delta = 10^5$ in practice. We note that it is not necessary to bound d_b and d_h, since it follows from (4.3b) that if $\boldsymbol{w}$ is fixed, then an *a priori* upper bound on $|d_b|$ and $|d_h|$ exists.

Thus the subproblem that we solve in our new algorithm is

$$\mathrm{QP}_k: \quad \begin{aligned} &\underset{\boldsymbol{d}}{\text{minimize}} && \frac{1}{2}\pi_k \boldsymbol{d}_{\boldsymbol{w}}^T \boldsymbol{d}_{\boldsymbol{w}} - d_h && (4.4\text{a}) \\ &\text{subject to} && AV^T(\boldsymbol{w}_k + \boldsymbol{d}_{\boldsymbol{w}}) + \boldsymbol{a}(b_k + d_b) - \boldsymbol{e}(h_k + d_h) \geq \boldsymbol{0} && (4.4\text{b}) \\ &&& -\boldsymbol{w}_k^T \boldsymbol{d}_{\boldsymbol{w}} = 0 && (4.4\text{c}) \\ &&& \|\boldsymbol{d}_{\boldsymbol{w}}\|_\infty \leq \Delta. && (4.4\text{d}) \end{aligned}$$

Since $\boldsymbol{d} = \boldsymbol{0}$ is feasible in QP_k, and $\boldsymbol{d}_{\boldsymbol{w}}$ is bounded, there always exists a solution. In practice we check $\|\boldsymbol{d}\| < \tau_d$ for some fixed small tolerance $\tau_d > 0$.

We shall denote the outcome of applying the resulting correction $\boldsymbol{d}$ by

$$(\boldsymbol{w}^\circ, b^\circ, h^\circ) = (\boldsymbol{w}_k + \boldsymbol{d}_{\boldsymbol{w}}, b_k + d_b, h_k + d_h). \tag{4.5}$$

We now differ slightly from the standard SQP algorithm by rescaling these values to obtain the next iterate, as

$$\begin{pmatrix} \boldsymbol{w}_{k+1} \\ b_{k+1} \\ h_{k+1} \end{pmatrix} = \frac{1}{\|\boldsymbol{w}^\circ\|} \begin{pmatrix} \boldsymbol{w}^\circ \\ b^\circ \\ h^\circ \end{pmatrix}, \tag{4.6}$$

which ensures that the new iterate is feasible in (2.2). The standard choice in SQP for the multiplier π_k would be $\pi_k = h_k$ by virtue of (3.6). However, if $h_k < 0$ this would result in QP$_k$ being a non-convex QP. Thus we choose

$$\pi_k = \max\{h_k, 0\} \tag{4.7}$$

as the multiplier estimate for (4.4a). For $h_k \leq 0$ we then note that (4.4) is in fact a linear programming (LP) calculation. Moreover, in the initial case (4.1), we shall see in Section 4.1 that the solution can be solved directly rather than by using an LP solver.

There are two particularly unusual and attractive features of the new algorithm. First we shall prove in Theorem 4.1 below that h_k is strictly monotonically increasing whilst $\boldsymbol{d_w} \neq \mathbf{0}$. Thus we have not needed to provide any globalization strategy to enforce this condition. Even more surprisingly, we have found that the sequence of iterates *terminates* at the solution after a finite number of iterations. We have not yet been able to prove that this must happen. We had conjectured that termination would happen as soon as the correct active set had been located by the algorithm, but numerical evidence has disproved this conjecture. A negative feature of the SP should also be noted, that being a non-convex NLP, there is no guarantee that all solutions are global solutions. Indeed, we were able to construct a case with a local but not global solution $\boldsymbol{w}_*$, and our SQP-like algorithm could be made to converge to this solution by choosing $\boldsymbol{w}_0$ close to $\boldsymbol{w}_*$. In practice, however, we have not recognized any other instances of this behaviour.

Before proving Theorem 4.1 we note that QP$_k$ (4.4) can equivalently be expressed as

$$\underset{\boldsymbol{w} \in \mathcal{W}_k, b, h}{\text{maximize}} \quad h - \frac{1}{2}\pi_k \boldsymbol{w}^T \boldsymbol{w} \tag{4.8a}$$

$$\text{subject to} \quad AV^T \boldsymbol{w} + \boldsymbol{a}b - \boldsymbol{e}h \geq \mathbf{0}, \tag{4.8b}$$

where

$$\mathcal{W}_k = \{\boldsymbol{w} \mid \boldsymbol{w}_k^T \boldsymbol{w} = 1, \ \|\boldsymbol{w} - \boldsymbol{w}_k\|_\infty \leq \Delta\}. \tag{4.9}$$

Hence the solution of QP$_k$ can also be expressed as

$$h^\circ = \max_{\boldsymbol{w} \in \mathcal{W}_k} \max_b \min_{i=1,\ldots,m} a_i(\boldsymbol{v}_i^T \boldsymbol{w} + b). \tag{4.10}$$

Since $\boldsymbol{w}^\circ$ is the maximizer over $\mathcal{W}_k$ it follows that

$$h^\circ = \max_b \min_{i=1,\ldots,m} a_i(\boldsymbol{v}_i^T \boldsymbol{w}^\circ + b) \tag{4.11}$$

and b° is the maximizer over b. We also observe from $\boldsymbol{w}_k^T \boldsymbol{d_w} = 0$, $\boldsymbol{w}_k^T \boldsymbol{w}_k = 1$ and $\boldsymbol{w}^\circ = \boldsymbol{w}_k + \boldsymbol{d_w}$ by Pythagoras' theorem that $\|\boldsymbol{w}^\circ\| \geq 1$, and if $\boldsymbol{d_w} \neq \mathbf{0}$, that

$$\|\boldsymbol{w}^\circ\| > 1. \tag{4.12}$$

Dividing through (4.11) by $\|\boldsymbol{w}^\circ\|$ yields

$$h_{k+1} = \max_b \min_{i=1,\dots,m} a_i\big(\boldsymbol{v}_i^T \boldsymbol{w}_{k+1} + b/\|\boldsymbol{w}^\circ\|\big), \tag{4.13}$$

and $b_{k+1} = b^\circ/\|\boldsymbol{w}^\circ\|$ is the maximizer over b. This provides an inductive proof of the result that

$$h_k = \max_b \min_{i=1,\dots,m} a_i\big(\boldsymbol{v}_i^T \boldsymbol{w}_k + b\big), \tag{4.14}$$

since we establish this result for $k = 0$ in solving (4.1).

Theorem 4.1. If at the kth iteration $\boldsymbol{d_w} = \boldsymbol{0}$ does not provide a solution of QP$_k$, then $h_{k+1} > h_k$.

Remark 4.2. Essentially we are assuming that if $\boldsymbol{d_w} = \boldsymbol{0}$ in any solution to QP$_k$, then the algorithm terminates. Otherwise we can assume that both $\boldsymbol{d_w} \neq \boldsymbol{0}$ and $h^\circ > h_k$ when proving the theorem.

Proof. First we consider the case that $h_k = \pi_k > 0$. We shall define

$$f^\circ = \max_{\boldsymbol{w}\in\mathcal{W}_k} \left(\max_b \min_{i=1,\dots,m} a_i\big(\boldsymbol{v}_i^T \boldsymbol{w} + b\big) - \frac{1}{2}\pi_k \boldsymbol{w}^T \boldsymbol{w} \right) \tag{4.15}$$

and note as in Section 2 that f° solves the problem

$$\begin{aligned} &\underset{\boldsymbol{w}\in\mathcal{W}_k, b, f}{\text{maximize}} \quad f \\ &\text{subject to } \; a_i\big(\boldsymbol{v}_i^T \boldsymbol{w} + b\big) - \frac{1}{2}\,\pi_k \boldsymbol{w}^T \boldsymbol{w} \geq f \qquad i = 1,\dots,m. \end{aligned} \tag{4.16}$$

If we substitute $f = h - \frac{1}{2}\pi_k \boldsymbol{w}^T \boldsymbol{w}$, we note that this becomes the problem

$$\begin{aligned} &\underset{\boldsymbol{w}\in\mathcal{W}_k, b, h}{\text{maximize}} \quad h \\ &\text{subject to } \; a_i\big(\boldsymbol{v}_i^T \boldsymbol{w} + b\big) \geq h \qquad i = 1,\dots,m, \end{aligned} \tag{4.17}$$

which is solved by $\boldsymbol{w}^\circ$, b° and h°. Thus we can identify $f^\circ = h^\circ - \frac{1}{2}\pi_k \boldsymbol{w}^{\circ T} \boldsymbol{w}^\circ$ and assert that

$$\begin{aligned} h^\circ = &\frac{1}{2}\,\pi_k (\boldsymbol{w}^\circ)^T \boldsymbol{w}^\circ \\ &+ \max_{\boldsymbol{w}\in\mathcal{W}_k} \left(\max_b \min_{i=1,\dots,m} a_i(\boldsymbol{v}_i^T \boldsymbol{w} + b) - \frac{1}{2}\,\pi_k \boldsymbol{w}^T \boldsymbol{w} \right). \end{aligned} \tag{4.18}$$

But $\boldsymbol{w}_k \in \mathcal{W}_k$, so it follows that

$$\begin{aligned} h^\circ \geq &\frac{1}{2}\,\pi_k (\boldsymbol{w}^\circ)^T \boldsymbol{w}^\circ \\ &+ \left(\max_b \min_{i=1,\dots,m} a_i(\boldsymbol{v}_i^T \boldsymbol{w}_k + b) \right) - \frac{1}{2}\,\pi_k \boldsymbol{w}_k^T \boldsymbol{w}_k. \end{aligned} \tag{4.19}$$

Using $\boldsymbol{w}_k^T \boldsymbol{w}_k = 1$, $\pi_k = h_k$ and the induction hypothesis (4.14), it follows that

$$h^\circ \geq \frac{1}{2} h_k (\boldsymbol{w}^\circ)^T \boldsymbol{w}^\circ + h_k - \frac{1}{2} h_k = \frac{1}{2} h_k \big(\|\boldsymbol{w}^\circ\|^2 + 1 \big). \tag{4.20}$$

Hence

$$h_{k+1} = h^\circ / \|\boldsymbol{w}^\circ\| \geq \frac{1}{2} h_k \big(\|\boldsymbol{w}^\circ\| + \|\boldsymbol{w}^\circ\|^{-1} \big). \tag{4.21}$$

Finally, from $\boldsymbol{d_w} \neq \mathbf{0}$ and (4.12) it follows that $h_{k+1} > h_k$ in this case.

In the case $h_k = 0$, we note that $h^\circ \geq h_k$ by virtue of (4.10) and the fact that $\boldsymbol{w}_k \in \mathcal{W}_k$ in (4.14). But if $h^\circ = h_k$, then $\boldsymbol{w}_k$ solves (4.4), which implies that $\boldsymbol{d_w} = \mathbf{0}$ in (4.4c), which is a contradiction. Thus $h_{k+1} = h^\circ / \|\boldsymbol{w}^\circ\| > 0$ and $h_k = 0$ so $h_{k+1} > h_k$, which completes the case $h_k = 0$.

Finally, we consider the case $h_k < 0$. It follows directly from $h^\circ > h_k$ and (4.12) that

$$h_{k+1} = \frac{h^\circ}{\|\boldsymbol{w}^\circ\|} > \frac{h_k}{\|\boldsymbol{w}^\circ\|} > h_k,$$

and the proof is complete. □

4.1. Solution boundedness and starting point

We mentioned that (4.3b) implies the boundedness of d_b and d_h, and hence of b and h. However, it is interesting to see directly how b and h are bounded for $\boldsymbol{w} \in \mathcal{W}_k$. Let $\mathcal{P} = \{i \mid a_i = +1\}$, $\mathcal{M} = \{i \mid a_i = -1\}$ and assume that they are both non-empty (otherwise we would not have a binary classification problem). For *fixed* $\boldsymbol{w} \in \mathcal{W}_k$ the solution of QP$_k$ for b and h is given by

$$\begin{aligned} h &= \max_b \min_i a_i \big(\boldsymbol{v}_i^T \boldsymbol{w} + b \big) \\ &= \max_b \min \Big\{ b + \min_{i \in \mathcal{P}} \big(\boldsymbol{v}_i^T \boldsymbol{w} \big), \ -b + \min_{i \in \mathcal{M}} \big(-\boldsymbol{v}_i^T \boldsymbol{w} \big) \Big\} \\ &= \max_b \min \big\{ b + \alpha, \ -b + \beta \big\}, \end{aligned} \tag{4.22}$$

where $\alpha = \min_{i \in \mathcal{P}} \big(\boldsymbol{v}_i^T \boldsymbol{w} \big)$ and $\beta = \min_{i \in \mathcal{M}} \big(-\boldsymbol{v}_i^T \boldsymbol{w} \big)$ depend only on the fixed $\boldsymbol{w}$. Now

$$b + \alpha \geq -b + \beta \quad \Leftrightarrow \quad b \geq \frac{1}{2}(\beta - \alpha)$$

and we have two cases: (i) if $b \geq (\beta - \alpha)/2$, then the minimum is given by $-b + \beta$, and the maximum over b is then obtained when $b = (\beta - \alpha)/2$, that is, $h = (\alpha - \beta)/2 + \beta = (\alpha + \beta)/2$; (ii) if $b \leq (\beta - \alpha)/2$, then the minimum is given by $b + \alpha$, and the maximum over b is obtained again when $b = (\beta - \alpha)/2$, which gives $h = (\beta - \alpha)/2 + \alpha = (\alpha + \beta)/2$. Thus, for any fixed $\boldsymbol{w} \in \mathcal{W}_k$ the solution of the max min problem (4.22) is given by

$$h = (\alpha + \beta)/2 \quad \text{and} \quad b = (\beta - \alpha)/2. \tag{4.23}$$

Since $\boldsymbol{w} \in \mathcal{W}_k$ is bounded, there exist bounds on α and β by continuity, and hence on b and h. Moreover, the equations (4.23) directly provide the solution of (4.1) for any fixed $\boldsymbol{w}$, so we do not need to solve the LP problem.

Also, we have already mentioned that we can start the algorithm from a normalized random vector $\boldsymbol{w}_0$. However, a better initial estimate can be obtained by choosing the normalized vector $\boldsymbol{w}_0$ joining the two nearest points of the opposite classes. Once we have chosen $\boldsymbol{w}_0$, we can compute α and β. Hence, our starting point for the SQP algorithm is given by $\boldsymbol{w}_0$, b_0, h_0, with b_0, h_0 as in (4.23). As we shall see later in the next section, this choice will also provide a natural way to initialize in the algorithm in the case of nonlinear SVMs.

5. Nonlinear SVMs

In practice it is very rare that the points $\boldsymbol{v}_i$ are adequately separated by a hyperplane $\boldsymbol{w}_*^T\boldsymbol{v} + b_* = 0$. An example which we use below is one where points in $\mathbb{R}^2$ are classified according to whether they lie in a black square or a white square of a chessboard. The *nonlinear SVM* technique aims to handle the situation by mapping $\boldsymbol{v}$ nonlinearly into a higher dimension space (the so-called *feature space*), in which a satisfactory separation can be obtained. Thus we are given some real functions

$$\boldsymbol{\phi}(\boldsymbol{v}) = \big(\phi_1(\boldsymbol{v}),\, \phi_2(\boldsymbol{v}), \ldots, \phi_N(\boldsymbol{v})\big)^T, \tag{5.1}$$

presently finite in number, and we solve the standard problem with the matrix

$$\Phi(\boldsymbol{v}) = \big[\boldsymbol{\phi}(\boldsymbol{v}_1) \quad \boldsymbol{\phi}(\boldsymbol{v}_2) \quad \cdots \quad \boldsymbol{\phi}(\boldsymbol{v}_m)\big], \tag{5.2}$$

replacing the matrix V. The solution values $\boldsymbol{w}_*$, b_*, h_* then provide a classification function

$$f(\boldsymbol{v}) = \boldsymbol{w}_*^T\boldsymbol{\phi}(\boldsymbol{v}) + b_* \tag{5.3}$$

which maps the optimal separating hyperplane in feature space, back into $\mathbb{R}^n$. The multipliers $\boldsymbol{x}_*$ obtained in the feature space also allow us to express

$$f(\boldsymbol{v}) = \frac{1}{h}\bigg(\sum_i (\boldsymbol{x}_*)_i a_i \boldsymbol{\phi}(\boldsymbol{v}_i)^T\boldsymbol{\phi}(\boldsymbol{v})\bigg) + b_* \tag{5.4}$$

when $h_* > 0$, and we notice that the sum only needs to be computed over the support vectors, by virtue of (3.4) and (3.5); see, for instance, Cucker and Zhou (2007) for a treatment of this subject from an approximation theory viewpoint.

Also, when $h_* > 0$, we can solve the SP by the transformation leading to the convex dual (2.7) in which $Q = A\Phi^T\Phi A$. The matrix Q is necessarily positive semidefinite, and if it is positive definite then the dual has a unique solution, and hence also the primal and the SP.

This development has suggested another approach in which a kernel function $\mathcal{K}(\boldsymbol{t}, \boldsymbol{y})$ is chosen, which can implicitly be factored into the infinite-dimensional scalar product $\sum_{i=1}^{\infty} \boldsymbol{\phi}_i(\boldsymbol{t})\boldsymbol{\phi}_i(\boldsymbol{y})$, in which the functions $\boldsymbol{\phi}_i(\cdot)$ are known to exist but may not be readily available;[2] see, for example, Shawe-Taylor and Cristianini (2004) and Schölkopf and Smola (2002). One of these kernel functions is the *Gaussian kernel*

$$\mathcal{K}(\boldsymbol{t}, \boldsymbol{y}) = \exp\bigl(-\|\boldsymbol{t} - \boldsymbol{y}\|^2/(2\sigma^2)\bigr). \tag{5.5}$$

An $m \times m$ matrix K with elements $K_{ij} = \mathcal{K}(\boldsymbol{v}_i, \boldsymbol{v}_j)$ may be computed from the $\boldsymbol{v}_i$, and K is positive semidefinite. For some kernels, such as the Gaussian kernel, K is always positive definite when the points $\boldsymbol{v}_i$ are distinct. Then $Q = AKA$ and the dual problem (2.7) may be solved to determine $\boldsymbol{x}_*$. Then the classification may be expressed as

$$f(\boldsymbol{v}) = \sum_{i=1}^{m} (\boldsymbol{x}_*)_i a_i \mathcal{K}(\boldsymbol{v}_i, \boldsymbol{v}). \tag{5.6}$$

Although the primal solution $\boldsymbol{w}_*$ and the map $\boldsymbol{\phi}(\boldsymbol{v})$ may be infinite in dimension, and hence not computable, the dual problem can always be attempted and, when Q is positive definite, always has a unique solution. Because of such considerations most existing research and software are dual-based.

In practice, however, m is likely to be extremely large, and Q is a dense matrix, so solving the dual is extremely challenging computationally. Also Q may be numerically singular, even for quite small values of m. It seems not to be well known that primal algorithms based on a kernel function are also practicable. Ignoring numerical considerations for the present, the approach is to calculate full-rank exact factors

$$K = U^T U \tag{5.7}$$

of the kernel matrix K, where $\operatorname{rank}(U) = N$ may be smaller than m. Then the SP is solved with U replacing V. The key step in determining the classification function $f(\boldsymbol{v})$ for arbitrary $\boldsymbol{v}$ is to find the least-squares solution of the system

$$U\boldsymbol{\theta} = \bigl(\mathcal{K}(\boldsymbol{v}_1, \boldsymbol{v}), \ldots, \mathcal{K}(\boldsymbol{v}_m, \boldsymbol{v})\bigr)^T. \tag{5.8}$$

[2] The fundamental hypothesis is that we can find a *Mercer's kernel*, that is, a symmetric and positive semidefinite function $\mathcal{K} : \mathcal{X} \times \mathcal{X} \to \mathbb{R}$, where $\mathcal{X}$ is a compact metric space. It is well known that, given such a symmetric and positive semidefinite function $\mathcal{K}$, there exists exactly one Hilbert space of functions $\mathcal{H}_{\mathcal{K}}$ such that: (i) $\mathcal{K}_x = \mathcal{K}(x, \cdot) \in \mathcal{H}_{\mathcal{K}}$, $\forall x \in \mathcal{X}$; (ii) $\operatorname{span}\{\mathcal{K}_x \,|\, x \in \mathcal{X}\}$ is dense in $\mathcal{H}_{\mathcal{K}}$; (iii) $f(x) = \langle \mathcal{K}_x, f \rangle_{\mathcal{H}_{\mathcal{K}}}$ $\forall f \in \mathcal{H}_{\mathcal{K}}$ and $\forall x \in \mathcal{X}$; (iv) the functions in $\mathcal{H}_{\mathcal{K}}$ are continuous on $\mathcal{X}$ with a bounded inclusion $\mathcal{H}_{\mathcal{K}} \subset \mathcal{C}^0(\mathcal{X})$. Property (iii) is known as the *reproducing property* of the kernel $\mathcal{K}$ and the function space $\mathcal{H}_{\mathcal{K}}$ is called the *reproducing kernel Hilbert space* (RKHS).

Then the classification function is

$$f(\boldsymbol{v}) = \boldsymbol{w}_*^T \boldsymbol{\theta} + b_*, \tag{5.9}$$

where $\boldsymbol{w}_*$ and b_* are obtained from the SP solution. It is readily observed when $\boldsymbol{v} = \boldsymbol{v}_j$ that $f(\boldsymbol{v}_j)$ is identical to the value obtained from (5.6) based on the dual solution. Also, when U is square ($N = m$) and non-singular, the same outcome as from (5.6) is obtained for all $\boldsymbol{v}$.

Because K is often nearly singular, it is even more effective and practical to consider calculating *low-rank approximate factors* $K \approx U^T U$ in which U has full rank ($\text{rank}(U) = N$) but $N < \text{rank}(K)$; see, for instance, Fine and Scheinberg (2001), Williams and Seeger (2001), Drineas and Mahoney (2005), Keerthi and DeCoste (2005) and Kulis, Sustik and Dhillon (2006). This enables the effects of ill-conditioning to be avoided, and is computationally attractive since the work scales up as $N^2 m$, as against m^3 for some dual-based methods. An efficient technique for calculating a suitable U is partial Cholesky factorization with diagonal pivoting (Goldfarb and Scheinberg 2004). This can be described in Matlab-like notation by

```
Initialize d=diag(K); U=[];
while 1
   i=argmax(d); if d(i)<=tol, break; end;
   s=sqrt(d(i)); u=(K(i,:)-U(i,:)'*U)/s;                (5.10)
   U=[U;u]; d=d-u.^2;
end.
```

The size of N is determined by the size of `tol`. It is advantageous that only the diagonal elements and one column of K on each iteration are needed, and not the entire matrix.

An important observation when using approximate Cholesky factors is the following. Let `pivots` denote the indices of the diagonal pivots used in calculating U. If $i \in$ `pivots` then row and column i of $K - U^T U$ are zero. If the SP is solved using the factor U, it then follows that the resulting value of $f(\boldsymbol{v}_i)$ agrees exactly with that which would be obtained by solving the SP with the exact U. Thus data points $\boldsymbol{v}_i$, $i \in$ `pivots`, are correctly classified when approximate factors are used. It is important that the support vectors arising from solving the SP are correctly classified, whereas errors in $f(\boldsymbol{v}_i)$ for $i \notin$ `pivots` may be tolerated. We have therefore experimented with an algorithm based on the following:

1 initialize `pivots` by a small number, $N = 10$ say, of diagonal pivots;

2 solve the SP using the resulting U and let `svs` denote the resulting set of support vectors;

3 set `newpivots = setdiff(pivots,svs)`;

4 finish if `newpivots` is empty;
5 set `pivots = pivots` $\cup$ `newpivots`, extend the factor U appropriately, and go to step 2.

We refer to the iterations of the scheme as *outer iterations.* Iterations of the SQP-like algorithm used to solve the SP are *inner iterations.*

When this algorithm terminates, `svs` $\subset$ `pivots`, and any classification errors are in data points of non-pivots. We have found on some smaller problems that the correct set of support vectors is identified more quickly, and with smaller values of N, than with diagonal pivoting. For some larger problems, we found that adding all the new pivots in step 5 could lead to difficulties in solving the SP due to ill-conditioning. It is therefore better to use diagonal pivoting amongst the new pivots, and not to add any new pivot for which d_i (see (5.10)) is very small. In fact the 2-norm of row i of $K - U^T U$ may be bounded by $(d_i \|\boldsymbol{d}\|_1)^{1/2}$, which enables a good estimate of the classification error of any data point to be made (see also Woodsend and Gondzio (2007*b*) for another way to choose the pivots). Moreover, the initialization step 1 in the previous algorithm can be fruitfully substituted by the following:

1 initialize `pivots` by $\{i_0, j_0\}$, where $i_0 \in \mathcal{P}$ and $j_0 \in \mathcal{M}$ are the indices used to compute α and β in (4.23).

The matrices U generated by the algorithm are most often *dense* (few zero elements), so the QPs are best handled by a dense matrix QP solver. We have used the BQPD code by Fletcher (1996–2007). It is important that the number of QP variables does not become too large, although the code can tolerate larger numbers of constraints. This equates to U having many more columns than rows, as is the case here. We also make frequent use of *warm starts*, allowing the QP solution to be found quickly starting from the active set of the previous QP calculation.

We are continuing to experiment with algorithms of this type and the interplay between classification error, choice of pivots and the effects of ill-conditioning. For example, one idea to be explored is the possibility of detecting pivots that are not support vectors, in order to reduce the size of U, promote speed-up, and improve conditioning.

6. Numerical experience

In this section we first give examples that illustrate some features of the primal-based low-rank SQP approach for nonlinear SVMs described in the previous section. We show on a small problem how this readily finds the solution, whereas there are potential difficulties when using the L_1QD formulation (2.9). We use a scaled-up version of this problem to explore some other aspects of the low-rank algorithm relating to computational efficiency.

Finally we outline preliminary numerical experience on some well-known data sets available on-line and often used for algorithm comparison. We do not need to use data pre-processing in our experiments (see the final discussion for more details).

6.1. A 3 × 3 chessboard problem

We have seen in Section 2.2 that even in (nonlinearly) separable cases, the 'solution' obtained from the L_1QD may not agree with the correct solution of the SP, because an insufficiently large value of the penalty parameter c has been used. We illustrate this feature using the small 3×3 chessboard-like example with 120 data points, shown in Figure 6.1(a). We use a Gaussian kernel with $\sigma = 1$. The solution computed by the low-rank SQP algorithm of Section 5 is shown in Figure 6.1(b).

Six outer iterations are required and each one terminates after 2 to 5 inner iterations. No pivots are thresholded out in these calculations. There are 16 data points recognized as support vectors, which are highlighted with thick squares around their symbols, and the contours of $f(\boldsymbol{v}) = +h$, 0, $-h$ are plotted, the thicker line being the zero contour. The solution has $h_* =$ 7.623918E$-$3, showing that the two classes are nonlinearly separable by the Gaussian, as the theory predicts. The number of pivots required is 30, and

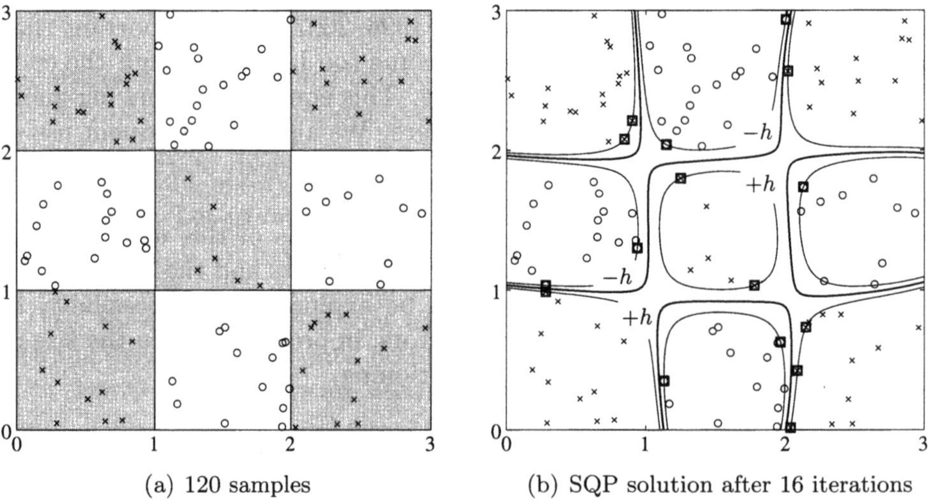

(a) 120 samples (b) SQP solution after 16 iterations

Figure 6.1. (a) 3×3 chessboard test problem. (b) Solution of the 3×3 problem computed by the low-rank SQP-like algorithm with Gaussian kernel ($\sigma = 1$): there are 16 support vectors, $b \approx 9.551186\text{E}{-2}$ and $h \approx 7.623918\text{E}{-3}$, showing that the two classes are separable by the given Gaussian.

this is also the final rank of U, showing that it is not necessary to factorize the whole of K to identify the exact solution. It is also seen that the zero contour quite closely matches the ideal chessboard classification shown in Figure 6.1(a).

To compare this with the usual SVM training approach based on the L_1QD (2.9), we solve this problem for a range of values of the penalty parameter c, whose choice is a major issue when attempting a solution; see, for example, Cristianini and Shawe-Taylor (2000), Schölkopf and Smola (2002), Cucker and Smale (2002) and Shawe-Taylor and Cristianini (2004). As the theory predicts, the SP solution can be recovered by choosing a sufficiently large c. Here the threshold is $c \geq 5000$ as in Figure 6.2(f). As can be seen in Figure 6.2, the dual solution is strongly sub-optimal even for quite large values of c. Such values might well have been chosen when using some of the usual heuristics, such as cross-validation, for the penalty parameter.

The behaviour of Lagrange multipliers is also of some interest. For the SP solution there are 16 non-zero multipliers corresponding to the 16 support vectors. For the L_1QD solutions, there are many more non-zero multipliers when $c < 5000$. We continue to refer to these as SVs in Figure 6.2 and Table 6.1. Multipliers on their upper bound (that is, $\boldsymbol{x}_i = c$, with $\boldsymbol{x}_i$ being computed from the L_1QD problem) are referred to as BSVs in the table. These data points violate the constraints that the distance from the optimal separating hypersurface is at least h. Some of these have a distance whose sign is opposite to their label. These points we refer to as being *misclassified*. By increasing c we have fewer multipliers that meet their upper bound, thus removing misclassifications: finally the remaining non-zero multipliers indicate the 'true' support vectors and the optimal separation is identified (see Table 6.1 and again Figure 6.2). These results agree with those of other well-known dual-based reference software such as SVM^{light}, LIBSVM and GPDT (see Section 6.3 for more information).

Table 6.1. Results for the direct solution of the L_1QD problem on the 3×3 chessboard data set.

c	SVs	BSVs	b	h	f	miscl.
0.5	109	103	7.876173E−2	1.041818E−1	−4.606665E+1	30
2.0	93	81	4.671941E−1	5.933552E−2	−1.420171E+2	25
20.0	58	45	3.031417E−2	2.487729E−2	−8.079118E+2	20
50.0	47	34	3.737594E−2	1.798236E−2	−1.546240E+3	14
1000.0	23	5	7.992453E−2	9.067501E−3	−6.081276E+3	15
5000.0	16	0	9.551186E−2	7.623918E−3	−8.602281E+3	0

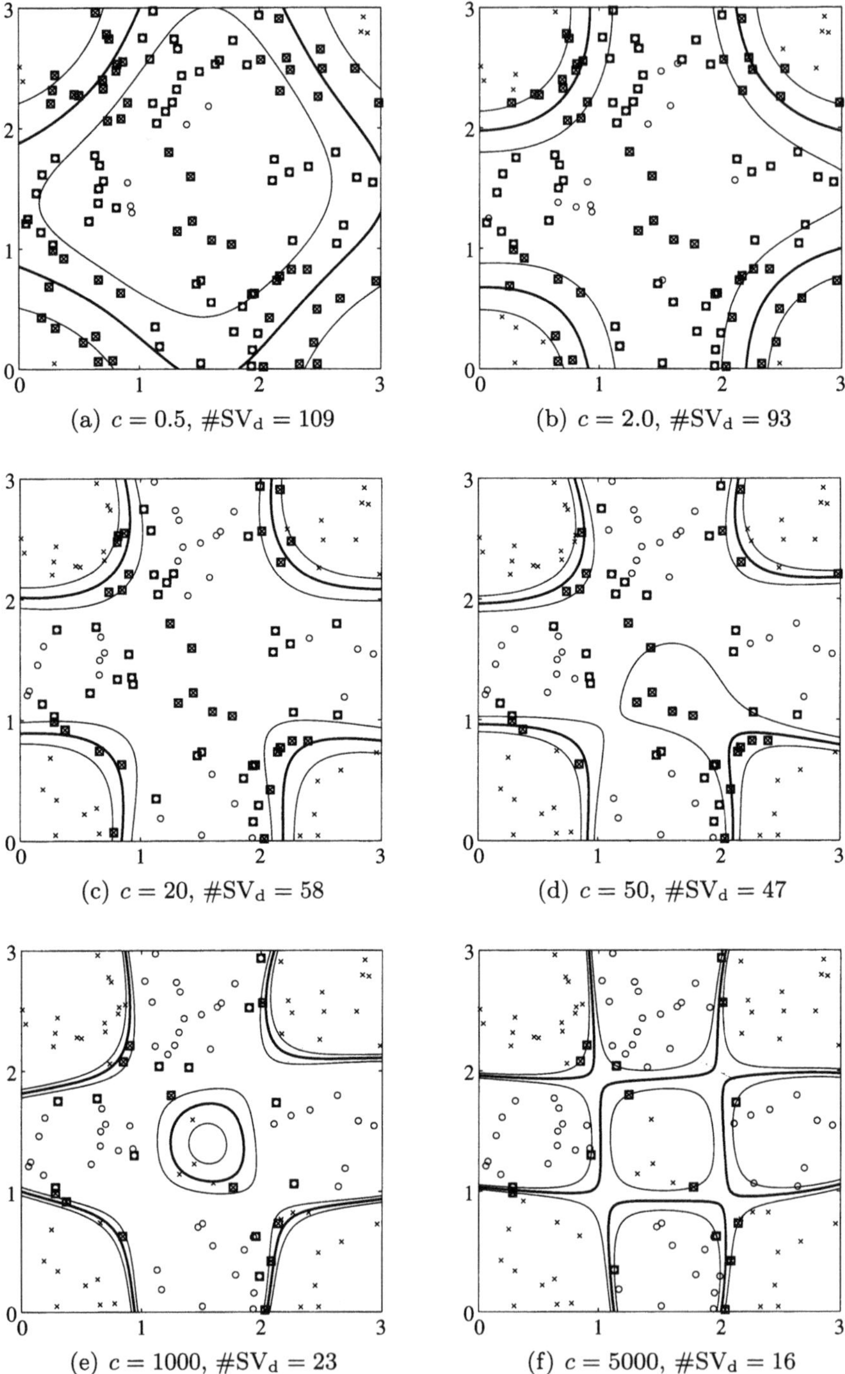

(a) $c = 0.5$, $\#\mathrm{SV_d} = 109$

(b) $c = 2.0$, $\#\mathrm{SV_d} = 93$

(c) $c = 20$, $\#\mathrm{SV_d} = 58$

(d) $c = 50$, $\#\mathrm{SV_d} = 47$

(e) $c = 1000$, $\#\mathrm{SV_d} = 23$

(f) $c = 5000$, $\#\mathrm{SV_d} = 16$

Figure 6.2. Different 'solutions' for the 3×3 problem computed by directly solving the QP dual with Gaussian kernel ($\sigma = 1$).

Also for $c < 5000$, the $f(\boldsymbol{v}) = 0$ contour agrees much less well with the true one obtained by the low-rank SQP-like algorithm. We conclude that in problems for which a satisfactory exact separation can be expected, solving the SP, for example by means of the low-rank algorithm, is likely to be preferable.

6.2. An 8×8 chessboard problem

To illustrate some other features we have used a scaled-up 8×8 chessboard problem with an increasing number m of data points. Again, we use a Gaussian kernel with $\sigma = 1$. The case $m = 1200$ is illustrated in Figure 6.3(a), and the solution of the SP by the low-rank SQP algorithm (Section 5) in Figure 6.3(b).

For the solution of this larger problem we set additional parameters in our low-rank SQP-like algorithm. In particular, in addition to the upper bound $\Delta = 10^5$ for constraint (4.4d), the tolerances `tol` in (5.10) for accepting a new pivot and τ_d for stopping the inner iterations at step 2 come into play. For this test problem we set them as `tol` $= 10^{-9}$ and $\tau_d = 10^{-6}$, respectively. Points recognized as support vectors are again highlighted with thick squares.

Only 9 outer iterations are required by this method. If we increase m to 12000 as shown in Table 6.2, the number of outer iterations is only 10. This is consistent with the fact that the additional points entered into the data set do not contain much more information about the intrinsic probability distribution, which is unknown to the trainer (see the final discussion for more details). From a numerical viewpoint, this is captured by the number of 'relevant' pivots selected for the construction of the low-rank approximation of the kernel matrix K. Actually, the additional information only

Table 6.2. Nonlinear training results of the SQP-like algorithm on the 8×8 cells chessboard data set. Here, $\mathrm{it}_{\mathrm{out}}$ are the outer iterations, $\mathrm{it}_{\mathrm{in}}$ the inner iterations, rank(U) and SVs are the final rank of the approximate Hessian factor and the final number of support vectors detected, respectively. For the inner iterations we report the total amount (Tot), the minimum (Min) the maximum (Max) and the average (Avg) counts over the whole run.

m	$\mathrm{it}_{\mathrm{out}}$	$\mathrm{it}_{\mathrm{in}}$				rank(U)	b	h	SVs
		Tot	Min	Max	Avg				
1200	9	60	2	16	6.7	268	1.2015E−3	1.1537E−4	123
12000	10	70	3	17	7.0	327	2.0455E−3	1.1653E−5	173

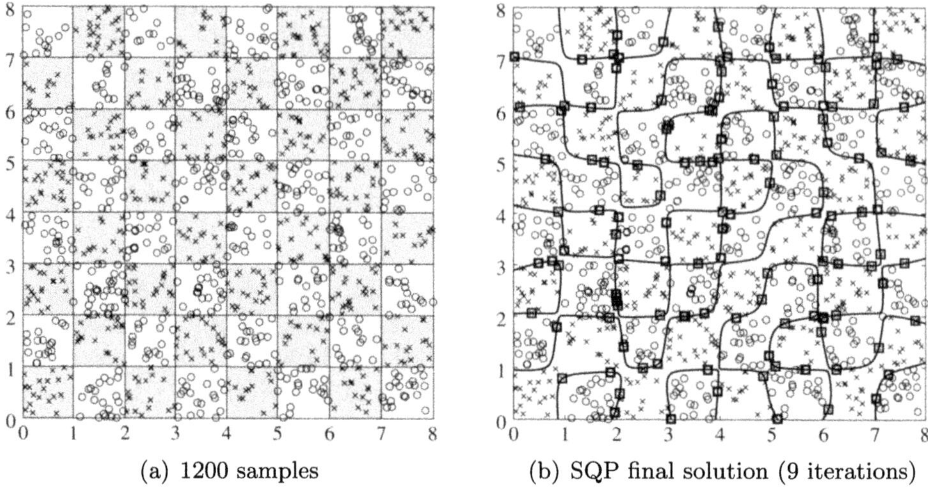

(a) 1200 samples (b) SQP final solution (9 iterations)

Figure 6.3. (a) 8×8 chessboard test problem. (b) Solution computed by the low-rank SQP-like algorithm.

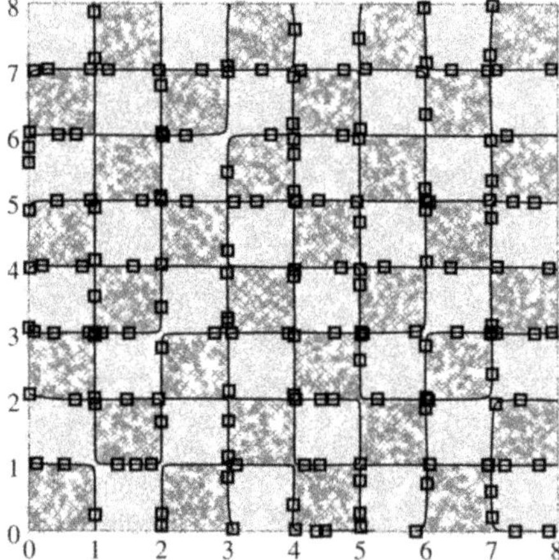

Figure 6.4. The reconstruction of the 8×8 chessboard with 12000 samples by the SQP-like algorithm. Only the few support vectors are highlighted; the positive class is in dark grey and the negative class is in light grey. The additional data allow much better approximation of the real separating hypersurface by the computed nonlinear solution.

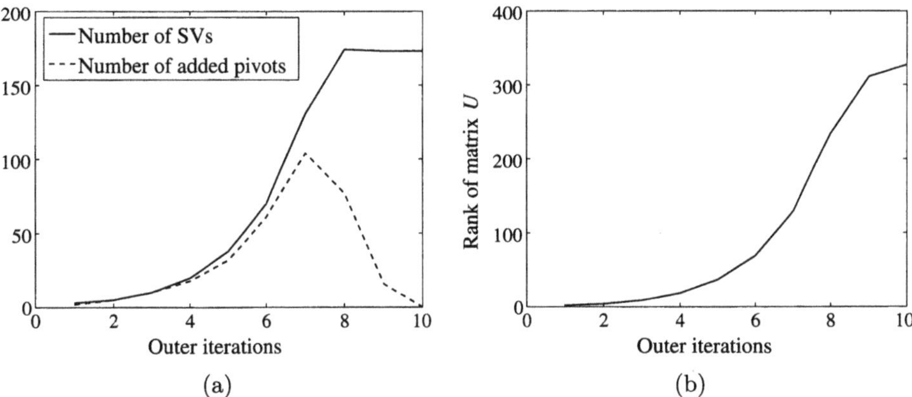

Figure 6.5. Identification performances on the chessboard case with 12000 samples. (a) Number of support vectors identified and number of new pivots to add to the model. (b) Rank of factor U.

affects minor details of the model, such as the positioning and accuracy of edges and corners (see Figure 6.4).

The ability of the low-rank algorithm to accumulate relevant information can be observed by how it builds up the approximate factor U. The plots in Figure 6.5 show that the pivot choice strategy promotes those columns which are most meaningful for the kernel matrix K. In the later iterations, fewer relevant pivots are detected, and are often thresholded out, so that fewer changes to the rank of U occur. This all contributes to the rapid identification of the exact solution. The threshold procedure is also an important feature for controlling ill-conditioning caused by the near singularity of K.

6.3. Larger data sets

We have also performed preliminary numerical experiments on some well-known data sets available on-line and often used for algorithm comparison.

We will report the training performance in terms of number of SQP iterations (itn), separability (h), displacement (b) and support vectors (SV) on the training sets, compared with those given by three other reference softwares that use the classical approach (that is, solving the L_1-dual): SVM^{light} by T. Joachims (Joachims 1999), LIBSVM by Chan and C. J. Lin (Chang and Lin 2001, Bottou and Lin 2007) and GPDT by T. Serafini, G. Zanghirati and L. Zanni (Serafini *et al.* 2005, Zanni *et al.* 2006). We have used a Fortran 77 code that is not yet optimized for performance: since the competing codes are all highly optimized C or C++ codes, we do not compare the computational time.

Data sets and program settings

The Web data sets are derived from a text categorization problem: each set is based on a 300-word dictionary that gives training points with 300 binary features. Each training example is related to a text document and the ith feature is 1 only if the ith dictionary keyword appears at least once in the document. The features are very sparse and there are replicated points (that is, some examples appear more than once, possibly with both positive and negative labels). The data set is available from the J. Platt's SMO page at www.research.microsoft.com/jplatt/smo.html.

Each one of the dual-based software codes implements a problem decomposition technique to solve the standard L_1-dual, in which the solution of the dual QP is sought by solving a sequence of smaller QP subproblems of the same form as the whole dual QP. After selecting the linear kernel, one is required to set some parameters to obtain satisfactory performance. These parameters are the dual upper bound c in (2.9), the subproblem size $\mathtt{q}$ for the decomposition technique, the maximum number $\mathtt{n}$ ($\leq \mathtt{q}$) of variables that can be changed from one decomposition iteration to the next, the stopping tolerance $\mathtt{e}$ (10^{-3} in all cases) and the amount of 'cache memory' $\mathtt{m}$ (300 MB in all cases). Note that the program cache memory affects only the computing time. We set $\mathtt{q} = 800$ and $\mathtt{n} = 300$ on all GPDT runs, whilst with SVM^{light} we set $\mathtt{q} = 50$, $\mathtt{n} = 10$ for the sets web1a to web3a and $\mathtt{q} = 80$, $\mathtt{n} = 20$ in the other cases. Since LIBSVM implements an SMO approach where $\mathtt{q} = 2$, it does not have these parameters.

Results

The results reported in Table 6.3 clearly show how our SQP approach is able to detect whether or not the classes are separable, in very few iterations. It can be seen that all the sets appear to be linearly weakly separable, with the OSH going through the origin: checking the data sets, this result is consistent with the fact that the origin appears many times as a training example, but with both the negative and the positive label. No other data points seem to be replicated in this way.

Moreover, from Table 6.4 we can see an interesting confirmation of the previous results for these data sets: using the Gaussian kernel the points of the two classes can be readily separated (we use $\sigma = \sqrt{10}$ as it is standard in the literature), and during the iterations an increasing number of support vectors is temporarily found whilst approaching the solution. Nevertheless in the final step, when the solution is located, only two support vectors are identified, which is consistent with the fact the the origin is labelled in both ways. Hence these data sets remain weakly separable, likewise with the linear classifier. We do not report numbers for the two larger data sets because our non-optimized implementation ran for too long, but partial results seem to confirm the same behaviour (in the full case, after 11 outer

Table 6.3. Linear training results on the Web data sets. For the dual-based codes $c = 1.0$ is used.

Web data set	size	b by J. Platt's document	SQP-like method itn	h	b	SVs
1a	2477	1.08553	1	0.47010E−19	0.47010E−19	8
2a	3470	1.10861	1	−0.69025E−30	−0.20892E−29	4
3a	4912	1.06354	1	−0.23528E−29	−0.41847E−29	6
4a	7366	1.07142	1	0.30815E−32	−0.20184E−30	2
5a	9888	1.08431	1	0.36543E−18	0.18272E−17	84
6a	17188	1.02703	1	−0.26234E−18	0.26234E−18	117
7a	24692	1.02946	1	−0.17670E−27	0.62793E−26	174
full	49749	1.03446	1	−0.34048E−18	−0.13268E−16	176

Web data set	GPDT itn	b	SVs	BSVs	SVM^{light} itn	b	SVs	BSVs
1a	6	1.0845721	173	47	245	1.0854851	171	47
2a	6	1.1091638	231	71	407	1.1089056	223	71
3a	6	1.0630223	279	106	552	1.0629392	277	104
4a	9	1.0710564	382	166	550	1.0708609	376	165
5a	10	1.0840553	467	241	583	1.0838959	465	244
6a	14	1.0710564	746	476	1874	1.0271027	759	479
7a	20	1.0295484	1011	682	2216	1.0297588	978	696
full	25	1.0295484	1849	1371	3860	1.0343829	1744	1395

Web data set	LIBSVM itn	b	SVs	BSVs
1a	3883	1.085248	169	47
2a	5942	1.108462	220	73
3a	8127	1.063190	275	112
4a	14118	1.070732	363	172
5a	30500	1.083892	455	247
6a	41092	1.026754	729	485
7a	105809	1.029453	968	702
full	112877	1.034590	1711	1423

Table 6.4. Nonlinear training results on the Web data set. For the dual-based codes $c = 1.0$ is used. Here $r_U = \text{rank}(U)$.

m	it_{out}	it_{in}				r_U	b	h	SVs
		Tot	Min	Max	Avg				
2477	7	19	2	3	2.7	88	1.5583E−15	7.9666E−18	2
3470	8	27	2	6	3.4	148	-1.7167E−03	1.6152E−17	2
4912	9	30	2	6	3.3	155	1.8521E−16	-2.0027E−18	2
7366	8	28	2	6	3.5	163	-4.0004E−18	0.0000E+00	2
9888	9	37	2	10	4.1	307	-6.1764E−17	4.4907E−17	2
17188	10	43	2	9	4.3	323	8.1184E−14	-1.2713E−13	2

iterations we observed that $h = -6.626\text{E}{-16}$ with only 4 support vectors and $\text{rank}(U) = 911$).

We have also performed some tests with subsets of the well-known MNIST data set of handwritten digits (LeCun and Cortes 1998, LeCun, Bottou, Bengio and Haffner 1998; www.research.att.com/~yann/ocr/mnist): this is actually a multiclass data set with ten classes, each one containing 6000 images (28×28 pixels wide) of the same digit, handwritten by different people. We constructed two binary classification problems by training a Gaussian SVM to recognize the digit '8' from the other 'not-8' digits. The problems have $m = 400$ and $m = 800$: they are obtained by randomly selecting 200 (400) examples of digit 8 and 200 (400) from the remaining not-8 digits. Examples of '8' are given class +1. The vectors $\boldsymbol{v}$ in these cases have 784-dimensional integer components ranging from 0 to 255 (grey levels), with an average of 81% zeros. Setting $\sigma = 1800$ (as is usual in the literature), our algorithm was able to compute the solution of both problems in a few outer iterations (12 and 13, respectively), detecting the usual number of support vectors (175 and 294, respectively) and significantly fewer pivots (183 and 311, respectively) than the number of examples. In fact only a handful of pivots are not support vectors, which is a clear justification of our approach.

7. Uncertain and mislabelled data

We begin this section with a small example that highlights potential difficulties in finding a meaningful classification for situations in which there are uncertain or mislabelled data points. This indicates that much more thought needs to be given in regard to what type of optimization problems it is best to solve. We make some suggestions as to future directions of research.

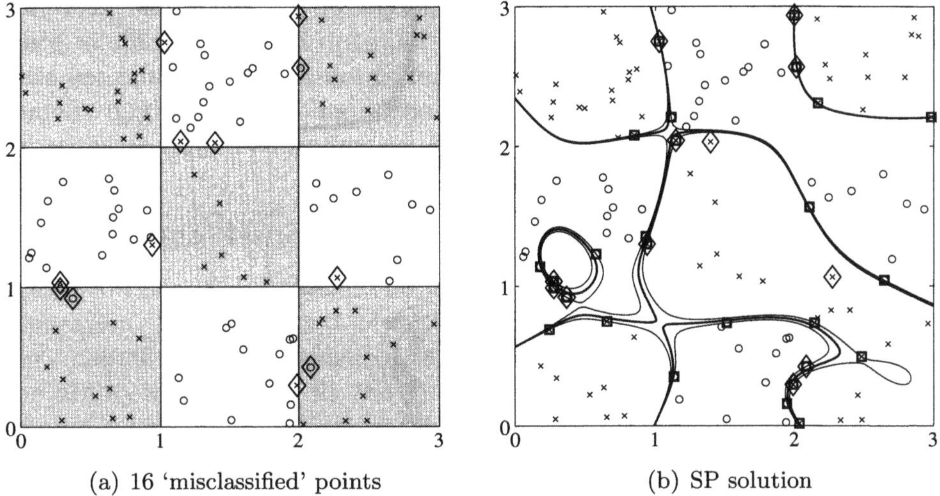

(a) 16 'misclassified' points

(b) SP solution

Figure 7.1. (a) 3×3 chessboard test problem where the labels of 12 points (marked by the diamonds) have been deliberately changed. (b) The corresponding SP solution.

We consider the same 3×3 chessboard problem as in Section 6.1, but we mislabel some points close to the grid lines as in Figure 7.1(a) (marked by diamonds). Again using a Gaussian kernel with $\sigma = 1$, we then solve the SP and find a separable solution shown in Figure 7.1(b).

Now we see a less satisfactory situation: in order to obtain separation, the zero contour of the classification function becomes very contorted in the vicinity of some of the mislabelled points. Also the separation is very small (very small h_*) and more support vectors (27) are needed.

Solving the L_1QD problem with $c = 5 \cdot 10^4$ produces the outcome of Figure 7.2, and we see that the zero contour is less contorted and some data points with L_1 errors, including some misclassifications, are indicated. This outcome is likely to be preferable to a user than the exact separable solution. Thus we need to give serious thought to how to deal with data in which there may be mislabellings, or if the labelling is uncertain. Yet we still find some aspects of the current approach based on L_1QD to be unsatisfactory. Choosing the parameter c gives a relative weighting to $\frac{1}{2}\boldsymbol{w}^T\boldsymbol{w}$ and the L_1 error in the constraints, which seems to us to be rather artificial. Moreover, choosing c is often a difficult task and can easily result in a bad outcome. Of course the method does provide 'solutions' which allow misclassified data, but it is not always clear that these provide a meaningful classification, or even that mislabelled data are correctly identified.

We therefore suggest alternative directions of research in which L_1 solution may be used. For example we may start by solving the SP problem to

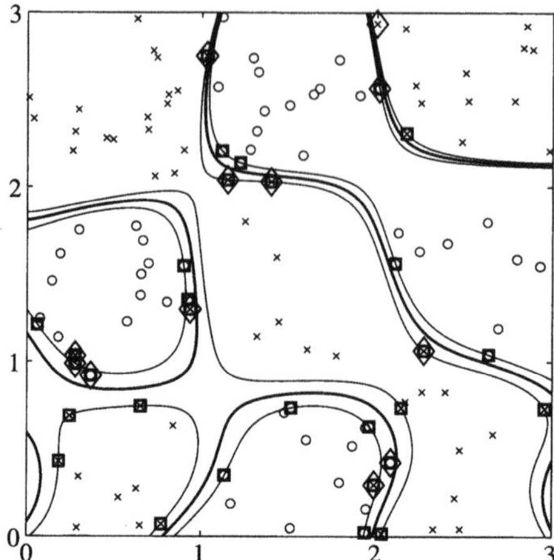

Figure 7.2. Solution of the 3 × 3 chessboard mislabelled problem computed by L_1QD with $c = 5 \cdot 10^4$.

decide if the problem is separable (we know this will be so for the Gaussian kernel) and find the optimum value h_*. If there are indications that this solution is undesirable (*e.g.*, h_* very small, multipliers very large), then we might ask the user to supply a margin $\hat{h} > h_*$ and solve the NLP problem

$$\begin{aligned} \underset{\boldsymbol{w},b,\boldsymbol{\xi}}{\text{minimize}} \quad & \boldsymbol{e}^T\boldsymbol{\xi} \\ \text{subject to} \quad & AV^T\boldsymbol{w} + \boldsymbol{a}b + \boldsymbol{\xi} - \boldsymbol{e}\hat{h} \geq \boldsymbol{0} \\ & \boldsymbol{w}^T\boldsymbol{w} = 1 \\ & \boldsymbol{\xi} \geq \boldsymbol{0}. \end{aligned} \tag{7.1}$$

The term $\boldsymbol{e}\hat{h}$ renders the constraint set infeasible, and we find the best L_1 relaxation as defined by $\boldsymbol{\xi}$. There is no reference to $\boldsymbol{w}$ in the objective function, which is desirable. We can devise an SLP algorithm to solve this problem in a similar way to Sections 2 and 5. A disadvantage to this approach is that the Jacobian matrix will contain a large dense block (U in Section 5) but also a sparse block (I) that multiplies $\boldsymbol{\xi}$. Thus an LP solver with a suitably flexible data structure to accommodate this system is likely to be important.

We also point out another feature that is worthy of some thought. In the Web problems there exist identical data points which are labelled in a contradictory way (both +1 and −1 labels). This suggests that, rather than ignore these points, we should use them to fix b, and then solve a maximal margin problem over the remaining points, but with b being fixed.

Some other data sets give rise to solutions of the SP in which all the data points are support vectors. Such large-scale problems are seriously intractable, in that the full-rank factors of K may be required to find the solution. Any practical method for computing the solution is likely to declare many misclassified data points, and one then has to ask whether the data set is at all able to determine a meaningful classification. A possible remedy might be to give more thought to choosing parameters in the kernel function, as described in Section 8.

8. Additional issues

We have not yet mentioned the issue of the *probabilistic nature* of binary separation. In the context of Machine Learning, the effectiveness of a binary classifier is also measured in regard to additional aspects other than the ability to correctly fit the given set of data. The most important of these aspects is *generalization*, that is, the ability of the classifier to correctly recognize previously *unseen* points, that is, points not involved in the training problem; see, for instance, Vapnik (1998, 1999), Herbrich (2002), Schölkopf and Smola (2002) and Shawe-Taylor and Cristianini (2004). This ability can be empirically evaluated on some test sets, but for many algorithms there exist theoretical probabilistic upper bounds; see, for instance, Agarwal and Niyogi (2009). Indeed, one of the main assumptions in the *supervised learning* context is that the observed data are samples of a probability distribution which is *fixed*, but *unknown*. However, once the problem formulation has been chosen, the generalization properties can still depend on the algorithm to be used for the problem solution as long as this solution is approximated to a low accuracy, as is common in the Machine Learning context; see, for instance, the discussion in Bennett and Parrado-Hernández (2006).

Another well-known consideration, relevant from the numerical viewpoint, is that badly scaled data easily lead to instability and ill-conditioning of the problem being solved, most often in the cases of a highly nonlinear kernel and/or large numbers of data points, because of accumulation of round-off errors. For these reasons, before attempting the training process it is often advisable to *scale* the data into some range, such as $[-1,+1]$; see, for example, Chang and Lin (2001). Moreover, other kinds of data *pre-processing* are sometimes used to facilitate the training process, such as the removal of duplicated data points, no matter if they have the same or different labels, to meet the requirement of L_1QD convergence theorems; see, for instance, Lin (2001*a*, 2001*b*, 2002), Hush and Scovel (2003) and Hush *et al.* (2006). However, we have not used such pre-processing in our experiments.

It is also worthwhile to describe our experience in choosing the parameter σ for the Gaussian kernel, since a sensible choice may be the difference between a successful solution, and the algorithm failing. We have observed in the MNIST subsets that setting $\sigma = 1$ gives solutions in which #SVs $= m$.

This makes the problems intractable, and also suggests that the classification will be unsatisfactory because all points are on the boundary of the classification. Moreover, we have observed that only two pivots per outer iteration are detected, which leads to a large total number of outer iterations. We also observed similar behaviour on other test sets when σ is badly chosen. A possible explanation may be that the effective range of influence of the Gaussians is not well matched to the typical separation of the data. Too small a value of σ leads to almost no overlap between the Gaussians, and too large a value leads to conflict between all points in the data set, both of which may be undesirable. This is clearly a problem-dependent issue: for instance, in the chessboard problems $\sigma = 1$ seems to be about the correct radius in which the Gaussians have influence.

9. Conclusion

Much has been written over the years on the subject of binary separation and SVMs. Theoretical and probabilistic properties have been investigated and software codes for solving large-scale systems have been developed. Our aim in this paper has been to look afresh at the foundations of this vast body of work, based as it is to a very large extent on a formulation involving the convex L_1-dual QP (2.9). We are particularly concerned about its inability to directly solve simple linear problems when the data are separable, particularly when weakly separable or nearly so.

Now (2.9) arises, as we have explained, from a transformation of a more fundamental *standard problem* (2.2), which is an NLP problem. When the problem is separable, an equivalent convex QP problem (2.7) is obtained. In order to get answers for non-separable problems, L_1 penalties are introduced, leading to the dual (2.9). Our aim has been to consider solving the (primal) SP directly. In the past the nonlinear constraint $\boldsymbol{w}^T\boldsymbol{w} = 1$ has been a disincentive to this approach. We have shown that an SQP-like approach is quite effective, and has some unusual features (for NLP) in that feasibility is maintained, monotonic progress to the solution is proved, and termination at the solution is observed.

However, all this material is relevant to the context of separation by a linear hyperplane. In practice this is not likely to be satisfactory, and we have to take on nonlinear SVM ideas. We have described how this ingenious concept arises and we have shown how it readily applies in a primal setting, as against the more common dual setting. Our attention has focused only on the use of a Gaussian kernel, although other kernels are possible and have been used. For the Gaussian kernel with distinct data, a separable solution can always be obtained. At first sight this seems to suggest that it would be quite satisfactory to solve problems like the SP (2.2) or the convex QPs (2.6) and (2.8), which do not resort to the use of penalties. This is true for 'good data' for which an exact separation might be expected. However, if the training set contains instances which are of uncertain labelling or are mislabelled, then the exact separation provides a very contorted separation surface and is undesirable. Now the L_1QD approach is successful insofar as it provides answers which identify certain points as being misclassified or of uncertain classification. Our concern is that these answers are obtained by optimizing a function which weights the L_1 penalties relative to the term $\frac{1}{2}\boldsymbol{w}^T\boldsymbol{w}$, which has arisen as an artifact of the transformation of the SP to the CQP in Section 2, and hence seems (at least from the numerical optimization point of view) artificial and lacking in meaning (although Vapnik (1999) assigns a probabilistic meaning in his *structural risk minimization* theory). Choice of the weighting parameter seems to need an *ad hoc* process, guided primarily by the extent to which the resulting separation looks 'reasonable'. Moreover, for very large problems, the L_1QD problem can only be solved approximately, adding another degree of uncertainty. At least the primal approach has the benefit of finding feasible approximate solutions. We hope to address these issues in future work, for example by using L_1 penalties in different ways as in (7.1).

For very large SVM problems (the usual case) the kernel matrix K is a huge dense matrix, and this presents a serious computational challenge when developing software. In some way or another the most meaningful information in K must (possibly implicitly) be extracted. Our approach has been via the use of low-rank Cholesky factors, $K \approx U^T U$, which is particularly beneficial in a primal context, leading to a significant reduction in the number of primal variables. We have suggested a method of choosing pivots which has proved effective when the problems are not too large. However, when the factor U tends towards being rank-deficient, we see signs of ill-conditioning becoming apparent. Again, we hope to address these issues in future work.

Acknowledgements

The authors are extremely grateful to Professor Luca Zanni and Dr Thomas Serafini of the University of Modena and Reggio-Emilia (Italy) for valuable discussions and suggestions.

REFERENCES

S. Agarwal and P. Niyogi (2009), 'Generalization bounds for ranking algorithms via algorithmic stability', *J. Mach. Learn. Res.* **10**, 441–474.

K. P. Bennett and E. Parrado-Hernández (2006), 'The interplay of optimization and machine learning research', *J. Mach. Learn. Res.* **7**, 1265–1281.

A. Bordes, S. Ertekin, J. Weston and L. Bottou (2005), 'Fast kernel classifiers with online and active learning', *J. Mach. Learn. Res.* **6**, 1579–1619.

B. E. Boser, I. Guyon and V. N. Vapnik (1992), A training algorithm for optimal margin classifiers. In *Proc. 5th Annual ACM Workshop on Computational Learning Theory* (D. Haussler, ed.), ACM Press, Pittsburgh, pp. 144–152.

L. Bottou and C.-J. Lin (2007), Support vector machine solvers. In *Large Scale Kernel Machines* (L. Bottou, O. Chapelle, D. DeCoste and J. Weston, eds), The MIT Press, pp. 301–320.

C. J. Burges and B. Schölkopf (1997), Improving the accuracy and speed of support vector machines. In *Advances in Neural Information Processing Systems*, Vol. 9, The MIT Press, pp. 375–381.

C. J. C. Burges (1998), 'A tutorial on support vector machines for pattern recognition', *Data Min. Knowl. Discovery* **2**, 121–167.

A. Caponnetto and L. Rosasco (2004), Non standard support vector machines and regularization networks. Technical report DISI-TR-04-03, University of Genoa, Italy.

B. Catanzaro, N. Sundaram and K. Keutzer (2008), Fast support vector machine training and classification on graphics processors. In *Proc. 25th International Conference on Machine Learning, Helsinki, Finland*, pp. 104–111.

C.-C. Chang and C.-J. Lin (2001), LIBSVM: A library for support vector machines. www.csie.ntu.edu.tw/~cjlin/libsvm

C.-C. Chang, C.-W. Hsu and C.-J. Lin (2000), 'The analysis of decomposition methods for support vector machines', *IEEE Trans. Neural Networks* **11**, 1003–1008.

E. Chang, K. Zhu, h. Wang, H. Bai, J. Li, Z. Qiu and H. Cui (2008), Parallelizing support vector machines on distributed computers. In *Advances in Neural Information Processing Systems*, Vol. 20, The MIT Press, pp. 257–264.

O. Chapelle (2007), 'Training a support vector machine in the primal', *Neural Comput.* **19**, 1155–1178.

P.-H. Chen, R.-E. Fan and C.-J. Lin (2006), 'A study on SMO-type decomposition methods for support vector machines', *IEEE Trans. Neural Networks* **17**, 893–908.

R. Collobert and S. Bengio (2001), 'SVMTorch: Support vector machines for large-scale regression problems', *J. Mach. Learn. Res.* **1**, 143–160.

N. Cristianini and J. Shawe-Taylor (2000), *An Introduction to Support Vector Machines and other Kernel-Based Learning Methods*, Cambridge University Press.

F. Cucker and S. Smale (2001), 'On the mathematical foundations of learning', *Bull. Amer. Math. Soc.* **39**, 1–49.

F. Cucker and S. Smale (2002), 'Best choices for regularization parameter in learning theory: On the bias-variance problem', *Found. Comput. Math.* **2**, 413–428.

F. Cucker and D. X. Zhou (2007), *Learning Theory: An Approximation Theory Viewpoint*, Cambridge University Press.

E. De Vito, L. Rosasco, A. Caponnetto, U. De Giovannini and F. Odone (2005), 'Learning from examples as an inverse problem', *J. Mach. Learn. Res.* **6**, 883–904.

E. De Vito, L. Rosasco, A. Caponnetto, M. Piana and A. Verri (2004), 'Some properties of regularized kernel methods', *J. Mach. Learn. Res.* **5**, 1363–1390.

J.-X. Dong, A. Krzyzak and C. Y. Suen (2003), A fast parallel optimization for training support vector machine. In *Proc. 3rd International Conference on Machine Learning and Data Mining* (P. Perner and A. Rosenfeld, eds), Vol. 2734 of *Lecture Notes in Artificial Intelligence*, Springer, pp. 96–105.

J.-X. Dong, A. Krzyzak and C. Y. Suen (2005), 'Fast SVM training algorithm with decomposition on very large data sets', *IEEE Trans. Pattern Anal. Mach. Intelligence* **27**, 603–618.

P. Drineas and M. W. Mahoney (2005), 'On the Nyström method for approximating a Gram matrix for improved kernel-based learning', *J. Mach. Learn. Res.* **6**, 2153–2175.

I. Durdanovic, E. Cosatto and H.-P. Graf (2007), Large-scale parallel SVM implementation. In *Large Scale Kernel Machines* (L. Bottou, O. Chapelle, D. DeCoste and J. Weston, eds), The MIT Press, pp. 105–138.

T. Evgeniou, M. Pontil and T. Poggio (2000), 'Regularization networks and support vector machines', *Adv. Comput. Math.* **13**, 1–50.

R.-E. Fan, K.-W. Chang, C.-J. Hsieh, X.-R. Wang and C.-J. Lin (2008), 'LIBLINEAR: A library for large linear classification', *J. Mach. Learn. Res.* **9**, 1871–1874.

M. C. Ferris and T. S. Munson (2002), 'Interior-point methods for massive support vector machines', *SIAM J. Optim.* **13**, 783–804.

S. Fine and K. Scheinberg (2001), 'Efficient SVM training using low-rank kernel representations', *J. Mach. Learn. Res.* **2**, 243–264.

S. Fine and K. Scheinberg (2002), INCAS: An incremental active set method for SVM. Technical report, IBM Research Labs, Haifa, Israel.

R. Fletcher (1987), *Practical Methods of Optimization*, 2nd edn, Wiley, Chichester.

R. Fletcher (1996–2007), BQPD: Linear and quadratic programming solver. www-new.mcs.anl.gov/otc/Guide/SoftwareGuide/Blurbs/bqpd.html

V. Franc and S. Sonnenburg (2008*a*), LIBOCAS: Library implementing OCAS solver for training linear SVM classifiers from large-scale data. cmp.felk.cvut.cz/~xfrancv/ocas/html

V. Franc and S. Sonnenburg (2008*b*), Optimized cutting plane algorithm for support vector machines. In *Proc. 25th International Conference on Machine Learning, Helsinki, Finland*, Vol. 307, ACM Press, New York, pp. 320–327.

E. M. Gertz and J. D. Griffin (2005), Support vector machine classifiers for large data sets. Technical report, Mathematics and Computer Science Division, Argonne National Laboratory, USA.

D. Goldfarb and K. Scheinberg (2004), 'A product-form cholesky factorization method for handling dense columns in interior point methods for linear programming', *Math. Program. Ser. A* **99**, 1–34.

H. P. Graf, E. Cosatto, L. Bottou, I. Dourdanovic and V. N. Vapnik (2005), Parallel support vector machines: The Cascade SVM. In *Advances in Neural Information Processing Systems* (L. Saul, Y. Weiss and L. Bottou, eds), Vol. 17, The MIT Press, pp. 521–528.

P. J. F. Groenen, G. Nalbantov and J. C. Bioch (2007), Nonlinear support vector machines through iterative majorization and I-splines. In *Advances in Data Analysis*, Studies in Classification, Data Analysis, and Knowledge Organization, Springer, pp. 149–161.

P. J. F. Groenen, G. Nalbantov and J. C. Bioch (2008), 'SVM-Maj: A majorization approach to linear support vector machines with different hinge errors', *Adv. Data Analysis and Classification* **2**, 17–43.

T. Hastie, R. Tibshirani and J. Friedman (2001), *The Elements of Statistical Learning: Data Mining, Inference, and Prediction*, Springer.

R. Herbrich (2002), *Learning Kernel Classifiers. Theory and Algorithms*, The MIT Press.

D. Hush and C. Scovel (2003), 'Polynomial-time decomposition algorithms for support vector machines', *Machine Learning* **51**, 51–71.

D. Hush, P. Kelly, C. Scovel and I. Steinwart (2006), 'QP algorithms with guaranteed accuracy and run time for support vector machines', *J. Mach. Learn. Res.* **7**, 733–769.

T. Joachims (1999), Making large-scale SVM learning practical. In *Advances in Kernel Methods: Support Vector Learning* (B. Schölkopf, C. J. C. Burges and A. Smola, eds), The MIT Press, pp. 169–184.

T. Joachims (2006), Training linear SVMs in linear time. In *Proc. 12th ACM SIGKDD International Conference on Knowledge Discovery and Data Mining, Philadelphia*, ACM Press, New York, pp. 217–226.

S. S. Keerthi and D. M. DeCoste (2005), 'A modified finite Newton method for fast solution of large-scale linear SVMs', *J. Mach. Learn. Res.* **6**, 341–361.

S. S. Keerthi and E. G. Gilbert (2002), 'Convergence of a generalized SMO algorithm for SVM classifier design', *Machine Learning* **46**, 351–360.

S. S. Keerthi, O. Chapelle and D. M. DeCoste (2006), 'Building support vector machines with reduced classifier complexity', *J. Mach. Learn. Res.* **7**, 1493–1515.

S. S. Keerthi, S. K. Shevade, C. Bhattacharyya and K. R. K. Murthy (2001), 'Improvements to Platt's SMO algorithm for SVM classifier design', *Neural Comput.* **13**, 637–649.

B. Kulis, M. Sustik and I. Dhillon (2006), Learning low-rank kernel matrices. In *Proc. 23rd International Conference on Machine Learning: ICML*, pp. 505–512.

Y. LeCun and C. Cortes (1998), The MNIST database of handwritten digits. www.research.att.com/~yann/ocr/mnist

Y. LeCun, L. Bottou, Y. Bengio and P. Haffner (1998), 'Gradient-based learning applied to document recognition', **86**, 2278–2324.

Y.-J. Lee and O. L. Mangasarian (2001*a*), RSVM: Reduced support vector machines. In *Proc. 1st SIAM International Conference on Data Mining, Chicago, April 5-7, 2001*, SIAM, Philadelphia, pp. 1–16.

Y.-J. Lee and O. L. Mangasarian (2001*b*), 'SSVM: A smooth support vector machine for classification', *Comput. Optim. Appl.* **20**, 5–22.

C.-J. Lin (2001*a*), Linear convergence of a decomposition method for support vector machines. Technical report, Department of Computer Science and Information Engineering, National Taiwan University, Taipei, Taiwan.

C.-J. Lin (2001*b*), 'On the convergence of the decomposition method for support vector machines', *IEEE Trans. Neural Networks* **12**, 1288–1298.

C.-J. Lin (2002), 'Asymptotic convergence of an SMO algorithm without any assumptions', *IEEE Trans. Neural Networks* **13**, 248–250.

O. L. Mangasarian (2000), Generalized support vector machines. In *Advances in Large Margin Classifiers*, The MIT Press, pp. 135–146.

O. L. Mangasarian (2002), 'A finite Newton method for classification', *Optim. Methods Software* **17**, 913–939.

O. L. Mangasarian (2006), 'Exact 1-norm support vector machines via unconstrained convex differentiable minimization', *J. Mach. Learn. Res.* **7**, 1517–1530.

O. L. Mangasarian and D. R. Musicant (2001), 'Lagrangian support vector machines', *J. Mach. Learn. Res.* **1**, 161–177.

E. Osuna, R. Freund and F. Girosi (1997), Training support vector machines: An application to face detection. In *Proc. IEEE Conference on Computer Vision and Pattern Recognition: CVPR97*, IEEE Computer Society, New York, pp. 130–136.

J. C. Platt (1998), Fast training of support vector machines using sequential minimal optimization. In *Advances in Kernel Methods: Support Vector Learning* (B. Schölkopf, C. Burges and A. Smola, eds), The MIT Press, pp. 185–210.

J. C. Platt (1999), Using analytic QP and sparseness to speed training of support vector machines. In *Advances in Neural Information Processing Systems* (M. Kearns *et al.*, eds), Vol. 11, The MIT Press, pp. 557–563.

M. Prato, L. Zanni and G. Zanghirati (2007) 'On recent machine learning algorithms for brain activity interpretation', *Applied Computational Electromagnetics Society Journal* **22**, 1939–1946.

K. Scheinberg (2006), 'An efficient implementation of an active set method for SVMs', *J. Mach. Learn. Res.* **7**, 2237–2257.

B. Schölkopf and A. J. Smola (2002), *Learning with Kernels*, The MIT Press.

T. Serafini and L. Zanni (2005), 'On the working set selection in gradient projection-based decomposition techniques for support vector machines', *Optim. Methods Software* **20**, 583–596.

T. Serafini, G. Zanghirati and L. Zanni (2005), 'Gradient projection methods for quadratic programs and applications in training support vector machines', *Optim. Methods Software* **20**, 353–378.

J. Shawe-Taylor and N. Cristianini (2004), *Kernel Methods for Pattern Analysis*, Cambridge University Press.

J. A. K. Suykens, T. Van Gestel, J. De Brabanter, B. De Moor and J. Vandewalle (2002), *Least Squares Support Vector Machines*, World Scientific, Singapore.

C. H. Teo, Q. V. Le, A. Smola and S. Vishwanathan (2009), BMRM: Bundle methods for regularized risk minimization.
users.rsise.anu.edu.au/~chteo/BMRM.html

I. W. Tsang, J. T. Kwok and P.-M. Cheung (2005), 'Core vector machines: Fast SVM training on very large data sets', *J. Mach. Learn. Res.* **6**, 363–392.

V. N. Vapnik (1998), *Statistical Learning Theory*, Wiley, New York.

V. N. Vapnik (1999), *The Nature of Statistical Learning Theory*, Information Science and Statistics, Springer.

C. K. Williams and M. Seeger (2001), Using the Nyström method to speed up kernel machines. In *Advances in Neural Information Processing Systems*, Vol. 13, The MIT Press, pp. 682–688.

K. Woodsend and J. Gondzio (2007*a*), Exploiting separability in large scale support vector machine training. Technical report MS-07-002, The University of Edinburgh.

K. Woodsend and J. Gondzio (2007*b*), Parallel support vector machine training with nonlinear kernels. Technical report MS-07-007, The University of Edinburgh.

K. Woodsend and J. Gondzio (2009), 'Hybrid MPI/OpenMP parallel linear support vector machine training', *J. Mach. Learn. Res.* **20**, 1937–1953.

M. Wyganowski (2008), Classification algorithms on the cell processor. PhD thesis, Kate Gleason College of Engineering, Rochester Institute of Technology, Rochester, NY, USA. http://hdl.handle.net/1850/7767

L. Zanni (2006), 'An improved gradient projection-based decomposition technique for support vector machines', *Comput. Management Sci.* **3**, 131–145.

L. Zanni, T. Serafini and G. Zanghirati (2006), 'Parallel software for training large-scale support vector machines on multiprocessors systems', *J. Mach. Learn. Res.* **7**, 1467–1492. http://dm.unife.it/gpdt

Acta Numerica (2010), pp. 159–208
doi:10.1017/S0962492910000036

Computing matrix functions

Nicholas J. Higham and Awad H. Al-Mohy
School of Mathematics,
University of Manchester,
Manchester, M13 9PL, UK
E-mail: higham@maths.man.ac.uk, almohy@maths.man.ac.uk

The need to evaluate a function $f(A) \in \mathbb{C}^{n\times n}$ of a matrix $A \in \mathbb{C}^{n\times n}$ arises in a wide and growing number of applications, ranging from the numerical solution of differential equations to measures of the complexity of networks. We give a survey of numerical methods for evaluating matrix functions, along with a brief treatment of the underlying theory and a description of two recent applications. The survey is organized by classes of methods, which are broadly those based on similarity transformations, those employing approximation by polynomial or rational functions, and matrix iterations. Computation of the Fréchet derivative, which is important for condition number estimation, is also treated, along with the problem of computing $f(A)b$ without computing $f(A)$. A summary of available software completes the survey.

CONTENTS

1. Introduction

Matrix functions are as old as matrix algebra itself. In his memoir that initiated the study of matrix algebra, Cayley (1858) treated matrix square roots. The theory of matrix functions was subsequently developed by many mathematicians over the ensuing 100 years. Today, functions of matrices are widely used in science and engineering and are of growing interest, due to the succinct way they allow solutions to be expressed and recent advances in numerical algorithms for computing them. New applications are regularly being found, but the archetypal application of matrix functions is in the solution of differential equations. Early recognition of the important role of the matrix exponential in this regard can be found in the book *Elementary Matrices and Some Applications to Dynamics and Differential Equations* by aerospace engineers Frazer, Duncan and Collar (1938), which was 'the first book to treat matrices as a branch of applied mathematics' (Collar 1978).

This article provides a survey of numerical methods for computing matrix functions and is organized as follows. Section 2 describes some key elements of the theory of matrix functions. Two recent applications, to networks and roots of transition matrices, are described in Section 3. The following three sections describe methods grouped by type: those based on similarity transformations, those employing polynomial or rational approximations, and matrix iterations. Section 7 treats computation of the Fréchet derivative by five different approaches and explains how to estimate the condition number of a matrix function. The problem of computing the action of $f(A)$ on a vector is treated in Section 8, while Section 9 describes available software. An appendix gives some new results on the comparison between truncated Taylor series and Padé approximants within the scaling and squaring method for the matrix exponential.

Throughout, $\|\cdot\|$ denotes an arbitrary matrix norm unless otherwise stated. A flop is a floating point operation: $+$, $-$, $*$ or $/$. The unit roundoff is denoted by u and has the value $u = 2^{-53} \approx 1.1 \times 10^{-16}$ in IEEE double precision arithmetic. We write $\widetilde{\gamma}_k := cku/(1 - cku)$ with c a small integer constant.

2. Theory

We are concerned with functions mapping $\mathbb{C}^{n\times n}$ to $\mathbb{C}^{n\times n}$ that are defined in terms of an underlying scalar function f. Thus, for example, $\det(A)$, the adjugate (or adjoint) matrix, and a matrix polynomial such as $p(X) = AX^2+BX+C$ (where all matrices are $n\times n$) are not matrix functions in the sense considered here, and elementwise evaluations such as $A \mapsto (\cos(a_{ij}))$ also are not of the required form.

The functions of a matrix in which we are interested can be defined in various ways. The multiplicity of definitions caused some confusion in the early years of the subject, until Rinehart (1955) showed all the definitions to be equivalent, modulo technical assumptions. We give two definitions, both of which are very useful in developing the theory.

2.1. Definitions

Definition 2.1. (Jordan form definition of $f(A)$) Let $A \in \mathbb{C}^{n\times n}$ have the Jordan canonical form $Z^{-1}AZ = J_A = \operatorname{diag}\big(J_1(\lambda_1), J_2(\lambda_2), \dots, J_p(\lambda_p)\big)$, where Z is non-singular,

$$J_k(\lambda_k) = \begin{bmatrix} \lambda_k & 1 & & \\ & \lambda_k & \ddots & \\ & & \ddots & 1 \\ & & & \lambda_k \end{bmatrix} \in \mathbb{C}^{m_k\times m_k}, \tag{2.1}$$

and $m_1 + m_2 + \cdots + m_p = n$. Then

$$f(A) := Zf(J_A)Z^{-1} = Z\operatorname{diag}(f(J_k(\lambda_k)))Z^{-1}, \tag{2.2}$$

where

$$f(J_k(\lambda_k)) := \begin{bmatrix} f(\lambda_k) & f'(\lambda_k) & \dots & \dfrac{f^{(m_k-1)}(\lambda_k)}{(m_k-1)!} \\ & f(\lambda_k) & \ddots & \vdots \\ & & \ddots & f'(\lambda_k) \\ & & & f(\lambda_k) \end{bmatrix}. \tag{2.3}$$

When the function f is multivalued and A has a repeated eigenvalue occurring in more than one Jordan block (*i.e.*, A is derogatory), we will take the same branch for f and its derivatives in each Jordan block. This gives a *primary matrix function.* If different branches are taken for the same eigenvalue in two different Jordan blocks then a *non-primary matrix function* is obtained. We will be concerned here only with primary matrix functions, and it is these that are needed in most applications. For more on non-primary matrix functions see Higham (2008, Section 1.4).

Definition 2.2. (polynomial interpolation definition of $f(A)$) Let $\lambda_1, \dots, \lambda_s$ denote the distinct eigenvalues of $A \in \mathbb{C}^{n\times n}$ and let n_i be the index of λ_i, that is, the order of the largest Jordan block in which λ_i appears. Then $f(A) := r(A)$, where r is the unique Hermite interpolating polynomial of degree less than $\sum_{i=1}^{s} n_i$ that satisfies the interpolation conditions

$$r^{(j)}(\lambda_i) = f^{(j)}(\lambda_i), \qquad j = 0\colon n_i - 1, \quad i = 1\colon s. \tag{2.4}$$

In both these definitions the values $f^{(j)}(\lambda_i)$ appearing in (2.4) are assumed to exist, in which case f is said to be *defined on the spectrum of A*.

A proof of the equivalence of these two definitions can be found in Higham (2008, Theorem 1.12). The equivalence is easily demonstrated for the

$m_k \times m_k$ Jordan block $J_k(\lambda_k)$ in (2.1). The polynomial satisfying the interpolation conditions (2.4) is then

$$r(t) = f(\lambda_k) + (t-\lambda_k)f'(\lambda_k) + \frac{(t-\lambda_k)^2}{2!}f''(\lambda_k) + \cdots + \frac{(t-\lambda_k)^{m_k-1}}{(m_k-1)!}f^{(m_k-1)}(\lambda_k),$$

which is just the first m_k terms of the Taylor series of f about λ_k (assuming the Taylor series exists). Hence, from Definition 2.2,

$$\begin{aligned} f(J_k(\lambda_k)) &= r(J_k(\lambda_k)) \\ &= f(\lambda_k)I + (J_k(\lambda_k)-\lambda_k I)f'(\lambda_k) + \frac{(J_k(\lambda_k)-\lambda_k I)^2}{2!}f''(\lambda_k) + \cdots \\ &\quad + \frac{(J_k(\lambda_k)-\lambda_k I)^{m_k-1}}{(m_k-1)!}f^{(m_k-1)}(\lambda_k). \end{aligned}$$

The matrix $(J_k(\lambda_k)-\lambda_k I)^j$ is zero except for 1s on the jth superdiagonal. This expression for $f(J_k(\lambda_k))$ is therefore equal to that in (2.3).

2.2. *Properties*

One of the most important basic properties is that $f(A)$ is a polynomial in $A \in \mathbb{C}^{n\times n}$, which is immediate from Definition 2.2. However, the coefficients of that polynomial depend on A. This property is not surprising in view of the Cayley–Hamilton theorem, which says that any matrix satisfies its own characteristic equation: $q(A) = 0$, where $q(t) = \det(tI - A)$ is the characteristic polynomial. The theorem implies that the nth power of A, and inductively all higher powers, are expressible as a linear combination of $I, A, \ldots, A^{n-1}$. Thus any power series in A can be reduced to a polynomial in A of degree at most $n-1$ (with coefficients depending on A).

Other important properties are collected in the next result, for a proof of which see Higham (2008, Theorem 1.13).

Theorem 2.3. Let $A \in \mathbb{C}^{n\times n}$ and let f be defined on the spectrum of A. Then

(a) $f(A)$ commutes with A;
(b) $f(A^T) = f(A)^T$;
(c) $f(XAX^{-1}) = Xf(A)X^{-1}$;
(d) the eigenvalues of $f(A)$ are $f(\lambda_i)$, where the λ_i are the eigenvalues of A;
(e) if $A = (A_{ij})$ is block triangular then $F = f(A)$ is block triangular with the same block structure as A, and $F_{ii} = f(A_{ii})$;
(f) if $A = \mathrm{diag}(A_{11}, A_{22}, \ldots, A_{mm})$ is block diagonal then

$$f(A) = \mathrm{diag}\big(f(A_{11}), f(A_{22}), \ldots, f(A_{mm})\big).$$

It is often convenient to represent a matrix function as a power series or Taylor series. The next result explains when such a series converges (Higham 2008, Theorem 4.7).

Theorem 2.4. (convergence of matrix Taylor series) Suppose f has a Taylor series expansion

$$f(z) = \sum_{k=0}^{\infty} a_k (z-\alpha)^k \qquad \left(a_k = \frac{f^{(k)}(\alpha)}{k!} \right) \tag{2.5}$$

with radius of convergence r. If $A \in \mathbb{C}^{n\times n}$ then $f(A)$ is defined and is given by

$$f(A) = \sum_{k=0}^{\infty} a_k (A-\alpha I)^k \tag{2.6}$$

if and only if each of the distinct eigenvalues $\lambda_1, \ldots, \lambda_s$ of A satisfies one of the conditions

(a) $|\lambda_i - \alpha| < r$,
(b) $|\lambda_i - \alpha| = r$ and the series for $f^{(n_i-1)}(\lambda)$ (where n_i is the index of λ_i) is convergent at the point $\lambda = \lambda_i$, $i = 1\colon s$.

Four books treat the theory of matrix functions in detail and should be consulted for more information: Gantmacher (1959, Chapter 5), Horn and Johnson (1991, Chapter 6), Lancaster and Tismenetsky (1985, Chapter 9), and Higham (2008).

2.3. Particular functions

We now turn to the definitions of some specific functions. For functions having a power series with an infinite radius of convergence, the matrix function can be defined by evaluating the power series at a matrix, by Theorem 2.4. Thus the matrix exponential is given by

$$\mathrm{e}^A = I + A + \frac{A^2}{2!} + \frac{A^3}{3!} + \cdots \tag{2.7}$$

and the matrix cosine and sine by

$$\cos(A) = I - \frac{A^2}{2!} + \frac{A^4}{4!} - \frac{A^6}{6!} + \cdots,$$
$$\sin(A) = A - \frac{A^3}{3!} + \frac{A^5}{5!} - \frac{A^7}{7!} + \cdots.$$

A natural question is to what extent scalar functional relations generalize to the matrix case. For example, are $(\mathrm{e}^A)^2 = \mathrm{e}^{2A}$, $\mathrm{e}^{\mathrm{i}A} = \cos(A) + \mathrm{i}\sin(A)$, and $\cos(2A) = 2\cos(A)^2 - I$ valid equalities for all A? The answer is yes for these particular examples. More generally, scalar identities in a single variable remain true with a matrix argument provided that all terms are defined and that the functions involved are single-valued. For multivalued functions such as the logarithm additional conditions may be needed to ensure the matrix identity is valid; an example is given below. The relevant general results can be found in Higham (2008, Section 1.3). Relations involving more than one variable do not usually generalize to matrices; for example, $e^{A+B} \neq e^A e^B$ in general.

An important function is the pth root of a matrix, where initially we assume p is a positive integer. For $A \in \mathbb{C}^{n\times n}$ we say X is a pth root of A if $X^p = A$. Note that defining pth roots implicitly via this equation gives a wider class of matrices than Definitions 2.1 and 2.2. For example, $\begin{bmatrix} 0 & 1 \\ 0 & 0 \end{bmatrix}^2 = \begin{bmatrix} 0 & 0 \\ 0 & 0 \end{bmatrix}$, but Definitions 2.1 and 2.2 provide only one square root of $\begin{bmatrix} 0 & 0 \\ 0 & 0 \end{bmatrix}$, namely itself. If A is singular with a defective zero eigenvalue then there are no primary pth roots, since the pth root function is not defined on the spectrum of A, but there may be non-primary pth roots (albeit not obtainable from Definitions 2.1 or 2.2). Conditions for the existence of pth roots of singular matrices are non-trivial (Psarrakos 2002). We will concentrate here on the non-singular case. Here there are always at least p pth roots and there are infinitely many if A is derogatory (that is, some eigenvalue appears in more than one Jordan block).

In the common case where A has no eigenvalues on $\mathbb{R}^-$, the closed negative real axis, there is a distinguished root that is real when A is real (Higham 2008, Theorem 7.2).

Theorem 2.5. (principal pth root) Let $A \in \mathbb{C}^{n\times n}$ have no eigenvalues on $\mathbb{R}^-$. There is a unique pth root X of A all of whose eigenvalues lie in the segment $\{ z : -\pi/p < \arg(z) < \pi/p \}$, and it is a primary matrix function of A. We refer to X as the *principal pth root* of A and write $X = A^{1/p}$. If A is real then $A^{1/p}$ is real.

In particular, *the principal square root* $A^{1/2}$ of a matrix A with no eigenvalues on $\mathbb{R}^-$ is the unique square root all of whose eigenvalues lie in the open right half-plane.

A logarithm of $A \in \mathbb{C}^{n\times n}$ can be defined as any matrix X such that $e^X = A$. A singular matrix has no logarithms, but for a non-singular matrix there are infinitely many. Indeed if $e^X = A$ then $e^{X+2\pi kiI} = A$ for all integers k. In practice it is usually the principal logarithm, defined in the next result, that is of interest (Higham 2008, Theorem 1.31).

Theorem 2.6. (principal logarithm) Let $A \in \mathbb{C}^{n\times n}$ have no eigenvalues on $\mathbb{R}^-$. There is a unique logarithm X of A all of whose eigenvalues lie in the strip $\{ z : -\pi < \operatorname{Im}(z) < \pi \}$. We refer to X as the *principal logarithm* of A and write $X = \log(A)$. If A is real then its principal logarithm is real.

From this point on, log always denotes the principal logarithm.

Care is needed in checking functional identities involving the logarithm. While the relation $\exp(\log(A)) = A$ is always true (for any logarithm, not just the principal one), by definition of the logarithm, the relation $\log(\exp(A)) = A$ holds if and only if $|\operatorname{Im}(\lambda_i)| < \pi$ for every eigenvalue λ_i of A (Higham 2008, Problem 1.39).

For $A \in \mathbb{C}^{n\times n}$ with no eigenvalues on $\mathbb{R}^-$, the logarithm provides a convenient way to define A^α for arbitrary real α: as $A^\alpha = \mathrm{e}^{\alpha \log(A)}$. From the relation in the previous paragraph it follows that

$$\log(A^\alpha) = \alpha \log(A), \qquad \alpha \in [-1, 1]. \tag{2.8}$$

The matrix sign function, introduced by Roberts in 1971 (Roberts 1980), corresponds to the choice

$$f(z) = \operatorname{sign}(z) = \begin{cases} 1, & \operatorname{Re} z > 0, \\ -1, & \operatorname{Re} z < 0 \end{cases}$$

in Definitions 2.1 and 2.2, for which $f^{(k)}(z) = 0$ for $k \geq 1$. Thus the matrix sign function is defined only for matrices $A \in \mathbb{C}^{n\times n}$ having no pure imaginary eigenvalues. If we arrange the Jordan canonical form as $A = Z \operatorname{diag}(J_1, J_2) Z^{-1}$, where the eigenvalues of $J_1 \in \mathbb{C}^{p\times p}$ lie in the open left half-plane and those of $J_2 \in \mathbb{C}^{q\times q}$ lie in the open right half-plane, then

$$\operatorname{sign}(A) = Z \begin{bmatrix} -I_p & 0 \\ 0 & I_q \end{bmatrix} Z^{-1}. \tag{2.9}$$

Another useful representation is (Higham 1994)

$$\operatorname{sign}(A) = A(A^2)^{-1/2}, \tag{2.10}$$

which generalizes the scalar formula $\operatorname{sign}(z) = z/(z^2)^{1/2}$.

3. Applications

Matrix functions are useful in a wide variety of applications. We describe just two recent ones here; more can be found in Higham (2008, Chapter 2).

3.1. Networks

Consider a network representing interactions between pairs of entities in a system. In recent years much work has focused on identifying computable measures that quantify characteristics of the network. Many measures are available in the literature, and they are typically expressed in terms of the network's associated undirected graph G with n nodes. The adjacency matrix $A \in \mathbb{R}^{n\times n}$ of the graph has (i, j) element equal to 1 if nodes i and j are connected and 0 otherwise. Assume $a_{ii} \equiv 0$, so that there are no loops in the graph. A walk of length m between two nodes i and j is an ordered list of nodes $i, k_1, k_2, \ldots, k_{m-1}, j$ such that successive nodes in the list are connected; the nodes need not be distinct and any of them may be i or j. When $i = j$ the walk starts and ends at the same node and is called closed. The walk is a path if all the nodes in the walk are distinct. Assume that the graph is connected, so that a path exists between any two distinct nodes. It is a standard fact in graph theory that the (i,j) element of A^m is the number of different walks, if $i \neq j$, or closed walks, if $i = j$, of length m between nodes i and j. A variety of measures have been built by combining different walk lengths into a single number. Estrada and Rodríguez-Velázquez (2005*b*) define the *subgraph centrality* of node i – a measure of its 'well-connectedness' – by

$$SC_i = \left(I + A + \frac{A^2}{2!} + \frac{A^3}{3!} + \cdots\right)_{ii} = (\mathrm{e}^A)_{ii}.$$

By combining walks of all possible lengths connecting node i to itself, and applying a weighting that decreases rapidly with the walk length, the subgraph centrality aims to capture the participation of the node in question in all subgraphs in the network. The sum of all subgraph centralities of the nodes in the graph is the *Estrada index*: $\mathrm{trace}(\mathrm{e}^A)$. Based on similar reasoning, Estrada and Hatano (2008) define the *communicability* between nodes i and j – a measure of how easy it is for 'information' to pass from node i to node j (and a generalization of the notion of shortest path between the nodes) – by

$$C_{ij} = \left(I + A + \frac{A^2}{2!} + \frac{A^3}{3!} + \cdots\right)_{ij} = (\mathrm{e}^A)_{ij}.$$

Finally, the *betweenness* of node r is defined in Estrada, D. J. Higham and Hatano (2009) by

$$\frac{1}{(n-1)^2 - (n-1)} \sum_{\substack{i,j \\ i \neq j, i \neq r, j \neq r}} \frac{(\mathrm{e}^A - \mathrm{e}^{A - E_r})_{ij}}{(\mathrm{e}^A)_{ij}},$$

where E_r is zero except in row and column r, where it agrees with A. The betweenness measures the relative change in communicability when node r is removed from the network. Experiments in the papers cited above show that these three measures can provide useful information about practically occurring networks that is not revealed by most other measures. In this description A is symmetric, but these concepts can be extended to directed graphs, for which the adjacency matrix is non-symmetric. Of course, the matrix exponential owes its appearance to the choice of weights in the sums over walklengths. Other weights could be chosen, resulting in different matrix functions in the definitions; see Estrada and D. J. Higham (2008).

When the elements of A not only indicate the existence of a link between nodes i and j but also assign a positive weight to a link, it is natural to normalize these definitions. Crofts and D. J. Higham (2009) generalize the definition of communicability to

$$C_{ij} = \mathrm{e}^{D^{-1/2}AD^{-1/2}},$$

where $D = \mathrm{diag}(d_i)$ and $d_i = \sum_{k=1}^{n} a_{ik}$ is the generalized degree of node i. They show how this communicability measure is a useful tool in clustering patients with brain disorders.

Finally, we note that Estrada and Rodríguez-Velázquez (2005*a*) propose $\beta(A) = \mathrm{trace}(\cosh(A))/\,\mathrm{trace}(\mathrm{e}^A)$ as a measure of how close a graph is to being bipartite: $\beta(A) \le 1$ with $\beta(A) = 1$ if and only if the graph G is bipartite.

3.2. Roots of transition matrices

A transition matrix is a stochastic matrix: a square matrix with nonnegative entries and row sums equal to 1. In credit risk, a transition matrix records the probabilities of a firm's transition from one credit rating to another over a given time interval (Jarrow, Lando and Turnbull 1997). The shortest period over which a transition matrix can be estimated is typically one year, and annual transition matrices can be obtained from rating agencies such as Moody's Investors Service and Standard & Poor's. However, for valuation purposes, a transition matrix for a period shorter than one year is usually needed. A short-term transition matrix can be obtained by computing a root of an annual transition matrix. A six-month transition matrix, for example, is a square root of the annual transition matrix. This property has led to interest in the finance literature in the computation or approximation of roots of transition matrices (Israel, Rosenthal and Wei 2001, Kreinin and Sidelnikova 2001). Exactly the same mathematical problem arises in Markov models of chronic diseases, where the transition matrix is built from observations of the progression in patients of a disease through different severity states. Again, the observations are at an interval longer than the short time intervals required for study and the need for a matrix root arises (Charitos, de Waal and van der Gaag 2008). An early discussion of this problem, which identifies the need for roots of transition matrices in models of business and trade, is that of Waugh and Abel (1967).

These applications require a stochastic root of a given stochastic matrix A, that is, a stochastic matrix X such that $X^p = A$, where p is typically an integer, but could be rational. A number of questions arise: does such a root exist; if so, how can one be computed; and what kind of approximation should be used if a stochastic root does not exist. These are investigated in Higham and Lin (2009) and the references therein.

More generally, matrix roots A^α with a real α arise in, for example, fractional differential equations (Ilić, Turner and Simpson 2009), discrete representations of norms corresponding to finite element discretizations of fractional Sobolev spaces (Arioli and Loghin 2009), and the computation of geodesic-midpoints in neural networks (Fiori 2008).

In the next three sections we consider methods based on three general approaches: similarity transformations, polynomial or rational approximations, and matrix iterations. We do not consider methods based on Definitions 2.1 or 2.2 because neither definition leads to an efficient and numerically reliable method in general.

4. Similarity transformations

The use of similarity transformations, considered in this section, rests on the identity $f(XAX^{-1}) = Xf(A)X^{-1}$ from Theorem 2.3(c). The aim is to choose X so that f is more easily evaluated at the matrix $B = XAX^{-1}$ than at A. When A is diagonalizable, B can be taken to be diagonal, and evaluation of $f(B)$ is then trivial.

In finite precision arithmetic, this approach is reliable only if X is well-conditioned, that is, if the condition number $\kappa(X) = \|X\|\|X^{-1}\|$ is not too large. Ideally, X will be unitary, so that in the 2-norm $\kappa_2(X) = 1$. For Hermitian A, or more generally normal A, the spectral decomposition $A = QDQ^*$ with Q unitary and D diagonal is always possible, and if this decomposition can be computed then the formula $f(A) = Qf(D)Q^*$ provides an excellent way of computing $f(A)$.

For general A, if X is restricted to be unitary then the furthest that A can be reduced is to Schur form: $A = QTQ^*$, where Q is unitary and T is upper triangular. This decomposition is computed by the QR algorithm. The problem is now reduced to that of evaluating f at a triangular matrix. The following result gives an explicit formula for the evaluation.

Theorem 4.1. (function of triangular matrix) Let $T \in \mathbb{C}^{n\times n}$ be upper triangular and suppose that f is defined on the spectrum of T. Then $F = f(T)$ is upper triangular with $f_{ii} = f(t_{ii})$ and

$$f_{ij} = \sum_{(s_0,\ldots,s_k)\in S_{ij}} t_{s_0,s_1} t_{s_1,s_2} \ldots t_{s_{k-1},s_k} f[\lambda_{s_0},\ldots,\lambda_{s_k}], \tag{4.1}$$

where $\lambda_i = t_{ii}$, S_{ij} is the set of all strictly increasing sequences of integers that start at i and end at j, and $f[\lambda_{s_0},\ldots,\lambda_{s_k}]$ is the kth-order divided difference of f at $\lambda_{s_0},\ldots,\lambda_{s_k}$.

Proof. See Davis (1973), Descloux (1963), or Van Loan (1975). □

While theoretically interesting, the formula (4.1) is of limited computational interest due to its exponential cost in n. However, the case $n = 2$ is worth noting. For $\lambda_1 \neq \lambda_2$ we have

$$f\left(\begin{bmatrix} \lambda_1 & t_{12} \\ 0 & \lambda_2 \end{bmatrix}\right) = \begin{bmatrix} f(\lambda_1) & t_{12}\dfrac{f(\lambda_2) - f(\lambda_1)}{\lambda_2 - \lambda_1} \\ 0 & f(\lambda_2) \end{bmatrix}. \tag{4.2}$$

When $\lambda_1 = \lambda_2 = \lambda$ we have, using a standard relation for confluent divided differences (Higham 2008, Section B.16),

$$f\left(\begin{bmatrix} \lambda & t_{12} \\ 0 & \lambda \end{bmatrix}\right) = \begin{bmatrix} f(\lambda) & t_{12}f'(\lambda) \\ 0 & f(\lambda) \end{bmatrix}. \tag{4.3}$$

(This is a special case of (7.10) below.)

A much better way to compute $f(T)$ is from a recurrence of Parlett (1976). From Theorem 2.3 we know that $F = f(T)$ is upper triangular with diagonal elements $f(t_{ii})$ and that it commutes with T. The elements in the strict upper triangle are determined by solving the equation $FT = TF$ in an appropriate order.

Algorithm 4.2. (Parlett recurrence) Given an upper triangular $T \in \mathbb{C}^{n\times n}$ with distinct diagonal elements and a function f defined on the spectrum of T, this algorithm computes $F = f(T)$ using Parlett's recurrence.

for $j = 1\colon n$
  $f_{jj} = f(t_{jj})$
  for $i = j-1\colon -1\colon 1$
    $f_{ij} = t_{ij}\dfrac{f_{ii} - f_{jj}}{t_{ii} - t_{jj}} + \left(\sum_{k=i+1}^{j-1} f_{ik}t_{kj} - t_{ik}f_{kj}\right) \Big/ (t_{ii} - t_{jj})$
  end
end

Cost: $2n^3/3$ flops.

The recurrence breaks down when $t_{ii} = t_{jj}$ for some $i \neq j$. In this case, T can be regarded as a block triangular matrix $T = (T_{ij})$, with square diagonal blocks, possibly of different sizes. Then $F = (F_{ij})$ has the same block triangular structure by Theorem 2.3(e) and by equating (i, j) blocks in $TF = FT$ we obtain

$$T_{ii}F_{ij} - F_{ij}T_{jj} = F_{ii}T_{ij} - T_{ij}F_{jj} + \sum_{k=i+1}^{j-1} (F_{ik}T_{kj} - T_{ik}F_{kj}), \qquad i < j. \tag{4.4}$$

The Sylvester equation (4.4) is non-singular precisely when T_{ii} and T_{jj} have no eigenvalue in common. Assuming that this property holds for all i and j we obtain the following algorithm.

Algorithm 4.3. (block Parlett recurrence) Given an upper triangular matrix $T = (T_{ij}) \in \mathbb{C}^{n\times n}$ partitioned in block $m \times m$ form with no two diagonal blocks having an eigenvalue in common, and a function f defined on the spectrum of T, this algorithm computes $F = f(T)$ using the block form of Parlett's recurrence.

```
for j = 1: m
    F_jj = f(T_jj)
    for i = j − 1: −1: 1
        Solve for F_ij the Sylvester equation (4.4).
    end
end
```

Cost: Dependent on the block sizes and f.

The problems of how to evaluate $f(T_{jj})$ and how to achieve a blocking with the desired properties are considered in the next section.

4.1. Schur–Parlett algorithm

In order to use the block Parlett recurrence we need to reorder and partition the matrix T so that no two diagonal blocks have an eigenvalue in common; here, reordering means applying a unitary similarity transformation to permute the diagonal elements whilst preserving triangularity. But in doing the reordering and defining the block structure we also need to take into account the difficulty of evaluating f at the diagonal blocks T_{ii} and the propagation of errors in the recurrence.

Consider first the evaluation of $f(T_{ii})$. For notational simplicity, let $T \in \mathbb{C}^{n\times n}$ play the role of T_{ii}. Assume that derivatives of f are available and that f has a Taylor series with an infinite radius of convergence. Then we can evaluate $f(T)$ from the Taylor series. Writing

$$T = \sigma I + M, \qquad \sigma = \mathrm{trace}(T)/n, \tag{4.5}$$

we evaluate f about the mean, σ, of the eigenvalues:

$$f(T) = \sum_{k=0}^{\infty} \frac{f^{(k)}(\sigma)}{k!} M^k. \tag{4.6}$$

If the eigenvalues of T are sufficiently close then the powers of M can be expected to decay quickly after the $(n-1)$st, and so a suitable truncation of (4.6) should yield good accuracy; indeed, in the special case where T has only one distinct eigenvalue ($t_{ii} \equiv \sigma$), M is nilpotent and $M^n = 0$.

By bounding the truncation error in the Taylor series we can construct the following algorithm that adaptively chooses the number of terms in order to achieve the desired accuracy; see Davies and Higham (2003) or Higham (2008, Section 9.1) for the details of the derivation.

Algorithm 4.4. (evaluate function of atomic block) Given a triangular matrix $T \in \mathbb{C}^{n\times n}$ whose eigenvalues $\lambda_1, \dots, \lambda_n$ are 'close,' a function f having a Taylor series with an infinite radius of convergence, and the ability to evaluate derivatives of f, this algorithm computes $F = f(T)$ using a truncated Taylor series.

$\sigma = n^{-1}\sum_{i=1}^{n} \lambda_i$, $M = T - \sigma I$, tol $= u$
$\mu = \|y\|_\infty$, where y solves $(I - |N|)y = e$ and N is the strictly upper triangular part of T. % $\mu = \|(I - |N|)^{-1}\|_\infty$
$F_0 = f(\sigma)I_n$
$P = M$
for $s = 1{:}\infty$
    $F_s = F_{s-1} + f^{(s)}(\sigma)P$
    $P = PM/(s+1)$
    if $\|F_s - F_{s-1}\|_F \le \text{tol}\|F_s\|_F$
        % Successive terms are close so check a truncation error bound.
        Estimate or bound $\Delta = \max_{0\le r\le n-1} \omega_{s+r+1}/r!$, where $\omega_k = \max\{\, |f^{(k)}(t_{ii})| : i = 1{:}n \,\}$.
        if $\mu\Delta\|P\|_F \le \text{tol}\|F_s\|_F$, $F = F_s$, quit, end
    end
end

Algorithm 4.4 costs $O(n^4)$ flops, since even if T has constant diagonal, so that M is nilpotent with $M^n = 0$, the algorithm may need to form the first $n-1$ powers of M. However, n here is the size of a block, and in most cases the blocks will be of much smaller dimension than the original matrix. Also, M is an upper triangular matrix, so forming all the powers $M^2, \dots, M^{n-1}$ costs $n^4/3$ flops – a factor 6 less than the flop count for multiplying full matrices.

Now we consider how to reorder and block the Schur factor T. We wish the Sylvester equations (4.4) to be well-conditioned, so that the equations are solved accurately and errors do not grow substantially within the recurrence. The conditioning of (4.4) is measured[1] by $\operatorname{sep}(T_{ii}, T_{jj})^{-1}$, where

$$\operatorname{sep}(T_{ii}, T_{jj}) = \min_{X\neq 0} \frac{\|T_{ii}X - XT_{jj}\|_F}{\|X\|_F}$$

is the separation of the diagonal blocks T_{ii} and T_{jj}. The separation is expensive to compute or estimate accurately, but we can approximate it by a lower bound that is cheap to evaluate:

$$\operatorname{sep}(T_{ii}, T_{jj})^{-1} \approx \frac{1}{\min\{\, |\lambda - \mu| : \lambda \in \Lambda(T_{ii}),\ \mu \in \Lambda(T_{jj}) \,\}}, \tag{4.7}$$

[1] In fact, this is a commonly used upper bound for the (interesting part of the) true condition number (Higham 2002, Section 10.3).

where $\Lambda(\cdot)$ denotes the spectrum. A second goal of the reordering is to produce diagonal blocks that have close eigenvalues, so that Algorithm 4.4 is efficient.

Denote the reordered upper triangular matrix by $\widetilde{T} = U^*TU = (\widetilde{T}_{ij})$, where U is unitary. A reasonable way to satisfy the above requirements for the diagonal blocks is to ensure that

(1) *separation between blocks*:

$$\min\{\, |\lambda - \mu| : \lambda \in \Lambda(\widetilde{T}_{ii}),\ \mu \in \Lambda(\widetilde{T}_{jj}),\ i \neq j \,\} > \delta,$$

(2) *separation within blocks*: for every block $\widetilde{T}_{ii}$ with dimension bigger than 1, for every $\lambda \in \Lambda(\widetilde{T}_{ii})$ there is a $\mu \in \Lambda(\widetilde{T}_{ii})$ with $\mu \neq \lambda$ such that $|\lambda - \mu| \leq \delta$.

Here, $\delta > 0$ is a blocking parameter. These conditions produce a blocking for which the spectra of different diagonal blocks are at least distance δ apart (thus ensuring that the right-hand side of (4.7) is at most δ^{-1}) while the eigenvalues within each diagonal block are close, in the sense that every eigenvalue is at most distance δ from some other eigenvalue. How to obtain such a reordering is described in Davies and Higham (2003) and Higham (2008, Section 9.3).

The overall algorithm is as follows.

Algorithm 4.5. (Schur–Parlett algorithm) Given $A \in \mathbb{C}^{n\times n}$, a function f having a Taylor series with an infinite radius of convergence, and the ability to evaluate derivatives of f, this algorithm computes $F = f(A)$.

1 Compute the Schur decomposition $A = QTQ^*$.
2 If T is diagonal, $F = f(T)$, goto line 10, end
3 Reorder and partition T and update Q to satisfy the conditions above with $\delta = 0.1$.
 % Now $A = QTQ^*$ is our reordered Schur decomposition,
 % with block $m \times m$ T.
4 for $j = 1\colon m$
5 Use Algorithm 4.4 to evaluate $F_{ii} = f(T_{ii})$.
6 for $i = j-1\colon -1\colon 1$
7   Solve the Sylvester equation (4.4) for F_{ij}.
8 end
9 end
10 $F = QFQ^*$

Cost: Roughly between $28n^3$ flops and $n^4/3$ flops, and dependent greatly on the eigenvalue distribution of A.

Algorithm 4.5 is the best available method for general functions f and it usually performs in a forward stable manner, that is, the forward error is usually bounded by a modest multiple of $\operatorname{cond}(f, A)u$, where the condition number $\operatorname{cond}(f, A)$ is defined in Section 7.1. However, the algorithm can be unstable, and empirically this seems most likely for matrices having large Jordan blocks. Instability can usually be cured by increasing δ to 0.2, but Davies and Higham (2003) have shown experimentally that the algorithm can be unstable for all choices of δ.

4.2. Schur method for matrix roots

For some functions it is possible to compute $f(T)$ by a different method than the Parlett recurrence. The main cases of practical interest are pth roots. If X is a square root of the upper triangular matrix T then the diagonal of X is readily determined and the superdiagonal elements can be obtained from the equation $X^2 = T$. The corresponding Schur method, due to Björck and Hammarling (1983), can be arranged as follows.

Algorithm 4.6. (Schur method for square root) Given a non-singular $A \in \mathbb{C}^{n\times n}$ this algorithm computes $X = \sqrt{A}$ via a Schur decomposition, where $\sqrt{\cdot}$ denotes any primary square root.

Compute a (complex) Schur decomposition $A = QTQ^*$.
for $j = 1{:}\, n$
  $u_{jj} = \sqrt{t_{jj}}$
  for $i = j-1{:}\, -1{:}\, 1$

$$u_{ij} = \frac{t_{ij} - \sum_{k=i+1}^{j-1} u_{ik}u_{kj}}{u_{ii} + u_{jj}}$$

  end
end
$X = QUQ^*$

Cost: $28\frac{1}{3}n^3$ flops.

Note that Algorithm 4.6 breaks down if $u_{ii} = -u_{jj}$ for some i and j, which can happen only if T has two equal diagonal elements that are mapped to different square roots – in other words, the algorithm is attempting to compute a non-primary square root. The algorithm can compute any primary square root by a suitable choice of the square roots of the diagonal elements. The computed square root $\widehat{X}$ satisfies

$$\widehat{X}^2 = A + \Delta A, \qquad \|\Delta A\|_F \le \widetilde{\gamma}_{n^3}\|\widehat{X}\|_F^2,$$

which is as good a residual bound as the rounded exact square root satisfies and so is essentially optimal.

For computing real square roots of real matrices it is more appropriate to employ the real Schur decomposition and thereby work entirely in real arithmetic; see Higham (1987) or Higham (2008, Section 6.2).

For pth roots, more complicated recurrences can be used to solve $X^p = T$, as shown by Smith (2003). The overall Schur algorithm has a cost of $O(pn^3)$ flops. If $p = p_1p_2 \dots p_t$ is composite, the pth root can be computed by successively computing the p_1th, p_2th, ..., p_tth roots, with a computational saving. Greco and Iannazzo (2010) show how to use the binary representation of p to compute X at a cost of $O(n^2p + n^3 \log_2 p)$ flops, which achieves further savings when p is large and not highly composite.

4.3. Block diagonalization

If we are willing to use non-unitary transformations then we can go beyond the Schur form to block diagonal form to obtain $A = XDX^{-1}$, where D is block diagonal. Such a form can be obtained by first computing the Schur form and then eliminating off-diagonal blocks by solving Sylvester equations (Bavely and Stewart 1979, Golub and Van Loan 1996, Section 7.6.3, Lavallée, Malyshev and Sadkane 1997). In order to guarantee a well-conditioned X a bound must be imposed on the condition numbers of the individual transformations, and this bound will be a parameter in the algorithm. The attraction of block diagonal form is that computing $f(A)$ reduces to computing $f(D)$, and hence to computing $f(D_{ii})$ for each diagonal block D_{ii}, and the D_{ii} are triangular if obtained as indicated above. However, evaluating $f(D_{ii})$ is still a non-trivial calculation because Algorithms 4.2 and 4.4 may not be applicable. The Schur–Parlett method and the block diagonalization method are closely related. Both employ a Schur decomposition, both solve Sylvester equations, and both must compute $f(T_{ii})$ for triangular blocks T_{ii}. Parlett and Ng (1985, Section 5) show that the two methods are mathematically equivalent, differing only in the order in which two commuting Sylvester operators are applied.

5. Polynomial and rational approximations

A natural way to approximate $f(A)$ is to mimic what is often done for scalar functions: to approximate $f(A)$ by $r(A)$ where r is some suitable polynomial or rational approximation to f. From scalar approximation theory we may know some region of $\mathbb{C}$ in which $f(z) \approx r(z)$ is a good approximation. However, if the spectrum of A lies in this region there is no guarantee that the matrix approximation $f(A) \approx r(A)$ is equally good. Indeed if A is diagonalizable with $A = ZDZ^{-1}$, then $f(A) - r(A) = Z(f(D) - r(D))Z^{-1}$, so that $\|f(A) - r(A)\| \le \kappa(Z)\|f(D) - r(D)\|$. Hence the error in the matrix approximation is potentially as much as $\kappa(Z)$ times larger than the error in the scalar approximation. If A is normal, so that we can take $\kappa_2(Z) = 1$, then the scalar and matrix approximation problems are essentially the same. But for non-normal matrices, achieving a good approximation requires more than simply approximating well at the eigenvalues.

A general framework for approximating $f(A)$ is as follows.

Framework 5.1. (for approximating $f(A)$)

(1) Choose a suitable rational approximation r and a transformation function g and set $A \leftarrow g(A)$.
(2) Compute $X = r(A)$ by some appropriate scheme.
(3) Apply transformations to X that undo the effect of the initial transformation on A.

The purpose of step (1) is to transform A so that $f(A) \approx r(A)$ is a sufficiently good approximation. Both g and r may depend on A. In the following subsections we describe how this can be done for some specific functions f.

For a polynomial approximation at step (2) the natural choice when f has a Taylor series $f(x) = \sum_{i=0}^{\infty} a_i x^i$ is the truncated Taylor series

$$T_k(A) = \sum_{i=0}^{k} a_i A^i. \tag{5.1}$$

Among rational approximations the Padé approximants prove to be particularly useful. Recall that $r_{km}(x) = p_{km}(x)/q_{km}(x)$ is a $[k/m]$ Padé approximant of f if p_{km} and q_{km} are polynomials of degree at most k and m, respectively, $q_{km}(0) = 1$, and

$$f(x) - r_{km}(x) = O(x^{k+m+1}) \tag{5.2}$$

(Brezinski and Van Iseghem 1995, Baker and Graves-Morris 1996). Thus a Padé approximant reproduces as many terms as possible of the Taylor series about the origin. If a $[k/m]$ Padé approximant exists then it is unique.

For the evaluation at step (2) there are many possibilities and which is best depends very much on r. Consider, first, the case of polynomial r. For a matrix argument, evaluation by Horner's method is not of optimal efficiency for polynomials of degree 4 and higher. More efficient alternatives are based on explicitly forming certain matrix powers, as is done in a general method of Paterson and Stockmeyer (1973) and a variant of Van Loan (1979). For rational $r = p/q$, the possibilities are more diverse:

(1) Evaluate $p(A)$ and $q(A)$ and then solve the multiple right-hand side system $q(A)r(A) = p(A)$.
(2) Evaluate $r(A)$ from a continued fraction representation of r in either top-down or bottom-up fashion.
(3) Evaluate $r(A)$ from a partial fraction representation, ideally with linear factors but possibly admitting higher-degree factors in order to keep the coefficients real.

The choice of scheme should take into account numerical stability and will, in general, depend on f. For details of all the above schemes see Higham (2008, Sections 4.2, 4.4.3).

We now focus on some specific transcendental functions of interest.

5.1. *Matrix exponential*

For Hermitian A, best L_∞ approximations to e^x can be employed, for which matrix level error bounds follow immediately from error bounds at the scalar level, as noted at the start of this section. However, we concentrate in this section on general matrices.

Truncating the Taylor series is one way to obtain an approximation. Padé approximation is particularly attractive because the $[k/m]$ Padé approximants to the exponential function are known explicitly for all k and m:

$$p_{km}(x) = \sum_{j=0}^{k} \frac{(k+m-j)!\,k!}{(k+m)!\,(k-j)!}\frac{x^j}{j!}, \quad q_{km}(x) = \sum_{j=0}^{m} \frac{(k+m-j)!\,m!}{(k+m)!\,(m-j)!}\frac{(-x)^j}{j!}. \tag{5.3}$$

Of course, the truncated Taylor series is just the $[k/0]$ Padé approximant. Among Padé approximants the diagonal approximants, $r_m \equiv r_{mm}$, are preferred, because they are more efficient (Moler and Van Loan 1978) and they have advantageous stability properties (Varga 2000, Chapter 8).

The standard way in which to use these approximations is within the scaling and squaring method. This method scales the matrix according to $A \leftarrow 2^{-s}A$, with s chosen so that $\|A\|$ is of order 1, evaluates $X = r_m(A)$ for some suitable diagonal Padé approximant r_m, then undoes the effect of the scaling by repeated squaring: $X \leftarrow X^{2^s}$. The method originates with Lawson (1967). Moler and Van Loan[2] (1978) give a backward error analysis that shows that $r_m(2^{-s}A)^{2^s} = e^{A+E}$ with an explicit bound on E expressed in terms of $\|A\|$. Here, E is measuring the effect of truncation errors in the Padé approximant and exact arithmetic is assumed. Based on this analysis they give a table indicating the optimal choice of s and m for a given $\|A\|$ and ϵ, where $\|E\| \le \epsilon\|A\|$ is required, and also conclude from it that Padé approximants are more efficient than Taylor series. The analysis in Moler and Van Loan (1978) led to implementations taking a fixed choice of m and s. For example, the function `expm` in version 7.1 and earlier of MATLAB used $m = 6$ and scaled so that $\|2^{-s}A\|_\infty \le 0.5$. However, a more efficient and more accurate (in floating point arithmetic) algorithm can be obtained by using higher-degree approximants, with the choice of m and s determined from sharper truncation error bounds.

[2] This classic paper was reprinted with an update in Moler and Van Loan (2003).

We need the following result of Al-Mohy and Higham (2009*b*), which is a slightly sharper version of a result of Higham (2005). Let $\nu_m = \min\{\,|t| : q_m(t) = 0\,\}$ and

$$\Omega_m = \{\, X \in \mathbb{C}^{n\times n} : \rho(\mathrm{e}^{-X} r_m(X) - I) < 1 \text{ and } \rho(X) < \nu_m \,\},$$

where ρ denotes the spectral radius.

Lemma 5.1. For $A \in \Omega_m$ we have $r_m(2^{-s}A)^{2^s} = \mathrm{e}^{A+\Delta A}$, where $\Delta A = h_{2m+1}(2^{-s}A)$ and $h_{2m+1}(x) = \log(\mathrm{e}^{-x} r_m(x))$. Moreover,

$$\frac{\|\Delta A\|}{\|A\|} \le \frac{\sum_{k=2m+1}^{\infty} |c_k|\, \|2^{-s}A\|^k}{\|2^{-s}A\|}, \tag{5.4}$$

where $h_{2m+1}(x) = \sum_{k=2m+1}^{\infty} c_k x^k$.

The bound in the lemma can be evaluated for a given value of $\|A\|$ by determining the coefficients c_k symbolically and then summing a suitable number of the terms in the numerator at high precision. Using a zero-finder we can therefore determine, for a range of m, the largest value of $\|2^{-s}A\|$, denoted by θ_m, such that the bound is no larger than $u = 2^{-53} \approx 1.1 \times 10^{-16}$, the unit roundoff for IEEE double precision arithmetic; some of these constants are given in Table 5.1. By taking account of the cost of evaluating $r_m(2^{-s}A)^{2^s}$, we can determine the optimal choice of s and m for a given $\|A\|$. The analysis also needs to ensure that the effect of rounding errors in the evaluation of r_m is not significant. The ideas above were used by Higham (2005) to derive the following algorithm.

Algorithm 5.2. (scaling and squaring) This algorithm evaluates the matrix exponential $X = \mathrm{e}^A$ of $A \in \mathbb{C}^{n\times n}$ using the scaling and squaring method. It uses the constants θ_m given in Table 5.1. The algorithm is intended for IEEE double precision arithmetic.

for $m = [3\ 5\ 7\ 9]$
    if $\|A\|_1 \le \theta_m$, evaluate $X = r_m(A)$, quit, end
end
$A \leftarrow A/2^s$ with $s \ge 0$ a minimal integer such that $\|A/2^s\|_1 \le \theta_{13}$ (*i.e.*, $s = \lceil \log_2(\|A\|_1/\theta_{13}) \rceil$).
Evaluate $X = r_{13}(A)$.
$X \leftarrow X^{2^s}$ by repeated squaring.

The details of how to evaluate r_m can be found in Higham (2005), Higham (2008, Section 10.3), or Higham (2009).

Although the derivation of Algorithm 5.2 is based on minimizing the cost of the computation, it is a welcome side effect that the accuracy is usually also improved compared with the previous choices of s and m. The reason is that the most dangerous phase of the algorithm – whose effect on the overall

Table 5.1. Constants θ_m needed in Algorithm 5.2.

m	θ_m
3	1.495585217958292e-2
5	2.539398330063230e-1
7	9.504178996162932e-1
9	2.097847961257068e0
13	5.371920351148152e0

numerical stability is still not well understood – is the s squarings, and the use of high Padé degrees up to $m = 13$ together with sharper truncation error bounds tends to produce smaller values of s and hence reduce the number of squarings.

The long-standing preference of Padé approximants over Taylor series within the scaling and squaring method needs revisiting with the aid of the sharper bound of Lemma 5.1. This is done in the Appendix, where we find that Padé approximants remain preferable.

We now reconsider the choice of s. Figure 5.1 shows what happens when we force Algorithm 5.2 to choose a particular value of s. For two different matrices we plot the relative error $\|X - \widehat{X}\|_1/\|X\|_1$, where X is an accurate approximation to e^A computed at high precision in MATLAB using the Symbolic Math Toolbox. For the first matrix, `-magic(6)^2` in MATLAB, we see in the first plot of Figure 5.1 that the relative error is of order 1 with no scaling ($s = 0$) and it reaches a minimum of 2.2×10^{-13} for $s = 11$. Algorithm 5.2 chooses $s = 12$, which is nearly optimal as regards the forward error. However, for the matrix

$$A = \begin{bmatrix} -1 & -1 & -10^4 & -10^4 \\ -1 & -1 & -10^4 & -10^4 \\ 0 & 0 & -1 & -1 \\ 0 & 0 & -1 & -1 \end{bmatrix}, \tag{5.5}$$

we see in the second plot of Figure 5.1 that the error grows mostly monotonically with s. The value $s = 12$ chosen by Algorithm 5.2 is much larger than the optimal value $s = 0$; thus Algorithm 5.2 computes a less accurate result than the optimum and at significantly greater cost. An explanation for this behaviour can be seen from the following MATLAB computation:

```
for i = 1:10, fprintf('%9.1e ', norm(A^i)^(1/i)), end
2.0e+004  2.8e+002  6.2e+001  2.8e+001  1.7e+001  1.3e+001
9.8e+000  8.2e+000  7.1e+000  6.3e+000
```

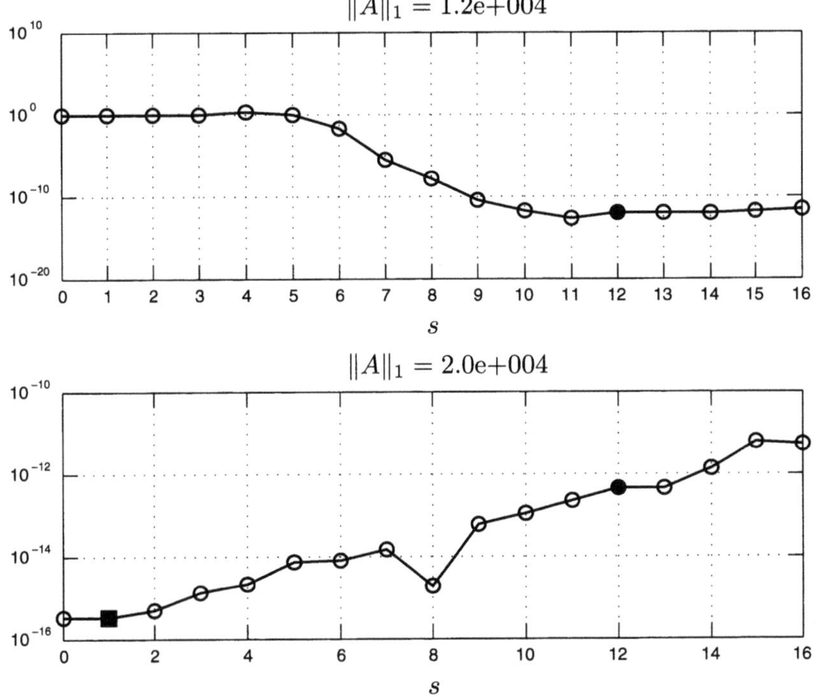

Figure 5.1. Scaling argument s versus relative error of scaling and squaring method, for A given by `-magic(6)^2` (*top*) and (5.5) (*bottom*). The symbol ● marks the value of s chosen by Algorithm 5.2; ■ marks the value given by the algorithm of Al-Mohy and Higham (2009*b*).

While $\|A\|_2$ is of order 10^4, the successive powers of A are smaller than might be expected, or, in other words, the inequalities $\|A^i\|_2 \leq \|A\|_2^i$ are very weak. Since the derivation of Algorithm 5.2 is based on bounding a power series by making use of such inequalities, it is not surprising that the algorithm can make a conservative choice of s. To do better, it is necessary to take account of the behaviour of the powers of A.

Notice that the matrix A in (5.5) is block 2×2 block upper triangular with both diagonal blocks having norm of order 1, while the off-diagonal block has norm of order 10^4. In this situation, Algorithm 5.2 is forced to take a large s in order to bring the overall matrix norm down to $\theta_{13} \approx 5.3$, even though this is not necessary as regards computing the exponentials of of the diagonal blocks. That it might not be necessary to let the off-diagonal blocks determine s can be seen from the formula (see, *e.g.*, Higham (2008, Problem 10.12), Van Loan (1978))

$$\exp\left(\begin{bmatrix} A_{11} & A_{12} \\ 0 & A_{22} \end{bmatrix}\right) = \begin{bmatrix} e^{A_{11}} & \int_0^1 e^{A_{11}(1-s)} A_{12} \, e^{A_{22}s} \, ds \\ 0 & e^{A_{22}} \end{bmatrix}. \tag{5.6}$$

Since A_{12} enters only *linearly* in the expression for e^A it is reasonable to argue that its norm should not influence s.

This phenomenon, whereby a much larger s is chosen than necessary in order to approximate e^A to the desired accuracy, was identified by Kenney and Laub (1998), and later by Dieci and Papini (2000), and is referred to as overscaling.

Al-Mohy and Higham (2009*b*) derive a new scaling and squaring algorithm that exploits the quantities $d_k = \|A^k\|^{1/k}$ for a few values of k. The key idea is to bound the truncation error for the Padé approximant using these values. Specifically, it can be shown that with h_{2m+1} defined as in Lemma 5.1 we have

$$\|h_{2m+1}(A)\| \le \widetilde{h}_{2m+1}\big(\max(d_p, d_{p+1})\big) \quad \text{if } 2m+1 \ge p(p-1),$$

where $\widetilde{h}_{2m+1}(x) = \sum_{k=2m+1}^{\infty} |c_k|\, x^k$. The algorithm re-uses the θ_m values in Table 5.1, but the scaling is now chosen so that $\max(d_k, d_{k+1})$, rather than $\|A\|$, is bounded by θ_m (for some appropriate k and m). In addition to computing d_k for values of k for which A^k must in any case be formed, the algorithm also estimates d_k for a few additional k using the matrix norm estimator of Higham and Tisseur (2000). The algorithm of Al-Mohy and Higham (2009*b*) also incorporates an improvement for the triangular case; the need for exponentials of triangular matrices arises, for example, in the solution of radioactive decay equations (Morai and Pacheco 2003, Yuan and Kernan 2007). In the squaring phase the diagonal and first superdiagonal of $X_i = (r_m(T))^{2^i}$ (where T is the scaled triangular matrix) are computed from exact formulae for the corresponding elements of $\mathrm{e}^{2^i T}$, which both gives a more accurate diagonal and first superdiagonal and reduces the propagation of errors in the squaring recurrence. The new algorithm generally provides accuracy at least as good as Algorithm 5.2 at no higher cost, and for matrices that are triangular or cause overscaling it usually yields significant improvements in accuracy, cost, or both. Figure 5.1 shows that for the matrix (5.5) the new algorithm chooses an almost optimal value of s, yielding a much more accurate solution than Algorithm 5.2.

5.2. Matrix logarithm

For the matrix logarithm we have the expansion

$$\log(I+X) = X - \frac{X^2}{2} + \frac{X^3}{3} - \frac{X^4}{4} + \cdots, \qquad \rho(X) < 1. \tag{5.7}$$

In order to use Framework 5.1 we need to choose the initial transformation g to bring A close to the identity. This is achieved by repeatedly taking square roots until $A^{1/2^k}$ is sufficiently close to I. By (2.8), the square roots are undone by scalar multiplication, so the overall approximation is $2^k r(A^{1/2^k} - I)$

for some suitable rational approximation $r(x)$ to $\log(1+x)$. This approximation was originally used by Briggs for computing logarithms of scalars (Goldstine 1977, Phillips 2000). For matrices, the approximation leads to the inverse scaling and squaring method, which was proposed by Kenney and Laub (1989*a*).

As for the exponential, the preferred approximations are diagonal Padé approximants. The diagonal Padé approximants r_m to $\log(1+x)$ are known explicitly in two different forms. They have the continued fraction expansion

$$r_m(x) = \cfrac{c_1 x}{1 + \cfrac{c_2 x}{1 + \cfrac{c_3 x}{1 + \cdots + \cfrac{c_{2m-2} x}{1 + \cfrac{c_{2m-1} x}{1 + c_{2m} x}}}}}, \tag{5.8a}$$

$$c_1 = 1, \quad c_{2j} = \frac{j}{2(2j-1)}, \quad c_{2j+1} = \frac{j}{2(2j+1)}, \quad j = 1, 2, \ldots, \tag{5.8b}$$

which is the truncation of an infinite continued fraction for $\log(1+x)$. They can also be represented in linear partial fraction form

$$r_m(x) = \sum_{j=1}^{m} \frac{\alpha_j^{(m)} x}{1 + \beta_j^{(m)} x}, \tag{5.9}$$

where the $\alpha_j^{(m)}$ are the weights and the $\beta_j^{(m)}$ are the nodes of the m-point Gauss–Legendre quadrature rule on $[0,1]$, all of which are real.

Analogously to the scaling and squaring method for the exponential, early algorithms used a fixed Padé degree and a fixed condition $\|A^{1/2^k} - I\| \le \theta$ for determining how many square roots to take. Kenney and Laub (1989*a*) take $m = 8$ and $\theta = 0.25$, while Dieci, Morini and Papini (1996) take $m = 9$ and $\theta = 0.35$. The use of a degree m dependent on $\|A\|$ originates with Cheng, Higham, Kenney and Laub (2001), who exploit the following result of Kenney and Laub (1989*b*).

Theorem 5.3. For $\|X\| < 1$ and any subordinate matrix norm,

$$\|r_m(X) - \log(I+X)\| \le \big|r_m(-\|X\|) - \log(1 - \|X\|)\big|. \tag{5.10}$$

This result says that the error in the matrix Padé approximation is bounded by the error in the scalar approximation at minus the norm of the matrix. After each square root is taken, this bound can be used to check whether the approximation error is guaranteed to be sufficiently small for a particular m. In designing an algorithm it is necessary to consider whether, when (5.10) is satisfied for some allowable m, it is worth taking another square root in the hope that the extra cost will be outweighed by being able to take a smaller m. Guidance in answering this question comes from the approximation, valid for large enough k,

$$\|I - A^{1/2^{k+1}}\| \approx \frac{1}{2}\|I - A^{1/2^k}\|, \tag{5.11}$$

which follows from $\bigl(I - A^{1/2^{k+1}}\bigr)\bigl(I + A^{1/2^{k+1}}\bigr) = I - A^{1/2^k}$.

Unlike for the exponential, the details of an inverse scaling and squaring algorithm depend on whether the matrix A is full or triangular, because the appropriate method for computing the square roots (and hence the cost of the square root stage) depends on the structure. If A is triangular, or an initial Schur factorization is computed to reduce to the triangular case, then the Schur method (Algorithm 4.6) can be used, otherwise a variant of the Newton iteration is appropriate; see Section 6.1. For evaluating the Padé approximant the best compromise between speed and accuracy turns out to be the partial fraction form (Higham 2001).

Cheng *et al.* (2001) give an inverse scaling and squaring algorithm based on the above ideas that accepts as input a parameter specifying the desired accuracy and computes square roots using the product form of the Denman–Beavers iteration (6.9), with the number of iterations used to compute each square root carefully chosen to minimize the overall cost. Higham (2008, Section 11.5) gives algorithms for both the full and triangular cases in which the parameters used in the algorithm's logic are precomputed, analogously as for Algorithm 5.2, rather than computed at run-time as in Cheng *et al.* (2001).

5.3. Trigonometric functions

We can apply Framework 5.1 to trigonometric functions by scaling $A \leftarrow 2^{-s}A$ in step (1) and using the appropriate double angle formulas in step (3). This idea was first proposed by Serbin and Blalock (1980) for the matrix cosine, using $\cos(2A) = 2\cos(A)^2 - I$.

It is not known whether Padé approximants r_{km} to the matrix cosine and sine exist for all degrees k and m, though they can be determined symbolically for the range of degrees of practical interest. Higham and Smith (2003) develop an algorithm that chooses s so that $\|2^{-s}A\|_\infty \le 1$, approximates $\cos(A) \approx r_8(2^{-s}A)$, and then uses the double angle recurrence. Hargreaves and Higham (2005) extend this approach by choosing the degree of the Padé approximant and the amount of scaling based on truncation error bounds expressed in terms of $\|A^2\|^{1/2}$ instead of $\|A\|$, thus using a more rudimentary version of the approach used for the exponential by Al-Mohy and Higham (2009*b*). They also derive an algorithm for computing both $\cos(A)$ and $\sin(A)$, by adapting the ideas developed for the cosine and intertwining the cosine and sine double angle recurrences.

So far there is little work on algorithms for other trigonometric functions. Some ideas concerning the tangent and inverse tangent for Hermitian matrices are given by Cheng, Higham, Kenney and Laub (2000).

6. Matrix iterations

Matrix roots, the matrix sign function, and the unitary polar factor (which we will not consider in this paper) are all amenable to computation by matrix iterations of the form

$$X_{k+1} = g(X_k), \tag{6.1}$$

where g is some nonlinear, usually rational, function. The starting matrix X_0 is almost always A in practice, and indeed in some cases A does not appear in the iteration itself, that is, g is independent of A. Such iterations are attractive because they are easy to implement, requiring just the basic building blocks of matrix multiplication and the solution of multiple right-hand side linear systems. It might appear that convergence analysis is straightforward, reducing to the scalar case that is usually well understood, but this is not necessarily so. Moreover, the numerical stability of matrix iterations in finite precision arithmetic is a subtle issue, with small changes in the form of the iteration sometimes greatly changing the stability.

6.1. Matrix sign function and square root

Two fundamental iterations are the Newton iteration for the matrix sign function,

$$X_{k+1} = \frac{1}{2}(X_k + X_k^{-1}), \qquad X_0 = A, \tag{6.2}$$

and the Newton iteration for the matrix square root,

$$X_{k+1} = \frac{1}{2}(X_k + X_k^{-1}A), \qquad X_0 = A. \tag{6.3}$$

The term 'Newton iteration' needs explaining. Iteration (6.2) is precisely Newton's method applied to the equation $X^2 = I$, assuming that the iterates are uniquely defined (Higham 2008, Problem 5.8). Likewise, iteration (6.3) is Newton's method applied to $X^2 = A$ provided the iterates are uniquely defined, and the proof (Higham 2008, Lemma 6.8) relies on the fact that $X_k A = AX_k$ for all k. Note, however, that if we select an arbitrary X_0 in (6.3) then the iteration is in general no longer equivalent to Newton's method, because the iterates X_k will not now commute with A.

For $A \in \mathbb{C}^{n\times n}$ having no pure imaginary eigenvalues, the iterates (6.2) converge quadratically to $\mathrm{sign}(A)$. For $A \in \mathbb{C}^{n\times n}$ with no eigenvalues on $\mathbb{R}^-$, the X_k from (6.3) converge quadratically to the principal square root, $A^{1/2}$. These results can be proved at the matrix level without using a

Table 6.1. Iteration functions $f_{\ell m}$ from the Padé family (6.4).

	$m=0$	$m=1$	$m=2$
$\ell=0$	x	$\dfrac{2x}{1+x^2}$	$\dfrac{8x}{3+6x^2-x^4}$
$\ell=1$	$\dfrac{x}{2}(3-x^2)$	$\dfrac{x(3+x^2)}{1+3x^2}$	$\dfrac{4x(1+x^2)}{1+6x^2+x^4}$
$\ell=2$	$\dfrac{x}{8}(15-10x^2+3x^4)$	$\dfrac{x}{4}\dfrac{(15+10x^2-x^4)}{1+5x^2}$	$\dfrac{x(5+10x^2+x^4)}{1+10x^2+5x^4}$

transformation to Jordan form to introduce the scalar case. For the sign iteration the convergence reduces to the fact that $G^k \to 0$ as $k \to \infty$ if the spectral radius $\rho(G) < 1$. The convergence of the square root iteration can be shown to be equivalent to that of the sign iteration for $X_0 = A^{1/2}$. See Higham (2008, Theorems 5.6, 6.9) for details.

The Newton iteration for $\operatorname{sign}(A)$, originally proposed by Roberts (1980), is the inverse of a member of an infinite Padé family of iterations that have some remarkable properties. The (ℓ, m) iteration is

$$X_{k+1} = X_k p_{\ell m}(I - X_k^2)\, q_{\ell m}(I - X_k^2)^{-1} =: f_{\ell m}(X_k), \qquad X_0 = A, \quad (6.4)$$

where $r_{\ell m}(\xi) = p_{\ell m}(\xi)/q_{\ell m}(\xi)$ is the $[\ell/m]$ Padé approximant to $h(\xi) = (1-\xi)^{-1/2}$. The appearance of h arises from the relation $\operatorname{sign}(z) = z/(z^2)^{1/2} = z/(1-(1-z^2))^{1/2} = zh(1-z^2)$.

Table 6.1 shows the first nine iteration functions $f_{\ell m}$ from this family. For $\ell = m$ and $\ell = m-1$, the polynomials $xp_{\ell m}(1-x^2)$ and $q_{\ell m}(1-x^2)$ are, respectively, the odd and even parts of $(1+x)^{\ell+m+1}$ (Kenney and Laub 1991*b*), which provides an easy way to generate the iteration functions. Note that f_{01} gives the iteration

$$X_{k+1} = 2X_k(I + X_k^2)^{-1}, \qquad X_0 = A, \quad (6.5)$$

which generates matrices that are the inverses of the those from (6.2), while f_{10} gives the Newton–Schulz iteration

$$X_{k+1} = \frac{1}{2}X_k(3I - X_k^2), \qquad X_0 = A. \quad (6.6)$$

This latter iteration can be derived from the Newton iteration (6.2) by approximating the inverse therein by one step of the Newton–Schulz iteration for the matrix inverse with X_k as starting value.

The following result of Kenney and Laub (1991*b*) describes the convergence properties.

Theorem 6.1. (convergence of Padé iterations) Let $A \in \mathbb{C}^{n\times n}$ have no pure imaginary eigenvalues and let $S = \operatorname{sign}(A)$. Consider the iteration (6.4) with $\ell + m > 0$ and any subordinate matrix norm.

(a) For $\ell \geq m-1$, if $\|I - A^2\| < 1$ then $X_k \to S$ as $k \to \infty$ and $\|I - X_k^2\| < \|I - A^2\|^{(\ell+m+1)^k}$.

(b) For $\ell = m-1$ and $\ell = m$,

$$(S - X_k)(S + X_k)^{-1} = \left[(S - A)(S + A)^{-1}\right]^{(\ell+m+1)^k}$$

and hence $X_k \to S$ as $k \to \infty$.

Theorem 6.1 shows that the iterations with $\ell = m-1$ and $\ell = m$ are globally convergent (that is, convergent for any A), while those with $\ell \geq m+1$ have local convergence, the convergence rate being $\ell + m + 1$ in every case.

Other interesting properties of the Padé family can be found in Higham (2008, Theorem 5.9).

The Padé iterations for the sign function have analogues for the square root that can be derived using the relation (Higham 1997)

$$\operatorname{sign}\left(\begin{bmatrix} 0 & A \\ I & 0 \end{bmatrix}\right) = \begin{bmatrix} 0 & A^{1/2} \\ A^{-1/2} & 0 \end{bmatrix}. \tag{6.7}$$

By applying any sign iteration to the matrix $\left[\begin{smallmatrix} 0 & A \\ I & 0 \end{smallmatrix}\right]$, using (6.7), and then reading off the (1,2) and (2,1) blocks, a coupled iteration for the square root is obtained. For example, if we start with the Newton iteration (6.2), we obtain

$$\begin{aligned} X_{k+1} &= \frac{1}{2}(X_k + Y_k^{-1}), \qquad & X_0 &= A, \\ Y_{k+1} &= \frac{1}{2}(Y_k + X_k^{-1}), \qquad & Y_0 &= I. \end{aligned} \tag{6.8}$$

This iteration was originally derived by Denman and Beavers (1976). It is easy to show that $Y_k \equiv A^{-1}X_k$, and so X_k satisfies (6.3). A general result of Higham, Mackey, Mackey and Tisseur (2005, Theorem 4.5) shows that for essentially any sign iteration this approach produces a coupled iteration with matrices X_k converging to $A^{1/2}$ and Y_k converging to $A^{-1/2}$, both with the same order of convergence as the original sign iteration. Thus the convergence is quadratic for (6.8). A variant of (6.8) that trades a matrix inverse for a multiplication is obtained by setting $M_k = X_k Y_k$ in (6.8):

$$\begin{aligned} X_{k+1} &= \frac{1}{2}X_k(I + M_k^{-1}), \qquad & X_0 &= A, \\ M_{k+1} &= \frac{1}{2}\left(I + \frac{M_k + M_k^{-1}}{2}\right), \qquad & M_0 &= A. \end{aligned} \tag{6.9}$$

At first sight, it may be unclear why (6.8) or (6.9) might be preferred to the Newton iteration (6.3), since they require $4n^3$ flops per iteration versus $8n^3/3$ flops for (6.3). The answer is that the stability properties of the iterations are very different.

In general, stability can be defined as follows for a matrix iteration (6.1).

Definition 6.2. (stability) Consider an iteration $X_{k+1} = g(X_k)$ with a fixed point X. Assume that g is Fréchet-differentiable[3] at X. The iteration is *stable in a neighbourhood of* X if the Fréchet derivative $L_g(X)$ has bounded powers, that is, there exists a constant c such that $\|L_g^i(X)\| \le c$ for all $i > 0$.

Note that stability concerns behaviour close to convergence and so is an asymptotic property. The motivation for the definition is that L_g determines how a perturbation E_k in X_k is propagated through the iteration, and this perturbation will have a bounded effect near X if the iteration is stable. Although it is an asymptotic property, this notion proves to be a good predictor of the overall numerical stability of a matrix iteration. It is sometimes the case that L_g is idempotent, that is, $L_g^2(X,E) = L_g(X, L_g(X,E)) = L_g(X,E)$, in which case stability is immediate. This notion of stability was introduced by Cheng *et al.* (2001) and developed further by Higham (2008, Section 4.9.4).

For the Newton iteration (6.3) we have $L_g(X,E) = (E - X^{-1}EX^{-1}A)/2$, and, at the fixed point, $L_g(A^{1/2},E) = (E - A^{-1/2}EA^{1/2})/2$. The eigenvalues of $L_g(A^{1/2})$ can be shown to be

$$\mu_{ij} = \frac{1}{2}(1 - \lambda_i^{1/2}\lambda_j^{-1/2}), \qquad i,j = 1\colon n,$$

where the λ_i are the eigenvalues of A, and the μ_{ij} will be within the unit circle (which is necessary for stability) only for very well-behaved matrices A. Thus in general the iteration is unstable, and this is easily demonstrated experimentally. The instability was first identified by Laasonen (1958). Higham (1986*b*) explained the instability and derived the stability condition $|\mu_{ij}| < 1$.

It is perhaps surprising that the process of rewriting the Newton iteration in the coupled form given by the Denman–Beavers iteration (6.8) stabilizes it. The iteration function is now

$$G(X,Y) = \frac{1}{2}\begin{bmatrix} X + Y^{-1} \\ Y + X^{-1} \end{bmatrix}.$$

[3] See Section 7 for the definition of Fréchet derivative.

At the limit $X = A^{1/2}$, $Y = A^{-1/2}$ we have

$$L_g(A^{1/2}, A^{-1/2}; E, F) = \frac{1}{2}\begin{bmatrix} E - A^{1/2}FA^{1/2} \\ F - A^{-1/2}EA^{-1/2} \end{bmatrix}, \tag{6.10}$$

and it is easy to see that $L_g(A^{1/2}, A^{-1/2})$ is idempotent. Hence the Denman–Beavers iteration is stable. The Denman–Beavers iteration is just one of several ways of rewriting the Newton iteration as a coupled iteration, all of which are stable; see Higham (2008, Chapter 6) for more details.

The iterations described above are all of limited use in their basic forms. If we take $A = \theta \in \mathbb{R}$ for some $\theta \gg 1$, then the sign iteration (6.2) approximately effects $x \leftarrow x/2$ in the early stages, so many steps are needed before the asymptotic quadratic convergence comes into effect. In practice the iterations are used in conjunction with scaling. Usually, scaling consists of replacing X_k by $\mu_k X_k$ for some $\mu_k > 0$ in the formula for X_{k+1} (with a corresponding change for the other iterate in a coupled iteration), though more general scalings with more than one parameter can also be considered. Standard scalings for the Newton sign iteration are:

$$\text{determinantal scaling} \quad \mu_k = |\det(X_k)|^{-1/n}, \tag{6.11}$$

$$\text{spectral scaling} \quad \mu_k = \sqrt{\rho(X_k^{-1})/\rho(X_k)}, \tag{6.12}$$

$$\text{norm scaling} \quad \mu_k = \sqrt{\|X_k^{-1}\|/\|X_k\|}. \tag{6.13}$$

Each of these is in some way trying to bring X_k closer to $\operatorname{sign}(X_k) = \operatorname{sign}(A)$. Experiments show that there is no clear best scaling, but spectral scaling does have a finite termination property when A has only real eigenvalues (Barraud 1979, Section 4, Higham 2008, Theorem 5.11). Corresponding scalings for the square root iterations can be derived by using the connections with the sign function in Higham (2008, Theorem 6.9) and (6.7).

A number of linearly convergent iterations have been investigated in the literature, mostly for the square root and usually with structured matrices in mind. We mention three of them. First, with $A \equiv I - C$, is the binomial iteration,

$$X_{k+1} = \frac{1}{2}(C + X_k^2), \qquad X_0 = 0, \tag{6.14}$$

so called because it is essentially a convenient way of evaluating the binomial expansion $(I - C)^{1/2} = \sum_{j=0}^{\infty} \binom{\frac{1}{2}}{j}(-C)^j \equiv I - P$. The iterates X_k converge to $I - P$ if the eigenvalues of C lie in the main cardioid of the Mandelbrot set stretched by a factor 4 (Higham 2008, Theorem 6.14). If C has non-negative elements ($C \geq 0$) and spectral radius less than 1 then the convergence is monotonic from below in the elementwise ordering. An important class of matrices satisfying the latter condition after scaling is the non-singular

M-matrices, which are the non-singular $A \in \mathbb{R}^{n\times n}$ such that $A = sI - B$, where $B \geq 0$ and $s > \rho(B)$. If we write $A = s(I - C)$ with $C = s^{-1}B \geq 0$ then $\rho(C) < 1$ and so the binomial iteration converges monotonically when applied to $I - C$.

Next, the Pulay iteration (Pulay 1966) writes $A^{1/2} = D^{1/2} + B$ with D diagonal and positive definite ($D = \mathrm{diag}(A)$ being the natural choice if it is positive definite) and computes B as the limit of the B_k from the iteration

$$D^{1/2}B_{k+1} + B_{k+1}D^{1/2} = A - D - B_k^2, \qquad B_0 = 0. \tag{6.15}$$

Finally, the Visser iteration (Visser 1937, Elsner 1970) has the form

$$X_{k+1} = X_k + \alpha(A - X_k^2), \qquad X_0 = (2\alpha)^{-1}I. \tag{6.16}$$

Both iterations are forms of modified Newton iterations with the Fréchet derivative approximated by a constant. A sufficient condition for the convergence of the Pulay iteration is that $A^{1/2} - D^{1/2}$ is small compared with $D^{1/2}$ (Higham 2008, Theorem 6.15). The Visser iteration is related to the binomial iteration; convergence to $A^{1/2}$ is guaranteed if the eigenvalues of $I - 4\alpha^2 A$ lie in the cardioid referred to above.

For Hermitian positive definite matrices a very good way to compute the principal square root without computing the spectral decomposition is as follows (Higham 1986*a*).

Algorithm 6.3. Given a Hermitian positive definite matrix $A \in \mathbb{C}^{n\times n}$ this algorithm computes $H = A^{1/2}$.

1 $A = R^*R$ (Cholesky factorization).
2 Compute the Hermitian polar factor H of R by applying (Higham 2008, Alg. 8.20) to R (exploiting the triangularity of R).

Cost: Up to about $15\frac{2}{3}n^3$ flops.

The algorithm used in step 2 of Algorithm 6.3 employs a (scaled) Newton iteration for the unitary polar factor that is closely related to the Newton iteration (6.2) for the matrix sign function.

6.2. *Matrix pth root*

The Newton iteration for the principal pth root of A analogous to (6.3) is

$$X_{k+1} = \frac{1}{p}\big[(p-1)X_k + X_k^{1-p}A\big], \qquad X_0 = I. \tag{6.17}$$

As in the $p = 2$ case, this iteration is unstable except for A with spectrum clustered around 1. The convergence properties for $p > 2$, however, are much more complicated than for $p = 2$; for convergence, the eigenvalues of A must lie in regions of the complex plane having a fractal structure. Indeed, these regions are convergence regions for the map $x_{k+1} = [(p-1)x_k + x_k^{1-p}a]/p$, which is a classic example for the analysis of rational iterations via the theory of Julia sets (Peitgen, Jürgens and Saupe 1992, Schroeder 1991).

Iannazzo (2006) identifies a set $\{ z \in \mathbb{C} : \operatorname{Re} z > 0 \text{ and } |z| \le 1 \} \cup \mathbb{R}^+$ such that if the eigenvalues of A lie in the set then the iteration (6.17) converges quadratically to $A^{1/p}$. He suggests an algorithm that uses an initial square root followed by a scaling such that a coupled, stable form of iteration (6.17) converges for the preprocessed matrix. Instead of applying Newton's method to $X^p = A$, which gives (6.17), we can apply it to $X^{-p} = A$, which leads to the inverse Newton iteration

$$X_{k+1} = \frac{1}{p}\big[(p+1)X_k - X_k^{p+1}A\big], \qquad X_0 = c^{-1}I. \tag{6.18}$$

Here, $c > 0$ is a parameter, and Guo and Higham (2006) show that the iteration converges quadratically to $A^{-1/p}$ if all the eigenvalues of A are in the set

$$\operatorname{conv}\{ \{ z : |z - c^p| \le c^p \}, (p+1)c^p \} \setminus \{ 0, (p+1)c^p \}, \tag{6.19}$$

where conv is the convex hull. One useful practical conclusion that can be drawn is that if A is stochastic and strictly row-diagonally dominant then iteration (6.18) with $c = 1$ converges; this is relevant for the application described in Section 3.2. It is necessary to rewrite (6.18) in a coupled form for stability:

$$\begin{aligned} X_{k+1} &= X_k\left(\frac{(p+1)I - M_k}{p}\right), & X_0 &= \frac{1}{c}I, \\ M_{k+1} &= \left(\frac{(p+1)I - M_k}{p}\right)^p M_k, & M_0 &= \frac{1}{c^p}A; \end{aligned} \tag{6.20}$$

we have $X_k \to A^{-1/p}$ and $M_k \to I$. Guo and Higham combine this iteration with an initial Schur reduction to triangular form followed by the computation of a sufficient number of square roots, computed using Algorithm 4.6, so that fast convergence is expected for (6.20), if $A^{-1/p}$ is required, or a variant of (6.20) given by Guo and Higham, if $A^{1/p}$ is required.

Guo (2009, Theorem 5) obtains a more useful convergence result for (6.17) than Iannazzo's by proving convergence when the eigenvalues of A lie in the set $\{ z \in \mathbb{C} : |z - 1| \le 1 \}$, and he gives an analogue of Guo and Higham's algorithm based on the coupled version of (6.17). Guo (2009, Theorem 13) also shows that if $A = I - B$ with $\rho(B) < 1$ then X_k from (6.17) satisfies $X_k = \sum_{i=0}^{\infty} c_i^{(k)} B^i$, where $c_i^{(k)} = b_i$, $i = 0\colon 2^k - 1$ and $(1-x)^{1/p} = \sum_{i=0}^{\infty} b_i x^i$. Thus k steps of the Newton iteration reproduce 2^k terms of the binomial series. He also obtains an analogous result for (6.18).

A Padé family of iterations for the pth root is investigated by Laszkiewicz and Ziętak (2009). A variety of other methods exist for computing pth roots; see Bini, Higham and Meini (2005).

7. Fréchet derivative

The Fréchet derivative of a matrix function $f : \mathbb{C}^{n\times n} \to \mathbb{C}^{n\times n}$ at $A \in \mathbb{C}^{n\times n}$ is a linear mapping

$$\begin{array}{ccc} \mathbb{C}^{n\times n} & \xrightarrow{L_f(A)} & \mathbb{C}^{n\times n} \\ E & \longmapsto & L_f(A,E) \end{array}$$

such that

$$f(A+E) - f(A) - L_f(A,E) = o(\|E\|) \tag{7.1}$$

for all $E \in \mathbb{C}^{n\times n}$. Thus it describes the first-order effect on f of perturbations in A. We note that a sufficient condition for the Fréchet derivative to be defined is that f is $2n-1$ times continuously differentiable on an open subset of $\mathbb{R}$ or $\mathbb{C}$ containing the spectrum of A (Higham 2008, Theorem 3.8).

7.1. Condition number

The norm of the Fréchet derivative,

$$\|L_f(A)\| := \max_{\|Z\|=1} \|L_f(A,Z)\|, \tag{7.2}$$

appears in an explicit expression for the condition number of the matrix function f at A (Higham 2008, Theorem 3.1):

$$\operatorname{cond}(f,A) := \lim_{\epsilon\to 0} \sup_{\|E\|\le\epsilon\|A\|} \frac{\|f(A+E)-f(A)\|}{\epsilon\|f(A)\|} = \frac{\|L_f(A)\|\,\|A\|}{\|f(A)\|}. \tag{7.3}$$

Ideally, along with $f(A)$ we would like to produce an estimate of $\operatorname{cond}(f,A)$. This can be done by converting the problem to one of matrix norm estimation. Since L_f is a linear operator,

$$\operatorname{vec}(L_f(A,E)) = K(A)\operatorname{vec}(E) \tag{7.4}$$

for some $K(A) \in \mathbb{C}^{n^2\times n^2}$ that is independent of E. We refer to $K(A)$ as the Kronecker form of the Fréchet derivative. To estimate $\|L_f(A)\|_F$ we can apply the power method to $K(A)$, since $\|L_f(A)\|_F = \|K(A)\|_2$. The resulting algorithm can be written entirely in terms of $L_f(A)$ and $L_f^\star(A)$, the adjoint of $L_f(A)$ defined with respect to the inner product $\langle X, Y\rangle = \operatorname{trace}(Y^*X)$. When $A \in \mathbb{R}^{n\times n}$ and $f : \mathbb{R}^{n\times n} \to \mathbb{R}^{n\times n}$, the adjoint is given by $L_f^\star(A) = L_f(A^T)$. In the complex case, $L_f^\star(A) = L_{\overline{f}}(A^*)$, where $\overline{f}(z) := \overline{f(\overline{z})}$, so that if f has a power series representation then $\overline{f}$ is obtained by conjugating the coefficients.

Algorithm 7.1. (power method on Fréchet derivative) Given $A \in \mathbb{C}^{n\times n}$ and the Fréchet derivative L_f of a function f, this algorithm uses the power method to produce an estimate $\gamma \le \|L_f(A)\|_F$.

```
Choose a non-zero starting matrix Z_0 ∈ C^{n×n}
for k = 0: ∞
    W_{k+1} = L_f(A, Z_k)
    Z_{k+1} = L_f^⋆(A, W_{k+1})
    γ_{k+1} = ||Z_{k+1}||_F / ||W_{k+1}||_F
    if converged, γ = γ_{k+1}, quit, end
end
```

A random Z_0 is a reasonable choice. However, our preference is to apply instead of the power method the block 1-norm estimator of Higham and Tisseur (2000), available as `normest1` in MATLAB. This produces a quantity γ with $\gamma \le \|K(A)\|_1$, where $\|K(A)\|_1 \in \left[n^{-1}\|L_f(A)\|_1, n\|L_f(A)\|_1\right]$.

Both the above approaches require the ability to evaluate $L_f(A, E)$ and $L_f^\star(A, E)$. In the rest of this section we discuss several general approaches.

7.2. Power series

When f has a power series expansion the Fréchet derivative can be expressed as a related series expansion (Higham 2008, Problem 3.6)

Theorem 7.2. Let f have the power series expansion $f(x) = \sum_{k=0}^{\infty} a_k x^k$ with radius of convergence r. Then, for $A, E \in \mathbb{C}^{n\times n}$ with $\|A\| < r$,

$$L_f(A, E) = \sum_{k=1}^{\infty} a_k \sum_{j=1}^{k} A^{j-1} E A^{k-j}. \tag{7.5}$$

The next theorem, from Al-Mohy and Higham (2009*a*), gives a recurrence that can be used to evaluate (7.5), as well as a useful bound on $\|L_f(A)\|$.

Theorem 7.3. Under the assumptions of Theorem 7.2,

$$L_f(A, E) = \sum_{k=1}^{\infty} a_k M_k, \tag{7.6}$$

where $M_k = L_{x^k}(A, E)$ satisfies the recurrence

$$M_k = M_{\ell_1} A^{\ell_2} + A^{\ell_1} M_{\ell_2}, \qquad M_1 = E, \tag{7.7}$$

with $k = \ell_1 + \ell_2$ and ℓ_1 and ℓ_2 positive integers. In particular,

$$M_k = M_{k-1} A + A^{k-1} M_1, \qquad M_1 = E. \tag{7.8}$$

In addition,

$$\|f(A)\| \le \widetilde{f}(\|A\|), \qquad \|L_f(A)\| \le \widetilde{f}'(\|A\|), \tag{7.9}$$

where $\widetilde{f}(x) = \sum_{k=0}^{\infty} |a_k| x^k$.

Proof. Since the power series can be differentiated term-by-term within its radius of convergence, we have

$$L_f(A,E) = \sum_{k=1}^{\infty} a_k M_k, \qquad M_k = L_{x^k}(A,E).$$

The product rule for Fréchet derivatives (Higham 2008, Theorem 3.3) yields

$$M_k = L_{x^k}(A,E) = L_{x^{\ell_1}}(A,E)A^{\ell_2} + A^{\ell_1}L_{x^{\ell_2}}(A,E) = M_{\ell_1}A^{\ell_2} + A^{\ell_1}M_{\ell_2}.$$

Taking $\ell_1 = k-1$ and $\ell_2 = 1$ gives (7.8). It is straightforward to see that $\|f(A)\| \le \widetilde{f}(\|A\|)$. Taking norms in (7.5) gives

$$\|L_f(A,E)\| \le \|E\| \sum_{k=1}^{\infty} k|a_k|\,\|A\|^{k-1} = \|E\|\widetilde{f}'(\|A\|),$$

and maximizing over all non-zero E gives $\|L_f(A)\| \le \widetilde{f}'(\|A\|)$. □

7.3. Block triangular matrix formula

If f is $2n-1$ times continuously differentiable on an open subset of $\mathbb{R}$ or $\mathbb{C}$ containing the spectrum of $A \in \mathbb{C}^{n\times n}$ then (Higham 2008, Section 3.2)

$$f\left(\begin{bmatrix} A & E \\ 0 & A \end{bmatrix}\right) = \begin{bmatrix} f(A) & L_f(A,E) \\ 0 & f(A) \end{bmatrix}. \tag{7.10}$$

Thus $L_f(A,E)$ can be obtained by evaluating f at the $2n \times 2n$ matrix $\left[\begin{smallmatrix} A & E \\ 0 & A \end{smallmatrix}\right]$ and reading off the (1,2) block. This approach is pragmatic, but for an $O(n^3)$ method its cost is up to 8 times the cost of evaluating $f(A)$ alone, this multiplier being mitigated by the block triangular, block Toeplitz structure of the argument. A drawback noted by Al-Mohy and Higham (2009*a*) is that since $L_f(A,\alpha E) = \alpha L_f(A,E)$ the norm of E can be chosen at will, but the choice may affect the accuracy of the algorithm used to evaluate (7.10), and it is difficult to know what is the optimal choice.

We illustrate the use of (7.10) for the matrix square root by applying the Denman–Beavers iteration (6.8) to $\widetilde{A} = \left[\begin{smallmatrix} A & E \\ 0 & A \end{smallmatrix}\right]$. Iterates $\widetilde{X}_k$ and $\widetilde{Y}_k$ are produced for which

$$\widetilde{X}_k = \begin{bmatrix} X_k & F_k \\ 0 & X_k \end{bmatrix}, \qquad \widetilde{Y}_k = \begin{bmatrix} Y_k & G_k \\ 0 & Y_k \end{bmatrix},$$

where X_k and Y_k satisfy (6.8) and

$$\begin{aligned} F_{k+1} &= \frac{1}{2}(F_k - Y_k^{-1}G_kY_k^{-1}), \qquad F_0 = E, \\ G_{k+1} &= \frac{1}{2}(G_k - X_k^{-1}F_kX_k^{-1}), \qquad G_0 = 0. \end{aligned} \tag{7.11}$$

From (7.10) we conclude that

$$\lim_{k\to\infty} F_k = L_{x^{1/2}}(A,E), \qquad \lim_{k\to\infty} G_k = L_{x^{-1/2}}(A,E). \tag{7.12}$$

The iteration (7.11) is due to Al-Mohy and Higham (2009*a*), who derive it in this way.

7.4. *Differentiating an algorithm*

If we have an algorithm for computing $f(A)$ then we might expect that differentiating it will provide an algorithm for computing the Fréchet derivative. We describe two situations where this idea proves useful.

Framework 5.1 uses a rational approximation $f(A) \approx r(A)$. Obviously, we can approximate $L_f(A,E)$ by $L_r(A,E)$, where the accuracy of this approximation remains to be investigated. By Fréchet-differentiating Framework 5.1 we obtain an algorithm for simultaneously computing $f(A)$ and $L_f(A,E)$.

Framework 7.1. Framework for approximating $f(A)$ and $L_f(A,E)$.

(1) Choose a suitable rational approximation r and a transformation function g and set $A \leftarrow g(A)$.
(2) Transform $E \leftarrow L_g(A,E)$ (since $L_{f\circ g} = L_f(g(A), L_g(A,E))$ by the chain rule for Fréchet derivatives (Higham 2008, Theorem 3.4)).
(3) Compute $r(A)$ and $L_r(A,E)$ by some appropriate scheme.
(4) Apply transformations to $r(A)$ and $L_r(A,E)$ that undo the effect of the initial transformation on A.

The natural way to obtain $L_r(A,E)$ at step (3) is by differentiating the scheme for r, which can be done with the aid of the following lemma from Al-Mohy and Higham (2009*a*) if the numerator and denominator polynomials are explicitly computed.

Lemma 7.4. The Fréchet derivative L_{r_m} of the rational function $r_m(x) = p_m(x)/q_m(x)$ satisfies

$$q_m(A)L_{r_m}(A,E) = L_{p_m}(A,E) - L_{q_m}(A,E)r_m(A). \tag{7.13}$$

Proof. Applying the Fréchet derivative product rule (Higham 2008, Theorem 3.3) to $q_m r_m = p_m$ gives

$$L_{p_m}(A,E) = L_{q_m r_m}(A,E) = L_{q_m}(A,E)r_m(A) + q_m(A)L_{r_m}(A,E),$$

which rearranges to the result. □

It can be shown (Al-Mohy and Higham 2009*a*, Theorem 4.1) that, for polynomials p and a class of schemes for evaluating $p(A)$ that contains all schemes of practical interest, the cost of evaluating $p(A)$ and $L_p(A,E)$ together is at most three times the cost of evaluating $p(A)$ alone.

Framework 7.1 has been used in conjunction with Algorithm 5.2 by Al-Mohy and Higham (2009*a*) to develop a scaling and squaring algorithm that computes e^A and $L_{\exp}(A, E)$ at about three times the cost of computing e^A alone. It improves on an earlier 'Kronecker–Sylvester scaling and squaring algorithm' of Kenney and Laub (1998) that is significantly more expensive and uses complex arithmetic even when A is real.

The second use of differentiation is to differentiate a matrix iteration. For example, we can differentiate the Newton iteration (6.2) for the matrix sign function to obtain

$$Y_{k+1} = \frac{1}{2}(Y_k - X_k^{-1} Y_k X_k^{-1}), \qquad Y_0 = E, \tag{7.14}$$

where X_k is defined by (6.2). The following result of Al-Mohy and Higham (2009*a*) shows that under reasonable assumptions this procedure will always produce an iteration having the required derivative as a fixed point.

Theorem 7.5. Let f and g be $2n-1$ times continuously differentiable on an open subset $\mathcal{D}$ of $\mathbb{R}$ or $\mathbb{C}$. Suppose that for any matrix $X \in \mathbb{C}^{n\times n}$ whose spectrum lies in $\mathcal{D}$, g has the fixed point $f(X)$, that is, $f(X) = g(f(X))$. Then, for any such X, L_g at $f(X)$ has the fixed point $L_f(X, E)$ for all E.

Theorem 7.5 does not readily lead to a convergence result. If we derive (7.14) by instead applying the Newton iteration to $\left[\begin{smallmatrix} A & E \\ 0 & A \end{smallmatrix}\right]$ (as was done by Mathias (1996)) then from (7.10) it is easy to see that $\lim_{k\to\infty} Y_k = L_{\text{sign}}(A, E)$. Iteration (7.14) is due to Kenney and Laub (1991*a*).

7.5. *Finite differences*

A natural approach is to approximate the Fréchet derivative by the finite difference

$$L_f(A, E) \approx \frac{f(A + hE) - f(A)}{h}, \tag{7.15}$$

for a suitably chosen h. Two types of errors affect this approximation: truncation errors caused by taking a finite h, and rounding errors in floating point arithmetic caused by subtracting two nearly equal matrices that are both contaminated by error. A standard argument based on balancing bounds for the two types of error leads to the choice (Higham 2008, Section 3.4)

$$h = \left(\frac{u\|f(A)\|}{\|E\|^2}\right)^{1/2}, \tag{7.16}$$

for which the overall error has a bound of order $u^{1/2}\|f(A)\|^{1/2}\|E\|$. The conclusion is that subtractive cancellation in floating point arithmetic limits the smallest relative error that can be obtained to order $u^{1/2}$.

7.6. Complex step approximation

Assume that $f : \mathbb{R}^{n\times n} \to \mathbb{R}^{n\times n}$ and $A, E \in \mathbb{R}^{n\times n}$. Replacing E by $\mathrm{i}hE$ in (7.1), where $\mathrm{i} = \sqrt{-1}$, and using the linearity of L_f, we obtain

$$f(A + \mathrm{i}hE) - f(A) - \mathrm{i}hL_f(A, E) = o(h).$$

Thus

$$f(A) \approx \operatorname{Re} f(A + \mathrm{i}hE), \tag{7.17}$$

$$L_f(A, E) \approx \operatorname{Im} \frac{f(A + \mathrm{i}hE)}{h}, \tag{7.18}$$

so that with one function evaluation at a complex argument we can approximate both f and $L_f(A, E)$. The approximation (7.18) is known as the complex step approximation; it has been known for some time in the scalar case (Squire and Trapp 1998, Giles, Duta, Müller and Pierce 2003, Martins, Sturdza and Alonso 2003, Shampine 2007), and was proposed for matrices by Al-Mohy and Higham (2010). In the latter paper it is shown that for analytic f the error in the approximations (7.17) and (7.18) is $O(h^2)$ and that the same is true for the matrix sign function.

An important advantage of the complex step approximation over the finite difference approximation (7.15) is that h is not restricted by floating point arithmetic considerations. Indeed practical experience reported in the papers cited above has demonstrated the ability of the approximation to produce accurate approximations in the scalar case even with h as small as 10^{-100}, which is the value used in software at the National Physical Laboratory according to Cox and Harris (2004). Al-Mohy and Higham (2010) show experimentally that when used in conjunction with condition estimation (see Section 7.1) the complex step approximation leads to significantly more reliable condition estimates than the finite difference approximation.

One caveat is that the underlying method for evaluating f must not employ complex arithmetic, which rules out methods based on the (complex) Schur form. The reason is that since the Fréchet derivative is assumed real, if the evaluation introduces a non-trivial imaginary part at any point then that term must be subject to massive subtractive cancellation in order for a final imaginary part of $O(h)$ to be produced. Looked at another way, the complex step approximation is essentially carrying out a form of automatic differentiation with h acting as a symbolic variable, and the introduction of pure imaginary numbers within the evaluation disturbs this process.

8. The $f(A)b$ problem

In many applications, especially those originating from partial differential equations and those with a large, sparse matrix A, it is the action of $f(A)$ on a vector, $f(A)b$, that is required and not $f(A)$. Of course, this is analogous to the requirement to solve a linear $Ax = b$ without computing A^{-1} (though the inverse function is distinguished from the other functions considered in this paper in that we rarely need to compute it explicitly). We discuss two general approaches.

8.1. *Quadrature and rational approximation*

If we have an integral representation $f(A) = \int r(A,t)\,\mathrm{d}t$, where r is a rational function of A, then we can apply quadrature and approximate $f(A)b$ by $\sum_i r(A,t_i)b$. Depending on how r is represented, the evaluation of this approximation reduces to solving one or more linear systems with coefficient matrices that are polynomials in A. This, of course, is equivalent to approximating $f(A)$ by the rational function $\sum_i r(A,t_i)$. Two notable examples of integral representations are (see, *e.g.*, Higham (2008))

$$\log(A) = \int_0^1 (A-I)\big[t(A-I)+I\big]^{-1}\,\mathrm{d}t, \tag{8.1}$$

$$\operatorname{sign}(A) = \frac{2}{\pi}A\int_0^\infty (t^2I + A^2)^{-1}\,\mathrm{d}t, \tag{8.2}$$

for which appropriate Gaussian quadrature rules lead to Padé approximants (for the details for (8.1), see Dieci *et al.* (1996, Theorem 4.3)).

More generally, for analytic functions f we can employ the Cauchy integral formula

$$f(A) = \frac{1}{2\pi\mathrm{i}}\int_\Gamma f(z)\,(zI-A)^{-1}\,\mathrm{d}z, \tag{8.3}$$

where Γ is a closed contour that lies in the region of analyticity of f and winds once around the spectrum in the anticlockwise direction. This formula is equivalent to the definitions given in Section 2.1 (Horn and Johnson 1991, Theorem 6.2.28). From (8.3) we have

$$f(A)b = \frac{1}{2\pi\mathrm{i}}\int_\Gamma f(z)\,(zI-A)^{-1}b\,\mathrm{d}z, \tag{8.4}$$

and so quadrature will reduce to solving linear systems with $zI - A$. Any method based on (8.4) will need to be specialized to particular classes of f and matrices A, since the selection of the contour Γ will be crucial to the efficiency and reliability of the method. Davies and Higham (2005) show that simply taking Γ to be a circle enclosing the spectrum is not generally a good choice. Hale, Higham and Trefethen (2008) develop methods for functions such as the square root and the logarithm with singularities in $(-\infty, 0]$ and for A with eigenvalues on or near the positive real axis. Their key idea is to use conformal mappings to transform the integral into one for which the repeated trapezium rule converges very quickly. We mention just one simple idea used therein, and will not illustrate the conformal mappings that are the most important part of the technique. For the square root, we can rewrite the problem as $A \cdot A^{-1}f(A)$ and change variables to $w = z^{1/2}$, so that (8.3) becomes

$$A^{1/2} = \frac{A}{2\pi \mathrm{i}} \int_{\Gamma_z} z^{-1/2} (zI - A)^{-1} \, \mathrm{d}z = \frac{A}{\pi \mathrm{i}} \int_{\Gamma_w} (w^2 I - A)^{-1} \, \mathrm{d}w. \tag{8.5}$$

The transformed integrand is analytic at the origin and hence easier to handle. A natural question is what rational approximation these methods produce. For f the square root and A with positive real eigenvalues, method 3 in Hale *et al.* (2008) produces a certain best rational approximation discovered by Zolotarev in 1877. The methods in Hale *et al.* (2008) are not restricted to Hermitian matrices but they do require estimates of the spectrum of A.

8.2. Krylov methods

The most studied methods for the $f(A)b$ problem are those based on Krylov subspaces. By computing a sequence of matrix–vector products with A they aim to reduce the problem to one of the same form but with a much smaller matrix. This is done by projecting the problem onto a Krylov subspace

$$\mathcal{K}_k(A, b) = \operatorname{span}\{b, Ab, \dots, A^{k-1}b\}.$$

If the Arnoldi process (Saad 2003, Section 6.3) with matrix A and starting vector $q_1 = b/\|b\|_2$ completes k steps then we have

$$AQ_k = Q_k H_k + h_{k+1,k} q_{k+1} e_k^T, \tag{8.6}$$

where $Q_k = [q_1, \dots, q_k]$ has orthonormal columns and $H_k = (h_{ij})$ is $k \times k$ upper Hessenberg. The columns of Q_k form an orthonormal basis for the Krylov subspace $\mathcal{K}_k(A, q_1)$. We can then approximate $f(A)b$ by

$$\begin{aligned} f_k &:= \|b\|_2 Q_k f(H_k) e_1, \\ &= Q_k f(H_k) Q_k^* b. \end{aligned} \tag{8.7}$$

The evaluation of f is carried out on the $k \times k$ matrix H_k, where $k \ll n$ in practice, and can be done by any available method. Effectively, we are evaluating f on the smaller Krylov subspace $\mathcal{K}_k(A, q_1)$ and then expanding the result back onto the original space $\mathbb{C}^n$. This procedure can be viewed as a form of model order reduction (Antoulas 2005, Frommer and Simoncini 2008*a*).

Few convergence results or error bounds are available for (8.7). However, two results of Saad (1992) provide fundamental insight into this approximation.

Lemma 8.1. Let $A \in \mathbb{C}^{n \times n}$ and Q_k, H_k be the result of k steps of the Arnoldi process on A with starting vector q_1. Then, for any polynomial p_j

of degree $j \le k-1$ we have

$$p_j(A)q_1 = Q_k p_j(H_k)e_1.$$

Theorem 8.2. Let Q_k, H_k be the result of k steps of the Arnoldi process on $A \in \mathbb{C}^{n\times n}$ with starting vector $q_1 = b/\|b\|_2$. Then

$$\|b\|_2 Q_k f(H_k)e_1 = \widetilde{p}_{k-1}(A)b,$$

where $\widetilde{p}_{k-1}$ is the unique polynomial of degree at most $k-1$ that interpolates f on the spectrum of H_k (that is, in the sense of (2.4)).

The lemma says that the approximation (8.7) is exact if f is a sufficiently low-degree polynomial. The theorem shows that the approximation (8.7) is an exact approximation not for f but for a Hermite interpolating polynomial based on the spectrum of H_k.

The use of Krylov methods for the $f(A)b$ problem is an active area of research. We will not try to give a summary, but instead point out a few very recent contributions, namely Afanasjew, Eiermann, Ernst and Güttel (2008), Grimm and Hochbruck (2008), Frommer and Simoncini (2008*b*), and Popolizio and Simoncini (2008).

9. The software scene

In this final section we give an outline of available software for computing matrix functions.

9.1. MATLAB

MATLAB has a number of functions for computing $f(A)$. Function `funm` implements the Schur–Parlett algorithm (Algorithm 4.5), and so is applicable to general functions having a Taylor series with an infinite radius of convergence. When invoked for the exponential it evaluates the exponential of any 2×2 diagonal blocks T_{ii} using an explicit formula (Higham 2008, Section 10.4.3) that, unlike the general formula (4.2), avoids cancellation in floating point arithmetic. Function `sqrtm` implements the Schur method for the matrix square root, Algorithm 4.6. Function `expm` implements Algorithm 5.2, the scaling and squaring algorithm. Function `logm` implements a specialized version of the Schur–Parlett algorithm in which $\log(T_{ii})$ is evaluated by an explicit formula (Higham 2008, Section 11.6.2) if T_{ii} is 2×2 or by the inverse scaling and squaring algorithm if T_{ii} has larger dimension.

The Symbolic Math Toolbox has two relevant functions, which are contained in the MuPAD engine (in MATLAB R2008b the default engine was changed from Maple to MuPAD). The function `numeric::expMatrix` (The MathWorks 2009*a*) can use hardware floating point arithmetic or variable precision software floating point arithmetic to compute e^A or $e^A b$. By default a Taylor series is used, apparently without scaling and squaring. Other options are diagonalization for diagonalizable matrices, interpolation (the form of interpolating polynomial is not specified), and a Krylov method for $e^A b$ only. The function `numeric::fMatrix` (The MathWorks 2009*b*) computes $f(A)$ for a general function f but requires that A is diagonalizable.

9.2. *Octave*

GNU Octave (Octave 2009) is a free 'MATLAB-like' system. It includes a function `expm` that implements Ward's (1977) version of the scaling and squaring method (which uses fixed Padé degree $m = 8$, with scaling so that $\|2^{-s}A\|_1 \le 1$), as well as a function `thfm` for evaluating trigonometric and hyperbolic functions. The latter function expresses a variety of trigonometric and hyperbolic functions in terms of the exponential, and inverse trigonometric and inverse hyperbolic functions in terms of the logarithm and square root. For example, it evaluates $\cos(A) = (e^{iA} + e^{-iA})/2$ (or as $\operatorname{Re} e^{iA}$ when A is real) and $\arctan(A) = -(i/2)\log((I + iA)(I - iA)^{-1})$. Some of these formulas are of uncertain numerical reliability and need careful numerical stability analysis before they can be recommended for use; *cf.* the analysis for the scalar case in Bradford, Corless, Davenport, Jeffrey and Watt (2002) and Kahan (1987).

9.3. *Other software*

Algorithm 5.2 is used by the MatrixExp function of Mathematica for matrices of machine numbers and by the NAG Library routine `F01ECF` (from Mark 22).

The Matrix Function Toolbox (Higham) contains over 40 MATLAB functions implementing many of the algorithms described in Higham (2008).

Sidje (1998) provides a package called Expokit containing MATLAB and Fortran codes for computing e^A and $e^A b$.

Koikari (2009) gives Fortran 95 code for computing the ψ functions by scaling and squaring or by a block Schur–Parlett algorithm, and the EXPINT package of Berland, Skaflestad and Wright (2007) also contains a function based on scaling and squaring for evaluating the ψ functions. These functions are defined by $\psi_k(z) = \sum_{j=0}^{\infty} z^j/(j+k)!$, $k = 0, 1, 2, \ldots$ and play an important role within exponential integrators (Hochbruck and Ostermann 2010).

Acknowledgements

We thank Lijing Lin for her helpful comments on a draft manuscript. This work was supported in part by EPSRC grant EP/E050441/1 (CICADA: Centre for Interdisciplinary Computational and Dynamical Analysis).

Appendix: Cost of Padé versus Taylor approximants within the scaling and squaring method

In this appendix we compare the efficiency of diagonal Padé approximants and Taylor approximants within the scaling and squaring method for the matrix exponential, based on the use of refined backward error bounds in both cases.

For $A \in \mathbb{C}^{n\times n}$ we use the Paterson–Stockmeyer scheme (see Paterson and Stockmeyer (1973), Higham (2008, Section 4.2)) in order to evaluate $T_m(A) = \sum_{k=0}^{m} A^k/k!$ as

$$T_m(A) = \sum_{k=0}^{\ell} g_k(A)(A^\tau)^k, \quad \ell = \lfloor m/\tau \rfloor, \tag{A.1}$$

where $1 \le \tau \le m$ is an integer and

$$g_k(A) = \begin{cases} \sum_{i=1}^{\tau} A^{\tau-i}/(\tau k + \tau - i)!, & k = 0 : \ell - 1, \\ \sum_{i=\ell\tau}^{m} A^{i-\ell\tau}/i!, & k = \ell. \end{cases}$$

Horner's rule is used on (A.1). This scheme evaluates $T_m(A)$ with a number of matrix multiplications equal to

$$\widetilde{\pi}_m = \ell + \tau - 1 - \phi(m,\tau), \qquad \phi(m,\tau) = \begin{cases} 1, & \text{if } \tau \mid m, \\ 0, & \text{otherwise.} \end{cases} \tag{A.2}$$

The choice $\tau = \sqrt{m}$ approximately minimizes this quantity (Higham 2008, Section 4.2), so we take τ either $\lfloor \sqrt{m} \rfloor$ or $\lceil \sqrt{m}\, \rceil$ since both yield the same operation count (Hargreaves 2005, Theorem 1.7.4).

Lemma 5.1 is applicable with trivial modifications to any rational approximation to e^x, not just diagonal Padé approximants, so we can replace r_m therein by T_m. Thus, with $h_m(x) = \log(\mathrm{e}^{-x}T_m(x)) = \sum_{k=m+1}^{\infty} c_k x^k$ in Lemma 5.1, we calculate the parameters

$$\widetilde{\theta}_m = \max\{\, \theta : \widetilde{h}_m(\theta)/\theta \le u = 2^{-53}\,\}, \tag{A.3}$$

where $\widetilde{h}_m(x) = \sum_{k=m+1}^{\infty} |c_k| x^k$, using the techniques described just after Lemma 5.1. Then we know that $T_m(2^{-s}A)^{2^s}$ has backward error at most $u = 2^{-53}$ for $\|2^{-s}A\| \le \widetilde{\theta}_m$. We select s as the smallest non-negative integer such that $2^{-s}\|A\| \le \widetilde{\theta}_m$, which is given by $s = \max(\lceil \log_2(\|A\|/\widetilde{\theta}_m)\rceil, 0)$. Then the number of matrix multiplications required to evaluate $T_m(2^{-s}A)^{2^s}$ is

$$\underbrace{\lfloor m/\lceil \sqrt{m}\,\rceil \rfloor + \lceil \sqrt{m}\,\rceil - 1 - \phi(m, \lceil \sqrt{m}\,\rceil)}_{\widetilde{\pi}_m} + \max(\lceil \log_2(\|A\|/\widetilde{\theta}_m)\rceil, 0). \tag{A.4}$$

When $s > 0$ we are interested in the m that minimizes the cost. To obtain

Table A.1. The number of matrix products $\widetilde{\pi}_m$ in (A.2) needed for the Paterson–Stockmeyer scheme, $\widetilde{\theta}_m$ defined by (A.3), and C_m from (A.5).

m	$\widetilde{\theta}_m$	$\widetilde{\pi}_m$	C_m	m	$\widetilde{\theta}_m$	$\widetilde{\pi}_m$	C_m	m	$\widetilde{\theta}_m$	$\widetilde{\pi}_m$	C_m
1	2.29e-16	0	53.00	11	2.14e-1	5	8.22	21	1.62	8	8.30
2	2.58e-8	1	27.21	12	3.00e-1	5	7.74	22	1.82	8	8.14
3	1.39e-5	2	19.14	13	4.00e-1	6	8.32	23	2.01	8	7.99
4	3.40e-4	2	14.52	14	5.14e-1	6	7.96	24	2.22	8	7.85
5	2.40e-3	3	12.70	15	6.41e-1	6	7.64	25	2.43	8	7.72
6	9.07e-3	3	10.79	16	7.81e-1	6	7.36	26	2.64	9	8.60
7	2.38e-2	4	10.39	17	9.31e-1	7	8.10	27	2.86	9	8.48
8	5.00e-2	4	9.32	18	1.09	7	7.87	28	3.08	9	8.38
9	8.96e-2	4	8.48	19	1.26	7	7.67	29	3.31	9	8.27
10	1.44e-1	5	8.79	20	1.44	7	7.48	30	3.54	9	8.18

a suitable measure of the cost we ignore the constant terms in (A.4) (since they are common to each m) and consider

$$C_m = \lfloor m/\lceil\sqrt{m}\,\rceil\rfloor + \lceil\sqrt{m}\,\rceil - \phi(m,\lceil\sqrt{m}\,\rceil) - \log_2(\widetilde{\theta}_m). \tag{A.5}$$

We tabulate $\widetilde{\theta}_m$, $\widetilde{\pi}_m$, and C_m, for $m = 1\colon 30$, in Table A.1, and find that $m = 16$ is the global minimizer of C_m, which suggests using $T_{16}(2^{-s}A)^{2^s}$ to approximate e^A when $\|A\| \geq \widetilde{\theta}_{16} \approx 0.78$. Corresponding analysis was done for Padé approximants by Higham (2005) and we use the number of matrix multiplications π_m from Higham (2005, Table 2.2), as well as the θ_i in Table 5.1.

Now we compare the cost of Taylor and Padé approximants. Assume first that $\|A\| \geq \theta_{13} \approx 5.4$. Computing $T_{16}(2^{-s}A)$ requires six matrix multiplications, and so the overall cost from (A.4) of approximating e^A is $c_T := 6 + s$, while Algorithm 5.2, which chooses a non-negative integer t so that $\frac{1}{2}\theta_{13} < \|2^{-t}A\| \leq \theta_{13}$, computes $r_{13}(2^{-t}A)^{2^t}$ with cost $c_P := 6+4/3+t$, where the term $4/3$ accounts for the solution of the multiple right-hand side linear system for the Padé approximant. Since $\frac{1}{2}\theta_{13} < 4\widetilde{\theta}_{16}$, there are two cases to consider. First, when $\|2^{-t}A\| \in (4\widetilde{\theta}_{16}, \theta_{13}]$ we have $2^{-t-3}\|A\| \leq \frac{1}{8}\theta_{13} < \widetilde{\theta}_{16}$ and hence $s = t+3$. Therefore, $1 < c_T/c_P = (9+t)/(7\frac{1}{3}+t) \leq 27/22$. Secondly, when $\|2^{-t}A\| \in (\frac{1}{2}\theta_{13}, 4\widetilde{\theta}_{16}]$ we have $2^{-t-2}\|A\| \leq \widetilde{\theta}_{16}$ and hence $s = t+2$. Therefore, $1 < c_T/c_P = (8+t)/(7\frac{1}{3}+t) \leq 12/11$.

A remaining question is whether when $\|A\| < \theta_{13}$ a Taylor series can be more efficient than a Padé approximant. The answer can be seen from Figure A.1, where '○' indicates the points $(\widetilde{\theta}_m, \widetilde{\pi}_m)$, $m = 4, 6, 9, 12, 16, 20, 25, 30$, and '□' indicates the points $(\theta_m, \pi_m + 4/3)$, $m = 3, 5, 7, 9, 13$. Notice that the dotted curve, which represents the cost of Taylor series, lies below the

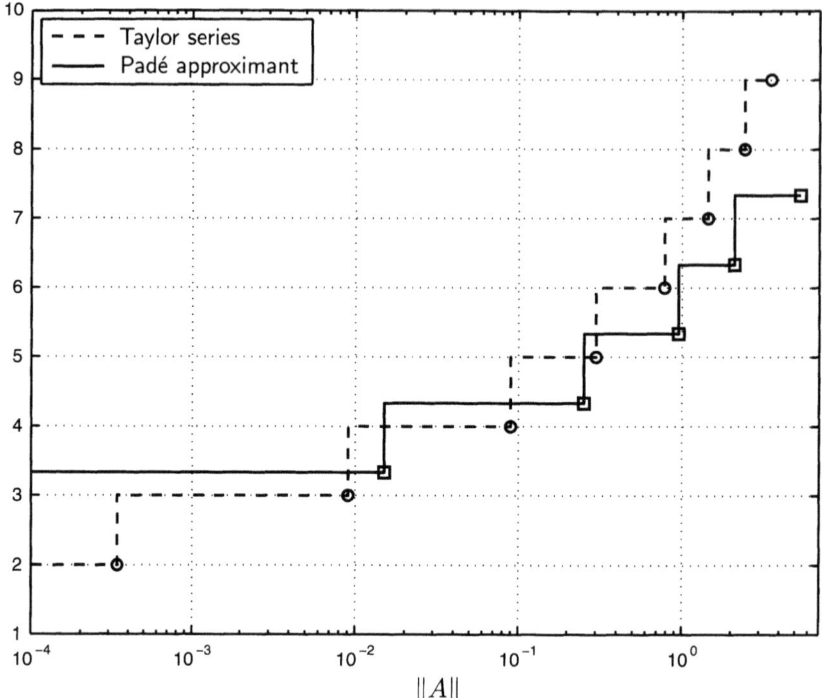

Figure A.1. $\|A\|$ versus cost in equivalent matrix multiplications of evaluating Taylor and Padé approximants to e^A in double precision.

solid curve in three intervals: $[0, \widetilde{\theta}_6]$, $(\theta_3, \widetilde{\theta}_9]$, and $(\theta_5, \widetilde{\theta}_{12}]$. Therefore, it is more efficient to use $T_m(A)$ rather Algorithm 5.2 if $\|A\|$ lies in any of these intervals.

We conclude that any algorithm based on Taylor series will cost up to 23% more than the Padé approximant-based Algorithm 5.2 and cannot have a lower cost for $\|A\| > \widetilde{\theta}_{12}$. Moreover the Taylor series requires a larger amount of scaling (since we are scaling to reduce $\|A\|$ below 0.78 instead of 5.4), which is undesirable from the point of view of possible numerical instability in the squaring phase.

We repeated the analysis for single precision: $u = 2^{-24} \approx 6.0 \times 10^{-8}$. For Padé approximants the optimal degree is now $m = 7$ with $\theta_7 \approx 3.9$ (Higham 2005), while for Taylor series it is $m = 9$ with $\widetilde{\theta}_9 \approx 0.78$. The conclusion is similar to that for double precision arithmetic: Padé approximation is more efficient than the Taylor series for $\|A\| > \widetilde{\theta}_9$ (up to 31% more efficient), and only for certain intervals of smaller $\|A\|$ is the Taylor series the more efficient.

In summary, Padé approximants are preferable to truncated Taylor series within the scaling and squaring method in both single and double precision, due to their greater efficiency and the lesser amount of scaling that they require.

REFERENCES

M. Afanasjew, M. Eiermann, O. G. Ernst and S. Güttel (2008), 'Implementation of a restarted Krylov subspace method for the evaluation of matrix functions', *Linear Algebra Appl.* **429**, 2293–2314.

A. H. Al-Mohy and N. J. Higham (2009*a*), 'Computing the Fréchet derivative of the matrix exponential, with an application to condition number estimation', *SIAM J. Matrix Anal. Appl.* **30**, 1639–1657.

A. H. Al-Mohy and N. J. Higham (2009*b*), 'A new scaling and squaring algorithm for the matrix exponential', *SIAM J. Matrix Anal. Appl.* **31**, 970–989.

A. H. Al-Mohy and N. J. Higham (2010), 'The complex step approximation to the Fréchet derivative of a matrix function', *Numer. Algorithms* **53**, 133–148.

A. C. Antoulas (2005), *Approximation of Large-Scale Dynamical Systems*, SIAM, Philadelphia, PA, USA.

M. Arioli and D. Loghin (2009), 'Discrete interpolation norms with applications', *SIAM J. Numer. Anal.* **47**, 2924–2951.

G. A. Baker, Jr. and P. Graves-Morris (1996), *Padé Approximants*, Vol. 59 of *Encyclopedia of Mathematics and its Applications*, second edn, Cambridge University Press, Cambridge, UK.

A. Y. Barraud (1979), 'Investigations autour de la fonction signe d'une matrice application a l'équation de Riccati', *RAIRO Automatique/Systems Analysis and Control* **13**, 335–368.

C. A. Bavely and G. W. Stewart (1979), 'An algorithm for computing reducing subspaces by block diagonalization', *SIAM J. Numer. Anal.* **16**, 359–367.

H. Berland, B. Skaflestad and W. Wright (2007), 'EXPINT: A MATLAB package for exponential integrators', *ACM Trans. Math. Software* **33**, #4.

D. A. Bini, N. J. Higham and B. Meini (2005), 'Algorithms for the matrix pth root', *Numer. Algorithms* **39**, 349–378.

Å. Björck and S. Hammarling (1983), 'A Schur method for the square root of a matrix', *Linear Algebra Appl.* **52/53**, 127–140.

R. J. Bradford, R. M. Corless, J. H. Davenport, D. J. Jeffrey and S. M. Watt (2002), 'Reasoning about the elementary functions of complex analysis', *Annals of Mathematics and Artificial Intelligence* **36**, 303–318.

C. Brezinski and J. Van Iseghem (1995), A taste of Padé approximation. In *Acta Numerica*, Vol. 4, Cambridge University Press, pp. 53–103.

A. Cayley (1858), 'A memoir on the theory of matrices', *Philos. Trans. Roy. Soc. London* **148**, 17–37.

T. Charitos, P. R. de Waal and L. C. van der Gaag (2008), 'Computing short-interval transition matrices of a discrete-time Markov chain from partially observed data', *Statistics in Medicine* **27**, 905–921.

S. H. Cheng, N. J. Higham, C. S. Kenney and A. J. Laub (2000), Return to the middle ages: A half-angle iteration for the logarithm of a unitary matrix. In *Proc. 14th International Symposium of Mathematical Theory of Networks and Systems, Perpignan, France.* CD ROM.

S. H. Cheng, N. J. Higham, C. S. Kenney and A. J. Laub (2001), 'Approximating the logarithm of a matrix to specified accuracy', *SIAM J. Matrix Anal. Appl.* **22**, 1112–1125.

A. R. Collar (1978), 'The first fifty years of aeroelasticity', *Aerospace* (*Royal Aeronautical Society Journal*) **5**, 12–20.

M. G. Cox and P. M. Harris (2004), Numerical analysis for algorithm design in metrology, Software Support for Metrology Best Practice Guide No. 11, National Physical Laboratory, Teddington, UK.

J. J. Crofts and D. J. Higham (2009), 'A weighted communicability measure applied to complex brain networks', *J. Roy. Soc. Interface* **6**, 411–414.

P. I. Davies and N. J. Higham (2003), 'A Schur–Parlett algorithm for computing matrix functions', *SIAM J. Matrix Anal. Appl.* **25**, 464–485.

P. I. Davies and N. J. Higham (2005), Computing $f(A)b$ for matrix functions f. In *QCD and Numerical Analysis III* (A. Boriçi, A. Frommer, B. Joó, A. Kennedy and B. Pendleton, eds), Vol. 47 of *Lecture Notes in Computational Science and Engineering*, Springer, Berlin, pp. 15–24.

C. Davis (1973), 'Explicit functional calculus', *Linear Algebra Appl.* **6**, 193–199.

E. D. Denman and A. N. Beavers, Jr. (1976), 'The matrix sign function and computations in systems', *Appl. Math. Comput.* **2**, 63–94.

J. Descloux (1963), 'Bounds for the spectral norm of functions of matrices', *Numer. Math.* **15**, 185–190.

L. Dieci and A. Papini (2000), 'Padé approximation for the exponential of a block triangular matrix', *Linear Algebra Appl.* **308**, 183–202.

L. Dieci, B. Morini and A. Papini (1996), 'Computational techniques for real logarithms of matrices', *SIAM J. Matrix Anal. Appl.* **17**, 570–593.

L. Elsner (1970), 'Iterative Verfahren zur Lösung der Matrizengleichung $X^2 - A = 0$', *Buletinul Institutului Politehnic din Iasi* **xvi**, 15–24.

E. Estrada and N. Hatano (2008), 'Communicability in complex networks', *Phys. Review E* **77**, 036111.

E. Estrada and D. J. Higham (2008), Network properties revealed through matrix functions. Mathematics Research Report 17, University of Strathclyde, Scotland, UK. To appear in *SIAM Rev.*

E. Estrada and J. A. Rodríguez-Velázquez (2005*a*), 'Spectral measures of bipartivity in complex networks', *Phys. Review E* **72**, 046105.

E. Estrada and J. A. Rodríguez-Velázquez (2005*b*), 'Subgraph centrality in complex networks', *Phys. Review E* **71**, 056103.

E. Estrada, D. J. Higham and N. Hatano (2009), 'Communicability betweenness in complex networks', *Physica A* **388**, 764–774.

S. Fiori (2008), 'Leap-frog-type learning algorithms over the Lie group of unitary matrices', *Neurocomputing* **71**, 2224–2244.

R. A. Frazer, W. J. Duncan and A. R. Collar (1938), *Elementary Matrices and Some Applications to Dynamics and Differential Equations*, Cambridge University Press, Cambridge, UK. 1963 printing.

A. Frommer and V. Simoncini (2008*a*), Matrix functions. In *Model Order Reduction: Theory, Research Aspects and Applications* (W. H. A. Schilders, H. A. van der Vorst and J. Rommes, eds), Springer, Berlin, pp. 275–303.

A. Frommer and V. Simoncini (2008*b*), 'Stopping criteria for rational matrix functions of Hermitian and symmetric matrices', *SIAM J. Sci. Comput.* **30**, 1387–1412.

F. R. Gantmacher (1959), *The Theory of Matrices*, Vol. one, Chelsea, New York.

M. B. Giles, M. C. Duta, J.-D. Müller and N. A. Pierce (2003), 'Algorithm developments for discrete adjoint methods', *AIAA Journal* **4**, 198–205.

H. H. Goldstine (1977), *A History of Numerical Analysis from the 16th through the 19th Century*, Springer, New York.

G. H. Golub and C. F. Van Loan (1996), *Matrix Computations*, third edn, Johns Hopkins University Press, Baltimore, MD, USA.

F. Greco and B. Iannazzo (2010), 'A binary powering algorithm for computing primary matrix roots', *Numer. Algorithms*. To appear.

V. Grimm and M. Hochbruck (2008), 'Rational approximation to trigonometric operator', *BIT* **48**, 215–229.

C.-H. Guo (2009), 'On Newton's method and Halley's method for the principal pth root of a matrix', *Linear Algebra Appl.* **432**, 1905–1922.

C.-H. Guo and N. J. Higham (2006), 'A Schur–Newton method for the matrix pth root and its inverse', *SIAM J. Matrix Anal. Appl.* **28**, 788–804.

N. Hale, N. J. Higham and L. N. Trefethen (2008), 'Computing A^α, $\log(A)$ and related matrix functions by contour integrals', *SIAM J. Numer. Anal.* **46**, 2505–2523.

G. Hargreaves (2005), Topics in matrix computations: Stability and efficiency of algorithms. PhD thesis, University of Manchester, Manchester, England.

G. I. Hargreaves and N. J. Higham (2005), 'Efficient algorithms for the matrix cosine and sine', *Numer. Algorithms* **40**, 383–400.

N. J. Higham (1986*a*), 'Computing the polar decomposition: With applications', *SIAM J. Sci. Statist. Comput.* **7**, 1160–1174.

N. J. Higham (1986*b*), 'Newton's method for the matrix square root', *Math. Comp.* **46**, 537–549.

N. J. Higham (1987), 'Computing real square roots of a real matrix', *Linear Algebra Appl.* **88/89**, 405–430.

N. J. Higham (1994), 'The matrix sign decomposition and its relation to the polar decomposition', *Linear Algebra Appl.* **212/213**, 3–20.

N. J. Higham (1997), 'Stable iterations for the matrix square root', *Numer. Algorithms* **15**, 227–242.

N. J. Higham (2001), 'Evaluating Padé approximants of the matrix logarithm', *SIAM J. Matrix Anal. Appl.* **22**, 1126–1135.

N. J. Higham (2002), *Accuracy and Stability of Numerical Algorithms*, second edn, SIAM, Philadelphia, PA, USA.

N. J. Higham (2005), 'The scaling and squaring method for the matrix exponential revisited', *SIAM J. Matrix Anal. Appl.* **26**, 1179–1193.

N. J. Higham (2008), *Functions of Matrices: Theory and Computation*, SIAM, Philadelphia, PA, USA.

N. J. Higham (2009), 'The scaling and squaring method for the matrix exponential revisited', *SIAM Rev.* **51**, 747–764.

N. J. Higham, 'The Matrix Function Toolbox'. http://www.ma.man.ac.uk/~higham/mftoolbox.

N. J. Higham and L. Lin (2009), On pth roots of stochastic matrices. MIMS EPrint 2009.21, Manchester Institute for Mathematical Sciences, The University of Manchester, UK.

N. J. Higham and M. I. Smith (2003), 'Computing the matrix cosine', *Numer. Algorithms* **34**, 13–26.

N. J. Higham and F. Tisseur (2000), 'A block algorithm for matrix 1-norm estimation, with an application to 1-norm pseudospectra', *SIAM J. Matrix Anal. Appl.* **21**, 1185–1201.

N. J. Higham, D. S. Mackey, N. Mackey and F. Tisseur (2005), 'Functions preserving matrix groups and iterations for the matrix square root', *SIAM J. Matrix Anal. Appl.* **26**, 849–877.

M. Hochbruck and A. Ostermann (2010), Exponential integrators. In *Acta Numerica*, Vol. 19, Cambridge University Press, pp. 209–286.

R. A. Horn and C. R. Johnson (1991), *Topics in Matrix Analysis*, Cambridge University Press, Cambridge, UK.

B. Iannazzo (2006), 'On the Newton method for the matrix Pth root', *SIAM J. Matrix Anal. Appl.* **28**, 503–523.

M. Ilić, I. W. Turner and D. P. Simpson (2009), 'A restarted Lanczos approximation to functions of a symmetric matrix', *IMA J. Numer. Anal.* Advance Access published on June 17, 2009. doi:10.1093/imanum/drp003.

R. B. Israel, J. S. Rosenthal and J. Z. Wei (2001), 'Finding generators for Markov chains via empirical transition matrices, with applications to credit ratings', *Mathematical Finance* **11**, 245–265.

R. A. Jarrow, D. Lando and S. M. Turnbull (1997), 'A Markov model for the term structure of credit risk spreads', *Rev. Financial Stud.* **10**, 481–523.

W. Kahan (1987), Branch cuts for complex elementary functions, or Much Ado About Nothing's sign bit. In *The State of the Art in Numerical Analysis* (A. Iserles and M. J. D. Powell, eds), Oxford University Press, pp. 165–211.

C. S. Kenney and A. J. Laub (1989*a*), 'Condition estimates for matrix functions', *SIAM J. Matrix Anal. Appl.* **10**, 191–209.

C. S. Kenney and A. J. Laub (1989*b*), 'Padé error estimates for the logarithm of a matrix', *Internat. J. Control* **50**, 707–730.

C. S. Kenney and A. J. Laub (1991*a*), 'Polar decomposition and matrix sign function condition estimates', *SIAM J. Sci. Statist. Comput.* **12**, 488–504.

C. S. Kenney and A. J. Laub (1991*b*), 'Rational iterative methods for the matrix sign function', *SIAM J. Matrix Anal. Appl.* **12**, 273–291.

C. S. Kenney and A. J. Laub (1998), 'A Schur–Fréchet algorithm for computing the logarithm and exponential of a matrix', *SIAM J. Matrix Anal. Appl.* **19**, 640–663.

S. Koikari (2009), 'Algorithm 894: On a block Schur–Parlett algorithm for φ-functions based on the sep-inverse estimate', *ACM Trans. Math. Software* **36**, #12.

A. Kreinin and M. Sidelnikova (2001), 'Regularization algorithms for transition matrices', *Algo Research Quarterly* **4**, 23–40.

P. Laasonen (1958), 'On the iterative solution of the matrix equation $AX^2 - I = 0$', *Math. Tables Aids Comp.* **12**, 109–116.

P. Lancaster and M. Tismenetsky (1985), *The Theory of Matrices*, second edn, Academic Press, London.

B. Laszkiewicz and K. Ziętak (2009), 'A Padé family of iterations for the matrix sector function and the matrix pth root', *Numer. Linear Algebra Appl.* **16**, 951–970.

P.-F. Lavallée, A. Malyshev and M. Sadkane (1997), Spectral portrait of matrices by block diagonalization. In *Numerical Analysis and its Applications* (L. Vulkov, J. Waśniewski and P. Yalamov, eds), Vol. 1196 of *Lecture Notes in Computer Science*, Springer, Berlin, pp. 266–273.

J. D. Lawson (1967), 'Generalized Runge-Kutta processes for stable systems with large Lipschitz constants', *SIAM J. Numer. Anal.* **4**, 372–380.

J. R. R. A. Martins, P. Sturdza and J. J. Alonso (2003), 'The complex-step derivative approximation', *ACM Trans. Math. Software* **29**, 245–262.

R. Mathias (1996), 'A chain rule for matrix functions and applications', *SIAM J. Matrix Anal. Appl.* **17**, 610–620.

The MathWorks (2009*a*), 'numeric::expMatrix: The exponential of a matrix'. http://www.mathworks.com/access/helpdesk/help/toolbox/mupad/numeric/expMatr.html, retrieved on September 29, 2009.

The MathWorks (2009*b*), 'numeric::fMatrix: Functional calculus for numerical square matrices'. http://www.mathworks.com/access/helpdesk/help/toolbox/mupad/numeric/fMatrix.html, retrieved on September 29, 2009.

C. B. Moler and C. F. Van Loan (1978), 'Nineteen dubious ways to compute the exponential of a matrix', *SIAM Rev.* **20**, 801–836.

C. B. Moler and C. F. Van Loan (2003), 'Nineteen dubious ways to compute the exponential of a matrix, twenty-five years later', *SIAM Rev.* **45**, 3–49.

L. Morai and A. F. Pacheco (2003), 'Algebraic approach to the radioactive decay equations', *Amer. J. Phys.* **71**, 684–686.

Octave Version 3.2.2 (2009). http://www.octave.org.

B. N. Parlett (1976), 'A recurrence among the elements of functions of triangular matrices', *Linear Algebra Appl.* **14**, 117–121.

B. N. Parlett and K. C. Ng (1985), Development of an accurate algorithm for $\exp(Bt)$. Technical Report PAM-294, Center for Pure and Applied Mathematics, University of California, Berkeley. Fortran program listings are given in an appendix with the same report number printed separately.

M. S. Paterson and L. J. Stockmeyer (1973), 'On the number of nonscalar multiplications necessary to evaluate polynomials', *SIAM J. Comput.* **2**, 60–66.

H.-O. Peitgen, H. Jürgens and D. Saupe (1992), *Fractals for the Classroom, Part Two: Complex Systems and Mandelbrot Set*, Springer, New York.

G. M. Phillips (2000), *Two Millennia of Mathematics: From Archimedes to Gauss*, Springer, New York.

M. Popolizio and V. Simoncini (2008), 'Acceleration techniques for approximating the matrix exponential operator', *SIAM J. Matrix Anal. Appl.* **30**, 657–683.

P. J. Psarrakos (2002), 'On the mth roots of a complex matrix', *Electron. J. Linear Algebra* **9**, 32–41.

P. Pulay (1966), 'An iterative method for the determination of the square root of a positive definite matrix', *Z. Angew. Math. Mech.* **46**, 151.

R. F. Rinehart (1955), 'The equivalence of definitions of a matric function', *Amer. Math. Monthly* **62**, 395–414.

J. D. Roberts (1980), 'Linear model reduction and solution of the algebraic Riccati equation by use of the sign function', *Internat. J. Control* **32**, 677–687. First issued as report CUED/B-Control/TR13, Department of Engineering, University of Cambridge, 1971.

Y. Saad (1992), 'Analysis of some Krylov subspace approximations to the matrix exponential operator', *SIAM J. Numer. Anal.* **29**, 209–228.

Y. Saad (2003), *Iterative Methods for Sparse Linear Systems*, second edn, SIAM, Philadelphia, PA, USA.

M. Schroeder (1991), *Fractals, Chaos, Power Laws: Minutes from an Infinite Paradise*, W. H. Freeman, New York.

S. M. Serbin and S. A. Blalock (1980), 'An algorithm for computing the matrix cosine', *SIAM J. Sci. Statist. Comput.* **1**, 198–204.

L. F. Shampine (2007), 'Accurate numerical derivatives in MATLAB', *ACM Trans. Math. Software.* #26, 17 pp.

R. B. Sidje (1998), 'Expokit: A software package for computing matrix exponentials', *ACM Trans. Math. Software* **24**, 130–156.

R. B. Sidje, 'Expokit'. http://www.maths.uq.edu.au/expokit, retrieved October 8, 2009.

M. I. Smith (2003), 'A Schur algorithm for computing matrix pth roots', *SIAM J. Matrix Anal. Appl.* **24**, 971–989.

W. Squire and G. Trapp (1998), 'Using complex variables to estimate derivatives of real functions', *SIAM Rev.* **40**, 110–112.

C. F. Van Loan (1975), A study of the matrix exponential. Numerical Analysis Report No. 10, University of Manchester, Manchester, UK. Reissued as MIMS EPrint 2006.397, Manchester Institute for Mathematical Sciences, The University of Manchester, UK, November 2006.

C. F. Van Loan (1978), 'Computing integrals involving the matrix exponential', *IEEE Trans. Automat. Control* **AC-23**, 395–404.

C. F. Van Loan (1979), 'A note on the evaluation of matrix polynomials', *IEEE Trans. Automat. Control* **AC-24**, 320–321.

R. S. Varga (2000), *Matrix Iterative Analysis*, second edn, Springer, Berlin.

C. Visser (1937), 'Note on linear operators', *Proc. Kon. Akad. Wet. Amsterdam* **40**, 270–272.

R. C. Ward (1977), 'Numerical computation of the matrix exponential with accuracy estimate', *SIAM J. Numer. Anal.* **14**, 600–610.

F. V. Waugh and M. E. Abel (1967), 'On fractional powers of a matrix', *J. Amer. Statist. Assoc.* **62**, 1018–1021.

D. Yuan and W. Kernan (2007), 'Explicit solutions for exit-only radioactive decay chains', *J. Appl. Phys.* **101**, 094907 1–12.

Acta Numerica (2010), pp. 209–286
doi:10.1017/S0962492910000048

Exponential integrators

Marlis Hochbruck
Karlsruher Institut für Technologie,
Institut für Angewandte und Numerische Mathematik,
D-76128 Karlsruhe,
Germany
E-mail: marlis.hochbruck@kit.edu

Alexander Ostermann
Institut für Mathematik,
Universität Innsbruck,
A-6020 Innsbruck,
Austria
E-mail: alexander.ostermann@uibk.ac.at

In this paper we consider the construction, analysis, implementation and application of exponential integrators. The focus will be on two types of stiff problems. The first one is characterized by a Jacobian that possesses eigenvalues with large negative real parts. Parabolic partial differential equations and their spatial discretization are typical examples. The second class consists of highly oscillatory problems with purely imaginary eigenvalues of large modulus. Apart from motivating the construction of exponential integrators for various classes of problems, our main intention in this article is to present the mathematics behind these methods. We will derive error bounds that are independent of stiffness or highest frequencies in the system.

Since the implementation of exponential integrators requires the evaluation of the product of a matrix function with a vector, we will briefly discuss some possible approaches as well. The paper concludes with some applications, in which exponential integrators are used.

CONTENTS

1. Introduction

Exponential integrators constitute an interesting class of numerical methods for the time integration of stiff systems of differential equations, that is,

$$u'(t) = F\big(t, u(t)\big), \quad u(0) = u_0. \tag{1.1}$$

In this survey article we will mainly consider two types of stiff problems. The first one is characterized by a Jacobian that possesses eigenvalues with large negative real parts. For these problems, the usual definition of stiffness applies which states that a differential equation is *stiff* whenever the implicit Euler method works (tremendously) better than the explicit Euler method. The reason for this behaviour lies in the different linear stability properties of the two methods. All available explicit integrators (with the exception of Runge–Kutta–Chebyshev methods) have a relatively small linear stability domain in the complex left half-plane, and this is the reason why explicit methods require unrealistically small step sizes for integrating stiff problems.

The second class of stiff problems considered in this survey consists of highly oscillatory problems with purely imaginary eigenvalues of large modulus. Again, explicit methods lack stability and are forced to use tiny time steps. For this class, however, the implicit Euler scheme does not perform well either. At first glance this behaviour is puzzling since the method has the required stability properties. A closer look reveals that the step size reduction is forced by accuracy requirements: the method tends to resolve all the oscillations in the solution, hence its numerical inefficiency.

The basic idea behind exponential integrators is to identify a prototypical differential equation that has stiffness properties similar to those of the underlying equation (1.1) and that can be solved exactly. This prototypical equation is often found by linearizing (1.1) at a certain state w. For autonomous systems, this yields

$$v'(t) + Av(t) = g(v(t)), \quad v(0) = u_0 - w, \tag{1.2}$$

with $A = -DF(w)$ and $v(t) = u(t) - w$. This linearization procedure gives rise to a semilinear equation with a comparably small nonlinear remainder g, if $u(t)$ is close to w. The linear part of (1.2),

$$v'(t) + Av(t) = 0, \quad v(0) = v_0, \tag{1.3}$$

can then serve as the prototypical equation with exact solution

$$v(t) = \mathrm{e}^{-tA} v_0. \tag{1.4}$$

If, for example, A is symmetric positive definite or skew-Hermitian with eigenvalues of large modulus, the exponential e^{-tA} enjoys favourable properties such as uniform boundedness, independent of the time step t, in contrast to the propagator $I - tA$ of the explicit Euler method. For oscillatory problems, the exponential contains the full information on linear oscillations, in contrast to the propagator $(I + tA)^{-1}$ of the implicit Euler method.

The numerical scheme for the full equation (1.1) or (1.2) is constructed by incorporating the exact propagator of (1.3) in an appropriate way. In the above situation, this can be achieved by considering the corresponding Volterra integral equation

$$u(t) = \mathrm{e}^{-tA}u_0 + \int_0^t \mathrm{e}^{-(t-\tau)A} g(u(\tau))\,\mathrm{d}\tau \tag{1.5}$$

instead of (1.2). This representation of the exact solution is also called *variation-of-constants formula.*

The simplest numerical method for (1.5) is obtained by interpolating the nonlinearity at the known value $g(u_0)$ only, leading to the exponential Euler approximation

$$u_1 = \mathrm{e}^{-hA}u_0 + h\varphi_1(-hA)g(u_0). \tag{1.6}$$

Here h denotes the step size and φ_1 is the entire function

$$\varphi_1(z) = \frac{\mathrm{e}^z - 1}{z}.$$

Obviously, method (1.6) makes use of the matrix exponential of A and a related function, hence its name 'exponential integrator'.

In the early days of stiff problems, the direct approximation of the matrix exponential and the related φ_1-function was not regarded as practical for large matrices. For this reason, the functions arising were approximated by rational (Padé) approximations which resulted in implicit or semi-implicit Runge–Kutta methods, Rosenbrock methods or W-schemes, just to mention a few well-established methods for stiff problems. The view, however, has changed as new methods for computing or approximating the product of a matrix exponential function with a vector have become available.

Apart from motivating the construction of exponential integrators for various classes of problems, our main intention in this article is to present the mathematics behind exponential integrators. We will derive error bounds that are independent of stiffness or highest frequencies in the system. In order to achieve this aim, we will distinguish – as already mentioned – between systems that admit smooth solutions and systems with highly oscillatory solutions. We also emphasize the analytic conditions that underlie the numerical schemes as well as the limitations of the given error bounds. We hope that this will help in choosing the right integrator for a particular application.

In this survey, we will concentrate on the convergence properties of exponential integrators for finite times. We will not discuss further important properties such as long-term behaviour or geometric properties of the discrete flow. For trigonometric integrators, such questions are fully addressed in Hairer, Lubich and Wanner (2006).

Whenever we speak about the order of convergence in this article, we refer to the so-called *stiff order*. Recall that a method has stiff order p (for a particular class of problems) if the global error on finite time intervals can be uniformly bounded by Ch^p, where C is a constant that depends on the length of the time interval and on the chosen class of problems, *but neither* on the stiffness nor on the step size h. The stiff order of a method must not be confused with its *classical order*. The classical order shows up in the non-stiff case for sufficiently small step sizes and is always an upper bound to the stiff order.

For stiff problems with smooth solutions, standard techniques from the theory of stiff differential equations can be used. The local error is determined by inserting the exact solution into the numerical scheme and Taylor-expanding it to determine the defects. Such a Taylor expansion is possible for smooth solutions with bounded derivatives of reasonable size and leads to error bounds in terms of the solution. The linearization A has to fulfil certain conditions to guarantee the stability of the error recursion. A prominent example of this class of problems is that of parabolic partial differential equations, either considered as abstract evolution equations in an appropriate Banach space or their spatial discretizations which result in large systems of stiff ordinary differential equations. For the latter, it is vital to obtain temporal error bounds that are basically independent of the spatial mesh width. In Section 2, we will discuss this approach for various classes of exponential one step and multistep methods.

Problems with highly oscillatory solutions are discussed in Section 3. For these problems, Taylor series expansion of the exact solution is not a viable option. Completely new techniques for constructing efficient methods and for proving error bounds have to be devised. In this section we discuss the construction and error analysis for Magnus integrators for first-order problems, and for trigonometric integrators for second-order problems. In order to avoid resonances, particular filter functions are required. Adiabatic integrators for singularly perturbed problems are briefly discussed as well.

The implementation of exponential integrators often requires the evaluation of the product of a matrix function with a vector. Many different approaches to evaluating this action in an efficient way have been proposed in the literature. In Section 4, we review Chebyshev methods, Krylov subspace methods, interpolation methods based on Leja points, and contour integral methods. For problems with dense matrices, we refer to the review by Higham and Al-Mohy (2010) and to the monograph by Higham (2008). Finally, we give some hints on the mathematical software.

Section 5 is devoted to applications of exponential integrators in science and technology. Exponential integrators, often in combination with splitting methods, have a long tradition in quantum dynamics and chemistry. New applications in mathematical finance and regularization of ill-posed problems, just to mention two fields, have emerged recently. We will briefly discuss such applications in this section.

We conclude this review in Section 6 with some comments on how the concept of exponential integrators has developed historically.

Having given this summary, we briefly discuss related topics that are, however, *not* included in this survey. As already mentioned, many interesting classes of methods for parabolic problems have their origin in the early days of exponential integrators. As the evaluation of the exponential of a large matrix was not regarded as practical at that time, methods were developed using rational approximations instead. Prominent classes are semi-implicit Runge–Kutta methods, Rosenbrock methods and W-methods. These methods will not be discussed further in this survey; we refer to the textbooks by Hairer and Wanner (1996) and Strehmel and Weiner (1992).

Splitting methods are in many respects competitive with exponential integrators; sometimes they are also used in combination with them. We do not discuss splitting methods here, but refer to the survey article of McLachlan and Quispel (2002). Methods for ordinary differential equations on manifolds and Lie group methods will also not be further discussed here. We refer to the review article by Iserles, Munthe-Kaas, Nørsett and Zanna (2000).

Some notation

We end this Introduction with some words on the notation employed. In order to stress the fundamental difference between smooth and highly oscillatory solutions, we have used different letters to denote them. Smooth solutions in Section 2 are generally denoted by u, whereas the letter ψ is reserved for highly oscillatory solutions of first-order equations in Section 3. Finally, (q, p) usually denotes the solution of a second-order problem, rewritten as a first-order system. The first component q is called the position and p is the momentum.

The end of a proof is marked by $\square$, the end of an example by $\diamond$, and the end of an assumption by $\circ$. The conjugate transpose of a matrix A is denoted by A^*.

Throughout the paper, $C > 0$ will denote a generic constant.

2. Parabolic problems, smooth solutions

In this section, we consider semilinear problems of the form

$$u'(t) + Au(t) = g(t, u(t)), \quad u(t_0) = u_0. \tag{2.1}$$

We are most interested in parabolic problems, which can be written in the form (2.1) either as an abstract ordinary differential equation on a suitable function space or as a system of ordinary differential equations in $\mathbb{R}^n$ or $\mathbb{C}^n$ stemming from a suitable spatial discretization. Throughout the section, we will only consider problems with temporally smooth solutions, so that we can always expand the solution in a Taylor series.

After motivating the analytical framework for our error analysis, we will first treat exponential quadrature rules for linear problems. The main part of this section is devoted to exponential Runge–Kutta methods, exponential Rosenbrock methods, exponential multistep methods, exponential general linear methods, and Magnus integrators for semilinear problems (2.1).

2.1. Preliminaries

In order to motivate the analytical framework of our error analysis, we start with a simple example.

Example 2.1. We consider the heat equation in one space dimension,

$$U_t(t,x) = U_{xx}(t,x), \quad U(0,x) = U_0(x), \quad x \in \Omega = (0,\pi), \tag{2.2}$$

subject to homogeneous Dirichlet boundary conditions,

$$U(t,0) = U(t,\pi) = 0.$$

We assume that the initial function satisfies $U_0 \in L^2(\Omega)$. In this case, it is well known that the solution of (2.2) is given by

$$U(x,t) = \sum_{k=1}^{\infty} \mu_k \mathrm{e}^{-k^2 t} \sin kx,$$

where the numbers μ_k are the Fourier coefficients of the initial function U_0,

$$\mu_k = \frac{2}{\pi} \int_0^{\pi} U_0(x) \sin(kx)\, \mathrm{d}x.$$

The assumption $U_0 \in L^2(\Omega)$ is equivalent to $(\mu_k)_k \in \ell^2$.

For an abstract formulation of (2.2), we consider the linear (differential) operator A defined by

$$(Av)(x) = -v_{xx}(x).$$

Obviously, A is an unbounded operator and not defined for all $v \in L^2(\Omega)$. In order to model homogeneous Dirichlet boundary conditions, we consider A on the domain

$$D(A) = H^2(\Omega) \cap H_0^1(\Omega), \tag{2.3}$$

where H^2 and H_0^1 denote the familiar Sobolev spaces. In one space dimension, functions in $D(A)$ are continuously differentiable and vanish on the boundary of Ω.

This operator has a complete system of orthogonal eigenfunctions $\sin(kx)$ corresponding to the eigenvalues k^2, $k = 1, 2, \ldots$. Due to the isomorphism between L^2 and ℓ^2, the operator A induces a corresponding operator on ℓ^2.

An application of A to an L^2-function corresponds to the multiplication of its Fourier coefficients by k^2 in ℓ^2. Given a function $\phi : \mathbb{C} \to \mathbb{C}$, this allows us to define the operator

$$\phi(tA) : L^2(\Omega) \to L^2(\Omega), \quad t \geq 0,$$

given by

$$\phi(tA)v = \sum_{k=1}^{\infty} \nu_k \phi(k^2 t) \sin kx \quad \text{for } v(x) = \sum_{k=1}^{\infty} \nu_k \sin kx.$$

Clearly, any function $\phi(\xi)$ bounded for $\xi \geq 0$ defines a bounded operator $\phi(tA)$ for $t \geq 0$. For instance, choosing $\phi(\xi) = \mathrm{e}^{-\xi}$ defines the bounded exponential operator

$$\mathrm{e}^{-tA} : L^2(\Omega) \to L^2(\Omega).$$

This operator has the properties of a semigroup, namely $\mathrm{e}^{-0A} = I$ and

$$\mathrm{e}^{-tA}\mathrm{e}^{-sA} = \mathrm{e}^{-(t+s)A}, \quad t, s \geq 0. \tag{2.4}$$

Moreover, it satisfies

$$\|\mathrm{e}^{-tA}\|_{L^2(\Omega)\leftarrow L^2(\Omega)} \leq \mathrm{e}^{-t} \leq 1, \quad t \geq 0,$$

and

$$\|(tA)^\gamma \mathrm{e}^{-tA}\|_{L^2(\Omega)\leftarrow L^2(\Omega)} = \sup_{k\geq 1}(tk^2)^\gamma \mathrm{e}^{-tk^2} \leq C(\gamma), \quad \gamma, t \geq 0. \tag{2.5}$$

Using the above notation, we can now formulate the heat equation as an abstract ordinary differential equation on the Hilbert space $X = L^2(\Omega)$ by defining $u(t)$ as the function that maps x to $U(t, x)$:

$$u(t) = [x \mapsto U(t, x)].$$

Problem (2.2) then reads

$$u' + Au = 0, \quad u(0) = u_0,$$

and its solution is given by

$$u(t) = \mathrm{e}^{-tA} u_0, \quad t \geq 0,$$

which looks formally like the familiar matrix exponential in $\mathbb{R}^n$.

For homogeneous Neumann boundary conditions, the operator A has to be considered on the domain

$$D(A) = \{v \in H^2(\Omega) \mid v_x(0) = v_x(1) = 0\}.$$

This operator has the complete system of orthogonal eigenfunctions $\cos(kx)$, $k = 0, 1, 2, \ldots$, corresponding to the eigenvalues k^2. Functions of this operator can again be defined with the help of Fourier series. $\diamond$

The above example motivates the following more general assumption. Background information on semigroups can be found in the textbooks by Henry (1981), Pazy (1992), Lunardi (1995), and Engel and Nagel (2000).

Assumption 2.2. Let X be a Banach space with norm $\|\cdot\|$. We assume that A is a linear operator on X and that $(-A)$ is the infinitesimal generator of a strongly continuous semigroup e^{-tA} on X. ○

In particular, this assumption implies that there exist constants C and ω such that

$$\|\mathrm{e}^{-tA}\|_{X\leftarrow X} \le C\,\mathrm{e}^{\omega t}, \quad t \ge 0. \tag{2.6}$$

Our error analysis will make use of this estimate only.

Example 2.3. Readers unfamiliar with functional analysis may want to think of $X = \mathbb{R}^n$ or $X = \mathbb{C}^n$. In this case the linear operator can be represented by an $n \times n$ matrix, and e^{-tA} is just the well-known matrix exponential function. It is important to note that condition (2.6) holds with $\omega = 0$ if the field of values of A is contained in the right complex half-plane. For instance, if A is Hermitian positive semidefinite or skew-Hermitian, then $C = 1$ and $\omega = 0$ hold in the Euclidean norm, independently of the dimension n. If A stems from a spatial discretization of a partial differential equation, then using Assumption 2.2 will yield temporal convergence results that are independent of the spatial mesh. ◇

2.2. Linear problems; exponential quadrature

In this section, we will derive error bounds for exponential Runge–Kutta discretizations of linear parabolic problems,

$$u'(t) + Au(t) = f(t), \quad u(0) = u_0, \tag{2.7}$$

with a time-invariant operator A. The solution of (2.7) at time

$$t_{n+1} = t_n + h_n, \quad t_0 = 0, \quad n = 0, 1, \ldots$$

is given by the variation-of-constants formula

$$u(t_{n+1}) = \mathrm{e}^{-h_n A} u(t_n) + \int_0^{h_n} \mathrm{e}^{-(h_n-\tau)A} f(t_n + \tau)\,\mathrm{d}\tau. \tag{2.8}$$

A first scheme is obtained by approximating the function f within the integral by its interpolation polynomial in certain non-confluent quadrature nodes $c_1, \ldots, c_s$. This yields the *exponential quadrature rule*

$$u_{n+1} = \mathrm{e}^{-h_n A} u_n + h_n \sum_{i=1}^{s} b_i(-h_n A) f(t_n + c_i h_n) \tag{2.9a}$$

with weights

$$b_i(-hA) = \int_0^1 \mathrm{e}^{-h(1-\theta)A} \ell_i(\theta)\,\mathrm{d}\theta. \tag{2.9b}$$

Here, ℓ_i are the familiar Lagrange interpolation polynomials

$$\ell_i(\theta) = \prod_{\substack{m=1 \\ m\neq i}}^{s} \frac{\theta - c_m}{c_i - c_m}, \qquad i = 1, \ldots, s.$$

Obviously, the weights $b_i(z)$ are linear combinations of the entire functions

$$\varphi_k(z) = \int_0^1 \mathrm{e}^{(1-\theta)z} \frac{\theta^{k-1}}{(k-1)!}\,\mathrm{d}\theta, \qquad k \geq 1. \tag{2.10}$$

These functions satisfy $\varphi_k(0) = 1/k!$ and the recurrence relation

$$\varphi_{k+1}(z) = \frac{\varphi_k(z) - \varphi_k(0)}{z}, \qquad \varphi_0(z) = \mathrm{e}^z. \tag{2.11}$$

Assumption 2.2 enables us to define the operators

$$\varphi_k(-hA) = \int_0^1 \mathrm{e}^{-h(1-\theta)A} \frac{\theta^{k-1}}{(k-1)!}\,\mathrm{d}\theta, \qquad k \geq 1.$$

The following lemma turns out to be crucial.

Lemma 2.4. Under Assumption 2.2, the operators $\varphi_k(-hA)$, $k = 1, 2, \ldots$ are bounded on X.

Proof. The boundedness simply follows from the estimate

$$\|\varphi_k(-hA)\| \leq \int_0^1 \|\mathrm{e}^{-h(1-\theta)A}\| \frac{\theta^{k-1}}{(k-1)!}\,\mathrm{d}\theta$$

and the bound (2.6) on the semigroup. □

Example 2.5. For $s = 1$ we have

$$\begin{aligned} u_{n+1} &= \mathrm{e}^{-h_n A} u_n + h_n \varphi_1(-h_n A) f(t_n + c_1 h_n) \\ &= u_n + h_n \varphi_1(-h_n A)\big(f(t_n + c_1 h_n) - A u_n\big). \end{aligned}$$

The choice $c_1 = 0$ yields the exponential Euler quadrature rule, while $c_1 = 1/2$ corresponds to the exponential midpoint rule. ◇

Example 2.6. For $s = 2$ we obtain the weights

$$\begin{aligned} b_1(z) &= \frac{c_2}{c_2 - c_1}\varphi_1(z) - \frac{1}{c_2 - c_1}\varphi_2(z), \\ b_2(z) &= -\frac{c_1}{c_2 - c_1}\varphi_1(z) + \frac{1}{c_2 - c_1}\varphi_2(z). \end{aligned}$$

The choice $c_1 = 0$ and $c_2 = 1$ yields the exponential trapezoidal rule. ◇

As a generalization to the exponential quadrature rules of collocation type (2.9) considered so far, we will now consider methods where the weights $b_i(-hA)$ do not necessarily satisfy condition (2.9b). This more general class of methods,

$$u_{n+1} = \mathrm{e}^{-h_n A} u_n + h_n \sum_{i=1}^{s} b_i(-h_n A) f(t_n + c_i h_n), \tag{2.12}$$

only requires the weights $b_i(-hA)$ to be uniformly bounded in $h \geq 0$.

In order to analyse (2.12), we expand the right-hand side of (2.8) in a Taylor series with remainder in integral form:

$$\begin{aligned} u(t_{n+1}) &= \mathrm{e}^{-h_n A} u(t_n) + \int_0^{h_n} \mathrm{e}^{-(h_n-\tau)A} f(t_n + \tau)\,\mathrm{d}\tau \\ &= \mathrm{e}^{-h_n A} u(t_n) + h_n \sum_{k=1}^{p} \varphi_k(-h_n A) h_n^{k-1} f^{(k-1)}(t_n) \qquad (2.13) \\ &\quad + \int_0^{h_n} \mathrm{e}^{-(h_n-\tau)A} \int_0^{\tau} \frac{(\tau-\xi)^{p-1}}{(p-1)!} f^{(p)}(t_n + \xi)\,\mathrm{d}\xi\,\mathrm{d}\tau. \end{aligned}$$

This has to be compared with the Taylor series of the numerical solution (2.12):

$$\begin{aligned} u_{n+1} &= \mathrm{e}^{-h_n A} u_n + h_n \sum_{i=1}^{s} b_i(-h_n A) f(t_n + c_i h_n) \\ &= \mathrm{e}^{-h_n A} u_n + h_n \sum_{i=1}^{s} b_i(-h_n A) \sum_{k=0}^{p-1} \frac{h_n^k c_i^k}{k!} f^{(k)}(t_n) \qquad (2.14) \\ &\quad + h_n \sum_{i=1}^{s} b_i(-h_n A) \int_0^{c_i h_n} \frac{(c_i h_n - \tau)^{p-1}}{(p-1)!} f^{(p)}(t_n + \tau)\,\mathrm{d}\tau. \end{aligned}$$

Obviously the error $e_n = u_n - u(t_n)$ satisfies

$$e_{n+1} = \mathrm{e}^{-h_n A} e_n - \delta_{n+1} \tag{2.15}$$

with

$$\delta_{n+1} = \sum_{j=1}^{p} h_n^j \psi_j(-h_n A) f^{(j-1)}(t_n) + \delta_{n+1}^{(p)}, \tag{2.16}$$

where

$$\psi_j(-h_n A) = \varphi_j(-h_n A) - \sum_{i=1}^{s} b_i(-h_n A) \frac{c_i^{j-1}}{(j-1)!} \tag{2.17}$$

and

$$\delta_{n+1}^{(p)} = \int_0^{h_n} e^{-(h_n-\tau)A} \int_0^{\tau} \frac{(\tau-\xi)^{p-1}}{(p-1)!} f^{(p)}(t_n+\xi)\, d\xi\, d\tau - h_n \sum_{i=1}^{s} b_i(-h_nA) \int_0^{c_ih_n} \frac{(c_ih_n-\tau)^{p-1}}{(p-1)!} f^{(p)}(t_n+\tau)\, d\tau.$$

The coefficients (2.17) of the low-order terms in (2.16) being zero turn out to be the desired *order conditions* of the exponential quadrature rule (2.12).

We are now ready to state our convergence result.

Theorem 2.7. Let Assumption 2.2 be fulfilled and let $f^{(p)} \in L^1(0,T)$. For the numerical solution of (2.7), consider the exponential quadrature rule (2.12) with uniformly bounded weights $b_i(-hA)$ for $h \geq 0$. If the method satisfies the order conditions

$$\psi_j(-hA) = 0, \quad j = 1, \ldots, p, \tag{2.18}$$

then it converges with order p. More precisely, the error bound

$$\|u_n - u(t_n)\| \leq C \sum_{j=0}^{n-1} h_j^p \int_{t_j}^{t_{j+1}} \|f^{(p)}(\tau)\|\, d\tau$$

then holds, uniformly on $0 \leq t_n \leq T$, with a constant C that depends on T, but is independent of the chosen step size sequence.

Proof. Solving the error recursion (2.15) yields the estimate

$$\|e_n\| \leq \sum_{j=0}^{n-1} \|e^{-(t_n-t_j)A}\|\, \|\delta_j^{(p)}\|.$$

The desired bound follows from the stability bound (2.6) and the assumption on the weights. □

Corollary 2.8. Let Assumption 2.2 be fulfilled. Then the exponential quadrature rule (2.9) satisfies the order conditions up to order s. It is thus convergent of order s.

Proof. The weights b_i of the exponential quadrature rule (2.9) satisfy the order conditions (2.18) for $p = s$ by construction. The boundedness of the weights follows from Lemma 2.4. □

Theorem 2.7 is not yet optimal for methods whose *underlying* quadrature rule $(b_i(0), c_i)$ satisfies additional order conditions, *e.g.*, $\psi_{s+1}(0) = 0$; see Hochbruck and Ostermann (2005*b*, Section 3) for details.

2.3. Exponential Runge–Kutta methods

For the numerical solution of semilinear problems (2.1), we proceed analogously to the construction of Runge–Kutta methods. We start from the variation-of-constants formula

$$u(t_n + h_n) = \mathrm{e}^{-h_n A} u(t_n) + \int_0^{h_n} \mathrm{e}^{-(h_n - \tau)A} g\big(t_n + \tau, u(t_n + \tau)\big)\, \mathrm{d}\tau. \quad (2.19)$$

Since, in contrast to linear problems, the integral now contains the unknown function u, we use (2.19) with h_n replaced by $c_i h_n$ to define internal stages. This leads to the following general class of one-step methods:

$$u_{n+1} = \chi(-h_n A) u_n + h_n \sum_{i=1}^{s} b_i(-h_n A) G_{ni}, \quad (2.20a)$$

$$U_{ni} = \chi_i(-h_n A) u_n + h_n \sum_{j=1}^{s} a_{ij}(-h_n A) G_{nj}, \quad (2.20b)$$

$$G_{nj} = g(t_n + c_j h_n, U_{nj}). \quad (2.20c)$$

Here, the method coefficients χ, χ_i, a_{ij}, and b_i are constructed from exponential functions or (rational) approximations of such functions evaluated at the matrix or operator $(-h_n A)$. For consistency reasons, we always assume that $\chi(0) = \chi_i(0) = 1$.

It seems worth mentioning that (2.20) reduces to a Runge–Kutta method with coefficients $b_i = b_i(0)$ and $a_{ij} = a_{ij}(0)$ if we consider the (formal) limit $A \to 0$. The latter method will henceforth be called the *underlying Runge–Kutta method*, while (2.20) will be referred to as an *exponential Runge–Kutta method*. Throughout this section, we suppose that the underlying Runge–Kutta method satisfies

$$\sum_{j=1}^{s} b_j(0) = 1, \qquad \sum_{j=1}^{s} a_{ij}(0) = c_i, \quad i = 1, \ldots, s,$$

which makes it invariant under the transformation of (2.1) to autonomous form.

A desirable property of numerical methods is that they preserve equilibria $u^\star$ of the autonomous problem

$$u'(t) + Au(t) = g(u(t)).$$

Requiring $U_{ni} = u_n = u^\star$ for all i and $n \geq 0$ immediately yields necessary and sufficient conditions. It turns out that the coefficients of the method have to satisfy

$$\sum_{j=1}^{s} b_j(z) = \frac{\chi(z) - 1}{z}, \qquad \sum_{j=1}^{s} a_{ij}(z) = \frac{\chi_i(z) - 1}{z}, \quad i = 1, \ldots, s. \quad (2.21)$$

Table 2.1. The exponential Runge–Kutta method (2.20) in tableau form. For χ, χ_i being the standard exponential functions (2.22), we omit the last column.

$$\begin{array}{c|cccc|c} c_1 & & & & & \chi_1(-hA) \\ c_2 & a_{21}(-hA) & & & & \chi_2(-hA) \\ \vdots & \vdots & \ddots & & & \vdots \\ c_s & a_{s1}(-hA) & \dots & a_{s,s-1}(-hA) & & \chi_s(-hA) \\ \hline & b_1(-hA) & \dots & b_{s-1}(-hA) & b_s(-hA) & \chi(-hA) \end{array}$$

Without further mention, we will assume throughout this section that these conditions are fulfilled. We mainly consider methods with

$$\chi(z) = \mathrm{e}^z \quad \text{and} \quad \chi_i(z) = \mathrm{e}^{c_i z}, \quad 1 \le i \le s. \tag{2.22}$$

For this choice, the simplifying assumptions (2.21) read

$$\sum_{j=1}^{s} b_j(z) = \varphi_1(z), \qquad \sum_{j=1}^{s} a_{ij}(z) = c_i \varphi_1(c_i z), \quad i = 1, \dots, s. \tag{2.23}$$

Our main interest lies in *explicit* methods for which, due to $c_1 = 0$,

$$\chi_1(z) = 1 \quad \text{and} \quad a_{ij}(z) = 0, \quad 1 \le i \le j \le s. \tag{2.24}$$

The general construction principle together with an error analysis is presented in Hochbruck and Ostermann (2005*b*). Explicit methods of the form (2.20) have already been proposed by Friedli (1978). He also presented non-stiff order conditions, which, however, are not sufficient to analyse parabolic problems. Methods of the form (2.20) with rational functions χ and χ_i in place of (2.22) have been proposed and analysed by Strehmel and Weiner (1987, 1992).

As usual, we represent the coefficients of (2.20) in a Butcher tableau: see Table 2.1. With the help of (2.21), the functions χ and χ_i can be eliminated in (2.20). The numerical scheme then takes the form

$$u_{n+1} = u_n + h_n \sum_{i=1}^{s} b_i(-h_n A)(G_{ni} - Au_n), \tag{2.25a}$$

$$U_{ni} = u_n + h_n \sum_{j=1}^{s} a_{ij}(-h_n A)(G_{nj} - Au_n). \tag{2.25b}$$

Conditions (2.21) also imply that we can restrict ourselves to autonomous problems,

$$u'(t) + Au(t) = g(u(t)), \quad u(0) = u_0, \tag{2.26}$$

since all methods satisfying (2.21) are invariant under the transformation of (2.1) to autonomous form.

In order to study interesting examples, we have to take care of the norms in which we derive error estimates. Our analysis will make use of the variation-of-constants formula and this will involve bounds of terms of the form

$$\mathrm{e}^{-t_n A}\big(g(u(t_n)) - g(u_n)\big).$$

Bounding g in the L^2-norm, for example, would exclude even very simple functions, such as polynomials of degree higher than one. Therefore, we have to refine our framework.

Assumption 2.9. Let X be a Banach space with norm $\|\cdot\|$. We assume that A is a linear operator on X and that $(-A)$ is the infinitesimal generator of an analytic semigroup e^{-tA} on X. ○

In a similar way to our motivating Example 2.1, generators of analytic semigroups allow us to define fractional powers of the operator. To be more precise, if A satisfies Assumption 2.9, then there exists an $\omega \in \mathbb{R}$ such that $(A + \omega I)^\alpha$ is defined for real powers α; see Henry (1981) and Pazy (1992). Writing

$$Au + g(u) = (A + \omega I)u + \big(g(u) - \omega u\big)$$

enables us to set $\omega = 0$ without loss of generality. Then Assumption 2.9 implies that there exist constants $C = C(\gamma)$ such that

$$\|\mathrm{e}^{-tA}\|_{X \leftarrow X} + \|t^\gamma A^\gamma \mathrm{e}^{-tA}\|_{X \leftarrow X} \le C, \quad \gamma, t \ge 0. \tag{2.27}$$

With this bound at hand, we can estimate the difference of g in a weaker norm:

$$\|\mathrm{e}^{-t_n A} A^\alpha \cdot A^{-\alpha}\big(g(u(t_n)) - g(u_n)\big)\| \le C t_n^{-\alpha} \|A^{-\alpha}\big(g(u(t_n)) - g(u_n)\big)\|.$$

This observation leads to the following framework, where our main assumption on the nonlinearity g is that of Henry (1981) and Pazy (1992).

Assumption 2.10. For $0 \le \alpha < 1$, let

$$V = \{v \in X \mid A^\alpha v \in X\}$$

be a Banach space with norm $\|v\|_V = \|A^\alpha v\|$. We assume that $g : [0,T] \times V \to X$ is locally Lipschitz-continuous in a strip along the exact solution u. Thus, there exists a real number $L = L(R,T)$ such that, for all $t \in [0,T]$,

$$\|g(t,v) - g(t,w)\| \le L\|v - w\|_V \tag{2.28}$$

if $\max\big(\|v - u(t)\|_V, \|w - u(t)\|_V\big) \le R$. ○

Example 2.11. It is well known that reaction–diffusion equations fit into this abstract framework, as well as the incompressible Navier–Stokes equations in two and three space dimensions; see, *e.g.*, Henry (1981, Chapter 3) and Lunardi (1995, Section 7.3). ◇

Convergence of the exponential Euler method

We next study the convergence of the simplest method of the class (2.20).

Example 2.12. For $s = 1$, the only reasonable choice is the exponential version of Euler's method. Applied to (2.1), it has the form

$$u_{n+1} = \mathrm{e}^{-h_n A} u_n + h_n \varphi_1(-h_n A) g(t_n, u_n). \tag{2.29}$$

The Butcher tableau of the method reads

$$\begin{array}{c|c} 0 & \\ \hline & \varphi_1 \end{array}. \qquad \diamond$$

In order to simplify the notation, we set

$$f(t) = g(t, u(t)),$$

and we consider constant step sizes $h_n = h$ only.

Our proofs are heavily based on the representation of the exact solution by the variation-of-constants formula (2.19), which coincides with (2.8) in our notation. We expand f in a Taylor series with remainder in integral form to receive

$$f(t_n + \tau) = f(t_n) + \int_0^\tau f'(t_n + \sigma)\,\mathrm{d}\sigma. \tag{2.30}$$

Inserting the exact solution into the numerical scheme yields

$$u(t_{n+1}) = \mathrm{e}^{-hA} u(t_n) + h\varphi_1(-hA) f(t_n) + \delta_{n+1} \tag{2.31}$$

where, by (2.13), the defect is given by

$$\delta_{n+1} = \int_0^h \mathrm{e}^{-(h-\tau)A} \int_0^\tau f'(t_n + \sigma)\,\mathrm{d}\sigma\,\mathrm{d}\tau. \tag{2.32}$$

For this defect we have the following estimate.

Lemma 2.13. Let the initial value problem (2.1) satisfy Assumption 2.9. Furthermore, let $0 < \beta \le 1$ and $A^{\beta-1} f' \in L^\infty(0, T; V)$. Then

$$\Big\| \sum_{j=0}^{n-1} \mathrm{e}^{-jhA} \delta_{n-j} \Big\|_V \le Ch \sup_{0 \le t \le t_n} \| A^{\beta-1} f'(t) \|_V \tag{2.33}$$

holds with a constant C, uniformly in $0 \le t_n \le T$.

Proof. We denote the supremum by

$$M := \sup_{0 \le t \le t_n} \| A^{\beta-1} f'(t) \|_V,$$

and write

$$e^{-jhA}\delta_{n-j} = e^{-jhA} \int_0^h e^{-(h-\tau)A} A^{1-\beta} \int_0^\tau A^{\beta-1} f'(t_{n-j-1}+\sigma)\,d\sigma\,d\tau.$$

Using the stability bound (2.27), the term for $j = 0$ can be bounded in V by

$$CM \int_0^h \tau(h-\tau)^{\beta-1}\,d\tau \le CMh^{1+\beta},$$

whereas the remaining sum is bounded by

$$\left\| \sum_{j=1}^{n-1} e^{-jhA}\delta_{n-j} \right\|_V \le CMh^2 \sum_{j=1}^{n-1} (jh)^{\beta-1} \le CMh \int_0^{t_{n-1}} t^{\beta-1}\,dt \le CMh.$$

This proves the desired estimate. □

For the exponential Euler method, we have the following convergence result.

Theorem 2.14. Let the initial value problem (2.1) satisfy Assumptions 2.9 and 2.10, and consider for its numerical solution the exponential Euler method (2.29). Further assume that $f : [0,T] \to X$ is differentiable and that $\beta \in (0,1]$ can be chosen such that $A^{\beta-1}f' \in L^\infty(0,T;V)$. Then, the error bound

$$\|u_n - u(t_n)\|_V \le C\,h \sup_{0\le t\le t_n} \|A^{\beta-1}f'(t)\|_V$$

holds uniformly in $0 \le nh \le T$. The constant C depends on T, but it is independent of n and h.

Proof. The exponential Euler method satisfies the error recursion

$$e_{n+1} = e^{-hA}e_n + h\varphi_1(-hA)\big(g(t_n,u_n) - f(t_n)\big) - \delta_{n+1} \tag{2.34}$$

with defect δ_{n+1} defined in (2.32). Solving this recursion yields

$$e_n = h\sum_{j=0}^{n-1} e^{-(n-j-1)hA}\varphi_1(-hA)\big(g(t_j,u_j) - f(t_j)\big) - \sum_{j=0}^{n-1} e^{-jhA}\delta_{n-j}.$$

Using (2.27), Assumption 2.10 and Lemma 2.13, we may estimate this in V by

$$\|e_n\|_V \le Ch\sum_{j=0}^{n-2} t_{n-j-1}^{-\alpha}\|e_j\|_V + Ch^{1-\alpha}\|e_{n-1}\|_V + Ch \sup_{0\le t\le t_n} \|A^{\beta-1}f'(t)\|_V.$$

The application of Lemma 2.15 concludes the proof. □

In the previous proof we used the following standard Discrete Gronwall Lemma. For its proof see Dixon and McKee (1986), for instance.

Lemma 2.15. (Discrete Gronwall Lemma) For $h > 0$ and $T > 0$, let $0 \le t_n = nh \le T$. Further assume that the sequence of non-negative numbers ε_n satisfies the inequality

$$\varepsilon_n \le ah \sum_{\nu=1}^{n-1} t_{n-\nu}^{-\rho} \varepsilon_\nu + b\, t_n^{-\sigma}$$

for $0 \le \rho, \sigma < 1$ and $a, b \ge 0$. Then the estimate

$$\varepsilon_n \le Cb\, t_n^{-\sigma}$$

holds, where the constant C depends on ρ, σ, a, and on T.

Convergence of the higher-order exponential Runge–Kutta methods

The convergence analysis of higher-order methods turns out to be much more complicated than that for the exponential Euler scheme, due to the low order of the internal stages. Moreover, it requires additional assumptions on the nonlinearity g.

Assumption 2.16. We assume that (2.1) possesses a sufficiently smooth solution $u : [0, T] \to V$ with derivatives in V, and that $g : [0, T] \times V \to X$ is sufficiently often Fréchet-differentiable in a strip along the exact solution. All occurring derivatives are assumed to be uniformly bounded. ○

Note that, under the above assumption, the composition

$$f : [0, T] \to X : t \mapsto f(t) = g\big(t, u(t)\big)$$

is a smooth mapping, too. This will be used frequently.

As usual for the analysis of stiff problems with smooth solutions, and as we have seen for the exponential Euler method, we start by inserting the exact solution into the numerical scheme. This yields

$$u(t_n + c_i h) = \mathrm{e}^{-c_i hA} u(t_n) + h \sum_{j=1}^{i-1} a_{ij}(-hA) f(t_n + c_j h) + \Delta_{ni}, \qquad (2.35\text{a})$$

$$u(t_{n+1}) = \mathrm{e}^{-hA} u(t_n) + h \sum_{i=1}^{s} b_i(-hA) f(t_n + c_i h) + \delta_{n+1}, \qquad (2.35\text{b})$$

with defect δ_{n+1} given in (2.16), and

$$\Delta_{ni} = \sum_{j=1}^{q_i} h^j \psi_{j,i}(-hA) f^{(j-1)}(t_n) + \Delta_{ni}^{(q_i)}, \qquad (2.36)$$

where

$$\psi_{j,i}(-hA) = \varphi_j(-c_i hA) c_i^j - \sum_{k=1}^{i-1} a_{ik}(-hA) \frac{c_k^{j-1}}{(j-1)!} \qquad (2.37)$$

and

$$\Delta_{ni}^{(q_i)} = \int_0^{c_i h} \mathrm{e}^{-(c_i h - \tau)A} \int_0^{\tau} \frac{(\tau - \sigma)^{q_i - 1}}{(q_i - 1)!} f^{(q_i)}(t_n + \sigma)\, \mathrm{d}\sigma\, \mathrm{d}\tau$$
$$- h \sum_{k=1}^{i-1} a_{ik}(-hA) \int_0^{c_k h} \frac{(c_k h - \sigma)^{q_i - 1}}{(q_i - 1)!} f^{(q_i)}(t_n + \sigma)\, \mathrm{d}\sigma.$$

Note that $\psi_{1,i} = 0$ due to the simplifying assumptions (2.23), but – as for all explicit Runge–Kutta schemes – it is not possible to achieve $\psi_{2,i} = 0$ for all i. This implies that the internal stages are of order one only, which makes the construction of higher-order methods quite involved. We refer to Hochbruck and Ostermann (2005*a*) for details, and only state the main result here.

Theorem 2.17. Let the initial value problem (2.1) satisfy Assumptions 2.9 and 2.16 with $V = X$ and consider for its numerical solution an explicit exponential Runge–Kutta method (2.20) satisfying (2.22)–(2.24). For $2 \le p \le 4$, assume that the order conditions of Table 2.2 hold up to order $p - 1$ and that $\psi_p(0) = 0$. Further assume that the remaining conditions of order p hold in a weaker form with $b_i(0)$ instead of $b_i(-hA)$ for $2 \le i \le s$. Then the numerical solution u_n satisfies the error bound

$$\|u_n - u(t_n)\| \le C\, h^p$$

uniformly in $0 \le nh \le T$. The constant C depends on T, but it is independent of n and h. □

We now consider some particular examples. We start with second-order methods.

Example 2.18. Second-order methods require two internal stages at least. For two stages, the order conditions taken from Table 2.2 are

$$b_1(-hA) + b_2(-hA) = \varphi_1(-hA), \tag{2.38a}$$
$$b_2(-hA)c_2 = \varphi_2(-hA), \tag{2.38b}$$
$$a_{21}(-hA) = c_2\varphi_1(-c_2hA). \tag{2.38c}$$

A straightforward elimination leads to the following one-parameter family of exponential Runge–Kutta methods:

$$\begin{array}{c|cc} 0 & & \\ c_2 & c_2\,\varphi_{1,2} & \\ \hline & \varphi_1 - \frac{1}{c_2}\varphi_2 & \frac{1}{c_2}\varphi_2 \end{array}\,. \tag{2.39}$$

Here and in the following Butcher tableaux, we use the short notation

$$\varphi_{j,k} = \varphi_j(-c_k hA), \quad \varphi_j = \varphi_j(-hA).$$

Table 2.2. Stiff order conditions for explicit exponential Runge–Kutta methods for $\alpha = 0$. Here J and K denote arbitrary bounded operators on X. The functions ψ_i and $\psi_{k,\ell}$ are defined in (2.17) and (2.37), respectively.

Number	Order	Order condition
1	1	$\psi_1(-hA) = 0$
2	2	$\psi_2(-hA) = 0$
3	2	$\psi_{1,i}(-hA) = 0$
4	3	$\psi_3(-hA) = 0$
5	3	$\sum_{i=1}^{s} b_i(-hA)J\psi_{2,i}(-hA) = 0$
6	4	$\psi_4(-hA) = 0$
7	4	$\sum_{i=1}^{s} b_i(-hA)J\psi_{3,i}(-hA) = 0$
8	4	$\sum_{i=1}^{s} b_i(-hA)J\sum_{j=2}^{i-1} a_{ij}(-hA)J\psi_{2,j}(-hA) = 0$
9	4	$\sum_{i=1}^{s} b_i(-hA)c_iK\psi_{2,i}(-hA) = 0$

It is also possible to omit the function φ_2 by weakening condition (2.38b) to $b_2(0)c_2 = \varphi_2(0) = \frac{1}{2}$. This yields another one-parameter family of methods:

$$\begin{array}{c|cc} 0 & & \\ c_2 & c_2\,\varphi_{1,2} & \\ \hline & \left(1 - \frac{1}{2c_2}\right)\varphi_1 & \frac{1}{2c_2}\varphi_1 \end{array} . \tag{2.40}$$

The most attractive choice here is $c_2 = \frac{1}{2}$, which yields $b_1 = 0$.

Methods (2.39) and (2.40) have already been proposed by Strehmel and Weiner (1992, Example 4.2.2) in the context of adaptive Runge–Kutta methods, where the functions φ_j are usually approximated by certain rational functions. It is shown in Strehmel and Weiner (1992, Section 4.5.3) that both methods are B-consistent of order one. ◇

The construction of various families of third-order methods can be found in Hochbruck and Ostermann (2005*a*). In addition, a discussion of the convergence of some related three-stage methods which can be found in the literature is presented there. Among these are the three-stage adaptive Runge–Kutta method of Strehmel and Weiner (1992, Example 4.5.4), the ETD3RK scheme by Cox and Matthews (2002) and the ETD2CF3 method, which is a variant of the commutator-free Lie group method CF3 of Celledoni, Marthinsen and Owren (2003).

We next discuss in more detail some four-stage methods that have recently been discussed in the literature.

Example 2.19. Cox and Matthews (2002) proposed the following exponential variant of the classical Runge–Kutta method:

$$\begin{array}{c|cccc} 0 & & & & \\ \frac{1}{2} & \frac{1}{2}\varphi_{1,2} & & & \\ \frac{1}{2} & 0 & \frac{1}{2}\varphi_{1,3} & & \\ 1 & \frac{1}{2}\varphi_{1,3}(\varphi_{0,3}-1) & 0 & \varphi_{1,3} & \\ \hline & \varphi_1 - 3\varphi_2 + 4\varphi_3 & 2\varphi_2 - 4\varphi_3 & 2\varphi_2 - 4\varphi_3 & 4\varphi_3 - \varphi_2 \end{array} . \tag{2.41}$$

This method satisfies conditions 1–4 of Table 2.2, the weakened but sufficient condition 6 ($\psi_4(0) = 0$), but not conditions 5, 7, 8 and 9. However, it satisfies a weakened form of conditions 5 and 9 (because $\psi_{2,2}(0)+\psi_{2,3}(0) = 0$ and $\psi_{2,4}(0) = 0$), and a very weak form of conditions 7 and 8 (where all arguments are evaluated for $A = 0$). In the worst case, this leads to an order reduction to order two only.

The method by Krogstad (2005) is given by

$$\begin{array}{c|cccc} 0 & & & & \\ \frac{1}{2} & \frac{1}{2}\varphi_{1,2} & & & \\ \frac{1}{2} & \frac{1}{2}\varphi_{1,3} - \varphi_{2,3} & \varphi_{2,3} & & \\ 1 & \varphi_{1,4} - 2\varphi_{2,4} & 0 & 2\varphi_{2,4} & \\ \hline & \varphi_1 - 3\varphi_2 + 4\varphi_3 & 2\varphi_2 - 4\varphi_3 & 2\varphi_2 - 4\varphi_3 & -\varphi_2 + 4\varphi_3 \end{array} . \tag{2.42}$$

This method satisfies conditions 1–5 and 9 of Table 2.2, the weakened but sufficient condition 6 ($\psi_4(0) = 0$), but not conditions 7 and 8, which are only satisfied in a very weak form (where all arguments are evaluated for $A = 0$). In the worst case, this leads to an order reduction to order three.

Strehmel and Weiner (1992, Example 4.5.5) suggested the scheme

$$\begin{array}{c|cccc} 0 & & & & \\ \frac{1}{2} & \frac{1}{2}\varphi_{1,2} & & & \\ \frac{1}{2} & \frac{1}{2}\varphi_{1,3} - \frac{1}{2}\varphi_{2,3} & \frac{1}{2}\varphi_{2,3} & & \\ 1 & \varphi_{1,4} - 2\varphi_{2,4} & -2\varphi_{2,4} & 4\varphi_{2,4} & \\ \hline & \varphi_1 - 3\varphi_2 + 4\varphi_3 & 0 & 4\varphi_2 - 8\varphi_3 & -\varphi_2 + 4\varphi_3 \end{array} . \tag{2.43}$$

This method satisfies the conditions of Table 2.2 in exactly the same way as Krogstad's method. It thus converges in our situation, with order three in the worst case. Strehmel and Weiner (1992) proved that the method is B-consistent of order two. ◇

The question whether it is possible to modify the coefficients of the above four-stage methods in such a way that they have order four for semilinear parabolic problems was answered in the negative by Hochbruck and Ostermann (2005*a*). There, the following five-stage method was constructed:

$$
\begin{array}{c|ccccc}
0 & & & & & \\
\frac{1}{2} & \frac{1}{2}\varphi_{1,2} & & & & \\
\frac{1}{2} & \frac{1}{2}\varphi_{1,3} - \varphi_{2,3} & \varphi_{2,3} & & & \\
1 & \varphi_{1,4} - 2\varphi_{2,4} & \varphi_{2,4} & \varphi_{2,4} & & \\
\frac{1}{2} & \frac{1}{2}\varphi_{1,5} - 2a_{5,2} - a_{5,4} & a_{5,2} & a_{5,2} & \frac{1}{4}\varphi_{2,5} - a_{5,2} & \\
\hline
 & \varphi_1 - 3\varphi_2 + 4\varphi_3 & 0 & 0 & -\varphi_2 + 4\varphi_3 & 4\varphi_2 - 8\varphi_3
\end{array}
$$

with

$$a_{5,2} = \frac{1}{2}\varphi_{2,5} - \varphi_{3,4} + \frac{1}{4}\varphi_{2,4} - \frac{1}{2}\varphi_{3,5}.$$

Under the assumptions of Theorem 2.17, this method has order four.

2.4. *Exponential Rosenbrock methods*

We now turn to the time discretization of (possibly abstract) differential equations in autonomous form,

$$u'(t) = F\big(u(t)\big), \quad u(t_0) = u_0. \tag{2.44}$$

The numerical schemes considered are based on a continuous linearization of (2.44) along the numerical solution. For a given point u_n in the state space, this linearization is

$$u'(t) = J_n u(t) + g_n\big(u(t)\big), \tag{2.45a}$$

$$J_n = \mathrm{D}F(u_n) = \frac{\partial F}{\partial u}(u_n), \quad g_n\big(u(t)\big) = F\big(u(t)\big) - J_n u(t), \tag{2.45b}$$

with J_n denoting the Jacobian of F and g_n the nonlinear remainder evaluated at u_n, respectively. The numerical schemes will make *explicit* use of these quantities.

Let u_n denote the numerical approximation to the solution of (2.44) at time t_n. Its value at t_0 is given by the initial condition. Applying an explicit exponential Runge–Kutta scheme (2.35) to (2.45), we obtain the following class of *exponential Rosenbrock methods*:

$$U_{ni} = \mathrm{e}^{c_i h_n J_n} u_n + h_n \sum_{j=1}^{i-1} a_{ij}(h_n J_n) g_n(U_{nj}), \quad 1 \le i \le s, \tag{2.46a}$$

$$u_{n+1} = \mathrm{e}^{h_n J_n} u_n + h_n \sum_{i=1}^{s} b_i(h_n J_n) g_n(U_{ni}). \tag{2.46b}$$

Methods of this form were proposed by Hochbruck and Ostermann (2006).

Without further mention, we will assume that the methods fulfil the simplifying assumptions (2.23). Therefore, $c_1 = 0$ and consequently $U_{n1} = u_n$. Methods that satisfy the simplifying assumptions possess several interesting features. In particular, they preserve equilibria of (2.44), they have small defects, which in turn leads to simple order conditions for stiff problems (see Table 2.3 below), they allow a reformulation for efficient implementation, and they can easily be extended to non-autonomous problems.

Example 2.20. The well-known exponential Rosenbrock–Euler method is given by

$$\begin{aligned} u_{n+1} &= \mathrm{e}^{h_n J_n} u_n + h_n \varphi_1(h_n J_n) g_n(u_n) \\ &= u_n + h_n \varphi_1(h_n J_n) F(u_n). \end{aligned} \tag{2.47}$$

It is computationally attractive since it requires only one matrix function per step. ◇

The first exponential integrator based on the local linearization (2.45) was proposed by Pope (1963). Tokman (2006) pursued the same approach leading to her EPI time integration. In her paper, multistep methods were also constructed.

Related approaches include the local linearization schemes by Ramos and García-López (1997) and De la Cruz, Biscay, Carbonell, Ozaki and Jimenez (2007). Implicit methods combined with fixed-point iteration and Krylov subspace approximations have been proposed by Friesner, Tuckerman, Dornblaser and Russo (1989).

For the implementation of an exponential Rosenbrock method it is crucial to approximate the application of matrix functions to vectors efficiently. We therefore suggest expressing the vectors $g_n(U_{nj})$ as

$$g_n(U_{nj}) = g_n(u_n) + D_{nj}, \quad 2 \le j \le s.$$

A similar idea was used in Tokman (2006). Due to the simplifying assumptions (2.23), the method (2.46) takes the equivalent form

$$U_{ni} = u_n + c_i h_n \varphi_1(c_i h_n J_n) F(u_n) + h_n \sum_{j=2}^{i-1} a_{ij}(h_n J_n) D_{nj}, \tag{2.48a}$$

$$u_{n+1} = u_n + h_n \varphi_1(h_n J_n) F(u_n) + h_n \sum_{i=2}^{s} b_i(h_n J_n) D_{ni}. \tag{2.48b}$$

Hence, each stage of the method consists of a perturbed exponential Rosenbrock–Euler step (2.47).

The main motivation for this reformulation is that the vectors D_{ni} are expected to be small in norm. When computing the application of matrix functions to these vectors with some Krylov subspace method, this should be possible in a low-dimensional subspace. Consequently, only one computationally expensive Krylov approximation will be required in each time step, namely that involving $F(u_n)$. A similar idea has also been used in Hochbruck, Lubich and Selhofer (1998) to make the code exp4 efficient.

The proposed method can easily be extended to non-autonomous problems,

$$u' = F(t, u), \quad u(t_0) = u_0, \tag{2.49}$$

by rewriting the problem in autonomous form,

$$U' = \mathcal{F}(U), \quad U = \begin{bmatrix} t \\ u \end{bmatrix}, \quad \mathcal{F}(U) = \begin{bmatrix} 1 \\ F(t, u) \end{bmatrix}, \tag{2.50a}$$

with Jacobian

$$\mathcal{J}_n = \begin{bmatrix} 0 & 0 \\ v_n & J_n \end{bmatrix}, \quad v_n = \frac{\partial F}{\partial t}(t_n, u_n), \quad J_n = \frac{\partial F}{\partial u}(t_n, u_n). \tag{2.50b}$$

This transformation is standard for Rosenbrock methods as well (see Hairer and Wanner (1996)), but it changes a linear non-autonomous problem into a nonlinear one.

In order to apply our method to the autonomous system (2.50), we have to compute matrix functions of $\mathcal{J}_n$. Using Cauchy's integral formula and exploiting the special structure of $\mathcal{J}_n$, we get

$$\varphi_k(h_n \mathcal{J}_n) = \begin{bmatrix} \varphi_k(0) & 0 \\ h_n \varphi_{k+1}(h_n J_n) v_n & \varphi_k(h_n J_n) \end{bmatrix}.$$

In our formulation, we will again work with the smaller quantities

$$D_{nj} = g_n(t_n + c_j h_n, U_{nj}) - g_n(t_n, u_n), \tag{2.51}$$

where

$$g_n(t, u) = F(t, u) - J_n u - v_n t.$$

Applying method (2.48) to the autonomous formulation (2.50), we get

$$\begin{aligned} U_{ni} &= u_n + h_n c_i \varphi_1(c_i h_n J_n) F(t_n, u_n) \\ &\quad + h_n^2 c_i^2 \varphi_2(c_i h_n J_n) v_n + h_n \sum_{j=2}^{i-1} a_{ij}(h_n J_n) D_{nj}, \end{aligned} \tag{2.52a}$$

$$\begin{aligned} u_{n+1} &= u_n + h_n \varphi_1(h_n J_n) F(t_n, u_n) \\ &\quad + h_n^2 \varphi_2(h_n J_n) v_n + h_n \sum_{i=2}^{s} b_i(h_n J_n) D_{ni}. \end{aligned} \tag{2.52b}$$

This is the format of an exponential Rosenbrock method for non-autonomous problems (2.49).

Example 2.21. The exponential Rosenbrock–Euler method for non-autonomous problems is given by

$$u_{n+1} = u_n + h_n\varphi_1(h_nJ_n)F(u_n) + h_n^2\varphi_2(h_nJ_n)v_n, \tag{2.53}$$

with J_n and v_n defined in (2.50b). This scheme was proposed by Pope (1963). ◇

In the subsequent analysis, we restrict our attention to autonomous semilinear problems,

$$u'(t) = F\big(u(t)\big), \quad F(u) = -Au + g(u), \quad u(t_0) = u_0. \tag{2.54}$$

This implies that (2.45b) takes the form

$$J_n = -A + \frac{\partial g}{\partial u}(u_n), \quad g_n\big(u(t)\big) = g\big(u(t)\big) - \frac{\partial g}{\partial u}(u_n)u(t). \tag{2.55}$$

We suppose that A satisfies Assumption 2.2. Our main hypothesis on the nonlinearity g is Assumption 2.16 with $V = X$. The latter assumption implies that the Jacobian

$$J = J(u) = \mathrm{D}F(u) = \frac{\partial F}{\partial u}(u)$$

satisfies the Lipschitz condition

$$\|J(u) - J(v)\|_{X\leftarrow X} \le C\|u - v\| \tag{2.56}$$

in a neighbourhood of the exact solution.

Convergence of higher-order exponential Rosenbrock methods

Theorem 2.22. Suppose the initial value problem (2.54) satisfies Assumptions 2.2 and 2.16 with $V = X$. Consider for its numerical solution an explicit exponential Rosenbrock method (2.46) that fulfils the order conditions of Table 2.3 up to order p for some $2 \le p \le 4$. Further, let the step size sequence h_j satisfy the condition

$$\sum_{k=1}^{n-1}\sum_{j=0}^{k-1} h_j^{p+1} \le C_{\mathrm{H}} \tag{2.57}$$

with a constant C_{H} that is uniform in $t_0 \le t_n \le T$. Then, for C_{H} sufficiently small, the numerical method converges with order p. In particular, the numerical solution satisfies the error bound

$$\|u_n - u(t_n)\| \le C\sum_{j=0}^{n-1} h_j^{p+1} \tag{2.58}$$

uniformly on $t_0 \le t_n \le T$. The constant C is independent of the chosen step size sequence satisfying (2.57).

Table 2.3. Stiff order conditions for exponential Rosenbrock methods applied to autonomous problems.

Number	Order condition	Order
1	$\sum_{i=1}^{s} b_i(z) = \varphi_1(z)$	1
2	$\sum_{j=1}^{i-1} a_{ij}(z) = c_i\varphi_1(c_i z), \quad 2 \le i \le s$	2
3	$\sum_{i=2}^{s} b_i(z)c_i^2 = 2\varphi_3(z)$	3
4	$\sum_{i=2}^{s} b_i(z)c_i^3 = 6\varphi_4(z)$	4

Proof. We present the proof for the exponential Rosenbrock–Euler method only. The proof for higher-order methods can be found in Hochbruck, Ostermann and Schweitzer (2009*c*).

To simplify the notation, we write $f_n(t) = g_n\big(u(t)\big)$. Inserting the exact solution into the numerical scheme yields

$$u(t_{n+1}) = \mathrm{e}^{h_n J_n} u(t_n) + h_n \varphi_1(h_n J_n) f_n(t_n) + \delta_{n+1} \tag{2.59}$$

with defect δ_{n+1}. Next, by using the variation-of-constants formula (2.8), we obtain

$$u(t_{n+1}) = \mathrm{e}^{h_n J_n} u(t_n) + \int_0^{h_n} \mathrm{e}^{(h_n-\tau)J_n} f_n(t_n+\tau)\,\mathrm{d}\tau.$$

Taylor expansion of f within the integral yields

$$f_n(t_n+\tau) = f_n(t_n) + \tau f_n'(t_n) + \int_0^{\tau} (\tau-\xi) f_n''(t_n+\xi)\,\mathrm{d}\xi,$$

where, by definition of g_n,

$$f_n'(t_n) = \left(\frac{\partial g}{\partial u}(u(t_n)) - \frac{\partial g}{\partial u}(u_n)\right) u'(t_n). \tag{2.60}$$

We thus have

$$\delta_{n+1} = h_n^2 \varphi_2(h_n J_n) f_n'(t_n) + \int_0^{h_n} \mathrm{e}^{(h_n-\tau)J_n} \int_0^{\tau} (\tau-\xi) f_n''(t_n+\xi)\,\mathrm{d}\xi\,\mathrm{d}\tau,$$

and this gives the bound

$$\|\delta_{n+1}\| \le C\big(h_n^2 \|e_n\| + h_n^3\big). \tag{2.61}$$

Subtracting (2.59) from the numerical solution leads to the error recursion

$$e_{n+1} = \mathrm{e}^{h_n J_n} e_n + h_n \varrho_n - \delta_{n+1}, \qquad e_0 = 0, \tag{2.62}$$

with

$$\varrho_n = \varphi_1(h_n J_n)\big(g_n(u_n) - f_n(t_n)\big).$$

Using the Lipschitz condition (2.56) in

$$g_n(u_n) - f_n(t_n) = \int_0^1 \left(\frac{\partial g_n}{\partial u}(u(t_n) + \theta e_n) - \frac{\partial g_n}{\partial u}(u_n) \right) e_n \, \mathrm{d}\theta,$$

we obtain

$$\|g_n(u_n) - f_n(t_n)\| \leq C \|e_n\|^2. \tag{2.63}$$

Solving the error recursion (2.62) and using $e_0 = 0$, we obtain

$$e_n = \sum_{j=0}^{n-1} h_j \, \mathrm{e}^{h_{n-1} J_{n-1}} \cdots \mathrm{e}^{h_{j+1} J_{j+1}} (\varrho_j - h_j^{-1} \delta_{j+1}). \tag{2.64}$$

Employing (2.61) and (2.63), we obtain the bound

$$\|\varrho_j\| + h_j^{-1} \|\delta_{j+1}\| \leq C \big(h_j \|e_j\| + \|e_j\|^2 + h_j^2 \big). \tag{2.65}$$

Inserting this into (2.64) and using the stability estimate of Hochbruck *et al.* (2009*c*, Theorem 3.7), we have

$$\|e_n\| \leq C \sum_{j=0}^{n-1} h_j \big(\|e_j\|^2 + h_j \|e_j\| + h_j^2 \big). \tag{2.66}$$

The constant in this estimate is uniform as long as

$$\sum_{j=1}^{n-1} \|e_j\| \leq C_{\mathtt{A}} \tag{2.67}$$

holds uniformly on $t_0 \leq t_n \leq T$. The application of a version of the Discrete Gronwall Lemma (Lemma 2.15) for variable step sizes (see, *e.g.*, Emmrich (2005)) to (2.66) then shows the desired bound (2.58).

It still remains to verify that condition (2.67) holds with a uniform bound $C_{\mathtt{A}}$. This now follows recursively from (2.58) and our assumption on the step size sequence (2.57) with $C_{\mathtt{H}}$ sufficiently small. □

The well-known exponential Rosenbrock–Euler method (2.47) obviously satisfies condition 1 of Table 2.3, while condition 2 is void. Therefore, it is second-order convergent for problems satisfying our analytic framework. A possible error estimator for (2.47) is described in Caliari and Ostermann (2009).

From the order conditions of Table 2.3, we constructed pairs of embedded methods of order three and four in Hochbruck *et al.* (2009*c*). For our variable step size implementation, we consider (2.46b) together with an embedded approximation

$$\widehat{u}_{n+1} = \mathrm{e}^{h_n J_n} u_n + h_n \sum_{i=1}^{s} \widehat{b}_i(h_n J_n) \, g_n(U_{ni}), \tag{2.68}$$

which relies on the same stages U_{ni}.

Example 2.23. The scheme exprb32 consists of a third-order exponential Rosenbrock method with a second-order error estimator (the exponential Rosenbrock–Euler method). Its coefficients are

$$\begin{array}{c|cc} 0 & & \\ 1 & \varphi_1 & \\ \hline & \varphi_1 - 2\varphi_3 & 2\varphi_3 \\ & \varphi_1 & \end{array} .$$

The additional bottom line in the Butcher tableau contains the weight $\widehat{b}_1$ of the embedded method (2.68). ◇

Example 2.24. The scheme exprb43 is a fourth-order method with a third-order error estimator. Its coefficients are

$$\begin{array}{c|ccc} 0 & & & \\ \frac{1}{2} & \frac{1}{2}\varphi_1(\frac{1}{2}\,\cdot) & & \\ 1 & 0 & \varphi_1 & \\ \hline & \varphi_1 - 14\varphi_3 + 36\varphi_4 & 16\varphi_3 - 48\varphi_4 & -2\varphi_3 + 12\varphi_4 \\ & \varphi_1 - 14\varphi_3 & 16\varphi_3 & -2\varphi_3 \end{array} .$$

◇

Note that the internal stages of the above methods are just exponential Rosenbrock–Euler steps. This leads to simple methods that can be implemented cheaply.

Evidently, the order conditions of Table 2.3 imply that the weights of any third-order method have to depend on φ_3, whereas those of any fourth-order method depend on φ_3 and φ_4 (in addition to φ_1).

Example 2.25. Hochbruck *et al.* (1998) proposed the following class of exponential integrators, which uses the φ_1-function only:

$$\begin{aligned} k_i &= \varphi_1(\gamma h J_n)\Big(J_n u_n + g_n(U_{ni}) + h J_n \sum_{j=1}^{i-1} \beta_{ij} k_j \Big), \quad i = 1, \ldots, s, \\ U_{ni} &= u_n + h \sum_{j=1}^{i-1} \alpha_{ij} k_j, \\ u_{n+1} &= u_n + h \sum_{i=1}^{s} b_i k_i. \end{aligned}$$

Here, $\gamma, \alpha_{ij}, \beta_{ij}, b_i$, are scalar coefficients that determine the method. This method is of the form (2.20), with χ_i, $i = 1, \ldots, s$ being linear combinations of exponential functions and with $\chi(z) = \mathrm{e}^z$. The method exp4 proposed by Hochbruck *et al.* (1998) can be interpreted as a three-stage exponential Rosenbrock-type method, which is of classical order four. However, for an efficient implementation, it should be written as a seven-stage method, which uses only three function evaluations. Although our theory does not fully cover this class of methods, it is obvious from the order conditions that methods using the φ_1-function only cannot have order larger than two. ◇

2.5. Exponential multistep methods

After the discussion of exponential one-step methods, we now turn to exponential multistep methods, which were first introduced by Certaine (1960) and in a more systematic way by Nørsett (1969). The idea of Nørsett was to generalize explicit Adams methods such that they become A-stable and thus suited to stiff problems. Given approximations $u_j \approx u(t_j)$, we replace the nonlinearity in (2.1) by its interpolation polynomial at the points

$$\big(t_{n-k+1}, g(t_{n-k+1}, u_{n-k+1})\big), \ldots, \big(t_n, g(t_n, u_n)\big),$$

that is,

$$g\big(t_n + \theta h, u(t_n + \theta h)\big) \approx \sum_{j=0}^{k-1} (-1)^j \binom{-\theta}{j} \nabla^j G_n, \quad G_j = g(t_j, u_j).$$

Here, $\nabla^j G_n$ denotes the jth backward difference defined recursively by

$$\nabla^0 G_n = G_n, \quad \nabla^{j+1} G_n = \nabla^j G_n - \nabla^j G_{n-1}, \quad j = 1, 2, \ldots.$$

By inserting this interpolation polynomial into the variation-of-constants formula (2.19), we obtain an explicit exponential Adams method,

$$u_{n+1} = \mathrm{e}^{-hA} u_n + h \sum_{j=0}^{k-1} \gamma_j(-hA) \nabla^j G_n, \tag{2.69a}$$

with weights

$$\gamma_j(z) = (-1)^j \int_0^1 \mathrm{e}^{(1-\theta)z} \binom{-\theta}{j} \, \mathrm{d}\theta. \tag{2.69b}$$

The weights satisfy the recurrence relation

$$\begin{aligned} \gamma_0(z) &= \varphi_1(z), \\ z\gamma_k(z) + 1 &= \sum_{j=0}^{k-1} \frac{1}{k-j} \gamma_j(z); \end{aligned}$$

see Nørsett (1969, (2.9), (2.10)). For $A = 0$, these exponential Adams methods reduce to the well-known explicit Adams methods; see, *e.g.*, Hairer, Nørsett and Wanner (1993, Chapter III).

For an efficient implementation, (2.69) should be reformulated as a corrected exponential Euler step, that is,

$$u_{n+1} = u_n + h\varphi_1(-hA)(G_n - Au_n) + h \sum_{j=1}^{k-1} \gamma_j(-hA) \nabla^j G_n. \tag{2.70}$$

Our error analysis below shows that $\|\nabla^j G_n\| = \mathcal{O}(h^j)$ for sufficiently smooth solutions, so it can be expected that Krylov approximations become cheaper with increasing j; see also Tokman (2006).

Example 2.26. For $k = 1$ we obtain the exponential Euler method (2.29), while for $k = 2$ we have

$$u_{n+1} = u_n + h\varphi_1(-hA)(G_n - Au_n) + h\varphi_2(-hA)(G_n - G_{n-1}), \qquad (2.71)$$

which will be seen to be second-order convergent. ◇

Cox and Matthews (2002) consider the same class of methods, while Calvo and Palencia (2006) constructed and analysed k-step methods, where the variation-of-constants formula is taken over an interval of length kh instead of h. In contrast to Adams methods, their methods have all parasitic roots on the unit circle. A variety of explicit and implicit schemes is given in Beylkin, Keiser and Vozovoi (1998). Related methods using rational approximation of the arising matrix functions are presented in Lambert and Sigurdsson (1972) and Verwer (1976).

Theorem 2.27. Let the initial value problem (2.1) satisfy Assumptions 2.9 and 2.10, and consider for its numerical solution the k-step exponential Adams method (2.69). For $f(t) = g\big(t, u(t)\big)$ assume that $f \in C^k([0,T], X)$. Then, if

$$\|u_j - u(t_j)\|_V \le c_0 h^k, \quad j = 1, \ldots, k-1,$$

the error bound

$$\|u_n - u(t_n)\|_V \le C \cdot h^k \sup_{0 \le t \le t_n} \|f^{(k)}(t)\|$$

holds uniformly in $0 \le nh \le T$. The constant C depends on T, but it is independent of n and h.

Proof. We will present the proof for the method (2.71), *i.e.*, for $k = 2$ only. The proof for $k > 2$ is analogous.

Inserting the exact solution into the numerical scheme yields

$$u(t_{n+1}) = \mathrm{e}^{-hA}u(t_n) + h\Big(\big(\varphi_1(-hA) + \varphi_2(-hA)\big)f(t_n) - \varphi_2(-hA)f(t_{n-1})\Big) + \delta_{n+1}$$

with defect δ_{n+1}. The interpolation error is given by

$$f(t_n + \theta h) = G_n + \theta \nabla G_n + \frac{1}{2}h^2\theta(\theta+1)f''(\zeta(\theta))$$

for certain intermediate times $\zeta(\theta) \in [t_{n-1}, t_{n+1}]$. Hence, by the variation-of-constants formula (2.8), the defect satisfies

$$\delta_{n+1} = \frac{1}{2}h^3 \int_0^1 \mathrm{e}^{-h(1-\theta)A}\theta(\theta+1)f''(\zeta(\theta))\,\mathrm{d}\theta$$

and is bounded by

$$\|\delta_{n+1}\| \le Ch^3M, \quad \|\delta_{n+1}\|_V \le Ch^{2-\alpha}M, \quad M = \sup_{0\le t\le t_{n+1}} \|f''(t)\|.$$

This yields the error recursion

$$\begin{aligned} e_{n+1} = \mathrm{e}^{-hA}e_n + h\Big(&\big(\varphi_1(-hA)+\varphi_2(-hA)\big)\big(G_n - f(t_n)\big) \\ &- \varphi_2(-hA)\big(G_{n-1} - f(t_{n-1})\big)\Big) - \delta_{n+1}. \end{aligned}$$

From (2.27) we obtain the estimate

$$\begin{aligned} \|e_{n+1}\|_V &\le C\|e_1\|_V + Ch\sum_{j=0}^{n} \frac{1}{((n+1-j)h)^\alpha}\big(\|e_j\|_V + h^2M\big) \\ &\le C\big(\|e_1\|_V + h^2M\big), \end{aligned}$$

where the last inequality follows from Gronwall's Lemma 2.15. □

2.6. Linearized exponential multistep methods

In the same way as for the derivation of exponential Rosenbrock methods in Section 2.4, one can also apply exponential multistep methods (2.69) to the locally linearized equation (2.45). This results in *linearized exponential multistep methods.*

Example 2.28. A particular two-step method of this type has been proposed in Tokman (2006, (39)):

$$u_{n+1} = u_n + h\varphi_1(hJ_n)F(u_n) - h\tfrac{2}{3}\varphi_2(hJ_n)\big(g_n(u_n) - g_n(u_{n-1})\big).$$

It can be interpreted as a perturbed exponential Rosenbrock–Euler step. ◇

Motivated by this example, we consider the variant

$$u_{n+1} = u_n + h\varphi_1(hJ_n)F(u_n) - 2h\varphi_3(hJ_n)\big(g_n(u_n) - g_n(u_{n-1})\big). \tag{2.72}$$

For $J_n = 0$, both methods coincide. The variant (2.72), however, has better convergence properties.

In the subsequent error analysis, we restrict our attention again to autonomous semilinear problems of the form (2.1). This implies that (2.45b) takes the form

$$J_n = -A + \frac{\partial g}{\partial u}(u_n), \quad g_n\big(u(t)\big) = g\big(u(t)\big) - \frac{\partial g}{\partial u}(u_n)u(t). \tag{2.73}$$

Theorem 2.29. Let the initial value problem (2.1) satisfy Assumptions 2.9, 2.10, and 2.16 and consider for its numerical solution the linearized exponential two-step method (2.72). For $f(t) = g\big(u(t)\big)$ assume that $f \in C^3([0,T],X)$. Then, if

$$\|u_1 - u(t_1)\|_V \le c_0 h^3,$$

the error bound

$$\|u_n - u(t_n)\|_V \le C \cdot h^3 \sup_{0\le t\le t_n} \|f^{(3)}(t)\|$$

holds uniformly in $0 \le nh \le T$. The constant C depends on T, but it is independent of n and h.

Proof. Let $f_n(t) = g_n\big(u(t)\big)$. The exact solution has the expansion

$$\begin{aligned} u(t_{n+1}) = u(t_n) &+ h\varphi_1(hJ_n)F\big(u(t_n)\big) + h^2\varphi_2(hJ_n)f_n'(t_n) \\ &+ h^3\varphi_3(hJ_n)f_n''(t_n) + h^4\varrho_3, \end{aligned}$$

with a remainder ϱ_3 as in (2.13) for $p = 3$. This remainder satisfies

$$\|\varrho_3\| \le C, \quad \|\varrho_3\|_V \le Ch^{-\alpha}. \tag{2.74}$$

Inserting the exact solution into the numerical scheme yields

$$\begin{aligned} u(t_{n+1}) = u(t_n) &+ h\varphi_1(hJ_n)F\big(u(t_n)\big) \\ &- 2h\varphi_3(hJ_n)\big(f_n(t_n) - f_n(t_{n-1})\big) + \delta_{n+1}, \end{aligned}$$

with defect

$$\delta_{n+1} = h^2\big(\varphi_2(hJ_n) + 2\varphi_3(hJ_n)\big)f_n'(t_n) + h^4\widetilde{\varrho}_3$$

and a remainder $\widetilde{\varrho}_3$ again satisfying (2.74). Due to (2.60), the defects are thus bounded by

$$\|\delta_{n+1}\| \le C\big(h^2\|e_n\|_V + h^4\big), \quad \|\delta_{n+1}\|_V \le C\big(h^{2-\alpha}\|e_n\|_V + h^{4-\alpha}\big),$$

where $e_n = u_n - u(t_n)$ denotes the error. This error satisfies the recursion

$$\begin{aligned} e_{n+1} = \mathrm{e}^{hJ_n}e_n &+ h\big(\varphi_1(hJ_n) - 2\varphi_3(hJ_n)\big)\big(g_n(u_n) - f_n(t_n)\big) \\ &+ 2h\varphi_3(hJ_n)\big(g_n(u_{n-1}) - f_n(t_{n-1})\big) - \delta_{n+1}, \end{aligned}$$

so that, by (2.27) (for $-J_n$ instead of A) and by the stability result of Hochbruck *et al.* (2009*c*, Appendix A),

$$\|e_n\|_V \le C\|e_1\|_V + Ch\sum_{j=0}^{n-1} \frac{1}{t_{n-j}^{\alpha}}\big(\|e_j\|_V + h^3\big).$$

The stated error bound then follows from a variant of the Discrete Gronwall Lemma (Lemma 2.15). □

Remark 2.30. Higher-order linearized exponential multistep methods can be constructed systematically following this approach. We will discuss details elsewhere.

2.7. Exponential general linear methods

In this section we study explicit exponential general linear methods for the autonomous semilinear problem (2.1). For given starting values $u_0, u_1, \ldots, u_{k-1}$, the numerical approximation u_{n+1} at time t_{n+1}, $n \geq k-1$, is given by the recurrence formula

$$\begin{aligned} u_{n+1} = \mathrm{e}^{-hA} u_n &+ h \sum_{i=1}^{s} b_i(-hA)\, g(t_n + c_i h, U_{ni}) \\ &+ h \sum_{\ell=1}^{k-1} v_\ell(-hA)\, g(t_{n-\ell}, u_{n-\ell}). \end{aligned} \tag{2.75a}$$

The internal stages U_{ni}, $1 \leq i \leq s$, are defined by

$$\begin{aligned} U_{ni} = \mathrm{e}^{-c_i hA} u_n &+ h \sum_{j=1}^{i-1} a_{ij}(-hA)\, g(t_n + c_j h, U_{nj}) \\ &+ h \sum_{\ell=1}^{k-1} w_{i\ell}(-hA)\, g(t_{n-\ell}, u_{n-\ell}). \end{aligned} \tag{2.75b}$$

The coefficient functions $a_{ij}(-hA)$, $w_{i\ell}(-hA)$, $b_i(-hA)$ and $v_\ell(-hA)$ are linear combinations of the exponential and related φ-functions.

The preservation of equilibria of (2.1) is guaranteed under the conditions

$$\begin{aligned} &\sum_{i=1}^{s} b_i(-hA) + \sum_{\ell=1}^{k-1} v_\ell(-hA) = \varphi_1(-hA), \\ &\sum_{j=1}^{i-1} a_{ij}(-hA) + \sum_{\ell=1}^{k-1} w_{i\ell}(-hA) = c_i\, \varphi_1(-c_i hA), \qquad 1 \leq i \leq s. \end{aligned} \tag{2.76}$$

Moreover, these conditions ensure the equivalence of our numerical methods for autonomous and non-autonomous problems. We tacitly assume (2.76) to be satisfied and further suppose $w_{1\ell} = 0$, which implies $c_1 = 0$ and thus $U_{n1} = u_n$.

The explicit exponential Runge–Kutta methods considered in Section 2.3 are contained in the method class (2.75) when setting $k = 1$. The exponential Adams methods of Section 2.5 result from (2.75) for the special case of a single stage $s = 1$.

For the analysis of exponential general linear methods, one proceeds as usual, by inserting the exact solution into the numerical scheme and

Table 2.4. The exponential general linear method (2.75) in tableau form.

$$
\begin{array}{c|cccc|ccc}
c_1 & & & & & & & \\
c_2 & a_{21}(-hA) & & & & w_{21}(-hA) & \dots & w_{2,k-1}(-hA) \\
\vdots & \vdots & \ddots & & & \vdots & & \vdots \\
c_s & a_{s1}(-hA) & \dots & a_{s,s-1}(-hA) & & w_{s1}(-hA) & \dots & w_{s,k-1}(-hA) \\
\hline
 & b_1(-hA) & \dots & b_{s-1}(-hA) & b_s(-hA) & v_1(-hA) & \dots & v_{k-1}(-hA)
\end{array}
$$

providing bounds for the defects by Taylor expansion within the variation-of-constants formula. If we denote the defects for the internal stages U_{ni} by Δ_{ni} and the defect of the new approximation u_{n+1} by δ_{n+1}, then the numerical scheme (2.75) is said to be of *stage order* q if $\Delta_{ni} = \mathcal{O}(h^{q+1})$ for $1 \le i \le s$ and of *quadrature order* p if $\delta_{n+1} = \mathcal{O}(h^{p+1})$. To achieve stage order q,

$$
\begin{aligned}
c_i^m \varphi_m(c_i z) = \sum_{j=1}^{i-1} \frac{c_j^{m-1}}{(m-1)!} a_{ij}(z) \\
+ \sum_{\ell=1}^{k-1} \frac{(-\ell)^{m-1}}{(m-1)!} w_{i\ell}(z), \quad 1 \le i \le s,
\end{aligned} \tag{2.77a}
$$

has to be satisfied for $1 \le m \le q$, and to achieve quadrature order p,

$$
\varphi_m(z) = \sum_{i=1}^{s} \frac{c_i^{m-1}}{(m-1)!} b_i(z) + \sum_{\ell=1}^{k-1} \frac{(-\ell)^{m-1}}{(m-1)!} v_\ell(z) \tag{2.77b}
$$

must hold for $1 \le m \le p$.

The following result from Ostermann, Thalhammer and Wright (2006) shows that for parabolic problems it suffices to satisfy, instead of (2.77b) for $m = p$, the weakened quadrature order conditions

$$
\frac{1}{p} = \sum_{i=1}^{s} c_i^{p-1} b_i(0) + \sum_{\ell=1}^{k-1} (-\ell)^{p-1} v_\ell(0), \tag{2.78}
$$

to obtain the full convergence order p, that is, the condition where $m = p$ is fulfilled for $z = 0$ only. A similar weakened condition appeared in Theorem 2.17 for exponential Runge–Kutta methods.

Theorem 2.31. Let Assumptions 2.9 and 2.16 be satisfied. Furthermore, suppose that the order conditions (2.77) are valid for $m = 1, \dots, p-1$ and that (2.78) holds. Moreover, let f be p-times differentiable with $f^{(p)} \in L^\infty(0,T;X)$ and $A^\beta f^{(p-1)}(t) \in X$ for some $0 \le \beta \le \alpha$ and for all $0 \le t \le T$.

Then, we have the bound

$$\|u(t_n) - u_n\|_V \le C \sum_{m=0}^{k-1} \|u(t_m) - u_m\|_V + Ch^{p-\alpha+\beta} \sup_{0\le t\le t_n} \|A^\beta f^{(p-1)}(t)\| \\ + Ch^p \sup_{0\le t\le t_n} \|f^{(p)}(t)\|, \quad t_k \le t_n \le T,$$

uniformly on $0 \le nh \le T$, with a constant $C > 0$ that is independent of n and h.

Lawson and generalized Lawson methods

A different approach to constructing higher-order exponential methods for semilinear problems (2.1) was proposed by Lawson (1967) and generalized later by Krogstad (2005). Krogstad called the methods (*generalized*) *integrated factor methods*. The idea is to use an appropriate transformation of variables, which is motivated by the Adams method (2.69) with a step size τ:

$$u(t_n + \tau) = \mathrm{e}^{-\tau A} v(\tau) + \tau \sum_{j=0}^{k-1} \gamma_j(-\tau A)\nabla^j G_n, \tag{2.79}$$

with weights (2.69b). Let

$$p(\tau) = \sum_{j=0}^{k-1} (-1)^j \binom{-\tau/h}{j} \nabla^j G_n \tag{2.80}$$

denote the interpolation polynomial through $(t_{n-k+1}, G_{n-k+1}), \ldots, (t_n, G_n)$, evaluated at $t_n + \tau$. Then, by definition of γ_j, we have

$$\tau \sum_{j=0}^{k-1} \gamma_j(-\tau A)\nabla^j G_n = \int_0^\tau \mathrm{e}^{-(\tau-\sigma)A} p(\sigma)\,\mathrm{d}\sigma. \tag{2.81}$$

Differentiating (2.79) with respect to τ by using (2.81), and inserting the result into (2.1), gives the following differential equation for v:

$$v'(\tau) = \mathrm{e}^{\tau A}\big(g(t_n + \tau, u(t_n + \tau)) - p(\tau)\big), \tag{2.82}$$

which is discretized with an explicit Runge–Kutta method or an explicit multistep method, and the result is transformed back to the original variables via (2.79). In the case of Runge–Kutta methods, the resulting methods will be exponential general linear methods. For their analysis see Section 2.7 and Ostermann *et al.* (2006).

At first glance, this looks strange due to the appearance of $\mathrm{e}^{\tau A}$, which is not defined for semigroups. However, this is only a formal transformation and the factor $\mathrm{e}^{\tau A}$ will eventually disappear. For Runge–Kutta methods, this requires the nodes $c_i \in [0, 1]$.

Example 2.32. The method of Lawson (1967) consists of taking $k = 0$ (so the sum in (2.79) is empty) and solving (2.82) with Runge's method. This yields the exponential Runge–Kutta method

$$\begin{array}{c|cc} 0 & & \\ \frac{1}{2} & \frac{1}{2}\mathrm{e}^{-h/2A} & \\ \hline & 0 & \mathrm{e}^{-h/2A} \end{array},$$

which is first-order accurate by Theorem 2.17. Ehle and Lawson (1975) also considered $k = 0$, but used higher-order Runge–Kutta methods implemented via rational approximations of the matrix functions. ◇

Example 2.33. Taking $k = 1$ and Runge's method gives

$$\begin{array}{c|cc} 0 & & \\ \frac{1}{2} & \frac{1}{2}\varphi_{1,2} & \\ \hline & \varphi_1 - \mathrm{e}^{-h/2A} & \mathrm{e}^{-h/2A} \end{array},$$

which is a second-order exponential Runge–Kutta method by Theorem 2.17. For $k = 2$ and Runge's method we obtain

$$\begin{array}{c|cc|c} 0 & & & \\ \frac{1}{2} & \frac{1}{2}(\varphi_{1,2} + \varphi_{2,2}) & & -\frac{1}{2}\varphi_{2,2} \\ \hline & \varphi_1 + \varphi_2 - \frac{3}{2}\mathrm{e}^{-h/2A} & \mathrm{e}^{-h/2A} & -\varphi_2 + \frac{1}{2}\mathrm{e}^{-h/2A} \end{array},$$

which is of second order, by Theorem 2.31. ◇

Example 2.34. The following scheme, with $k = 2$ and the classical fourth-order Runge–Kutta method, was proposed by Krogstad (2005):

$$\begin{array}{c|cccc|c} 0 & & & & & \\ \frac{1}{2} & \frac{1}{2}\varphi_{1,2} + \frac{1}{4}\varphi_{2,2} & & & & -\frac{1}{4}\varphi_{2,2} \\ \frac{1}{2} & \frac{1}{2}\varphi_{1,3} + \frac{1}{4}\varphi_{2,3} - \frac{3}{4}I & \frac{1}{2}I & & & -\frac{1}{4}\varphi_{2,3} + \frac{1}{4}I \\ 1 & \varphi_{1,4} + \varphi_{2,4} - \frac{3}{2}\varphi_{0,2} & 0 & \varphi_{0,2} & & -\varphi_{2,4} + \frac{1}{2}\varphi_{0,2} \\ \hline & \varphi_1 + \varphi_2 - \varphi_{0,2} - \frac{1}{3}I & \frac{1}{3}\varphi_{0,2} & \frac{1}{3}\varphi_{0,2} & \frac{1}{6}I & -\varphi_2 + \frac{1}{3}\varphi_{0,2} + \frac{1}{6}I \end{array}.$$

This scheme satisfies (2.77) for $m = 1, 2$ and (2.78) for $p = 3$. According to Theorem 2.31, it is of order three. ◇

2.8. Magnus integrators

A simple exponential integrator for the non-autonomous linear problem

$$u'(t) = A(t)u(t) + b(t), \quad 0 < t \le T, \qquad u(0) = u_0 \tag{2.83}$$

can be obtained by freezing A and b at $t_{n+1/2}$. This yields the numerical scheme

$$u_{n+1} = \mathrm{e}^{hA_{n+1/2}}u_n + h\varphi_1(hA_{n+1/2})b_{n+1/2}, \quad n \ge 0, \tag{2.84}$$

with

$$A_{n+1/2} = A\left(t_n + \frac{h}{2}\right), \quad b_{n+1/2} = b\left(t_n + \frac{h}{2}\right).$$

The method belongs to the class of Magnus integrators. For $b = 0$ it was studied for time-dependent Schrödinger equations in Hochbruck and Lubich (2003); see Section 3.1 below.

For the analysis of scheme (2.84), stability bounds of the form

$$\left\| \prod_{i=m}^{n} \mathrm{e}^{hA_{i+1/2}} \right\|_{X \leftarrow X} \leq C \tag{2.85}$$

are essential. We omit the details and refer to González, Ostermann and Thalhammer (2006), where the Magnus integrator (2.84) is shown to be second-order convergent under appropriate regularity assumptions. An extension of this theory to quasi-linear parabolic problems can be found in González and Thalhammer (2007). Higher-order methods for the special case $b = 0$ are analysed in Thalhammer (2006).

3. Highly oscillatory problems

In this section, we discuss numerical methods for problems having solutions that are highly oscillatory in time. Obviously, the techniques of the previous section, which always used Taylor expansion of the exact solution, are no longer applicable since higher-order time derivatives will now have large norms. In order to obtain error estimates that are useful for practical applications, it is crucial to base the analysis on assumptions related to the particular problem, for instance the assumption that the energy of a physical problem remains bounded.

We will treat first- and second-order differential equations such as Schrödinger equations with time-dependent Hamiltonian, Newtonian equations of motion, and semilinear wave equations. In order to keep this review to a reasonable length, among the many important properties of numerical methods for highly oscillatory problems, we will only address their finite-time error analysis. In particular, we will show that such problems can be solved numerically so that the error of the numerical solution is bounded independently of the highest frequencies arising in the problem. An overview of various principles for the construction of such integrators is given by Cohen, Jahnke, Lorenz and Lubich (2006). We refer to Hairer *et al.* (2006) and references therein for a detailed study of further properties of the discrete flow, such as long-time behaviour and energy conservation.

The numerical integration of highly oscillatory functions is a strongly related problem. We refer to Iserles and Nørsett (2004) for these quadrature methods. Related numerical integrators for highly oscillatory ordinary differential equations are discussed in Iserles (2002*a*, 2002*b*); an application to electronic engineering is given in Condon, Deaño and Iserles (2009).

3.1. Magnus integrators

This subsection is devoted to the construction of numerical integrators for linear differential equations of the form

$$\psi'(t) = A(t)\psi(t), \quad \psi(0) = \psi_0, \tag{3.1}$$

with a time-dependent, skew-Hermitian matrix $A(t) = -A(t)^*$. We follow an approach due to Magnus (1954). In the context of geometric integration, such integrators were studied in Budd and Iserles (1999) and Iserles and Nørsett (1999). An extensive review with many references on Magnus methods is given by Blanes, Casas, Oteo and Ros (2009).

Without loss of generality, we scale the initial state $\psi(0) = \psi_0$ such that

$$\|\psi_0\| = 1.$$

The assumption that A is skew-Hermitian implies $\|\psi(t)\| = 1$ for all t. Moreover, we will use the following short notation:

$$\psi_n(\tau) = \psi(t_n + \tau), \quad A_n(\tau) = A(t_n + \tau).$$

The idea of Magnus consists in writing the solution of (3.1) as

$$\psi_n(\tau) = \exp(\Omega_n(\tau))\, \psi_n(0), \quad n = 0, 1, \ldots, \tag{3.2}$$

for suitable matrices $\Omega_n(\tau)$, which are determined by differentiating (3.2),

$$\psi_n'(\tau) = \operatorname{dexp}_{\Omega_n(\tau)}(\Omega_n'(\tau))\, \psi_n(0).$$

Here, the dexp operator can be expressed by

$$\operatorname{dexp}_\Omega(B) = \varphi_1(\operatorname{ad}_\Omega)(B) = \sum_{k\geq 0} \frac{1}{(k+1)!} \operatorname{ad}_\Omega^k(B), \tag{3.3}$$

where $\operatorname{ad}_\Omega^k$ is defined recursively by

$$\begin{aligned} \operatorname{ad}_\Omega(B) &= [\Omega, B] = \Omega B - B\Omega, \\ \operatorname{ad}_\Omega^{k+1}(B) &= \operatorname{ad}_\Omega(\operatorname{ad}_\Omega^k(B)), \quad k = 1, 2, \ldots. \end{aligned}$$

Hence, ψ defined in (3.2) solves (3.1) if

$$A_n(\tau) = \operatorname{dexp}_{\Omega_n(\tau)}(\Omega_n'(\tau)), \quad \Omega_n(0) = 0. \tag{3.4}$$

In order to obtain an explicit differential equation for Ω_n, we have to invert the $\operatorname{dexp}_{\Omega_n(\tau)}$ operator. Unfortunately, this operator cannot be guaranteed to be invertible unless $\|\Omega_n(\tau)\| < \pi$. In this case, with β_k denoting the kth Bernoulli number, the series

$$\operatorname{dexp}^{-1}_{\Omega_n(\tau)}(A_n(\tau)) = \sum_{k\geq 0} \frac{\beta_k}{k!} \operatorname{ad}^k_{\Omega_n(\tau)}(A_n(\tau)) \tag{3.5}$$

converges. Taking the first terms of (3.5) yields

$$\Omega_n'(\tau) = A_n(\tau) - \frac{1}{2}[\Omega_n(\tau), A_n(\tau)] + \frac{1}{12}[\Omega_n(\tau), [\Omega_n(\tau), A_n(\tau)]] + \cdots .$$

Picard iteration yields the *Magnus expansion*:

$$\begin{aligned}\Omega_n(h) = &\int_0^h A_n(\tau)\,\mathrm{d}\tau - \frac{1}{2}\int_0^h \left[\int_0^\tau A_n(\sigma)\,\mathrm{d}\sigma, A_n(\tau)\right]\mathrm{d}\tau \\ &+ \frac{1}{4}\int_0^h \left[\int_0^\tau \left[\int_0^\sigma A_n(\mu)\,\mathrm{d}\mu, A_n(\sigma)\right]\mathrm{d}\sigma, A_n(\tau)\right]\mathrm{d}\tau \\ &+ \frac{1}{12}\int_0^h \left[\int_0^\tau A_n(\sigma)\,\mathrm{d}\sigma, \left[\int_0^\tau A_n(\mu)\,\mathrm{d}\mu, A_n(\tau)\right]\right]\mathrm{d}\tau + \cdots .\end{aligned} \tag{3.6}$$

Numerical methods are obtained by truncating this series and approximating the integrals by quadrature rules. If, for instance, $A(t)$ is locally replaced by an interpolation polynomial, then the integrals in the Magnus expansion can be computed analytically. Denoting the result by Ω_n, we obtain

$$\psi_{n+1} = \exp(\Omega_n)\psi_n, \quad \Omega_n \approx \Omega_n(h) \tag{3.7}$$

as a numerical approximation to $\psi(t_{n+1})$ at $t_{n+1} = t_n + h$.

Example 3.1. The simplest method is obtained by truncating the series after the first term and using the midpoint rule for approximating the integral. This yields the exponential midpoint rule

$$\psi_{n+1} = \mathrm{e}^{hA(t_n+h/2)}\psi_n, \quad n = 0, 1, 2, \ldots, \tag{3.8}$$

which corresponds to the choice

$$\Omega_n = hA_n(h/2) = hA(t_n + h/2) \tag{3.9}$$

in (3.7). ◇

Example 3.2. Using two terms of the series and the two-point Gauss quadrature rule with nodes $c_{1,2} = 1/2 \mp \sqrt{3}/6$ yields the fourth-order scheme with

$$\Omega_n = \frac{h}{2}\big(A_n(c_1h) + A_n(c_2h)\big) + \frac{\sqrt{3}h^2}{12}[A_n(c_2h), A_n(c_1h)]. \tag{3.10}$$

This method requires two evaluations of A and one commutator in each time step. ◇

High-order interpolatory Magnus integrators require the computation of many commutators per step. Blanes, Casas and Ros (2002) constructed (non-interpolatory) Magnus integrators for which the number of commutators could be reduced significantly.

While the Magnus series approach nicely motivates the numerical schemes, it is not at all useful for studying the convergence of Magnus integrators in the case of large $\|hA(t)\|$. The reason is – as mentioned before – that dexp_{Ω_n} need not be invertible and the Magnus expansion need not converge; see Moan and Niesen (2008) for a discussion of the convergence properties of the Magnus expansion. Nevertheless, in important practical applications, Magnus integrators work extremely well even with step sizes for which $\|hA(t)\|$ is large.

For the analysis of the exponential midpoint rule (3.8), we use the following assumption.

Assumption 3.3. We assume that $A(t)$ is skew-Hermitian and that the mth derivative satisfies

$$\|A^{(m)}(t)\| \leq M_m, \quad m = 1, 2, \ldots, p$$

for an appropriate integer $p \geq 1$. ○

The following theorem was given by Hochbruck and Lubich (1999*b*).

Theorem 3.4.

(i) If Assumption 3.3 is satisfied with $p = 1$, then the error of the method (3.8) is bounded by

$$\|\psi_n - \psi(t_n)\| \leq C\,h$$

for $0 \leq t_n \leq T$. Here, C depends only on M_1 and T.

(ii) Let Assumption 3.3 hold with $p = 2$ and let the Hermitian matrix

$$H(t) = -\mathrm{i}A(t)$$

be positive definite. Further assume that there exists $0 < \alpha \leq 1$ such that

$$\|H(t)^{\alpha}\psi(t)\| \leq C_{\alpha}, \tag{3.11}$$

$$\|H(t)^{\alpha}A'(t)\psi(t)\| \leq C_{\alpha} \tag{3.12}$$

for $0 \leq t \leq T$. Then the error of the method (3.8) is bounded by

$$\|\psi_n - \psi(t_n)\| \leq C\,h^{1+\alpha}$$

for $0 \leq t_n \leq T$. Here, C only depends on C_{α}, M_1, M_2, and T.

The assumption of a positive definite $H(t)$ is not essential. If the eigenvalues of $H(t)$ are bounded from below by $-\kappa$, then the result still holds when $H(t)$ is replaced by $H(t) + (\kappa + 1)I$ in (3.11) and (3.12).

Proof of Theorem 3.4. **(i)** We insert the exact solution into the numerical scheme and obtain

$$\psi(t_{n+1}) = \mathrm{e}^{hA_n(h/2)}\psi(t_n) + \delta_{n+1} \tag{3.13}$$

with defect δ_{n+1}. By writing (3.1) as

$$\psi'(t) = A(t)\psi(t) = A_n(h/2)\psi(t) + \big(A(t) - A_n(h/2)\big)\psi(t)$$

and applying the variation-of-constants formula (2.8), we find

$$\delta_{n+1} = \int_0^h \mathrm{e}^{(h-\tau)A_n(h/2)}\big(A_n(\tau) - A_n(h/2)\big)\psi(t_n+\tau)\,\mathrm{d}\tau. \tag{3.14}$$

By assumption, we have $\|\delta_{n+1}\| \le CM_1h^2$. Denote the error as usual by $e_n = \psi_n - \psi(t_n)$. Subtracting (3.13) from the numerical scheme (3.8) yields the error recursion

$$e_{n+1} = \mathrm{e}^{hA_n(h/2)}e_n - \delta_{n+1}.$$

Solving this recursion and using $e_0 = 0$ gives

$$\|e_n\| \le \sum_{j=1}^{n} \|\delta_j\| \le CM_1Th.$$

(ii) Using once more the variation-of-constants formula for $\psi(t_n+\tau)$ in (3.14) yields

$$\delta_{n+1} = \int_0^h \mathrm{e}^{(h-\tau)A_n(h/2)}(\tau - h/2)A_n'(h/2)\mathrm{e}^{\tau A_n(h/2)}\psi(t_n)\,\mathrm{d}\tau + \mathcal{O}(h^3),$$

where the constant hidden in the $\mathcal{O}(h^3)$ term depends only on M_2.

In the defect δ_{n+1}, we write $\psi(t_n) = H_{n+1/2}^{-\alpha}H_{n+1/2}^{\alpha}\psi(t_n)$, where $H_{n+1/2} = H(t_n+h/2)$. From $H(t) = -\mathrm{i}A(t)$ we obtain the bound

$$\|(\mathrm{e}^{sA_n(h/2)} - I)H_n^{-\alpha}\| \le \max_{x>0}|(\mathrm{e}^{-\mathrm{i}sx} - 1)/x^\alpha| \le C\,|s|^\alpha.$$

By condition (3.11) we have

$$\int_0^h \mathrm{e}^{(h-\tau)A_n(h/2)}(\tau - h/2)A_n'(h/2)(\mathrm{e}^{\tau A_n(h/2)} - I)H_{n+1/2}^{-\alpha}H_{n+1/2}^{\alpha}\psi(t_n)\,\mathrm{d}\tau = \mathcal{O}(h^{2+\alpha}),$$

and therefore the defect can be written as

$$\delta_{n+1} = \int_0^h \mathrm{e}^{(h-\tau)A_n(h/2)}(\tau - h/2)\,\mathrm{d}\tau\, A_n'(h/2)\psi(t_n) + \mathcal{O}(h^{2+\alpha}).$$

Using condition (3.12) and, once more, the same argument to eliminate the factor $e^{(h-\tau)A_n(h/2)}$, we obtain

$$\delta_{n+1} = \int_0^h (\tau - h/2)\,d\tau\, A_n'(h/2)\psi(t_n) + \mathcal{O}(h^{2+\alpha}) = \mathcal{O}(h^{2+\alpha})$$

because the integral vanishes. This proves the desired estimate. □

Remark 3.5. Under certain non-resonance conditions and conditions on the smoothness of the eigen-decompositions of $A(t)$, second-order error bounds can be proved: see Hochbruck and Lubich (1999*b*, Theorem 2.1) for details.

The convergence behaviour of higher-order Magnus integrators was explained in Hochbruck and Lubich (2003). Our presentation closely follows this paper.

Assumption 3.6. We assume that there exist constants $M_q > 0$ and a symmetric, positive definite matrix D such that

$$A(t) = -i(D^2 + V(t)) \tag{3.15}$$

with $\|V^{(q)}(t)\| \le M_q$ for $q = 0, 1, 2, \ldots$. Moreover, we assume that there exists a constant $K > 0$ such that, for all vectors v and for all step sizes $h > 0$, the commutator bound

$$\|[A(\tau), A(\sigma)]v\| \le Kh\,\|Dv\| \quad \text{for } |\tau - \sigma| \le h \tag{3.16}$$

holds. ○

Without loss of generality, we further assume

$$\|v\| \le \|Dv\| \quad \text{for all } v. \tag{3.17}$$

This can be achieved by shifting D by the identity matrix.

Example 3.7. We consider the linear Schrödinger equation on a d-dimensional torus $\mathcal{T}$ with time-dependent potential,

$$i\psi' = H(t)\psi = -\Delta\psi + V(t)\psi.$$

Here Δ denotes the Laplacian and $V(t)$ is a bounded and smooth multiplication operator. By setting $D^2 = -\Delta + I$, we obtain

$$\|Dv\|^2 = \int_{\mathcal{T}} |\nabla v|^2\,dx + \int_{\mathcal{T}} v^2\,dx,$$

so that $\|Dv\|$ is the familiar H^1 Sobolev norm of v. In the spatially discretized case, $\|Dv\|$ can be viewed as a discrete Sobolev norm. For a space discretization with minimal grid spacing Δx, we note that $\|D^2\| \sim \Delta x^{-2}$ and $\|D\| \sim \Delta x^{-1}$.

In the continuous case, the commutator bound (3.16) is valid since $[D^2, V]$ is a *first-order* differential operator. For a spectral discretization the commutator bound is proved, uniformly in the discretization parameter, in Jahnke and Lubich (2000, Lemma 3.1). ◇

The idea of Hochbruck and Lubich (2003) was to interpret the numerical approximation as the exact solution of a modified problem. Truncation of the Magnus expansion (3.6) amounts to using a modified $\widetilde{\Omega}_n$ instead of Ω_n in (3.2), *i.e.*,

$$\widetilde{\psi}_n(\tau) = \exp(\widetilde{\Omega}_n(\tau))\psi(t_n). \tag{3.18}$$

For the exponential midpoint rule we truncate the series after the first term and obtain

$$\widetilde{\Omega}_n(\tau) = \int_0^\tau A_n(\sigma)\,\mathrm{d}\sigma, \quad 0 \le \tau \le h. \tag{3.19}$$

By differentiating (3.18), we obtain the approximate solution $\widetilde{\psi}_n$ as the solution of the modified differential equation

$$\widetilde{\psi}'_n(\tau) = \widetilde{A}_n(\tau)\widetilde{\psi}_n(\tau), \quad \widetilde{\psi}_n(0) = \psi_n(0) = \psi(t_n), \tag{3.20}$$

with

$$\widetilde{A}_n(\tau) = \operatorname{dexp}_{\widetilde{\Omega}_n(\tau)}(\widetilde{\Omega}'_n(\tau)).$$

Note that the truncated Magnus series $\widetilde{\Omega}_n(\tau)$ and the modified operator $\widetilde{A}_n(\tau)$ are skew-Hermitian if $A(t)$ is skew-Hermitian. As the following lemma shows, a bound on $\widetilde{A}_n(\tau) - A_n(\tau)$ then immediately leads to a local error bound.

Lemma 3.8. Let ψ be a solution of (3.1) with skew-Hermitian $A(t)$, and $\widetilde{\psi}$ a solution of (3.20). Then their difference is bounded by

$$\|\widetilde{\psi}_n(\tau) - \psi_n(\tau)\| \le Ch^3 \max_{0\le\sigma\le h} \|D\psi_n(\sigma)\|, \quad 0 \le \tau \le h.$$

Proof. For $E_n(\tau) = \widetilde{A}_n(\tau) - A_n(\tau)$, we write (3.1) as

$$\psi'_n(\tau) = A_n(\tau)\psi_n(\tau) = \widetilde{A}_n(\tau)\psi_n(\tau) - E_n(\tau)\psi_n(\tau)$$

and subtract it from (3.20). This shows that the error $\widetilde{e}_n(\tau) = \widetilde{\psi}_n(\tau) - \psi_n(\tau)$ satisfies

$$\widetilde{e}'_n(\tau) = \widetilde{A}_n(\tau)\widetilde{e}_n(\tau) + E_n(\tau)\psi_n(\tau), \quad \widetilde{e}_n(0) = 0. \tag{3.21}$$

Taking the inner product with $\widetilde{e}_n$ on both sides of (3.21) and noting that $\widetilde{A}_n$ is skew-Hermitian, we have

$$\operatorname{Re}\langle \widetilde{e}'_n, \widetilde{e}_n\rangle = \operatorname{Re}\langle E_n\,\psi_n, \widetilde{e}_n\rangle \le \|E_n\,\psi_n\|\,\|\widetilde{e}_n\|.$$

On the other hand, we have

$$\operatorname{Re}\langle \widetilde{e}'_n, \widetilde{e}_n\rangle = \frac{1}{2}\frac{\mathrm{d}}{\mathrm{d}\tau}\|\widetilde{e}_n\|^2 = \|\widetilde{e}_n\|\,\frac{\mathrm{d}}{\mathrm{d}\tau}\|\widetilde{e}_n\| \le \|E_n\,\psi_n\|\;\|\widetilde{e}_n\|.$$

Integrating the inequality proves the bound

$$\|\widetilde{\psi}_n(\tau) - \psi_n(\tau)\| \le \int_0^\tau \|\big(\widetilde{A}_n(\sigma) - A_n(\sigma)\big)\psi_n(\sigma)\|\,\mathrm{d}\sigma.$$

For $\widetilde{\Omega}_n = \widetilde{\Omega}_n(h)$, we next use the expansion

$$\widetilde{A}_n = \operatorname{dexp}_{\widetilde{\Omega}_n}(\widetilde{\Omega}'_n) = \widetilde{\Omega}'_n + \varphi_2(\operatorname{ad}_{\widetilde{\Omega}_n})(\operatorname{ad}_{\widetilde{\Omega}_n}(\widetilde{\Omega}'_n))$$

and $A_n = \widetilde{\Omega}'_n$ by (3.19). It was shown in Hochbruck and Lubich (2003, Lemma 5.1) that the remainder can be bounded by

$$\|\varphi_2(\operatorname{ad}_{\widetilde{\Omega}_n})(\operatorname{ad}_{\widetilde{\Omega}_n}(\widetilde{\Omega}'_n))\psi_n(\sigma)\| \le Ch^2\,\|D\psi_n(\sigma)\|, \quad 0 \le \tau \le h. \tag{3.22}$$

This finally proves the desired result. □

Theorem 3.9. Let Assumption 3.6 hold. If $A(t)$ satisfies the commutator bound (3.16), then the error of the exponential midpoint rule (3.8) is bounded by

$$\|\psi_n - \psi(t_n)\| \le Ch^2\,t_n \max_{0\le t\le t_n}\|D\psi(t)\|.$$

The constant C depends only on M_m for $m \le 2$ and on K. In particular, C is independent of n, h, and $\|D\|$.

Proof. Inserting the exact solution ψ of (3.1) into the numerical scheme yields

$$\psi(t_{n+1}) = \exp(\Omega_n)\psi(t_n) + \delta_{n+1}, \tag{3.23}$$

with defect

$$\delta_{n+1} = \psi(t_{n+1}) - \exp(\widetilde{\Omega}_n)\psi(t_n) + \exp(\widetilde{\Omega}_n)\psi(t_n) - \exp(\Omega_n)\psi(t_n).$$

Since the midpoint rule is of order two and due to $\|A''(t)\| \le M_2$, the quadrature error is bounded by

$$\|\widetilde{\Omega}_n - \Omega_n\| = \left\|\int_0^h \big(A_n(\tau) - A_n(h/2)\big)\,\mathrm{d}\tau\right\| \le \frac{1}{24}M_2h^3,$$

and this leads immediately to

$$\|\mathrm{e}^{\widetilde{\Omega}_n} - \mathrm{e}^{\Omega_n}\| = \left\|\int_0^1 \mathrm{e}^{(1-s)\Omega_n}(\widetilde{\Omega}_n - \Omega_n)\mathrm{e}^{s\widetilde{\Omega}_n}\,\mathrm{d}s\right\| \le \frac{1}{24}M_2h^3. \tag{3.24}$$

By Lemma 3.8, estimates (3.22) and (3.24), and assumption (3.17), we obtain

$$\|\delta_{n+1}\| \le Ch^3 \max_{t_n\le t\le t_{n+1}}\|D\psi(t)\|.$$

Subtracting (3.23) from (3.7) leads to the error recursion

$$e_{n+1} = \exp(\Omega_n)e_n - \delta_{n+1}$$

for $e_n = \psi_n - \psi(t_n)$, and thus

$$\|e_n\| \le \sum_{j=1}^{n} \|\delta_j\|. \tag{3.25}$$

This proves the stated error estimate. □

The following error bound of Hochbruck and Lubich (2003, Theorem 3.2) holds for pth-order interpolatory Magnus integrators, *i.e.*, those based on a pth-order truncation of the Magnus series and polynomial interpolation of $A(t)$ at the nodes of a pth-order quadrature formula.

Theorem 3.10. Let Assumption 3.6 hold and assume that for a method of classical order p, which contains commutator products of $A(t_n+c_jh)$ with r factors, the commutator bounds

$$\left\| \left[A(\tau_k), \left[\dots, \left[A(\tau_1), \frac{\mathrm{d}^m}{\mathrm{d}t^m}V(\tau_0)\right]\right]\dots\right]v\right\| \le K\,\|D^k v\| \tag{3.26}$$

hold for arbitrary times τ_j for $0 \le m \le p$ and $k+1 \le rp$. Then the pth-order interpolatory Magnus integrators satisfy the error bound

$$\|\psi_n - \psi(t_n)\| \le Ch^p\, t_n \max_{0\le t\le t_n} \|D^{p-1}\psi(t)\|,$$

for time steps h satisfying $h\|D\| \le c$. The constant C depends only on M_m for $m \le p$, on K, c, and on p. In particular, C is independent of n, h, and $\|D\|$ as long as $h\|D\| \le c$.

In Example 3.2, we have $r = 2$ for order $p = 4$. For all Magnus methods proposed in the literature, $r \le p-1$ holds.

Adiabatic integrators for time-dependent Schrödinger equations

Next we consider the related singularly perturbed problem

$$\psi'(t) = \frac{1}{\varepsilon}A(t)\psi(t), \quad 0 < \varepsilon \ll 1. \tag{3.27}$$

We still impose Assumption 3.3. For background information about the adiabatic theory in quantum dynamics we refer to Teufel (2003).

Solving this problem by the exponential midpoint rule would require time steps of size $h = \mathcal{O}(\varepsilon)$ to achieve reasonable accuracy. Moreover, the convergence bounds of Section 3.1 are not practical here, due to the large factor $1/\varepsilon$ multiplying the derivatives of $A(t)$.

A different approach to the solution of such problems is based on adiabatic transformations: see Jahnke and Lubich (2003), Jahnke (2004), Hairer *et al.*

(2006, Chapter XIV) and references therein. The main idea is to represent the solution in a rotating frame of eigenvectors of A. Let

$$-\mathrm{i}A(t) = H(t) = Q(t)\Lambda(t)Q(t)^*, \quad \Lambda(t) = \mathrm{diag}(\lambda_k(t)) \tag{3.28}$$

be the eigen-decomposition of the real symmetric matrix $H(t)$. Here, $\lambda_k(t)$ are the eigenvalues and $Q(t)$ is the orthogonal matrix of eigenvectors of $H(t)$ depending smoothly on t (unless, possibly, crossings of eigenvalues occur). Then the adiabatic transformation is defined by

$$\eta(t) = \mathrm{e}^{-\frac{\mathrm{i}}{\varepsilon}\Phi(t)}Q(t)^*\psi(t), \tag{3.29a}$$

where

$$\Phi(t) = \mathrm{diag}(\phi_j(t)) = \int_0^t \Lambda(s)\,\mathrm{d}s. \tag{3.29b}$$

Differentiation with respect to t leads to the differential equation

$$\eta'(t) = \mathrm{e}^{-\frac{\mathrm{i}}{\varepsilon}\Phi(t)}W(t)\mathrm{e}^{\frac{\mathrm{i}}{\varepsilon}\Phi(t)}\eta(t), \quad W(t) = Q'(t)^*Q(t). \tag{3.30}$$

The orthogonality of Q ensures that W is a skew-symmetric matrix for all t. Note that the assumption on the smoothness of the eigenvectors guarantees that $\|\eta'(t)\|$ is bounded. The quantum-adiabatic theorem of Born and Fock (1928) yields that the solution of (3.30) satisfies

$$\eta(t) = \eta(0) + \mathcal{O}(\varepsilon) \tag{3.31}$$

uniformly on bounded time intervals if the eigenvalues are separated from each other by a constant δ, which is independent of ε, *i.e.*, if

$$|\lambda_j(t) - \lambda_k(t)| \geq \delta, \tag{3.32}$$

and if $\|\Lambda'(t)\| \leq C$, $\|Q'(t)\| \leq C$ for some constant C for $0 \leq t \leq T$. Quantities satisfying (3.31) are called *adiabatic invariants.*

We consider (3.30) with the bounded matrix

$$B(t) = \mathrm{e}^{-\frac{\mathrm{i}}{\varepsilon}\Phi(t)}W(t)\mathrm{e}^{\frac{\mathrm{i}}{\varepsilon}\Phi(t)}.$$

Since B is highly oscillatory, the exponential midpoint rule should not be applied directly. Instead, an averaging process should first be applied. We will explain the basic idea presented by Jahnke and Lubich (2003) and Jahnke (2004) for the construction of such methods. For the simplest scheme – which, however, is not of practical interest – the phase is approximated by

$$\Phi(t_n + \tau) \approx \Phi(t_n + h/2) + (h/2 - \tau)\Lambda(t_n + h/2) =: \widetilde{\Phi}_{n+1/2}(\tau),$$

with an error of size $\mathcal{O}(h^2)$ for $0 \leq \tau \leq h$. This approximation is inserted into $B(t)$, the slow variable W is approximated by its value at the midpoint, and the time average is taken over a time step. This results in the following approximation:

$$B(t_n+h/2) \approx \widetilde{B}_{n+1/2} = \frac{1}{h}\int_0^h \mathrm{e}^{-\frac{\mathrm{i}}{\varepsilon}\widetilde{\Phi}_{n+1/2}(\tau)} W(t_n+h/2)\mathrm{e}^{\frac{\mathrm{i}}{\varepsilon}\widetilde{\Phi}_{n+1/2}(\tau)}\,\mathrm{d}\tau. \quad (3.33)$$

The integration can be done exactly, leading to

$$\widetilde{B}_{n+1/2} = E(\Phi(t_{n+1/2})) \bullet \mathcal{I}(\Lambda(t_{n+1/2})) \bullet W(t_{n+1/2}),$$

where $\bullet$ denotes the entrywise product of matrices. The matrices E and $\mathcal{I}$ are defined by

$$\begin{aligned} E(\Phi)_{jk} &= \mathrm{e}^{-\frac{\mathrm{i}}{\varepsilon}(\phi_j-\phi_k)}, \\ \mathcal{I}(\Lambda)_{jk} &= \int_{-h/2}^{h/2} \mathrm{e}^{-\frac{\mathrm{i}}{\varepsilon}\tau(\lambda_j-\lambda_k)}\,\mathrm{d}\tau = \operatorname{sinc}\Big(\frac{h}{2\varepsilon}(\lambda_j-\lambda_k)\Big). \end{aligned}$$

The averaged adiabatic exponential midpoint rule finally reads

$$\eta_{n+1} = \mathrm{e}^{h\widetilde{B}_{n+1/2}}\eta_n, \quad n = 0,1,2,\ldots. \quad (3.34)$$

Unfortunately, the error of this scheme is $\mathcal{O}(\min\{\varepsilon, h\})$ only, which is not satisfactory because η is an adiabatic invariant.

Practically relevant schemes use better approximations of the phase and of the factor $W(s)$, and are combined with recursive usage of the variation-of-constants formula or higher-order Magnus integrators. Such methods are presented and analysed in Jahnke and Lubich (2003) and Jahnke (2003, 2004).

3.2. Second-order differential equations

We now consider the second-order differential equation

$$q''(t) = f(q(t)), \quad q(0) = q_0, \; q'(0) = p_0, \quad (3.35a)$$

where we assume that the force f can be decomposed into fast and slow forces:

$$f = f_{\text{fast}} + f_{\text{slow}}. \quad (3.35b)$$

In practice, the fast forces are often cheap to evaluate while slow forces are expensive to compute. Therefore, we are interested in constructing methods that use only one evaluation of the slow force per time step.

We start with the semilinear problem where f_{fast} is a linear force and where, for simplicity, the slow force is denoted by g:

$$q''(t) = -\Omega^2 q(t) + g(q(t)), \quad q(0) = q_0, \; q'(0) = p_0. \quad (3.36)$$

Ω is assumed to be a symmetric positive definite matrix. We are interested in the case where high frequencies occur, *i.e.*, $\|\Omega\| \gg 1$.

The variation-of-constants formula (2.19) applied to the equivalent first-order system

$$\begin{bmatrix} q(t) \\ p(t) \end{bmatrix}' = \begin{bmatrix} 0 & I \\ -\Omega^2 & 0 \end{bmatrix} \begin{bmatrix} q(t) \\ p(t) \end{bmatrix} + \begin{bmatrix} 0 \\ g(q(t)) \end{bmatrix} \tag{3.37}$$

yields

$$\begin{bmatrix} q(t) \\ p(t) \end{bmatrix} = R(t\Omega) \begin{bmatrix} q(0) \\ p(0) \end{bmatrix} + \int_0^t \begin{bmatrix} \Omega^{-1} \sin((t-s)\Omega) \\ \cos((t-s)\Omega) \end{bmatrix} g(q(s))\, \mathrm{d}s, \tag{3.38}$$

with

$$R(t\Omega) = \exp\left(t \begin{bmatrix} 0 & I \\ -\Omega^2 & 0 \end{bmatrix} \right) = \begin{bmatrix} \cos(t\Omega) & \Omega^{-1}\sin(t\Omega) \\ -\Omega \sin(t\Omega) & \cos(t\Omega) \end{bmatrix}. \tag{3.39}$$

Numerical methods can be constructed by approximating the integral appropriately, either by using a standard quadrature rule or by approximating the function g and integrating exactly. We consider explicit schemes only.

Example 3.11. Gautschi (1961) developed a number of trigonometric multistep methods that use the above variation-of-constants formula (3.38). A symmetric two-step scheme is obtained by replacing $g(q(s))$ by $g(q_n)$ to approximate $q(t_n \pm h)$. Adding and subtracting yield

$$q_{n+1} - 2\cos(h\Omega) q_n + q_{n-1} = h^2 \operatorname{sinc}^2\left(\tfrac{h}{2}\Omega\right) g(q_n),$$

where $\operatorname{sinc}\xi = \sin\xi/\xi$. Approximations to the momenta can be obtained via

$$p_{n+1} - p_{n-1} = 2h \operatorname{sinc}(h\Omega)(-\Omega^2 q_n + g(q_n)).$$

The same approximation of g within the integral (3.38) leads to the one-step scheme

$$\begin{bmatrix} q_{n+1} \\ p_{n+1} \end{bmatrix} = R(h\Omega) \begin{bmatrix} q_n \\ p_n \end{bmatrix} + \frac{h}{2} \begin{bmatrix} h \operatorname{sinc}^2\left(\frac{h}{2}\Omega\right) \\ 2\operatorname{sinc}(h\Omega) \end{bmatrix} g(q_n).$$

However, we would like to stress that the two schemes are not equivalent. ◇

Example 3.12. Deuflhard (1979) refined Gautschi's method by using the trapezoidal rule to approximate the integrals. He suggested the two-step scheme

$$q_{n+1} - 2\cos(h\Omega) q_n + q_{n-1} = h^2 \operatorname{sinc}(h\Omega)\, g(q_n)$$

and its one-step formulation

$$\begin{bmatrix} q_{n+1} \\ p_{n+1} \end{bmatrix} = R(h\Omega) \begin{bmatrix} q_n \\ p_n \end{bmatrix} + \frac{h}{2} \begin{bmatrix} h \operatorname{sinc}(h\Omega) g(q_n) \\ \cos(h\Omega) g(q_n) + g(q_{n+1}) \end{bmatrix}.$$

In contrast to Gautschi's method, this scheme is symmetric as a one- and a two-step method. ◇

If $\|\Omega\| \gg 1$, the solution of (3.36) will have highly oscillatory components. The pointwise evaluation of the right-hand side g in combination with large time steps is not advisable. Much better behaviour can be expected if g is evaluated at an averaged value of q. Clearly, there is a lot of freedom in choosing such averages. García-Archilla, Sanz-Serna and Skeel (1998) propose solving an auxiliary problem,

$$y'' = -\Omega^2 y, \quad y(0) = q, \; y'(0) = 0,$$

and computing the averaged solutions over a time step of length h,

$$a(q) = \frac{1}{h}\int_0^h y(\tau)\,\mathrm{d}\tau = \operatorname{sinc}(h\Omega)q. \tag{3.40}$$

The examples above and the averaging approach suggest the following class of two-step methods for the solution of (3.36):

$$q_{n+1} - 2\cos(h\Omega)q_n + q_{n-1} = h^2\psi(h\Omega)\,g(\phi(h\Omega)q_n). \tag{3.41}$$

The schemes are symmetric if the filter functions ψ and ϕ are even, which we will assume henceforth.

We also consider the corresponding class of one-step schemes given by

$$\begin{bmatrix} q_{n+1} \\ p_{n+1} \end{bmatrix} = R(h\Omega)\begin{bmatrix} q_n \\ p_n \end{bmatrix} + \frac{h}{2}\begin{bmatrix} h\Psi g(\Phi q_n) \\ \Psi_0 g(\Phi q_n) + \Psi_1 g(\Phi q_{n+1}) \end{bmatrix}, \tag{3.42}$$

where

$$\Phi = \phi(h\Omega), \quad \Psi = \psi(h\Omega), \quad \Psi_0 = \psi_0(h\Omega), \quad \Psi_1 = \psi_1(h\Omega)$$

with suitable functions ϕ, ψ, ψ_0 and ψ_1; see Hairer *et al.* (2006, Chapter XIII). The one-step method (3.42) is symmetric if and only if

$$\psi(\xi) = \psi_1(\xi)\operatorname{sinc}\xi, \quad \psi_0(\xi) = \psi_1(\xi)\cos\xi. \tag{3.43}$$

A symmetric one-step scheme is equivalent to its corresponding two-step scheme for appropriate initial values.

Finite-time error analysis

We now study the error of methods of the class (3.42) over a finite time interval $[0, T]$ for problems whose solutions satisfy a finite-energy condition. Our presentation is based on Grimm and Hochbruck (2006).

Assumption 3.13. We assume that there exists a constant $K > 0$ such that the solution of (3.37) satisfies the *finite-energy condition*

$$\frac{1}{2}\|p(t)\|^2 + \frac{1}{2}\|\Omega q(t)\|^2 \le K^2 \tag{3.44}$$

uniformly for $0 \le t \le T$. ○

The even analytic functions defining the integrator (3.42) are assumed to be bounded on the non-negative real axis, *i.e.*,

$$\max_{\xi\geq 0}\left|\chi(\xi)\right| \leq M_1, \quad \chi = \phi,\, \psi,\, \psi_0,\, \psi_1, \tag{3.45}$$

for some constant M_1. Moreover, we assume $\phi(0) = 1$ and thus the existence of a constant M_2 such that

$$\max_{\xi\geq 0}\left|\frac{\phi(\xi)-1}{\xi}\right| \leq M_2. \tag{3.46}$$

In addition, we assume

$$\max_{\xi\geq 0}\left|\frac{1}{\sin\frac{\xi}{2}}\left(\operatorname{sinc}^2\frac{\xi}{2} - \psi(\xi)\right)\right| \leq M_3 \tag{3.47}$$

and

$$\max_{\xi\geq 0}\left|\frac{1}{\xi\sin\frac{\xi}{2}}(\operatorname{sinc}\xi - \chi(\xi))\right| \leq M_4, \quad \chi = \phi,\, \psi_0,\, \psi_1. \tag{3.48}$$

The assumptions made so far are necessary to prove second-order error bounds for the positions $q_n \approx q(t_n)$. In order to verify first-order error bounds for the momenta p, we assume that

$$\max_{\xi\geq 0}\left|\xi\,\psi(\xi)\right| \leq M_5, \quad \max_{\xi\geq 0}\left|\frac{\xi}{\sin\frac{\xi}{2}}\left(\operatorname{sinc}^2\frac{\xi}{2} - \psi(\xi)\right)\right| \leq M_6. \tag{3.49}$$

Clearly, the constants M_1 to M_6 only depend on the choice of the analytic functions. It is easy to find analytic functions for which

$$M := \max_{i=1,\ldots,6} M_i$$

is a small constant.

Example 3.14. The method of Gautschi (1961) described in Example 3.11 uses the filters

$$\phi(\xi) = 1, \quad \psi(\xi) = \operatorname{sinc}^2(\tfrac{1}{2}\xi), \quad \psi_0(\xi) = 2\operatorname{sinc}\xi, \quad \psi_1(\xi) = 0.$$

Hence, condition (3.48) is not satisfied for ψ_0 and ψ_1. For the method of Deuflhard (1979), we have

$$\phi(\xi) = 1, \quad \psi(\xi) = \operatorname{sinc}\xi, \quad \psi_0(\xi) = \cos\xi, \quad \psi_1(\xi) = 1.$$

These methods do not use an inner filter ϕ, and thus suffer from resonances.

Filter functions that fulfil (3.45)–(3.49) have been proposed in the literature. García-Archilla *et al.* (1998) suggested the choice

$$\phi(\xi) = \operatorname{sinc}\xi, \quad \psi(\xi) = \operatorname{sinc}^2\xi, \tag{3.50}$$

and Grimm and Hochbruck (2006) suggested

$$\phi(\xi) = \operatorname{sinc}\xi, \quad \psi(\xi) = \operatorname{sinc}^3\xi. \qquad \diamond$$

The following theorem of Grimm and Hochbruck (2006) states that suitably chosen filter functions lead to second-order errors in the positions q and first-order error bounds in the momenta p.

Theorem 3.15. In (3.36), let Ω be an arbitrary symmetric positive semi-definite matrix and let the solution satisfy Assumption 3.13. Moreover, suppose that g and the derivatives g_q and g_{qq} are bounded in the Euclidean norm or the norms induced by the Euclidean norm, respectively. If the even analytic functions of the scheme (3.42) satisfy (3.45)–(3.48), then

$$\|q(t_n) - q_n\| \leq h^2 C, \quad 0 \leq t_n = nh \leq T.$$

The constant C only depends on $T, K, M_1, \ldots, M_4$, $\|g\|, \|g_q\|$, and $\|g_{qq}\|$.
If, in addition, (3.49) holds, then

$$\|p(t_n) - p_n\| \leq h\,\widetilde{C}, \quad 0 \leq t_n = nh \leq T.$$

The constant $\widetilde{C}$ only depends on T, K, M, $\|g\|$, $\|g_q\|$, and $\|g_{qq}\|$.

Proof. Substitution of the exact solution into the integration scheme (3.42) gives

$$\begin{bmatrix} q(t_{n+1}) \\ p(t_{n+1}) \end{bmatrix} = R(h\Omega) \begin{bmatrix} q(t_n) \\ p(t_n) \end{bmatrix} + \begin{bmatrix} \frac{1}{2}h^2 \Psi g(\Phi q(t_n)) \\ \frac{h}{2}\big(\Psi_0 g(\Phi q(t_n)) + \Psi_1 g(\Phi q(t_{n+1}))\big) \end{bmatrix} + \begin{bmatrix} \delta_{n+1} \\ \delta'_{n+1} \end{bmatrix},$$

with the defects δ_{n+1} and δ'_{n+1}. Subtraction of equation (3.42) and summation leads to

$$\begin{bmatrix} e_{n+1} \\ e'_{n+1} \end{bmatrix} = R(h\Omega)^{n+1} \begin{bmatrix} e_0 \\ e'_0 \end{bmatrix} + \sum_{j=0}^{n} R(h\Omega)^{n-j} \begin{bmatrix} \frac{1}{2}h^2 \Psi F_j e_j \\ \frac{1}{2}h\Psi_0 F_j e_j + \frac{1}{2}h\Psi_1 F_{j+1} e_{j+1} \end{bmatrix} + \begin{bmatrix} \Delta_{n+1} \\ \Delta'_{n+1} \end{bmatrix}, \tag{3.51}$$

where $e_n := q(t_n) - q_n$ and $e'_n := p(t_n) - p_n$,

$$F_n := \int_0^1 g_q\big(\Phi(q_n + \theta e_n)\big)\, \mathrm{d}\theta\, \Phi, \quad \|F_n\| \leq M_1 \|g_q\|,$$

and

$$\begin{bmatrix} \Delta_n \\ \Delta'_n \end{bmatrix} = \sum_{j=1}^{n} R\big((n-j)h\Omega\big) \begin{bmatrix} \delta_j \\ \delta'_j \end{bmatrix}.$$

Unfortunately, the defects δ_j are of size $\mathcal{O}(h^2)$ and the defects δ'_j are of size $\mathcal{O}(h)$ only. Bounding the sums in a standard way would yield first-order error bounds for the positions and nothing for the velocities. The key point of the proof is now to write

$$\delta_{n+1} = h^2 \cdot (\text{highly oscillatory function}) \cdot g(\Phi q_n) + \mathcal{O}(h^3),$$

and to use a similar but more complicated form for $h\delta'_{n+1}$. Such expressions are derived in Grimm and Hochbruck (2006, Lemmas 1 and 2). Using summation by parts, the sums of highly oscillatory functions are bounded independently of n as long as no resonances occur, while the differences of the g-function yield an additional factor of h due to the smoothness of g. In general, resonant frequencies will deteriorate the convergence and this is exactly the place where the filter functions come into play. They are constructed in such a way that possible resonances disappear. The analysis is detailed in Grimm and Hochbruck (2006, Lemmas 3–6). We then have

$$\|\Delta_n\| \le Ch^2 \quad \text{and} \quad \|\Delta'_n\| \le Ch,$$

with a constant C independent of n as long as $0 \le nh \le T$. Due to $e_0 = e'_0 = 0$, the recursion (3.51) reads

$$e_n = h \sum_{j=1}^{n-1} L_j e_j + \Delta_n,$$

where

$$L_j := \frac{1}{2}\Big(h\cos\big((n-j)h\Omega\big)\,\Psi + (n-j)h\ \mathrm{sinc}\big((n-j)h\Omega\big)\,\Psi_0 + (n+1-j)h\ \mathrm{sinc}\big((n+1-j)h\Omega\big)\,\Psi_1\Big)F_j.$$

This yields

$$\|L_j\| \le \frac{3}{2}TM_1^2\,\|g_q\|,$$

so that $\|e_n\| \le Ch^2$ follows from Gronwall's Lemma 2.15. Assumption (3.49) and the recursion for e'_n finally show $\|e'_n\| \le Ch$. □

Example 3.16. A similar result for two-step methods (3.41) was proved in Hochbruck and Lubich (1999*c*). There, it was shown that the choice

$$\psi(\xi) = \mathrm{sinc}^2\frac{\xi}{2}, \quad \phi(\xi) = \left(1 + \frac{1}{3}\mathrm{sinc}^2\frac{\xi}{2}\right)\mathrm{sinc}\,\xi$$

yields a method with a small error constant. The conjecture that the error constant is uniform in the number of time steps and the dimension of the problem was proved elegantly in Grimm (2005*a*). ◇

In Grimm (2002, 2005*b*) the error analysis is extended to more general problems, where $\Omega = \Omega(t,q)$ varies smoothly, *i.e.*, the first and the second partial derivatives of $\Omega(t,q)$ are bounded.

Impulse and mollified impulse method

Writing (3.35) as a first-order differential equation, and using Strang splitting applied to

$$\begin{bmatrix} q' \\ p' \end{bmatrix} = \begin{bmatrix} p \\ f_{\text{fast}}(q) \end{bmatrix} + \begin{bmatrix} 0 \\ f_{\text{slow}}(q) \end{bmatrix},$$

yields the *impulse method*, by Grubmüller, Heller, Windemuth and Schulten (1991) and Tuckerman, Berne and Martyna (1992). Following García-Archilla *et al.* (1998), we state it in the following form:

$$\begin{aligned}
\text{kick} \quad & p_n^+ = p_n + \frac{h}{2} f_{\text{slow}}(q_n), \\
\text{oscillate} \quad & \text{solve } q'' = f_{\text{fast}}(q) \text{ with initial values } (q_n, p_n^+) \text{ over} \\
& \text{a time step of length } h \text{ to obtain } (q_{n+1}, p_{n+1}^-), \\
\text{kick} \quad & p_{n+1} = p_{n+1}^- + \frac{h}{2} f_{\text{slow}}(q_{n+1}).
\end{aligned}$$

For linear fast forces (3.36), the 'oscillate' step can be computed exactly, leading to

$$\begin{bmatrix} q_{n+1} \\ p_{n+1}^- \end{bmatrix} = R(h\Omega) \begin{bmatrix} q_n \\ p_n^+ \end{bmatrix}.$$

The method is then precisely the scheme of Deuflhard (1979), described in Example 3.12 above. For more general forces f, the 'oscillate' step can be done numerically using smaller time steps. This is affordable since the fast forces are usually cheap to evaluate.

It was observed in Biesiadecki and Skeel (1993) that the impulse method also suffers from resonances, which may occur if eigenvalues of $h\Omega$ are integer multiples of π. They can be circumvented by suitable averaging. When the slow force has a potential U, $f_{\text{slow}} = -\nabla U$, García-Archilla *et al.* (1998) suggest replacing $U(q)$ by an averaged potential $\overline{U}(q) = U(a(q))$. This leads to a mollified force:

$$\overline{f}_{\text{slow}}(q) = a'(q)^T f_{\text{slow}}(a(q)). \tag{3.52}$$

The same technique can then be used in the general case where f does not have a potential. The *mollified impulse method* of García-Archilla *et al.* (1998) consists of using the averaged slow force $\overline{f}_{\text{slow}}$ instead of f_{slow} in the impulse method.

To compute the mollifier, we solve the auxiliary problem

$$y'' = f_{\text{fast}}(y), \quad y(0) = q, \ y'(0) = 0,$$

together with the variational equation

$$Y'' = f'_{\text{fast}}(y(t))Y, \quad Y(0) = I, \ Y'(0) = 0,$$

and compute the averaged solutions over a time step of length h,

$$a(q) = \frac{1}{h}\int_0^h y(\tau)\,\mathrm{d}\tau, \quad a'(q) = \frac{1}{h}\int_0^h Y(\tau)\,\mathrm{d}\tau.$$

Example 3.17. For linear forces as in (3.36), we obtain the averaging function (3.40) and thus $a'(q) = \mathrm{sinc}(h\Omega)$. The mollified impulse method then reads:

$$\begin{aligned} &\text{kick} \quad p_n^+ = p_n + \frac{h}{2}\,\mathrm{sinc}(h\Omega) g\big(\mathrm{sinc}(h\Omega) q_n\big), \\ &\text{oscillate} \quad \begin{bmatrix} q_{n+1} \\ p_{n+1}^- \end{bmatrix} = R(h\Omega)\begin{bmatrix} q_n \\ p_n^+ \end{bmatrix}, \\ &\text{kick} \quad p_{n+1} = p_{n+1}^- + \frac{h}{2}\,\mathrm{sinc}(h\Omega) g\big(\mathrm{sinc}(h\Omega) q_{n+1}\big), \end{aligned}$$

and this is equivalent to (3.42) with filter (3.50). Thus the convergence is covered by Theorem 3.15. Different proofs have been given by García-Archilla *et al.* (1998) and by Hochbruck and Lubich (1999*c*). ◇

Multiple time-stepping

Yet another way to treat different scales in the general case (3.35) is motivated by the following relation satisfied by the exact solution:

$$q(t+h) - 2q(t) + q(t-h) = h^2 \int_{-1}^{1} (1-|\theta|) f(q(t+\theta h))\,\mathrm{d}\theta;$$

see Hairer *et al.* (2006, Chapter VIII.4). The force is approximated by

$$f(q(t_n+\theta h)) \approx f_{\mathrm{fast}}(y(\theta h)) + f_{\mathrm{slow}}(q_n),$$

where $y(\tau)$ is a solution of the auxiliary problem

$$y''(\tau) = f_{\mathrm{fast}}(y(\tau)) + f_{\mathrm{slow}}(q_n), \quad y(0) = q_n,\ y'(0) = p_n. \tag{3.53}$$

This yields

$$\int_{-1}^{1} (1-|\theta|) f(q_n + y(\theta h))\,\mathrm{d}\theta = \frac{1}{h^2}\big(y(h) - 2y(0) + y(-h)\big). \tag{3.54}$$

For the velocities we proceed analogously, starting from

$$y'(t+h) - y'(t-h) = h\int_{-1}^{1} f(q(t+\theta h))\,\mathrm{d}\theta.$$

The auxiliary problem (3.53) can either be solved exactly or by a standard method such as the Störmer–Verlet scheme with a smaller step size on the time interval $[-h, h]$. Denoting the approximations by $y_{n\pm1} \approx y(\pm h)$, this yields the following symmetric two-step method of Hochbruck and Lubich (1999*c*):

$$q_{n+1} - 2q_n + q_{n-1} = y_{n+1} - 2y_n + y_{n-1},$$
$$p_{n+1} - p_{n-1} = y'_{n+1} - y'_{n-1}.$$

To obtain a one-step scheme, we solve

$$y''(\tau) = f_{\text{fast}}(y(\tau)) + f_{\text{slow}}(q_n), \quad y(0) = q_n, \ y'(0) = 0. \tag{3.55}$$

Note that, for linear fast forces, the averaged force (3.54) is independent of $y'(0)$, as can be seen from (3.38). Moreover, the solution of (3.55) is even, $y(-\tau) = y(\tau)$, which means that the integration has to be done for $\tau \in [0, h]$ only. Hence the averaged force (3.54) can be computed from the solution of (3.55) via

$$\widetilde{f}_n = \frac{2}{h^2}(y(h) - q_n).$$

This leads to the scheme

$$\begin{aligned} p_{n+1/2} &= p_n + \frac{h}{2}\widetilde{f}_n, \\ q_{n+1} &= q_n + hp_{n+1/2}, \\ p_{n+1} &= p_{n+1/2} + \frac{h}{2}\widetilde{f}_{n+1}; \end{aligned}$$

see Hairer *et al.* (2006, Chapter VIII.4). The variables p_n can be interpreted as averaged velocities

$$p_n = \frac{q_{n+1} - q_{n-1}}{2h} \approx \frac{q(t_{n+1}) - q(t_{n-1})}{2h} = \frac{1}{2}\int_{-1}^{1} q'(t_n + \theta h)\,\mathrm{d}\theta.$$

Methods using averaged forces have been used for applications in quantum-classical molecular dynamics. We refer to the review by Cohen *et al.* (2006) for details and further references.

Remark 3.18. The geometric properties of the methods considered above have recently been intensively studied. It is straightforward to verify that symmetric one-step methods (3.42) are symplectic if and only if

$$\psi(\xi) = \operatorname{sinc}(\xi)\phi(\xi).$$

In Hairer and Lubich (2000), however, long-time near-energy conservation for linear problems has been proved under the condition

$$\psi(\xi) = \operatorname{sinc}^2(\xi)\phi(\xi),$$

which cannot be satisfied by a symplectic method. For an overview of geometric properties and, in particular, the techniques of modulated Fourier expansions, we refer to Hairer *et al.* (2006) and Hairer and Lubich (2009).

Adiabatic integrators

Finally, we consider the singularly perturbed second-order differential equation

$$q''(t) + \frac{1}{\varepsilon^2}\Omega(t)^2 q(t) = \frac{1}{\varepsilon^2} f(t), \tag{3.56}$$

with real symmetric positive definite matrix $\Omega(t)$. We start with the homogeneous problem and use the notation

$$\psi(t) = \begin{bmatrix} q(t) \\ p(t) \end{bmatrix}.$$

Rewriting (3.56) as a first-order system reads

$$\psi'(t) = \frac{1}{\varepsilon} A(t)\psi(t) + G(t)\psi(t),$$

with

$$A(t) = \begin{bmatrix} 0 & \Omega(t) \\ -\Omega(t) & 0 \end{bmatrix}, \quad G(t) = \begin{bmatrix} 0 & 0 \\ 0 & -\Omega(t)^{-1}\Omega'(t) \end{bmatrix}.$$

Except for a perturbation which is bounded independently of ε, this perturbed first-order problem is of the form (3.27). Next we diagonalize A to obtain (3.28). Note that the eigenvalues of H occur in pairs of $\pm\omega_j$, where $\omega_j > 0$ are the eigenvalues of Ω. Using the adiabatic transformation (3.29) leads to

$$\eta'(t) = \mathrm{e}^{-\frac{\mathrm{i}}{\varepsilon}\Phi(t)} W(t) \mathrm{e}^{\frac{\mathrm{i}}{\varepsilon}\Phi(t)} \eta(t) + \mathrm{e}^{-\frac{\mathrm{i}}{\varepsilon}\Phi(t)} \widetilde{G}(t) \mathrm{e}^{\frac{\mathrm{i}}{\varepsilon}\Phi(t)} \eta(t), \tag{3.57}$$

with

$$W(t) = Q'(t)^* Q(t) \quad \text{and} \quad \widetilde{G}(t) = Q(t)^* G(t) Q(t).$$

If, in addition to (3.32), we assume that the frequencies are bounded away from zero,

$$|\lambda_j| \geq \frac{\delta}{2},$$

and that Q depends smoothly on t, (3.57) is a linear differential equation with bounded operator. This allows us to apply a Magnus-type integrator, for instance a variant of the exponential midpoint rule (3.34). For details, we refer to Lorenz, Jahnke and Lubich (2005).

The inhomogeneity f in (3.56) can be handled by yet another transformation:

$$w(t) = q(t) - \Omega(t)^{-2} f(t) + \varepsilon^2 \Omega(t)^{-2} \frac{\mathrm{d}^2}{\mathrm{d}t^2}\big(\Omega(t)^{-2} f(t)\big).$$

Differentiating twice shows that w satisfies

$$w''(t) + \frac{1}{\varepsilon^2}\Omega(t)^2 w(t) = \varepsilon^2 \frac{\mathrm{d}^2}{\mathrm{d}t^2}\left(\Omega(t)^{-2} \frac{\mathrm{d}^2}{\mathrm{d}t^2}\big(\Omega(t)^{-2} f(t)\big)\right).$$

The term on the right-hand side leads to an error of size $\mathcal{O}(\varepsilon^4)$ in the solution, and hence can be omitted if an approximation with error $\mathcal{O}(h^2)$ is required for step sizes $h > \varepsilon$. Then the equation for w is again of the type (3.56).

Adiabatic integrators for mechanical systems with time-dependent frequencies are studied in more detail in Hairer *et al.* (2006, Chapter XIV.2). In particular, an error analysis for the mollified impulse method for solving singularly perturbed second-order problems is presented there.

4. Implementation issues

The implementation of exponential integrators requires the approximation of products of matrix functions with vectors, that is,

$$\phi(A)v, \quad A \in \mathbb{R}^{d\times d}, \quad v \in \mathbb{R}^d.$$

Clearly, the efficiency of these integrators strongly depends on the numerical linear algebra used to compute these approximations. Standard methods such as diagonalization or Padé approximation will only be useful if the dimension of the matrix is not too large; see Moler and Van Loan (2003) and Higham (2008). We refer to the review by Higham and Al-Mohy (2010) in this volume and concentrate here on large-scale problems.

For simplicity, we always scale v to have unit norm.

4.1. Chebyshev methods

If the matrix A is Hermitian or skew-Hermitian for which *a priori* information about an enclosure of spectrum is available, then a very efficient way of approximating $\phi(A)v$ for some vector v is to use a truncated Chebyshev series. For simplicity, we only consider the exponential function here.

It is well known that a smooth function $\phi : [-1,1] \to \mathbb{C}$ can be expanded in a Chebyshev series:

$$\phi(\xi) = c_0 + 2\sum_{k=1}^{\infty} c_k T_k(\xi).$$

Here, T_k denotes the kth Chebyshev polynomial, which is defined by

$$T_k(\xi) = \cos(k \arccos \xi), \quad \xi \in [-1,1]$$

and

$$c_k = \frac{1}{\pi}\int_{-1}^{1} T_k(\xi)\phi(\xi)\frac{1}{\sqrt{1-\xi^2}}\,\mathrm{d}\xi = \frac{1}{\pi}\int_0^{\pi} \cos(k\theta)\phi(\cos\theta)\,\mathrm{d}\theta. \tag{4.1}$$

Chebyshev polynomials satisfy the recursion

$$T_{k+1}(\xi) = 2\xi T_k(\xi) - T_{k-1}(\xi), \quad k = 1,2,\ldots$$

initialized by $T_0(\xi) = 1$ and $T_1(\xi) = \xi$.

An approximation of ϕ can be obtained by truncating the Chebyshev series of ϕ after m terms,

$$S_m\phi(\xi) = c_0 + 2\sum_{k=1}^{m-1} c_k T_k(\xi).$$

Note that S_m is a polynomial of degree $m-1$.

For real ω and $\phi(\xi) = \mathrm{e}^{-\omega\xi}$, we have by (4.1)

$$c_k = \mathrm{i}^k J_k(\mathrm{i}\omega) = I_k(-\omega),$$

where J_k denotes the kth Bessel function and I_k the kth modified Bessel function, while for $\phi(\xi) = \mathrm{e}^{\mathrm{i}\omega\xi}$ we get

$$c_k = \mathrm{i}^k J_k(\omega).$$

If A is Hermitian with eigenvalues in $[a,b] \subset \mathbb{R}$, then a linear transformation to the interval $[-1,1]$ yields the approximation

$$S_m\mathrm{e}^{-hA} = \mathrm{e}^{-h(a+b)/2}\left(c_0 + 2\sum_{k=1}^{m-1} c_k T_k\left(\frac{2}{b-a}\left(A - \frac{a+b}{2}I\right)\right)\right),$$

with

$$c_k = I_k\left(-h\frac{b-a}{2}\right).$$

On the other hand, if A is skew-Hermitian with eigenvalues contained in the interval $\mathrm{i}[a,b]$ on the imaginary axis, then

$$S_m\mathrm{e}^{hA} = \mathrm{e}^{\mathrm{i}h(a+b)/2}\left(c_0 + 2\sum_{k=1}^{m-1} c_k T_k\left(\frac{-2}{b-a}\left(\mathrm{i}A + \frac{a+b}{2}I\right)\right)\right),$$

with

$$c_k = \mathrm{i}^k J_k\left(h\frac{b-a}{2}\right).$$

The approximation $S_m\mathrm{e}^{\pm hA}v$ can be computed efficiently with the Clenshaw algorithm.

For the approximation error in the case of Hermitian A, the following result was given by Stewart and Leyk (1996, Section 4); see also Bergamaschi and Vianello (2000).

Theorem 4.1. Let A be a Hermitian positive semi-definite matrix with eigenvalues in the interval $[0,\rho]$. Then the error of the mth Chebyshev approximation of e^{-hA}, *i.e.*, $\varepsilon_m := \|\mathrm{e}^{-hA} - S_m\mathrm{e}^{-hA}\|$, is bounded by

$$\varepsilon_m \le 2\exp\left(-\frac{\beta m^2}{h\rho}\right)\left(1 + \sqrt{\frac{h\rho\pi}{4\beta}}\right) + \frac{2\delta^{h\rho}}{1-\delta}, \quad 0 < m \le h\rho,$$

where $\beta = 2/(1+\sqrt{5}) \approx 0.618$ and $\delta = \mathrm{e}^{\beta}/(2+\sqrt{5}) \approx 0.438$.

Remark 4.2. The bounds given in this theorem are not optimal. One can obtain sharper bounds (similar to Theorem 4.4) by combining Bernstein's theorem (see Lubich (2008, Theorem III.2.1)) with the techniques used in the proof of Hochbruck and Lubich (1997, Theorem 2). We do not elaborate on this here.

For skew-Hermitian matrices, the subsequent convergence result is given in Lubich (2008, Theorem III.2.4).

Theorem 4.3. Let A be a skew-Hermitian matrix with eigenvalues in the interval $\mathrm{i}[a,b]$. Then the error of the mth Chebyshev approximation of e^{hA}, *i.e.*, $\varepsilon_m := \|\mathrm{e}^{hA} - S_m \mathrm{e}^{hA}\|$, is bounded by

$$\varepsilon_m \le 4\left(\exp\left(1-\left(\frac{\omega}{2m}\right)^2\right)\frac{\omega}{2m}\right)^m, \qquad \omega = h\frac{b-a}{2} \le m.$$

For general matrices, one has to use truncated Faber series instead of Chebyshev series. This technique has been employed by Knizhnerman (1991) and Moret and Novati (2001) for the exponential function.

As mentioned at the start of this subsection, Chebyshev methods require *a priori* information on the spectrum. In contrast, Krylov subspace methods, which we consider next, work without any *a priori* information on the field of values of the matrix. Moreover, they exploit clustering of the spectrum and take advantage of particular properties of the vector b.

4.2. Krylov subspace methods

We assume that the reader is familiar with the basics of solving a linear system with a Krylov subspace method; see, for instance, Saad (2003). The mth Krylov subspace with respect to a vector $b \in \mathbb{C}^d$ and a matrix $A \in \mathbb{C}^{d\times d}$ will be denoted by

$$\mathcal{K}_m(A,b) = \operatorname{span}\{b, Ab, \ldots, A^{m-1}b\}.$$

Without loss of generality we scale b such that $\|b\| = 1$.

There are several different ways to derive Krylov approximations to the product of a matrix function with a vector. Here, we present an approach motivated by the Cauchy integral formula:

$$\phi(A)b = \frac{1}{2\pi \mathrm{i}}\int_\Gamma \phi(\lambda)(\lambda I - A)^{-1}b\,\mathrm{d}\lambda = \frac{1}{2\pi \mathrm{i}}\int_\Gamma \phi(\lambda)x(\lambda)\,\mathrm{d}\lambda. \tag{4.2}$$

The curve Γ is a suitable contour surrounding the field of values $\mathcal{F}(A)$ of the matrix A and ϕ is assumed to be analytic in a neighbourhood of $\mathcal{F}(A)$. The integrand contains, for each $\lambda \in \Gamma$, the solution of a linear system of equations:

$$(\lambda I - A)x(\lambda) = b. \tag{4.3}$$

A Krylov subspace method for approximating the solution of (4.3) first constructs a basis of the Krylov subspace $\mathcal{K}_m(\lambda I - A, b)$. Fortunately, since $\mathcal{K}_m(A,b) = \mathcal{K}_m(\lambda I - A, b)$ for each $\lambda \in \mathbb{C}$, the same Krylov subspace can be used for all λ.

The Arnoldi method constructs an orthonormal basis $V_m \in \mathbb{C}^{d\times m}$ of $\mathcal{K}_m(A,b)$ and an unreduced upper Hessenberg matrix $H_m \in \mathbb{C}^{m\times m}$ satisfying the standard Krylov recurrence formula

$$AV_m = V_m H_m + h_{m+1,m} v_{m+1} e_m^T, \quad V_m^* V_m = I_m.$$

Here e_m denotes the mth unit vector in $\mathbb{C}^d$. For the shifted systems, we thus have the relation

$$(\lambda I - A)V_m = V_m(\lambda I - H_m) - h_{m+1,m} v_{m+1} e_m^T.$$

A Galerkin approximation to the solution of (4.3) is defined by requiring that the residual

$$r_m(\lambda) = b - (\lambda I - A)x_m(\lambda)$$

is orthogonal to $\mathcal{K}_m(A,b)$. The orthogonality of V_m leads to the approximation

$$x_m(\lambda) = V_m(\lambda I - H_m)^{-1} e_1. \tag{4.4}$$

Note that the Galerkin approximation exists for each m, since by assumption Γ surrounds the field of values of A and thus also $\mathcal{F}(H_m) \subset \mathcal{F}(A)$. An approximation to $\phi(A)b$ is obtained by replacing $x(\lambda) \approx x_m(\lambda)$ in (4.2). This yields

$$\begin{aligned} \phi(A)b &\approx \frac{1}{2\pi \mathrm{i}} \int_\Gamma \phi(\lambda) x_m(\lambda)\,\mathrm{d}\lambda \\ &= \frac{1}{2\pi \mathrm{i}} \int_\Gamma \phi(\lambda) V_m (\lambda I - H_m)^{-1} e_1\,\mathrm{d}\lambda \\ &= V_m \phi(H_m) e_1, \end{aligned} \tag{4.5}$$

since V_m is independent of λ.

To summarize, the Krylov approximation of $\phi(A)b$ involves two steps. The first is the computation of the basis, for instance using the Arnoldi method, and the second is the computation of $\phi(H_m)e_1$, which can be done by standard methods such as diagonalization or Padé approximation (Higham 2008, Higham and Al-Mohy 2010), or methods based on contour integrals (López-Fernández 2009).

The convergence of Krylov subspace approximations has been studied extensively: see, for instance, Druskin and Knizhnerman (1991), Knizhnerman (1991, 1992), Gallopoulos and Saad (1992), Druskin and Knizhnerman (1994, 1995) and Saad (1992). A remarkable property is that the convergence is always superlinear. However, the number of iterations to reach the regime of superlinear convergence depends on the geometry of the field of values of the matrix.

The following convergence result for the Hermitian case was given by Hochbruck and Lubich (1997, Theorem 2).

Theorem 4.4. Let A be a Hermitian positive semi-definite matrix with eigenvalues in the interval $[0,\rho]$. Then the error in the Arnoldi approximation of $\mathrm{e}^{-hA}v$, *i.e.*, $\varepsilon_m := \|\mathrm{e}^{-hA}v - V_m \mathrm{e}^{-hH_m} e_1\|$, is bounded as follows:

$$\varepsilon_m \le \left(3\frac{\rho h}{m^2} + 4\frac{\sqrt{\rho h}}{m}\right)\mathrm{e}^{-\beta m^2/(\rho h)}, \qquad \sqrt{\rho h} \le m \le \frac{1}{2}\rho h,$$

$$\varepsilon_m \le \left(\frac{20}{\rho h} + \frac{6\sqrt{\pi}}{\sqrt{\rho h}}\right)\mathrm{e}^{(\rho h)^2/(16m)}\mathrm{e}^{-\rho h/2}\left(\frac{\mathrm{e}\rho h}{4m}\right)^m, \qquad \frac{1}{2}\rho h \le m,$$

where $\beta > 0.92$.

For the skew-Hermitian case, the regime of superlinear convergence behaviour starts much later, namely with $m \ge \|hA\|$ instead of $m \ge \sqrt{\|hA\|}$ for Hermitian matrices. The result is taken from Hochbruck and Lubich (1997, Theorem 4) in the refined version of Lubich (2008, Theorem III.2.10).

Theorem 4.5. Let A be a skew-Hermitian matrix with eigenvalues contained in an interval $\mathrm{i}[a,b]$ of the imaginary axis. Then, for $\rho = (b-a)/2$, the error in the Arnoldi approximation of $e^{hA}v$ is bounded by

$$\varepsilon_m \le 8\,\mathrm{e}^{-(\rho h)^2/(4m)}\left(\frac{\mathrm{e}\rho h}{2m}\right)^m, \qquad m \ge \rho h.$$

For $\rho h > 1$ we have

$$\varepsilon_m \le \frac{1}{3}\left(\frac{8}{\rho h} + \frac{11\sqrt{2}}{\sqrt{\rho h}}\right)\mathrm{e}^{-(\rho h)^2/(4m)}\left(\frac{\mathrm{e}\rho h}{2m}\right)^m, \qquad m \ge \rho h.$$

Additional results for matrices with field of values contained in a disc or a wedge-shaped set can be found in Hochbruck and Lubich (1997). Here we only note that the onset of superlinear convergence behaviour starts with

$$m \ge (h\|A\|)^\alpha, \qquad \alpha = \frac{\pi}{2}\frac{1}{\pi - \theta},$$

where θ denotes the angle between negative real axis and the boundary of the wedge at the origin. The situation in Theorem 4.4 corresponds to $\theta = 0$, while Theorem 4.4 treats the case $\theta = \pi/2$.

Variants of Krylov subspace methods for the evaluation of matrix functions are an active field of research. Some recent advances such as restarted Krylov subspace methods are given by Eiermann and Ernst (2006) and Niehoff (2007).

4.3. Leja interpolation

Yet a different approach for constructing polynomial approximations is to use interpolation. Since interpolation is ill-conditioned, in general, it is essential to choose the nodes carefully. Caliari, Vianello and Bergamaschi (2004) proposed the use of Leja points for the approximation of the function φ_1 of a matrix whose field of values is contained in an ellipse with foci $(a, 0)$ and $(b, 0)$ on the real axis. In the following, we summarize the basic features of the method. For more details and further references, the reader is referred to Caliari *et al.* (2004). An extension to φ_k for $k > 1$ is given in Caliari (2007).

A sequence of Leja points $\{z_i\}$ is defined recursively, usually starting from $|z_0| = \max\{|a|, |b|\}$, in such a way that the $(m+1)$st point z_m satisfies

$$\prod_{i=0}^{m-1} |z_m - z_i| = \max_{z\in[a,b]} \prod_{i=0}^{m-1} |z - z_i|.$$

In practice, Leja points can be extracted from a sufficiently dense set of uniformly distributed points on $[a, b]$. From the definition, it is clear that, in contrast to interpolation in Chebyshev nodes, it is possible to increase the interpolation degree by simply adding new nodes of the same sequence.

Leja points guarantee maximal and superlinear convergence of the interpolant on every ellipse of the confocal family. Therefore, they also ensure superlinear convergence of the corresponding matrix polynomials, provided that the spectrum (or the field of values) of the matrix is contained in one of the above ellipses.

As for Chebyshev approximation, interpolation in Leja points requires *a priori* information about the field of values of A, *e.g.*, by using Gerschgorin's discs. For stability reasons, the method is applied to a scaled and shifted function. Let c and γ be defined such that $[a, b] = [c - 2\gamma, c + 2\gamma]$. We interpolate $\varphi_k(h(c+\gamma\xi))$ at the Leja points $\{\xi_i\}$, $\xi_0 = 2$, of the reference interval $[-2, 2]$. The latter can be computed once and for all. The matrix Newton polynomial $p_m(hA)v$ of degree m, which approximates $\varphi_k(hA)v$, is then

$$\begin{aligned} p_m(hA)v &= p_{m-1}(hA)v + d_m q_m, \\ q_m &= \big((A - cI)/\gamma - \xi_{m-1}I\big)q_{m-1}, \end{aligned} \tag{4.6a}$$

where

$$p_0(hA)v = d_0 q_0, \quad q_0 = v, \tag{4.6b}$$

and $\{d_i\}_{i=0}^m$ are the divided differences of the function $\varphi_k(h(c + \gamma\xi))$ at the points $\{\xi_i\}$. The accurate evaluation of divided differences is considered in Caliari (2007). The method is computationally attractive because of the underlying two-term recurrence. When the expected degree m for convergence is too large, an efficient sub-stepping procedure can be used: see Caliari and Ostermann (2009).

4.4. Contour integrals

A major disadvantage of polynomial approximations is that the required degree of the polynomials grows with the norm of A. If linear systems with coefficient matrix $zI + hA$ can be solved efficiently, rational approximations constitute a good alternative. The most famous example is the approximation of the exponential function on the negative real line, which converges geometrically with rate $1/9.28903\ldots$; see Schmelzer and Trefethen (2007) for a near-best approximation based on the Carathéodory–Fejér procedure, and references therein.

Among the many different approaches, we mention rational Krylov methods. Some recent references are Moret and Novati (2004), Hochbruck and van den Eshof (2006), Frommer and Simoncini (2008) and Grimm and Hochbruck (2008). Extended Krylov subspace methods are investigated in Knizhnerman and Simoncini (2009). It is beyond the scope of this review paper to give a complete list.

Other rational approaches are based on representing the matrix function as an appropriate contour integral: see, *e.g.*, Schmelzer and Trefethen (2007), Trefethen, Weideman and Schmelzer (2006), Schädle, López-Fernández and Lubich (2006), to mention just a few recent articles. Here we present the approach of López-Fernández, Palencia and Schädle (2006), which is based on the inversion of the Laplace transform. Let

$$\Sigma_\delta = \{z \in \mathbb{C} \mid |\arg(-z)| \le \delta\} \cup \{0\}, \quad 0 < \delta < \pi/2,$$

be a sector in the left complex half-plane, and let

$$\Phi : \mathbb{C} \setminus \Sigma_\delta \to X$$

be a holomorphic function satisfying

$$\|\Phi(z)\| \le \frac{M}{|z|} \tag{4.7}$$

for some constant $M > 0$. It is well known that, in this case, Φ is the Laplace transform of

$$\varphi(t) = \frac{1}{2\pi \mathrm{i}} \int_\Gamma \mathrm{e}^{tz} \Phi(z) \,\mathrm{d}z, \quad |\arg t| \le \pi/2 - \delta. \tag{4.8}$$

Our aim is to reconstruct φ from a few evaluations of $\Phi(z)$ on a certain contour Γ.

For a given δ, we select parameters $d > 0$, $\alpha > 0$ satisfying

$$0 < \alpha - d < \alpha + d < \pi/2 - \delta. \tag{4.9}$$

To invert the Laplace transform (4.8), we use the contour formed by the left branch of a hyperbola,

$$\Gamma = \{\lambda T(x) + \gamma \mid x \in \mathbb{R}\},$$

where $\lambda > 0$ and

$$T(x) = 1 - \sin(\alpha + \mathrm{i}x).$$

Inserting this contour into (4.8), we obtain

$$\varphi(t) = \int_{-\infty}^{\infty} G(t,x)\,\mathrm{d}x, \tag{4.10}$$

where the function

$$G(t,w) = -\frac{\lambda T'(w)}{2\pi\,\mathrm{i}} \mathrm{e}^{t(\lambda T(w)+\gamma)} \Phi(\lambda T(w) + \gamma)$$

is holomorphic on the horizontal strip $\{w \in \mathbb{C} \mid |\mathrm{Im}\, w| < d\}$.

For the numerical approximation $\varphi(n,t)$ of (4.10), we consider the quadrature rule

$$\varphi(n,t) = h \sum_{\ell=-n}^{n} G(t, x_\ell), \tag{4.11}$$

with nodes $x_\ell = \ell h$. This approximation can be rewritten in terms of Φ as

$$\varphi(n,t) = h \sum_{\ell=-n}^{n} w_\ell \mathrm{e}^{t z_\ell} \Phi(z_\ell),$$

with weights

$$w_\ell = \frac{h\lambda}{2\pi\,\mathrm{i}} T'(x_\ell)$$

and nodes $z_\ell = \lambda T(x_\ell)$.

For this approximation, López-Fernández *et al.* (2006) proved the following error bound.

Theorem 4.6. Let Φ satisfy (4.7) and let α, d be chosen according to (4.9). Then, for $t_0 > 0$, $\Lambda \geq 1$, $0 < \theta < 1$, $n \geq 1$, and

$$a(\theta) = \operatorname{arccosh}\left(\frac{\Lambda}{(1-\theta)\sin\alpha}\right),$$

the choice

$$h = \frac{a(\theta)}{n}, \qquad \lambda = \frac{2\pi d n(1-\theta)}{t_0 \Lambda a(\theta)}$$

yields the uniform estimate

$$\|\varphi(t) - \varphi(n,t)\| \leq \frac{4M}{\pi} \psi(\alpha, d, \lambda, t_0) \frac{\varepsilon_n(\theta)^\theta}{1 - \varepsilon_n(\theta)},$$

for $t_0 \le t \le \Lambda t_0$, where

$$\psi(\alpha, d, \lambda, t_0) = \sqrt{\frac{1+\sin(\alpha+d)}{1-\sin(\alpha+d)}}\Big(1 + \big|\log\big(1 - \mathrm{e}^{-\lambda t_0 \sin(\alpha-d)}\big)\big|\Big)$$

and

$$\varepsilon_n(\theta) = \exp\Big(-\frac{2\pi d n}{a(\theta)}\Big).$$

López-Fernández (2009) suggests applying these techniques for evaluating the φ-functions of generators of analytic semigroups. For this purpose, we write

$$\varphi_k(-hA) = \int_0^1 \mathrm{e}^{-h(1-\theta)A}\frac{\theta^{k-1}}{(k-1)!}\,\mathrm{d}\theta = \mathcal{L}^{-1}\big(\mathcal{L}(\chi(\cdot, hA))\mathcal{L}(\varrho_k)\big)(1),$$

where $\mathcal{L}$ denotes the Laplace transform and

$$\chi(\theta, hA) = \mathrm{e}^{-h\theta A}, \quad \rho_k(\theta) = \frac{\theta^{k-1}}{(k-1)!}.$$

This shows at once that

$$\Phi_k(z, -hA) = \frac{1}{z^k}(zI + hA)^{-1}.$$

If A is a matrix and v is a vector of appropriate dimension, the evaluation of $\Phi_k(z_\ell, -hA)v$ requires the solution of a linear system of equations.

4.5. Hints on mathematical software

Most of the mathematical software for exponential integrators is still in its infancy. Although there exist many experimental codes, only a few of these programs are sufficiently developed and documented for general use. The following packages, however, are well established and much used.

The exponential integrator **exp4**,[1] by Hochbruck *et al.* (1998), is a variable step size implementation of an exponential Rosenbrock-type method, described in Example 2.25. Implementations in both MATLAB and C are available. The code has an option for dense matrices, where the required matrix functions are computed exactly. For large sparse matrices, however, Krylov subspace methods are employed. The software is well tested and has been successfully used in various applications.

In contrast to the integrator **exp4**, the MATLAB package EXPINT,[2] by Berland, Skaflestad and Wright (2007*b*), is more of a developing and testing tool for exponential integrators. It is a constant step size implementation of various integrators from the literature, and it is supplemented

[1] http://www.am.uni-duesseldorf.de/en/Research/04_Software.php
[2] http://www.math.ntnu.no/num/expint/matlab.php

by some numerical examples for the purpose of testing and comparison. The φ-functions are evaluated by an extension of the well-known scaling and squaring method for the matrix exponential function (Moler and Van Loan 2003, Higham 2008).

The next two packages can be used to compute the product of the matrix exponential and the matrix φ_1-function with a vector, respectively. In this capacity, they are useful building blocks for exponential integrators. Moreover, they can be used for solving first-order linear systems

$$u'(t) + Au(t) = b.$$

One of these packages is Expokit,[3] by Sidje (1998). It is available in MATLAB and FORTRAN. The computation of the above-mentioned matrix functions is based on Krylov subspace methods. The other package is SPARSKIT,[4] by Saad (1994). Designed as a basic toolkit for sparse matrix computations, its routines EXPPRO and PHIPRO provide the product of the matrix exponential and the matrix φ_1-function with a vector, respectively. The computations are again based on Krylov subspace methods.

Finally, we mention the recent MATLAB function `phipm`,[5] by Niesen and Wright (2009). It provides the action of matrix φ-functions, computed by Krylov subspace methods.

5. Applications

In recent years, exponential integrators have been employed in various large-scale computations. Here we will discuss some typical applications that illustrate the computational benefits of exponential integrators. The given list is by no means exhaustive, and is steadily growing.

Nonlinear partial differential equations with constant coefficients on rectangular domains can be spatially discretized by spectral methods. Integrating the resulting semidiscrete ordinary differential equations with constant step sizes allows us to compute the arising matrix functions once and for all at the beginning of the time integration. Such an approach has been advocated by Kassam and Trefethen (2005), and was successfully used in Klein (2008).

5.1. Reaction–advection–diffusion equations

Time-dependent partial differential equations that couple reaction terms with advection and diffusion terms form an important class of (semi)linear parabolic equations. For a recent review on numerical methods for such equations, we refer to the textbook by Hundsdorfer and Verwer (2007).

[3] http://www.maths.uq.edu.au/expokit/
[4] http://www-users.cs.umn.edu/~saad/software/SPARSKIT/sparskit.html
[5] http://www.amsta.leeds.ac.uk/~jitse/software.html

Exponential integrators for reaction–diffusion systems are used in Friesner *et al.* (1989). For linear advection–diffusion problems, Bergamaschi, Caliari and Vianello (2004) and Caliari *et al.* (2004) consider a finite element discretization with mass lumping. The arising linear differential equation is integrated exactly with an exponential integrator. The required φ_1-function of the non-symmetric stiffness matrix is approximated by interpolation at Leja points. A comparison with the Crank–Nicolson method is given for problems in two and three space dimensions. Bergamaschi, Caliari, Martínez and Vianello (2005) and Martínez, Bergamaschi, Caliari and Vianello (2009) describe parallel implementations of the same approach. Numerical comparisons for a reaction–advection–diffusion equation in two dimensions are given in Caliari and Ostermann (2009). Krylov and Leja approximations for large-scale matrix exponentials are compared in Bergamaschi, Caliari, Martínez and Vianello (2006).

5.2. Mathematical finance

Option pricing in mathematical finance is yet another source of parabolic (integro-)differential equations. Tangman, Gopaul and Bhuruth (2008) consider European, barrier and butterfly spread options for the Black–Scholes model and Merton's jump diffusion model. In that paper, the performance of exponential integrators is compared with the traditional Crank–Nicolson method. American options are considered by Rambeerich, Tangman, Gopaul and Bhuruth (2009), and Gondal (2010).

Lee and Sheen (2009) consider a contour integral approach for the solution of the Black–Scholes equations. A hyperbolic contour is used for the numerical inversion of the Laplace transform. In't Hout and Weideman (2009) extend this approach to the Heston model in two dimensions. They consider a parabolic contour and compare the performance of the resulting method with conventional ADI splitting methods.

5.3. Classical and quantum–classical molecular dynamics

Molecular dynamics is concerned with the simulation of long-range interaction of molecules. The different time scales involved, together with high oscillations, make these simulations very time-consuming, even on supercomputers. The challenges arising in computational molecular biophysics are well documented in the survey paper by Schlick, Skeel, Brunger, Kalé, Board, Hermans and Schulten (1999).

The mollified impulse method (see Section 3.2) was an important step towards more efficient integrators that allow longer time steps. More details and some applications are given in Izaguirre, Reich and Skeel (1999), Ma, Izaguirre and Skeel (2003) and Ma and Izaguirre (2003*a*, 2003*b*).

Exponential integrators have also been proposed for the time integration of mixed quantum–classical models: see Nettesheim, Bornemann, Schmidt and Schütte (1996), Schütte and Bornemann (1999), Nettesheim and Schütte (1999), and the collection by Deuflhard, Hermans, Leimkuhler, Mark, Reich and Skeel (1999). These models describe a small but important part of the system by quantum mechanics while the majority of atoms is described by classical mechanics. The quantum–classical molecular dynamics (QCMD) model is a famous example, where a singularly perturbed Schrödinger equation is coupled nonlinearly to classical Newtonian equations of motion. Due to the different time scales in the classical and the quantum evolution, the solutions are typically highly oscillatory. Hochbruck and Lubich (1999*b*) propose and analyse a variety of methods in which the action of the exponential of the Hamiltonian to a wave function is approximated by the Lanczos algorithm. Methods for more general problems, including the Car–Parinello equations of *ab initio* molecular dynamics, are presented in Hochbruck and Lubich (1999*a*). The methods allow the use of step sizes that are much larger than the inverse of the largest frequency in the system.

Numerical integrators for quantum dynamics close to the adiabatic limit are studied in Jahnke and Lubich (2003) and Jahnke (2004). The main idea here is to apply a clever transformation of variables (*cf.* (3.29)). For the reformulated problem, time-reversible numerical integrators have been proposed, which can use much larger step sizes than standard schemes. A generalization of these techniques to QCMD models can be found in Jahnke (2003) and to linear second-order differential equations with time-varying eigen-decompositions in Lorenz *et al.* (2005).

5.4. Schrödinger equations

Krylov subspace methods and Chebyshev approximations of the matrix exponential operator have a long tradition in computational chemistry and physics: see, *e.g.*, Nauts and Wyatt (1983), Tal-Ezer and Kosloff (1984), Park and Light (1986), Tal-Ezer, Kosloff and Cerjan (1992), Kosloff (1994) and Peskin, Kosloff and Moiseyev (1994).

Berland, Owren and Skaflestad (2006) compare the performance of Lawson methods and exponential Runge–Kutta methods for nonlinear Schrödinger equations, and Berland, Islas and Schober (2007*a*) use exponential integrators for the cubic nonlinear Schrödinger equation with periodic boundary conditions. Celledoni, Cohen and Owren (2008) derive and study symmetric exponential integrators and present some results on the cubic nonlinear Schrödinger equation.

5.5. Maxwell equations

Botchev, Harutyunyan and van der Vegt (2006) suggest using a Gautschi-type method (see Section 3.2) combined with Krylov approximations for the solution of a finite element discretization of linear three-dimensional

Maxwell equations. The paper also includes an analysis of the Krylov approximation error and of the dispersion of the numerical solution. Moreover, comparisons with a standard leap-frog method are reported. Later, Botchev, Faragó and Horváth (2009) presented further comparisons of this method with a variety of splitting methods, and in Verwer and Botchev (2009), the exponential trapezoidal rule described in Example 2.6 is considered.

Tokman and Bellan (2002) present a three-dimensional magnetohydrodynamic (MHD) model for the description of the evolution of coronal magnetic arcades in response to photospheric flows, which is able to reproduce a number of features characteristic of coronal mass ejection observations. For the time integration, the scheme `exp4` by Hochbruck *et al.* (1998), which is a particular variant of the exponential Rosenbrock-type method discussed in Example 2.25, is used.

Karle, Schweitzer, Hochbruck, Laedke and Spatschek (2006) and Karle, Schweitzer, Hochbruck and Spatschek (2008) suggest using Gautschi-type integrators for the simulation of nonlinear wave motion in dispersive media. The model derived in these papers applies to laser propagation in a relativistic plasma. For the one-dimensional problem, the matrix functions are evaluated by fast Fourier transformations, while for a two-dimensional problem this is no longer efficient if the laser pulse propagates within the plasma. The authors propose the use of two-dimensional fast Fourier transformations only in vacuum, where the integrator can take very large time steps. Within the plasma, where smaller time steps are required, a physically motivated spatial splitting is used for the implementation. Comparisons with a standard leap-frog integrator are reported and an implementation on parallel computers is discussed.

Another application is the simulation of electromagnetic wave propagation in optical and photonic systems. Niegemann, Tkeshelashvili and Busch (2007) suggest using an exponential quadrature rule (see Section 2.2) for the solution of linear Maxwell's equations that uses only matrix exponentials instead of linear combinations of φ-functions as in (2.9). In contrast to the standard finite difference time-domain (FDTD) method, their exponential quadrature rule allows the handling of lossy and anisotropic materials, as well as advanced boundary conditions, such as perfectly matched layers or auxiliary differential equations. For the implementation, Krylov subspace methods are used.

Since many complex optical and photonic systems lead to nonlinear problems, Pototschnig, Niegemann, Tkeshelashvili and Busch (2009) generalize this approach to Maxwell's equations with general nonlinear polarizations. As concrete examples, they investigate instantaneous Kerr nonlinearities, and give an overview of the treatment of coupled systems dynamics. They discuss Lawson methods (see Section 2.7) and a particular case of exponential Rosenbrock methods from Hochbruck *et al.* (1998) (see Section 2.4).

5.6. *Regularization of ill-posed problems*

Asymptotic regularization is a well-established tool for treating nonlinear ill-posed problems. Tautenhahn (1994) analysed the convergence of this approach for linear and nonlinear problems.

For its numerical realization, an appropriate numerical method for solving the Showalter differential equation is required. Applications of standard integration schemes yield well-known regularization methods. For example, the explicit Euler method and the linearly implicit Euler method are equivalent to the Landweber method and the Levenberg–Marquardt regularization, respectively. Hochbruck, Hönig and Ostermann (2009*a*, 2009*b*) present a variable step size analysis for the exponential Euler method. Optimal convergence rates are achieved under suitable assumptions on the initial error.

6. Historical remarks

We conclude this review with some comments on how the concept of exponential integrators has developed historically. However, it is not our intention to provide a full account of the history of exponential integrators. Relevant comments with appropriate references have already been given at various locations in the main text. Additional historical details can be found in Minchev and Wright (2005).

To the best of our knowledge, exponential integrators were first considered by Hersch (1958). The starting point of his investigation was the observation that standard difference schemes for the numerical solution of differential equations do not give the exact solution, in general, even if the differential equation is simple and can be solved by elementary methods. He then proposed new schemes that are *exact* for linear problems with constant coefficients. His main interest in this work was the study of improved schemes for boundary value and eigenvalue problems with variable coefficients in one or several spatial variables.

Certaine (1960) used the variation-of-constants formula to derive exponential multistep methods for the numerical solution of semilinear initial value problems (2.1). He suggested using the exponential Euler scheme (2.29) as a predictor for implicit exponential one- and two-step methods. Pope (1963) considered general nonlinear problems (2.44) and suggested linearizing them in each time step. The simplest scheme making use of this linearization is the exponential Rosenbrock–Euler method (2.53) for non-autonomous problems.

Gautschi (1961) proposed the use of trigonometric polynomials for the construction of time integration schemes for second-order problems. Probably the most famous example is the two-step method of Example 3.11. Related methods were constructed by Deuflhard (1979), using a theoretical approach to extrapolation methods. Example 3.12 contains his two-step method for semilinear second-order differential equations.

Nørsett (1969) used an idea similar to that of Certaine (1960) to construct explicit exponential Adams methods for semilinear problems (2.1). These methods are discussed in Section 2.5. A different idea due to Lawson (1967) consists in using an appropriate transformation of variables. Here, the simplest scheme is obtained by using the exponential function as a transformation: see Example 2.32. The hope is that the transformed equation will be non-stiff and can be discretized efficiently with an explicit Runge–Kutta or multistep method. The resulting approximations are then transformed back to the original variables. The idea was further developed in Ehle and Lawson (1975).

The first exponential Runge–Kutta methods (2.20) were constructed by Lawson (1967) and Ehle and Lawson (1975), with coefficients being exponential functions. Higher-order methods, however, require more φ-functions. Such methods were first proposed by Friedli (1978).

The term 'exponential integrators' was coined in the seminal paper by Hochbruck, Lubich and Selhofer (1998), which proposed an implementation of the code `exp4` (which belongs to the class of Rosenbrock-type methods; see Section 2.4), including adaptive time-stepping combined with Krylov subspace approximations of the φ_1-functions. The class of methods had already been introduced in Hochbruck and Lubich (1997), but it turned out that a naive implementation of exponential integrators would not be efficient in general. The integrator `exp4` used a clever construction which minimizes the numerical linear algebra costs. In this way, the first exponential integrator was obtained that was competitive for certain time-dependent partial differential equations.

The paper by Hochbruck *et al.* (1998) led to a revival of exponential integrators and initiated various activities in different directions on the construction, analysis, implementation, and application of such methods. These activities can be divided into four different groups: integrators for problems with temporally smooth solutions, integrators for problems with highly oscillatory solutions, improvements of the numerical linear algebra, and integrators designed for a specific application. In fact, these four topics correspond to the four main chapters of this review paper.

Acknowledgements

We thank the members of our groups for their careful reading of this paper and helpful comments. Moreover, we are grateful to Christian Lubich, Valeria Simoncini, and Nick Higham for their comments on an earlier version of this paper.

This work was supported by the Deutsche Forschungsgemeinschaft via SFB TR 18.

REFERENCES

L. Bergamaschi and M. Vianello (2000), 'Efficient computation of the exponential operator for large, sparse, symmetric matrices', *Numer. Linear Algebra Appl.* **7**, 27–45.

L. Bergamaschi, M. Caliari and M. Vianello (2004), The ReLPM exponential integrator for FE discretizations of advection–diffusion equations. In *Computational Science: ICCS 2004*, Vol. 3039 of *Lecture Notes in Computer Science*, Springer, pp. 434–442.

L. Bergamaschi, M. Caliari, A. Martínez and M. Vianello (2005), A parallel exponential integrator for large-scale discretizations of advection–diffusion models. In *Recent Advances in Parallel Virtual Machine and Message Passing Interface*, Vol. 3666 of *Lecture Notes in Computer Science*, Springer, pp. 483–492.

L. Bergamaschi, M. Caliari, A. Martínez and M. Vianello (2006), Comparing Leja and Krylov approximations of large scale matrix exponentials. In *Computational Science: ICCS 2006*, Vol. 3994 of *Lecture Notes in Computer Science*, Springer, pp. 685–692.

H. Berland, A. L. Islas and C. M. Schober (2007*a*), 'Conservation of phase space properties using exponential integrators on the cubic Schrödinger equation', *J. Comput. Phys.* **225**, 284–299.

H. Berland, B. Owren and B. Skaflestad (2006), 'Solving the nonlinear Schrödinger equation using exponential integrators', *Model. Identif. Control* **27**, 201–217.

H. Berland, B. Skaflestad and W. M. Wright (2007*b*), 'EXPINT: A MATLAB package for exponential integrators', *ACM Trans. Math. Software* **33**, 4:1–4:17.

G. Beylkin, J. M. Keiser and L. Vozovoi (1998), 'A new class of time discretization schemes for the solution of nonlinear PDEs', *J. Comput. Phys.* **147**, 362–387.

J. J. Biesiadecki and R. D. Skeel (1993), 'Dangers of multiple time step methods', *J. Comput. Phys.* **109**, 318–328.

S. Blanes, F. Casas and J. Ros (2002), 'High order optimized geometric integrators for linear differential equations', *BIT* **42**, 262–284.

S. Blanes, F. Casas, J. Oteo and J. Ros (2009), 'The Magnus expansion and some of its applications', *Physics Reports* **470**, 151–238.

M. Born and V. Fock (1928), 'Beweis des Adiabatensatzes', *Z. Phys. A Hadrons and Nuclei* **51**, 165–180.

M. A. Botchev, I. Faragó and R. Horváth (2009), 'Application of operator splitting to the Maxwell equations including a source term', *Appl. Numer. Math.* **59**, 522–541.

M. A. Botchev, D. Harutyunyan and J. J. W. van der Vegt (2006), 'The Gautschi time stepping scheme for edge finite element discretizations of the Maxwell equations', *J. Comput. Phys.* **216**, 654–686.

C. Budd and A. Iserles (1999), 'On the solution of linear differential equations in Lie groups', *Philos. Trans. Royal Soc. A* **357**, 946–956.

M. Caliari (2007), 'Accurate evaluation of divided differences for polynomial interpolation of exponential propagators', *Computing* **80**, 189–201.

M. Caliari and A. Ostermann (2009), 'Implementation of exponential Rosenbrock-type integrators', *Appl. Numer. Math.* **59**, 568–581.

M. Caliari, M. Vianello and L. Bergamaschi (2004), 'Interpolating discrete advection–diffusion propagators at Leja sequences', *J. Comput. Appl. Math.* **172**, 79–99.

M. P. Calvo and C. Palencia (2006), 'A class of explicit multistep exponential integrators for semilinear problems', *Numer. Math.* **102**, 367–381.

E. Celledoni, D. Cohen and B. Owren (2008), 'Symmetric exponential integrators with an application to the cubic Schrödinger equation', *Found. Comp. Math.* **8**, 303–317.

E. Celledoni, A. Marthinsen and B. Owren (2003), 'Commutator-free Lie group methods', *Future Generation Computer Systems* **19**, 341–352.

J. Certaine (1960), The solution of ordinary differential equations with large time constants. In *Mathematical Methods for Digital Computers*, Wiley, pp. 128–132.

D. Cohen, T. Jahnke, K. Lorenz and C. Lubich (2006), Numerical integrators for highly oscillatory Hamiltonian systems: A review. In *Analysis, Modeling and Simulation of Multiscale Problems* (A. Mielke, ed.), Springer, pp. 553–576.

M. Condon, A. Deaño and A. Iserles (2009), 'On highly oscillatory problems arising in electronic engineering', *Mathematical Modelling and Numerical Analysis* **43**, 785–804.

S. M. Cox and P. C. Matthews (2002), 'Exponential time differencing for stiff systems', *J. Comput. Phys.* **176**, 430–455.

H. De la Cruz, R. J. Biscay, F. Carbonell, T. Ozaki and J. Jimenez (2007), 'A higher order local linearization method for solving ordinary differential equations', *Appl. Math. Comput.* **185**, 197–212.

P. Deuflhard (1979), 'A study of extrapolation methods based on multistep schemes without parasitic solutions', *Z. Angew. Math. Phys.* **30**, 177–189.

P. Deuflhard, J. Hermans, B. Leimkuhler, A. Mark, S. Reich and R. D. Skeel, eds (1999), *Algorithms for Macromolecular Modelling*, Vol. 4 of *Lecture Notes in Computational Science and Engineering*, Springer.

J. Dixon and S. McKee (1986), 'Weakly singular discrete Gronwall inequalities', *Z. Angew. Math. Mech.* **66**, 535–544.

V. L. Druskin and L. A. Knizhnerman (1991), 'Error bounds in the simple Lanczos procedure for computing functions of symmetric matrices and eigenvalues', *Comput. Math. Math. Phys.* **31**, 20–30.

V. L. Druskin and L. A. Knizhnerman (1994), 'On application of the Lanczos method to solution of some partial differential equations', *J. Comput. Appl. Math.* **50**, 255–262.

V. L. Druskin and L. A. Knizhnerman (1995), 'Krylov subspace approximation of eigenpairs and matrix functions in exact and computer arithmetic', *Numer. Linear Algebra Appl.* **2**, 205–217.

B. L. Ehle and J. D. Lawson (1975), 'Generalized Runge–Kutta processes for stiff initial-value problems', *J. Inst. Math. Appl.* **16**, 11–21.

M. Eiermann and O. G. Ernst (2006), 'A restarted Krylov subspace method for the evaluation of matrix functions', *SIAM J. Numer. Anal.* **44**, 2481–2504.

E. Emmrich (2005), 'Stability and error of the variable two-step BDF for semilinear parabolic problems', *J. Appl. Math. Comput.* **19**, 33–55.

K.-J. Engel and R. Nagel (2000), *One-Parameter Semigroups for Linear Evolution Equations*, Vol. 194 of *Graduate Texts in Mathematics*, Springer.

A. Friedli (1978), Verallgemeinerte Runge–Kutta Verfahren zur Lösung steifer Differentialgleichungssysteme. In *Numerical Treatment of Differential Equations* (R. Burlirsch, R. Grigorieff and J. Schröder, eds), Vol. 631 of *Lecture Notes in Mathematics*, Springer, pp. 35–50.

R. A. Friesner, L. S. Tuckerman, B. C. Dornblaser and T. V. Russo (1989), 'A method for exponential propagation of large systems of stiff nonlinear differential equations', *J. Sci. Comput.* **4**, 327–354.

A. Frommer and V. Simoncini (2008), Matrix functions. In *Model Order Reduction: Theory, Research Aspects and Applications* (W. H. Schilders and H. A. van der Vorst, eds), Mathematics in Industry, Springer, pp. 275–304.

E. Gallopoulos and Y. Saad (1992), 'Efficient solution of parabolic equations by Krylov approximation methods', *SIAM J. Sci. Statist. Comput.* **13**, 1236–1264.

B. García-Archilla, J. M. Sanz-Serna and R. D. Skeel (1998), 'Long-time-step methods for oscillatory differential equations', *SIAM J. Sci. Comput.* **20**, 930–963.

W. Gautschi (1961), 'Numerical integration of ordinary differential equations based on trigonometric polynomials', *Numer. Math.* **3**, 381–397.

M. A. Gondal (2010), 'Exponential Rosenbrock integrators for option pricing', *J. Comput. Appl. Math.* **234**, 1153–1160.

C. González and M. Thalhammer (2007), 'A second-order Magnus-type integrator for quasi-linear parabolic problems', *Math. Comp.* **76**, 205–231.

C. González, A. Ostermann and M. Thalhammer (2006), 'A second-order Magnus-type integrator for nonautonomous parabolic problems', *J. Comput. Appl. Math.* **189**, 142–156.

V. Grimm (2002), Exponentielle Integratoren als Lange-Zeitschritt-Verfahren für oszillatorische Differentialgleichungen zweiter Ordnung. Dissertation, Heinrich-Heine Universität Düsseldorf.

V. Grimm (2005*a*), 'A note on the Gautschi-type method for oscillatory second-order differential equations', *Numer. Math.* **102**, 61–66.

V. Grimm (2005*b*), 'On error bounds for the Gautschi-type exponential integrator applied to oscillatory second-order differential equations', *Numer. Math.* **100**, 71–89.

V. Grimm and M. Hochbruck (2006), 'Error analysis of exponential integrators for oscillatory second-order differential equations', *J. Phys. A* **39**, 5495–5507.

V. Grimm and M. Hochbruck (2008), 'Rational approximation to trigonometric operators', *BIT* **48**, 215–229.

H. Grubmüller, H. Heller, A. Windemuth and K. Schulten (1991), 'Generalized Verlet algorithm for efficient molecular dynamics simulations with long-range interactions', *Molecular Simulation* **6**, 121–142.

E. Hairer and C. Lubich (2000), 'Long-time energy conservation of numerical methods for oscillatory differential equations', *SIAM J. Numer. Anal.* **38**, 414–441.

E. Hairer and C. Lubich (2009), Oscillations over long times in numerical Hamiltonian systems. In *Highly Oscillatory Problems* (E. H. B. Engquist, A. Fokas and A. Iserles, eds), Vol. 366 of *London Mathematical Society Lecture Notes*, Cambridge University Press, pp. 1–24.

E. Hairer and G. Wanner (1996), *Solving Ordinary Differential Equations II: Stiff and Differential-Algebraic Problems*, Vol. 14 of *Springer Series in Computational Mathematics*, 2nd edn, Springer.

E. Hairer, C. Lubich and G. Wanner (2006), *Geometric Numerical Integration, Structure-Preserving Algorithms for Ordinary Differential Equations*, Vol. 31 of *Springer Series in Computational Mathematics*, Springer.

E. Hairer, S. P. Nørsett and G. Wanner (1993), *Solving Ordinary Differential Equations I: Nonstiff Problems*, Vol. 8 of *Springer Series in Computational Mathematics*, 2nd edn, Springer.

D. Henry (1981), *Geometric Theory of Semilinear Parabolic Equations*, Vol. 840 of *Lecture Notes in Mathematics*, Springer.

J. Hersch (1958), 'Contribution à la méthode des équations aux différences', *Z. Angew. Math. Phys.* **9**, 129–180.

N. J. Higham (2008), *Functions of Matrices: Theory and Computation*, SIAM.

N. J. Higham and A. H. Al-Mohy (2010), Computing matrix functions. In *Acta Numerica*, Vol. 19, Cambridge University Press, pp. 159–208.

M. Hochbruck and C. Lubich (1997), 'On Krylov subspace approximations to the matrix exponential operator', *SIAM J. Numer. Anal.* **34**, 1911–1925.

M. Hochbruck and C. Lubich (1999*a*), A bunch of time integrators for quantum/classical molecular dynamics. In Deuflhard *et al.* (1999), pp. 421–432.

M. Hochbruck and C. Lubich (1999*b*), 'Exponential integrators for quantum–classical molecular dynamics', *BIT* **39**, 620–645.

M. Hochbruck and C. Lubich (1999*c*), 'A Gautschi-type method for oscillatory second-order differential equations', *Numer. Math.* **83**, 403–426.

M. Hochbruck and C. Lubich (2003), 'On Magnus integrators for time-dependent Schrödinger equations', *SIAM J. Numer. Anal.* **41**, 945–963.

M. Hochbruck and A. Ostermann (2005*a*), 'Explicit exponential Runge–Kutta methods for semilinear parabolic problems', *SIAM J. Numer. Anal.* **43**, 1069–1090.

M. Hochbruck and A. Ostermann (2005*b*), 'Exponential Runge–Kutta methods for parabolic problems', *Appl. Numer. Math.* **53**, 323–339.

M. Hochbruck and A. Ostermann (2006), 'Explicit integrators of Rosenbrock-type', *Oberwolfach Reports* **3**, 1107–1110.

M. Hochbruck and J. van den Eshof (2006), 'Preconditioning Lanczos approximations to the matrix exponential', *SIAM J. Sci. Comput.* **27**, 1438–1457.

M. Hochbruck, M. Hönig and A. Ostermann (2009*a*), 'A convergence analysis of the exponential Euler iteration for nonlinear ill-posed problems', *Inverse Problems* **25**, 075009.

M. Hochbruck, M. Hönig and A. Ostermann (2009*b*), 'Regularization of nonlinear ill-posed problems by exponential integrators', *Mathematical Modelling and Numerical Analysis* **43**, 709–720.

M. Hochbruck, C. Lubich and H. Selhofer (1998), 'Exponential integrators for large systems of differential equations', *SIAM J. Sci. Comput.* **19**, 1552–1574.

M. Hochbruck, A. Ostermann and J. Schweitzer (2009*c*), 'Exponential Rosenbrock-type methods', *SIAM J. Numer. Anal.* **47**, 786–803.

W. Hundsdorfer and J. G. Verwer (2007), *Numerical Solution of Time-Dependent Advection–Diffusion–Reaction Equations*, Vol. 33 of *Springer Series in Computational Mathematics*, corrected 2nd printing, Springer.

K. J. In't Hout and J. A. C. Weideman (2009), Appraisal of a contour integral method for the Black–Scholes and Heston equations. Technical report, Department of Mathematics and Computer Science, University of Antwerp.

A. Iserles (2002*a*), 'On the global error of discretization methods for highly-oscillatory ordinary differential equations', *BIT* **42**, 561–599.

A. Iserles (2002*b*), 'Think globally, act locally: Solving highly-oscillatory ordinary differential equations', *Appl. Numer. Math.* **43**, 145–160.

A. Iserles and S. P. Nørsett (1999), 'On the solution of linear differential equations in Lie groups', *Philos. Trans. Royal Soc. A* **357**, 983–1019.

A. Iserles and S. Nørsett (2004), 'On quadrature methods for highly oscillatory integrals and their implementation', *BIT* **44**, 755–772.

A. Iserles, H. Z. Munthe-Kaas, S. P. Nørsett and A. Zanna (2000), Lie-group methods. In *Acta Numerica*, Vol. 9, Cambridge University Press, pp. 215–365.

J. A. Izaguirre, S. Reich and R. D. Skeel (1999), 'Longer time steps for molecular dynamics', *J. Chem. Phys.* **110**, 9853–9864.

T. Jahnke (2003), Numerische Verfahren für fast adiabatische Quantendynamik. PhD thesis, Eberhard-Karls-Universität, Tübingen, Germany.

T. Jahnke (2004), 'Long-time-step integrators for almost-adiabatic quantum dynamics', *SIAM J. Sci. Comput.* **25**, 2145–2164.

T. Jahnke and C. Lubich (2000), 'Error bounds for exponential operator splittings', *BIT* **40**, 735–744.

T. Jahnke and C. Lubich (2003), 'Numerical integrators for quantum dynamics close to the adiabatic limit', *Numer. Math.* **94**, 289–314.

C. Karle, J. Schweitzer, M. Hochbruck and K.-H. Spatschek (2008), 'A parallel implementation of a two-dimensional fluid laser-plasma integrator for stratified plasma-vacuum systems', *J. Comput. Phys.* **227**, 7701–7719.

C. Karle, J. Schweitzer, M. Hochbruck, E. W. Laedke and K.-H. Spatschek (2006), 'Numerical solution of nonlinear wave equations in stratified dispersive media', *J. Comput. Phys.* **216**, 138–152.

A.-K. Kassam and L. N. Trefethen (2005), 'Fourth-order time-stepping for stiff PDEs', *SIAM J. Sci. Comput.* **26**, 1214–1233.

C. Klein (2008), 'Fourth order time-stepping for low dispersion Korteweg–de Vries and nonlinear Schrödinger equations', *Electron. Trans. Numer. Anal.* **29**, 116–135.

L. A. Knizhnerman (1991), 'Computation of functions of unsymmetric matrices by means of Arnoldi's method', *J. Comput. Math. Math. Phys.* **31**, 5–16 (in the Russian issue).

L. A. Knizhnerman (1992), 'Error bounds in Arnoldi's method: The case of a normal matrix', *Comput. Math. Math. Phys.* **32**, 1199–1211.

L. Knizhnerman and V. Simoncini (2009), 'A new investigation of the extended Krylov subspace method for matrix function evaluations', *Numer. Linear Algebra Appl.* In press.

R. Kosloff (1994), 'Propagation methods for quantum molecular dynamics', *Annu. Rev. Phys. Chem.* **45**, 145–178.

S. Krogstad (2005), 'Generalized integrating factor methods for stiff PDEs', *J. Comput. Phys.* **203**, 72–88.

J. D. Lambert and S. T. Sigurdsson (1972), 'Multistep methods with variable matrix coefficients', *SIAM J. Numer. Anal.* **9**, 715–733.

J. D. Lawson (1967), 'Generalized Runge–Kutta processes for stable systems with large Lipschitz constants', *SIAM J. Numer. Anal.* **4**, 372–380.

H. Lee and D. Sheen (2009), 'Laplace transformation method for the Black–Scholes equations', *Int. J. Numer. Anal. Model.* **6**, 642–659.

M. López-Fernández (2009), On the implementation of exponential methods for semilinear parabolic equations. Technical report, Instituto de Ciencias Matemáticas, Madrid, Spain.

M. López-Fernández, C. Palencia and A. Schädle (2006), 'A spectral order method for inverting sectorial Laplace transforms', *SIAM J. Numer. Anal.* **44**, 1332–1350.

K. Lorenz, T. Jahnke and C. Lubich (2005), 'Adiabatic integrators for highly oscillatory second-order linear differential equations with time-varying eigendecomposition', *BIT* **45**, 91–115.

C. Lubich (2008), *From Quantum to Classical Molecular Dynamics: Reduced Models and Numerical Analysis*, Zurich Lectures in Advanced Mathematics, European Mathematical Society (EMS).

A. Lunardi (1995), *Analytic Semigroups and Optimal Regularity in Parabolic Problems*, Vol. 16 of *Progress in Nonlinear Differential Equations and their Applications*, Birkhäuser.

Q. Ma and J. A. Izaguirre (2003*a*), Long time step molecular dynamics using targeted Langevin stabilization. In *SAC '03: Proc. 2003 ACM Symposium on Applied Computing*, ACM, New York, pp. 178–182.

Q. Ma and J. A. Izaguirre (2003*b*), 'Targeted mollified impulse: A multiscale stochastic integrator for long molecular dynamics simulations', *Multiscale Model. Simul.* **2**, 1–21.

Q. Ma, J. A. Izaguirre and R. D. Skeel (2003), 'Verlet-I/R-RESPA/impulse is limited by nonlinear instabilities', *SIAM J. Sci. Comput.* **24**, 1951–1973.

W. Magnus (1954), 'On the exponential solution of differential equations for a linear operator', *Comm. Pure Appl. Math.* **7**, 649–673.

A. Martínez, L. Bergamaschi, M. Caliari and M. Vianello (2009), 'A massively parallel exponential integrator for advection–diffusion models', *J. Comput. Appl. Math.* **231**, 82–91.

R. I. McLachlan and G. R. W. Quispel (2002), Splitting methods. In *Acta Numerica*, Vol. 11, Cambridge University Press, pp. 341–434.

B. V. Minchev and W. Wright (2005), A review of exponential integrators for first order semi-linear problems. Preprint, NTNU Trondheim.

P. C. Moan and J. Niesen (2008), 'Convergence of the Magnus series', *Found. Comput. Math.* **8**, 291–301.

C. Moler and C. Van Loan (2003), 'Nineteen dubious ways to compute the exponential of a matrix, twenty-five years later', *SIAM Rev.* **45**, 3–49.

I. Moret and P. Novati (2001), 'An interpolatory approximation of the matrix exponential based on Faber polynomials', *J. Comput. Appl. Math.* **131**, 361–380.

I. Moret and P. Novati (2004), 'RD-rational approximations of the matrix exponential', *BIT* **44**, 595–615.

A. Nauts and R. E. Wyatt (1983), 'New approach to many-state quantum dynamics: The recursive-residue-generation method', *Phys. Rev. Lett.* **51**, 2238–2241.

P. Nettesheim and C. Schütte (1999), Numerical integrators for quantum–classical molecular dynamics. In Deuflhard *et al.* (1999), pp. 396–411.

P. Nettesheim, F. A. Bornemann, B. Schmidt and C. Schütte (1996), 'An explicit and symplectic integrator for quantum–classical molecular dynamics', *Chem. Phys. Lett.* **256**, 581–588.

J. Niegemann, L. Tkeshelashvili and K. Busch (2007), 'Higher-order time-domain simulations of Maxwell's equations using Krylov-subspace methods', *J. Comput. Theor. Nanoscience* **4**, 627–634.

J. Niehoff (2007), Projektionsverfahren zur Approximation von Matrixfunktionen mit Anwendungen auf die Implementierung exponentieller Integratoren. Dissertation, Heinrich-Heine Universität Düsseldorf, Mathematisches Institut.

J. Niesen and W. Wright (2009), A Krylov subspace algorithm for evaluating the φ-functions appearing in exponential integrators. Preprint: arXiv:0907.4631v1.

S. P. Nørsett (1969), An A-stable modification of the Adams-Bashforth methods. In *Conference on the Numerical Solution of Differential Equations*, Vol. 109 of *Lecture Notes in Mathematics*, Springer, pp. 214–219.

A. Ostermann, M. Thalhammer and W. M. Wright (2006), 'A class of explicit exponential general linear methods', *BIT* **46**, 409–431.

T. J. Park and J. C. Light (1986), 'Unitary quantum time evolution by iterative Lanczos reduction', *J. Chem. Phys.* **85**, 5870–5876.

A. Pazy (1992), *Semigroups of Linear Operators and Applications to Partial Differential Equations*, Vol. 44 of *Applied Mathematical Sciences*, corrected 2nd printing, Springer.

U. Peskin, R. Kosloff and N. Moiseyev (1994), 'The solution of the time dependent Schrödinger equation by the (t, t') method: The use of global polynomial propagators for time dependent Hamiltonians', *J. Chem. Phys.* **100**, 8849–8855.

D. A. Pope (1963), 'An exponential method of numerical integration of ordinary differential equations', *Comm. Assoc. Comput. Mach.* **6**, 491–493.

M. Pototschnig, J. Niegemann, L. Tkeshelashvili and K. Busch (2009), 'Time-domain simulations of nonlinear Maxwell equations using operator-exponential methods', *IEEE Trans. Antenn. Propag.* **57**, 475–483.

N. Rambeerich, D. Y. Tangman, A. Gopaul and M. Bhuruth (2009), 'Exponential time integration for fast finite element solutions of some financial engineering problems', *J. Comput. Appl. Math.* **224**, 668–678.

J. I. Ramos and C. M. García-López (1997), 'Piecewise-linearized methods for initial-value problems', *Appl. Math. Comput.* **82**, 273–302.

Y. Saad (1992), 'Analysis of some Krylov subspace approximations to the matrix exponential operator', *SIAM J. Numer. Anal.* **29**, 209–228.

Y. Saad (1994), SPARSKIT: A basic tool kit for sparse matrix computations, version 2. Technical report, Department of Computer Science and Engineering, University of Minnesota.

Y. Saad (2003), *Iterative Methods for Sparse Linear Systems*, 2nd edn, SIAM.

A. Schädle, M. López-Fernández and C. Lubich (2006), 'Fast and oblivious convolution quadrature', *SIAM J. Sci. Comput.* **28**, 421–438.

T. Schlick, R. D. Skeel, A. T. Brunger, L. V. Kalé, J. A. Board, J. Hermans and K. Schulten (1999), 'Algorithmic challenges in computational molecular biophysics', *J. Comput. Phys.* **151**, 9–48.

T. Schmelzer and L. N. Trefethen (2007), 'Evaluating matrix functions for exponential integrators via Carathéodory-Fejér approximation and contour integrals', *Electron. Trans. Numer. Anal.* **29**, 1–18.

C. Schütte and F. A. Bornemann (1999), 'On the singular limit of the quantum–classical molecular dynamics model', *SIAM J. Appl. Math.* **59**, 1208–1224.

R. B. Sidje (1998), 'Expokit: A software package for computing matrix exponentials', *ACM Trans. Math. Software* **24**, 130–156.

D. E. Stewart and T. S. Leyk (1996), 'Error estimates for Krylov subspace approximations of matrix exponentials', *J. Comput. Appl. Math.* **72**, 359–369.

K. Strehmel and R. Weiner (1987), 'B-convergence results for linearly implicit one step methods', *BIT* **27**, 264–281.

K. Strehmel and R. Weiner (1992), *Linear-implizite Runge–Kutta Methoden und ihre Anwendungen*, Vol. 127 of *Teubner-Texte zur Mathematik*, Teubner.

H. Tal-Ezer and R. Kosloff (1984), 'An accurate and efficient scheme for propagating the time-dependent Schrödinger equation', *J. Chem. Phys.* **81**, 3967–3971.

H. Tal-Ezer, R. Kosloff and C. Cerjan (1992), 'Low-order polynomial approximation of propagators for the time-dependent Schrödinger equation', *J. Comput. Phys.* **100**, 179–187.

D. Y. Tangman, A. Gopaul and M. Bhuruth (2008), 'Exponential time integration and Chebychev discretisation schemes for fast pricing of options', *Appl. Numer. Math.* **58**, 1309–1319.

U. Tautenhahn (1994), 'On the asymptotical regularization of nonlinear ill-posed problems', *Inverse Problems* **10**, 1405–1418.

S. Teufel (2003), *Adiabatic Perturbation Theory in Quantum Dynamics*, Vol. 1821 of *Lecture Notes in Mathematics*, Springer.

M. Thalhammer (2006), 'A fourth-order commutator-free exponential integrator for nonautonomous differential equations', *SIAM J. Numer. Anal.* **44**, 851–864.

M. Tokman (2006), 'Efficient integration of large stiff systems of ODEs with exponential propagation iterative (EPI) methods', *J. Comput. Phys.* **213**, 748–776.

M. Tokman and P. M. Bellan (2002), 'Three-dimensional model of the structure and evolution of coronal mass ejections', *Astrophys. J.* **567**, 1202–1210.

L. N. Trefethen, J. A. C. Weideman and T. Schmelzer (2006), 'Talbot quadratures and rational approximations', *BIT* **46**, 653–670.

M. Tuckerman, B. J. Berne and G. J. Martyna (1992), 'Reversible multiple time scale molecular dynamics', *J. Chem. Phys.* **97**, 1990–2001.

J. Verwer (1976), 'On generalized linear multistep methods with zero-parasitic roots and an adaptive principal root', *Numer. Math.* **27**, 143–155.

J. G. Verwer and M. A. Botchev (2009), 'Unconditionally stable integration of Maxwell's equations', *Linear Algebra Appl.* **431**, 300–317.

Acta Numerica (2010), pp. 287–449
doi:10.1017/S096249291000005X

Verification methods: Rigorous results using floating-point arithmetic

Siegfried M. Rump
Institute for Reliable Computing,
Hamburg University of Technology,
Schwarzenbergstraße 95, 21071 Hamburg, Germany
and
Visiting Professor at Waseda University,
Faculty of Science and Engineering,
3–4–1 Okubo, Shinjuku-ku, Tokyo 169–8555, Japan
E-mail: rump@tu-harburg.de

A classical mathematical proof is constructed using pencil and paper. However, there are many ways in which computers may be used in a mathematical proof. But 'proof by computer', or even the use of computers in the course of a proof, is not so readily accepted (the December 2008 issue of the *Notices of the American Mathematical Society* is devoted to formal proofs by computer).

In the following we introduce verification methods and discuss how they can assist in achieving a mathematically rigorous result. In particular we emphasize how floating-point arithmetic is used.

The goal of verification methods is ambitious. For a given problem it is proved, with the aid of a computer, that there exists a (unique) solution of a problem within computed bounds. The methods are constructive, and the results are rigorous in every respect. Verification methods apply to data with tolerances as well, in which case the assertions are true for all data within the tolerances.

Non-trivial problems have been solved using verification methods. For example, Tucker (1999) received the 2004 EMS prize awarded by the European Mathematical Society for giving 'a rigorous proof that the Lorenz attractor exists for the parameter values provided by Lorenz. This was a long-standing challenge to the dynamical system community, and was included by Smale in his list of problems for the new millennium. The proof uses computer estimates with rigorous bounds based on higher dimensional interval arithmetics.' Also, Sahinidis and Tawaralani (2005) received the 2006 Beale–Orchard–Hays Prize for their package BARON, which 'incorporates techniques from automatic differentiation, interval arithmetic, and other areas to yield an automatic, modular, and relatively efficient solver for the very difficult area of global optimization.'

This review article is devoted to verification methods and consists of three parts of similar length. In Part 1 the working tools of verification methods are discussed, in particular floating-point and interval arithmetic; my findings in Section 1.5 (Historical remarks) seem new, even to experts in the field.

In Part 2, the development and limits of verification methods for finite-dimensional problems are discussed in some detail. In particular, we discuss how verification is *not* working. For example, we give a probabilistic argument that the so-called interval Gaussian elimination (IGA) does not work even for (well-conditioned) random matrices of small size. Verification methods are discussed for problems such as dense systems of linear equations, sparse linear systems, systems of nonlinear equations, semi-definite programming and other special linear and nonlinear problems, including M-matrices, finding simple and multiple roots of polynomials, bounds for simple and multiple eigenvalues or clusters, and quadrature. The necessary automatic differentiation tools to compute the range of gradients, Hessians, Taylor coefficients, and slopes are also introduced.

Concerning the important area of optimization, Neumaier (2004) gave in his *Acta Numerica* article an overview on global optimization and constraint satisfaction methods. In view of the thorough treatment there, showing the essential role of interval methods in this area, we restrict our discussion to a few recent, complementary issues.

Finally, in Part 3, verification methods for infinite-dimensional problems are presented, namely two-point boundary value problems and semilinear elliptic boundary value problems.

Throughout the article, many examples of the *inappropriate* use of interval operations are given. In the past such examples contributed to the dubious reputation of interval arithmetic (see Section 1.3), whereas they are, in fact, simply a misuse.

One main goal of this review article is to introduce the principles of the design of verification algorithms, and how these principles differ from those for traditional numerical algorithms (see Section 1.4).

Many algorithms are presented in executable MATLAB/INTLAB code, providing the opportunity to test the methods directly. INTLAB, the MATLAB toolbox for reliable computing, was, for example, used by Bornemann, Laurie, Wagon and Waldvogel (2004) in the solution of half of the problems of the SIAM 10×10-digit challenge by Trefethen (2002).

CONTENTS

PART ONE
Fundamentals

1. Introduction

It is not uncommon for parts of mathematical proofs to be carried out with the aid of computers. For example,

- the non-existence of finite projective planes of order 10 by Lam, Thiel and Swiercz (1989),
- the existence of Lyons' simple group (of order $2^8 \cdot 3^7 \cdot 5^6 \cdot 7 \cdot 11 \cdot 31 \cdot 37 \cdot 67$) by Gorenstein, Lyons and Solomon (1994),
- the uniqueness of Thompson's group and the existence of O'Nan's group by Aschbacher (1994)

were proved with substantial aid of digital computers. Those proofs have in common that they are based on integer calculations: no 'rounding errors' are involved. On the other hand,

- the proof of universality for area-preserving maps (Feigenbaum's constant) by Eckmann, Koch and Wittwer (1984),
- the verification of chaos in discrete dynamical systems by Neumaier and Rage (1993),
- the proof of the double-bubble conjecture by Hass, Hutchings and Schlafly (1995),
- the verification of chaos in the Lorenz equations by Galias and Zgliczynski (1998),
- the proof of the existence of eigenvalues below the essential spectrum of the Sturm–Liouville problem by Brown, McCormack and Zettl (2000)

made substantial use of proof techniques based on floating-point arithmetic (for other examples see Frommer (2001)). We do not want to philosophize to what extent such proofs are rigorous, a theme that even made it into *The New York Times* (Browne 1988). Assuming a computer is *working according to its specifications*, the aim of this article is rather to present methods providing rigorous results which, in particular, use floating-point arithmetic.

We mention that there are possibilities for performing an entire mathematical proof by computer (where the ingenuity is often with the programmer). There are many projects in this direction, for example proof assistants like *Coq* (Bertot and Castéran 2004), theorem proving programs such as *HOL* (Gordon 2000), combining theorem provers and interval arithmetic (Daumas, Melquiond and Muñoz 2005, Hölzl 2009), or the ambitious project FMathL by Neumaier (2009), which aims to formalize mathematics in a very general way.

A number of proofs for non-trivial mathematical theorems have been carried out in this way, among them the Fundamental Theorem of Algebra, the impossibility of trisecting a 60° angle, the prime number theorem, and Brouwer's Fixed-Point Theorem.

Other proofs are routinely performed by computers, for example, integration in finite terms: for a given function using basic arithmetic operations and elementary standard functions, Risch (1969) gave an algorithm for deciding whether such an integral exists (and eventually computing it). This algorithm is implemented, for example, in Maple (2009) or Mathematica (2009) and solves any such problem in finite time. In particular the proof of non-existence of an integral in closed form seems appealing.

1.1. Principles of verification methods

The methods described in this article, which are called *verification methods* (or *self-validating methods*), are of quite a different nature. It will be discussed how *floating-point arithmetic* can be used in a rigorous way. Typical problems to be solved include the following.

- Is a given matrix non-singular? (See Sections 1.6, 10.5, 10.8.)
- Compute error bounds for the minimum value of a function $f : D \subseteq \mathbb{R}^n \to \mathbb{R}$. (See Sections 8, 11.6.)
- Compute error bounds for a solution of a system of nonlinear equations $f(x) = 0$. (See Section 13.)
- Compute error bounds for the solution of an ordinary or partial differential equation. (See Sections 15, 16.)

Most verification methods rely on a good initial approximation. Although the problems (and their solutions) are of different nature, they are solved by the following common

> DESIGN PRINCIPLE OF VERIFICATION METHODS:
> Mathematical theorems are formulated whose assumptions are verified with the aid of a computer. (1.1)

A verification method is the interplay between mathematical theory and practical application: a major task is to derive the theorems and their assumptions in such a way that the verification is likely to succeed. Mostly those theorems are sufficient criteria: if the assumptions are satisfied, the assertions are true; if not, nothing can be said.[1] The verification of the assumptions is based on *estimates* using floating-point arithmetic. Following Hadamard, a problem is said to be well-posed if it has a unique solution which depends continuously on the input data. A verification method *solves*

[1] In contrast, methods in computer algebra (such as Risch's algorithm) are *never-failing*: the correct answer will be computed in finite time, and the maximally necessary computing time is estimated depending on the input.

a problem by *proving* existence and (possibly local) uniqueness of the solution. Therefore, the inevitable presence of rounding errors implies the

$$\text{SOLVABILITY PRINCIPLE OF VERIFICATION METHODS:} \\ \text{Verification methods solve well-posed problems.} \tag{1.2}$$

As a typical example, there are efficient verification methods to prove the non-singularity of a matrix (see Section 1.6); however, the proof of singularity is outside the scope of verification methods because in every open neighbourhood of a matrix there is a non-singular matrix.

There are partial exceptions to this principle, for example for integer input data or, in semidefinite programming, if not both the primal and dual problem have distance zero to infeasibility, see Section 14.2. After regularizing an ill-posed problem, the resulting well-posed problem can be treated by verification algorithms.

There will be many examples of these principles throughout the paper. The ambitious goal of verification methods is to produce *rigorous error bounds*, correct in a mathematical sense – taking into account all possible sources of errors, in particular rounding errors.

Furthermore, the goal is to derive verification algorithms which are, for certain classes of problems, not much slower than the best numerical algorithms, say by at most an order of magnitude. Note that comparing computing times is not really fair because the two types of algorithms deliver results of different nature.

Part 1 of this review article, Sections 1 to 9, introduces tools for verification methods. From a mathematical point of view, much of this is rather trivial. However, we need this bridge between mathematics and computer implementation to derive successful verification algorithms.

Given that numerical algorithms are, in general, very reliable, one may ask whether it is necessary to compute verified error bounds for numerical problems. We may cite William (Vel) Kahan, who said that 'numerical errors are rare, rare enough not to worry about all the time, yet not rare enough to ignore them.' Moreover, problems are not restricted to numerical problems: see the short list at the beginning.

Besides this I think it is at the core of mathematics to produce true results. Nobody would take it seriously that Goldbach's conjecture is likely to be true because in trillions of tests no counter-example was found.

Verification methods are to be sharply distinguished from approaches *increasing* reliability. For example, powerful stochastic approaches have been developed by La Porte and Vignes (1974), Vignes (1978, 1980), Stewart (1990) and Chatelin (1988). Further, Demmel *et al.* (2004) have proposed a very well-written linear system solver with improved iterative refinement, which proves reliable in millions of examples. However, none of these approaches claims to produce always true results.

As another example, I personally believe that today's standard function libraries produce floating-point approximations accurate to at least the second-to-last bit. Nevertheless, they cannot be used 'as is' in verification methods because there is no *proof* of that property.[2] In contrast, basic floating-point operations $+, -, \cdot, /, \sqrt{\cdot}$, according to IEEE 754, are *defined precisely*, and are accurate to the last bit.[3] Therefore verification methods willingly use floating-point arithmetic, not least because of its tremendous speed.

1.2. Well-known pitfalls

We need designated tools for a mathematically rigorous verification. For example, it is well known that a small residual of some approximate solution is not sufficient to prove that the true solution is anywhere near it. Similarly, a solution of a discretized problem need not be near a solution of the continuous problem.

Consider, for example, Emden's equation $-\Delta u = u^2$ with Dirichlet boundary conditions on a rectangle with side lengths f and $1/f$, which models an actively heated metal band. It is theoretically known, by the famous result by Gidas, Ni and Nirenberg (1979) on symmetries of positive solutions to semilinear second-order elliptic boundary value problems, that there is a unique non-trivial centro-symmetric solution. However, the discretized equation, dividing the edges into 64 and 32 intervals for $f = 4$, has the solution shown in Figure 1.1. The height of the peak is normed to 4 in the figure; the true height is about 278. The norm of the residual divided by the norm of the solution is about $4 \cdot 10^{-12}$.

Note this is a *true* solution of the discretized equation computed by a verification method described in Section 13. Within the computed (narrow) bounds, this solution of the nonlinear system is unique. The nonlinear system has other solutions, among them an approximation to the solution of the exact equation.

The computed true solution in Figure 1.1 of the discretized equation is far from symmetric, so according to the theoretical result it cannot be near the solution of the continuous problem. Methods for computing rigorous inclusions of infinite-dimensional problems will be discussed in Sections 15 and 16.

It is also well known that if a computation yields similar approximations in various precisions, this approximation need not be anywhere near the true

[2] In Section 7 we briefly discuss how to take advantage of the fast floating-point standard functions.

[3] Admittedly assuming that the actual implementation follows the specification, a principal question we will address again.

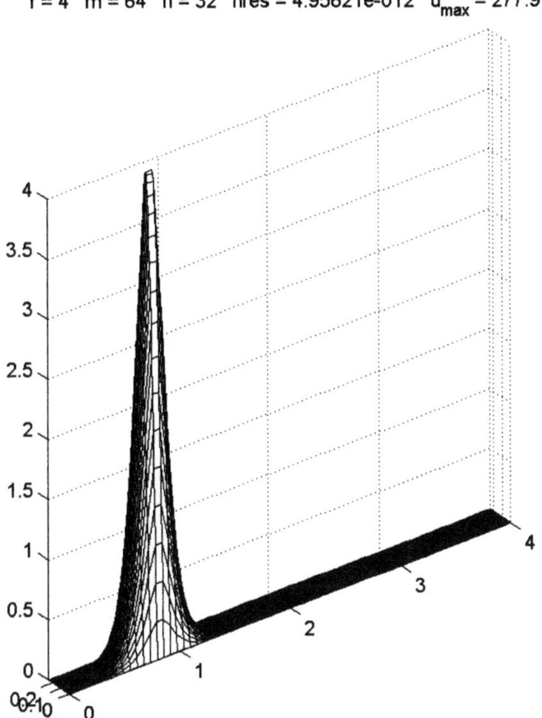

Figure 1.1. True solution of discretized, but spurious solution of the continuous Emden equation $-\Delta u = u^2$.

solution. To show this I constructed the arithmetic expression (Rump 1994)

$$f = 333.75b^6 + a^2(11a^2b^2 - b^6 - 121b^4 - 2) + 5.5b^8 + \frac{a}{2b}, \tag{1.3}$$

with $a = 77617$ and $b = 33096$, in the mid-1980s for the arithmetic on IBM S/370 mainframes. In single, double and extended precision[4] corresponding to about 8, 17 and 34 decimal digits, respectively, the results are

$$\begin{array}{ll} \text{single precision} & f \approx 1.172603\cdots \\ \text{double precision} & f \approx 1.1726039400531\cdots \\ \text{extended precision} & f \approx 1.172603940053178\cdots, \end{array} \tag{1.4}$$

whereas the true value is $f = -0.827386\cdots = a/2b - 2$.

The true sum of the main part in (1.3) (everything except the last fraction) is -2 and subject to heavy cancellation. Accidentally, in all precisions the floating-point sum of the main term cancels to 0, so the computed result is just $a/2b$. Further analysis has been given by Cuyt, Verdonk, Becuwe and Kuterna (2001) and Loh and Walster (2002).

[4] Multiplications are carried out to avoid problems with exponential and logarithm.

Almost all of today's architectures follow the IEEE 754 arithmetic standard. For those the arithmetic expression

$$f = 21 \cdot b \cdot b - 2 \cdot a \cdot a + 55 \cdot b \cdot b \cdot b \cdot b - 10 \cdot a \cdot a \cdot b \cdot b + \frac{a}{2b} \tag{1.5}$$

with $a = 77617$ and $b = 33096$ yields the same (wrong) results as in (1.4) when computing in single, double and extended precision (corresponding to about 7, 16 and 19 decimal digits precision, respectively). The reason is the same as for (1.3), and the true value is again

$$f = -0.827386\cdots = \frac{a}{2b} - 2.$$

1.3. The dubious reputation of interval arithmetic

One of the tools we use is interval arithmetic, which bears a dubious reputation. Some exponents in the interval community contributed to a history of overselling intervals as a panacea for problems in scientific computing. In fact, like almost every tool, interval arithmetic is no panacea: if used in a way it should not be used, the results may be useless. And there has been a backlash: the result is that interval techniques have been under-utilized.

It is quite natural that claims were criticized, and the criticism was justified. However, only the critique of the claims and of inappropriate use of interval arithmetic is appropriate; the extension of the criticism to interval arithmetic as a whole is understandable, but overstated.

One of the aims of this article is to show that by using interval arithmetic *appropriately*, certain non-trivial problems can be solved (see the abstract); for example, in Bornemann *et al.* (2004), half of the problems of the SIAM 10×10-digit challenge by Trefethen (2002) were (also) solved using INTLAB.

Consider the following well-known fact:

The (real or complex) roots of $x^2 + px + q = 0$ are $x_{1,2} = -\frac{p}{2} \pm \sqrt{\frac{p^2}{4} - q}$. (1.6)

It is also well known that, although mathematically correct, this innocent formula may produce poor approximations if not used or programmed appropriately. For example, for $p = 10^8$ and $q = 1$, the MATLAB statements

```
>> p=1e8; q=1; x1tilde=-p/2+sqrt(p^2/4-q)
```

produce

```
x1tilde =
 -7.4506e-009
```

The two summands $-p/2$ and $+\sqrt{p^2/4 - q}$ are almost equal in magnitude, so although using double-precision floating-point arithmetic corresponding to 16 decimal figures of precision, the cancellation leaves no correct digits. A naive user may trust the result because it comes without warning.

All this is well known. It is also well known that one should compute the root of smaller magnitude by $q/(-p/2-\sqrt{p^2/4-q})$ (for positive p). Indeed

```
>> p=1e8; q=1; x1=q/(-p/2-sqrt(p^2/4-q))                    (1.7)
```

produces correctly

```
x1 =
 -1.0000e-008
```

We want to stress explicitly that we think that, in this example, interval arithmetic is by no means better than floating-point arithmetic: an inclusion of x_1 computed by (1.6) yields the wide interval $[-1.491, -0.745]\cdot 10^{-8}$. This would be a typical example of inappropriate use of interval arithmetic.[5] But (1.7) yields a good, though not optimal inclusion for positive p.

We also want to stress that we are not arguing against the use of floating-point arithmetic or even trying to imply that floating-point calculations are not trustworthy per se. On the contrary, every tool should be used appropriately. In fact, verification methods depend on floating-point arithmetic to derive rigorous results, and we will use its speed to solve larger problems.

Because it is easy to use interval arithmetic inappropriately, we find it necessary to provide the reader with an easy way to check the claimed properties and results. Throughout the article we give many examples using INTLAB, developed and written by Rump (1999*a*), the MATLAB (2004) toolbox for reliable computing. INTLAB can be downloaded freely for academic purposes, and it is entirely written in MATLAB. For an introduction, see Hargreaves (2002). Our examples are given in executable INTLAB code; we use Version 6.0.

Recently a book by Moore, Kearfott and Cloud (2009) on verification methods using INTLAB appeared; Rohn (2009*b*) gives a large collection of verification algorithms written in MATLAB/INTLAB.

1.4. Numerical methods versus verification methods

A main purpose of this article is to describe how verification methods work and, in particular, how they are designed. There is an essential difference from numerical methods. Derivations and estimates valid over the field of real numbers frequently carry over, in some way, to approximate methods using floating-point numbers.

Sometimes care is necessary, as for the *pq*-formula (1.6) or when solving least-squares problems by normal equations; but usually one may concentrate on the real analysis, in both senses of the term. As a rule of thumb,

[5] One might think that this wide interval provides information on the sensitivity of the problem. This conclusion is not correct because the wide intervals may as well be attributed to some overestimation by interval arithmetic (see Sections 5, 6, 8).

straight replacement of real operations by floating-point operations is not unlikely to work.

For verification methods it is almost the other way around, and many examples of this will be given in this article. Well-known terms such as 'convergence' or even 'distributivity' have to be revisited. As a rule of thumb, straight replacement of real operations by interval operations is likely *not* to work.

We also want to stress that interval arithmetic is a convenient *tool* for implementing verification methods, but is by no means mandatory. In fact, in Sections 3 or 10.8.1 fast and rigorous methods will be described using solely floating-point arithmetic (with rounding to nearest mode). However, a rigorous analysis is sometimes laborious, whereas interval methods offer convenient and simple solutions; see, for example, the careful analysis by Viswanath (1999) to bound his constant concerned with random Fibonacci sequences, and the elegant proof by Oliveira and de Figueiredo (2002) using interval operations. In Section 9.2 we mention possibilities other than traditional intervals for computing with sets.

1.5. Historical remarks

Historically, the development of verification methods has divided into three major steps.

First, interval operations were defined in a number of papers such as Young (1931), Dwyer (1951) and Warmus (1956), but without proper rounding of endpoints and without any applications.

A second, major step was to *solve problems* using interval arithmetic. In an outstanding paper, his Master's Thesis at the University of Tokyo, Sunaga (1956) introduced:

- the interval lattice, the law of subdistributivity, differentiation, gradients, the view of intervals as a topological group, *etc.*,
- infimum–supremum and midpoint-radius arithmetic theoretically and with outward rounded bounds, real and complex including multi-dimensional intervals,
- the inclusion property (5.16), the inclusion principle (5.17) and the subset property,
- the centred form (11.10) as in Section 11.6, and subdivision to improve range estimation,
- the interval Newton procedure (Algorithm 6.1),
- verified interpolation for accurate evaluation of standard functions,
- fast implementation of computer arithmetic by redundant number representation, *e.g.*, addition in two cycles (rediscovered by Avizienis (1961)),
- inclusion of definite integrals using Simpson's rule as in Section 12, and
- the solution of ODEs with stepwise refinement.

His thesis is (handwritten) in Japanese and difficult to obtain. Although Sunaga (1958) summarized some of his results in English (see also Markov and Okumura (1999)), the publication was in an obscure journal and his findings received little attention.

Interval arithmetic became popular through the PhD thesis by Moore (1962), which is the basis for his book (Moore 1966). Overestimation by interval operations was significantly reduced by preconditioning proposed by Hansen and Smith (1967).

Sunaga finished his thesis on February 29, 1956. As Moore (1999) states that he conceived of interval arithmetic and some of its ramifications in the spring of 1958, the priority goes to Sunaga.

Until the mid-1970s, however, either known error estimates (such as for Simpson's rule) were computed with rigour, or assumed inclusions were refined, such as by Krawczyk (1969*a*). However, much non-trivial mathematics was developed: see Alefeld and Herzberger (1974).

The existence tests proposed by Moore (1977) commence the third and major step from interval to verification methods. Now fixed-point theorems such as Brouwer's in finite dimensions or Schauder's in infinite dimensions are used to certify that a solution to a problem exists within given bounds. For the construction of these bounds an iterative scheme was introduced in Rump (1980), together with the idea to include not the solution itself but the error with respect to an approximate solution (see Section 10.5 for details). These techniques are today standard in all kinds of verification methods.

An excellent textbook on verification methods is Neumaier (1990). Moreover, the introduction to numerical analysis by Neumaier (2001) is very much in the spirit of this review article: along with the traditional material the tools for rigorous computations and verification methods are developed. Based on INTLAB, some alternative verification algorithms for linear problems are described in Rohn (2005).

Besides INTLAB, which is free for academic use, ACRITH (1984) and ARITHMOS (1986) are commercial libraries for algorithms with result verification. Both implement the algorithms in my habilitation thesis (Rump 1983).

1.6. A first simple example of a verification method

A very simple first example illustrates the DESIGN PRINCIPLE (1.1) and the SOLVABILITY PRINCIPLE (1.2), and some of the reasoning behind a verification method.

Theorem 1.1. Let matrices $A, R \in \mathbb{R}^{n\times n}$ be given, and denote by I the $n \times n$ identity matrix. If the spectral radius $\varrho(I - RA)$ of $I - RA$ is less than 1, then A is non-singular.

Proof. If A is singular, then $I - RA$ has an eigenvalue 1, a contradiction. □

If the assumption $\varrho(I - RA) < 1$ is satisfied, then it is satisfied for all matrices in a small neighbourhood of A. This corresponds to the SOLVABILITY PRINCIPLE (1.2). Note that verification of singularity of a matrix is an ill-posed problem: an arbitrarily small change of the input data may change the answer.

Theorem 1.1 is formulated in such way that its assumption $\varrho(I-RA) < 1$ is likely to be satisfied for an approximate inverse R of A calculated in floating-point arithmetic. This is the interplay between the mathematical theorem and the practical application.

Note that in principle the matrix R is arbitrary, so neither the 'quality' of R as an approximate inverse nor its non-singularity need to be checked. The only assumption to be verified is that an upper bound of $\varrho(I - RA)$ is less than one. If, for example, $\|I - RA\|_\infty < 1$ is rigorously verified, then Theorem 1.1 applies. One way to achieve this is to estimate the individual error of each floating-point operation; this will be described in the next section.

Also note that all numerical experience should be used to design the mathematical theorem, the assumptions to be verified, and the way to compute R. In our particular example it is important to calculate R as a 'left inverse' of A: see Chapter 13 in Higham (2002) (for the drastic difference of the residuals $\|I - RA\|$ and $\|I - AR\|$ see the picture on the front cover of his book). For the left inverse it is proved in numerical analysis that, even for ill-conditioned matrices, it is likely that $\varrho(I - RA) < 1$ is satisfied.

Given that this has been done with care one may ask: What is the validity and the value of such a proof? Undoubtedly the computational part lacks beauty, surely no candidate for 'the Book'. But is it rigorous?

Furthermore, a proof assisted by a computer involves many components such as the programming, a compiler, the operating system, the processor itself, and more. A purely mathematical proof also relies on trust, however, but at least every step *can* be checked.

The trust in the correctness of a proof assisted by a computer can be increased by extensive testing. Verification algorithms allow a kind of testing which is hardly possible for numerical algorithms. Suppose a problem is constructed in such a way that the true solution is π. An approximate solution $p = 3.14159265358978$ would hardly give any reason to doubt, but the wrong 'inclusion' $[3.14159265358978, 3.14159265358979]$ would reveal an error. In INTLAB we obtain a correct inclusion of π by

```
>> P = 4*atan(intval(1))
intval P =
[   3.14159265358979,    3.14159265358980]
```

or simply by `P = intval('pi')`.

1 000 000	
− 999 999.9	3
0.1	

Figure 1.2. Erroneous result on pocket calculators.

The deep mistrust of 'computer arithmetic' as a whole is nourished by examples such as the following. For decades, floating-point arithmetic on digital computers was not precisely defined. Even worse, until very recent times the largest computers fell into the same trap as practically all cheap pocket calculators[6] still do today: calculating 1 000 000–999 999.93 results in 0.1 rather than 0.07 on an 8-decimal-digit calculator. This is because both summands are moved into the 8-digit accumulator, so that the last digit 3 of the second input disappears.

The reason is that the internal accumulator has no extra digit, and the effect is catastrophic: the relative error of a single operation is up to 100%. For other examples and details see, for example, the classical paper by Goldberg (1991).

It is a trivial statement that the floating-point arithmetic we are dealing with has to be precisely defined for any rigorous conclusion. Unfortunately, this was not the case until 1985. Fortunately this is now the case with the IEEE 754 arithmetic standard.

The floating-point arithmetic of the vast majority of existing computers follows this standard, so that the result of every (single) floating-point operation is *precisely defined.* Note that for composite operations, such as dot products, intermediate results are sometimes accumulated in some extended precision, so that floating-point approximations may differ on different computers. However, the error estimates to be introduced in the next section remain valid. This allows mathematically rigorous conclusions, as will be shown in the following.

2. Floating-point arithmetic

A floating-point operation approximates the given real or complex operations. For simplicity we restrict the following discussion to the reals. An excellent and extensive treatment of various aspects of floating-point arithmetic is Muller *et al.* (2009). For another, very readable discussion see Overton (2001).

Let a finite subset $\mathbb{F} \subseteq \mathbb{R} \cup \{-\infty, +\infty\}$ be given, where $\infty \in \mathbb{F}$ and $-\mathbb{F} = \mathbb{F}$. We call the elements of $\mathbb{F}$ *floating-point numbers.* Moreover, set $\texttt{realmax} := \max\{|f| : f \in \mathbb{F} \cap \mathbb{R}\}$.

[6] 8 or 10 or 12 decimal digits without exponent.

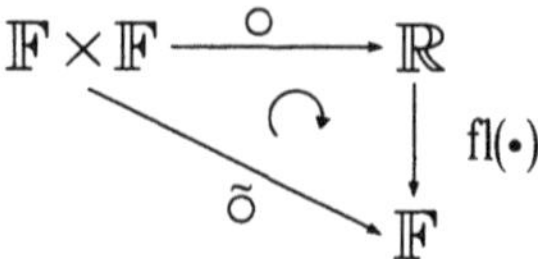

Figure 2.1. Definition of floating-point arithmetic through rounding.

For $a, b \in \mathbb{F}$, a floating-point operation $a \,\tilde{\circ}\, b : \mathbb{F} \times \mathbb{F} \to \mathbb{F}$ with $\circ \in \{+, -, \cdot, /\}$ should approximate the real result $a \circ b$. Using a rounding $\mathrm{fl} : \mathbb{R} \to \mathbb{F}$, a natural way to define floating-point operations is

$$a \,\tilde{\circ}\, b := \mathrm{fl}(a \circ b) \quad \text{for all} \quad a, b \in \mathbb{F}. \tag{2.1}$$

In other words, the diagram in Figure 2.1 commutes. Note that

$$a \,\tilde{\circ}\, b = a \circ b \quad \text{whenever} \quad a \circ b \in \mathbb{F}.$$

There are several ways to define a rounding $\mathrm{fl}(\cdot)$. One natural way is *rounding to nearest*, satisfying

$$|\mathrm{fl}(x) - x| = \min\{|f - x| : f \in \mathbb{F}\} \quad \text{for all } x \in \mathbb{R} \text{ with } |x| \le \texttt{realmax}. \tag{2.2}$$

Real numbers larger than **realmax** in absolute value require some special attention. Besides this, such a rounding is optimal in terms of approximating a real number by a floating-point number. Note that the only freedom is the result of $\mathrm{fl}(x)$ for x being the midpoint between two adjacent floating-point numbers. This ambiguity is fixed by choosing the floating-point number with zero last bit in the mantissa.[7]

This defines uniquely the result of all floating-point operations in rounding to nearest mode, and (2.1) together with (2.2) is exactly the definition of floating-point arithmetic in the IEEE 754 standard in rounding to nearest mode. For convenience we denote the result of a floating-point operation by $\mathrm{fl}(a \circ b)$. In terms of minimal relative error the definition is best possible.

In the following we use only IEEE 754 double-precision format, which corresponds to a relative rounding error unit of $\mathbf{u} := 2^{-53}$. It follows for operations $\circ \in \{+, -, \cdot, /\}$ that

$$a \circ b = \mathrm{fl}(a \circ b) \cdot (1 + \varepsilon) + \delta \quad \text{with} \quad |\varepsilon| \le \mathbf{u}, |\delta| \le \frac{1}{2}\eta, \tag{2.3}$$

where $\eta := 2^{-1074}$ denotes the underflow unit. Furthermore, we always have $\varepsilon\delta = 0$ and, since $\mathbb{F} = -\mathbb{F}$, taking an absolute value causes no rounding error.

For addition and subtraction the estimates are particularly simple, because the result is exact if underflow occurs:

$$a \circ b = \mathrm{fl}(a \circ b) \cdot (1 + \varepsilon) \quad \text{with} \quad |\varepsilon| \le \mathbf{u} \quad \text{for} \quad \circ \in \{+, -\}. \tag{2.4}$$

[7] Called 'rounding tie to even'.

Such inequalities can be used to draw rigorous conclusions, for example, to verify the assumption of Theorem 1.1.[8]

In order to verify $\|I - RA\|_\infty < 1$ for matrices $R, A \in \mathbb{F}^{n\times n}$, (2.3) can be applied successively. Such estimates, however, are quite tedious, in particular if underflow is taken into account. Much simplification is possible, as described in Higham (2002), (3.2) and Lemma 8.2, if underflow is neglected, as we now see.

Theorem 2.1. Let $x, y \in \mathbb{F}^n$ and $c \in \mathbb{F}$ be given, and denote by $\mathrm{fl}(\sum x_i)$ and $\mathrm{fl}(c - x^T y)$ the floating-point evaluation of $\sum_{i=1}^n x_i$ and $c - x^T y$, respectively, in any order. Then

$$|\mathrm{fl}(\sum x_i) - \sum x_i| \le \gamma_{n-1} \sum |x_i|, \tag{2.5}$$

where $\gamma_n := \frac{n\mathrm{u}}{1-n\mathrm{u}}$. Furthermore, provided no underflow has occurred,

$$|\mathrm{fl}(c - x^T y) - (c - x^T y)| \le \gamma_n |x|^T |y|, \tag{2.6}$$

where the absolute value is taken entrywise.

To obtain a computable bound for the right-hand side of (2.6), we abbreviate $e := |x|^T|y|$, $\tilde{e} := \mathrm{fl}(|x|^T|y|)$ and use (2.6) to obtain

$$|e| \le |\tilde{e}| + |\tilde{e} - e| \le |\tilde{e}| + \gamma_n |e|, \tag{2.7}$$

and therefore

$$|\mathrm{fl}(c - x^T y) - (c - x^T y)| \le \gamma_n |e| \le \frac{\gamma_n}{1-\gamma_n}|\tilde{e}| = \frac{1}{2}\gamma_{2n} \cdot \mathrm{fl}(|x|^T|y|). \tag{2.8}$$

This is true provided no over- or underflow occurs. To obtain a computable bound, the error in the floating-point evaluation of γ_n has to be estimated as well. Denote

$$\tilde{\gamma}_n := \mathrm{fl}(n \cdot \mathrm{u}/(1 - n \cdot \mathrm{u})).$$

Then for $n\mathrm{u} < 1$ neither $N := \mathrm{fl}(n \cdot \mathrm{u})$ nor $\mathrm{fl}(1 - N)$ causes a rounding error, and (2.3) implies

$$\gamma_n \le \tilde{\gamma}_{n+1}.$$

Therefore, with (2.8), estimating the error in the multiplication by e and observing that division by 2 causes no rounding error, we obtain

$$|\mathrm{fl}(c - x^T y) - (c - x^T y)| \le \mathrm{fl}\left(\frac{1}{2}\tilde{\gamma}_{2n+2} \cdot \tilde{e}\right) \quad \text{with } \tilde{e} = \mathrm{fl}(|x|^T|y|). \tag{2.9}$$

[8] The following sample derivation of floating-point estimates serves a didactic purpose; in Section 5 we show how to avoid (and improve) these tedious estimates.

Applying this to each component of $I - RA$ gives the entrywise inequality

$$|\mathrm{fl}(I - RA) - (I - RA)| \le \mathrm{fl}\left(\left(\frac{1}{2}\tilde{\gamma}_{2n+2}\right) \cdot E\right) =: \beta \quad \text{with} \quad E := \mathrm{fl}(|R||A|). \tag{2.10}$$

Note that the bound is valid for all $R, A \in \mathbb{F}^{n\times n}$ with $(2n+2)\mathrm{u} < 1$ provided that no underflow has occurred. Also note that β is a computable matrix of floating-point entries.

It is clear that one can continue like this, eventually obtaining a computable and rigorous upper bound for $\bar{c} := \|I - RA\|_\infty$. Note that $\bar{c} < 1$ *proves* (R and) A to be non-singular. The proof is rigorous provided that $(2n+2)\mathrm{u} < 1$, that no underflow occurred, and that the processor, compiler, operating system, and all components involved work to their specification.

These rigorous derivations of error estimates for floating-point arithmetic are tedious. Moreover, for each operation the worst-case error is (and has to be) assumed. Both will be improved in the next section.

Consider the model problem with an $n \times n$ matrix A_n based on the following pattern for $n = 3$:

$$A_n = \begin{pmatrix} 1 & \frac{1}{4} & \frac{1}{7} \\ \frac{1}{2} & \frac{1}{5} & \frac{1}{8} \\ \frac{1}{3} & \frac{1}{6} & \frac{1}{9} \end{pmatrix} \quad \text{for} \quad n = 3. \tag{2.11}$$

Now the application of Theorem 1.1 is still possible but even more involved since, for example, $\frac{1}{3}$ is not a floating-point number. Fortunately there is an elegant way to obtain rigorous and sharper error bounds using floating-point arithmetic, even for this model problem with input data not in $\mathbb{F}$.

Before we come to this we discuss a quite different but very interesting method for obtaining not only rigorous but exact results in floating-point arithmetic.

3. Error-free transformations

In the following we consider solely rounding to nearest mode, and we assume that no overflow occurs. As we have seen, the result of every floating-point operation is uniquely defined by (2.1). This not only allows error estimates such as (2.3), but it can be shown that the error of every floating-point operation is itself a floating-point number:[9]

$$x = \mathrm{fl}(a \circ b) \quad \Rightarrow \quad x + y = a \circ b \quad \text{with} \quad y \in \mathbb{F} \tag{3.1}$$

for $a, b \in \mathbb{F}$ and $\circ \in \{+, -, \cdot\}$. Remarkably, the error y can be calculated using only basic floating-point operations. The following algorithm by Knuth (1969) does it for addition.

[9] For division, $q, r \in \mathbb{F}$ for $q := \mathrm{fl}(a/b)$ and $a = qb + r$, and for the square root $x, y \in \mathbb{F}$ for $x = \mathrm{fl}(\sqrt{a})$ and $a = x^2 + y$.

Algorithm 3.1. Error-free transformation of the sum of two floating-point numbers:

```
function [x, y] = TwoSum(a, b)
    x = fl(a + b)
    z = fl(x − a)
    e = fl(x − z)
    f = fl(b − z)
    y = fl(fl(a − e) + f)
```

Theorem 3.2. For any two floating-point numbers $a, b \in \mathbb{F}$, the computed results x, y of Algorithm 3.1 satisfy

$$x = \mathrm{fl}(a + b) \quad \text{and} \quad x + y = a + b. \tag{3.2}$$

Algorithm 3.1 needs six floating-point operations. It was shown by Dekker (1971) that the same can be achieved in three floating-point operations if the input is sorted by absolute value.

Algorithm 3.3. Error-free transformation of the sum of two sorted floating-point numbers:

```
function [x, y] = FastTwoSum(a, b)
    x = fl(a + b)
    y = fl(fl(a − x) + b)
```

Theorem 3.4. The computed results x, y of Algorithm 3.3 satisfy (3.2) for any two floating-point numbers $a, b \in \mathbb{F}$ with $|a| \geq |b|$.

One may prefer Dekker's Algorithm 3.3 with a branch and three operations. However, with today's compiler optimizations we note that Knuth's Algorithm 3.1 with six operations is often faster: see Section 9.1.

Error-free transformations are a very powerful tool. As Algorithms 3.1 and 3.3 transform a pair of floating-point numbers into another pair, Algorithm 3.5 transforms a vector of floating-point numbers into another vector without changing the sum.

Algorithm 3.5. Error-free vector transformation for summation:

```
function q = VecSum(p)
    π_1 = p_1
    for i = 2 : n
        [π_i, q_{i−1}] = TwoSum(p_i, π_{i−1})
    end for
    q_n = π_n
```

As in Algorithm 3.1, the result vector q splits into an approximation q_n of $\sum p_i$ and into error terms $q_{1...n-1}$ without changing the sum.

We use the convention that for floating-point numbers p_i, $\mathrm{fl}_\uparrow\left(\sum p_i\right)$ denotes their recursive floating-point sum starting with p_1.

Theorem 3.6. For a vector $p \in \mathbb{F}^n$ of floating-point numbers let $q \in \mathbb{F}^n$ be the result computed by Algorithm 3.5. Then

$$\sum_{i=1}^{n} q_i = \sum_{i=1}^{n} p_i. \tag{3.3}$$

Moreover,

$$q_n = \mathrm{fl}_\uparrow\left(\sum_{i=1}^{n} p_i\right) \quad \text{and} \quad \sum_{i=1}^{n-1} |q_i| \le \gamma_{n-1} \sum_{i=1}^{n} |p_i|, \tag{3.4}$$

where $\gamma_k := k\mathbf{u}/(1-k\mathbf{u})$ using the relative rounding error unit $\mathbf{u}$.

Proof. By Theorem 3.2 we have $\pi_i + q_{i-1} = p_i + \pi_{i-1}$ for $2 \le i \le n$. Summing up yields

$$\sum_{i=2}^{n} \pi_i + \sum_{i=1}^{n-1} q_i = \sum_{i=2}^{n} p_i + \sum_{i=1}^{n-1} \pi_i \tag{3.5}$$

or

$$\sum_{i=1}^{n} q_i = \pi_n + \sum_{i=1}^{n-1} q_i = \sum_{i=2}^{n} p_i + \pi_1 = \sum_{i=1}^{n} p_i. \tag{3.6}$$

Moreover, $\pi_i = \mathrm{fl}(p_i + \pi_{i-1})$ for $2 \le i \le n$, and $q_n = \pi_n$ proves $q_n = \mathrm{fl}_\uparrow\left(\sum p_i\right)$. Now (2.3) and (2.5) imply

$$|q_{n-1}| \le \mathbf{u}|\pi_n| \le \mathbf{u}(1+\gamma_{n-1}) \sum_{i=1}^{n} |p_i|.$$

It follows by an induction argument that

$$\sum_{i=1}^{n-1} |q_i| \le \gamma_{n-2} \sum_{i=1}^{n-1} |p_i| + \mathbf{u}(1+\gamma_{n-1}) \sum_{i=1}^{n} |p_i| \le \gamma_{n-1} \sum_{i=1}^{n} |p_i|. \tag{3.7}$$

□

This is a rigorous result using only floating-point computations. The vector p is transformed into a new vector of $n-1$ error terms $q_1, \ldots, q_{n-1}$ together with the floating-point approximation q_n of the sum $\sum p_i$.

A result 'as if' computed in quadruple precision can be achieved by adding the error terms in floating-point arithmetic.

Theorem 3.7. For a vector $p \in \mathbb{F}^n$ of floating-point numbers, let $q \in \mathbb{F}^n$ be the result computed by Algorithm 3.5. Define

$$e := \mathrm{fl}\left(\sum_{i=1}^{n-1} q_i\right), \tag{3.8}$$

where the summation can be executed in any order. Then

$$\left|q_n + e - \sum_{i=1}^{n} p_i\right| \le \gamma_{n-2}\gamma_{n-1} \sum_{i=1}^{n} |p_i|. \tag{3.9}$$

Further,

$$|\tilde{s} - s| \le \mathbf{u}|s| + \gamma_n^2 \sum_{i=1}^{n} |p_i|, \tag{3.10}$$

for $\tilde{s} := \mathrm{fl}(q_n + e)$ and $s := \sum_{i=1}^{n} p_i$.

Proof. Using (3.3), (2.5) and (3.4), we find

$$\left|q_n + e - \sum_{i=1}^{n} p_i\right| = \left|e - \sum_{i=1}^{n-1} q_i\right| \le \gamma_{n-2} \sum_{i=1}^{n-1} |q_i| \le \gamma_{n-2}\gamma_{n-1} \sum_{i=1}^{n} |p_i| \tag{3.11}$$

and, for some $|\varepsilon| \le \mathbf{u}$,

$$|\mathrm{fl}(q_n + e) - s| = |\varepsilon s + (q_n + e - s)(1 + \varepsilon)| \le \mathbf{u}|s| + \gamma_n^2 \sum_{i=1}^{n} |p_i| \tag{3.12}$$

by (2.4) and (3.11). □

Putting things together, we arrive at an algorithm using solely double-precision floating-point arithmetic but achieving a result of quadruple-precision quality.

Algorithm 3.8. Approximating some $\sum p_i$ in double-precision arithmetic with quadruple-precision quality:

function res $=$ Sums(p)
    $\pi_1 = p_1;\ e = 0$
    for $i = 2 : n$
        $[\pi_i, q_{i-1}] = \mathrm{TwoSum}(p_i, \pi_{i-1})$
        $e = \mathrm{fl}(e + q_{i-1})$
    end for
    res $= \mathrm{fl}(\pi_n + e)$

This algorithm was given by Neumaier (1974) using FastTwoSum with a branch. At the time he did not know Knuth's or Dekker's error-free transformation, but derived the algorithm in expanded form. Unfortunately, his paper is written in German and did not receive a wide audience.

Note that the pair (π_n, e) can be used as a result representing a kind of simulated quadruple-precision number. This technique is used today in the XBLAS library (see Li *et al.* (2002)).

Similar techniques are possible for the calculation of dot products. There are error-free transformations for computing $x + y = a \cdot b$, again using only floating-point operations. Hence a dot product can be transformed into a sum without error provided no underflow has occurred.

Once dot products can be computed with quadruple-precision quality, the residual of linear systems can be improved substantially, thus improving the quality of inclusions for the solution. We come to this again in Section 10.4. The precise dot product was, in particular, popularized by Kulisch (1981).

4. Directed roundings

The algebraic properties of $\mathbb{F}$ are very poor. In fact it can be shown under general assumptions that for a floating-point arithmetic neither addition nor multiplication can be associative.

As well as rounding to nearest mode, IEEE 754 allows other rounding modes,[10] $\text{fldown}, \text{flup} : \mathbb{R} \to \mathbb{F}$, namely rounding towards $-\infty$ mode and rounding towards $+\infty$ mode:

$$\begin{aligned} \text{fldown}(a \circ b) &:= \max\{f \in \mathbb{F} : f \leq a \circ b\}, \\ \text{flup}(a \circ b) &:= \min\{f \in \mathbb{F} : a \circ b \leq f\}. \end{aligned} \tag{4.1}$$

Note that the inequalities are always valid for all operations $\circ \in \{+, -, \cdot, /\}$, including possible over- or underflow, and note that $\pm\infty$ may be a result of a floating-point operation with directed rounding mode. It follows for all $a, b \in \mathbb{F}$ and $\circ \in \{+, -, \cdot, /\}$ that

$$\text{fldown}(a \circ b) = \text{flup}(a \circ b) \quad \Longleftrightarrow \quad a \circ b \in \mathbb{F}, \tag{4.2}$$

a nice mathematical property. On most computers the operations with directed roundings are particularly easy to execute: the processor can be set into a specific rounding mode such as to nearest, towards $-\infty$ or towards $+\infty$, so that *all* subsequent operations are executed in this rounding mode until the next change.

Let $x, y \in \mathbb{F}^n$ be given. For $1 \leq i \leq n$ we have

$$s_i := \text{fldown}(x_i \cdot y_i) \leq x_i \cdot y_i \leq \text{flup}(x_i \cdot y_i) =: t_i.$$

Note that $s_i, t_i \in \mathbb{F}$ but $x_i \cdot y_i \in \mathbb{R}$ and, in general, $x_i \cdot y_i \notin \mathbb{F}$. It follows that

$$d_1 := \text{fldown}\Big(\sum s_i\Big) \leq x^T y \leq \text{flup}\Big(\sum t_i\Big) =: d_2, \tag{4.3}$$

where $\text{fldown}(\sum s_i)$ indicates that all additions are performed with rounding towards $-\infty$ mode. The summations may be executed in any order.

[10] IEEE 754 also defines rounding towards zero. This is the *only* rounding mode available on cell processors, presumably because it is fast to execute and avoids overflow.

The function setround in INTLAB performs the switching of the rounding mode, so that for two column vectors $\mathtt{x}, \mathtt{y} \in \mathbb{F}^n$ the MATLAB/INTLAB code

```
setround(-1)                % rounding downwards mode
d1 = x'*y
setround(1)                 % rounding upwards mode
d2 = x'*y
```

calculates floating-point numbers $\mathtt{d1}, \mathtt{d2} \in \mathbb{F}$ with $\mathtt{d1} \le x^T y \le \mathtt{d2}$. Note that these inequalities are rigorous, including the possibility of over- or underflow. This can be applied directly to verify the assumptions of Theorem 1.1. For given $A \in \mathbb{F}^{n\times n}$, consider the following algorithm.

Algorithm 4.1. Verification of the non-singularity of the matrix A:

```
R = inv(A);                 % approximate inverse
setround(-1)                % rounding downwards mode
C1 = R*A-eye(n);            % lower bound for RA-I
setround(1)                 % rounding upwards mode
C2 = R*A-eye(n);            % upper bound for RA-I
C = max(abs(C1),abs(C2));   % upper bound for |RA-I|
c = norm( C , inf );        % upper bound for ||RA-I||_inf
```

We claim that $\mathtt{c} < 1$ proves that A and R are non-singular.

Theorem 4.2. Let matrices $A, R \in \mathbb{F}^{n\times n}$ be given, and let c be the quantity computed by Algorithm 4.1. If $\mathtt{c} < 1$, then A (and R) are non-singular.

Proof. First note that

$$\mathrm{fldown}(a-b) \le a-b \le \mathrm{flup}(a-b) \quad \text{for all} \quad a, b \in \mathbb{F}.$$

Combining this with (4.3) and observing the rounding mode, we obtain

$$\mathtt{C1} \le RA - I \le \mathtt{C2}$$

using entrywise inequalities. Taking absolute value and maximum do not cause a rounding error, and observing the rounding mode when computing the ∞-norm together with $\varrho(I-RA) \le \|I-RA\|_\infty = \|\,|RA-I|\,\|_\infty \le \|\mathtt{C}\|_\infty$ proves the statement. □

Theorem 4.2 is a very simple first example of a verification method. According to the DESIGN PRINCIPLE OF VERIFICATION METHODS (1.1), $\mathtt{c} < 1$ is verified with the aid of the computer. Note that this proves non-singularity of all matrices within a small neighbourhood $U_\epsilon(A)$ of A.

According to the SOLVABILITY PRINCIPLE OF VERIFICATION METHODS (1.2), it is possible to verify non-singularity of a given matrix, but not singularity. An arbitrarily small perturbation of the input matrix may change the answer from yes to no. The verification of singularity excludes the use of estimates: it is only possible using exact computation.

Note that there is a trap in the computation of C1 and C2: Theorem 4.2 is not valid when replacing $\mathtt{R} * \mathtt{A} - \mathtt{eye(n)}$ by $\mathtt{eye(n)} - \mathtt{R} * \mathtt{A}$ in Algorithm 4.1. In the latter case the multiplication and subtraction must be computed in opposite rounding modes to obtain valid results.

This application is rather simple, avoiding tedious estimates. However, it does not yet solve our model problem (2.11) where the matrix entries are not floating-point numbers. An elegant solution for this is interval arithmetic.

5. Operations with sets

There are many possibilities for defining operations on sets of numbers, most prominently the power set operations. Given two sets $X, Y \subseteq \mathbb{R}$, the operations $\circ : \mathbb{PR} \times \mathbb{PR} \to \mathbb{PR}$ with $\circ \in \{+, -, \cdot, /\}$ are defined by

$$X \circ Y := \{x \circ y : x \in X, y \in Y\}, \tag{5.1}$$

where $0 \notin Y$ is assumed in the case of division. The input sets X, Y may be interpreted as *available information* of some quantity. For example, $\pi \in [3.14, 3.15]$ and $e \in [2.71, 2.72]$, so, with only this information at hand,

$$d := \pi - e \in [0.42, 0.44] \tag{5.2}$$

is true and the best we can say. The operations are optimal, the result is the minimum element in the infimum lattice $\{\mathbb{PR}, \cap, \cup\}$. However, this may no longer be true when it comes to composite operations and if operations are executed one after the other. For example,

$$d + e \in [3.13, 3.16] \tag{5.3}$$

is again true and the best we can say using only the information $d \in [0.42, 0.44]$; using the definition of d reveals $d + e = \pi$.

5.1. *Interval arithmetic*

General sets are hardly representable. The goal of implementable operations suggests restricting sets to intervals. Ordinary intervals are not the only possibility: see Section 9.2.

Denote[11] the set of intervals $\{[\underline{x}, \overline{x}] : \underline{x}, \overline{x} \in \mathbb{R}, \underline{x} \le \overline{x}\}$ by $\mathbb{IR}$. Then, provided $0 \notin Y$ in the case of division, the result of the power set operation $X \circ Y$ for $X, Y \in \mathbb{IR}$ is again an interval, and we have

$$[\underline{x}, \overline{x}] \circ [\underline{y}, \overline{y}] := [\min(\underline{x} \circ \underline{y}, \underline{x} \circ \overline{y}, \overline{x} \circ \underline{y}, \overline{x} \circ \overline{y}),\ \max(\underline{x} \circ \underline{y}, \underline{x} \circ \overline{y}, \overline{x} \circ \underline{y}, \overline{x} \circ \overline{y})]. \tag{5.4}$$

For a practical implementation it becomes clear that in most cases it can be decided *a priori* which pair of the bounds $\underline{x}, \overline{x}, \underline{y}, \overline{y}$ lead to the lower and

[11] A standardized interval notation is proposed by Kearfott *et al.* (2005).

upper bound of the result. In the case of addition and subtraction this is particularly simple, namely

$$\begin{aligned} X + Y &= [\underline{x} + \underline{y}, \overline{x} + \overline{y}], \\ X - Y &= [\underline{x} - \overline{y}, \overline{x} - \underline{y}]. \end{aligned} \tag{5.5}$$

For multiplication and division there are case distinctions for X and Y depending on whether they are entirely positive, entirely negative, or contain zero. In all but one case, namely where both X and Y contain zero for multiplication, the pair of bounds is *a priori* clear. In that remaining case $0 \in X$ and $0 \in Y$ we have

$$[\underline{x}, \overline{x}] \cdot [\underline{y}, \overline{y}] := [\min(\underline{x} \circ \overline{y}, \overline{x} \circ \underline{y}),\ \max(\underline{x} \circ \underline{y}, \overline{x} \circ \overline{y})]. \tag{5.6}$$

As before, we assume from now on that a denominator interval does not contain zero. We stress that *all* results remain valid without this assumption; however, various statements become more difficult to formulate without giving substantial new information.

We do not discuss complex interval arithmetic in detail. Frequently complex discs are used, as proposed by Sunaga (1958). They were also used, for example, by Gargantini and Henrici (1972) to enclose roots of polynomials. The implementation is along the lines of real interval arithmetic. It is included in INTLAB.

5.2. Overestimation

A measure for accuracy or overestimation by interval operations is the diameter $d(X) := \overline{x} - \underline{x}$. Obviously,

$$d(X + Y) = d(X) + d(Y); \tag{5.7}$$

the diameter of the sum is the sum of the diameters. However,

$$d(X - Y) = (\overline{x} - \underline{y}) - (\underline{x} - \overline{y}) = d(X) + d(Y), \tag{5.8}$$

so the diameter of the sum and the difference of intervals cannot be smaller than the minimum of the diameters of the operands. In particular,

$$d(X - X) = 2 \cdot d(X). \tag{5.9}$$

This effect is not due to interval arithmetic but occurs in power set operations as well: see (5.2) and (5.3).

This can also be seen when writing an interval X in Gaussian notation $x \pm \Delta x$ as a number with a tolerance:

$$\begin{aligned} (x \pm \Delta x) + (y \pm \Delta y) &= (x + y) \pm (\Delta x + \Delta y), \\ (x \pm \Delta x) - (y \pm \Delta y) &= (x - y) \pm (\Delta x + \Delta y). \end{aligned} \tag{5.10}$$

That means the absolute errors add, implying a large relative error if the result is small in absolute value. This is exactly the case when catastrophic cancellation occurs. In contrast, neglecting higher-order terms,

$$(x \pm \Delta x) \cdot (y \pm \Delta y) \sim x \cdot y \pm (\Delta x \cdot |y| + |x|\Delta y),$$
$$\frac{x \pm \Delta x}{y \pm \Delta y} = \frac{(x \pm \Delta x) \cdot (y \mp \Delta y)}{y^2 - \Delta y^2} \sim \frac{x}{y} \pm \frac{\Delta x \cdot |y| + |x|\Delta y}{y^2}, \tag{5.11}$$

so that

$$\frac{\Delta(x \cdot y)}{|x \cdot y|} = \frac{\Delta(x/y)}{|x/y|} = \frac{\Delta x}{|x|} + \frac{\Delta y}{|y|}. \tag{5.12}$$

This means that for multiplication and division the relative errors add. Similarly, not much overestimation occurs in interval multiplication or division.

Demmel, Dumitriu, Holtz and Koev (2008) discuss in their recent *Acta Numerica* paper how to evaluate expressions to achieve accurate results in linear algebra. One major sufficient (but not necessary) condition is the NO INACCURATE CANCELLATION PRINCIPLE (NIC). It allows multiplications and divisions, but additions (subtractions) only on data with the same (different) sign, or on input data.

They use this principle to show that certain problems can be solved with high accuracy if the structure is taken into account. A distinctive example is the accurate computation of the smallest singular value of a Hilbert matrix of, say, dimension $n = 100$ (which is about 10^{-150}) by looking at it as a Cauchy matrix.

We see that the NO INACCURATE CANCELLATION PRINCIPLE (NIC) means precisely that replacement of every operation by the corresponding interval operation produces an accurate result. This is called 'naive interval arithmetic', and it is, in general, bound to fail (see Section 6).

In general, if no structure in the problem is known, the sign of summands cannot be predicted. This leads to the

UTILIZE INPUT DATA PRINCIPLE OF VERIFICATION METHODS:
Avoid re-use of computed data; use input data where possible. (5.13)

Our very first example in Theorem 1.1 follows this principle: the verification of $\varrho(I - RA) < 1$ is based mainly on the input matrix A.

Once again we want to stress that the effect of *data dependency* is not due to interval arithmetic but occurs with power set operations as well. Consider two enclosures $X = [3.14, 3.15]$ and $Y = [3.14, 3.15]$ for π. Then

$$X - Y = [-0.01, +0.01] \quad \text{or} \quad X/Y = [0.996, 1.004]$$

is (rounded to 3 digits) the best we can deduce using the given information; but when adding the information that both X and Y are inclusions for π, the results can be sharpened into 0 and 1, respectively.

5.3. *Floating-point bounds*

Up to now the discussion has been theoretical. In a practical implementation on a digital computer the bounds of an interval are floating-point numbers. Let $\mathbb{IF} \subset \mathbb{IR}$ denote the set $\{[\underline{x}, \overline{x}] : \underline{x}, \overline{x} \in \mathbb{F}, \underline{x} \leq \overline{x}\}$. We define a rounding $\diamond : \mathbb{IR} \to \mathbb{IF}$ by

$$\diamond([\underline{x}, \overline{x}]) := [\text{fldown}(\underline{x}), \text{flup}(\overline{x})], \tag{5.14}$$

and operations $\circledcirc : \mathbb{IF} \times \mathbb{IF} \to \mathbb{IF}$ for $X, Y \in \mathbb{IF}$ and $\circ \in \{+, -, \cdot, /\}$ by

$$X \circledcirc Y := \diamond(X \circ Y). \tag{5.15}$$

Fortunately this definition is straightforward to implement on today's computers using directed roundings, as introduced in Section 4. Basically, definitions (5.4) and (5.6) are used where the lower (upper) bound is computed with the processor switched into rounding downwards (upwards) mode. Note that the result is best possible.

There are ways to speed up a practical implementation of scalar interval operations. Fast C++ libraries for interval operations are PROFIL/BIAS by Knüppel (1994, 1998). Other libraries for interval operations include Intlib (Kearfott, Dawande, Du and Hu 1992) and C-XSC (Klatte *et al.* 1993). The main point for this article is that interval operations with floating-point bounds are rigorously implemented.

For vector and matrix operations a fast implementation is mandatory, as discussed in Section 9.1.

5.4. *Infinite bounds*

We defined $-\infty$ and $+\infty$ to be floating-point numbers. This is particularly useful for maintaining (5.4) and (5.14) without nasty exceptions in the case of overflow. Also, special operations such as division by a zero interval can be consistently defined, such as by Kahan (1968).

We feel this is not the main focus of interest, and requires too much detail for this review article. We therefore assume from now on that all intervals are finite. Once again, all results (for example, computed by INTLAB) remain rigorous, even in the presence of division by zero or infinite bounds.

5.5. *The inclusion property*

For readability we will from now on denote interval quantities by bold letters $\mathbf{X}, \mathbf{Y}, \ldots$. Operations between interval quantities are always interval operations, as defined in (5.15). In particular, an expression such as

$$\mathbf{R} = (\mathbf{X} + \mathbf{Y}) - \mathbf{X}$$

is to be understood as

$$\mathbf{Z} = \mathbf{X} + \mathbf{Y} \quad \text{and} \quad \mathbf{R} = \mathbf{Z} - \mathbf{X},$$

where both addition and subtraction are interval operations. If the order of execution is ambiguous, assertions are valid for any order of evaluation.

We will frequently demonstrate examples using INTLAB. Even without familiarity with the concepts of MATLAB it should not be difficult to follow our examples; a little additional information is given where necessary to understand the INTLAB code.

INTLAB uses an operator concept with a new data type `intval`. For $\mathtt{f}, \mathtt{g} \in \mathbb{F}$, the 'type casts' `intval(f)` or `infsup(f,g)` produce intervals $[\mathtt{f}, \mathtt{f}]$ or $[\mathtt{f}, \mathtt{g}]$, respectively. In the latter case $\mathtt{f} \leq \mathtt{g}$ is checked. Operations between interval quantities and floating-point quantities `f` are possible, where the latter are automatically replaced by the interval $[\mathtt{f}, \mathtt{f}]$. Such a quantity is called a *point interval.* With this the above reads

```
R = (X+Y)-X;
```

as executable INTLAB code, where `X` and `Y` are interval quantities.

The operator concept with the natural embedding of $\mathbb{F}$ into $\mathbb{IF}$ implies that an interval operation is applied if at least one of the operands is of type `intval`. Therefore,

```
X1 = 1/intval(3);
X2 = intval(1)/3;
X3 = intval(1)/intval(3);
```

all have the same result, namely $[\mathrm{fldown}(1/3), \mathrm{flup}(1/3)]$. The most important property of interval operations is the

INCLUSION PROPERTY:

$$\begin{aligned}&\text{Given } \mathbf{X}, \mathbf{Y} \in \mathbb{IF},\ \circ \in \{+, -, \cdot, /\} \text{ and any } x, y \in \mathbb{R} \\ &\text{with } \ x \in \mathbf{X}, y \in \mathbf{Y}, \ \text{ it is true that } x \circ y \in \mathbf{X} \circ \mathbf{Y}.\end{aligned} \tag{5.16}$$

For a given arithmetic expression $f(x_1, \ldots, x_n)$, we may replace each operation by its corresponding interval operation. Call that new expression $F(x_1, \ldots, x_n)$ the *natural interval extension* of f. Using the natural embedding $x_i \in \mathbf{X}_i$ with $\mathbf{X}_i := [x_i, x_i] \in \mathbb{IF}$ for $i \in \{1, \ldots, n\}$, it is clear that $f(x_1, \ldots, x_n) \in f(\mathbf{X}_1, \ldots, \mathbf{X}_n)$ for $x_i \in \mathbb{F}$. More generally, applying the INCLUSION PROPERTY (5.16) successively, we have the

INCLUSION PRINCIPLE:

$$x_i \in \mathbb{R}, \mathbf{X}_i \in \mathbb{IF} \text{ and } x_i \in \mathbf{X}_i \Longrightarrow f(x_1, \ldots, x_n) \in F(\mathbf{X}_1, \ldots, \mathbf{X}_n). \tag{5.17}$$

These remarkable properties, the INCLUSION PROPERTY (5.16) and the INCLUSION PRINCIPLE (5.17), due to Sunaga (1956, 1958) and rediscovered by Moore (1962), allow the estimation of the range of a function over a given domain in a simple and rigorous way. The result will always be true; however, much overestimation may occur (see Sections 6, 8 and 11.6).

Concerning infinite bounds, (5.17) is also true when using INTLAB; however, it needs some interpretation. An invalid operation such as division by two intervals containing zero leads to the answer NaN. This symbol *Not a Number* is the MATLAB representation for invalid operations such as $\infty - \infty$. An interval result NaN is to be interpreted as 'no information available'.

6. Naive interval arithmetic and data dependency

One may try to replace each operation in some algorithm by its corresponding interval operation to overcome the rounding error problem. It is a true statement that the true value of each (intermediate) result is included in the corresponding interval (intermediate) result. However, the direct and naive use of the INCLUSION PRINCIPLE (5.17) in this way will almost certainly fail by producing wide and useless bounds.

Consider $f(x) = x^2 - 2$. A Newton iteration $x_{k+1} = x_k - (x_k^2 - 2)/2x_k$ with starting value x_0 not too far from $\sqrt{2}$ will converge rapidly. After at most 5 iterations, any starting value in $[1, 2]$ produces a result with 16 correct figures, for example, in executable MATLAB code:

```
>> x=1; for i=1:5, x = x - (x^2-2)/(2*x), end
x = 1.500000000000000
x = 1.416666666666667
x = 1.414215686274510
x = 1.414213562374690
x = 1.414213562373095
```

Now consider a naive interval iteration starting with $\mathbf{X}_0 := [1.4, 1.5]$ in executable INTLAB code.

Algorithm 6.1. Naive interval Newton procedure:

```
>> X=infsup(1.4,1.5);
>> for i=1:5, X = X - (X^2-2)/(2*X), end
intval X =
[    1.3107,     1.5143]
intval X =
[    1.1989,     1.6219]
intval X =
[    0.9359,     1.8565]
intval X =
[    0.1632,     2.4569]
intval X =
[  -12.2002,     8.5014]
```

Rather than converging, the interval diameters increase, and the results are of no use. This is another typical example of inappropriate use of interval arithmetic. The reason for the behaviour is data dependency. This is not due to interval arithmetic, but, as has been noted before, the results would be the same when using power set operations. It corresponds to the rules of thumb mentioned in Section 1.4.

Instead of an inclusion of

$$\{x - (x^2 - 2)/(2x) : x \in \mathbf{X}_k\}, \tag{6.1}$$

naive interval arithmetic computes in the kth iteration an inclusion of

$$\{\xi_1 - (\xi_2^2 - 2)/(2\xi_3) : \xi_1, \xi_2, \xi_3 \in \mathbf{X}_k\}. \tag{6.2}$$

To improve this, the Newton iteration is to be redefined in an appropriate way, utilizing the strengths of interval arithmetic and diminishing weaknesses (see Moore (1966)).

Theorem 6.2. Let a differentiable function $f : \mathbb{R} \to \mathbb{R}$, $\mathbf{X} = [x_1, x_2] \in \mathbb{IR}$ and $\tilde{x} \in \mathbf{X}$ be given, and suppose $0 \notin f'(\mathbf{X})$. Using interval operations, define

$$N(\tilde{x}, \mathbf{X}) := \tilde{x} - f(\tilde{x})/f'(\mathbf{X}). \tag{6.3}$$

If $N(\tilde{x}, \mathbf{X}) \subseteq \mathbf{X}$, then $\mathbf{X}$ contains a unique root of f. If $N(\tilde{x}, \mathbf{X}) \cap \mathbf{X} = \emptyset$, then $f(x) \neq 0$ for all $x \in \mathbf{X}$.

Proof. If $N(\tilde{x}, \mathbf{X}) \subseteq \mathbf{X}$, then

$$x_1 \leq \tilde{x} - f(\tilde{x})/f'(\xi) \leq x_2 \tag{6.4}$$

for all $\xi \in \mathbf{X}$. Therefore $0 \notin f'(\mathbf{X})$ implies

$$(f(\tilde{x}) + f'(\xi_1)(x_1 - \tilde{x})) \cdot (f(\tilde{x}) + f'(\xi_2)(x_2 - \tilde{x})) \leq 0$$

for all $\xi_1, \xi_2 \in \mathbf{X}$, and in particular $f(x_1) \cdot f(x_2) \leq 0$. So there is a root of f in $\mathbf{X}$, which is unique because $0 \notin f'(\mathbf{X})$.

Suppose $\hat{x} \in \mathbf{X}$ is a root of f. By the Mean Value Theorem there exists $\xi \in X$ with $f(\tilde{x}) = f'(\xi)(\tilde{x} - \hat{x})$ or $\hat{x} = \tilde{x} - f(\tilde{x})/f'(\xi) \in N(\tilde{x}, \mathbf{X})$, and the result follows. □

For a univariate function f it is not difficult to certify that an interval contains a root of f. However, to verify that a certain interval does *not* contain a root is not that simple or obvious.

Theorem 6.2 and in particular (6.3) are suitable for application of interval arithmetic. Let $\mathbf{X}$ be given. The assumptions of Theorem 6.2 are verified as follows.

(1) Let $\mathbf{F}$ and $\mathbf{Fs}$ be interval extensions of f and f', respectively.
(2) If $\mathbf{Fs}(\mathbf{X})$ does not contain zero, then $f'(x) \neq 0$ for $x \in \mathbf{X}$.
(3) If $\tilde{x} - \mathbf{F}(\tilde{x})/\mathbf{Fs}(\mathbf{X}) \in \mathbf{X}$, then $\mathbf{X}$ contains a unique root of f.
(4) If $\{\tilde{x} - \mathbf{F}(\tilde{x})/\mathbf{Fs}(\mathbf{X})\} \cap \mathbf{X} = \emptyset$, then $\mathbf{X}$ contains no root of f.

Step (1) is directly solved by writing down the functions in INTLAB. We then define the computable function

$$\mathbf{N}(\tilde{x}, \mathbf{X}) := \tilde{x} - \mathbf{F}(\tilde{x})/\mathbf{Fs}(\mathbf{X}) : \mathbb{F} \times \mathbb{IF} \to \mathbb{IF}, \tag{6.5}$$

where all operations are interval operations with floating-point bounds. Then always $N(\tilde{x}, \mathbf{X}) \subseteq \mathbf{N}(\tilde{x}, \mathbf{X})$, so that the assumptions of Theorem 6.2 can be confirmed on the computer.

After verifying step (2) for the initial interval $\mathbf{X} := [1, 2]$, we obtain the following results:

```
>> X=infsup(1,2);
   for i=1:4, xs=intval(mid(X)); X = xs - (xs^2-2)/(2*X), end
intval X =
[   1.37499999999999,   1.43750000000001]
intval X =
[   1.41406249999999,   1.41441761363637]
intval X =
[   1.41421355929452,   1.41421356594718]
intval X =
[   1.41421356237309,   1.41421356237310]
```

Starting with the wide interval $[1, 2]$, an accurate inclusion of the root of f is achieved after 4 iterations. Note that the type cast of `xs` to type `intval` by `xs` $=$ `intval(mid(X))` is mandatory. Using `xs` $=$ `mid(X)` instead, the computation of `xs^2-2` would be performed in floating-point arithmetic. Moreover, as in Theorem 6.2, `xs` does not need to be the exact midpoint of `X`; only `xs` $\in$ `X` is required. This is true for the `mid`-function. Note that $[\tilde{x} - \mathbf{F}(\tilde{x})/\mathbf{F}(\mathbf{X})] \cap \mathbf{X}$ contains the root of f as well; in our example, however, it makes no difference.

Also note that all output of INTLAB is rigorous. This means that a *displayed* lower bound is less than or equal to the computed lower bound, and similarly for the upper bound.

For narrow intervals we have found another form of display useful:

```
>> format _; X=infsup(1,2);
   for i=1:4, xs=intval(mid(X)); X = xs - (xs^2-2)/(2*X), end
intval X =
[   1.37499999999999,   1.43750000000001]
intval X =
   1.414___________
intval X =
   1.41421356______
intval X =
   1.41421356237309
```

A true inclusion is obtained from the display by subtracting and adding 1 to the last displayed digit. For example, a true inclusion of the last iterate is

$$[1.41421356237308, 1.41421356237310].$$

This kind of display allows us to grasp easily the accuracy of the result. In this particular case it also illustrates the quadratic convergence.

7. Standard functions and conversion

The concepts of the previous section can be directly applied to the evaluation of standard functions. For example,

$$\sin(x) \in x - x^3/3! + x^5/5! + \frac{x^7}{7!}[-1, 1] \tag{7.1}$$

is a true statement for all $x \in \mathbb{R}$. A remarkable fact is that (7.1) can be applied to interval input as well:

$$\sin(x) \in \left\{ \mathbf{X} - \mathbf{X}^3/3! + \mathbf{X}^5/5! + \frac{\mathbf{X}^7}{7!}[-1, 1] \right\} \cap [-1, 1] \tag{7.2}$$

is a true statement for all $x \in \mathbf{X}$. Of course, the quality of such an inclusion may be weak. Without going into details, we mention that, among others, for

- exp, log, sqrt
- sin, cos, tan, cot and their inverse functions (7.3)
- sinh, cosh, tanh, coth and their inverse functions

interval extensions are implemented in INTLAB. Much care is necessary to choose the right approximation formulas, and in particular how to evaluate them. In consequence, the computed results are mostly accurate to the last digit. The algorithms are based on a table-driven approach carefully using the built-in functions at specific sets of points and using addition theorems for the standard functions. When no addition theorem is available (such as for the inverse tangent), other special methods have been developed.

Those techniques for standard functions are presented in Rump (2001*a*) with the implementational details. In particular a fast technique by Payne and Hanek (1983) was rediscovered to evaluate sine and cosine accurately for very large arguments. All standard functions in (7.3) are implemented in INTLAB also for complex (interval) arguments, for point and for interval input data. The implementation of the latter follows Börsken (1978).

Other techniques for the implementation of elementary standard functions were given by Braune (1987) and Krämer (1987). Interval implementations for some higher transcendental functions are known as well, for example for the Gamma function by Krämer (1991).

7.1. Conversion

Some care is necessary when converting real numbers into intervals. When executing the statement

```
X = intval(3.5)
```

MATLAB first converts the input '3.5' into the nearest floating-point number f, and then defines `X` to be the point interval $[f, f]$. In this case indeed $f = 3.5$, because 3.5 has a finite binary expansion and belongs to $\mathbb{F}$. However, the statement

```
X = intval(0.1)
```

produces a point interval *not* containing the real number $0.1 \notin \mathbb{F}$. The problem is that the routine `intval` receives as input the floating-point number nearest to the real number 0.1, because conversion of $0.1 \in \mathbb{R}$ into $\mathbb{F}$ has already taken place. To overcome this problem, the conversion has to be performed by INTLAB using

```
X = intval('0.1')
```

The result is an interval truly containing the real number 0.1. Similar considerations apply to transcendental numbers. For example,

```
>> Y = sin(intval(pi))
intval Y =
  1.0e-015 *
[   0.12246467991473,    0.12246467991474]
```

is an inclusion of $\sin(f)$, where f is the nearest floating-point number to π. Note that `Y` does not contain zero. The command

```
>> sin(intval('pi'))
intval ans =
  1.0e-015 *
[  -0.32162452993533,    0.12246467991474]
```

uses an inclusion of the transcendental number π as argument for the sine, hence the inclusion of the sine must contain zero. The statement `sin(4 * atan(intval(1)))` would also work.

8. Range of a function

The evaluation of the interval extension of a function results in an inclusion of the range of the function over the input interval. Note that this is achieved by a mere evaluation of the function, without further knowledge such as extrema, Lipschitz properties and the like. Sometimes it is possible to rearrange the function to improve this inclusion.

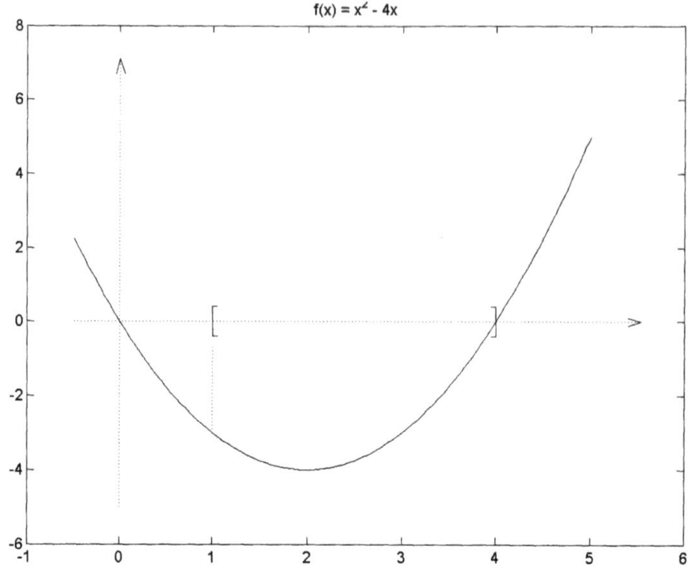

Figure 8.1. Graph of $f(x) = x^2 - 4x$.

8.1. Rearrangement of a function

Consider a simple model function

$$f(x) := x^2 - 4x \quad \text{over} \quad \mathbf{X} := [1, 4]. \tag{8.1}$$

From the graph of the function in Figure 8.1, the range is clearly $\{f(x) : x \in \mathbf{X}\} = [-4, 0]$. The straightforward interval evaluation yields

```
>> X = infsup(1,4); X^2-4*X
intval ans =
[  -15.0000,    12.0000]
```

Now $f(x) = x(x - 4)$, so

```
>> X = infsup(1,4); X*(X-4)
intval ans =
[  -12.0000,     0.0000]
```

is an inclusion as well. But also $f(x) = (x - 2)^2 - 4$, and this yields the exact range:

```
>> X = infsup(1,4); (X-2)^2-4
intval ans =
[   -4.0000,     0.0000]
```

Note that the real function $f : \mathbb{R} \to \mathbb{R}$ is manipulated, *not* the interval extension $\mathbf{F} : \mathbb{IF} \to \mathbb{IF}$. Manipulation of expressions including intervals should be done with great care; for example, only $(\mathbf{X}+\mathbf{Y})\cdot\mathbf{Z} \subseteq \mathbf{X}\cdot\mathbf{Y}+\mathbf{X}\cdot\mathbf{Z}$

is valid. Also note that the interval square function is the interval extension of $s(x) := x^2$ and is therefore evaluated by

$$\mathbf{X}^2 := [\mathbf{X} \cdot \mathbf{X}] \cap [0, \infty], \tag{8.2}$$

the bounds of which are easily calculated by a case distinction, whether $\mathbf{X}$ is completely positive, negative or contains zero.

It is not by accident that $f(x) = (x-2)^2 - 4$ yields the exact range of $f(\mathbf{X})$ (neglecting possible overestimation by directed rounding). In fact, it achieves this for every input interval $\mathbf{X}$. The reason is that the input interval occurs only once, so that there is no dependency. Because each individual operation computes the best-possible result, this is true for the final inclusion as well (apparently this was known to Sunaga (1956); it was formulated in Moore (1966, p. 11)).

Theorem 8.1. Let a function $f : \mathbb{R}^n \to \mathbb{R}$ be given, and let $\mathbf{F} : \mathbb{IR}^n \to \mathbb{IR}$ be its interval extension. If f is evaluated by some arithmetic expression and each variable x_i occurs at most once, then

$$\mathbf{F}(\mathbf{X}_1, \ldots, \mathbf{X}_n) = \{f(x_1, \ldots, x_n) : x_i \in \mathbf{X}_i \quad \text{for} \quad 1 \le i \le n\}, \tag{8.3}$$

for all $\mathbf{X}_1, \ldots, \mathbf{X}_n \in \mathbb{IF}$.

Unfortunately there seems to be no general recipe for how to rearrange a function in order to minimize overestimation. It may well be that in one expression the input variables occur less often than in another, and yet the overestimation is worse.

8.2. Oscillating functions

Consider more involved functions. One may expect that for functions with many extrema, an overestimation of the range is more likely. Consider

$$f(x) := \sin\left(\frac{2x^2}{\sqrt{\cosh x}} - x\right) - \operatorname{atan}(4x+1) + 1 \quad \text{over} \quad X := [0, 4]. \tag{8.4}$$

The graph of this function is shown in Figure 8.2. One verifies that the minimum and maximum are achieved near $x_1 = 0.408$ and $x_2 = 1.556$, so that

$$\{f(x) : x \in \mathbf{X}\} \subseteq [-0.2959, 0.5656] \tag{8.5}$$

is a true statement and best possible in 4 digits. The direct interval evaluation yields

```
>> f = inline('sin(2*x^2/sqrt(cosh(x))-x)-atan(4*x+1)+1');
   Y = f(infsup(0,4))
intval Y =
[   -1.5121,     1.2147]
```

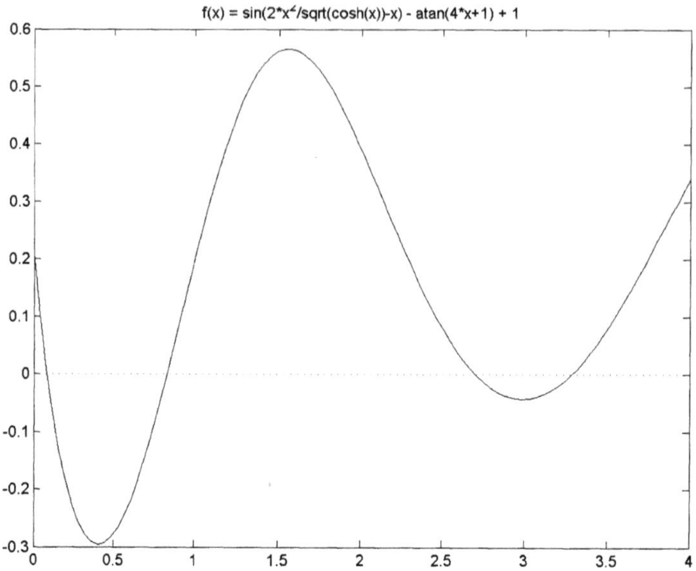

Figure 8.2. Graph of $f(x)$ as defined in (8.4).

which is a true result with some overestimation. Next we change the function by simply replacing the sine function by the hyperbolic sine:

$$g(x) := \sinh\left(\frac{2x^2}{\sqrt{\cosh x}} - x\right) - \operatorname{atan}(4x+1) + 1 \quad \text{over} \quad X := [0,4]. \quad (8.6)$$

The graph of this function is shown in Figure 8.3. One might think that the range of this function would be even simpler to estimate. The best-possible inclusion to 4 digits is

$$\{g(x) : x \in \mathbf{X}\} \subseteq [-0.2962, 6.7189], \quad (8.7)$$

whereas the interval evaluation yields the gross overestimate

```
>> g = inline('sinh(2*x^2/sqrt(cosh(x))-x)-atan(4*x+1)+1');
   format short e
   Y = g(infsup(0,4))
intval Y =
[ -2.7802e+001,  3.9482e+013]
```

Note that the interval evaluation of the standard functions is practically best possible. The reason is once again data dependency. The function

$$h_1(x) := \frac{2x^2}{\sqrt{\cosh x}}$$

is not too far from $h_2(x) := x$ over $[0,4]$, so that the range of the argument $h_3(x) := h_1(x) - x$ of the hyperbolic sine is $h_3([0,4]) \subseteq [-0.1271, 2.6737]$.

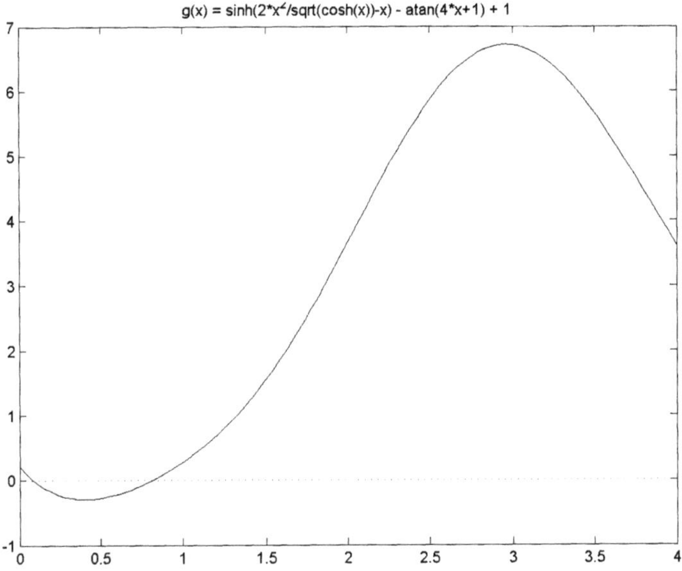

Figure 8.3. Graph of $g(x)$ as defined in (8.6).

However,

$$\left\{ \frac{2\xi_1^2}{\sqrt{\cosh \xi_2}} - \xi_3 : \xi_i \in [0, 4] \right\} = [-4, 32].$$

This overestimate is due to data dependency, and the result using power set operations is the same. Then this overestimate is amplified by the hyperbolic sine and produces the observed gross overestimate for the function g defined in (8.6).

In contrast, in the first function f defined in (8.4), the sine function reduces the overestimate, since the output of the interval sine function is always bounded by $[-1, 1]$. We come to this again in Section 11.6.

8.3. Improved range estimation

Let a real function $f : \mathbb{R} \to \mathbb{R}$ and its interval extension $\mathbf{F} : \mathbb{IF} \to \mathbb{IF}$ be given. For a given $\mathbf{X} = [a, b] \in \mathbb{IF}$, we always have $f(\mathbf{X}) \subseteq \mathbf{F}(\mathbf{X})$, so that $0 \notin \mathbf{F}(\mathbf{X})$ implies that f has no roots in $\mathbf{X}$. Applying this to an interval extension $\mathbf{Fs} : \mathbb{IF} \to \mathbb{IF}$ of f' means that f is monotone on $\mathbf{X}$ provided $0 \notin \mathbf{Fs}(\mathbf{X})$. In this case

$$0 \notin \mathbf{Fs}(\mathbf{X}) \quad \Rightarrow \quad f(\mathbf{X}) = \{f(x) : x \in \mathbf{X}\} = f(a) \,\underline{\cup}\, f(b). \tag{8.8}$$

Hence a straightforward way to improve the range estimation $f(\mathbf{X})$ is to apply (8.8) if $0 \notin \mathbf{Fs}(\mathbf{X})$ is true, and otherwise to apply a bisection scheme.

Table 8.1. Range inclusion for g as in (8.6), using bisection scheme and (8.8).

Bisection depth	Function evaluations	Range
4	31	$[-0.6415, 19.1872]$
5	53	$[-0.4864, 11.2932]$
6	73	$[-0.3919, 8.7007]$
7	93	$[-0.3448, 7.6441]$
8	115	$[-0.3206, 7.1664]$
9	137	$[-0.3084, 6.9389]$
10	159	$[-0.3023, 6.8280]$
true range		$[-0.2962, 6.7189]$
`fminbnd`	19	$]-0.2961, 6.7189[$

Doing this for the function g defined in (8.6) for different bisection depths yields the results displayed in Table 8.1.

In the second-to-last row the true range is displayed. So, as may be expected, with increasing bisection depth the range estimation improves at the price of more function evaluations.

Note that, despite possible overestimation, interval evaluation has two advantages. First, the estimates come without further knowledge of the function, and second, all range estimates are rigorous. Using pure floating-point arithmetic, this rigour is hardly achievable. We may use some nonlinear function minimization, for example the function `fminbnd` in MATLAB. According to the specification it attempts to find a local minimizer of a given function within a specified interval. Applying this to g and $-g$ yields the result displayed in the last row of Table 8.1. Obviously, with a few function evaluations the true range is found.

Note that in principle the local minimization approach finds an *inner* inclusion of the true range,[12] and it may occur that the global minimum within the given interval is missed. Applying the same scheme to the function f as in (8.4), we obtain the results displayed in Table 8.2.

Again, increasing bisection depth increases the accuracy of the range estimation. However, the minimization function `fminbnd` fails to find the global minimum near 0.4 and underestimates the true range. The displayed bounds are calculated using the default values. Setting the maximum number of function evaluations by `fminbnd` to 1000 and the termination tolerance on

[12] Provided function evaluations are rigorous.

Table 8.2. Range inclusion for f as in (8.4), using bisection scheme and (8.8).

Bisection depth	Function evaluations	Range
4	31	$[-1.0009, 0.6251]$
5	59	$[-0.5142, 0.5920]$
6	91	$[-0.3903, 0.5772]$
7	101	$[-0.3442, 0.5717]$
8	113	$[-0.3202, 0.5686]$
9	127	$[-0.3081, 0.5671]$
10	137	$[-0.3020, 0.5664]$
true range		$[-0.2959, 0.5656]$
`fminbnd`	18	$]-0.0424, 0.5656[$

the minimizer and the function value to 10^{-12} increases the number of function evaluations to 21 but produces the same result.

The approach described in this section (as well as the Newton iteration in Theorem 6.2) requires an interval extension of the derivative. It is not satisfactory for this to be provided by a user. In fact, the range of the first and higher derivatives in the univariate as well as the multivariate case can be computed automatically. This will be described in Section 11.

Moreover, in Section 11.6 derivatives and slopes will be used to improve the range estimation for narrow input intervals; see also Section 13.1. Further background information, together with a challenge problem, is given by Neumaier (2010).

9. Interval vectors and matrices

In order to treat multivariate functions, we need to define interval vectors and matrices. An interval vector $\mathbf{X}$ may be defined as an n-tuple with interval entries such that

$$\mathbf{X} := \{x \in \mathbb{R}^n : x_i \in \mathbf{X}_i \quad \text{for} \quad 1 \le i \le n\}.$$

Obviously an interval vector is the Cartesian product of (one-dimensional) intervals. There is another, equivalent definition using the partial ordering $\le$ on $\mathbb{R}^n$, namely $\mathbf{X} = [\underline{x}, \overline{x}]$ for $\underline{x}, \overline{x} \in \mathbb{R}^n$ with

$$\mathbf{X} = \{x \in \mathbb{R}^n : \underline{x} \le x \le \overline{x}\}, \tag{9.1}$$

where comparison of vectors and matrices is always to be understood componentwise. Both representations are useful. Since they are equivalent we use $\mathbb{IR}^n$ to denote interval vectors. The set of interval matrices $\mathbb{IR}^{n \times n}$

is defined similarly. The lower and upper bound of an interval quantity $\mathbf{X}$ is denoted by $\underline{\mathbf{X}}$ and $\overline{\mathbf{X}}$, respectively.

Interval operations extend straightforwardly to vectors and matrices by replacement of each individual (real) operation by its corresponding interval operation. For example,

$$\mathbf{A} = \begin{pmatrix} [-2,1] & -2 \\ [0,2] & [1,3] \end{pmatrix} \quad \text{and} \quad \mathbf{X} = \begin{pmatrix} [-3,2] \\ 4 \end{pmatrix} \quad \text{yield} \quad \mathbf{AX} = \begin{pmatrix} [-12,-2] \\ [-2,16] \end{pmatrix}. \tag{9.2}$$

As before, we use the natural embedding of $\mathbb{R}$ into $\mathbb{IR}$. This concept of point intervals extends to vectors and matrices as well, and the names point vector or point matrix are used, respectively. Note that for $\mathbf{A} \in \mathbb{IR}^{n\times n}$ and $\mathbf{X} \in \mathbb{IR}^n$ the inclusion property (5.16) implies

$$A \in \mathbf{A}, x \in \mathbf{X} \quad \Rightarrow \quad Ax \in \mathbf{AX}. \tag{9.3}$$

As before, we assume interval operations to be used if at least one operand is an interval quantity. Next consider two special cases. First, let an interval matrix $\mathbf{A} \in \mathbb{IR}^{n\times n}$ and a vector $x \in \mathbb{R}^n$ be given. Then Theorem 8.1 implies

$$\mathbf{A}x = \{Ax : A \in \mathbf{A}\}. \tag{9.4}$$

That means there is no overestimation; the interval operation and the power set operation coincide. The reason is that the components $\mathbf{A}_{ij}$ of $\mathbf{A}$, the only occurring intervals, are used only once in $\mathbf{A}x$.

Secondly, consider a real matrix $A \in \mathbb{R}^{n\times n}$ and an interval vector $\mathbf{X} \in \mathbb{IR}^n$. Now in the interval multiplication $A\mathbf{X}$ each interval component $\mathbf{X}_i$ is used n times, so in general the interval product $A\mathbf{X}$ will be an overestimation by the power set operation:

$$\{Ax : x \in \mathbf{X}\} \subseteq A\mathbf{X}. \tag{9.5}$$

However, for the computation of each *component* $(A\mathbf{X})_i$ of the result, each interval component $\mathbf{X}_j$ is used only once, so componentwise there is no overestimation. In other words, the resulting interval vector $A\mathbf{X}$ is narrowest possible, that is,

$$A\mathbf{X} = \bigcap \{\mathbf{Z} \in \mathbb{IR} \,:\, Ax \in \mathbf{Z}, x \in \mathbf{X}\} =: \operatorname{hull}\big(\{Ax : x \in X\}\big). \tag{9.6}$$

This is called the *interval hull.* Consider

$$A = \begin{pmatrix} 2 & -1 \\ 1 & 2 \end{pmatrix} \quad \text{and} \quad \mathbf{X} = \begin{pmatrix} [2,3] \\ [1,3] \end{pmatrix}. \tag{9.7}$$

Then Figure 9.1 illustrates the result of the power set operation $\{Ax : x \in \mathbf{X}\}$ and the interval operation $A\mathbf{X}$. The power set operation is a linear transformation of the axis-parallel rectangle $\mathbf{X}$.

This creates a pitfall. Consider a system of linear equations $Ax = b$ for $A \in \mathbb{R}^{n\times n}$ and $b, x \in \mathbb{R}^n$. Using an approximate solution $\tilde{x}$ computed

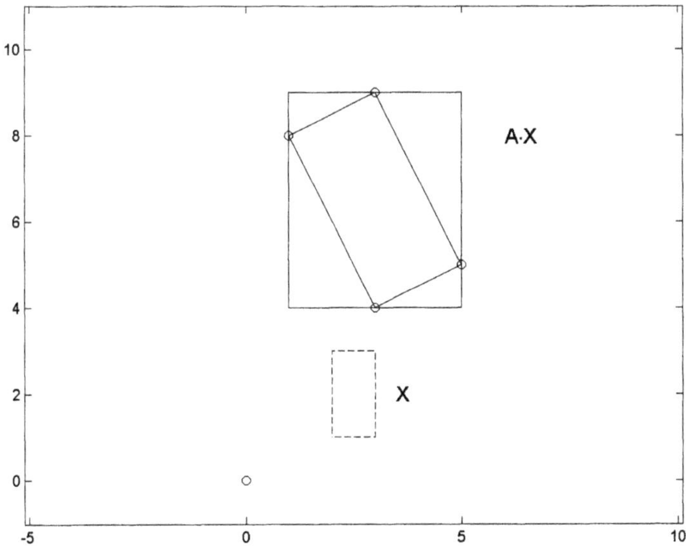

Figure 9.1. Power set and interval product for A and $\mathbf{X}$ as in (9.7).

by some standard algorithm, define an interval vector $\mathbf{X}$ containing $\tilde{x}$, for example $\mathbf{X} := [1-e, 1+e] \cdot \tilde{x}$ for some small $0 < e \in \mathbb{R}$. Because $A\mathbf{X}$ is narrowest possible, one might conclude that $\mathbf{X}$ contains the solution $A^{-1}b$ if $b \in A\mathbf{X}$. It is true that $Ax \in A\mathbf{X}$ for all $x \in \mathbf{X}$; however,

$$\forall\, y \in A\mathbf{X}\ \exists\, x \in \mathbf{X} \ :\ y = Ax \quad \text{is not true} \tag{9.8}$$

(see Figure 9.1 for the data in (9.7)). For example, $x := (4,8)^T \in A\mathbf{X} = ([1,5],[4,9])^T$, but $A^{-1}x = (3.2, 2.4)^T \notin \mathbf{X}$.

9.1. Performance aspects

The definition of interval vector and matrix operations by replacement of each individual (real) operation by its corresponding interval operation is theoretically useful, but it implies a severe performance impact. This is because on today's architectures dramatic speed-up is obtained by instruction-level parallelism and by avoiding cache-misses. On the contrary, branches, in particular, may slow down an algorithm significantly.

Our definition of the product of two interval matrices $\mathbf{A}, \mathbf{B} \in \mathbb{IF}^{n\times n}$, say, implies that in the inner loop an interval product $\mathbf{A}_{ik}\mathbf{B}_{kj}$ has to be computed. This requires at least two branches to decide whether the input intervals are non-negative, non-positive or contain zero, and two switches of the rounding mode to compute the lower and upper bound. In the worst case of two intervals both containing zero in the interior, four products must be computed in total.

Table 9.1. Performance in GFLOPs for LAPACK matrix multiplication (DGEMM), Gaussian elimination with partial pivoting (DGETRF) and with total pivoting (DGETC2); one branch is counted as one FLOP.

n	DGEMM	DGETRF	DGETC2
2000	121	68	0.20
4000	151	114	0.15
6000	159	128	0.12
8000	163	142	0.14
10000	166	150	0.11

But a pure counting of floating-point operations does not show how well a code can be optimized, and in particular it hides the disastrous effect of branches. As an example of how branches may slow down a computation we refer to the well-known LAPACK package described in Anderson *et al.* (1995). All routines are well written by the best experts in the field. We compare the routines DGEMM for multiplying two matrices, Gaussian elimination with partial pivoting DGETRF and Gaussian elimination with total pivoting DGETC2. For Table 9.1 we count each addition, multiplication and branch as 1 FLOP and display the GFLOPs achieved for all routines. The environment is a PC with four Quad-Core AMD Opteron 8393 SE with 3.1 GHz clock speed, thus 16 cores in total. Each core may execute up to four operations in parallel, so that the peak performance is $64 \times 3.1 = 198.4$ GFLOPs.[13]

Note that if all codes could be equally well optimized and there were no time penalty for branches, all GFLOP counts in Table 9.1 would be equal. However, we observe a slow-down by a factor well over 1000 for dimension $n = 10\,000$. Similarly a significant slow-down is observed when implementing the interval matrix product according to the theoretical definition by case distinctions.

For a fast implementation of interval vector and matrix operations, Rump (1999*b*) analysed that the conversion into midpoint-radius form is useful. We defined an interval by its left and right bound. Equivalently, we may use

$$\mathbf{X} = \langle m, r \rangle := \{x \in \mathbb{R} : m - r \leq x \leq m + r\}, \tag{9.9}$$

and similarly for vectors and matrices using the componentwise partial ordering in $\mathbb{R}^n$ and $\mathbb{R}^{n\times n}$. Sometimes the midpoint-radius is advantageous

[13] Thanks to Takeshi Ogita for performing the tests.

because the minimum radius is not restricted to the distance of adjacent floating-point numbers, but tiny radii are possible as well.

First, consider the product $\mathbf{A} \cdot B$ of an interval matrix $\mathbf{A} = [\underline{A}, \overline{A}] \in \mathbb{IF}^{m \times k}$ and a real matrix $B \in \mathbb{F}^{k \times n}$. Setting $mA := 0.5(\underline{A} + \overline{A})$ and $rA := 0.5(\overline{A} - \underline{A})$, a little thought reveals

$$\mathbf{A} \cdot B = \langle mA, rA \rangle \cdot B = \langle mA \cdot B,\ rA \cdot |B| \rangle. \tag{9.10}$$

For the computer implementation note that both the conversion of the interval matrix $\mathbf{A}$ into midpoint-radius form and the products $mA \cdot B$ and $rA \cdot |B|$ are subject to rounding errors. Fortunately this problem can be solved in a simple way.

Algorithm 9.1. Fast matrix multiplication $[\mathtt{Ainf}, \mathtt{Asup}] \cdot \mathtt{B}$:

```
setround(1)
mA = 0.5*(Ainf+Asup);
rA = mA - Ainf;
rC = rA*abs(B);
Csup = mA*B + rC;
setound(-1)
Cinf = mA*B - rC;
```

The elegant conversion from left and right bound to midpoint-radius form is due to Oishi (1998). Algorithm 9.1 ensures that, for the computed floating-point matrices mA and rA,

$$\mathtt{mA} - \mathtt{rA} \leq \mathtt{Ainf} \leq \mathtt{Asup} \leq \mathtt{mA} + \mathtt{rA}. \tag{9.11}$$

Moreover, $\mathtt{abs(B)} = |\mathtt{B}|$ because $\mathbb{F} = -\mathbb{F}$, and the setting of the rounding mode yields

$$0 \leq \mathtt{rA} \cdot |\mathtt{B}| \leq \mathtt{rC} \tag{9.12}$$

for the floating-point matrices rA, B and rC, so that

$$\mathtt{Cinf} \leq \mathtt{mA} \cdot |\mathtt{B}| - \mathtt{rC} \leq \mathtt{mA} \cdot |\mathtt{B}| + \mathtt{rC} \leq \mathtt{Csup} \quad \text{and} \quad \mathbf{A} \cdot B \subseteq [\mathtt{Cinf}, \mathtt{Csup}]. \tag{9.13}$$

Note that not only are various branches and switchings of the rounding mode omitted but the main work reduces to 3 multiplications of real matrices. Therefore, the very fast BLAS routines can be used, which are again much faster than an ordinary 3-loop implementation.

Secondly, consider the product $\mathbf{A} \cdot \mathbf{B}$ of two interval matrices $\mathbf{A} = [mA - rA, mA + rA] \in \mathbb{IF}^{m \times k}$ and $\mathbf{B} = [mB - rB, mB + rB] \in \mathbb{IF}^{k \times n}$. Again it is easy to see that

$$\begin{aligned} &\mathbf{A} \cdot \mathbf{B} \subseteq [mC - rC, mC + rC] \quad \text{for} \\ &\qquad mC := mA \cdot mB \quad \text{and} \quad rC := rA \cdot (|mB| + rB) + |mA| \cdot rB. \end{aligned} \tag{9.14}$$

Note that now there is some overestimation due to the fact that rB is used

Table 9.2. Computing time for real-interval and interval-interval matrix multiplication in C, floating-point multiplication normed to 1.

	Real × interval matrix		Interval × interval matrix	
Dimension	Traditional	(9.10)	Traditional	(9.14)
100	11.3	3.81	110.2	5.36
200	12.4	3.53	131.1	4.94
500	12.3	3.35	134.2	4.60
1000	20.1	3.25	140.0	4.45
1500	23.2	3.18	142.6	4.32
2000	23.6	3.14	144.6	4.25

twice in the computation of rC, and the product of the midpoints $mA \cdot mB$ is not the midpoint of the product $\mathbf{A} \cdot \mathbf{B}$. For example, $[1,3] \cdot [5,9] = [5,27]$ is the usual interval product identical to the power set operation, but $\langle 2,1\rangle \cdot \langle 7,2\rangle = \langle 14,13\rangle = [1,27]$ using (9.14).

However, something strange happens, namely that the *computed* quantities $[mC - rC, mC + rC]$ are sometimes more narrow than $\mathbf{A} \cdot \mathbf{B}$, contradicting (9.14). This is due to outward rounding, as in the following example:

$$[2.31, 2.33] \cdot [3.74, 3.76] = [8.6394, 8.7608] \subseteq [8.63, 8.77] \tag{9.15}$$

is the best-possible result in a 3-digit decimal arithmetic, whereas (9.14) yields

$$\begin{aligned} \langle 2.32, 0.01\rangle \cdot \langle 3.75, 0.01\rangle &= \langle 8.7, 0.01 \cdot (3.75 + 0.01) + 2.32 \cdot 0.01\rangle \\ &= \langle 8.7, 0.0608\rangle, \end{aligned} \tag{9.16}$$

which has a 13% smaller diameter than the result in (9.14).

An implementation of (9.14) along the lines of Algorithm 9.1 is straightforward. Note that the main costs are 4 real matrix multiplications, which are, as before, very fast using BLAS routines.

We first compare the new approaches in (9.10) and (9.14) with a C implementation along the traditional lines. With the time for one ordinary BLAS matrix multiplication in floating-point normed to 1, Table 9.2 displays the timing for different dimensions.[14] It clearly shows the advantage of the new approach, which almost achieves the theoretical ratio.

This method, as in (9.10) and (9.14), is in particular mandatory for a MATLAB implementation: although much effort has been spent on diminishing the effects of interpretation, loops may still slow down a computation

[14] Thanks to Viktor Härter for performing the tests.

Table 9.3. Computing time for interval matrix multiplication in INTLAB, floating-point multiplication normed to 1.

Dimension	3-loop	Rank-1 update	Mid-rad by (9.14)
50	610659.1	81.3	6.4
100		79.6	5.2
200		100.5	4.7
500		173.2	4.3
1000		261.7	4.1

significantly, in particular when user-defined variables such as intervals are involved. The fastest way to compute a tight inclusion without overestimation, as in (9.14), seems to be via rank-1 updates. Both possibilities, tightest and fast inclusions, are available in INTLAB via a system variable.[15]

Table 9.3 shows the timing for the product of two interval matrices for a standard 3-loop approach, for the tight inclusion using rank-1 updates and the fast inclusion via (9.14). The computing time for an ordinary product of two real matrices of the same size is again normed to 1.

As can be seen, the overhead for the tight inclusion increases with the dimension, whereas the time ratio for the fast inclusion approaches the theoretical factor 4. The standard 3-loop approach is unacceptably slow even for small dimensions.

As in the scalar case, interval matrices are a convenient way to deal with matrices not representable in floating-point arithmetic. For example, Algorithm 9.2 produces an inclusion A of the matrix in our model problem (2.11).

Algorithm 9.2. Computation of an interval matrix $\mathtt{A} \in \mathbb{IF}^{n\times n}$ containing the real matrix as in (2.11):

```
A = zeros(n);
for i=1:n, for j=1:n
  A(i,j) = intval(1)/(i+(j-1)*n);
end, end
```

Then, for example, the matrix product A ∗ A surely contains the square of the original matrix in (2.11). In Section 10 we discuss how to compute an inclusion of the solution of a system of linear equations, in particular when the matrix is not representable in floating-point arithmetic.

[15] This is done by `intvalinit('SharpIVmult')` and `intvalinit('FastIVmult')`, and applies only if both factors are thick interval quantities, *i.e.*, the lower and upper bound do not coincide.

9.2. Representation of intervals

The Cartesian product of a one-dimensional interval suggests itself as a representation of interval vectors. Although operations are easy to define and fast to execute, there are drawbacks.

In particular, there is no 'orientation': the boxes are always parallel to the axes which may cause overestimation. The simplest example is the 2-dimensional unit square rotated by 45: the best inclusion increases the radii by a factor $\sqrt{2}$. This is known as the *wrapping effect*, one of the major obstacles when integrating ODEs over a longer time frame.

As an alternative to interval vectors, Beaumont (2000) considers oblique boxes $Q\mathbf{X}$ for orthogonal Q, where the product is not executed but Q and $\mathbf{X}$ are stored separately.

Another approach involving ellipsoids is defined by Neumaier (1993) (see also Kreinovich, Neumaier and Xiang (2008)), where the ellipsoid is the image of the unit ball by a triangular matrix; for interesting applications see Ovseevich and Chernousko (1987).

Andrade, Comba and Stolfi (1994) define an affine arithmetic, apparently used in other areas, by representing an interval by $x_0 + \sum x_i \mathbf{E}_i$ for $x_i \in \mathbb{R}$ and $\mathbf{E}_i = [-1, 1]$. Often this seems efficient in combatting overestimation; see de Figueiredo and Stolfi (2004).

For each representation, however, advantages are counterbalanced by increased computational costs for interval operations.

PART TWO
Finite-dimensional problems

10. Linear problems

The solution of systems of linear equations is one of the most common of all numerical tasks. Therefore we discuss various aspects of how to compute verified error bounds, from dense systems to large systems, input data with tolerances, estimation of the quality to NP-hard problems. Almost-linear problems such as the algebraic eigenvalue problem are discussed in Section 13.4. We start by showing that a naive approach is bound to fail.

10.1. The failure of the naive approach: interval Gaussian elimination (IGA)

The most commonly used algorithm for a general dense linear system $Ax = b$ is factoring A by Gaussian elimination. From an LU decomposition, computed with partial pivoting, the solution can be directly computed.

Once again it is a true statement that replacement of each operation in this process by its corresponding interval operation produces a correct inclusion of the result – if not ended prematurely by division by an interval containing zero. This naive approach is likely to produce useless results for almost any numerical algorithm, and Gaussian elimination is no exception.

Unfortunately, it is still not fully understood why Gaussian elimination works so successfully in floating-point arithmetic. A detailed and very interesting argument was given by Trefethen and Schreiber (1990). They investigate in particular why partial pivoting suffices despite worst-case exponential growth factors. Their model can be used to shed some light on the failure of IGA.

Let $A^{(1)} := A \in \mathbb{R}^{n\times n}$ be given, and denote by $A^{(k)}$ the modified matrix before step k of Gaussian elimination (we assume the rows of A are already permuted so that the partial pivots are already in place). Then $A^{(k)} = L^{(k)} \cdot A^{(k-1)}$ with $L_{ik}^{(k)} = \varphi_i^{(k)} := -A_{ik}^{(k)}/A_{kk}^{(k)}$ for $k+1 \le i \le n$, or

$$A_{ij}^{(k+1)} = A_{ij}^{(k)} - \varphi_i^{(k)} A_{kj}^{(k)} \quad \text{for} \quad k+1 \le i,j \le n. \tag{10.1}$$

Now suppose Gaussian elimination is performed in interval arithmetic. For intervals $\mathbf{X}, \mathbf{Y} \in \mathbb{IR}$ it is easy to see that

$$\texttt{rad}(\mathbf{XY}) \ge \texttt{rad}(\mathbf{X}) \cdot |\texttt{mid}(\mathbf{Y})| + |\texttt{mid}(\mathbf{X})| \cdot \texttt{rad}(\mathbf{X})$$

(Neumaier 1990, Proposition 1.6.5).[16] Using this and (5.8), a very crude

[16] Note the similarity to $(uv)' = u'v + uv'$.

estimate for the radius of the new elements of the elimination is

$$\texttt{rad}(A_{ij}^{(k+1)}) \geq \texttt{rad}(A_{ij}^{(k)}) + |\varphi_i^{(k)}| \cdot \texttt{rad}(A_{kj}^{(k)}). \tag{10.2}$$

In matrix notation this takes the form

$$\texttt{rad}(A^{(k+1)}) \geq |L^{(k)}| \cdot \texttt{rad}(A^{(k)}),$$

valid for the first k rows of the upper triangle of $A^{(k+1)}$ and for the lower right square of $A^{(k+1)}$, *i.e.*, for indices $k+1 \leq i,j \leq n$. Finally we obtain

$$\texttt{rad}(U) \geq \text{upper triangle}\big[|L^{(n-1)}| \cdot \ldots \cdot |L^{(2)}| \cdot |L^{(1)}| \cdot \texttt{rad}(A)\big]. \tag{10.3}$$

A unit lower triangular matrix with non-zero elements below the diagonal only in column k is inverted by negating those elements. Hence $|L^{(k)}|^{-1} = \langle L^{(k)} \rangle$ using Ostrowski's comparison matrix,

$$\langle A \rangle_{ij} := \begin{cases} |A_{ij}| & \text{for } i = j, \\ -|A_{ij}| & \text{otherwise.} \end{cases} \tag{10.4}$$

It follows that

$$|L^{(n-1)}| \cdot \ldots \cdot |L^{(2)}| \cdot |L^{(1)}| = \big[\langle L^{(1)} \rangle \cdot \langle L^{(2)} \rangle \cdot \ldots \cdot \langle L^{(n-1)} \rangle\big]^{-1} = \langle L \rangle^{-1},$$

and (10.3) implies

$$\texttt{rad}(U) \geq \text{upper triangle}\big[\langle L \rangle^{-1} \cdot \texttt{rad}(A)\big]. \tag{10.5}$$

It is known that for a random matrix $\|\langle L \rangle^{-1}\| / \|L^{-1}\|$ is large. However, the L factor of Gaussian elimination is far from random. One reason is that L is generally well-conditioned, even for ill-conditioned A, despite the fact that Viswanath and Trefethen (1998) showed that random lower triangular matrices are generally ill-conditioned.

Adopting the analysis of Gaussian elimination by Trefethen and Schreiber (1990), for many classes of matrices we can expect the multipliers, *i.e.*, the elements of L, to have mean zero with standard deviation

$$\sigma(L_{ik}) \approx \frac{1}{W(m)}\left(1 - \frac{\sqrt{2/\pi}W(m)\mathrm{e}^{-W(m)^2/2}}{\mathrm{erf}(W(m)/\sqrt{2})}\right)^{1/2} \sim \left(\frac{1}{2\log(m\sqrt{2/\pi})}\right)^{1/2}, \tag{10.6}$$

where $m = n + 1 - k$ for partial pivoting, and $W(m)$ denotes the 'winner function' for which

$$W(m) \approx \alpha\left(1 - \frac{2\log\alpha}{1+\alpha^2}\right)^{1/2} + \mathcal{O}\left(\frac{1}{\log m}\right)$$

with $\alpha := \sqrt{2\log(m\sqrt{2/\pi})}$. The expected absolute value of the multipliers is about 0.6745 times the standard deviation. Hence we can compute a matrix $\mathcal{L}$ which, in general, is not too far from $|L|$, and can compute $\Phi := \|\langle \mathcal{L} \rangle^{-1}\|$.

Table 10.1. Lower bound for the amplification factor for interval Gaussian elimination (IGA).

	$n = 20$	$n = 50$	$n = 100$	$n = 150$	$n = 170$
Φ	$1.0 \cdot 10^2$	$1.6 \cdot 10^5$	$1.4 \cdot 10^{10}$	$7.6 \cdot 10^{14}$	$5.4 \cdot 10^{16}$
$\Vert\langle L\rangle^{-1}\Vert$	$2.6 \cdot 10^2$	$2.2 \cdot 10^6$	$4.6 \cdot 10^{13}$	$2.7 \cdot 10^{19}$	$5.3 \cdot 10^{21}$

After the first elimination step of interval Gaussian elimination, the inevitable presence of rounding errors produces an interval matrix $\mathbf{A}^{(2)}$ with $\mathtt{rad}(\mathbf{A}^{(2)}) \geq \mathrm{u}|A^{(2)}|$. Thus Φ is a lower bound for the amplification of radii in IGA.

Table 10.1 lists this lower bound Φ for different dimensions. In addition, $\Vert\langle L\rangle^{-1}\Vert$ for the factor L of some random matrix is displayed. For small dimension the amplification Φ exceeds u^{-1}, which means breakdown of IGA. These estimates are very crude; in practice the behaviour is even worse than the second line of Table 10.1, as explained in a moment.

The model does not apply to special matrices such as M-matrices or diagonally dominant matrices. In fact one can show that IGA does not break down for such matrices. However, for such matrices very fast methods are available to solve linear systems with rigour; see Section 10.9.

In what follows we monitor the practical behaviour of IGA. For general linear systems the condition number is usually reflected in the U factor. Thus it seems reasonable to monitor the relative error of the last component U_{nn}. We define the relative error of an interval $\mathbf{X} \in \mathbb{IF}$ by

$$\mathtt{relerr}(\mathbf{X}) := \begin{cases} \left|\frac{\mathtt{rad}(\mathbf{X})}{\mathtt{mid}(\mathbf{X})}\right| & \text{if } 0 \notin \mathbf{X}, \\ \mathtt{rad}(\mathbf{X}) & \text{otherwise.} \end{cases} \tag{10.7}$$

For randomly generated matrices with normally distributed random entries, Gaussian elimination with partial pivoting is applied to generate $\mathbf{L}$ and $\mathbf{U}$ factors. All operations are interval operations, and the pivot element is the one with largest *mignitude* $\mathrm{mig}(\mathbf{X}) := \min\{|x| : x \in \mathbf{X}\}$. This is about the best one can do.

In Table 10.2 we display the time ratio for a Gaussian elimination for a matrix of the same size in pure floating-point arithmetic, and the median relative error of $\mathbf{U}_{nn}$ over 100 samples. If some interval $\mathbf{U}_{ii}$ contains 0 we call that a failure. The percentage of failures is listed as well, and the time ratio and relative errors displayed are the means over successful examples. As can be seen, the relative error of $\mathbf{U}_{nn}$ increases with the dimension, and for dimensions about 60 to 70 the approach fails completely. Note that this is true even though random matrices are well known to be reasonably

Table 10.2. Results for interval Gaussian elimination (IGA) for random matrices and random orthogonal matrices.

Dimension	Random matrices			Random orthogonal matrices		
	Time ratio	relerr(U_{nn})	Failure %	Time ratio	relerr(U_{nn})	Failure %
10	173.9	3.4e-013	0	171.6	1.3e-014	0
20	320.5	2.7e-009	0	320.9	6.3e-012	0
30	432.3	2.1e-007	0	419.5	2.5e-009	0
40	482.2	1.0e-004	0	497.7	1.0e-006	0
50	407.0	4.9e-002	2	434.6	4.4e-004	0
60	454.0	6.6e-001	96	414.2	1.4e-001	4
70			100			100

well-conditioned. Moreover, not only do the results become useless, but the approach is also very slow due to many branches and rounding switches, as explained in Section 9.1.[17]

The reason is solely the number of consecutive interval operations, permanently violating the UTILIZE INPUT DATA PRINCIPLE (5.13). Data dependencies quickly produce wide and useless results. Interval arithmetic is not to blame for this: the result would be the same with power set operations applied in this way.

To confirm that even the mild condition numbers of random matrices do not contribute, similar data for randomly generated orthogonal matrices are also displayed in Table 10.2.

The picture changes when the input data have a specific structure, such as A being an M-matrix. Then no overestimation occurs, because the NO INACCURATE CANCELLATION PRINCIPLE (NIC) is satisfied. However, the time penalty persists, and the methods in Section 10.9 should be used.

To emphasize the point, we take a randomly generated 3×3 matrix and multiply it by another random matrix several times in interval arithmetic. Note that all matrices are point matrices (left and right bounds coincide), and that every factor is a new random matrix. All entries of the intermediate products are summed up, and the relative error of this sum is displayed. The following code produces a semi-logarithmic graph of the result.

Algorithm 10.1. Product of random matrices:

```
imax = 65; y = zeros(1,imax); A = intval(randn(3));
for i=1:imax, A=A*randn(3); y(i)=relerr(sum(A(:))); end
close, semilogy(1:imax,y)
```

[17] Note that the code was vectorized where possible.

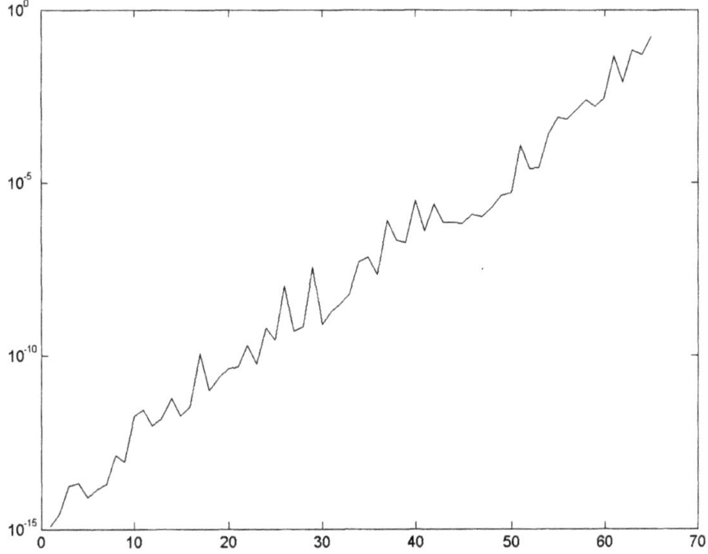

Figure 10.1. Overestimation by naive interval arithmetic and dependencies.

A typical graph of the result is shown in Figure 10.1. As can be seen, rounding errors add from step to step, resulting in an exponentially growing relative error. A rule of thumb roughly predicts an increase of the relative error by a factor 2^K after K iterations. Since $2^{48} \sim 10^{16}$ and double precision corresponds to about 16 decimal places of precision, a relative error 1 can be expected after some 50 iterations.

10.2. Partial pivoting

The method of choice for a dense system of linear equations is Gaussian elimination with partial pivoting. Although it is known that the growth factor may increase exponentially with the dimension, the corresponding examples seemed to be constructed and not occurring in practice. In 1993 Wright (1993) showed practical examples with exponential growth factor. For example, integrating

$$\dot{x} = x - 1 \quad \text{for} \quad x(0) = x(40)$$

and applying a trapezoid rule results in a linear system $Ax = b$, the solution of which is obviously the vector of ones. For given n, the following MATLAB code, given by Foster (1994),

```
T = 40; h = T/(n-1);
b = -(0:n-1)'*h;
A = - h * tril(ones(n),-1);
```

```
A = A + (1-h/2) * diag(ones(n,1));
A(:,1) = -(h/2)*ones(n,1);
A(1,1) = 1; A(:,n) = -ones(n,1);
A(n,n) = -(h/2);
```

computes A and b. Up to $n = 60$ the solution is very accurate, but for larger n the exponential increase of the growth factor produces meaningless results. For example, the components $\mathtt{x}_{62...65}$ of `x = A\b` for $n = 65$ are

```
-23.272727272727273
                  0
-93.090909090909093
  1.000000000000000
```

rather than all ones. The condition number of the matrix is less than 30, so the disastrous effect of rounding errors is only due to the large growth factor. Remarkably, no warning is given by MATLAB, so a user might trust the computed values. It is also well known that one extra residual iteration in working precision produces a backward stable result. Doing this, even for large dimension, produces accurate results.

There seems little potential for converting floating-point Gaussian elimination into a verification algorithm, because known error estimates need an upper bound for the condition number. There are condition estimators as well, but there is strong evidence that a reliable condition estimator costs as much as a reliable computation of A^{-1}. In fact Demmel, Diament and Malajovich (2001) have shown that reliably computing a bound for $\|A^{-1}\|$ has at least the cost of testing whether the product of two $n \times n$ matrices is zero, which in turn is believed to actually cost as much as computing the product.

10.3. Preconditioning

One way to follow the UTILIZE INPUT DATA PRINCIPLE (5.13) and to avoid successive operations on computed data is preconditioning by some approximate inverse R. This very important principle was proposed by Hansen and Smith (1967), analysed in Ris (1972), and its optimality in a certain sense shown by Neumaier (1984).

It is an unwritten rule in numerical analysis never to compute the inverse of a matrix, especially not to solve a linear system $Ax = b$. Indeed, the direct solution by Gaussian elimination is not only faster but also produces more accurate results than multiplication of b by a computed inverse. For the purpose of verification other rules apply, in particular to rely only on the input data if possible. There are other (and faster) verification approaches using factorizations of A; see Oishi and Rump (2002). However, preconditioning by some approximate inverse is of superior quality.

Given a linear system $Ax = b$ with non-singular A, some $R \in \mathbb{R}^{n \times n}$ and some $\tilde{x} \in \mathbb{R}^n$, there follows

$$\|A^{-1}b - \tilde{x}\| = \|\big(I - (I - RA)\big)^{-1} R(b - A\tilde{x})\| \leq \frac{\|R(b - A\tilde{x}\|}{1 - \|I - RA\|} \tag{10.8}$$

provided $\|I - RA\| < 1$. We stress that there are no mathematical assumptions on R: if A is too ill-conditioned and/or R is of poor quality, then $\|I - RA\| < 1$ is not satisfied. Moreover, if $\|I - RA\| < 1$ is satisfied, we conclude *a posteriori* that A (and R) is non-singular. This proves the following.

Theorem 10.2. Let $A, R \in \mathbb{R}^{n \times n}$ and $b, \tilde{x} \in \mathbb{R}^n$ be given, and denote by I the $n \times n$ identity matrix. If $\|I - RA\| < 1$ for some matrix norm, then A is non-singular and

$$\|A^{-1}b - \tilde{x}\| \leq \frac{\|R(b - A\tilde{x})\|}{1 - \|I - RA\|}. \tag{10.9}$$

In particular, the ∞-norm is useful because it directly implies componentwise error bounds on the solution, and it is easy to calculate. This theorem is especially suited to deriving error estimates using interval arithmetic.

10.4. Improved residual

Note that the quality of the bound in (10.9) depends directly on the size of the residual $\|b - A\tilde{x}\|$ and can be improved by some residual iteration on $\tilde{x}$. If dot products can be calculated accurately, an inclusion of the solution $A^{-1}b$ accurate to the last bit can be computed provided $\|I - RA\| < 1$.

This can be done[18] by multi-precision packages, such as, for example, the MPFR-package of Fousse *et al.* (2005), or based on 'error-free transformations', as discussed in Section 3. Recently the latter techniques were used to derive very fast algorithms for computing rigorous error bounds of sums and dot products of vectors of floating-point numbers to arbitrary accuracy, for example by Zielke and Drygalla (2003), Zhu, Yong and Zheng (2005), Rump, Ogita and Oishi (2008) and Rump (2009). The algorithms are particularly fast because there are no branches, and only floating-point operations in one working precision are used. INTLAB contains reference implementations, which however, severely suffer from interpretation overhead.

Some improvement can be achieved by the 'poor man's residual' algorithm `lssresidual`. It is based on the following algorithm by Dekker for splitting a floating-point number into some higher- and lower-order part.

[18] Sometimes it is preferable to solve the given linear system using a multi-precision package from the beginning.

Algorithm 10.3. Error-free splitting of a floating-point number into two parts:

$$\begin{aligned} &\text{function } [x, y] = \text{Split}(a) \\ &\quad c = \text{fl}(\varphi \cdot a) \qquad\qquad \% \ \varphi = 2^s + 1 \\ &\quad x = \text{fl}(c - \text{fl}(c - a)) \\ &\quad y = \text{fl}(a - x) \end{aligned}$$

As a result $a = x + y$ for all $a \in \mathbb{F}$, and in 53-bit precision x and y have at most $53 - s$ and $s - 1$ significant bits. In particular, for $s = 27$ a 53-bit number is split into two 26-bit parts, which can be multiplied in floating-point arithmetic without error. The trick is that the sign bit is used as an extra bit of information.

Algorithm 10.4. Improved computation of the residual of a linear system (poor man's residual):

```
function res = lssresidual(A,x,b)
   factor = 68719476737;    % heuristically optimal splitting 2^36+1
   C = factor*A;
   Abig = C - A;
   A1 = C - Abig;           % upper part of A, first 17 bits
   A2 = A - A1;             % A = A1+A2 exact splitting
   x = -x; y = factor*x;
   xbig = y - x;
   x1 = y - xbig;           % upper part of -x, first 17 bits
   x2 = x - x1;             % -x = x1+x2 exact splitting
   res = (A1*x1+b)+(A1*x2+A2*x);
```

This algorithm `lssresidual` splits `A` and `x` into $17 + 35$ bits, so that the product `A1` $*$ `x1` does not cause a rounding error if the elements of neither `A` nor `x` cover a wide exponent range.

`lssresidual` is a cheap way to improve the residual, and thus the quality of the solution of a linear system, also for numerical algorithms without verification. The command `intvalinit('ImprovedResidual')` sets a flag in INTLAB so that `verifylss` uses this residual improvement. It is applicable for a matrix residual $I - RA$ as well.

10.5. Dense linear systems

Before discussing other approaches for verification, we show the computed bounds by Theorem 10.2 for a linear system with matrix as in (2.11) of dimension $n = 9$. Note that, except for $n = 2$, these matrices are more ill-conditioned than the well-known and notoriously ill-conditioned Hilbert matrices. The right side is computed so that the solution $A^{-1}b$ is the vector of ones. This is done by the following executable code, first for the floating-point matrix nearest to the original matrix:

```
n = 9; A = 1./reshape(1:n^2,n,n); b = A*ones(n,1);
R = inv(A); xs = A\b;
d = norm(eye(n)-R*intval(A),inf);
if d<1
  midrad( xs , mag(norm(R*(b-A*intval(xs)),inf)/(1-d)) )
end
```

Note that the quantity d is estimated in interval arithmetic and is itself of type interval. The magnitude of an interval $\mathbf{X}$ is defined by $\text{mag}(\mathbf{X}) := \max\{|x| : x \in \mathbf{X}\} \in \mathbb{F}$. The result is as follows:

```
intval ans =
[    0.8429,    1.1571]
[    0.8429,    1.1571]
[    0.8429,    1.1571]
[    0.8428,    1.1569]
[    0.8441,    1.1582]
[    0.8384,    1.1526]
[    0.8514,    1.1656]
[    0.8353,    1.1495]
[    0.8455,    1.1596]
```

The quality of the inclusion seems poor, but it corresponds roughly to the condition number $\text{cond}(A) = 4.7 \cdot 10^{14}$. This is the largest dimension of the matrix as in (2.11), for which Theorem 10.2 is applicable. For $n = 10$ the condition number of A is $8.6 \cdot 10^{16}$.

For a completely rigorous result there are two flaws. First, as has been mentioned, we do not use the true, real matrices as in (2.11), but floating-point approximations. Second, the right-hand side b is computed in floating-point arithmetic as well, which means that the vector of ones is likely *not* the solution of the linear system. Both problems are easily solved by the following code:

```
n = 9; A = 1./intval(reshape(1:n^2,n,n)); b = A*ones(n,1);
R = inv(A.mid); xs = A.mid\b.mid;
d = norm(eye(n)-R*A,inf);
if d<1
  midrad( xs , mag(norm(R*(b-A*xs),inf)/(1-d)) )
end
```

Now the interval matrix A contains the original real matrix as in (2.11). The assertions of Theorem 10.2 are valid for all matrices within the interval matrix A, in particular for the original real matrix. This statement includes the non-singularity as well as the error bounds. The result is shown in Table 10.3; the true solution is again the vector of all ones.

Table 10.3. Inclusion by Theorem 10.2 for a 9×9 linear system with matrix as in (2.11).

```
intval ans =
[    0.6356,     1.3644]
[    0.6356,     1.3644]
[    0.6356,     1.3644]
[    0.6357,     1.3645]
[    0.6344,     1.3632]
[    0.6403,     1.3691]
[    0.6267,     1.3555]
[    0.6435,     1.3723]
[    0.6329,     1.3617]
```

The diameter is even worse because now the set of solutions $A^{-1}b$ is included for all $A \in \mathtt{A}$ and $b \in \mathtt{b}$. However, it may equally well be attributed to overestimation by interval arithmetic. But it is possible to estimate the overestimation, as will be described in Section 10.6.

A drawback to this approach is that $\|I - RA\| < 1$ is mandatory. In order to be easily and reliably computable there is not much choice for the matrix norm. If the matrix is badly scaled this causes problems. It is possible to incorporate some kind of scaling in the verification process. The corresponding inclusion theorem is based on a lemma by Rump (1980). The following simple proof is by Alefeld.

Lemma 10.5. Let $\mathbf{Z}, \mathbf{X} \in \mathbb{IR}^n$ and $\mathbf{C} \in \mathbb{IR}^{n \times n}$ be given. Assume

$$\mathbf{Z} + \mathbf{C}\mathbf{X} \subset \operatorname{int}(\mathbf{X}), \tag{10.10}$$

where $\operatorname{int}(\mathbf{X})$ denotes the interior of $\mathbf{X}$. Then, for every $C \in \mathbf{C}$ the spectral radius $\varrho(C)$ of C is less than one.

Proof. Let $C \in \mathbf{C}$ be given and denote $\mathbf{Z} = [\underline{\mathbf{Z}}, \overline{\mathbf{Z}}]$. Then, following (9.10) yields

$$\mathbf{Z} + C\mathbf{X} = C \cdot \operatorname{mid}(X) + [\underline{\mathbf{Z}} - |C| \cdot \operatorname{rad}(X), \overline{\mathbf{Z}} + |C| \cdot \operatorname{rad}(X)] \subset \operatorname{int}(X),$$

and therefore

$$|C| \cdot \operatorname{rad}(X) < \operatorname{rad}(X).$$

Since $\operatorname{rad}(X)$ is a positive real vector, Perron–Frobenius theory yields $\varrho(C) \leq \varrho(|C|) < 1$. □

Based on this we can formulate a verification method, *i.e.*, a theorem the assumptions of which can be verified on digital computers. The theorem was formulated by Krawczyk (1969a) as a refinement of a given inclusion of $A^{-1}b$, and modified as an existence test by Moore (1977). Our exposition follows Rump (1980).

Theorem 10.6. Let $A, R \in \mathbb{R}^{n\times n}$, $b \in \mathbb{R}^n$ and $\mathbf{X} \in \mathbb{IR}^n$ be given, and denote by I the $n \times n$ identity matrix. Assume

$$Rb + (I - RA)\mathbf{X} \subset \operatorname{int}(\mathbf{X}). \tag{10.11}$$

Then the matrices A and R are non-singular and $A^{-1}b \in Rb + (I - RA)\mathbf{X}$.

Proof. By Lemma 10.5 the spectral radius of $C := I - RA$ is less than one. Hence the matrices R and A are non-singular, and the iteration $x^{(k+1)} := z + Cx^{(k)}$ with $z := Rb$ converges to $(I - C)^{-1}z = (RA)^{-1}Rb = A^{-1}b$ for every starting point $x^{(0)} \in \mathbb{R}^n$. By (10.11), $x^{(k)} \in \mathbf{X}$ for any $k \geq 0$ if $x^{(0)} \in \mathbf{X}$, and the result follows. □

Condition (10.11) is the only one to be verified on the computer. To build an efficient verification algorithm, we need three improvements introduced in Rump (1980). First, condition (10.11) is still a sufficient criterion for given $\mathbf{X}$, and the problem remains of how to determine a suitable candidate. The proof suggests an interval iteration,

$$\mathbf{X}^{(k+1)} := Rb + (I - RA)\mathbf{X}^{(k)}, \tag{10.12}$$

so that with $\mathbf{X}^{(k+1)} \subseteq \operatorname{int}(\mathbf{X}^{(k)})$ the assumption of Theorem 10.6 is satisfied for $\mathbf{X} := \mathbf{X}^{(k)}$.

This is not sufficient, as trivial examples show. Consider a 1×1 linear system with $b = 0$, a poor 'approximate inverse' R so that $1 - RA = -0.6$ and a starting interval $\mathbf{X}^{(k)} := [-1, 2]$. Obviously the solution 0 is enclosed in every iterate, but $\mathbf{X}^{(k+1)}$ never belongs to $\operatorname{int}(\mathbf{X}^{(k)})$.

So, secondly, it needs a so-called *epsilon-inflation*, that is, to enlarge $\mathbf{X}^{(k)}$ intentionally before the next iteration. This was first used in Caprani and Madsen (1978); the term was coined and the properties of the method analysed in Rump (1980). One possibility is

$$\begin{aligned} \mathbf{Y} &:= \mathbf{X}^{(k)} \cdot [0.9, 1.1] + [-e, e], \\ \mathbf{X}^{(k+1)} &:= Rb + (I - RA)\mathbf{Y}, \end{aligned} \tag{10.13}$$

with some small constant e and to check $\mathbf{X}^{(k+1)} \subseteq \operatorname{int}(\mathbf{Y})$.

The third improvement is that it is often better to calculate an inclusion of the difference of the true solution to an approximate solution $\tilde{x}$. For linear systems this is particularly simple, by verifying

$$\mathbf{Y} := R(b - A\tilde{x}) + (I - RA)\mathbf{X} \subset \operatorname{int}(\mathbf{X}) \tag{10.14}$$

to prove $A^{-1}b \in \tilde{x} + \mathbf{Y}$. Nowadays these three improvements are standard tools for nonlinear and infinite-dimensional problems (see Sections 13, 15 and 16).

Table 10.4. Inclusion by Algorithm 10.7 for a 9×9 linear system with matrix as in (2.11).

```
intval X =
[    0.9999,     1.0001]
[    0.9999,     1.0001]
[    0.9998,     1.0002]
[    0.9955,     1.0050]
[    0.9544,     1.0415]
[    0.8363,     1.1797]
[    0.6600,     1.3095]
[    0.7244,     1.3029]
[    0.8972,     1.0935]
```

In summary, we obtain the following verification algorithm presented in Rump (1980) for general dense systems of linear equations with point or interval matrix and right-hand side.[19]

Algorithm 10.7. Verified bounds for the solution of a linear system:

```
function XX = VerifyLinSys(A,b)
  XX = NaN;                          % initialization
  R = inv(mid(A));                   % approximate inverse
  xs = R*mid(b);                     % approximate solution
  C = eye(dim(A))-R*intval(A);       % iteration matrix
  Z = R*(b-A*intval(xs));
  X = Z; iter = 0;
  while iter<15
    iter = iter+1;
    Y = X*infsup(0.9,1.1) + 1e-20*infsup(-1,1);
    X = Z+C*Y;                       % interval iteration
    if all(in0(X,Y)), XX = xs + X; return; end
  end
```

Theorem 10.8. If Algorithm 10.7 ends successfully for a given interval matrix $\mathbf{A} \in \mathbb{IR}^{n\times n}$ and interval right-hand side $\mathbf{b} \in \mathbb{IR}^n$, then the following is true. All matrices $A \in \mathbf{A}$ are non-singular, and the computed `XX` satisfies

$$\Sigma(\mathbf{A}, \mathbf{b}) := \{x \in \mathbb{R}^n : Ax = b \text{ for some } A \in \mathbf{A}, b \in \mathbf{b}\} \subseteq \mathtt{XX}. \tag{10.15}$$

Proof. The proof follows by applying Theorem 10.6 to fixed but arbitrary $A \in \mathbf{A}$ and $b \in \mathbf{b}$ and the INCLUSION PROPERTY (5.16). □

Note that the result is valid for a matrix right-hand side $\mathbf{b} \in \mathbb{IR}^{n\times k}$ as well. In particular, it allows one to compute an inclusion of the inverse of a

[19] The routine `in0(X, Y)` checks $\mathtt{X} \subset \mathrm{int}(\mathtt{Y})$ componentwise.

Table 10.5. Computing time of Algorithm 10.7 for dense point and interval linear systems, Gaussian elimination in floating-point arithmetic normed to 1.

Dimension	Point data	Interval data
100	7.3	8.2
200	6.4	8.2
500	6.8	9.9
1000	7.0	9.5
2000	7.7	10.4

matrix $A \in \mathbb{R}^{n\times n}$, and also of the set of inverses $\{A^{-1} : A \in \mathbf{A}\}$ including the proof of non-singularity of all $A \in \mathbf{A}$. It was shown by Poljak and Rohn (1993) that the latter is an NP-hard problem. Rohn (2009*a*) gives 40 necessary and sufficient criteria for non-singularity of an interval matrix.

Theorem 10.8 is superior to Theorem 10.2. For our model problem in (2.11) for $n = 9$ with the interval matrix $\mathbf{A}$ containing the true real matrix, for example, we obtain $\mathbf{C} := I - R\mathbf{A}$ with $\mathrm{norm}(\mathrm{mag}(\mathbf{C})) = 0.0651$ but $\varrho(\mathrm{mag}(\mathbf{C})) = 0.0081$. Accordingly the result of Algorithm 10.7, as displayed in Table 10.4, is superior to that in Table 10.3. Recall that the condition number $\|A^{-1}\|\|A\|$ of the midpoint matrix is again increased by taking the (narrowest) interval matrix including the matrix (2.11) for $n = 9$. For $n = 10$ we have $\mathrm{norm}(\mathrm{mag}(\mathbf{C})) = 10.29$ but $\varrho(\mathrm{mag}(\mathbf{C})) = 0.94$. Therefore Theorem 10.2 is not applicable, whereas with Theorem 10.8 an inclusion is still computed. Due to the extreme[20] condition number $8.2 \cdot 10^{16}$ of the midpoint matrix, however, the quality is very poor.

Epsilon-inflation is theoretically important as well. It can be shown that basically an inclusion will be computed by Algorithm 10.7 if and only if $\varrho(|I - RA|) < 1$. Note the similarity to the real iteration $x^{(k+1)} := Rb + (I - RA)x^{(k)}$, which converges if and only if $\varrho(I - RA) < 1$. In a practical application there is usually not too much difference between $\varrho(|I - RA|)$ and $\varrho(I - RA)$. Also note that the only matrix norms which can be easily estimated are the 1-norm, the ∞-norm and the Frobenius norm, for which the norm of C and $|C|$ coincides.

The main computational effort in Algorithm 10.7 is the computation of `R` and `C`. Using (9.10) and (9.14) this corresponds to the cost of 3 or 4 real matrix multiplications, respectively. The cost of Gaussian elimination is one third of one real matrix multiplication, so we can expect the verification algorithm to take 9 or 12 times the computing time of Gaussian elimination

[20] Computed with a symbolic package corresponding to infinite precision.

Table 10.6. Quality of inclusions $\mu/\text{cond}(A)$ of Algorithm 10.7 for point data.

	$n = 100$	$n = 200$	$n = 500$	$n = 1000$	$n = 2000$
$\text{cond}(A) = 10^{10}$	$6.7 \cdot 10^{-16}$	$2.1 \cdot 10^{-15}$	$6.3 \cdot 10^{-15}$	$2.0 \cdot 10^{-14}$	$6.0 \cdot 10^{-14}$
$\text{cond}(A) = 10^{11}$	$5.8 \cdot 10^{-16}$	$1.5 \cdot 10^{-15}$	$8.4 \cdot 10^{-15}$	$2.1 \cdot 10^{-14}$	$4.6 \cdot 10^{-14}$
$\text{cond}(A) = 10^{12}$	$8.1 \cdot 10^{-16}$	$1.9 \cdot 10^{-15}$	$7.7 \cdot 10^{-15}$	$1.5 \cdot 10^{-14}$	$5.0 \cdot 10^{-14}$
$\text{cond}(A) = 10^{13}$	$5.2 \cdot 10^{-16}$	$2.1 \cdot 10^{-15}$	$4.5 \cdot 10^{-15}$	$2.3 \cdot 10^{-14}$	$6.4 \cdot 10^{-14}$
$\text{cond}(A) = 10^{14}$	$8.4 \cdot 10^{-16}$	$3.0 \cdot 10^{-15}$	$7.5 \cdot 10^{-15}$	$9.6 \cdot 10^{-13}$	failed

for point or interval input matrices, respectively. This is confirmed by the measured computing time in Table 10.5, where the ratio is a little better than expected since BLAS routines for matrix multiplication are, in contrast to Gaussian elimination, likely to achieve nearly peak performance.

The sensitivity of the solution of a linear system is measured by the condition number. Computing in double precision corresponding to 16 decimal digits precision, for $\text{cond}(A) = 10^k$ we cannot expect more than about $16-k$ correct digits of the solution. What is true for the pure floating-point algorithm is also true for the verification process. If μ is the median of the relative errors of an inclusion computed by Algorithm 10.7, then the ratio $\mu/\text{cond}(A)$ should be constantly 10^{-16}. For randomly generated matrices with specified condition number, this is confirmed in Table 10.6.

The practical implementation `verifylss` in INTLAB of a verification algorithm for dense systems of linear equations is based on Algorithm 10.7.

10.6. Inner inclusion

The traditional condition number for a linear system $Ax = b$ is defined by

$$\kappa_{E,f}(A,x) := \lim_{\epsilon \to 0} \sup \left\{ \frac{\|\Delta x\|}{\epsilon \|x\|} : (A + \Delta A)(x + \Delta x) = b + \Delta b,\ \Delta A \in \mathbb{R}^{n \times n}, \right.$$
$$\left. \Delta b \in \mathbb{R}^n,\ \|\Delta A\| \le \epsilon \|E\|,\ \|\Delta b\| \le \epsilon \|f\| \right\}, \quad (10.16)$$

and it is known (Higham 2002) that

$$\kappa_{E,f}(A,x) = \|A^{-1}\|\,\|E\| + \frac{\|A^{-1}\|\,\|f\|}{\|x\|}. \quad (10.17)$$

This is the sensitivity of the solution with respect to infinitesimally small perturbations of the input data. The solution set $\Sigma(\mathbf{A}, \mathbf{b})$ in (10.15) characterizes the sensitivity for finite perturbations. As shown by Kreinovich, Lakeyev and Noskov (1993) and Rohn (1994), the computation of the interval hull of $\Sigma(A,b)$ is an NP-hard problem, even when computed only to a certain precision.

Algorithm 10.7 computes an inclusion $\mathbf{X}$ of the solution set, thus upper bounds for the sensitivity of $A^{-1}b$ with respect to finite perturbations of A and b. Estimating the amount of overestimation of $\mathbf{X}$ yields lower bounds for the sensitivity, and this is possible by 'inner' inclusions of $\Sigma(\mathbf{A},\mathbf{b})$.

For interval input data, besides the solution set defined in (10.15), the 'tolerable (or restricted) solution set' (see Shary (2002))

$$\Sigma_{\forall\exists}(\mathbf{A},\mathbf{b}) := \{x \in \mathbb{R} : \forall A \in \mathbf{A}\, \exists b \in \mathbf{b} \text{ with } Ax = b\} = \{x \in \mathbb{R}^n : \mathbf{A}x \subseteq \mathbf{b}\} \tag{10.18}$$

and the 'controllable solution set'

$$\Sigma_{\exists\forall}(\mathbf{A},\mathbf{b}) := \{x \in \mathbb{R} : \forall b \in \mathbf{b}\, \exists A \in \mathbf{A} \text{ with } Ax = b\} = \{x \in \mathbb{R}^n : \mathbf{A}x \supseteq \mathbf{b}\} \tag{10.19}$$

are also of interest. We restrict our attention to what is sometimes called the 'united solution set' (10.15).

An inclusion $\mathbf{X} \in \mathbb{IF}$ of the solution set defined in (10.15) computed by Algorithm 10.7 may be wide due to overestimation and/or due to the sensitivity of the problem. The best-possible computable inclusion $\mathbf{Y} \in \mathbb{IF}$ is

$$\mathbf{Y} := \bigcap \{\mathbf{Z} \in \mathbb{IF} : \Sigma(\mathbf{A},\mathbf{b}) \subseteq \mathbf{Z}\} = \mathrm{hull}(\Sigma(\mathbf{A},\mathbf{b})). \tag{10.20}$$

The following theorem shows how to estimate the overestimation of $\mathbf{X}$ with respect to $\mathbf{Y}$. Such 'inner' inclusions were introduced by Neumaier (1987, 1989). Here we follow Rump (1990), who showed how to obtain them practically without additional computational effort.

Theorem 10.9. Let $\mathbf{A} \in \mathbb{IR}^{n\times n}, \mathbf{b} \in \mathbb{IR}^n$, $\tilde{x} \in \mathbb{R}^n$ and $\mathbf{X} \in \mathbb{IR}^n$ be given. Define

$$\mathbf{Z} := R(\mathbf{b} - \mathbf{A}\tilde{x}) \quad \text{and} \quad \boldsymbol{\Delta} := (I - R\mathbf{A})\mathbf{X}, \tag{10.21}$$

and suppose

$$\mathbf{Z} + \boldsymbol{\Delta} \subset \mathrm{int}(\mathbf{X}). \tag{10.22}$$

Then R and every matrix $A \in \mathbf{A}$ is non-singular, and $\mathbf{Y}$ as defined in (10.20) satisfies

$$\tilde{x} + \underline{\mathbf{Z}} + \underline{\boldsymbol{\Delta}} \le \underline{\mathbf{Y}} \le \tilde{x} + \underline{\mathbf{Z}} + \overline{\boldsymbol{\Delta}} \quad \text{and} \quad \tilde{x} + \overline{\mathbf{Z}} + \underline{\boldsymbol{\Delta}} \le \overline{\mathbf{Y}} \le \tilde{x} + \overline{\mathbf{Z}} + \overline{\boldsymbol{\Delta}}. \tag{10.23}$$

Proof. Let $A \in \mathbf{A}$ and $b \in \mathbf{b}$ be given. Applying Theorem 10.6 to the linear system $Ay = b - A\tilde{x}$ implies that A and R are non-singular, and $y = A^{-1}b - \tilde{x} \in \mathbf{Z} + \boldsymbol{\Delta}$. Since A and b are chosen arbitrarily and $\mathbf{Z} + \boldsymbol{\Delta} = [\underline{\mathbf{Z}} + \underline{\boldsymbol{\Delta}}, \overline{\mathbf{Z}} + \overline{\boldsymbol{\Delta}}]$, this proves $\Sigma(\mathbf{A},\mathbf{b}) \subseteq \tilde{x} + \mathbf{Z} + \boldsymbol{\Delta}$ and the first and last inequality in (10.23). Using (10.22), it follows that $\Sigma(\mathbf{A},\mathbf{b}) - \tilde{x} \subseteq \mathbf{X}$.

Table 10.7. Ratio of diameter of inner and outer inclusions for $n = 100$ and interval data.

Width of input	$n = 50$	$n = 100$	$n = 200$	$n = 500$	$n = 1000$
10^{-12}	1.000	1.000	1.000	1.000	1.000
10^{-8}	1.000	1.000	1.000	1.000	1.000
10^{-6}	1.000	0.998	0.991	0.955	0.954
10^{-5}	0.996	0.992	0.974	0.710	0.641
10^{-4}	0.962	0.684	0.502	failed	failed

Furthermore,

$$\begin{aligned}
&\{\tilde{x} + R(b - A\tilde{x}) \,:\, A \in \mathbf{A}, b \in \mathbf{b}\} \\
&\quad = \{A^{-1}b - (I - RA)(A^{-1}b - \tilde{x}) \,:\, A \in \mathbf{A}, b \in \mathbf{b}\} \\
&\quad \subseteq \Sigma(\mathbf{A}, \mathbf{b}) - \{(I - RA)(A^{-1}b - \tilde{x}) \,:\, A \in \mathbf{A}, b \in \mathbf{b}\} \\
&\quad \subseteq \Sigma(\mathbf{A}, \mathbf{b}) - \{(I - RA_1)(A_2^{-1}b - \tilde{x}) \,:\, A_1, A_2 \in \mathbf{A}, b \in \mathbf{b}\} \\
&\quad \subseteq \Sigma(\mathbf{A}, \mathbf{b}) - (I - R\mathbf{A})(\Sigma(\mathbf{A}, \mathbf{b}) - \tilde{x}) \\
&\quad \subseteq \Sigma(\mathbf{A}, \mathbf{b}) - \mathbf{\Delta},
\end{aligned}$$

where all set operations, even on interval quantities, are power set operations. If $U \subseteq V$ for two sets $U, V \in \mathbb{R}^n$, then $\mathrm{hull}(U) \subseteq \mathrm{hull}(V)$, and (9.4) and (9.6) imply

$$\tilde{x} + [\underline{\mathbf{Z}}, \overline{\mathbf{Z}}] = \tilde{x} + \mathbf{Z} \subseteq \mathbf{Y} - \mathbf{\Delta} = [\underline{\mathbf{Y}} - \overline{\mathbf{\Delta}}, \overline{\mathbf{Y}} - \underline{\mathbf{\Delta}}]. \tag{10.24}$$

This finishes the proof. □

Special care is necessary when applying this on the computer because lower *and upper* bounds are required in (10.23) for $\underline{\mathbf{Z}}$ and for $\overline{\mathbf{Z}}$. Using directed roundings and (9.6), this is not difficult. Note that for the outer and the inner bounds of $\mathbf{Y}$ only an outer inclusion of $\mathbf{\Delta}$ is necessary. Therefore, the inner bounds in Theorem 10.9 come with an extra effort of $\mathcal{O}(n^2)$ operations, so almost for free.

For interval input $\mathbf{A}, \mathbf{b}$, Theorem 10.9 yields an inner inclusion $\tilde{x} + [\underline{\mathbf{Z}} + \overline{\mathbf{\Delta}}, \overline{\mathbf{Z}} + \underline{\mathbf{\Delta}}]$, provided the radius of $\mathbf{\Delta}$ is no larger than the radius of $\mathbf{Z}$. Since $\mathbf{X}$ is the potential inclusion of the distance of $\tilde{x}$ to the true solution, $\mathbf{\Delta}$ is the product of two small quantities and can be expected to be of small radius.

In Table 10.7 we display the median of the componentwise ratios of the inner inclusion to the radius of the outer inclusion $\tilde{x} + \mathbf{Z} + \mathbf{\Delta}$. The matrices are generated by `midrad(randn(n), r * rand(n))` for `r` as given in the first column. This corresponds approximately to the relative precision of the matrix components. The right-hand side is computed by `b = A * ones(n, 1)`.

Table 10.8. Comparison of Monte Carlo and verification method for $n = 100$.

$K = 10$	$K = 100$	$K = 1000$	$K = 10000$
0.084	0.153	0.215	0.266

For increasing dimensions the ratio of the radii of inner and outer inclusions is not too far from 1 if the relative precision of the input data is not too large. For a relative precision of 10^{-4} the verification algorithm fails for dimension 500 and 1000, presumably because singular matrices slip into the input interval matrix. For these cases the spectral radius of $\mathrm{mag}(I - R\mathbf{A})$ is 1.15 and 2.18, respectively.

For relative precision 10^{-6} the radius of the inner inclusion is at least 95% of the outer inclusion, *i.e.*, the computed inclusion $\mathbf{X}$ by Algorithm 10.7 is not too far from the best possible inclusion $\mathbf{Y}$ as in (10.20). A Monte Carlo approach underestimates $\mathbf{Y}$ in principle. For fixed dimension $n = 100$ we choose K vertex matrices A and right-hand sides b, solve the linear system $Ax = b$ and form the narrowest interval vector $\mathbf{Z}$ containing all those solutions. We compare this with the solution $\mathbf{X}$ computed by Algorithm 10.7. In Table 10.8 the median of the quotient of the radii of $\mathbf{Z}$ and $\mathbf{X}$ is displayed for different values of K.

The computational effort for the Monte Carlo method is about $K/10$ times as much as for verification; however, improvement is slowly increasing with K. Note that the result of Algorithm 10.7 is of high quality and proved to be correct.

However, there is a major drawback to verification routines applied to interval data. All interval input is regarded as *independent* of each other. Simple dependencies such as symmetry are often difficult to take account of in verification algorithms. For linear systems this is possible, even for general linear dependencies of the input data, however, at the cost of roughly doubling the computational effort (see Section 10.7). In contrast, rather general, also nonlinear, dependencies are easily taken into account with a Monte Carlo approach.

The solution set was characterized by Oettli and Prager (1964) as

$$x \in \Sigma(\mathbf{A}, \mathbf{b}) \quad \Leftrightarrow \quad |\mathrm{mid}(\mathbf{A})x - \mathrm{mid}(\mathbf{b})| \le \mathtt{rad}(\mathbf{A})|x| + \mathtt{rad}(\mathbf{b}).$$

Moreover, the shape of the intersection of $\Sigma(\mathbf{A}, \mathbf{b})$ with some orthant is precisely prescribed by rewriting the problem as an LP-problem, as follows.

Let $S \in \mathbb{R}^{n \times n}$ be a signature matrix, *i.e.*, a diagonal matrix with $|S_{ii}| = 1$ for $1 \le i \le n$. The set $\mathcal{O} := \{x \in \mathbb{R}^n : Sx \ge 0\}$ defines an orthant in $\mathbb{R}^n$.

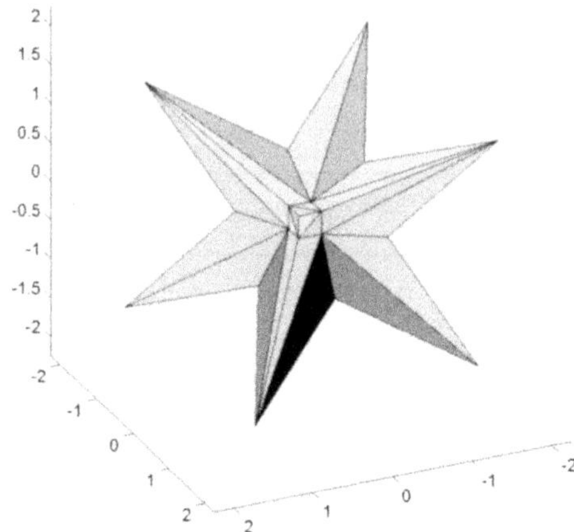

Figure 10.2. Solution set $\Sigma(\mathbf{A},\mathbf{b})$ for $\mathbf{A},\mathbf{b}$ as in (10.27).

For $\mathbf{A} := [mA - rA, mA + rA]$ and $x \in \mathcal{O}$ it follows that

$$\mathbf{A}x = [mA - rA, mA + rA]x = mA \cdot x + [-rA \cdot Sx, rA \cdot Sx] \qquad (10.25)$$

because $Sx \geq 0$. As explained in (9.4), there is no overestimation in the computation of $\mathbf{A}x$; interval and power set operations coincide. Therefore

$$\Sigma(\mathbf{A},\mathbf{b}) \cap \mathcal{O} = \{x \in \mathbb{R}^n : Sx \geq 0,\ (mA - rA \cdot S)x \leq \overline{\mathbf{b}},\ (mA + rA \cdot S)x \geq \underline{\mathbf{b}}\}. \qquad (10.26)$$

Hence the intersection of $\Sigma(\mathbf{A},\mathbf{b})$ with every orthant is obtained by solving some LP-problems, the number of which is polynomial in the dimension n. However, there are 2^n orthants, and $\Sigma(\mathbf{A},\mathbf{b})$ may have a non-empty intersection with each orthant although all matrices in $\mathbf{A}$ are non-singular. Here is the reason for the NP-hardness. Jansson (1997) gave an algorithm to compute exact bounds for $\Sigma(\mathbf{A},\mathbf{b})$ with computing time not necessarily exponential in the dimension, and Jansson and Rohn (1999) gave a similar algorithm for checking non-singularity of an interval matrix. Both algorithms, however, are worst-case exponential.

The INTLAB function `plotlinsol(A,b)` computes the graph of the solution set of a 2- or 3-dimensional interval linear system. For example (taken from Neumaier (1990, p. 96 *ff.*)),

$$\mathbf{A} = \begin{pmatrix} 3.5 & [0,2] & [0,2] \\ [0,2] & 3.5 & [0,2] \\ [0,2] & [0,2] & 3.5 \end{pmatrix} \quad \text{and} \quad \mathbf{b} = \begin{pmatrix} [-1,1] \\ [-1,1] \\ [-1,1] \end{pmatrix} \qquad (10.27)$$

produce the solution set $\Sigma(\mathbf{A},\mathbf{b})$, as shown in Figure 10.2. Because of the wide input intervals, the inclusion

```
intval X =
[   -8.6667,    8.6667]
[   -8.6667,    8.6667]
[   -8.6667,    8.6667]
```

computed by **verifylss** overestimates the interval hull $[-1.7648, 1.7648] \cdot (1,1,1)^T$ of the true solution set $\Sigma(\mathbf{A}, \mathbf{b})$, and the inner inclusion by Theorem 10.9 is empty. Note that the true solution set of the preconditioned linear system using the true inverse $R := \mathrm{mid}(A)^{-1}$ is

$$\mathrm{hull}\big(\Sigma(R\mathbf{A}, R\mathbf{b})\big) = [-8.6667, 8.6667] \cdot (1,1,1)^T,$$

so that the reason for the overestimation by **verifylss** is solely the preconditioning.

10.7. Data dependencies

In practice, matrices are often algebraically structured, for example symmetric or Toeplitz. Denote by M_n^{struct} a set of matrices such that $A, B \in M_n^{\mathrm{struct}} \subseteq \mathbb{R}^{n\times n}$ implies $A+B \in M_n^{\mathrm{struct}}$. Then Higham and Higham (1992*b*) generalize the usual condition number (10.16) into a structured condition number by

$$\kappa_{E,f}^{\mathrm{struct}}(A,x) := \tag{10.28}$$
$$\lim_{\epsilon\to 0} \sup\left\{ \frac{\|\Delta x\|}{\epsilon\|x\|} : (A+\Delta A)(x+\Delta x) = b + \Delta b,\ \Delta A \in M_n^{\mathrm{struct}},\right.$$
$$\left.\Delta b \in \mathbb{R}^n, \|\Delta A\| \le \epsilon\|E\|,\ \|\Delta b\| \le \epsilon\|f\| \right\},$$

where the spectral norm is used. It includes general matrices (no structure), *i.e.*, $M_n^{\mathrm{struct}} = \mathbb{R}^{n\times n}$.

For symmetric matrices, Bunch, Demmel and Van Loan (1989) showed that the unstructured and structured condition numbers are equal, *i.e.*, among the worst-case perturbations for A is a symmetric one; for many other structures Rump (2003*b*) showed that there is also no – or not much – difference. For some structures, however, such as Toeplitz matrices, Rump and Sekigawa (2009) showed that there exist $A_n \in M_n^{\mathrm{Toep}}$ and $b \in \mathbb{R}^n$ such that $\kappa_{A_n,b}^{\mathrm{Toep}}/\kappa_{A_n,b} < 2^{-n}$.

Note that this happens only for a specific right-hand side: the *matrix* condition number for general and structured perturbations is the same for many structures. More precisely, define

$$\kappa^{\mathrm{struct}}(A) := \tag{10.29}$$
$$\lim_{\epsilon\to 0} \sup\left\{ \frac{\|(A+\Delta A)^{-1} - A^{-1}\|}{\epsilon\|A^{-1}\|} :\ \Delta A \in M_n^{\mathrm{struct}}, \|\Delta A\| \le \epsilon\|A\| \right\}.$$

Then Rump (2003*b*) showed that

$$\kappa^{\text{struct}}(A) = \kappa(A) = \|A^{-1}\| \cdot \|A\| \tag{10.30}$$

for general, symmetric, persymmetric, skewsymmetric, Toeplitz, symmetric Toeplitz, Hankel, persymmetric Hankel, and circulant structures. Moreover, define the structured distance to the nearest singular matrix by

$$\delta^{\text{struct}}(A) := \min\{\alpha : \ \Delta A \in M_n^{\text{struct}}, \|\Delta A\| \le \alpha\|A\|, A + \Delta A \text{ singular}\}. \tag{10.31}$$

The (absolute) distance to the nearest singular matrix in the spectral norm is the smallest singular value, so that $\delta(A) = [\|A^{-1}\| \cdot \|A\|]^{-1} = \kappa(A)^{-1}$ for general perturbations. Rump (2003*b*) showed a remarkable generalization of this famous result by Eckart and Young (1936), namely

$$\delta^{\text{struct}}(A) = \kappa^{\text{struct}}(A)^{-1} \tag{10.32}$$

for all structures mentioned above.

As in Section 10.6, the sensitivity of a linear system with respect to finite structured perturbations can be estimated by inner inclusions. What is straightforward for Monte Carlo-like perturbations requires some effort in verification methods.

Let $\mathbf{A}$ be an interval matrix such that $\mathbf{A}_{ij} = \mathbf{A}_{ji}$ for all i, j. The *symmetric solution set* is defined by

$$\Sigma^{\text{sym}}(\mathbf{A}, \mathbf{b}) := \{x \in \mathbb{R}^n : \ Ax = b \quad \text{for} \quad A^T = A \in \mathbf{A}, b \in \mathbf{b}\}.$$

Due to Jansson (1991), inclusions of the symmetric solution set can be computed. Consider the following example taken from Behnke (1989):

$$\mathbf{A} = \begin{pmatrix} 3 & [1,2] \\ [1,2] & 3 \end{pmatrix} \quad \text{and} \quad \mathbf{b} = \begin{pmatrix} [10, 10.5] \\ [10, 10.5] \end{pmatrix}. \tag{10.33}$$

Computed inner and outer inclusions for the solution set $\Sigma(\mathbf{A}, \mathbf{b})$ and the symmetric solution set $\Sigma^{\text{sym}}(\mathbf{A}, \mathbf{b})$ are shown in Figure 10.3. To compute outer and inner inclusions for the symmetric solution we adapt Theorem 10.9. For data without dependencies we know that

$$\mathbf{b} - \mathbf{A}\tilde{x} = \{b - A\tilde{x} : \ A \in \mathbf{A}, b \in \mathbf{b}\}, \tag{10.34}$$

by (9.4), and

$$R(\mathbf{b} - \mathbf{A}\tilde{x}) = \text{hull}\big(\{R(b - A\tilde{x}) : \ A \in \mathbf{A}, b \in \mathbf{b}\}\big), \tag{10.35}$$

by (9.6), so $\mathbf{Z}$ is best possible. As we have seen, $\mathbf{\Delta}$ in (10.21) is usually small compared to $\mathbf{Z}$. So the main task is to obtain a sharp inclusion of

$$\mathbf{Z}^{\text{sym}} := \text{hull}\big(\{R(b - A\tilde{x}) : \ A^T = A \in \mathbf{A}, b \in \mathbf{b}\}\big). \tag{10.36}$$

In view of Theorem 8.1 we use the symmetry to rearrange the computation

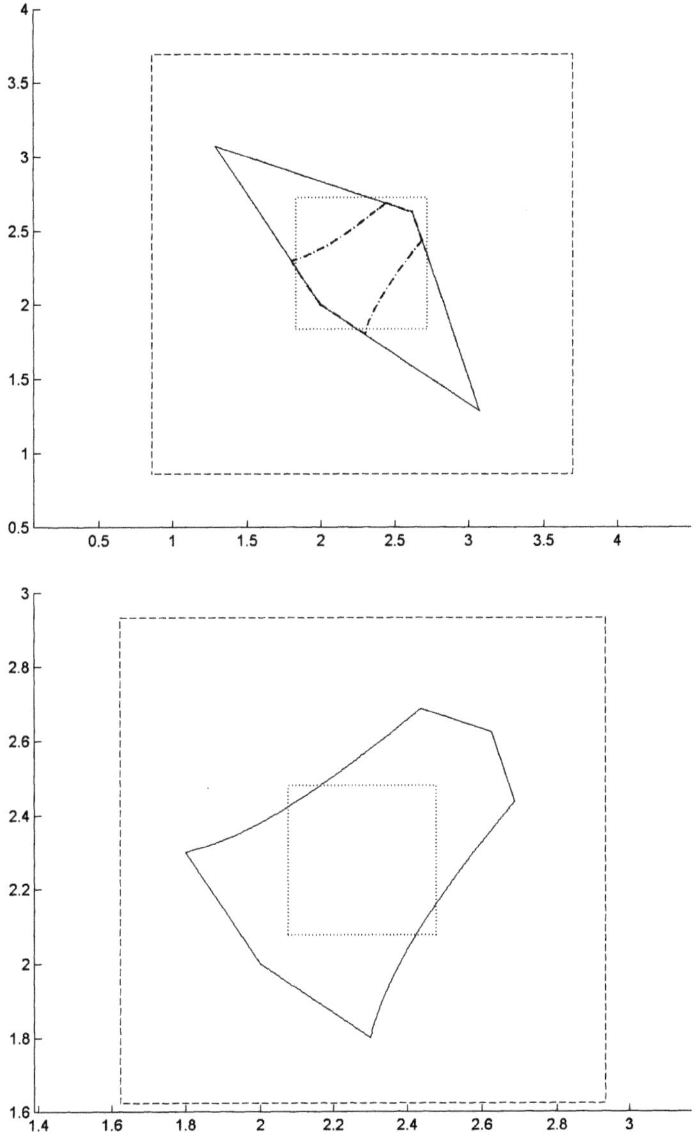

Figure 10.3. Inner and outer inclusions for the solution set $\Sigma(\mathbf{A}, \mathbf{b})$ and the symmetric solution set $\Sigma^{\mathrm{sym}}(\mathbf{A}, \mathbf{b})$ for $\mathbf{A}, \mathbf{b}$ as in (10.33).

so that each interval quantity occurs only once:

$$\big(R(b - A\tilde{x})\big)_i = \sum_{j=1}^{n} R_{ij}(b_j - A_{jj}\tilde{x}_j) - \sum_{\substack{j,k=1 \\ j<k}}^{n} (R_{ij}\tilde{x}_k + R_{ik}\tilde{x}_j)A_{jk}. \quad (10.37)$$

It follows that inserting $\mathbf{A}$ and $\mathbf{b}$ into (10.37) produces no overestimation, resulting in the componentwise (sharp) computation of $\mathbf{Z}_i^{\mathrm{sym}}$. Therefore we may proceed as in the proof of Theorem 10.9, to obtain the following.

Theorem 10.10. Let $\mathbf{A} \in \mathbb{IR}^{n\times n}, \mathbf{b} \in \mathbb{IR}^n$, $\tilde{x} \in \mathbb{R}^n$ and $\mathbf{X} \in \mathbb{IR}^n$ be given. Define

$$\mathbf{\Delta} := (I - R\mathbf{A})\mathbf{X}, \quad (10.38)$$

and let $\mathbf{Z}$ be computed componentwise by (10.37). Furthermore, suppose

$$\mathbf{Z} + \mathbf{\Delta} \subset \mathrm{int}(\mathbf{X}). \quad (10.39)$$

Then R and every matrix $A^T = A \in \mathbf{A}$ is non-singular, and

$$\mathbf{Y} := \mathrm{hull}\big(\Sigma^{\mathrm{sym}}(\mathbf{A}, \mathbf{b})\big)$$

satisfies

$$\tilde{x} + \underline{\mathbf{Z}} + \underline{\mathbf{\Delta}} \le \underline{\mathbf{Y}} \le \tilde{x} + \underline{\mathbf{Z}} + \overline{\mathbf{\Delta}} \quad \text{and} \quad \tilde{x} + \overline{\mathbf{Z}} + \underline{\mathbf{\Delta}} \le \overline{\mathbf{Y}} \le \tilde{x} + \overline{\mathbf{Z}} + \overline{\mathbf{\Delta}}. \quad (10.40)$$

Theorem 10.10 was used to compute the data in Figure 10.3. The idea of computing a sharp inclusion of $R(\mathbf{b} - \mathbf{A}\tilde{x})$ for symmetric $\mathbf{A}$ was extended by Rump (1994) to general linear dependencies in the matrix and the right-hand side as follows.

The following exposition borrows an idea by Higham and Higham (1992*a*), which they used to analyse the sensitivity of linear systems with data dependency. Denote by $\mathrm{vec}(A) \in \mathbb{R}^{mn}$ the columns of a matrix $A \in \mathbb{R}^{m\times n}$ stacked into one vector, and let $A \otimes B \in \mathbb{R}^{mp\times nq}$ be the Kronecker product of $A \in \mathbb{R}^{m\times n}$ and $B \in \mathbb{R}^{pq}$.

Let $\varphi : \mathbb{R}^k \to \mathbb{R}^{n\times n}$ be a matrix-valued function depending linearly on parameters $p \in \mathbb{R}^k$, so that $A(p) := \varphi(p)$ implies

$$A(p)_{ij} = [c^{(ij)}]^T p \quad (10.41)$$

for vectors $c^{(ij)} \in \mathbb{R}^k$. If $\Phi \in \mathbb{R}^{n^2\times k}$ denotes the matrix with $\{(i-1)n+j\}$th row $[c^{(ij)}]^T$, then

$$\mathrm{vec}(A(p)) = \Phi p. \quad (10.42)$$

The set $M_n^{\mathrm{struct}} := \{A(p) : p \in \mathbb{R}^k\}$ defines a set of structured matrices. Obviously, $A, B \in M_n^{\mathrm{struct}}$ implies $A + B \in M_n^{\mathrm{struct}}$, so that the sensitivity with respect to infinitesimally small structured perturbations is characterized by (10.28), and can be computed according to Higham and Higham (1992*b*).

Table 10.9. Number of parameters for different structures.

Structure	Number of parameters
symmetric	$n(n+1)/2$
real skew-symmetric	$n(n-1)/2$
Toeplitz	$2n-1$
symmetric Toeplitz	n
circulant	n

Next consider finite perturbations generalizing componentwise structured perturbations discussed in Rump (2003*c*).

For fixed dimension, the matrix Φ depends only on the given structure. For example, symmetry is modelled by Φ^{sym} with $k = n(n+1)/2$ rows and one non-zero entry 1 per row. For common structures the number k of parameters is shown in Table 10.9.

A right-hand side $b(p)$ linear depending on p is modelled in the same way by some matrix $\Psi \in \mathbb{R}^{n\times k}$ with $b(p) = \Psi p$. For an interval vector $\mathbf{p} \in \mathbb{R}^k$, define

$$A(\mathbf{p}) := \{A(p) : p \in \mathbf{p}\} \quad \text{and} \quad b(\mathbf{p}) := \{b(p) : p \in \mathbf{p}\}. \tag{10.43}$$

Then (9.6) implies

$$\Phi\mathbf{p} = \mathrm{hull}(A(\mathbf{p})) \quad \text{and} \quad \Psi\mathbf{p} = \mathrm{hull}(b(\mathbf{p})). \tag{10.44}$$

As for (10.37), one verifies, for $p \in \mathbb{R}^k$,

$$R\big(b(p) - A(p)\tilde{x}\big) = \big(R\Psi - (\tilde{x}^T \otimes R)\Phi\big)p. \tag{10.45}$$

Since $R\Psi - (\tilde{x}^T \otimes R)\Phi \in \mathbb{R}^{n\times k}$ is a point matrix, it follows by (9.4) that

$$\big(R\Psi - (\tilde{x}^T \otimes R)\Phi\big)\mathbf{p} = \mathrm{hull}\big\{R\big(b(p) - A(p)\tilde{x}\big) : p \in \mathbf{p}\big\}. \tag{10.46}$$

As for Theorems 10.9 and 10.10, this is the foundation for calculating outer and inner inclusions for systems of linear equations, with general linear dependencies in the coefficients of the matrix and the right-hand side.

Theorem 10.11. For $k, n \in \mathbb{N}$, let $\Phi \in \mathbb{R}^{k\times n^2}$, $\Psi \in \mathbb{R}^{k\times n}$, $\mathbf{p} \in \mathbb{IR}^k$, $\tilde{x} \in \mathbb{R}^n$ and $\mathbf{X} \in \mathbb{IR}^n$ be given. Define

$$\mathbf{Z} := \big(R\Psi - (\tilde{x}^T \otimes R)\Phi\big)\mathbf{p} \quad \text{and} \quad \mathbf{\Delta} := (I - R\mathbf{A})\mathbf{X} \tag{10.47}$$

and

$$A(\mathbf{p}) := \{A : \mathrm{vec}(A) = \Phi p,\ p \in \mathbf{p}\} \quad \text{and} \quad b(\mathbf{p}) := \{\Psi p : p \in \mathbf{p}\}. \tag{10.48}$$

Suppose

$$\mathbf{Z} + \mathbf{\Delta} \subset \operatorname{int}(\mathbf{X}). \tag{10.49}$$

Then R and every matrix $A \in A(\mathbf{p})$ is non-singular, and the hull of the structured solution set

$$\mathbf{Y} := \operatorname{hull}\big(\{x \in \mathbb{R}^n : Ax = b, A \in A(\mathbf{p}), b \in b(\mathbf{p})\}\big) \tag{10.50}$$

satisfies

$$\tilde{x} + \underline{\mathbf{Z}} + \underline{\mathbf{\Delta}} \le \underline{\mathbf{Y}} \le \tilde{x} + \underline{\mathbf{Z}} + \overline{\mathbf{\Delta}} \quad \text{and} \quad \tilde{x} + \overline{\mathbf{Z}} + \underline{\mathbf{\Delta}} \le \overline{\mathbf{Y}} \le \tilde{x} + \overline{\mathbf{Z}} + \overline{\mathbf{\Delta}}. \tag{10.51}$$

Note again that special care is necessary in the computation of $\mathbf{Z}$ in (10.47): inner bounds for $\mathbf{Z}$ are necessary for inner bounds of $\mathbf{Y}$, whereas for both outer and inner bounds of $\mathbf{Y}$, outer bounds for $\mathbf{\Delta}$ are sufficient.

The exact, in general, nonlinear shape (see Figure 10.3) of the solution set for symmetric and other linear dependencies has been characterized by Alefeld, Kreinovich and Mayer (1997, 2003).

10.8. Sparse linear systems

For general dense linear systems, verification algorithms generally compute an inclusion if Gaussian elimination in floating-point arithmetic provides at least one correct digit. The computing time for the verification algorithm is within an order of magnitude of Gaussian elimination. Unfortunately, for general sparse linear systems such an algorithm is not known, one major open problem of verification algorithms.

For sparse linear systems Algorithm 10.7 is, in general, not suitable because the inverse of a sparse matrix is likely to be dense. For symmetric positive definite matrices, verified error bounds can be computed effectively using the following methods by Rump (1994). But already for indefinite symmetric problems, the known approaches may fail for reasonable condition number and/or require extensive computational time. Moreover, the field of verification algorithms based on iterative methods is basically a blank slate.

For didactic purposes, we let there be given a symmetric positive definite matrix A. Later it is seen that this significant assumption is *proved a posteriori* by the proposed algorithm.

Let $A \in \mathbb{R}^{n\times n}$ and $b, \tilde{x} \in \mathbb{R}^n$ be given. As for (10.8) we note that

$$\|A^{-1}b - \tilde{x}\|_2 \le \|A^{-1}\|_2 \cdot \|b - A\tilde{x}\|_2 = \frac{\|b - A\tilde{x}\|_2}{\sigma_n(A)}, \tag{10.52}$$

where $\|\cdot\|_2$ denotes the spectral norm and $\sigma_1(A) \ge \cdots \ge \sigma_n(A)$ denote the singular values of A. Thus a lower bound on the smallest singular value of A yields a bound on the error of $\tilde{x}$. We use the following well-known perturbation result given by Wilkinson (1965, p. 99 *ff.*).

Lemma 10.12. Let the eigenvalues of a symmetric matrix A be denoted by $\lambda_1(A) \geq \cdots \geq \lambda_n(A)$. Let symmetric $A, E \in \mathbb{R}^{n\times n}$ be given. Then

$$\lambda_i(A) + \lambda_n(E) \leq \lambda_i(A+E) \leq \lambda_i(A) + \lambda_1(E) \quad \text{for } 1 \leq i \leq n. \tag{10.53}$$

Let $\alpha \in \mathbb{R}$ be an approximation of $\sigma_n(A)$, and set $\tilde{\alpha} := 0.9 \cdot \alpha$. Further, let $G \in \mathbb{R}^{n\times n}$ and define $E := A - \tilde{\alpha} I - G^T G$. Then Lemma 10.12 implies

$$\lambda_n(A - \tilde{\alpha} I) \geq \lambda_n(G^T G) + \lambda_n(E) \geq -\|E\|_2 \quad \text{or} \quad \lambda_n(A) \geq \tilde{\alpha} - \|E\|_2 := \mu. \tag{10.54}$$

If $\mu > 0$ it follows that A is symmetric positive definite and

$$\sigma_n(A) \geq \mu. \tag{10.55}$$

For practical applications, it suffices to compute an upper bound for $\|E\|_2$ which is readily obtained by evaluating $\|A - \tilde{\alpha} I - G^T G\|_1$ in interval arithmetic, or with the methods given in Section 10.11. The obvious choice for G is an approximate Cholesky factor of $A - \tilde{\alpha} I$. Note that only the input data is used corresponding to the UTILIZE INPUT DATA PRINCIPLE (5.13).

The Cholesky decomposition allows one to estimate all rounding errors effectively to obtain an *a priori* bound without calculating $A - \tilde{\alpha} I - G^T G$, resulting in an efficient verification method.

10.8.1. Verification of positive definiteness

Following Demmel (1989) we use *pure* floating-point arithmetic to verify positive definiteness of $A - \tilde{\alpha} I$ and to obtain a lower bound $\sigma_n(A) \geq \tilde{\alpha}$. It is another typical example of the design of a verification method following the DESIGN PRINCIPLE (DP) (1.1) and the UTILIZE INPUT DATA PRINCIPLE (5.13). It is much faster than (10.54), albeit with a more restricted range of applicability.

When executing Cholesky decomposition in floating-point arithmetic and, for the moment, barring underflow, successive applications of Theorem 2.1 imply, for the computed factor $\tilde{G}$,

$$\tilde{G}^T \tilde{G} = A + \Delta A \quad \text{with } |\Delta A| \leq \operatorname{diag}(\gamma_2, \ldots, \gamma_{n+1}) |\tilde{G}^T| |\tilde{G}|, \tag{10.56}$$

where $\gamma_k := k\mathbf{u}/(1 - k\mathbf{u})$ is defined as in Theorem 2.1. Cholesky decomposition is a very stable algorithm, and no pivoting is necessary, so the size of $|\Delta A|$ can be estimated by $\tilde{G}$ and the input data. If the Cholesky decomposition runs to completion, then $a_{jj} \geq 0$, and one can show

$$\|\tilde{g}_j\|_2^2 = \tilde{g}_j^T \tilde{g}_j \leq a_{jj} + \gamma_{j+1} |\tilde{g}_j^T| |\tilde{g}_j| = a_{jj} + \gamma_{j+1} \tilde{g}_j^T \tilde{g}_j, \tag{10.57}$$

where $\tilde{g}_j$ denotes the jth column of $\tilde{G}$. Therefore

$$\|\tilde{g}_j\|_2^2 \leq (1 - \gamma_{j+1})^{-1} a_{jj} =: d_j,$$

Table 10.10. Smallest singular value for which `isspd` verifies positive definiteness.

Full matrices			Sparse matrices		
Dimension	$\alpha/\|A\|_2$	Time (sec)	Dimension	$\alpha/\|A\|_2$	Time (sec)
100	$1.5 \cdot 10^{-13}$	0.003	5000	$3.5 \cdot 10^{-11}$	0.06
200	$5.9 \cdot 10^{-13}$	0.009	10000	$1.2 \cdot 10^{-10}$	0.22
500	$3.6 \cdot 10^{-12}$	0.094	20000	$5.1 \cdot 10^{-10}$	1.0
1000	$1.4 \cdot 10^{-11}$	0.38	50000	$3.2 \cdot 10^{-9}$	11.9
2000	$5.6 \cdot 10^{-11}$	2.1	100000	$9.1 \cdot 10^{-9}$	84

and setting $\overline{d}_j := (\gamma_{j+1}(1-\gamma_{j+1})^{-1}\,a_{jj})^{1/2}$ and $k := \min(i,j)$, we have

$$|\Delta a_{ij}| \le \gamma_{k+1}\,|\tilde{g}_i^T|\,|\tilde{g}_j| \le \gamma_{i+1}^{1/2}\,\|\tilde{g}_i\|_2\,\gamma_{j+1}^{1/2}\,\|\tilde{g}_j\|_2 \le \overline{d}_i\,\overline{d}_j, \tag{10.58}$$

so that $\overline{d} := (\overline{d}_1,\ldots,\overline{d}_n)$ implies

$$\|\Delta A\| \le \|\,|\Delta A|\,\| \le \|\overline{d}\,\overline{d}^T\| = \overline{d}^T\,\overline{d} = \sum_{j=1}^{n} \gamma_{j+1}(1-\gamma_{j+1})^{-1}\,a_{jj}. \tag{10.59}$$

Lemma 10.13. Let symmetric $A \in \mathbb{F}^{n\times n}$ be given, and suppose Cholesky decomposition, executed in floating-point arithmetic satisfying (2.3), runs to completion. Then, barring over- and underflow, the computed $\tilde{G}$ satisfies

$$\tilde{G}^T\tilde{G} = A + \Delta A \quad \text{with } \|\Delta A\| \le \sum_{j=1}^{n} \varphi_{j+1}a_{jj}, \tag{10.60}$$

where $\varphi_k := \gamma_k(1-\gamma_k)^{-1}$.

Note that corresponding to the SOLVABILITY PRINCIPLE OF VERIFICATION METHODS (1.2) positive definiteness (and indefiniteness) can be verified if the smallest singular value is not too small in magnitude. For a singular matrix, neither is possible in general.

With some additional work covering underflow, Lemma 10.13 leads to algorithm `isspd` in INTLAB: a quantity $\tilde{\alpha}$ based on $\sum_{j=1}^{n}\varphi_{j+1}a_{jj}$ in (10.60) and covering underflow can be computed *a priori*, and if the floating-point Cholesky decomposition applied to $A - \tilde{\alpha}I$ runs to completion, then A is proved to be positive definite.

For an interval input matrix $\mathbf{A}$, algorithm `isspd` verifies positive definiteness of every symmetric matrix within $\mathbf{A}$. But rather than working with the interval matrix, the following lemma is used.

Lemma 10.14. Let $\mathbf{A} = [M-R, M+R] \in \mathbb{IR}^{n\times n}$ with symmetric matrices $M, R \in \mathbb{R}^{n\times n}$ be given. If, for some $c \in \mathbb{R}$, $M - cI$ is positive definite for $\|R\|_2 \le c$, then every symmetric matrix $A \in \mathbf{A}$ is positive definite.

Table 10.11. Results and ratio of computing time for sparse linear systems.

	Double-precision residual		Improved residual	
Dimension	Inclusion of A_{11}^{-1}	Time ratio	Inclusion of A_{11}^{-1}	Time ratio
10000	0.302347266456_	5.0	0.3023472664558_	6.5
40000	0.30234727323__	4.7	0.302347273226_	6.3
250000	0.3023472737__	4.2	0.30234727367__	5.3

Moreover, for every positive vector $x \in \mathbb{R}^n$,

$$\|R\|_2 \le \sqrt{\mu} \quad \text{with} \quad \mu := \max_i \frac{(R^T(Rx))_i}{x_i}. \tag{10.61}$$

Proof. Lemma 10.12 implies that $|\sigma_i(A+\Delta)-\sigma_i(A)| \le \|\Delta\|_2$ for symmetric $A, \Delta \in \mathbb{R}^{n\times n}$, and $\|\Delta\|_2 \le \|\,|\Delta|\,\|_2$ proves that all $A \in \mathbf{A}$ are positive definite. Collatz (1942) showed that μ is an upper bound for the Perron root of (the non-negative matrix) $R^T R$, and the result follows. □

The applicability of Lemma 10.13 is tested by generating a numerically singular symmetric matrix A by `B = randn(n, n − 1); A = B * B'`; and finding the smallest $\alpha > 0$ for which positive definiteness of $A - \alpha I$ can be verified. The results for dense and sparse matrices (with about 5 non-zero elements per row) are displayed in Table 10.10; computations are performed on a 1.6 GHz laptop in MATLAB.

Estimating $\Delta A = \tilde{G}^T\tilde{G} - A$ in (10.60) using interval arithmetic takes more computing time, but positive definiteness can be verified for $\alpha/\|A\|_2$ almost of size $\mathbf{u}$.

10.8.2. Solution of sparse linear systems with s.p.d. matrix

As before, let $\alpha \in \mathbb{R}$ be an approximation of $\sigma_n(A)$, and denote $\tilde{\alpha} := 0.9{\cdot}\alpha$. If positive definiteness of $A-\tilde{\alpha}I$ can be verified (using `isspd`), then $\|A\|_2 \ge \tilde{\alpha}$ and (10.52) can be applied.

This algorithm is implemented in `verifylss` in INTLAB for sparse input data. For a five-point discretization A of the Laplace equation on the unit square, we compute A_{11}^{-1} and display the results in Table 10.11.

The time for the MATLAB call $A\backslash b$ is normed to 1, and the results are shown with residual computed in double precision and with the improved residual by Algorithm 10.4. For this well-conditioned linear system, the floating-point approximation is also accurate. For some test matrices from the Harwell–Boeing test case library, the numbers are given in Table 10.12 (the name of the matrix, the dimension, number of non-zero elements, time for MATLAB $A\backslash b$ normed to 1, and median relative error of the inclusion are displayed using the improved residual).

Table 10.12. Computational results for Harwell–Boeing test matrices.

Matrix	Dimension	Non-zero elements	Time ratio	Median relative error of inclusion
bcsstk15	3948	117816	7.69	$1.3 \cdot 10^{-5}$
bcsstk16	4884	290378	11.41	$2.9 \cdot 10^{-7}$
bcsstk17	10974	428650	9.37	$4.1 \cdot 10^{-5}$
bcsstk18	11948	149090	9.50	$5.0 \cdot 10^{-5}$
bcsstk21	3600	26600	7.00	$2.0 \cdot 10^{-11}$
bcsstk23	3134	45178	7.85	$1.8 \cdot 10^{-3}$
bcsstk24	3562	159910	15.02	$6.8 \cdot 10^{-5}$
bcsstk25	15439	252241	7.49	$1.3 \cdot 10^{-3}$
bcsstk28	4410	219024	13.13	$3.5 \cdot 10^{-7}$

For matrices with small bandwidth, however, the verification is significantly slower than the pure floating-point computation. For example, computing inclusions of the first column of A^{-1} for the classic second difference operator on n points gives the results in Table 10.13. For such special matrices, however, specialized verification methods are available.

However, no stable and fast verification method for reasonably ill-conditioned matrices is known for symmetric (or general) matrices. There are some methods based on an LDL^T-decomposition along the lines of (10.54) discussed in Rump (1994); however, the method is only stable with some pivoting, which may, in turn, produce significant fill-in. The problem may also be attacked using the normal equations $A^TAy = A^Tb$, but this squares the condition number. To be more specific, we formulate the following challenge.

Challenge 10.15. Derive a verification algorithm which computes an inclusion of the solution of a linear system with a general symmetric sparse matrix of dimension 10 000 with condition number 10^{10} in IEEE 754 double precision, and which is no more than 10 times slower than the best numerical algorithm for that problem.

For more challenges see Neumaier (2002).

10.9. Special linear systems

If the matrix of a linear system has special properties, adapted methods may be used. For example, the discretization of an elliptic partial differential equation leads in general to an M-matrix, *i.e.*, a matrix of the form $A := cI - B$ with non-negative B and $\varrho(B) < c$.

Table 10.13. Results and ratio of computing time for the classic second difference operator.

Dimension	Time ratio	Median relative error of inclusion
10000	79.6	$1.24 \cdot 10^{-7}$
30000	94.0	$2.99 \cdot 10^{-6}$
100000	92.8	$6.25 \cdot 10^{-5}$
300000	151.9	$6.03 \cdot 10^{-4}$
1000000	147.0	$1.86 \cdot 10^{-2}$

In this case it is well known that $A^{-1} > 0$ and thus $\|A^{-1}\|_\infty = \|A^{-1}e\|_\infty$ for the vector e of all ones. Hence for $\tilde{y} \in \mathbb{R}^n$ it follows that

$$\|A^{-1}\|_\infty = \|\tilde{y} + A^{-1}(e - A\tilde{y})\|_\infty \le \|\tilde{y}\| + \|A^{-1}\|_\infty \|e - A\tilde{y}\|_\infty$$

(see Ogita, Oishi and Ushiro (2001)), and therefore, along the lines of (10.8),

$$\|A^{-1}b - \tilde{x}\|_\infty \le \|A^{-1}\|_\infty \|b - A\tilde{x}\|_\infty \le \frac{\|\tilde{y}\|_\infty}{1 - \|e - A\tilde{y}\|_\infty} \|b - A\tilde{x}\|_\infty \quad (10.62)$$

for $\tilde{x} \in \mathbb{R}^n$. For given approximate solutions $\tilde{y}$ of $Ay = e$ and $\tilde{x}$ of $Ax = b$, this is an $\mathcal{O}(n^2)$ bound of good quality.

An interval matrix $\mathbf{A}$ is called an H-matrix if $\mathbf{A}v > 0$ for some positive vector v. Note that by (9.4) the interval and power set product $\mathbf{A}v$ coincide. Moreover, Ostrowski's comparison matrix $\langle \mathbf{A} \rangle \in \mathbb{R}^{n \times n}$ for interval matrices is defined by

$$\langle \mathbf{A} \rangle_{ij} := \begin{cases} \min\{|\alpha| : \alpha \in \mathbf{A}_{ij}\} & \text{for } i = j, \\ -\max\{|\alpha| : \alpha \in \mathbf{A}_{ij}\} & \text{otherwise.} \end{cases} \quad (10.63)$$

In fact, all matrices $A \in \langle \mathbf{A} \rangle$ are M-matrices. It follows that

$$|A^{-1}| \le \langle \mathbf{A} \rangle^{-1} \quad \text{for all} \quad A \in \mathbf{A} \quad (10.64)$$

(see Neumaier (1990)).

If the midpoint matrix of $\mathbf{A}$ is a diagonal matrix, *i.e.*, off-diagonal intervals are centred around zero, then the exact solution set can be characterized. This remarkable result is known as the Hansen–Bliek–Rohn–Ning–Kearfott–Neumaier enclosure of an interval linear system. Recall that for a general linear system with interval matrix $\mathbf{A}$, it is an NP-hard problem to compute narrow bounds for the solution set defined in (10.15) (Rohn and Kreinovich 1995).

Theorem 10.16. Let $\mathbf{A} \in \mathbb{IR}^{n\times n}$ be an H-matrix, and let $\mathbf{b} \in \mathbb{IR}^n$ be a right-hand side. Define

$$u := \langle \mathbf{A} \rangle^{-1} |\mathbf{b}| \in \mathbb{R}^n, \quad d_i := (\langle \mathbf{A} \rangle^{-1})_{ii} \in \mathbb{R}, \tag{10.65}$$

and

$$\alpha_i := \langle \mathbf{A} \rangle_{ii} - 1/d_i, \quad \beta_i := u_i/d_i - |\mathbf{b}_i|. \tag{10.66}$$

Then the solution set $\Sigma(\mathbf{A}, \mathbf{b}) = \{x \in \mathbb{R}^n : Ax = b \text{ for } A \in \mathbf{A}, b \in \mathbf{b}\}$ is contained in the interval vector $\mathbf{X}$ with components

$$\mathbf{X}_i := \frac{\mathbf{b}_i + [-\beta_i, \beta_i]}{\mathbf{A}_{ii} + [-\alpha_i, \alpha_i]}. \tag{10.67}$$

Moreover, if the midpoint matrix of $\mathbf{A}$ is diagonal, then

$$\text{hull}(\Sigma(\mathbf{A}, \mathbf{b})) = \mathbf{X}.$$

In practice, Theorem 10.16 is applied to a preconditioned linear system with matrix $R\mathbf{A}$ and right-hand side $R\mathbf{b}$, where R is an approximation of the inverse of $\text{mid}(\mathbf{A})$.

It is likely that the midpoint of $R\mathbf{A}$ is not far from the identity matrix. However, even using exact arithmetic with $R := \text{mid}(\mathbf{A})^{-1}$ so that $\text{mid}(R\mathbf{A}) = I$, the solution set $\Sigma(R\mathbf{A}, R\mathbf{b})$ is only a superset of $\Sigma(\mathbf{A}, \mathbf{b})$ due to data dependencies. So this approach offers no means to attack the original NP-hard problem.

This approach is slower than Algorithm 10.7; therefore it is used in `verifylss` in INTLAB as a 'second stage' if the verification using Algorithm 10.7 fails.

10.10. The determinant

As another example for the DESIGN PRINCIPLE OF VERIFICATION METHODS (1.1), consider the determinant of a matrix.

About the worst one can do is to perform Gaussian elimination in interval arithmetic (IGA). Instead one may proceed as follows:

```
[L,U,p] = lu(A,'vector');           % approximate factorization of A
Linv = inv(L); Uinv = inv(U);       % approximate preconditioners
C = Linv*(A(p,:)*intval(Uinv)); % inclusion of preconditioned matrix
D = prod(gershgorin(C));            % inclusion of det(Linv*A(p,:)*Uinv)
```

The first statement calculates approximate factors `L`, `U` such that `A(p,:)` $\approx$ `LU` for a permutation vector `p`. Multiplying `A(p,:)` from the left and right by approximate inverses of `L` and `U` results approximately in the identity matrix, the determinant of which can be estimated by the (complex interval) product of the Gershgorin circles.

Most of the computations are performed in floating-point arithmetic: only the computation of C and D is performed in interval arithmetic. A typical result for a 1000×1000 random matrix is

```
intval D =
<   1.00000000000008 +  0.00000000000000i,  0.00000062764860>
```

where the complex result is displayed by midpoint and radius. Since the input matrix was real, so must be the determinant resulting in

$$\det(\texttt{Linv} * \texttt{A}(\texttt{p},:) * \texttt{Uinv}) \in [0.99999937, 1.00000063].$$

Now $\det(\texttt{L}) = \det(\texttt{Linv}) = 1$ and $\det(A)$ (or, better, $\log\det(A)$ to avoid over- or underflow) is easily and accurately computed.

Note that the approach applies to interval matrices as well, by computing an approximate decomposition of the midpoint matrix and otherwise proceeding as before.

10.11. The spectral norm of a matrix

Any non-trivial vector $x \in \mathbb{R}^n$ yields a lower bound of $\|A\|_2$ by evaluating $\|Ax\|_2/\|x\|_2$ in interval arithmetic, so the best numerical algorithm at hand may be used to compute a suitable x. However, an upper bound seems non-trivial.

If A is symmetric and $\alpha \approx \|A\|_2$ is given, then Lemma 10.13 may be applied to verify that $\tilde{\alpha}I - A$ and $\tilde{\alpha}I + A$ are positive definite for some $\tilde{\alpha} > \alpha$, thus verifying $\|A\|_2 \leq \tilde{\alpha}$.

Let a general matrix A be given, together with an approximation $\alpha \approx \|A\|_2$. Using perturbation bounds similar to Lemma 10.12, it is not difficult to compute an inclusion of a singular value of A near α. However, there is no proof that this is the largest singular value.

But in this case there is no problem using $A^H A$ in the above approach, *i.e.*, verifying that $\tilde{\alpha}^2 I - A^H A$ is positive definite to prove $\|A\|_2 \leq \tilde{\alpha}$; the squared condition number has no numerical side effect.

The cost of a verified upper bound of $\|A\|_2$, however, is $\mathcal{O}(n^3)$, whereas a few power iterations on $A^H A$ require some $\mathcal{O}(n^2)$ operations and usually lead to an accurate approximation of $\|A\|_2$. For a verified upper bound, the standard estimations $\|A\|_2 \leq \sqrt{\|A\|_1 \|A\|_\infty}$ or $\|A\|_2 \leq \|A\|_F$ are sometimes weak. Also, $\|A\|_2 \leq \||A|\|_2$ together with (10.61) is often weak.

Challenge 10.17. Given $A \in \mathbb{R}^{n \times n}$, derive a verification algorithm to compute an upper bound for $\|A\|_2$ with about 1% accuracy in $\mathcal{O}(n^2)$ operations.

A verified lower bound is easy to compute in $\mathcal{O}(n^2)$ operations, but I find it hard to believe that there is such a method for an upper bound of $\|A\|_2$.

11. Automatic differentiation

For enclosing solutions of systems of nonlinear equations we need to approximate the Jacobian of nonlinear functions and to compute the range of the Jacobian over some interval.

The method of 'automatic differentiation' accomplishes this. Because this is mandatory for the following, we want to make at least a few remarks here. For a thorough discussion see Rall (1981), Corliss *et al.* (2002) and the *Acta Numerica* article by Griewank (2003).

11.1. Gradients

The method was found and forgotten several times, starting in the 1950s. When giving a talk on automatic differentiation in the 1980s, the audience would usually split in two groups, one not understanding or believing in the method and the other knowing it. The reason is that it seems, at first sight, not much more than the classical differentiation formulas $(uv)' = u'v + uv'$ or $g(f(x))' = g'(f(x))f'(x)$.

One way to understand it is similar to the concept of differential fields. Define a set $\mathcal{D}$ of pairs $(a, \alpha) \in \mathbb{R}^2$, and define operations $+, -, \cdot, /$ on $\mathcal{D}$ by

$$\begin{aligned} (a, \alpha) \pm (b, \beta) &:= (a + b, \alpha \pm \beta), \\ (a, \alpha) \cdot (b, \beta) &:= (a \cdot b, \alpha b + a\beta), \\ (a, \alpha)/(b, \beta) &:= (a/b, (\alpha - a\beta/b)/b), \end{aligned} \tag{11.1}$$

with non-zero denominator assumed in the case of division. Let a differentiable function $f : \mathbb{R} \to \mathbb{R}$ be given by means of an arithmetic expression $\mathfrak{A}$ in one independent variable x, so that $\mathfrak{A}(x) = f(x)$.

When replacing constants $c \in \mathbb{R}$ in $\mathfrak{A}$ by $(c, 0)$, then evaluating $\mathfrak{A}((\tilde{x}, 1))$ for some $\tilde{x} \in \mathbb{R}$ using (11.1) yields a pair (r, ρ) with property $f(\tilde{x}) = r$ and $f'(\tilde{x}) = \rho$, provided no division by 0 has occurred.

Using a programming language with an operator concept, the implementation is particularly simple, basically implementing (11.1). Standard functions are easily added, for example,

$$\begin{aligned} \mathrm{e}^{(b,\beta)} &:= (\mathrm{e}^b, \beta\,\mathrm{e}^b), \\ \cos(b, \beta) &:= (\cos(b), -\beta \sin(b)). \end{aligned} \tag{11.2}$$

Replacing all operations again by the corresponding interval operations and successively applying the inclusion property (5.16), it is clear that an inclusion of the range of a function and its derivative over some interval $\mathbf{X}$ is obtained as well. For example, for the function f given in (8.4), an inclusion of the range of f and f' over $\mathbf{X} := [2.4, 2.5]$ is obtained by

```
f = inline('sin(2*x^2/sqrt(cosh(x))-x)-atan(4*x+1)+1');
Y = f(gradientinit(intval('[2.4,2.5]')))
```

```
intval gradient value Y.x =
[   -0.2435,     0.3591]
intval gradient derivative(s) Y.dx =
[   -1.1704,    -0.0736]
```

The function **gradientinit(x)** initializes the gradient operator, *i.e.*, replaces constants c by $(c, 0)$ and replaces the argument x by $(x, 1)$. It follows that $-1.1704 \le f'(x) \le -0.0736$ for all $x \in [2.4, 2.5]$. Note that the notation `intval('[2.4,2.5]')` is necessary to obtain a true inclusion of $\mathbf{X}$ because $2.4 \notin \mathbb{F}$. As always, the inclusions are rigorous but may be subject to overestimation.

In an environment with operator concept such as MATLAB we can conveniently define a function and, depending on the input argument, obtain an approximation or an inclusion of the range of the function and the derivative. From the definitions (11.1) and (11.2) it is clear that the computing time for the function value and its derivative is less than about 5 times the computing time for only the function value.

Some care is necessary if a program contains branches. For example,

```
function y = f(x)
  if x==3, y=9; else y=x^2; end
```

is an unusual but correct implementation for the function $f(x) = x^2$. A straightforward automatic differentiation program will, however, deliver $f'(3) = 0$.

Applying the discussed principles to n independent variables $x_1, \ldots, x_n$ successively, an approximation and also an inclusion of the gradient or the Hessian of a function $f : \mathbb{R}^n \to \mathbb{R}$ is obtained. This is called the forward mode. The computing time for the gradient, however, is up to $5n$ times that for a function evaluation.

11.2. Backward mode

A major breakthrough of automatic differentiation was the so-called backward mode, in which the time to compute the function *and* gradient is not more than 5 times as much as only the evaluation of the function. This is independent of the dimension n. The idea is as follows.

Assume we are given an arithmetic expression $\mathfrak{A}$ to evaluate a function $f : \mathbb{R}^n \to \mathbb{R}$ at $x \in \mathbb{R}^n$. Denote the result of each intermediate operation in $\mathfrak{A}$ by x_i for $n+1 \le i \le n+m$, so that the evaluation consists of m steps and x_{n+m} is the final result $f(x)$. Collecting the initial vector x and the vector of intermediate results in one vector, the evaluation of $\mathfrak{A}(x)$ corresponds to successive computation of

$$y^{(k)} = \Phi_k(y^{(k-1)}) \quad \text{for} \quad k = 1, \ldots, m, \tag{11.3}$$

with the initial vector $y^{(0)} = (x, 0)^T \in \mathbb{R}^{n+m}$ and $x_{n+k} = e_{n+k}^T y^{(k)}$. The functions $\Phi_k : \mathbb{R}^{n+m} \rightarrow \mathbb{R}^{n+m}$ correspond to a basic arithmetic operation or standard function on already computed quantities. In other words, Φ_k depends only on components $1, \ldots, n+k-1$ of $y^{(k-1)}$. Therefore

$$f(x) = e_{n+m}^T \cdot \Phi_m \circ \ldots \circ \Phi_1\left(\begin{pmatrix} x \\ 0 \end{pmatrix}\right), \qquad (11.4)$$

with 0 denoting a vector of m zeros, and

$$\nabla f(x) = e_{n+m}^T \cdot \Phi_m'(y^{(m-1)}) \cdots \Phi_1'(y^{(0)}) \cdot \begin{pmatrix} I \\ 0 \end{pmatrix}, \qquad (11.5)$$

with 0 denoting the $m \times n$ zero matrix. The Jacobians $\Phi_k'(y^{(k-1)}) \in \mathbb{R}^{(n+m)\times(n+m)}$ have a very simple structure corresponding to the operations they are associated with. Note that m corresponds to the number of intermediate steps in the evaluation of $\mathfrak{A}$, and n corresponds to the number of unknowns.

One can see that the forward mode corresponds to the evaluation of (11.5) from right to left, so that each operation is a $(p \times p)$ times $(p \times n)$ matrix–matrix multiplication with $p := n + m$. However, when evaluating (11.5) from left to right, each operation is a $(p \times p)$ matrix multiplied by a p-vector from the left. This is the backward mode.

The remarkable speed of backward automatic differentiation comes at a price. While the forward mode is straightforward to implement with an operator concept at hand, the implementation of the backward mode is involved. However, packages are available that transform a computer program for evaluating a function into another program for evaluating the function together with the gradient and/or Hessian (in the fast backward mode); see, for example, Bischof, Carle, Corliss and Griewank (1991). For the range estimation of derivatives, however, forward differentiation sometimes gives much better enclosures than those from the backward mode.

11.3. Hessians

Automatic differentiation in forward mode for gradients and Hessians is implemented in INTLAB. Consider, for example, the function $f : \mathbb{R}^2 \rightarrow \mathbb{R}$:

$$f(x,y) := \exp(\sin(50x)) + \sin(60\exp(y)) + \sin(70\sin(x)) + \sin(\sin(80y)) \\ - \sin(10(x+y)) + (x^2+y^2)/4. \qquad (11.6)$$

It was Problem 4 of the 10×10-digit challenge by Trefethen (2002) to compute the global minimum of this function over $\mathbb{R}^2$. The INTLAB code

```
f = inline('exp(sin(50*x(1))) + sin(60*exp(x(2)))
      + sin(70*sin(x(1))) + sin(sin(80*x(2)))
      - sin(10*(x(1)+x(2))) + (x(1)^2+x(2)^2)/4');
X = verifynlss(f,[-0.02;0.2],'h')
Y = f(hessianinit(X))
```

evaluates the function value, gradient and Hessian of f over the given X, resulting in

```
intval X =
  -0.02440307969437
   0.21061242715535
intval Hessian value Y.x =
  -3.3068686474752_
intval Hessian first derivative(s) Y.dx =
  1.0e-011 *
  -0.0_____________    0.______________
intval Hessian second derivative(s) Y.hx =
  1.0e+003 *
   5.9803356010662_    0.09578721471459
   0.09578721471459    9.895778741947__
```

The call `X = verifynlss(f,[-0.02;0.2],'h')` proves that X is an inclusion of a stationary point (see Section 13), and by Gershgorin's theorem, every 2×2 matrix included in the Hessian is positive definite. This proves that X contains a strict local minimizer.

Consider the test problem

$$f(x) := \sum_{i=1}^{n-10} \frac{x_i}{x_{i+10}} + \sum_{i=1}^{n} (x-1)^2 - \sum_{i=1}^{n-1} x_i x_{i+1} = \text{Min!}, \tag{11.7}$$

with initial guess $\tilde{x} := (1, \ldots, 1)^T$. Given f implementing the function in (11.7), the code

```
n = 2000;
X = verifynlss(@f,ones(n,1),'h');
H = f(hessianinit(X));
isMin = isspd(H.hx)
```

computes for $n = 2000$ an inclusion X of a solution of the nonlinear system $\nabla f(x) = 0$, *i.e.*, of a stationary point $\hat{x}$ of f, to at least 10 decimal places. Using **isspd** as described in Section 10.8, the result $\texttt{isMin} = 1$ verifies that every matrix $A \in$ **H.hx** is positive definite, in particular the Hessian of f at $\hat{x}$. Therefore, f has a strict (local) minimum in X. The Hessian has bandwidth 10, and the Gershgorin circles contain 0.

11.4. Taylor coefficients

To obtain higher-order derivatives, Taylor coefficients can be computed along the previous lines. Here we follow Rall (1981); see also Moore (1966, Section 11). Let two functions $f, g : \mathbb{R} \to \mathbb{R}$ with $f, g \in \mathcal{C}^K$ be given, and denote their respective Taylor coefficients at some $\tilde{x} \in \mathbb{R}$ by

$$a_k := \frac{1}{k!} f^{(k)}(\tilde{x}) \quad \text{and} \quad b_k := \frac{1}{k!} g^{(k)}(\tilde{x}),$$

for $0 \le k \le K$. Then the Taylor coefficients of selected composite functions are as follows:

$$\begin{array}{ll} \text{operation} & \text{Taylor coefficient} \\ c = f \pm g & c_k = a_k \pm b_k \\ c = f \cdot g & c_k = \sum_{j=0}^{k} a_j b_{k-j} \\ c = f/g & c_k = \frac{1}{b_0}\left(a_k - \sum_{j=1}^{k} b_j c_{k-j}\right) \\ c = \exp(f) & c_k = \frac{1}{k}\sum_{j=1}^{k} j a_j c_{k-j}. \end{array} \tag{11.8}$$

As before, assume a function f is given by means of an arithmetic expression. Then, initializing constants c by $(c, 0, \ldots, 0)$ and the independent variable x by $(\tilde{x}, 1, 0, \ldots, 0)$, and replacement of each operation or standard function by the corresponding Taylor operation or standard function, result in a vector $(r_0, \ldots, r_K)$ of Taylor coefficients of f at $\tilde{x}$.

In INTLAB, Taylor operations for all functions listed in (7.3) are implemented. Again, replacement of the argument by an interval $\mathbf{X}$ computes inclusions of the Taylor coefficients $\frac{1}{k!}f^{(k)}(\tilde{x})$ for $\tilde{x} \in \mathbf{X}$. For example, for the function f given in (8.4), an inclusion of the range of the Taylor coefficients up to order 4 over $\mathbf{X} := [2.4, 2.5]$ is obtained by

```
f = inline('sin(2*x^2/sqrt(cosh(x))-x)-atan(4*x+1)+1');
Y = f(taylorinit(intval('[2.4,2.5]'),4))
intval Taylor value Y.t =
[   -0.2435,     0.3591]
[   -1.0692,    -0.1034]
[   -0.3933,     1.1358]
[   -0.5512,     1.7307]
[   -2.2706,     1.0008]
```

Note that the inclusion $[-1.0692, -0.1034]$ for $f'([2.4, 2.5])$ is narrower than before.

Along similar lines, 'automatic Lipschitz estimates' can also be defined.

11.5. Slopes

Yet another approach uses automatic slopes, introduced by Krawczyk and Neumaier (1985). For $f : \mathbb{R}^m \to \mathbb{R}^n$, let $\tilde{x} \in \mathbb{R}^m$ and $\mathbf{X} \in \mathbb{IR}^m$ be given. The triple $(\mathbf{C}, \mathbf{R}, \mathbf{S}) \in \mathbb{IR}^n \times \mathbb{IR}^n \times \mathbb{IR}^{n \times m}$ of 'centre, range and slope' is a slope expansion with respect to f, $\tilde{x}$ and $\mathbf{X}$ if $f(\tilde{x}) \in \mathbf{C}$, $\{f(x) : x \in \mathbf{X}\} \subseteq \mathbf{R}$, and

$$f(x) \in f(\tilde{x}) + \mathbf{S}(x - \tilde{x}) \quad \text{for all} \quad x \in \mathbf{X}. \tag{11.9}$$

Note that $\tilde{x} \in \mathbf{X}$ is not necessary. An automatic slope package, which is contained in INTLAB, initializes constants c by the point interval $(c, c, 0)$, and the ith independent variable by $(\tilde{x}_i, \mathbf{X}_i, e_i)$. This satisfies (11.9), and

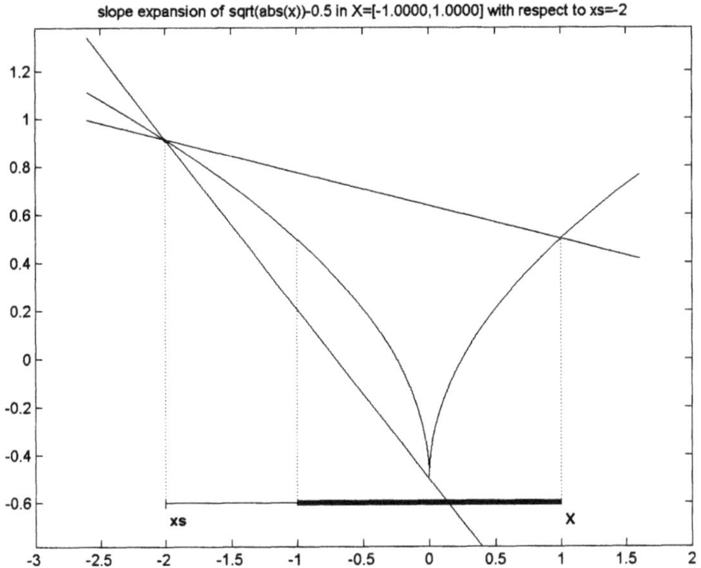

Figure 11.1. Visualization of slope for $f(x) = \sqrt{|x|} - 0.5$ and $\tilde{x} = -2$, $\mathbf{X} = [-1, 1]$.

automatic slopes are computed by defining arithmetic operations and standard functions appropriately.

As an example for $f(x) = \sqrt{|x|} - 0.5$ and $\tilde{x} = -2$, $\mathbf{X} = [-1, 1]$, the slope expansion is computed by

```
f=inline('sqrt(abs(x))-0.5'); y=f(slopeinit(-2,infsup(-1,1)))
slope intval center y.c =
    0.9142
slope intval range y.r =
[   -0.5000,     0.5000]
slope intval slope y.s =
[   -0.7072,    -0.1380]
```

and visualized in Figure 11.1.[21] The derivative over $\mathbf{X}$ is $[-\infty, \infty]$, but the slope is finite. Slopes offer some possibility for computing inclusions when the Jacobian contains singular matrices; the practical use, however, seems to be limited.

11.6. Range of a function

Gradients and slopes may be used to improve bounds for the range of a function over narrow intervals. For, given $f : \mathbb{R} \to \mathbb{R}$ and $\mathbf{X} = [m - r, m + r]$

[21] Generated by `slopeplot('sqrt(abs(x))-0.5',-2,infsup(-1,1),[],10000)`.

Table 11.1. Range estimation and overestimation for the function f as in (8.4) and $\text{mid}(\mathbf{X}) = 2$.

rad(**X**)	True range **R**	d(**X1**)/d(**R**)	d(**X2**)/d(**R**)	d(**X3**)/d(**R**)
10^{-6}	[0.3904, 0.3905]	5.86	1.00	1.00
10^{-4}	[0.3903, 0.3906]	5.86	1.00	1.00
10^{-2}	[0.3839, 0.3969]	5.86	1.21	1.03
0.1	[0.3241, 0.4520]	5.81	3.56	1.31
0.5	[0.0794, 0.5656]	4.22	18.1	3.16
1	[−0.0425, 0.5656]	3.49	148	3.49
2	[−0.2959, 0.5656]	3.17	2041	3.17

$\in \mathbb{IR}$, we obviously have

$$\text{Range}(f, \mathbf{X}) := \{f(x) : x \in \mathbf{X}\} \subseteq f(m) + f'(\mathbf{X})[-r, r]. \tag{11.10}$$

Hansen (1969) showed that the overestimation of this 'centred form' converges quadratically to zero for small radii of $\mathbf{X}$. However, the overestimation also increases quadratically for larger radii.

Another possibility for bounding $\text{Range}(f, \mathbf{X})$ uses slopes, as described in Section 11.5. The following code computes inclusions for $\text{Range}(f, \mathbf{X})$ by directly using interval operations, by the centred form (11.10) and by slopes:

```
X1 = f(X);                          % naive interval arithmetic
y = f(gradientinit(m));             % gradient inclusion of f(m)
Y = f(gradientinit(X));             % gradient inclusion of f'(X)
X2 = y.x + Y.dx*infsup(-r,r);       % inclusion by centred form
y = f(slopeinit(m,X));              % slope w.r.t. (m,X)
X3 = y.r;                           % inclusion by slopes
```

We compute the overestimation by means of the ratio of diameters of the computed inclusion and the true range for our model function f in (8.4). Table 11.1 displays the results for input interval $\mathbf{X}$ with $\text{mid}(\mathbf{X}) = 2$ and different radii $\text{rad}(\mathbf{X})$. For small radii both the centred form and the slope inclusion are almost optimal, whereas naive interval evaluation shows some overestimation. For larger radii it is the other way around for the centred form, whereas direct evaluation and the slope inclusion show moderate overestimation.

For arbitrarily wide input interval, the range of the sine function is always bounded by $[-1, 1]$. While oscillations may be a problem for numerical approximations, they can be advantageous for interval evaluation.

The picture changes completely when changing the first sine function in the definition (8.4) of f into the hyperbolic sine, which is then the function

Table 11.2. Range estimation and overestimation for the function g as in (8.6) and $\text{mid}(\mathbf{X}) = 2$.

$\text{rad}(\mathbf{X})$	True range $\mathbf{R}$	$\text{d}(\mathbf{X1})/\text{d}(\mathbf{R})$	$\text{d}(\mathbf{X2})/\text{d}(\mathbf{R})$	$\text{d}(\mathbf{X3})/\text{d}(\mathbf{R})$
10^{-6}	$[3.6643, 3.6644]$	6.33	1.00	1.00
10^{-4}	$[3.6639, 3.6649]$	6.33	1.00	1.00
10^{-2}	$[3.6166, 3.7122]$	6.34	1.18	1.04
0.1	$[3.1922, 4.1434]$	7.15	4.3	1.35
0.5	$[1.5495, 5.8700]$	89	778	4.28
1	$[0.2751, 6.7189]$	$5.6 \cdot 10^4$	$5.0 \cdot 10^6$	36.5
2	$[-0.2962, 6.7189]$	$5.6 \cdot 10^{12}$	$9.8 \cdot 10^{15}$	$3.3 \cdot 10^{11}$

defined in (8.6). From the graphs in Figure 8.2 and Figure 8.3 we know that both functions behave similarly over the interval $[0, 4]$. However, interval evaluation for the same data as in Table 11.1 for the function g produces the results shown in Table 11.2.

Clearly interval arithmetic is no panacea, particularly for wide input intervals. It is the main goal of verification methods to use appropriate mathematical tools to avoid such situations. For more details see Ratschek and Rokne (1984) or Neumaier (1990).

12. Quadrature

A direct application of the possibility of computing inclusions of Taylor coefficients is the inclusion of an integral. Often it suffices to merely implement a known error estimate using interval arithmetic. This is one (of the not so common) examples where naive evaluation in interval arithmetic is applicable. The following error estimate for Kepler's Faßregel (published in 1615, but commonly known as Simpson's [1710–1761] formula) was already known to James Stirling [1692–1770]; the computation of verified error bounds appeared in Sunaga (1956).

Theorem 12.1. For $[a, b] \in \mathbb{IR}$ let $f : [a, b] \to \mathbb{R}$ with $f \in C^4$ be given. For even $n \in \mathbb{N}$, define $x_i := a + ih$ for $0 \le i \le n$ and $h := (b - a)/n$. Then

$$\int_a^b f(x)\,\mathrm{d}x = \frac{h}{3}\big(f(x_0)+4f(x_1)+2f(x_2)+\cdots+4f(x_{n-1})+f(x_n)\big)-E \quad (12.1)$$

with

$$E := \frac{h^4}{180}(b-a)f^{(4)}(\xi) \quad \text{for some} \quad \xi \in [a, b]. \quad (12.2)$$

For a given function f, for a, b and some n the application is straightforward, for example by the following code.

Table 12.1. Integral of f and g as in (8.4) and (8.6).

	Approximation by quad		Verification by verifyquad	
Function	Approximation	Time (sec)	Inclusion	Time (sec)
f as in (8.4)	0.563112654015933	0.025	0.563112_{58}^{60}	0.410
g as in (8.6)	12.711373479890620	0.030	12.7113_{64}^{80}	0.363

Algorithm 12.2. Computation of X including $\int_a^b f(x)\,dx$:

```
D = b-intval(a);   H = D/n;
x = a + intval(0:n)/n*D;
w=2*ones(1,n+1); w(2:2:n)=4; w(1)=1; w(n+1)=1;
V = sum( H/3 * w .* f(intval(x)) );
Y = f(taylorinit(infsup(a,b),4));    % inclusion of approximation
E = H^4*D/180 * Y.t(4);              % error term
X = V - E;                           % inclusion of integral
```

A more sophisticated algorithm, verifyquad, is implemented in INTLAB based on a Romberg scheme with error term and automatic choice of n, depending on the behaviour of the function.

If the input function is well behaved, the verification algorithm is slower than an approximate routine. For example, the integral of f and g as defined in (8.4) and (8.6) over $[0,4]$ is approximated by the MATLAB routine quad and included by verifyquad using default parameters. The results are shown in Table 12.1.[22]

One of the main problems of a numerical integration routine, namely the stopping criterion, is elegantly solved by interval arithmetic: The verification algorithm stops (to increase n, for example) when the inclusion is (provably) good enough or does not improve. As an example, consider

$$\int_0^8 \sin(x + \mathrm{e}^x)\,\mathrm{d}x \tag{12.3}$$

with the results displayed in Table 12.2. Note that $\mathrm{e}^8 \approx 949\pi$.

Note that no digit of the MATLAB approximation is correct, and no warning is given by MATLAB. Presumably, some internal results in the MATLAB routine quad were sufficiently close to make quad stop without warning. As in (1.4) for the example in (1.3), even coinciding results computed in different ways give no guarantee of correctness.

[22] Note that the MATLAB routine quad requires the function to be specified in 'vectorized' form, *e.g.*, f as in (8.4) by

```
f=vectorize(inline('sin(2*x^2/sqrt(cosh(x))-x)-atan(4*x+1)+1'));
```

Table 12.2. Integral of f as in (12.3) using the MATLAB routine `quad`, Algorithm 12.2 and `verifyquad`.

		Result	Time (sec)
	`quad`	0.2511	1.77
Algorithm 12.2	$n = 2^{10}$	$[-0.47, 1.01]$	0.07
	$n = 2^{12}$	0.32_{2}^{9}	0.12
	$n = 2^{14}$	0.347_{39}^{42}	0.17
	$n = 2^{16}$	0.347400_{1}^{3}	0.60
	$n = 2^{18}$	0.347400172_{5}^{9}	2.26
	`verifyquad`	0.3474001_{6}^{8}	2.66

The presented Algorithm 12.2 is a first step. More serious algorithms handle singularities within the range of integration and may integrate through the complex plane: see Corliss and Rall (1987), Petras (2002), Okayama, Matsuo and Sugihara (2009) and Yamanaka, Okayama, Oishi and Ogita (2009).

13. Nonlinear problems

Let a nonlinear system $f(x) = 0$ with differentiable function $f : \mathbf{D} \to \mathbb{R}^n$ with $\mathbf{D} \in \mathbb{IR}^n$ be given. We assume a MATLAB program `f` be given such that `f(x)` evaluates $f(x)$. Using the INTLAB operators, according to the type of the argument `x`, an approximation or inclusion of the function value or gradient or Hessian is computed.

Let $\tilde{x} \in \mathbf{D}$ be given. Denote the Jacobian of f at x by $J_f(x)$. Then, by the n-dimensional Mean Value Theorem, for $x \in \mathbf{D}$, there exist $\xi_1, \ldots, \xi_n \in x \underline{\cup} \tilde{x}$, the convex union of x and $\tilde{x}$, with

$$f(x) = f(\tilde{x}) + \begin{pmatrix} \nabla f_1(\xi_1) \\ \cdots \\ \nabla f_n(\xi_n) \end{pmatrix} \big(x - \tilde{x}\big), \tag{13.1}$$

using the component functions $f_i : \mathbf{D}_i \to \mathbb{R}$. As is well known, the ξ_i cannot, in general, be replaced by a single ξ, so that the matrix in (13.1) is only row-wise equal to some Jacobian J_f of f.

For $X \in \mathbb{PR}^n$, recall that $\mathrm{hull}(X) \in \mathbb{IR}^n$ is defined by

$$\mathrm{hull}(X) := \bigcap \{\mathbf{Z} \in \mathbb{IR}^n : X \subseteq \mathbf{Z}\}. \tag{13.2}$$

For $x, \tilde{x} \in \mathbf{D}$, also $\mathbf{X} := \text{hull}(x \underline{\cup} \tilde{x}) \subseteq \mathbf{D}$, and the inclusion property (5.16) implies

$$\begin{pmatrix} \nabla f_1(\xi_1) \\ \cdots \\ \nabla f_n(\xi_n) \end{pmatrix} \in J_f(\mathbf{X}) \quad \text{with} \quad J_f(\mathbf{X}) := \text{hull}\{J_f(x) : x \in \mathbf{X}\} \tag{13.3}$$

for all $\xi_1, \ldots, \xi_n \in \mathbf{X}$. Therefore, using interval operations, the Mean Value Theorem can be written in an elegant way.

Theorem 13.1. Let there be given continuously differentiable $f : \mathbf{D} \to \mathbb{R}^n$ with $\mathbf{D} \in \mathbb{IR}^n$ and $x, \tilde{x} \in \mathbf{D}$. Then

$$f(x) \in f(\tilde{x}) + J_f(\mathbf{X})(x - \tilde{x}) \tag{13.4}$$

for $\mathbf{X} := \text{hull}(x \underline{\cup} \tilde{x})$.

This allows an interval Newton's method similar to the univariate version in Theorem 6.2. The proof is taken from Alefeld (1994).

Theorem 13.2. Let differentiable $f : \mathbf{X} \to \mathbb{R}^n$ with $\mathbf{X} \in \mathbb{IR}^n$ be given. Suppose all matrices $M \in J_f(\mathbf{X})$ are non-singular, and define, for some $\tilde{x} \in \mathbf{X}$,

$$N(\tilde{x}, \mathbf{X}) := \{\tilde{x} - M^{-1} f(\tilde{x}) : M \in J_f(\mathbf{X})\}. \tag{13.5}$$

If $N(\tilde{x}, \mathbf{X}) \subseteq \mathbf{X}$, then $\mathbf{X}$ contains a unique root $\hat{x}$ of f in $\mathbf{X}$. If $N(\tilde{x}, \mathbf{X}) \cap \mathbf{X} = \emptyset$, then $f(x) \neq 0$ for all $x \in \mathbf{X}$. Moreover, $\hat{x} \in N(\tilde{x}, \mathbf{X})$.

Proof. Using

$$f(x) - f(\tilde{x}) = \int_0^1 \frac{\mathrm{d}}{\mathrm{d}t} f(\tilde{x} + t(x - \tilde{x}))\, \mathrm{d}t,$$

it follows that

$$f(x) - f(\tilde{x}) = M_x(x - \tilde{x}) \quad \text{for} \quad M_x := \int_0^1 \frac{\partial f}{\partial x}(\tilde{x} + t(x - \tilde{x}))\, \mathrm{d}t \in J_f(\mathbf{X}) \tag{13.6}$$

for all $x \in \mathbf{X}$. The function

$$g(x) := \tilde{x} - M_x^{-1} f(\tilde{x}) : \mathbf{X} \to \mathbb{R}^n \tag{13.7}$$

is continuous, and by assumption $\{g(x) : x \in \mathbf{X}\} \subseteq \mathbf{X}$. Therefore Brouwer's Fixed-Point Theorem implies existence of $\hat{x} \in \mathbf{X}$ with

$$g(\hat{x}) = \hat{x} = \tilde{x} - M_{\hat{x}}^{-1} f(\tilde{x}) \in N(\tilde{x}, \mathbf{X}),$$

so that (13.6) implies $f(\hat{x}) = 0$. Furthermore, the root $\hat{x}$ is unique in $\mathbf{X}$ by the non-singularity of all $M \in J_f(\mathbf{X})$. Finally, if $f(y) = 0$ for $y \in \mathbf{X}$, then $-f(\tilde{x}) = M_y(y - \tilde{x})$ by (13.6), so that $y = \tilde{x} - M_y^{-1} f(\tilde{x}) \in N(\tilde{x}, \mathbf{X})$. ☐

To apply Theorem 13.2, first an inclusion $\mathbf{J} \in \mathbb{IF}^{n\times n}$ of $J_f(\mathbf{X})$ is computed by automatic differentiation. Then an inclusion of $\mathbf{\Delta}$ of the solution set $\Sigma(\mathbf{J}, f(\tilde{x}))$ defined in (10.15) is computed by a verification method described in Section 10. Note that this implies in particular that all matrices $M \in J_f(\mathbf{X})$ are non-singular. If $\tilde{x} - \mathbf{\Delta} \subseteq \mathbf{X}$, then Theorem 13.2 implies existence and uniqueness of a root $\hat{x}$ of f in $\mathbf{X}$ (and in $\tilde{x} - \mathbf{\Delta}$).

For a practical implementation, first an approximate solution $\tilde{x}$ should be improved by numerical means. The better $\tilde{x}$, the smaller is the residual $f(\tilde{x})$ and the more likely is an inclusion.

Next an interval refinement $\mathbf{X}^{(k+1)} := N(\tilde{x}, \mathbf{X}^{(k)}) \cap \mathbf{X}^{(k)}$ may be applied, starting with the first inclusion $\mathbf{X}^{(0)} := \mathbf{X}$.

However, this requires at each step the solution of an interval linear system. Therefore, it is in practice often superior to use the following modification (Rump 1983) of an operator given by Krawczyk (1969*a*).

Theorem 13.3. Let there be given continuously differentiable $f: D \to \mathbb{R}^n$ and $\tilde{x} \in \mathbb{R}^n$, $\mathbf{X} \in \mathbb{IR}^n$, $R \in \mathbb{R}^{n\times n}$ with $0 \in \mathbf{X}$ and $\tilde{x} + \mathbf{X} \subseteq D$. Suppose

$$S(\mathbf{X}, \tilde{x}) := -Rf(\tilde{x}) + \{I - RJ_f(\tilde{x} + \mathbf{X})\}\mathbf{X} \subseteq \operatorname{int}(\mathbf{X}). \tag{13.8}$$

Then R and all matrices $M \in J_f(\tilde{x} + \mathbf{X})$ are non-singular, and there is a unique root $\hat{x}$ of f in $\tilde{x} + S(\mathbf{X}, \tilde{x})$.

Proof. Define $g : \mathbf{X} \to \mathbb{R}^n$ by $g(x) := x - Rf(\tilde{x}+x)$. Then, using $\tilde{x} \in \tilde{x}+\mathbf{X}$ and Theorem 13.1 implies that

$$g(x) = x - R\big(f(\tilde{x}) + M_{\tilde{x}+x}x\big) = -Rf(\tilde{x}) + \{I - RM_{\tilde{x}+x}\}x \in S(\mathbf{X}, \tilde{x}) \subseteq \mathbf{X} \tag{13.9}$$

for $x \in \mathbf{X}$. By Brouwer's Fixed-Point Theorem there exists a fixed point $\hat{x} \in \mathbf{X}$ of g, so that $Rf(\tilde{x} + \hat{x}) = 0$. Moreover, $\hat{x} = g(\hat{x}) \in S(\mathbf{X}, \tilde{x})$ by (13.9). Now Lemma 10.5 applied to (13.8) implies that R and every matrix $M \in J_f(\tilde{x} + \mathbf{X})$ is non-singular, and therefore $f(\tilde{x} + \hat{x}) = 0$. The non-singularity of all $M \in J_f(\tilde{x} + \mathbf{X})$ implies that f is injective over $\tilde{x} + \mathbf{X}$. □

Much along the lines of Algorithm 10.7 with the improvements therein, the following algorithm computes an inclusion of a solution of the nonlinear system given by a function `f` near some approximation `xs`.

Algorithm 13.4. Verified bounds for the solution of a nonlinear system:

```
function XX = VerifyNonLinSys(f,xs)
  XX = NaN;                          % initialization
  y = f(gradientinit(xs));
  R = inv(y.dx);                     % approximate inverse of J_f(xs)
  Y = f(gradientinit(intval(xs)));
  Z = -R*Y.x;                        % inclusion of -R*f(xs)
  X = Z; iter = 0;
```

Table 13.1. Solution of (13.10) using MATLAB's `fsolve` and INTLAB's `verifynlss`.

	`fsolve`	`verifynlss`	
Dimension	Median relative error	Maximum relative error	Ratio computing time
50	$5.7 \cdot 10^{-14}$	$6.1 \cdot 10^{-16}$	1.10
100	$5.8 \cdot 10^{-10}$	$6.8 \cdot 10^{-16}$	0.70
200	$8.0 \cdot 10^{-8}$	$5.7 \cdot 10^{-16}$	0.40
500	$2.5 \cdot 10^{-9}$	$5.7 \cdot 10^{-16}$	0.15
1000	$2.1 \cdot 10^{-7}$	$8.4 \cdot 10^{-16}$	0.13
2000	$2.2 \cdot 10^{-8}$	$8.1 \cdot 10^{-16}$	0.11

```
while iter<15
  iter = iter+1;
  Y = hull( X*infsup(0.9,1.1) + 1e-20*infsup(-1,1) , 0 );
  YY = f(gradientinit(xs+Y));     % YY.dx inclusion of J_f(xs+Y)
  X = Z + (eye(n)-R*YY.dx)*Y;     % interval iteration
  if all(in0(X,Y)), XX = xs + X; return; end
end
```

As has been mentioned before, the initial approximation $\tilde{x}$ should be improved by some numerical algorithm. This is included in the algorithm `verifynlss` in INTLAB for solving systems of nonlinear equations. It has been used for the verified solution of the discretized Emden's equation shown in Figure 1.1, and the inclusion of a stationary point of f in (11.6) and (11.7).

The nonlinear function in Algorithm 13.4 may depend on parameters $p \in \mathbb{R}^k$, so that an inclusion $\mathbf{X} \in \mathbb{IR}^n$ of a solution of $f(p,x) = 0$ is computed. If an interval vector $\mathbf{p} \in \mathbb{IR}^k$ is specified for the parameter vector and an inclusion $\mathbf{X}$ is computed, then for all $p \in \mathbf{p}$ there exists a unique solution of $f(p,x) = 0$ in $\mathbf{X}$. This is an upper bound for the sensitivity of the solution with respect to finite perturbations of the parameters. Lower bounds for the sensitivity, *i.e.*, inner inclusions of the solution set as in Section 10.6, can be computed as well: see Rump (1990).

As another example, consider the discretization of

$$3\ddot{y}y + \dot{y}^2 = 0 \quad \text{with} \quad y(0) = 0,\, y(1) = 20, \tag{13.10}$$

given by Abbott and Brent (1975). The true solution is $y = 20x^{0.75}$. As initial approximation we use equally spaced points in $[0, 20]$. The results for different dimensions, comparing `fsolve` of the MATLAB R2009b Optimization toolbox and `verifynlss` of INTLAB, are displayed in Table 13.1.

The third column displays the maximum relative error of the inclusion by `verifynlss`. The inclusions are very accurate, so the relative error of the approximation by `fsolve` can be estimated.

It seems strange that the inclusion is more accurate than the approximation. However, `verifynlss` itself performs a (simplified) Newton iteration, which apparently works better than `fsolve`.

In this particular example the verification is even faster than the approximative routine, except for dimension $n = 50$. This, however, is by no means typical. Note that an inclusion of the solution of the discretized problem (13.10) is computed; verification algorithms for infinite-dimensional problems are discussed in Sections 15 and 16.

The interval Newton method in Theorem 13.2 requires the solution of a linear system with interval matrix. We note that assumption (13.8) of Theorem 13.3 can be verified by solving a linear system with point matrix and interval right-hand side. Define $[mC - rC, mC + rC] := J_f(\tilde{x} + \mathbf{X})$, assume mC is non-singular and $\mathbf{Y}$ is an inclusion of $\Sigma(mC, -f(\tilde{x}) + [-rC, rC]\mathbf{X})$. Then $\mathbf{Y} \subset \operatorname{int}(\mathbf{X})$ implies (13.8).

To see this, set $R := mC^{-1}$ and observe

$$-mC^{-1}f(\tilde{x}) + \{I - mC^{-1}[mC + \Delta]\}x = mC^{-1}\{-f(\tilde{x}) - \Delta x\}$$

for $x \in \mathbb{R}^n$ and $\Delta \in \mathbb{R}^{n \times n}$. Applying this to $x \in \mathbf{X}$ and using $|\Delta| \le rC$ proves the assertion. Note that the non-singularity of mC follows when computing $\mathbf{Y}$ by one of the algorithms in Section 10.

This little derivation is another example of how to prove a result when interval quantities are involved: the assertion is shown for fixed but arbitrary elements out of the interval quantities and the INCLUSION PRINCIPLE (5.17) is used.

Although Algorithm 13.4 works well on some problems, for simple circumstances it is bound to fail. For example, if $\mathbf{X}$ is wide enough that $J_f(\mathbf{X})$ or $J_f(\tilde{x} + \mathbf{X})$ contain a singular matrix, then Theorem 13.2 or Theorem 13.3 is not applicable, respectively, because the non-singularity of every such matrix is proved.

One possibility for attacking this is to use automatic slopes; however, the practical merit of bounding the solution of nonlinear problems seems limited. When interval bounds get wider, slopes and their second-order variants (see Kolev and Mladenov (1997)) become more and more useful.

13.1. Exclusion regions

Verification methods based on Theorem 13.3 prove that a certain interval vector $\mathbf{X} \in \mathbb{IR}^n$ contains a unique root of a nonlinear function $f : \mathbb{R}^n \to \mathbb{R}^n$. A common task, in particular in computer algebra, is to find *all* roots of f within some region. In this case a major problem is to prove that a certain region does *not* contain a root.

Jansson (1994) suggested the following. The box $\mathbf{X}$ is widened into a box $\mathbf{Y} \in \mathbb{IR}^n$ with $\mathbf{X} \subseteq \mathbf{Y}$. If, for example by Theorem 13.3, it can be verified that $\mathbf{Y}$ contains a unique root of f as well, then $f(x) \neq 0$ for all $x \in \mathbf{Y}\backslash\mathbf{X}$. This method proves to be useful for diminishing the so-called cluster effect (Kearfott 1997) in global optimization; see also Neumaier (2004). For verification of componentwise and affine invariant existence, uniqueness, and non-existence regions, see Schichl and Neumaier (2004).

13.2. Multiple roots I

Theorem 13.3 proves that the inclusion interval contains a unique root of the given function. This implies in particular that no inclusion is possible for a multiple root.

There are two ways to deal with a multiple root or a cluster of roots. First, it can be proved that a slightly perturbed problem has a true multiple root, where the size of the perturbation is estimated as well. Second, it can be proved that the original problem has k roots within computed bounds. These may be k simple roots, or one k-fold, or anything in-between. The two possibilities will be discussed in this and the following section.

The proof that the original problem has a k-fold root is possible in exact computations such as computer algebra systems, *e.g.*, Maple (2009), but it is outside the scope of verification methods for $k \geq 2$, by the SOLVABILITY PRINCIPLE OF VERIFICATION METHODS (1.2).

We first consider double roots. In the univariate case, for given $f : \mathbb{R} \to \mathbb{R}$ define

$$g(x, \varepsilon) := \begin{pmatrix} f(x) - \varepsilon \\ f'(x) \end{pmatrix} \quad \text{with Jacobian} \quad J_g(x, \varepsilon) = \begin{pmatrix} f'(x) & -1 \\ f''(x) & 0 \end{pmatrix}. \tag{13.11}$$

Then the function $F(x) := f(x) - \hat{\varepsilon}$ satisfies $F(\hat{x}) = F'(\hat{x}) = 0$ for a root $(\hat{x}, \hat{\varepsilon})^T \in \mathbb{R}^2$ of g. The nonlinear system $g(x, \varepsilon) = 0$ can be solved using Theorem 13.3, evaluating the derivatives by the methods described in Section 11.

In the multivariate case, following Werner and Spence (1984), let a function $f = (f_1, \ldots, f_n) : \mathbb{R}^n \to \mathbb{R}^n$ be given and let $\hat{x} \in \mathbb{R}^n$ be such that $f(\hat{x}) = 0$ and the Jacobian $J_f(\hat{x})$ of f at $\hat{x}$ is singular or almost singular. Adding a smoothing parameter ε, we define $F_\varepsilon : \mathbb{R}^{n+1} \to \mathbb{R}^n$ by

$$F_\varepsilon(x) = f(x) - \varepsilon e_k, \tag{13.12}$$

for fixed k, where e_k denotes the kth column of the $n \times n$ identity matrix. In addition, the Jacobian of f is forced to be singular by

$$J_f(x)y = 0 \tag{13.13}$$

for $y = (y_1, \ldots, y_{m-1}, 1, y_{m+1}, \ldots, y_n)^T$ and some fixed m. This defines a nonlinear system in $2n$ unknowns $g(x,\varepsilon,y)=0$ with

$$g(x,\varepsilon,y)=\begin{pmatrix}F_\varepsilon(x)\\ J_f(x)y\end{pmatrix}, \tag{13.14}$$

and again it can be solved using Theorem 13.3 by evaluating the derivatives using the methods described in Section 11. For $g(\hat{x},\hat{\varepsilon},\hat{y})=0$, it follows that $F_{\hat{\varepsilon}}(\hat{x})=0$ and $\det J_{F_{\hat{\varepsilon}}}(\hat{x})=0$ for the perturbed nonlinear function F_ε with $\varepsilon:=\hat{\varepsilon}$ in (13.12). The problem was also discussed by Kanzawa and Oishi (1999*b*, 1999*a*) and Rump and Graillat (2009), where some heuristics are given on the choice of k and m.

The case of multiple roots of univariate functions will first be explained for a triple root. Let a function $f:\mathbb{R}\to\mathbb{R}$ be given with $f^{(i)}(\tilde{x})\approx 0$ for $0\le i\le 2$ and $f'''(\tilde{x})$ not too small. Assume $\mathbf{X}$ is an inclusion of a root of f'' near $\tilde{x}$ computed by Algorithm 13.4, so that there is $\hat{x}\in\mathbf{X}$ with $f''(\hat{x})=0$. Moreover, $\hat{x}$ is a simple root of f'', so $f'''(\hat{x})\ne 0$.

Compute $\mathbf{E}_0,\mathbf{E}_1\in\mathbb{IR}$ with

$$f'(m)+f''(\mathbf{X})(\mathbf{X}-m)\subseteq\mathbf{E}_0 \quad\text{and}\quad f(m)+f'(\mathbf{X})(\mathbf{X}-m)-\mathbf{E}_0\mathbf{X}\subseteq\mathbf{E}_1 \tag{13.15}$$

for some $m\in\mathbf{X}$. Then $\hat{\varepsilon}_0:=f'(\hat{x})\in\mathbf{E}_0$ and $\hat{\varepsilon}_1:=f(\hat{x})-f'(\hat{x})\hat{x}\in\mathbf{E}_1$, and the function

$$F(x):=f(x)-\hat{\varepsilon}_0x-\hat{\varepsilon}_1 \tag{13.16}$$

satisfies

$$F(\hat{x})=F'(\hat{x})=F''(\hat{x})=0 \quad\text{and}\quad F'''(\hat{x})\ne 0, \tag{13.17}$$

so that $\hat{x}$ is a triple root of the perturbed function F in $\mathbf{X}$.

Following Rump and Graillat (2009), this approach can be applied to compute an inclusion of a k-fold root of a perturbed function.

Theorem 13.5. Let $f:\mathbb{R}\to\mathbb{R}$ with $f\in\mathcal{C}^{k+1}$ be given. Assume $\mathbf{X}\in\mathbb{IR}$ is an inclusion of a root $\hat{x}$ of $f^{(k-1)}$. Let $\mathbf{E}_j\in\mathbb{IR}$ be computed by

$$\mathbf{E}_j=f^{(k-2-j)}(m)+f^{(k-1-j)}(\mathbf{X})(\mathbf{X}-m)-\sum_{\nu=0}^{j-1}\frac{\mathbf{E}_\nu}{(j-\nu)!}\mathbf{X}^{j-\nu} \tag{13.18}$$

for some $m\in\mathbf{X}$. Then, for $0\le j\le k-2$ there exist $\hat{\varepsilon}_j\in\mathbf{E}_j$ with

$$f^{(k-2-j)}(\hat{x})=\hat{\varepsilon}_j+\sum_{\nu=0}^{j-1}\frac{\hat{\varepsilon}_\nu}{(j-\nu)!}\hat{x}^{j-\nu}. \tag{13.19}$$

Define $F:\mathbb{R}\to\mathbb{R}$ by

$$F(x):=f(x)-\sum_{\nu=0}^{k-2}\frac{\hat{\varepsilon}_\nu}{(k-2-\nu)!}\,x^{k-2-\nu}. \tag{13.20}$$

Table 13.2. Expansion of g_5, in total 7403 characters.

```
g_5 =
inline('1+40*sin(x)*cos(x^2/cosh(x)^(1/2))^2*atan(4*x+1)^3
                    ...  ...  ...
-240*cos(x)^2*sin(x^2/cosh(x)^(1/2))^2*cos(x^2/cosh(x)^(1/2))^2
  *sin(x)*atan(4*x+1)')
```

Table 13.3. Inclusions of a k-fold root of perturbed g_k by `verifynlss2` near $\tilde{x} = 0.82$.

k	$\mathbf{X}$	mag($\mathbf{E}_0$)	mag($\mathbf{E}_1$)	mag($\mathbf{E}_2$)	mag($\mathbf{E}_3$)
1	0.81990735694229_3^5				
2	0.819907356694_2^4	$1.28 \cdot 10^{-13}$			
3	0.81990735669_3^6	$1.30 \cdot 10^{-11}$	$1.24 \cdot 10^{-11}$		
4	0.819907358_6^8	$8.85 \cdot 10^{-10}$	$1.61 \cdot 10^{-9}$	$1.93 \cdot 10^{-9}$	
5	failed				

Then $F^{(j)}(\hat{x}) = 0$ for $0 \le j \le k-1$. If the inclusion $\mathbf{X}$ is computed by a verification method based on Theorem 13.3, then the multiplicity of the root $\hat{x}$ of F in $\mathbf{X}$ is exactly k.

The methods described in this section are implemented by `verifynlss2` in INTLAB. As an example, consider our model function g in (8.6). To obtain multiple roots, we define $g_k := g(x)^k$. Using the symbolic toolbox of MATLAB, the symbolic expression

```
expand((sinh(2*x^2/sqrt(cosh(x))-x)-atan(4*x+1)+1)^k)
```

is specified for g_k, that is, the expanded version of $g(x)^k$. For example, part of the expression for g_5 with a total length of the string of 7403 characters is shown in Table 13.2.

The call `verifynlss2(g_k, 0.82, k)` attempts to compute an inclusion $\mathbf{X}$ of a k-fold root of

$$G(x) := g(x) - \epsilon_0 - \epsilon_1 x - \cdots - \epsilon_{k-2} x^{k-2}$$

near $\tilde{x} = 0.82$ and inclusions $\mathbf{E}_\nu$ of the perturbations ϵ_ν for $0 \le \nu \le k-2$. Note that, compared with (13.20), the true coefficients $\epsilon_\nu = \hat{\varepsilon}_\nu (k-2-\nu)!$ of $x^{k-2-\nu}$ are included. Since, by Figure 8.3, there is a root of g near $\tilde{x} = 0.82$, so there is a k-fold root of g_k. The computational results are shown in Table 13.3.

The test functions are based on the function g in (8.6). When using f in (8.4) instead, the results are a little more accurate, and the inclusion of the 5-fold root does not fail. Note that the results are always verified to be correct; failing means that no rigorous statement can be made.

Another approach to multiple roots of nonlinear equations going beyond Theorem 13.5 was presented by Kearfott, Dian and Neumaier (2000). They derive an efficient way to compute the topological degree of nonlinear functions that can be extended to an analytic function (or to a function that can be approximated by an analytic function).

13.3. Multiple roots II

The second kind of verification method for a multiple or a cluster of roots is to verify that, counting multiplicities, there are exactly k roots within some interval. Note that this does not contradict the SOLVABILITY PRINCIPLE OF VERIFICATION METHODS (1.2).

Neumaier (1988) gives a sufficient criterion for the univariate case. He shows that if

$$\left|\mathrm{Re}\,\frac{f^{(k)}(z)}{k!}\right| > \sum_{i=0}^{k-1}\left|\frac{f^{(i)}(\tilde{z})}{i!}\right| r^{i-k} \tag{13.21}$$

is satisfied for all z in the disc $D(\tilde{z}, r)$, then f has exactly k roots in D.

We now discuss the approach by Rump and Oishi (2009). They showed how to omit the $(k-1)$st summand on the right of (13.21), and computed sharper expressions for the left-hand side. In particular, they gave a constructive scheme for how to find a suitable disc D.

Let a function $f : D_0 \to \mathbb{C}$ be given, which is analytic in the open disc D_0. Suppose some $\tilde{x} \in D_0$ is given such that $\tilde{x}$ is a numerically k-fold zero, *i.e.*,

$$f^{(\nu)}(\tilde{x}) \approx 0 \quad \text{for} \quad 0 \le \nu < k \tag{13.22}$$

and $f^{(k)}(\tilde{x})$ not too small. Note that the latter is not a mathematical assumption to be verified. For $z, \tilde{z} \in D_0$, denote the Taylor expansion by

$$f(z) = \sum_{\nu=0}^{\infty} c_\nu (z-\tilde{z})^\nu \quad \text{with} \quad c_\nu = \frac{1}{\nu!} f^{(\nu)}(\tilde{z}). \tag{13.23}$$

Let $X \subset D_0$ denote a complex closed disc near $\tilde{x}$ such that $f^{(k-1)}(\hat{x}) = 0$ for some $\hat{x} \in X$. It can be computed, for example, by `verifynlss` in INTLAB applied to $f^{(k-1)}(x) = 0$.

We aim to prove that some closed disc $Y \subset D_0$ with $X \subseteq Y$ contains exactly k roots of f. First f is expanded with respect to $\hat{x}$ and the series is split into

$$f(y) = q(y) + g(y)(y-\hat{x})^k \quad \text{and} \quad g(y) = c_k + e(y) \tag{13.24}$$

with

$$q(y) = \sum_{\nu=0}^{k-2} c_\nu (y-\hat{x})^\nu \quad \text{and} \quad e(y) = \sum_{\nu=k+1}^{\infty} c_\nu (y-\hat{x})^{\nu-k}. \tag{13.25}$$

Note that g is holomorphic in D_0, and that $c_{k-1} = 0$ by assumption. The minimum of $|g(y)|$ on Y can be estimated by the maximum of the remainder term $|e(y)|$. This is possible by the following, apparently not so well-known version of a complex Mean Value Theorem due to Darboux (1876).[23]

Theorem 13.6. Let holomorphic $f : D_0 \to \mathbb{C}$ in the open disc D_0 be given and $a, b \in D_0$. Then, for $1 \le p \le k+1$ there exists $0 \le \Theta \le 1$ and $\omega \in \mathbb{C}$, $|\omega| \le 1$ such that, for $h := b-a$ and $\xi := a + \Theta(b-a)$,

$$f(b) = \sum_{\nu=0}^{k} \frac{h^\nu}{\nu!} f^{(\nu)}(a) + \omega \frac{h^{k+1}}{k!} \frac{(1-\Theta)^{k-p+1}}{p} f^{(k+1)}(\xi). \tag{13.26}$$

The following proof is due to Bünger (2008).

Proof. Set $\ell := |b-a|$, which is non-zero without loss of generality, and define a function $g : [0,\ell] \to a \underline{\cup} b$ by $g(t) := a + t\frac{b-a}{\ell}$. Then

$$|g'(t)| = \frac{|b-a|}{\ell} \equiv 1.$$

For

$$F(x) := \sum_{\nu=0}^{k} \frac{(b-x)^\nu}{\nu!} f^{(\nu)}(x),$$

this means

$$\begin{aligned} F'(x) &= f'(x) + \sum_{\nu=1}^{k} -\frac{(b-x)^{\nu-1}}{(\nu-1)!} f^{(\nu)}(x) + \frac{(b-x)^\nu}{\nu!} f^{(\nu+1)}(x) \\ &= \frac{(b-x)^k}{k!} f^{(k+1)}(x). \end{aligned}$$

With this we use $|g'(t)| \equiv 1$ and $|b - g(t)| = \ell - t$ to obtain

$$\begin{aligned} |F(b) - F(a)| = |F(g(\ell)) - F(g(0))| &= \left| \int_0^\ell (F \circ g)'(t)\, \mathrm{dt} \right| \\ &\le \int_0^\ell |F'(g(t))|\, |g'(t)|\, \mathrm{dt} = \int_0^\ell \left| \frac{|b-g(t)|^k}{k!} \right| |f^{(k+1)}(g(t))|\, \mathrm{dt} \\ &= \int_0^\ell \frac{(\ell-t)^k}{k!p(\ell-t)^{p-1}} |f^{(k+1)}(g(t))| p(\ell-t)^{p-1}\, \mathrm{dt} \end{aligned}$$

[23] Thanks to Prashant Batra for pointing this out.

$$\begin{aligned}&\le \frac{(\ell-t^*)^{k-p+1}}{k!p}|f^{(k+1)}(g(t^*))|\int_0^\ell (-(\ell-t)^p)'\,\mathrm{d}t\\&=\frac{(\ell-t^*)^{k-p+1}\ell^p}{k!p}|f^{(k+1)}(g(t^*))|\end{aligned}$$

for $1 \le p \le k+1$ and some $t^* \in [0,\ell]$. For $\Theta := \frac{t^*}{\ell} \in [0,1]$, the last expression is equal to

$$\frac{\ell^{k+1}(1-\Theta)^{k-p+1}}{k!p}|f^{(k+1)}(a+\Theta(b-a))|,$$

so that there exists complex ω with $|\omega| \le 1$ and

$$\begin{aligned}f(b)-f(a)-\sum_{\nu=1}^{k}\frac{(b-a)^\nu}{\nu!}f^{(\nu)}(a)&=F(b)-F(a)\\&=-\omega\frac{(b-a)^{k+1}}{k!}\frac{(1-\Theta)^{k-p+1}}{p}|f^{(k+1)}(a+\Theta(b-a))|.\end{aligned}$$ □

Using Taylor coefficients as described in Section 11.4, an inclusion of $c_k = \frac{1}{k!}f^{(k)}(\hat{x})$ can be evaluated, and with Theorem 13.6 the remainder term $e(y)$ in (13.24) can be estimated as well.

Note that there is some freedom to choose p. The choice $p = k+1$ gives the traditional-looking expansion

$$f(b)=\sum_{\nu=0}^{k}\frac{h^\nu}{\nu!}f^{(\nu)}(a)+\omega\frac{h^{k+1}}{(k+1)!}f^{(k+1)}(\xi)\quad\text{with}\quad |\omega|\le 1,$$

so that

$$|e(y)| \le \frac{|b-a|}{(k+1)!}\max_{z\in\partial Y}|f^{(k+1)}(z)| \quad \forall\, y \in Y. \tag{13.27}$$

For $p = k$ the interval for Θ may be split to obtain $|e(y)| \le \max(\beta_1, \beta_2)$ with

$$\beta_1 := \frac{|b-a|}{k!}\max_{|y-\hat{x}|\le\frac{r}{2}}|f^{(k+1)}(y)| \tag{13.28}$$

and

$$\beta_2 := \frac{|b-a|}{2k!}\max_{|y-\hat{x}|\le r}|f^{(k+1)}(y)|, \tag{13.29}$$

where $r := \max_{y\in Y}|y-\hat{x}|$. By the definition (13.24) this gives a computable lower bound for $|g(y)|$.

Let a polynomial $P(z) \in \mathbb{C}[z]$ with $P(z) = \sum_{\nu=0}^{n} p_\nu z^\nu$ be given with $p_n \ne 0$. The Cauchy polynomial $C(P)$ with respect to P is defined by $C(P) := |p_n x^n| - \sum_{\nu=0}^{n-1}|p_\nu|x^\nu \in \mathbb{R}[x]$. By Descartes' rule of signs, $C(P)$ has exactly one non-negative root, called the Cauchy bound $\overline{C(P)}$. It is well known that the Cauchy bound is an upper bound for the absolute value of all (real and complex) roots of P:

$$P(z) = 0 \;\Rightarrow\; |z| \le \overline{C(P)}. \tag{13.30}$$

In fact, it is the best upper bound taking only the absolute values $|p_\nu|$ into account. Note that the leading coefficient p_n must be non-zero.

The Cauchy bound can be defined for interval polynomials as well. For $\mathfrak{P}(z) \in \mathbb{IK}[z]$ with $\mathfrak{P}(z) = \sum_{\nu=0}^{n} \mathfrak{p}_\nu z^\nu$, $\mathfrak{p}_\nu \in \mathbb{IK}$ and $\mathbb{K} \in \{\mathbb{R}, \mathbb{C}\}$, define

$$C(\mathfrak{P}) := \operatorname{mig}(\mathfrak{p}_n) x^n - \sum_{\nu=0}^{n-1} \operatorname{mag}(\mathfrak{p}_\nu) x^\nu \in \mathbb{R}[x], \tag{13.31}$$

where $\operatorname{mig}(\mathfrak{p}_n) := \min\{|\pi| : \pi \in \mathfrak{p}_n\}$ and $\operatorname{mag}(\mathfrak{p}_\nu) := \max\{|\pi| : \pi \in \mathfrak{p}_\nu\}$. Then the unique non-negative root $\overline{C(\mathfrak{P})}$ of $C(\mathfrak{P})$ is a root bound for all polynomials $P \in \mathfrak{P}$:

$$P \in \mathfrak{P} \;\text{ and }\; P(z) = 0 \;\Rightarrow\; |z| \le \overline{C(\mathfrak{P})}. \tag{13.32}$$

The Cauchy bound for real or complex interval polynomials is easily upper-bounded by applying a few Newton iterations on $C(\mathfrak{P})$ starting at some other traditional root bound. Note that the iteration converges quickly to $\overline{C(\mathfrak{P})}$.

Theorem 13.7. Let holomorphic $f : D_0 \to \mathbb{C}$ in the open disc D_0 and fixed $k \in \mathbb{N}$ be given, and closed discs $\mathbf{X}, \mathbf{Y} \subset D_0$ with $\mathbf{X} \subseteq \mathbf{Y}$. Assume there exists $\hat{x} \in \mathbf{X}$ with $f^{(k-1)}(\hat{x}) = 0$. Define $g(y)$ as in (13.24), and let $\mathbf{G} \in \mathbb{IC}$ be a complex disc with $g(y) \in \mathbf{G}$ for all $y \in Y$. Assume $0 \notin \mathbf{G}$, and define the interval polynomial

$$\mathfrak{P}(z) := q(z) + \mathbf{G} \cdot (z - \hat{x})^k \in \mathbb{IC}[z]. \tag{13.33}$$

Denote the closed complex disc with centre m and radius r by $D(m; r)$. Assume that the Cauchy bound $\overline{C(\mathfrak{P})}$ for $\mathfrak{P}$ satisfies

$$D(\hat{x}\,; \overline{C(\mathfrak{P})}) \subset \operatorname{int}(Y). \tag{13.34}$$

Then, counting multiplicities, there are exactly k roots of the function f in $D(\hat{x}\,; \overline{C(\mathfrak{P})})$.

Proof. Define the parametrized set of polynomials

$$P_y(z) := q(z) + g(y)(z - \hat{x})^k \in \mathbb{C}[z]. \tag{13.35}$$

Note that only the leading term depends on the parameter y. By definition (13.24) we have $f(y) = P_y(y)$. Moreover, $P_y \in \mathfrak{P}$ for all $y \in Y$, so that $g(y) \ne 0$ and (13.32) imply that $P_y(z) = 0$ is only possible for $z \in D(\hat{x}\,; \overline{C(\mathfrak{P})})$. Thus (13.34) implies for all $y \in Y$ that $P_y(z) \ne 0$ for all $z \in \partial Y$. Next define

$$P_{y,t}(z) := t \cdot q(z) + g(y)(z - \hat{x})^k \tag{13.36}$$

and the homotopy function

$$h_t(y) := P_{y,t}(y) = t \cdot q(y) + g(y)(y - \hat{x})^k. \tag{13.37}$$

Since q is a polynomial and g is holomorphic, all functions h_t are holomorphic as well. The definition of the Cauchy bound implies

$$\overline{C(P_{y,t})} \le \overline{C(P_y)} \le \overline{C(\mathfrak{P})} \tag{13.38}$$

for all $t \in [0,1]$ and all $y \in Y$. Thus definition (13.37) implies that for all $t \in [0,1]$ we have $h_t(y) \neq 0$ for all $y \in \partial Y$. We conclude that all holomorphic functions h_t must have the same number of roots in Y, in particular h_0 and h_1.

For $t = 0$ we have $h_0(y) = g(y)(y - \hat{x})^k$, which has exactly k roots in Y because $g(y) \neq 0$ for all $y \in Y$. Hence

$$h_1(y) = q(y) + g(y)(y - \hat{x})^k = P_y(y) = f(y)$$

must have exactly k roots in Y. By (13.38), for all $t \in [0,1]$ and all $y \in Y$, all roots of $P_{y,t}(z)$ lie in $D(\hat{x}\,;\overline{C(\mathfrak{P})})$, so in particular the roots of f. This concludes the proof. □

For the applicability of Theorem 13.7 in a verification method, note that the quality of the bound depends directly on the lower bound on $|g(\mathbf{Y})|$, which means by (13.24) on the lower bound of $c_k = \frac{1}{k!}f^{(k)}(\hat{x}) \in \frac{1}{k!}f^{(k)}(\mathbf{X})$.

The direct computation of an inclusion of $\frac{1}{k!}f^{(k)}(\mathbf{X})$ by interval arithmetic can be improved using the centred form

$$c_k \in \frac{1}{k!}f^{(k)}(\tilde{x}) + \frac{1}{(k+1)!}f^{(k+1)}(\mathbf{X}) \cdot (\mathbf{X} - \tilde{x}) \tag{13.39}$$

for $\tilde{x} \in \mathbf{X}$. A suitable choice is some $\tilde{x}$ near the midpoint of $\mathbf{X}$.

The problem remains to find a suitable inclusion interval Y. Note that the inclusion interval is necessarily complex: if the assumptions of Theorem 13.7 are satisfied for some function f, they are by continuity satisfied for a suitably small perturbation of f as well. But an arbitrary small perturbation of f may move a double real root into two complex roots. This is another example of the SOLVABILITY PRINCIPLE OF VERIFICATION METHODS (1.2).

Since $\hat{x} \in \mathbf{X}$ is necessary by assumption, a starting interval may be $\mathbf{Y}^0 := \mathbf{X}$. However, the sensitivity of a k-fold root is $\varepsilon^{1/k}$ for an ε-perturbation of the coefficients (see below). But the quality of the inclusion X of the *simple* root of $f^{(k-1)}$ can be expected to be nearly machine precision.

The polynomial in (13.33), $\mathfrak{P}_{\mathbf{Y}}$, say, depends on $\mathbf{Y}$. The main condition to check is (13.34). Thus a suitable candidate for a first inclusion interval is $\mathbf{Y}^1 := D(\hat{x}\,;\overline{C(\mathfrak{P}_{\mathbf{Y}^0})})$. This already defines an iteration scheme, where $\mathbf{Y}^{m+1} \subset \mathrm{int}(\mathbf{Y}^m)$ verifies the conditions of Theorem 13.7. Equipped with an epsilon-inflation as in (10.13), this is a suitable verification method for

Table 13.4. Inclusions of k roots of the original function g_k by `verifynlss2` near $\tilde{x} = 0.82$ and sensitivity of the root.

k	$\mathfrak{Re}(\mathbf{X})$	$\mathrm{rad}(\mathbf{X})$	Sensitivity
1	0.81990735694429_{3}^{5}	$6.32 \cdot 10^{-15}$	$2.14 \cdot 10^{-16}$
2	0.819907_{0}^{7}	$2.68 \cdot 10^{-7}$	$2.53 \cdot 10^{-8}$
3	0.819_{820}^{995}	$8.71 \cdot 10^{-5}$	$1.25 \cdot 10^{-5}$
4	0.8_{18}^{22}	$1.54 \cdot 10^{-3}$	$2.95 \cdot 10^{-4}$
5	failed		$1.96 \cdot 10^{-3}$

the inclusion of k roots of a univariate nonlinear function. It is implemented in `verifynlss2` in INTLAB.

For computational results we use the same function as in the previous section, namely $g_k := g(x)^k$ for our model function g in (8.6). Again the expanded function as in Table 13.2 is used. The results are displayed in Table 13.4.

Note that by construction, and according to the SOLVABILITY PRINCIPLE (1.2), the inclusion is a complex disc. In all examples in Table 13.4 the inclusion was the smallest disc including $\mathfrak{Re}(\mathbf{X})$, *i.e.*, the imaginary part of the midpoint was zero.

Although the inclusions in Table 13.4 may appear wide, we show that they are, when computing in double precision, almost best possible. Let analytic f be given with k-fold root $\hat{z}$. For given ϵ define $\tilde{f}(z) := f(z) - \epsilon$. By continuity, for sufficiently small ϵ there exists small h with $\tilde{f}(\hat{z}+h) = 0$, so that

$$0 = -\epsilon + c_k h^k + \mathcal{O}(h^{k+1}), \tag{13.40}$$

using the Taylor expansion $f(\hat{z}+h) = \sum c_\nu f^{(\nu)}(\hat{z})h^\nu$. Thus h, the sensitivity of the k-fold root $\hat{z}$, is of the order $(\epsilon/c_k)^{1/k}$ for small ϵ.

Let a function f be given by an arithmetic expression, such as g_5 in Table 13.2. As a course analysis, $f(\tilde{x})$ is evaluated as a floating-point sum $\mathrm{fl}(t_1 + \cdots + t_m)$ of some $t_\mu \in \mathbb{F}$, where $t_\mu := \mathrm{fl}(T_\mu(\tilde{x}))$ is the result of the floating-point evaluation of (possibly large) terms $T_\mu(\tilde{x})$.

Due to rounding errors, the accuracy of $\mathrm{fl}(t_1 + \cdots + t_m)$ is at best $\epsilon := \mathbf{u} \cdot \max_\mu |t_\mu|$ for the relative rounding error unit $\mathbf{u} = 2^{-53}$, ignoring possible sources of errors in the evaluation of the summands $T_\mu(\tilde{x})$.

Thus the inevitable presence of rounding errors creates the sensitivity $h = (\mathbf{u} \cdot \max_\mu |t_\mu|/c_k)^{1/k}$ of a k-fold root. This sensitivity is shown in Table 13.4, and it is not far from the radius of the inclusion. The analysis is confirmed by the floating-point evaluation of g_5 over $[0.816, 0.824]$, as shown in Figure 13.1.

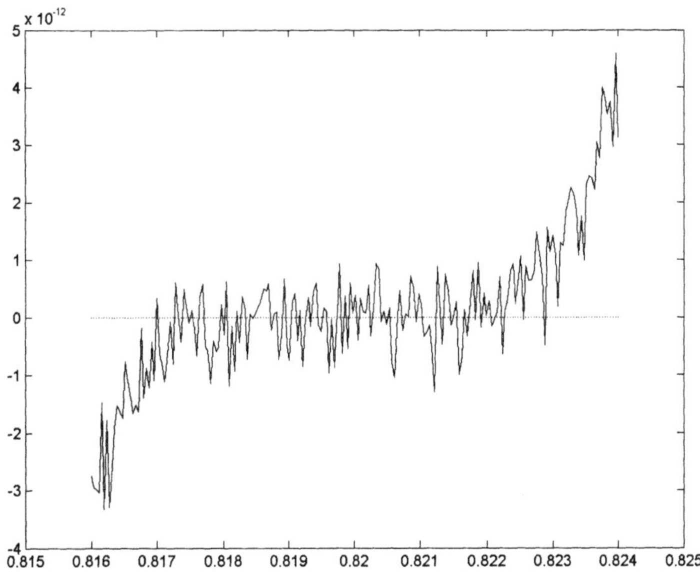

Figure 13.1. Floating-point evaluation of g_5 as in Table 13.2 on 200 equidistant mesh points in [0.816,0.824].

13.4. Simple and multiple eigenvalues

The methods discussed so far can be applied to a variety of particular problems. For example, Theorem 13.3 and Algorithm 13.4 imply a verification algorithm for simple roots of polynomials. For simple roots this is efficient where, of course, the derivative can be computed directly.

For multiple roots of polynomials, Neumaier (2003) and Rump (2003*a*) present a number of specific methods taking advantage of the special structure of the problem.

An eigenvector/eigenvalue pair (x, λ) can be written as a solution of the nonlinear system $Ax - \lambda x = 0$ together with some normalization: see Krawczyk (1969*b*). Again, multiple eigenvalues can be treated by the general approach as in the last section, but it is superior to take advantage of the structure of the problem. The approach in Rump (2001*b*) for non-self-adjoint matrices, to be described in the following, is a further example of how to develop verification methods. We mention that the methods apply *mutatis mutandis* to the generalized eigenvalue problem $Ax = \lambda Bx$.

Another example of the SOLVABILITY PRINCIPLE (1.2) is the following. Suppose $\mathbf{\Lambda} \in \mathbb{IR}$ and $\mathbf{X} \in \mathbb{IR}^k$ have been calculated by a verification algorithm such that $\mathbf{\Lambda}$ contains k not necessarily distinct eigenvalues $\lambda_1, \ldots, \lambda_k$ of a matrix $A \in \mathbb{R}^{n \times n}$, and the columns of $\mathbf{X}$ contain an inclusion of basis vectors of the corresponding invariant subspace. Then Rump and Zemke (2004) show under plausible assumptions that all eigenvalues λ_ν must have

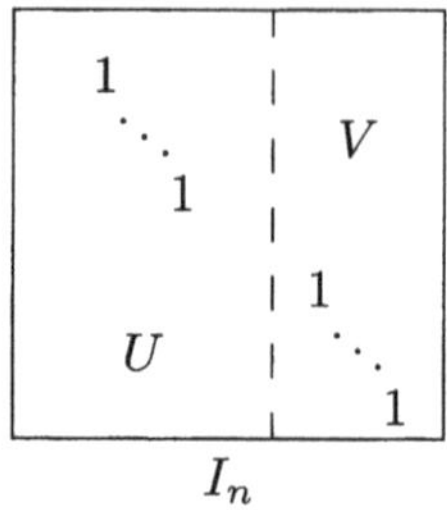

Figure 13.2. Partition of the identity matrix.

geometric multiplicity 1, *i.e.*, have, up to normalization, a unique eigenvector. This is even true for symmetric A. The reason is that for an eigenvalue λ of A of geometric multiplicity $m > 1$, for any eigenvector x in the m-dimensional eigenspace there exists an arbitrarily small perturbation $\tilde{A}$ of A such that λ is a simple eigenvalue of $\tilde{A}$ and x is (up to normalization) the unique eigenvector.

For $A \in \mathbb{K}^{n\times n}$, $\mathbb{K} \in \{\mathbb{R}, \mathbb{C}\}$, let $\tilde{X} \in \mathbb{K}^{n\times k}$ be an approximation to a k-dimensional invariant subspace corresponding to a multiple or a cluster of eigenvalues near some $\tilde{\lambda} \in \mathbb{K}$, such that $A\tilde{X} \approx \tilde{\lambda}\tilde{X}$. As always, there are no *a priori* assumptions on the quality of the approximations $\tilde{X}$ and $\tilde{\lambda}$.

The degree of arbitrariness is removed by freezing k rows of the approximation $\tilde{X}$. If the set of these rows is denoted by v, and by definition $u := \{1, \ldots, n\}\backslash v$, then denote by $U \in \mathbb{R}^{n\times(n-k)}$ the submatrix of the identity matrix with columns in u. Correspondingly, define $V \in \mathbb{R}^{n\times k}$ to comprise of the columns in v out of the identity matrix. Denoting the $n \times n$ identity matrix by I_n, we have $UU^T + VV^T = I_n$, and $V^T\tilde{X} \in \mathbb{K}^{k\times k}$ is the normalizing part of $\tilde{X}$. Note that $U^TU = I_{n-k}$ and $V^TV = I_k$. For example, for $u = \{1, \ldots, n-k\}$, $v = \{n-k+1, \ldots, n\}$ the situation is as in Figure 13.2.

For given $\tilde{X} \in \mathbb{K}^{n\times k}$ and $\tilde{\lambda} \in \mathbb{K}$, suppose

$$AY = YM \quad \text{for } Y \in \mathbb{K}^{n\times k},\ M \in \mathbb{K}^{k\times k}, \tag{13.41}$$

such that Y and $\tilde{X}$ coincide in the normalizing part of $\tilde{X}$: $V^TY = V^T\tilde{X}$. The unknown quantities U^TY and M are collected into $\hat{X} \in \mathbb{K}^{n\times k}$. In other words, $\hat{X}$ will be computed with $U^T\hat{X} = U^TY$ and $V^T\hat{X} = M$. Note that M is not assumed to be diagonal. For $u = \{1, \ldots, n-k\}, v = \{n-k+1, \ldots, n\}$ the situation is as in Figure 13.3. This implies the eigen-equation

$$A(UU^T\hat{X} + VV^T\tilde{X}) = (UU^T\hat{X} + VV^T\tilde{X})V^T\hat{X}, \tag{13.42}$$

such that, according to (13.41), $Y = UU^T\hat{X} + VV^T\tilde{X}$ and $M = V^T\hat{X}$. It can be shown that the following algorithm, Algorithm 13.8, converges quadratically under reasonable conditions.

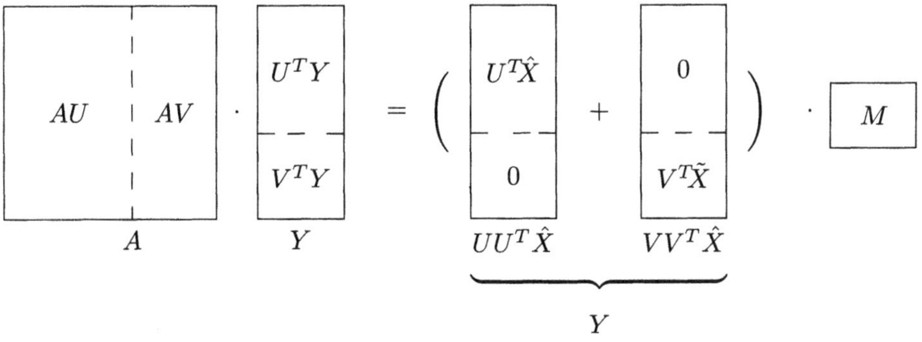

$$\underbrace{\left(\begin{array}{c|c} AU & AV \end{array}\right)}_{A} \cdot \underbrace{\begin{pmatrix} U^TY \\ \hline V^TY \end{pmatrix}}_{Y} = \underbrace{\left(\underbrace{\begin{pmatrix} U^T\hat{X} \\ \hline 0 \end{pmatrix}}_{UU^T\hat{X}} + \underbrace{\begin{pmatrix} 0 \\ \hline V^T\tilde{X} \end{pmatrix}}_{VV^T\hat{X}} \right)}_{Y} \cdot \boxed{M}$$

Figure 13.3. Nonlinear system for multiple or clusters of eigenvalues.

Algorithm 13.8. Newton-like iteration for eigenvalue clusters:

$X^0 := UU^T\tilde{X} + \tilde{\lambda}V$
for $\nu = 0, 1, \ldots$
  $\lambda_\nu := \mathrm{trace}(V^TX^\nu)/k$
  $C_\nu := (A - \lambda_\nu I_n)UU^T - (UU^TX^\nu + VV^T\tilde{X})V^T$
  $X^{\nu+1} := UU^TX^\nu + \lambda_\nu V - C_\nu^{-1} \cdot (A - \lambda_\nu I_n)(UU^TX^\nu + VV^T\tilde{X})$

Note that the scalar λ_ν is an approximation to the cluster, and is adjusted in every iteration to be the mean value of the eigenvalues of V^TX^ν. This iteration is the basis for the following verification method.

Theorem 13.9. Let $A \in \mathbb{K}^{n\times n}, \tilde{X} \in \mathbb{K}^{n\times k}, \tilde{\lambda} \in \mathbb{K}, R \in \mathbb{K}^{n\times n}$ and $X \in \mathbb{K}^{n\times k}$ be given, and let U, V partition the identity matrix as defined in Figure 13.2. Define

$$f(X) := -R(A\tilde{X} - \tilde{\lambda}\tilde{X}) + \{I - R((A - \tilde{\lambda}I)UU^T - (\tilde{X} + UU^T \cdot X)V^T)\} \cdot X. \tag{13.43}$$

Suppose

$$f(\mathbf{X}) \subseteq \mathrm{int}(\mathbf{X}). \tag{13.44}$$

Then there exists $\hat{M} \in \mathbb{K}^{k\times k}$ with $\hat{M} \in \tilde{\lambda}I_k + V^T\mathbf{X}$ such that the Jordan canonical form of $\hat{M}$ is identical to a $k \times k$ principal submatrix of the Jordan canonical form of A, and there exists $\hat{Y} \in \mathbb{K}^{n\times k}$ with $\hat{Y} \in \tilde{X} + UU^T\mathbf{X}$ such that $\hat{Y}$ spans the corresponding invariant subspace of A. We have $A\hat{Y} = \hat{Y}\hat{M}$.

Proof. The continuous mapping $f : \mathbb{K}^n \to \mathbb{K}^n$ defined by (13.43) maps by (13.44) the non-empty, convex and compact set $\mathbf{X}$ into itself. Therefore, Brouwer's Fixed-Point Theorem implies existence of a fixed point $\hat{X} \in \mathbb{K}^n$ with $f(\hat{X}) = \hat{X}$ and $\hat{X} \in \mathbf{X}$. Inserting in (13.43) yields

$$-R\{(A\tilde{X} - \tilde{\lambda}\tilde{X}) + (A - \tilde{\lambda}I)UU^T\hat{X} - (\tilde{X} + UU^T\hat{X})V^T\hat{X}\} = 0. \tag{13.45}$$

Furthermore, (13.43), (13.44) and Lemma 10.5 imply R and every matrix within $\mathbf{B} := (A - \tilde{\lambda}I)UU^T - (\tilde{X} + UU^T \cdot \mathbf{X})V^T \in \mathbb{IK}^{n \times n}$ to be non-singular. Collecting terms in (13.45) yields

$$A(\tilde{X} + UU^T\hat{X}) = (\tilde{X} + UU^T\hat{X})(\tilde{\lambda}I_k + V^T\hat{X})$$

or

$$A\hat{Y} = \hat{Y}\hat{M} \quad \text{for} \quad \hat{Y} := \tilde{X} + UU^T\hat{X} \quad \text{and} \quad \hat{M} := \tilde{\lambda}I_k + V^T\hat{X}.$$

Finally, $(A - \tilde{\lambda}I)UU^T - (\tilde{X} + UU^T\hat{X})V^T \in \mathbf{B}$ is non-singular and has k columns equal to $-\hat{Y}$. Therefore, $\hat{Y}$ has full rank and is a basis for an invariant subspace of A. For $\hat{M} = ZJZ^{-1}$ denoting the Jordan canonical form, $A\hat{Y} = \hat{Y}\hat{M}$ implies $A(\hat{Y}Z) = (\hat{Y}Z)J$. The theorem is proved. □

Note that Theorem 13.9 is applicable for $k = 1, \ldots, n$. For $k = 1$ we have the usual eigenvalue/eigenvector inclusion, basically corresponding to the application of Theorem 13.3 to $Ax - \lambda x = 0$, freezing some component of x. For $k = n$ the maximum spectral radius of $\tilde{\lambda}I + X$, $X \in \mathbf{X}$ is an inclusion of all eigenvalues.

For a practical implementation, $\tilde{X}$ and $\tilde{\lambda}$ are such that $A\tilde{X} \approx \tilde{\lambda}\tilde{X}$, and the matrix R serves as a preconditioner, with (13.43) indicating the obvious choice

$$R \approx \left((A - \tilde{\lambda}I)UU^T - \tilde{X}V^T\right)^{-1}.$$

As before, Theorem 13.9 computes an inclusion $\mathbf{X}$ for the error with respect to $\tilde{\lambda}$ and $\tilde{X}$. For an interval iteration with epsilon-inflation as in (10.13), an initial choice for $\mathbf{X}$ is a small superset of the correction term $-R(A\tilde{X} - \tilde{\lambda}\tilde{X})$.

It remains to compute an inclusion of the eigenvalue cluster, that is, an inclusion of the eigenvalues of $\hat{M}$. Using Gershgorin circles of $\tilde{\lambda}I_k + V^T\mathbf{X}$ would yield quite pessimistic bounds for defective eigenvalues.

For an interval matrix $\mathbf{C} \in \mathbb{K}^{k \times k}$, denote by $|\mathbf{C}| \in \mathbb{R}^{k \times k}$ the matrix of the entrywise maximum modulus of $\mathbf{C}$. Therefore, $|C_{ij}| \le (|\mathbf{C}|)_{ij}$ for every $C \in \mathbf{C}$. Then,

$$\begin{aligned} &\text{for } \; r := \varrho(|V^T\mathbf{X}|) \; \text{ there are } k \text{ eigenvalues of } A \\ &\text{in } \; U_r(\tilde{\lambda}) := \{z \in \mathbb{C} : |z - \tilde{\lambda}| \le r\}, \end{aligned} \tag{13.46}$$

where ϱ denotes the spectral radius, in this case the Perron root of $|V^T\mathbf{X}| \in \mathbb{R}^{k \times k}$, which can be estimated as in (10.61). As a matter of principle, the inclusion is complex.

To see (13.46), observe that for $\hat{M} = \tilde{\lambda}I_k + \tilde{M}$, for some $\tilde{M} \in V^T\mathbf{X}$, the eigenvalues of $\hat{M}$ are the eigenvalues of $\tilde{M}$ shifted by $\tilde{\lambda}$, and for any eigenvalue of μ of $\tilde{M}$, Perron–Frobenius theory implies $|\mu| \le \varrho(\tilde{M}) \le \varrho(|\tilde{M}|) \le \varrho(|V^T\mathbf{X}|) = r$. Using (13.46) is especially advantageous for defective eigenvalues.

The matrix $V^T\mathbf{X}$ basically contains error terms except for large off-diagonal quantities characterizing the Jordan blocks. If the error terms are of size ε and off-diagonal elements of size 1, the spectral radius of $|V^T\mathbf{X}|$ is of size $\varepsilon^{1/m}$, where m is the size of the largest Jordan block. Therefore, the radius of the inclusion is of size $\varepsilon^{1/m}$, which corresponds to the sensitivity of defective eigenvalues given by Wilkinson (1965, p. 81).

In turn, this implies that if the distance of an m-fold defective eigenvalue to the rest of the spectrum is of the order $\varepsilon^{1/m}$, then 'numerically' the cluster comprises at least $m+1$ eigenvalues, and for $k = m$ no inclusion is possible.

For the same reason the quality of a numerical algorithm to approximate an eigenvalue corresponding to a $k \times k$ Jordan block will be no better than $\varepsilon^{1/k}$. We demonstrate this with the following example.

Algorithm 13.10. Computing approximations and inclusions of eigenvalues of `A`:

```
n = 8; C = triu(tril(ones(n),1));
H = hadamard(n); A = H'*C*H/8,
[V,D] = eig(A);
e = diag(D), close, plot(real(e),imag(e),'*')
[L,X] = verifyeig(A,mean(e),V)
```

The matrix `C` is unit upper triangular with ones on the first upper diagonal, *i.e.*, one Jordan block to $\lambda = 1$. MATLAB seems to check for triangular matrices because `eig(C)` produces a vector of ones.

The 8×8 Hadamard matrix has integer entries, and $8^{-1}H^TH = I$, also in floating-point arithmetic for $8 = 2^3$. Hence `A` has the same Jordan structure as `C`, *i.e.*, one 8×8 Jordan block to the 8-fold eigenvalue 1.

The computed eigenvalues (in the diagonal of `D`) are shown as asterisks in Figure 13.4. Basically, they are on a circle around 1 with radius 0.1. MATLAB produces those values without error message.

The final statement in Algorithm 13.10 computes the inclusion $\mathcal{C} := \{z \in \mathbb{C} : |z - 1.0002| \leq 0.0355\}$ based on the mean of the eigenvalue approximations and the approximation `V` of the corresponding invariant subspace. Note that `V` is numerically singular because there is only one eigenvector to the 8-fold eigenvalue 1.

Since $\varepsilon^{1/8} = 0.0101$, the quality of $\mathcal{C}$ corresponds to the sensitivity of the 8-fold eigenvalue. Note that the floating-point approximations of the eigenvalue, which may be regarded as of poor quality, represent, in fact, a backward stable result: there is a small perturbation of the input matrix with true eigenvalues near the computed approximations.

It is proved that $\mathcal{C}$ contains 8 eigenvalues of `A`, and the analysis shows that there is not much room for improvement. We note that for $n = 16$ the same approach fails; no inclusion is computed.

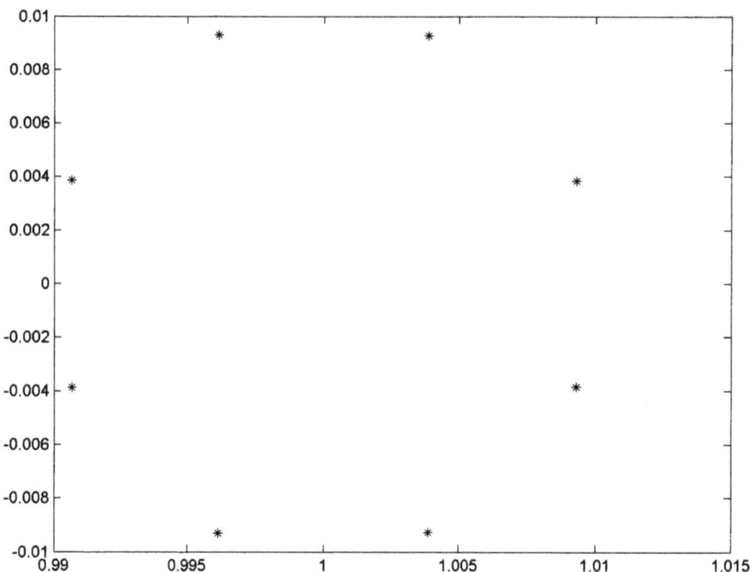

Figure 13.4. The computed eigenvalue approximations for A as in Algorithm 13.10.

Applying Theorem 13.9 to an interval matrix $\mathbf{A} \in \mathbb{K}^{n\times n}$ yields an inclusion of eigenvalues and eigenvectors of all $A \in \mathbf{A}$. In this case, as for general nonlinear equations, inner inclusions may be computed as well. Again this corresponds to bounds for the sensitivity with respect to finite perturbations of A.

Moreover, outer and inner inclusions for input matrices out of some structure may be computed as well. This corresponds to structured finite perturbations and generalizes the structured condition numbers for eigenvalues and pseudospectra discussed in Rump (2006).

14. Optimization

As mentioned in the abstract, Sahinidis and Tawaralani (2005) received the 2006 Beale–Orchard–Hays Prize for their global optimization package BARON, which 'incorporates techniques from automatic differentiation, interval arithmetic, and other areas to yield an automatic, modular, and relatively efficient solver for the very difficult area of global optimization' (from the laudatio).

In optimization some important tasks are treated successfully by interval methods which could hardly be solved in any other way. In particular, nonlinear terms in constraint propagation for branch and bound methods, the estimation of the range of a function, and verification of the non-existence of roots within a domain (see Section 13.1) are reserved to interval methods or related techniques from convex analysis.

Neumaier (2004), in his *Acta Numerica* article, gave a detailed overview on global optimization and constraint satisfaction methods. In view of the thorough treatment there, showing the essential role of interval methods in this area, we restrict our discussion to more recent, complementary issues.

14.1. Linear and convex programming

One might be inclined to presume that convex optimization problems are less affected by numerical problems. However, the NETLIB (2009) suite of linear optimization problems contains practical examples from various areas, and a study by Ordóñez and Freund (2003) revealed that 72% of these real-life problems are ill-conditioned; they show that many commercial solvers fail.

For mixed integer linear programming problems, preprocessing in particular may change the status of the linear program from feasible to infeasible, and *vice versa*. Jansson (2004*b*) and Neumaier and Shcherbina (2004) give methods describing how safe bounds for the solution of linear and mixed integer linear programming problems can be obtained with minimal additional computational effort (also, a simple example is given for which many commercial solvers fail). A generalization of their method will be described below.

In my experience, although straightforward, it is not easy to program a *robust* simplex algorithm. Even for small problems it is not unlikely that an incorrect branch will lead to a sub-optimal result or apparent infeasibility.

Much more demanding is the situation when applying local programming solvers. Tawaralani and Sahinidis (2004) pointed out that nonlinear programming solvers often fail even in solving convex problems. Due to this lack of reliability, as one consequence, they used in their global optimization package BARON only linear instead of nonlinear convex relaxations.

For linear programming, efficient verification methods have been implemented in LURUPA by Keil (2006). This is a C++ package based on PROFIL/BIAS by Knüppel (1994, 1998). It includes a routine for computing rigorous bounds for the condition number. Numerical results for the NETLIB (2009) lp-library, a collection of difficult-to-solve applications, can be found in Keil and Jansson (2006).

For convex optimization problems, too, some postprocessing of computed data allows one to compute rigorous bounds for the result with little additional effort.

14.2. Semidefinite programming

For this review article we restrict the exposition to semidefinite programming problems (SDP), and briefly sketch some promising results in this direction, following Jansson, Chaykin and Keil (2007). This class is rather extensive, since many nonlinear convex problems can be reformulated within this framework. For a general introduction to semidefinite programming, see Todd (2001).

Consider the *(primal) semidefinite program* in block diagonal form,

$$p^* := \min \sum_{j=1}^{n} \langle C_j, X_j \rangle \quad \text{such that} \quad \sum_{j=1}^{n} \langle A_{ij}, X_j \rangle = b_i \quad \text{for } i = 1, \ldots, m,$$
$$X_j \succeq 0 \quad \text{for } j = 1, \ldots, n, \tag{14.1}$$

where C_j, A_{ij}, and X_j are symmetric $s_j \times s_j$ matrices, $b \in \mathbb{R}^m$, and

$$\langle C, X \rangle = \text{trace}(C^T X) \tag{14.2}$$

denotes the inner product for the set of symmetric matrices. Moreover, $\succeq$ is the Löwner partial ordering, that is, $X \succeq Y$ if and only if $X - Y$ is positive semidefinite.

Semidefinite programming problems generalize linear programming problems as by $s_j = 1$ for $j = 1, \ldots, n$, in which case C_j, A_{ij} and X_j are real numbers. On the other hand, linear and semidefinite programming are special cases of conic programming. This is a universal form of convex programming, and refers to non-smooth problems with linear objective function, linear constraints, and variables that are restricted to a cone.

The Lagrangian dual of (14.1) is

$$d^* := \max b^T y \quad \text{such that} \quad \sum_{i=1}^{m} y_i A_{ij} \preceq C_j \quad \text{for } j = 1, \ldots, n, \tag{14.3}$$

where $y \in \mathbb{R}^m$, so that the constraints $\sum_{i=1}^{m} y_i A_{ij} \preceq C_j$ are linear matrix inequalities (LMI). We use the convention that $p^* = -\infty$ if (14.1) is unbounded and $p^* = \infty$ if (14.1) is infeasible, analogously for (14.3).

It is known (Vandenberghe and Boyd 1996, Ben-Tal and Nemirovskii 2001) that semidefinite programs satisfy weak duality $d^* \leq p^*$, which turns into strong duality if the so-called Slater constraint qualifications are satisfied.

Theorem 14.1. (Strong Duality Theorem)

(a) If the primal problem (14.1) is strictly feasible (*i.e.*, there exist feasible positive definite matrices X_j for $j = 1, \ldots, n$) and p^* is finite, then $p^* = d^*$ and the dual supremum is attained.

(b) If the dual problem (14.3) is strictly feasible (*i.e.*, there exists some $y \in \mathbb{R}^m$ such that $C_j - \sum_{i=1}^{m} y_i A_{ij}$ are positive definite for $j = 1, \ldots, n$) and d^* is finite, then $p^* = d^*$, and the primal infimum is attained.

In general, one of the problems (14.1) and (14.3) may have optimal solutions while its dual is infeasible, or the duality gap may be positive at optimality. This is in contrast to linear programming, where strong duality is fulfilled without any assumptions.

As has been pointed out by Neumaier and Shcherbina (2004), ill-conditioning is, for example, likely to take place in combinatorial optimization when branch-and-cut procedures sequentially generate linear or semidefinite programming relaxations. Rigorous results in combinatorial optimization can be obtained when solving the relaxations rigorously.

The following results generalize methods for linear programming by Jansson (2004*b*) and Neumaier and Shcherbina (2004). For convex programming problems that cannot be reformulated as semidefinite or conic problems, see Jansson (2004*a*). These techniques can also be generalized for computing rigorous error bounds for infinite-dimensional non-smooth conic optimization problems within the framework of functional analysis: see Jansson (2009).

The main goal is to obtain rigorous bounds by postprocessing already computed data.

Theorem 14.2. Let a semidefinite program (14.1) be given, let $\tilde{y} \in \mathbb{R}^m$, set

$$D_j := C_j - \sum_{i=1}^{m} \tilde{y}_i A_{ij} \quad \text{for } j = 1, \ldots, n, \tag{14.4}$$

and suppose that

$$\underline{d}_j \le \lambda_{\min}(D_j) \quad \text{for} \quad j = 1, \ldots, n. \tag{14.5}$$

Assume further that a primal feasible solution X_j of (14.1) is known, together with upper bounds

$$\lambda_{\max}(X_j) \le \overline{x}_j \quad \text{for} \quad j = 1, \ldots, n \tag{14.6}$$

for the maximal eigenvalues, where $\overline{x}_j$ may be infinite. If

$$\underline{d}_j \ge 0 \quad \text{for those } j \text{ with } \ \overline{x}_j = +\infty, \tag{14.7}$$

then, abbreviating $\underline{d}_j^- := \min(0, \underline{d}_j)$,

$$p^* \ge \inf\left\{ b^T \tilde{y} + \sum_{j=1}^{n} s_j \cdot \underline{d}_j^- \cdot \overline{x}_j \right\} \tag{14.8}$$

is satisfied, and the right-hand side of (14.8) is finite. Moreover, for every j with $\underline{d}_j \ge 0$,

$$\sum_{i=1}^{m} \tilde{y}_i A_{ij} - C_j \preceq 0.$$

Proof. Let $E, Y \in \mathbb{R}^{k \times k}$ be symmetric matrices with

$$\underline{d} \le \lambda_{\min}(E), \quad 0 \le \lambda_{\min}(Y), \quad \text{and} \quad \lambda_{\max}(Y) \le \overline{x}. \tag{14.9}$$

For the eigenvalue decomposition $E = Q\Lambda Q^T$, it follows that

$$\langle E, Y\rangle = \mathrm{trace}(Q\Lambda Q^T Y) = \mathrm{trace}(\Lambda Q^T Y Q) = \sum_{\nu=1}^{k} \lambda_\nu(E) e_\nu^T Q^T Y Q e_\nu.$$

Then (14.9) implies $0 \le e_\nu^T Q^T Y Q e_\nu \le \overline{x}$ and thus, using $\underline{d}^- := \min(0, \underline{d})$,

$$\langle E, Y\rangle \ge k \cdot \underline{d}^- \cdot \overline{x}. \tag{14.10}$$

The definitions (14.4) and (14.1) imply

$$\sum_{j=1}^{n} \langle C_j, X_j\rangle - b^T\tilde{y} = \sum_{j=1}^{n} \langle D_j, X_j\rangle,$$

and application of (14.10) yields

$$\sum_{j=1}^{n} \langle D_j, X_j\rangle \ge \sum_{j=1}^{n} s_j \cdot \underline{d}_j^- \cdot \overline{x}_j,$$

which proves inequality (14.8), and assumption (14.7) ensures a finite right-hand side. The last statement is an immediate consequence of $\lambda_{\min}(D_j) \ge \underline{d}_j \ge 0$. □

Note that Theorem 14.2 includes the case when no information on primal feasible solutions is available. In this case $\overline{x}_j = +\infty$ for all j.

The application of Theorem 14.2 is as follows. The lower bounds for the smallest eigenvalues as in (14.5) are calculated by the methods explained in Section 10.8.1. If (14.7) is satisfied, only (14.8) has to be evaluated. Otherwise, the constraints j violating (14.7) are relaxed by replacing C_j by $C_j - \epsilon_j I$. Then the dual optimal solution $y(\epsilon)$ satisfies the constraints

$$C_j - \sum_{i=1}^{m} y_i(\varepsilon)\, A_{ij} \succeq \varepsilon_j I,$$

increasing the minimal eigenvalues of the new defect

$$D_j(\varepsilon) := C_j - \sum_{i=1}^{m} y_i(\varepsilon) A_{ij}.$$

Some heuristic is applied to choose ϵ_j: see Jansson *et al.* (2007).

Algorithms for computing rigorous lower and upper bounds bounds for the optimal value, existence of optimal solutions and rigorous bounds for ϵ-optimal solutions as well as verified certificates of primal and dual infeasibility have been implemented in VSDP by Jansson (2006), a MATLAB toolbox for verified semidefinite programming solving based on INTLAB.

Numerical results for problems from the SDPLIB collection by Borchers (1999), a collection of large and real-life test problems, were reported by

Jansson (2009). The verification method in VSDP could compute for all problems a rigorous lower bound of the optimal value, and could verify the existence of strictly dual feasible solutions proving that all problems have a zero duality gap. A finite rigorous upper bound could be computed for all well-posed problems, with one exception (Problem hinf2). For all 32 ill-posed problems in the SDPLIB collection, VSDP computed the upper bound $\overline{f}_d = +\infty$ indicating the zero distance to primal infeasibility.

The package SDPT3 by Tütüncü, Toh and Todd (2003) (with default values), for example, gave warnings for 7 problems, where 2 warnings were given for well-posed problems. Hence, no warnings were given for 27 ill-posed problems with zero distance to primal infeasibility.

PART THREE
Infinite-dimensional problems

15. Ordinary differential equations

In the following we briefly summarize current verification methods for initial value problems; two-point boundary value problems will be discussed in the next section in more detail, because the generalization from finite-dimensional problems (Section 13) to the infinite-dimensional case is very natural.

Most approaches proceed iteratively, computing an inclusion $\mathbf{y}_{\nu+1}$ based on an inclusion $\mathbf{y}_{\nu}$. This verification is performed in two steps: an initial inclusion $\tilde{\mathbf{y}}_{\nu+1}$ followed by a refinement step. Because of the inevitable presence of rounding errors, the first inclusion $\mathbf{y}_1$ will have non-zero width, so that, at each step, an initial value problem has to be integrated over a set of initial values.

This leads to a fundamental problem, the *wrapping effect* already discussed in Section 9.2, similar to the widening of intervals in interval Gaussian elimination (see Section 10.1). To combat this, Moore (1966) proposed a moving coordinate system to reduce the local error. Eijgenraam (1981) proposed using higher-order approximations and inclusions. Instead of directly multiplying the transformation matrices, Lohner (1988) applies a QR-decomposition, an important advantage for long-term integration. His package AWA (Anfangswertaufgaben)[24] is well known.

Most verification methods for ODEs follow this two-step approach (see Nedialkov and Jackson (2000)), and much research has been done on improving the next initial inclusion (the first step), and improving the refinement. For the latter step, common techniques are Taylor series methods, as in AWA, constraint satisfaction methods, which are also used in global optimization (Schichl and Neumaier 2005), and the Hermite–Obreschkoff method by Nedialkov (1999). The latter is the basis for the software package VNODE. Note that these approaches contradict the UTILIZE INPUT DATA PRINCIPLE (5.13).

Finally we mention the one-phase method Cosy Infinity by Berz and Makino (1999), where the solution is directly modelled by an automatic Taylor expansion (see Section 11.4). Their code is not freely available.

However, we also mention the very detailed but also discouraging study by Bernelli azzera, Vasile, Massari and Di Lizia (2004) on the solution of space-related problems. The result is basically that long-term integration is an open problem. This was also formulated as a challenge in Neumaier (2002).

[24] Initial value problems.

15.1. Two-point boundary value problems

The ideas presented in the previous sections on finite-dimensional problems can be used to derive verification methods for infinite-dimensional problems by extending the tools used so far to Banach and Hilbert spaces.

In this section we briefly sketch a verification method for an ODE, whereas in the final section semilinear elliptic PDEs are considered in much more detail. Once more it demonstrates the principle of verification methods: to derive mathematical theorems, the assumptions of which can be verified with the aid of a computer. The result is a computer-assisted proof.

In particular, we are here concerned with a two-point boundary value problem of the form

$$\begin{aligned} -u'' &= ru^N + f, \quad 0 < x < 1, \\ u(0) &= u(1) = 0, \end{aligned} \tag{15.1}$$

for $N \in \mathbb{N}$, $r \in L^\infty(0,1)$ and $f \in L^2(0,1)$. We assume $N \geq 2$; the following is based on Nakao, Hashimoto and Watanabe (2005) and Takayasu, Oishi and Kubo (2009*a*) (the simpler linear case $N = 1$ is handled in Takayasu, Oishi and Kubo (2009*b*); see below).

The verification method transforms the problem into a nonlinear operator equation with a solution operator $\mathcal{K}$ to the Poisson equation

$$\begin{aligned} -u'' &= f, \qquad 0 < x < 1, \\ u(0) &= u(1) = 0. \end{aligned} \tag{15.2}$$

Let $(\cdot,\cdot)$ denote the inner product in $L^2(0,1)$, and let $H^m(0,1)$ denote the L^2-Sobolev space of order m. For

$$H_0^1(0,1) = \{u \in H^1(0,1) : u(0) = u(1) = 0 \text{ in the sense of traces}\}$$

with the inner product (u', v') and norm $\|u\|_{H_0^1} = \|u'\|_{L^2}$, the solution operator $\mathcal{K}$ will be regarded as a bounded linear operator from L^2 into $H^2 \cap H_0^1$.

Since the embedding $H^2 \hookrightarrow H^1$ is compact, $\mathcal{K}$ is also a compact linear operator from H_0^1 to H_0^1.

Recall that the eigenvalue problem

$$\begin{aligned} -u'' &= \lambda u, \qquad 0 < x < 1, \\ u(0) &= u(1) = 0, \end{aligned} \tag{15.3}$$

has eigenvalues $\lambda_k = k^2\pi^2$, so the minimal eigenvalue is $\lambda_{\min} = \pi^2$. The usual Rayleigh quotient estimate for $u \in H^2 \cap H_0^1$ of (15.3) implies

$$\lambda_{\min}(u,u) \leq (u',u') = (-u'',u) \leq \|u''\|_{L^2}\|u\|_{L^2},$$

hence

$$\|u\|_{L^2} \le \frac{1}{\pi}\|u'\|_{L^2} = \frac{1}{\pi}\|u\|_{H_0^1} \quad \text{and} \quad \|u\|_{H_0^1} \le \frac{1}{\pi}\|u''\|_{L^2}, \tag{15.4}$$

which means in particular

$$\|\mathcal{K}\|_{\mathcal{L}(L^2,H_0^1)} \le \frac{1}{\pi}. \tag{15.5}$$

Note that $\|u\|_{L^2} \le \frac{1}{\pi}\|u\|_{H_0^1}$ holds for all $u \in H_0^1$ since $H^2 \cap H_0^1$ is dense in H_0^1 (and the embedding $H^1 \hookrightarrow L^2$ is compact).

Recall that, for every $u \in H_0^1$,

$$|u(x)| = \left|\int_0^x u' \,\mathrm{d}t\right| \le \int_0^x |u'| \,\mathrm{d}t \quad \text{and} \quad |u(x)| = \left|\int_1^x u' \,\mathrm{d}t\right| \le \int_x^1 |u'| \,\mathrm{d}t,$$

which implies

$$2|u(x)| \le \int_0^1 |u'| \,\mathrm{d}t \le \|u'\|_{L^2} = \|u\|_{H_0^1},$$

and thus

$$\|u\|_\infty \le \frac{1}{2}\|u\|_{H_0^1}. \tag{15.6}$$

For numerical approximations we will use piecewise cubic finite element basis functions. For this purpose consider the cubic polynomials

$$\begin{aligned}
N_0(x) &= (1-x)^2(1+2x),\\
N_1(x) &= x(1-x)^2,\\
N_2(x) &= x^2(3-2x),\\
N_3(x) &= -x^2(1-x),
\end{aligned}$$

which form a basis of the vector space P_3 of all real polynomials of degree ≤ 3. Note that for $p \in P_3$ the corresponding dual basis $N_i^* \in P_3^*$, $0 \le i \le 3$, is given by

$$\begin{aligned}
N_0^*(p) &:= p(0),\\
N_1^*(p) &:= p'(0),\\
N_2^*(p) &:= p(1),\\
N_3^*(p) &:= p'(1),
\end{aligned}$$

so that $p = p(0)N_0 + p'(0)N_1 + p(1)N_2 + p'(1)N_3$.

Now let $n \in \mathbb{N}$, $h := 1/(n+1)$ and $x_i := ih$, $-1 \le i \le n+2$. Then $x_0, \dots, x_{n+1}$ is an equidistant partition of the interval $[0,1]$. The functions

$$\psi_{i,0}(x) = \begin{cases} N_2(\frac{x-x_{i-1}}{h}), & x \in [x_{i-1}, x_i] \cap [0,1], \\ N_0(\frac{x-x_i}{h}), & x \in \,]x_i, x_{i+1}] \cap [0,1], \\ 0, & \text{otherwise}, \end{cases} \qquad 0 \le i \le n+1,$$

$$\psi_{i,1}(x) = \begin{cases} hN_3(\frac{x-x_{i-1}}{h}), & x \in [x_{i-1}, x_i] \cap [0,1], \\ hN_1(\frac{x-x_i}{h}), & x \in \,]x_i, x_{i+1}] \cap [0,1], \\ 0, & \text{otherwise}, \end{cases} \qquad 0 \le i \le n+1,$$

are $2(n+2)$ linearly independent H^2-conforming finite element basis functions. We choose the ordering

$$(\psi_{0,0}, \psi_{0,1}, \psi_{1,0}, \psi_{1,1}, \dots, \psi_{n,0}, \psi_{n,1}, \psi_{n+1,1}, \psi_{n+1,0}) =: (\phi_0, \dots, \phi_{m+1}),$$

where $m := 2(n+1)$, and denote the spanned $(m+2)$-dimensional space by

$$\overline{X}_n := \operatorname{span}\{\phi_0, \dots, \phi_{m+1}\} \subset H^2. \tag{15.7}$$

The subspace spanned by the H_0^1-conforming finite element basis functions $(\phi_1, \dots, \phi_m)$ is denoted by

$$X_n := \operatorname{span}\{\phi_1, \dots, \phi_m\} \subset H^2 \cap H_0^1. \tag{15.8}$$

Note that the coefficients $v_0, \dots, v_{m+1}$ of a function

$$v = \sum_{i=0}^{m+1} v_i \phi_i \in \overline{X}_n$$

can be easily computed by $v_{2i} = v(x_i)$, $v_{2i+1} = v'(x_i)$, $i = 0, \dots, n$, and $v_m = v'(1)$, $v_{m+1} = v(1)$. The orthogonal projection $\mathcal{P}_n : H_0^1 \to X_n$ is defined by

$$(u' - (\mathcal{P}_n u)', \phi') = 0 \quad \text{for all } \ u \in H_0^1 \ \text{ and all } \ \phi \in X_n, \tag{15.9}$$

and the dual projection $\mathcal{Q}_n : H^2 \to \overline{X}_n$ is defined by

$$\mathcal{Q}_n(u) := \sum_{i=0}^{n} (u(x_i)\phi_{2i} + u'(x_i)\phi_{2i+1}) + u'(1)\phi_m + u(1)\phi_{m+1}, \tag{15.10}$$

where in (15.10), as usual, $u \in H^2$ is regarded as a member of $C^1([0,1])$ in which H^2 is compactly embedded. We recall basic estimates of the finite element projection errors. Even though well known in general, these estimates often do not occur explicitly for the one-dimensional case in standard textbooks. Thus, for completeness, the proofs are included.

Theorem 15.1. (FEM-projection error bounds) For $u \in H^2$ the following inequalities hold:

(a) $\|u - \mathcal{P}_n u\|_{H_0^1} \le \|u - \mathcal{Q}_n u\|_{H_0^1}$ if $u \in H^2 \cap H_0^1$,

(b) $\|u - \mathcal{Q}_n u\|_{L^2} \le \frac{h}{\pi}\|u - \mathcal{Q}_n u\|_{H_0^1}$,

(c) $\|u - \mathcal{Q}_n u\|_{H_0^1} \le \frac{h}{\pi}\|(u - \mathcal{Q}_n u)''\|_{L^2} \le \frac{h}{\pi}\|u''\|_{L^2}$,

(d) $\|u - \mathcal{Q}_n u\|_{H_0^1} \le \frac{h^2}{\pi^2}\|u'''\|_{L^2}$ if $u \in H^3 \cap H_0^1$.

Proof. **(a)** Since $u \in H^2 \cap H_0^1$, $\mathcal{Q}_n u \in X_n$ and therefore $\mathcal{P}_n u - \mathcal{Q}_n u \in X_n$. It follows from the definition of the orthogonal projection (15.9) that $((u - \mathcal{P}_n u)', (\mathcal{P}_n u - \mathcal{Q}_n u)') = 0$, which gives the Pythagorean identity

$$\|u - \mathcal{Q}_n u\|_{H_0^1}^2 = \|u - \mathcal{P}_n u\|_{H_0^1}^2 + \|\mathcal{P}_n u - \mathcal{Q}_n u\|_{H_0^1}^2 \ge \|u - \mathcal{P}_n u\|_{H_0^1}^2.$$

(b) The function $e := u - \mathcal{Q}_n u$ fulfils $e(x_i) = 0$ for each $i \in \{0, \dots, n+1\}$. Hence each function $e(x_i + hx)$, $i \in \{0, \dots, n\}$, belongs to H_0^1. Now (15.4) supplies

$$\begin{aligned}\int_0^1 e(x_i + hx)^2 \,\mathrm{d}x &= \|e(x_i + hx)\|_{L^2}^2 \le \frac{1}{\pi^2}\|(e(x_i + hx))'\|_{L^2}^2 \\ &= \frac{h}{\pi^2}\int_0^1 h e'(x_i + hx)^2 \,\mathrm{d}x = \frac{h}{\pi^2}\int_{x_i}^{x_{i+1}} e'(x)^2 \,\mathrm{d}x.\end{aligned}$$

Thus

$$\begin{aligned}\|e\|_{L^2}^2 &= \sum_{i=0}^{n}\int_{x_i}^{x_{i+1}} e(x)^2 \,\mathrm{d}x = h\sum_{i=0}^{n}\int_{x_i}^{x_{i+1}} e\left(x_i + h \cdot \frac{x - x_i}{h}\right)^2 \frac{1}{h}\,\mathrm{d}x \\ &= h\sum_{i=0}^{n}\int_0^1 e(x_i + hx)^2 \,\mathrm{d}x \le \frac{h^2}{\pi^2}\sum_{i=0}^{n}\int_{x_i}^{x_{i+1}} e'(x)^2 \,\mathrm{d}x = \frac{h^2}{\pi^2}\|e\|_{H_0^1}^2.\end{aligned}$$

(c) The function $d := (u - \mathcal{Q}_n u)'$ fulfils $d(x_i) = 0$ for each $i \in \{0, \dots, n+1\}$. Therefore $d \in H_0^1$ as $u, \mathcal{Q}_n u \in H^2$, and $\|d\|_{L^2} \le \frac{h}{\pi}\|d'\|_{L^2}$ follows by the same computation as in (b) with d instead of e. This is the first part,

$$\|u - \mathcal{Q}_n u\|_{H_0^1} \le \frac{h}{\pi}\|(u - \mathcal{Q}_n u)''\|_{L^2},$$

of the inequality in (c). Since $\mathcal{Q}_n u$ is a piecewise cubic polynomial,

$$(\mathcal{Q}_n u|_{[x_i, x_{i+1}]})'''' = 0$$

supplies

$$\begin{aligned}&\frac{1}{2}\left(-\|(u - \mathcal{Q}_n u)''\|_{L^2}^2 + \|u''\|_{L^2}^2 - \|(\mathcal{Q}_n u)''\|_{L^2}^2\right) \qquad (15.11)\\ &\quad = \int_0^1 (\mathcal{Q}_n u)''(u - \mathcal{Q}_n u)'' \,\mathrm{d}x\end{aligned}$$

$$= \sum_{i=0}^{n} [(\mathcal{Q}_n u)''(u - \mathcal{Q}_n u)']_{x_i}^{x_{i+1}} - [(\mathcal{Q}_n u)'''(u - \mathcal{Q}_n u)]_{x_i}^{x_{i+1}} + \int_{x_i}^{x_{i+1}} (\mathcal{Q}_n u)''''(\mathcal{Q}_n u)\,\mathrm{d}x = 0.$$

Thus

$$\|(u - \mathcal{Q}_n u)''\|_{L^2}^2 = \|u''\|_{L^2}^2 - \|(\mathcal{Q}_n u)''\|_{L^2}^2 \le \|u''\|_{L^2}^2,$$

giving the second part of the asserted inequality.

(d) If $u \in H^3 \cap H_0^1$, then using (15.11), the Cauchy–Schwarz inequality and (c) yield

$$\|(u - \mathcal{Q}_n u)''\|_{L^2}^2 = \int_0^1 u''(u - \mathcal{Q}_n u)''\,\mathrm{d}x = -\int_0^1 u'''(u - \mathcal{Q}_n u)'\,\mathrm{d}x \le \|u'''\|_{L^2}\|(u - \mathcal{Q}_n u)'\|_{L^2} \le \frac{h}{\pi}\|u'''\|_{L^2}\|(u - \mathcal{Q}_n u)''\|_{L^2}.$$

Therefore

$$\|(u - \mathcal{Q}_n u)''\|_{L^2} \le \frac{h}{\pi}\|u'''\|_{L^2},$$

which, inserted into (c), gives

$$\|u - \mathcal{Q}_n u\|_{H_0^1} \le \frac{h^2}{\pi^2}\|u'''\|_{L^2}.$$ □

Applying the solution operator $\mathcal{K}$ for (15.2), the problem (15.1) is transformed into $u = \mathcal{K}(ru^N + f)$, or equivalently into the nonlinear operator equation

$$F(u) := u - \mathcal{K}ru^N - \mathcal{K}f = 0. \tag{15.12}$$

Based on an approximate finite element solution $\hat{u}$, the Newton–Kantorovich theorem is applicable. Note that $\hat{u}$ can be computed by any standard (approximate) numerical method.

Theorem 15.2. (Newton–Kantorovich) Let F be given as in (15.12), and assume the Fréchet derivative $F'(\hat{u})$ is non-singular and satisfies

$$\|F'(\hat{u})^{-1}F(\hat{u})\|_{H_0^1} \le \alpha \tag{15.13}$$

for some positive α. Furthermore, assume

$$\|F'(\hat{u})^{-1}(F'(v) - F'(w))\|_{\mathcal{L}(H_0^1, H_0^1)} \le \omega\|v - w\|_{H_0^1} \tag{15.14}$$

for some positive ω and for all $v, w \in U_{2\alpha}(\hat{u}) := \{z \in H_0^1, \|z - \hat{u}\|_{H_0^1} < 2\alpha\}$. If

$$\alpha\omega \le \frac{1}{2}, \tag{15.15}$$

then $F(u) = 0$ has a solution u satisfying

$$\|u - \hat{u}\|_{H_0^1} \le \rho := \frac{1 - \sqrt{1 - 2\alpha\omega}}{\omega}. \tag{15.16}$$

Within this bound, the solution is unique.

Since $\mathcal{K} : H_0^1 \to H_0^1$ is a compact operator, $r \in L^\infty(0,1)$ and $\hat{u} \in X_n \subset L^\infty$, the operator

$$\mathcal{T} : H_0^1 \to H_0^1, \quad v \mapsto \mathcal{K} N r \hat{u}^{N-1} v \tag{15.17}$$

is compact as well. The verification method is based on the computation of constants C_1, C_2 and C_3 with

$$\|F'(\hat{u})^{-1}\|_{\mathcal{L}(H_0^1,H_0^1)} = \|(I - \mathcal{T})^{-1}\|_{\mathcal{L}(H_0^1,H_0^1)} \le C_1, \tag{15.18}$$

$$\|F(\hat{u})\|_{H_0^1} = \|\hat{u} - \mathcal{K} r \hat{u}^N - \mathcal{K} f\|_{H_0^1} \le C_2, \tag{15.19}$$

where I denotes the identity on H_0^1, and

$$\|F'(v) - F'(w)\|_{\mathcal{L}(H_0^1,H_0^1)} \le C_3 \|v - w\|_{H_0^1} \quad \text{for all } \; v, w \in U_{2\alpha}(\hat{u}). \tag{15.20}$$

If (15.15) is satisfied for

$$\alpha := C_1 C_2 \quad \text{and} \quad \omega := C_1 C_3, \tag{15.21}$$

then (15.16) follows. The main part, the estimation of C_1, uses the following result by Oishi (2000).

Theorem 15.3. Let $\mathcal{T} : H_0^1 \to H_0^1$ be a compact linear operator. Assume that $\mathcal{P}_n \mathcal{T}$ is bounded by

$$\|\mathcal{P}_n \mathcal{T}\|_{\mathcal{L}(H_0^1,H_0^1)} \le K, \tag{15.22}$$

that the difference between $\mathcal{T}$ and $\mathcal{P}_n \mathcal{T}$ is bounded by

$$\|\mathcal{T} - \mathcal{P}_n \mathcal{T}\|_{\mathcal{L}(H_0^1,H_0^1)} \le L, \tag{15.23}$$

and that the finite-dimensional operator $(I - \mathcal{P}_n \mathcal{T})|_{X_n} : X_n \to X_n$ is invertible with

$$\|(I - \mathcal{P}_n \mathcal{T})|_{X_n}^{-1}\|_{\mathcal{L}(H_0^1,H_0^1)} \le M. \tag{15.24}$$

If $(1 + MK)L < 1$, then the operator $(I - \mathcal{T}) : H_0^1 \to H_0^1$ is invertible and

$$\|(I - \mathcal{T})^{-1}\|_{\mathcal{L}(H_0^1,H_0^1)} \le \frac{1 + MK}{1 - (1 + MK)L}.$$

Proof. Using the finite-dimensional operator $(I - \mathcal{P}_n \mathcal{T})|_{X_n}^{-1} : X_n \to X_n$, one can show by direct computation that

$$(I - \mathcal{P}_n \mathcal{T})^{-1} = I + (I - \mathcal{P}_n \mathcal{T})|_{X_n}^{-1} \mathcal{P}_n \mathcal{T}.$$

The assumptions imply that this inverse operator is bounded by

$$\begin{aligned}\|(I-\mathcal{P}_n\mathcal{T})^{-1}\|_{\mathcal{L}(H_0^1,H_0^1)} &\le 1+\|(I-\mathcal{P}_n\mathcal{T})|_{X_n}^{-1}\|_{\mathcal{L}(H_0^1,H_0^1)}\cdot\|\mathcal{P}_n\mathcal{T}\|_{\mathcal{L}(H_0^1,H_0^1)}\\ &\le 1+MK.\end{aligned}$$

Moreover, for $\phi\in H_0^1$ and $\psi:=(I-\mathcal{T})\phi$, using

$$\phi=\phi-(I-\mathcal{P}_n\mathcal{T})^{-1}(I-\mathcal{T})\phi+(I-\mathcal{P}_n\mathcal{T})^{-1}\psi,$$

we have

$$\begin{aligned}\|\phi\|_{H_0^1} &\le \|I-(I-\mathcal{P}_n\mathcal{T})^{-1}(I-\mathcal{T})\|_{\mathcal{L}(H_0^1,H_0^1)}\|\phi\|_{H_0^1}+\|(I-\mathcal{P}_n\mathcal{T})^{-1}\psi\|_{H_0^1}\\ &\le \|(I-\mathcal{P}_n\mathcal{T})^{-1}\|_{\mathcal{L}(H_0^1,H_0^1)}\|\mathcal{T}-\mathcal{P}_n\mathcal{T}\|_{\mathcal{L}(H_0^1,H_0^1)}\|\phi\|_{H_0^1}+(1+MK)\|\psi\|_{H_0^1}\\ &\le (1+MK)L\|\phi\|_{H_0^1}+(1+MK)\|\psi\|_{H_0^1},\end{aligned} \tag{15.25}$$

and therefore

$$\|\phi\|_{H_0^1}\le\frac{1+MK}{1-(1+MK)L}\|(I-\mathcal{T})\phi\|_{H_0^1}.$$

This implies that $(I-\mathcal{T}):H_0^1\to H_0^1$ is injective, and by the Fredholm alternative it has a bounded inverse satisfying $\phi=(I-\mathcal{T})^{-1}\psi$. The theorem follows. □

To estimate C_1 by Theorem 15.3, three constants K, L and M are needed, which can be computed as follows. Using (15.4) and (15.5), we obtain for $u\in H_0^1$

$$\|\mathcal{P}_n\mathcal{T}u\|_{H_0^1}\le\|\mathcal{T}u\|_{H_0^1}\le\frac{1}{\pi}N\|r\hat{u}^{N-1}\|_\infty\|u\|_{L^2}\le\frac{1}{\pi^2}N\|r\hat{u}^{N-1}\|_\infty\|u\|_{H_0^1},$$

which implies that (15.22) holds for

$$K:=\frac{1}{\pi^2}N\|r\hat{u}^{N-1}\|_\infty. \tag{15.26}$$

Furthermore, applying the error bound

$$\|(I-\mathcal{P}_n)v\|_{H_0^1}\le\frac{h}{\pi}\|v''\|_{L^2}$$

for $v\in H^2\cap H_0^1$ (see Theorem 15.1(a), (c)), we obtain, using (15.4),

$$\|(\mathcal{T}-\mathcal{P}_n\mathcal{T})u\|_{H_0^1}\le\frac{h}{\pi}\|(\mathcal{T}u)''\|_{L^2}=\frac{h}{\pi}N\|r\hat{u}^{N-1}u\|_{L^2}\le\frac{h}{\pi^2}N\|r\hat{u}^{N-1}\|_\infty\|u\|_{H_0^1}.$$

Hence (15.23) holds for

$$L:=\frac{h}{\pi^2}N\|r\hat{u}^{N-1}\|_\infty=hK. \tag{15.27}$$

If $r \in W^{1,\infty}$, the L^∞-Sobolev space of order 1, then $\mathcal{T}u \in H^3 \cap H_0^1$, and using Theorem 15.1(a), (d) gives

$$\begin{aligned}
&\|(\mathcal{T} - \mathcal{P}_n\mathcal{T})u\|_{H_0^1} \\
&\quad \le \frac{h^2}{\pi^2}\|(\mathcal{T}u)'''\|_{L^2} = \frac{h^2}{\pi^2}N\|(r\hat{u}^{N-1}u)'\|_{L^2} \\
&\quad \le \frac{h^2}{\pi^2}N\left(\frac{1}{\pi}\|r'\hat{u}^{N-1} + (N-1)r\hat{u}^{N-2}\hat{u}'\|_\infty + \|r\hat{u}^{N-1}\|_\infty\right)\|u\|_{H_0^1},
\end{aligned}$$

and (15.23) holds for the $O(h^2)$ bound

$$L := \frac{h^2}{\pi^2}N\left(\frac{1}{\pi}\|r'\hat{u}^{N-1} + (N-1)r\hat{u}^{N-2}\hat{u}'\|_\infty + \|r\hat{u}^{N-1}\|_\infty\right). \tag{15.28}$$

Condition (15.24) requires us to calculate a constant M satisfying

$$\left\|(I - \mathcal{P}_n\mathcal{T})|_{X_n}^{-1}\left(\sum_{i=1}^m v_i\phi_i\right)\right\|_{H_0^1} \le M\left\|\sum_{i=1}^m v_i\phi_i\right\|_{H_0^1} \quad \text{for all } v_1,\dots,v_m \in \mathbb{R},$$

and thus

$$\sum_{i,j=1}^m v_iv_j\int_0^1 \psi_i'\psi_j'\,\mathrm{d}x \le M^2\sum_{i,j=1}^m v_iv_j\int_0^1 \phi_i'\phi_j'\,\mathrm{d}x, \tag{15.29}$$

where

$$\psi_i = \sum_{j=1}^m \alpha_{ij}\phi_j := (I - \mathcal{P}_n\mathcal{T})|_{X_n}^{-1}\phi_i, \quad \textit{i.e.}, \quad (I - \mathcal{P}_n\mathcal{T})\psi_i = \phi_i.$$

Taking inner H_0^1-products with ϕ_k, $k = 1,\dots,m$, the latter is equivalent to

$$\sum_{j=1}^m \alpha_{ij}\left[(\phi_j', \phi_k') - ((\mathcal{T}\phi_j)', \phi_k')\right] = (\phi_i', \phi_k'). \tag{15.30}$$

By partial integration,

$$((\mathcal{T}\phi_j)', \phi_k') = -((\mathcal{T}\phi_j)'', \phi_k) = N(r\hat{u}^{N-1}\phi_j, \phi_k).$$

Thus, defining $A = (\alpha_{ij})_{1\le i,j\le m}$ and

$$\begin{aligned}
R &:= ((\phi_j', \phi_k') - N(r\hat{u}^{N-1}\phi_j, \phi_k))_{1\le j,k\le m}, \\
G &:= ((\phi_i', \phi_k'))_{1\le i,k\le m},
\end{aligned} \tag{15.31}$$

(15.30) reads $AR = G$, whence $A = GR^{-1}$. Note that all integrals have to be bounded rigorously. This can be done, for example, using the INTLAB algorithm `verifyquad` described in Section 12. Also note that the invertibility of R, which is equivalent to the invertibility of the operator

$(I - \mathcal{P}_n\mathcal{T})|_{X_n}$, has to be verified. The methods described below ensure this. Now

$$(\psi_i', \psi_j') = \sum_{k,l=1}^{m} \alpha_{ik}\alpha_{jl}(\phi_k', \phi_l') = (AGA^T)_{ij} = (GR^{-1}GR^{-1}G)_{ij}.$$

Insertion into (15.29) gives the equivalent condition

$$v^T(GR^{-1}GR^{-1}G)v \le M^2 v^T G v \quad \text{for all} \quad v \in \mathbb{R}^m,$$

and thus (15.24) holds for

$$M := \sqrt{\lambda_{\max}}, \tag{15.32}$$

with $\lambda_{\max}$ denoting the largest (in absolute value) eigenvalue of the matrix eigenvalue problem

$$GR^{-1}GR^{-1}Gv = \lambda Gv. \tag{15.33}$$

Since M is equal to the spectral radius of $R^{-1}G$, an upper bound can be computed in INTLAB, for example, as follows. First, `C = verifylss(R,G)` (see Section 10.5) verifies that R is non-singular and computes an inclusion `C` of $R^{-1}G$, so that $M \le \|R^{-1}G\|_2 \le \sqrt{\|R^{-1}G\|_\infty \|R^{-1}G\|_1}$ and

$$\texttt{Mb} = \texttt{mag(sqrt(norm(C, inf) * norm(C, 1)))} \quad \text{implies} \quad M \le \texttt{Mb}. \tag{15.34}$$

Another way to estimate M is by a (verified) similarity transformation and Gershgorin circles as follows:

```
[V,D] = eig(mid(C));
Y = verifylss(V,C*V);
Mb = max(mag(gershgorin(Y)));
```

Again $M \le$ `Mb` holds. Neither approach uses the symmetry of R and G. Alternatively, note that for a complex eigenvalue/eigenvector pair (v, λ) of (15.33) it follows that

$$\overline{v}^T GR^{-1}GR^{-1}Gv = \lambda \overline{v}^T Gv. \tag{15.35}$$

But $\overline{v}^T Gv > 0$ because G is positive definite, and the left-hand side of (15.35) is real because $GR^{-1}GR^{-1}G$ is symmetric, so that λ must be real and non-zero. Moreover, λ is an eigenvalue of $(R^{-1}G)^2$, so $\lambda > 0$, and, because G is positive definite, $\lambda' G - GR^{-1}GR^{-1}G$ is positive semidefinite if and only if $\lambda' \ge \lambda_{\max}$. The latter bounds are often significantly superior to (15.34).

Thus, choosing λ' a little larger than an approximation of $\lambda_{\max}$, and verifying that $\lambda' G - GR^{-1}GR^{-1}G$ is positive semidefinite by algorithm `isspd` discussed in Section 10.8.1, proves $M \le \sqrt{\lambda'}$. We mention that the latter two approaches yield significantly better estimates of M than (15.34).

With K, L, M computed according to (15.26), (15.27) (or (15.28) if $r \in W^{1,\infty}$) and (15.32), a constant

$$C_1 := \frac{1+MK}{1-(1+MK)L} \tag{15.36}$$

satisfying (15.18) is obtained via Theorem 15.3, provided $(1+MK)L < 1$ (which requires h to be sufficiently small).

It follows from (15.5) that

$$\begin{aligned}\|\hat{u} - \mathcal{K}(r\hat{u}^N) - \mathcal{K}f\|_{H_0^1} &= \|\mathcal{K}(-\hat{u}'' - r\hat{u}^N - f)\|_{H_0^1} \\ &\le \frac{1}{\pi}\|\hat{u}'' + r\hat{u}^N + f\|_{L^2} =: C_2. \end{aligned} \tag{15.37}$$

The constant C_2 is expected to be small if the approximation $\hat{u}$ is sufficiently accurate.

In the following, another estimate for a defect bound $\|\hat{u} - \mathcal{K}(r\hat{u}^N + f)\|_{H_0^1} \le C_2$ is given, which naturally leads to a Newton method for finding $\hat{u}$. For simplicity we assume from now on that $r, f \in H^2$, so that the dual projection $\mathcal{Q}_n$ can be applied to these functions. The main idea is to replace $\|\hat{u} - \mathcal{K}(r\hat{u}^N + f)\|_{H_0^1}$ by a projected finite element defect, namely by $\|\hat{u} - \mathcal{P}_n\mathcal{K}\mathcal{Q}_n(r\hat{u}^N + f)\|_{H_0^1}$. Note that $r\hat{u}^N \in H^2$, since H^2 is an algebra in the one-dimensional case. Using the triangle inequality, Theorem 15.1(a), (d) and (15.5) imply

$$\begin{aligned}&\|\hat{u} - \mathcal{K}(r\hat{u}^N + f)\|_{H_0^1} \\ &\quad \le \|\hat{u} - \mathcal{P}_n\mathcal{K}\mathcal{Q}_n(r\hat{u}^N + f)\|_{H_0^1} + \|\mathcal{P}_n\mathcal{K}\mathcal{Q}_n(r\hat{u}^N + f) - \mathcal{K}\mathcal{Q}_n(r\hat{u}^N + f)\|_{H_0^1} \\ &\qquad + \|\mathcal{K}\mathcal{Q}_n(r\hat{u}^N + f) - \mathcal{K}(r\hat{u}^N + f)\|_{H_0^1} \\ &\quad = \|\hat{u} - \mathcal{P}_n\mathcal{K}\mathcal{Q}_n(r\hat{u}^N + f)\|_{H_0^1} + \|(\mathcal{P}_n - I)\mathcal{K}\mathcal{Q}_n(r\hat{u}^N + f)\|_{H_0^1} \\ &\qquad + \|\mathcal{K}(\mathcal{Q}_n - I)(r\hat{u}^N + f)\|_{H_0^1} \\ &\quad \le \|\hat{u} - \mathcal{P}_n\mathcal{K}\mathcal{Q}_n(r\hat{u}^N + f)\|_{H_0^1} + \frac{h^2}{\pi^2}\|(\mathcal{Q}_n(r\hat{u}^N + f))'\|_{L^2} \\ &\qquad + \frac{1}{\pi}\|(\mathcal{Q}_n - I)(r\hat{u}^N + f)\|_{L^2}.\end{aligned}$$

The three summands in the last line are abbreviated by

$$\begin{aligned}C'_{21} &:= \|\hat{u} - \mathcal{P}_n\mathcal{K}\mathcal{Q}_n(r\hat{u}^N + f)\|_{H_0^1}, \\ C_{22} &:= \frac{h^2}{\pi^2}\|(\mathcal{Q}_n(r\hat{u}^N + f))'\|_{L^2}, \\ C_{23} &:= \frac{1}{\pi}\|(\mathcal{Q}_n - I)(r\hat{u}^N + f)\|_{L^2}.\end{aligned}$$

Furthermore, define the mass and stiffness matrices

$$H := ((\phi_i, \phi_j))_{1\le i\le m,\ 0\le j\le m+1} \in \mathbb{R}^{m,m+2},$$
$$\overline{G} := ((\phi_i', \phi_j'))_{0\le i,j\le m+1} \in \mathbb{R}^{m+2,m+2}.$$

Recall that the symmetric positive definite matrix

$$G = ((\phi_i', \phi_j'))_{1\le i,j\le m} \in \mathbb{R}^{m,m},$$

defined in (15.31), is the inner submatrix of $\overline{G}$ of order m.

Let $\overline{u}_h = (u_0, \dots, u_{m+1}) \in \mathbb{R}^{m+2}$ denote the vector representation of $u \in \overline{X}_n$ with respect to the basis $(\phi_0, \dots, \phi_{m+1})$, and analogously let $u_h = (u_1, \dots, u_m) \in \mathbb{R}^m$ denote the vector representation of $u \in X_n$ with respect to the basis $(\phi_1, \dots, \phi_m)$ of X_n. For $v \in \overline{X}_n$ and $u := \mathcal{P}_n\mathcal{K}v \in X_n$, for each $i = 1, \dots, m$ the definition of the orthogonal projection (15.9) yields

$$\begin{aligned}(Gu_h)_i &= \sum_{j=1}^{m} u_j(\phi_i', \phi_j') = (\phi_i', u') = (\phi_i', (\mathcal{P}_n\mathcal{K}v)') = (\phi_i', (\mathcal{K}v)') \\ &= -(\phi_i, (\mathcal{K}v)'') = (\phi_i, v) = \sum_{j=0}^{m+1} v_j(\phi_i, \phi_j) = (H\overline{v}_h)_i.\end{aligned}$$

Hence $u_h = G^{-1}H\overline{v}_h$ shows that the matrix representation of $\mathcal{P}_n\mathcal{K}|_{\overline{X}_n} : \overline{X}_n \to X_n$ with respect to the bases $(\phi_0, \dots, \phi_{m+1})$ of $\overline{X}_n$ and $(\phi_1, \dots, \phi_m)$ of X_n is $G^{-1}H$. Now define $v := \mathcal{Q}_n(r\hat{u}^N + f)$. Then

$$\begin{aligned}C_{21}' &= \|\hat{u} - \mathcal{P}_n\mathcal{K}v\|_{H_0^1} = [(\hat{u}_h - G^{-1}H\overline{v}_h)^t G(\hat{u}_h - G^{-1}H\overline{v}_h)]^{\frac{1}{2}} \qquad (15.38)\\ &= [(G\hat{u}_h - H\overline{v}_h)^t G^{-1}(G\hat{u}_h - H\overline{v}_h)]^{\frac{1}{2}} \le \frac{1}{\sqrt{\lambda_{\min}(G)}}\|G\hat{u}_h - H\overline{v}_h\|_2,\end{aligned}$$

where $\lambda_{\min}(G)$ denotes the smallest eigenvalue of the symmetric positive definite matrix G. A positive lower bound $\texttt{lambda} \le \lambda_{\min}(G)$ can be estimated and verified using INTLAB, for example, by the following pattern:

```
lambda = 0.99*min(eig(mid(G)));
I_m = eye(m);
while not(isspd(G-lambda*I_m))
    lambda = 0.99*lambda;
end
```

Here `isspd` verifies that a symmetric matrix is positive definite; see Section 10.8.1. Note that the loop was never executed in our examples.

Having computed `lambda` successfully, define

$$C_{21} := \frac{1}{\sqrt{\texttt{lambda}}}\|G\hat{u}_h - H\overline{v}_h\|_2.$$

Then the constant

$$C_2 := C_{21} + C_{22} + C_{23} \tag{15.39}$$

is an upper bound for $\|\hat{u} - \mathcal{K}(r\hat{u}^N + f)\|_{H_0^1}$. Besides

$$C_{22} = \frac{h^2}{\pi^2}\sqrt{\overline{v}_h^t \overline{G}\overline{v}_h},$$

we also have that

$$C_{23} = \frac{1}{\pi}\|(\mathcal{Q}_n - I)(r\hat{u}^N + f)\|_{L^2} \le \frac{h^2}{\pi^3}\|(r\hat{u}^N + f)''\|_{L^2}$$

(see Theorem 15.1(b), (c)) is of order $O(h^2)$, so that in general C_2 cannot be expected to be of higher order.

Line (15.38) suggests a Newton method for determining an approximate solution $\hat{u}$, which is described below. Define

$$g : \mathbb{R}^m \to \mathbb{R}^m, \quad u \mapsto Gu - H\overline{v}_h,$$

where

$$v := \mathcal{Q}_n\left(r\left(\sum_{i=1}^m u_i\phi_i\right)^N + f\right) \in \overline{X}_n,$$

with vector representation $\overline{v}_h = (v_0, \ldots, v_{m+1}) \in \mathbb{R}^{m+2}$. For $\mathcal{Q}_n f,\ \mathcal{Q}_n r \in \overline{X}_n$ with vector representations

$$\begin{aligned}\overline{f}_h &= (f(0), f'(0), f(h), f'(h), \ldots, f(1-h), f'(1-h), f'(1), f(1))\\ &= (f_0, \ldots, f_{m+1}),\end{aligned}$$

and similarly $\overline{r}_h = (r_0, \ldots, r_{m+1})$, respectively, it follows, utilizing $N \ge 2$, for $i \in \{0, \ldots, m+1\}$, that

$$v_i = \begin{cases} f_i, & \text{if } i \in \{0, 1, m, m+1\},\\ r_i u_i^N + f_i, & \text{if } i \in \{2, \ldots, m-2\} \text{ is even},\\ r_i u_{i-1}^N + N r_{i-1} u_{i-1}^{N-1} u_i + f_i, & \text{if } i \in \{3, \ldots, m-1\} \text{ is odd}.\end{cases}$$

Hence the Jacobian J of g in $u \in \mathbb{R}^m$ reads, in MATLAB notation,

```
J = G;
for j = [2:2:m-2]
  J(:,j) = J(:,j)-N*(H(:,j)*r(j)*u(j)^(N-1)...
                   +H(:,j+1)*(r(j+1)*u(j)^(N-1)...
                   +(N-1)*r(j)*u(j)^(N-2)*u(j+1)));
end
for j = [3:2:m-1]
  J(:,j) = J(:,j)-N*H(:,j)*r(j-1)*u(j-1)^(N-1);
end
```

and a Newton step improves an approximation u into $u - du$ with $du := J^{-1}(Gu - H\overline{v}_h)$. In practice, of course, du is calculated as an approximate solution of the linear system with matrix J and right-hand side $Gu - H\overline{v}_h$.

Finally, to compute a constant C_3 satisfying (15.20), we use (15.5) and (15.6) to obtain, for $v, w, \varphi \in H_0^1$,

$$\begin{aligned}
&\|(F'(v) - F'(w))\varphi\|_{H_0^1} \\
&\quad = \|\mathcal{K}Nr(v^{N-1} - w^{N-1})\varphi\|_{H_0^1} \le \frac{N}{\pi}\|r(v^{N-1} - w^{N-1})\varphi\|_{L^2} \\
&\quad \le \frac{N}{\pi}\|r(v^{N-1} - w^{N-1})\|_{L^2}\|\varphi\|_\infty \le \frac{N}{2\pi}\|r(v^{N-1} - w^{N-1})\|_{L^2}\|\varphi\|_{H_0^1}.
\end{aligned}$$

If

$$\|v - \hat{u}\|_{H_0^1} < 2\alpha, \quad \|w - \hat{u}\|_{H_0^1} < 2\alpha,$$

and thus

$$\|v - \hat{u}\|_\infty < \alpha, \quad \|w - \hat{u}\|_\infty < \alpha$$

by (15.6), this means, using (15.4), that

$$\begin{aligned}
&\|F'(v) - F'(w)\|_{\mathcal{L}(H_0^1, H_0^1)} \\
&\quad \le \frac{N}{2\pi}\|r(v^{N-1} - w^{N-1})\|_{L^2} \\
&\quad = \frac{N}{2\pi}\|r(v^{N-2} + v^{N-3}w + \cdots + vw^{N-3} + w^{N-2})(v - w)\|_{L^2} \\
&\quad \le \frac{N}{2\pi}\|r\|_\infty(\|v\|_\infty^{N-2} + \|v\|_\infty^{N-3}\|w\|_\infty + \cdots + \|w\|_\infty^{N-2})\|v - w\|_{L^2} \\
&\quad \le \frac{N(N-1)}{2\pi^2}\|r\|_\infty(\|\hat{u}\|_\infty + \alpha)^{N-2}\|v - w\|_{H_0^1},
\end{aligned}$$

which gives

$$C_3 := \frac{N(N-1)}{2\pi^2}\|r\|_\infty(\|\hat{u}\|_\infty + \alpha)^{N-2}. \tag{15.40}$$

If $\alpha\omega = C_1^2 C_2 C_3 \le \frac{1}{2}$, then the Newton–Kantorovich theorem implies that there exists a unique solution $u \in \overline{U_\rho(\hat{u})}$ satisfying the two-point boundary value problem (15.1), with ρ defined in (15.16).[25]

As a numerical example, consider

$$\begin{aligned}
-u'' &= (x+2)^3 u^3 - \cos 2\pi x, \quad 0 < x < 1, \\
u(0) &= u(1) = 0.
\end{aligned} \tag{15.41}$$

[25] Even though excluded in the beginning, the linear case $N = 1$ can be treated in a similar but more direct way resulting in $\|u - \hat{u}\|_{H_0^1} \le \alpha = C_1 C_2$. The constant C_3 is not needed.

Table 15.1. Guaranteed error estimates.

		Number of grid points			
		65	129	257	513
u_{-1}	$\alpha\omega$	0.85	$1.88 \cdot 10^{-1}$	$4.56 \cdot 10^{-2}$	$1.14 \cdot 10^{-2}$
	ρ	–	$2.39 \cdot 10^{-3}$	$5.41 \cdot 10^{-4}$	$1.34 \cdot 10^{-4}$
u_0	$\alpha\omega$	$4.77 \cdot 10^{-5}$	$1.18 \cdot 10^{-5}$	$2.95 \cdot 10^{-6}$	$7.37 \cdot 10^{-7}$
	ρ	$1.11 \cdot 10^{-4}$	$2.79 \cdot 10^{-5}$	$6.97 \cdot 10^{-6}$	$1.74 \cdot 10^{-6}$
u_1	$\alpha\omega$	0.59	$1.34 \cdot 10^{-1}$	$3.27 \cdot 10^{-2}$	$8.19 \cdot 10^{-3}$
	ρ	–	$2.09 \cdot 10^{-3}$	$4.88 \cdot 10^{-4}$	$1.21 \cdot 10^{-4}$

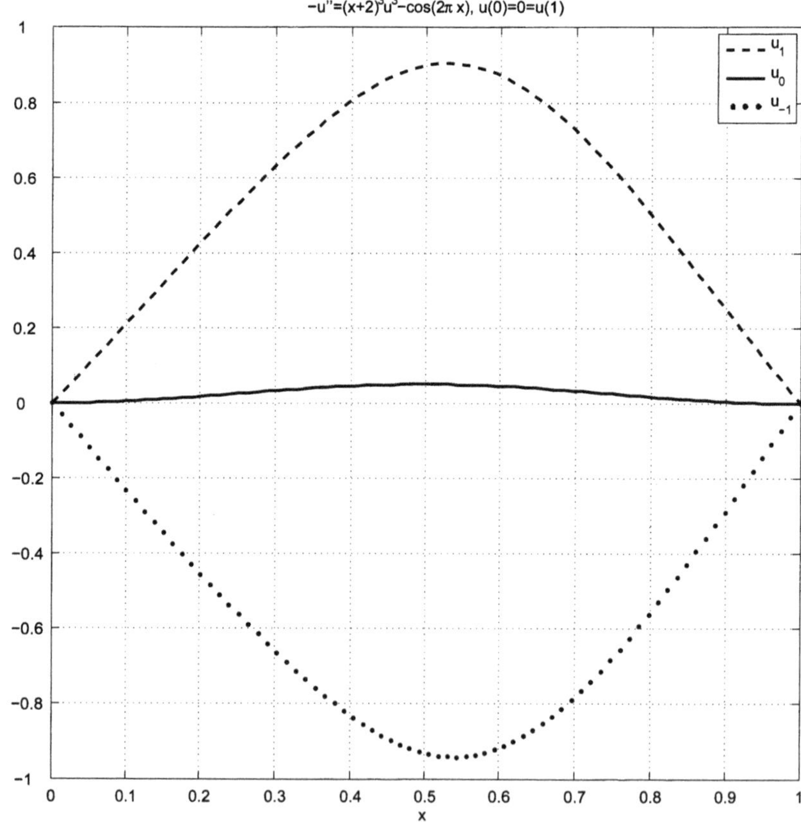

Figure 15.1. Verified inclusions for three distinct solutions of (15.41)

Three approximate solutions $\hat{u}_{-1}$, $\hat{u}_0$, $\hat{u}_1$ are obtained by some Newton iterates starting with $\hat{u}_{-1}^0 \equiv -1$, $\hat{u}_0^0 \equiv 0$, $\hat{u}_1^0 \equiv 1$, respectively. The computed bounds for $\alpha\omega$ and ρ are displayed in Table 15.1. If the number of grid points $n+1$ is at least 129, then $\alpha\omega < 0.5$ in all three cases. It follows that there exist solutions u_{-1}, u_0 and u_1 of (15.41) in the ball centred at $\hat{u}_i$ with corresponding radius ρ, *i.e.*, $\|u_i - \hat{u}_i\|_{H_0^1} \leq \rho$ and therefore

$$\|u_i - \hat{u}_i\|_\infty \leq \frac{1}{2}\rho \quad \text{for} \quad i = -1, 0, 1.$$

Since ρ is sufficiently small, these solutions are pairwise distinct. Note that $\|u_{-1}\|_\infty \approx 0.91$, $\|u_0\|_\infty \approx 0.05$ and $\|u_1\|_\infty \approx 0.95$, so that for all three solutions the relative error near the extremum is about 10^{-4}.

Figure 15.1 displays the $\|\cdot\|_\infty$-inclusions for these three solutions. The radius ρ is so small that the upper and lower bounds seem to lie on one single line. At first glance Figure 15.1 might suggest $u_{-1} = -u_1$, which is of course not the case, on checking (15.41) for symmetries. We do not know whether there are more solutions of (15.41) besides u_1, u_0, u_{-1}.

16. Semilinear elliptic boundary value problems (by Michael Plum, Karlsruhe)

In this final section we will describe in more detail a verification method for semilinear elliptic boundary value problems of the form

$$-\Delta u(x) + f(x, u(x)) = 0 \quad (x \in \Omega), \quad u(x) = 0 \quad (x \in \partial\Omega), \tag{16.1}$$

with $\Omega \subset \mathbb{R}^n$ denoting some given domain, and $f : \Omega \times \mathbb{R} \to \mathbb{R}$ some given nonlinearity.

Such problems have been (and still are) extensively studied in the differential equations literature, and they have a lot of applications, for instance in mathematical physics. Often they serve as model problems for more complex mathematical situations.

Starting perhaps with Picard's successive iterations at the end of the nineteenth century,various analytical methods and techniques have been (and are being) developed to study existence and multiplicity of solutions to problem (16.1), among them variational methods (including mountain pass methods), index and degree theory, monotonicity methods, fixed-point methods, and more. However, many questions remain open, offering opportunities for computer-assisted proofs and verification methods to supplement these purely analytical approaches.

As for finite-dimensional problems, we start with an approximate solution $\tilde{u}$ in some suitable function space and rewrite (16.1) as a boundary value problem for the error $v = u - \tilde{u}$. This is transformed into an equivalent fixed-point equation,

$$v \in X, \quad v = T(v), \tag{16.2}$$

in a Banach space X, and some fixed-point theorem is applied as in Sections 13 or 15.

In the finite-dimensional case, Brouwer's Fixed-Point Theorem was the easiest choice. Here we use its generalization to Banach spaces, that is, Schauder's Fixed-Point Theorem, provided that some compactness properties are available, or Banach's Fixed-Point Theorem if we are ready to accept an additional contraction condition. The existence of a solution v^* of (16.2) in some suitable set $V \subset X$ then follows, provided that

$$T(V) \subset V. \tag{16.3}$$

Consequently, $u^* := \tilde{u} + v^*$ is a solution of (16.1) (which gives the desired existence result), and $u^* \in \tilde{u} + V$ is an enclosure for u^*.

So the crucial condition to be verified, for some suitable set V, is (16.3). In the corresponding finite-dimensional situation, for given V the left-hand side of (16.3) was evaluated more or less in closed form, since there the operator $T : \mathbb{R}^n \to \mathbb{R}^n$ was composed of computable or enclosable terms such as the given nonlinear function, its Jacobian, and so forth.

In the infinite-dimensional situation, the operator T will in principle be built in a similar way, with the Jacobian now replaced by the elliptic differential operator L obtained by linearization of the left-hand side of (16.1). Evaluating or enclosing L^{-1} (as is needed if T is of Newton type, as in 13.2) is, however, not directly possible in general, and it is unclear how an approximation $C \approx L^{-1}$ should be computed such that $I - CL$ or $I - LC$ has a norm less than 1 (as in 13.3).

Therefore – choosing a Newton-type operator T – a normwise bound for L^{-1}, and also for the other ingredients of T, will be used instead. Again there is an analogy to the finite-dimensional case. If the system dimension is too large and hence the effort for enclosing – or even approximating – an inverse matrix A^{-1} is too high, a lower bound for the smallest singular value of A is used as in Section 10.8.1, which obviously corresponds to a norm bound for A^{-1}. If a norm ball V (centred at the origin) is chosen as a candidate for (16.3), then (16.3) results in an inequality involving the radius of V and the norm bounds indicated above. Since these bounds are computable, either directly or by additional computer-assisted means (such as the bound for $\|L^{-1}\|$), the resulting sufficient inequality for (16.3) can be checked.

It is important to remark that this inequality will indeed be satisfied if the approximate solution $\tilde{u}$ has been computed with sufficient accuracy, and if $\|L^{-1}\|$ is not too large (*i.e.*, if the situation is 'sufficiently non-degenerate'). Both conditions also appear in the finite-dimensional case.

We remark that Nagatou, Nakao and Yamamoto (1999), Nakao (1993) and Nakao and Yamamoto (1995) have chosen an approach avoiding the direct computation of a bound for $\|L^{-1}\|$. They use a finite-dimensional projection of L, which can be treated by the linear algebra verifying tools explained in the previous sections. However, the (infinite-dimensional) projection error also needs to be bounded in a suitable way, as in Nakao (1988) or Nakao *et al.* (2005), which is certainly possible for not too complicated domains Ω, but may be problematic, *e.g.*, for unbounded domains.[26]

16.1. Abstract formulation

In order to see the structural essence of the proposed enclosure methods for problem (16.1), and in particular its analogy to finite-dimensional problems, we first study (16.1) in the abstract form

$$\text{Find } u \in X \text{ satisfying } \mathcal{F}(u) = 0, \tag{16.4}$$

with $(X, \langle\cdot,\cdot\rangle_X)$ and $(Y, \langle\cdot,\cdot\rangle_Y)$ denoting two real Hilbert spaces, and $\mathcal{F} : X \to Y$ some Fréchet-differentiable mapping. Concrete choices of X and Y will be discussed in the next two subsections.

Let $\tilde{u} \in X$ denote some *approximate* solution to (16.4) (computed, *e.g.*, by numerical means), and denote by

$$L := \mathcal{F}'(\tilde{u}) : X \to Y, \quad x \mapsto L[x] \tag{16.5}$$

the Fréchet derivative of $\mathcal{F}$ at $\tilde{u}$, *i.e.*, $L \in \mathcal{B}(X, Y)$ (the Banach space of all bounded linear operators from X to Y) and

$$\lim_{\substack{h \in X \setminus \{0\} \\ h \to 0}} \frac{1}{\|h\|_X} \|\mathcal{F}(\tilde{u} + h) - \mathcal{F}(\tilde{u}) - L[h]\|_Y = 0.$$

Suppose that constants δ and K, and a non-decreasing function $g : [0, \infty) \to [0, \infty)$, are known such that

$$\|\mathcal{F}(\tilde{u})\|_Y \leq \delta, \tag{16.6}$$

i.e., δ bounds the *defect* (residual) of the approximate solution $\tilde{u}$ to (16.4),

$$\|u\|_X \leq K \, \|L[u]\|_Y \quad \text{for all} \quad u \in X, \tag{16.7}$$

i.e., K bounds the *inverse* of the linearization L,

$$\|\mathcal{F}'(\tilde{u} + u) - \mathcal{F}'(\tilde{u})\|_{\mathcal{B}(X,Y)} \leq g(\|u\|_X) \quad \text{for all} \quad u \in X, \tag{16.8}$$

i.e., g majorizes the modulus of continuity of $\mathcal{F}'$ at $\tilde{u}$, and

$$g(t) \to 0 \quad \text{as} \quad t \to 0 \tag{16.9}$$

(which, in particular, requires $\mathcal{F}'$ to be continuous at $\tilde{u}$).

The concrete computation of such δ, K, and g is the main challenge in our approach, with particular emphasis on K. We will however not address these

[26] Part of the following is taken from Plum (2008).

questions on the abstract level, but postpone them to the more specific case of the boundary value problem (16.1). For now, assume that (16.6)–(16.9) hold true.

In order to obtain a suitable fixed-point formulation (16.3) for our problem (16.4), the operator L must be *onto* because $L^{-1} : Y \to X$ will be used. (Note that L is one-to-one by (16.7).) There are two alternative ways to do this, both suited to the later treatment of problem (16.1).

(1) *The compact case.* Suppose that $\mathcal{F}$ admits a splitting

$$\mathcal{F} = L_0 + \mathcal{G}, \tag{16.10}$$

with a *bijective* linear operator $L_0 \in \mathcal{B}(X, Y)$, and a *compact* and Fréchet-differentiable operator $\mathcal{G} : X \to Y$, with compact Fréchet derivative $\mathcal{G}'(\tilde{u})$. Then the Fredholm alternative holds for the equation $L[u] = r$ (where $r \in Y$), and since L is one-to-one by (16.7), it is therefore onto.

(2) *The dual and symmetric case.* Suppose that $Y = X'$, the (topological) *dual* of X, *i.e.*, the space of all bounded linear functionals $l : X \to \mathbb{R}$. $X'(= \mathcal{B}(X, \mathbb{R}))$ is a Banach space endowed with the usual operator sup-norm. Indeed, this norm is generated by the inner product (which therefore makes X' a Hilbert space)

$$\langle r, s \rangle_{X'} := \langle \Phi^{-1}[r], \Phi^{-1}[s] \rangle_X \quad (r, s \in X'), \tag{16.11}$$

where $\Phi : X \to X'$ is the canonical isometric isomorphism given by

$$(\Phi[u])[v] := \langle u, v \rangle_X \ (u, v \in X). \tag{16.12}$$

To ensure that $L : X \to Y = X'$ is onto, we make the additional assumption that $\Phi^{-1}L : X \to X$ is *symmetric* with respect to $\langle \cdot, \cdot \rangle_X$, which by (16.12) amounts to the relation

$$(L[u])[v] = (L[v])[u] \quad \text{for all} \quad u, v \in X. \tag{16.13}$$

This implies the denseness of the range $(\Phi^{-1}L)(X) \subset X$: given any u in its orthogonal complement, we have, for all $v \in X$,

$$0 = \langle u, (\Phi^{-1}L)[v] \rangle_X = \langle (\Phi^{-1}L)[u], v \rangle_X,$$

and hence $(\Phi^{-1}L)[u] = 0$, which implies $L[u] = 0$ and thus $u = 0$ by (16.7).

Therefore, since Φ is isometric, the range $L(X) \subset X'$ is *dense*. To prove that L is onto, it remains to show that $L(X) \subset X'$ is *closed*. For this purpose, let $(L[u_n])_{n \in \mathbb{N}}$ denote some sequence in $L(X)$ converging to some $r \in X'$. Then (16.7) shows that $(u_n)_{n \in \mathbb{N}}$ is a Cauchy sequence in X. With $u \in X$ denoting its limit, the boundedness of L implies $L[u_n] \to L[u]$ $(n \to \infty)$. Thus, $r = L[u] \in L(X)$, proving closedness of $L(X)$.

We are now able to formulate and prove our main theorem, which is similar to the Newton–Kantorovich theorem.

Theorem 16.1. Let δ, K, g satisfy conditions (16.6)–(16.9). Suppose that some $\alpha > 0$ exists such that

$$\delta \leq \frac{\alpha}{K} - G(\alpha), \tag{16.14}$$

where $G(t) := \int_0^t g(s)\,\mathrm{d}s$. Moreover, suppose that:

(1) the *compact case* applies, or

(2) the *dual and symmetric case* applies, and we have the additional condition

$$Kg(\alpha) < 1. \tag{16.15}$$

Then, there exists a solution $u \in X$ of the equation $\mathcal{F}(u) = 0$ satisfying

$$\|u - \tilde{u}\|_X \leq \alpha. \tag{16.16}$$

Remark 1. Due to (16.9), $G(t) = \int_0^t g(s)\,\mathrm{d}s$ is *superlinearly* small as $t \to 0$. Therefore, the crucial condition (16.14) is indeed satisfied for some 'small' α if K is 'moderate' (*i.e.*, not too large) and δ is sufficiently small, which means, according to (16.6), that the approximate solution $\tilde{u}$ to problem (16.4) must be computed with *sufficient accuracy*, and (16.14) tells us *how* accurate the computation has to be.

Remark 2. To prove Theorem 16.1, the (abstract) *Green's operator* L^{-1} will be used to reformulate problem (16.4) as a fixed-point equation, and some fixed-point theorem will be applied. If the space X were finite-dimensional, *Brouwer's* Fixed-Point Theorem would be most suitable for this purpose. In the application to differential equation problems such as (16.1), however, X has to be infinite-dimensional, whence Brouwer's Theorem is not applicable. There are two choices.

(S) We use the generalization of Brouwer's theorem to infinite-dimensional spaces, *i.e.*, *Schauder's* Fixed-Point Theorem, which explicitly requires additional compactness properties (holding automatically in the finite-dimensional case). In our later application to (16.1), this compactness is given by compact embeddings of Sobolev function spaces, provided that the domain Ω is bounded (or at least has finite measure).

(B) We use *Banach's* Fixed-Point Theorem. No compactness is needed, but the additional contraction condition (16.15) is required. Due to (16.9), this condition is, however, not too critical if α (computed according to (16.14)) is 'small'. This option includes unbounded domains Ω.

Proof of Theorem 16.1. We rewrite problem (16.4) as

$$L[u-\tilde{u}] = -\mathcal{F}(\tilde{u}) - \big\{\mathcal{F}(u) - \mathcal{F}(\tilde{u}) - L[u-\tilde{u}]\big\},$$

which due to the bijectivity of L amounts to the equivalent fixed-point equation

$$v \in X, \quad v = -L^{-1}\big[\mathcal{F}(\tilde{u}) + \{\mathcal{F}(\tilde{u}+v) - \mathcal{F}(\tilde{u}) - L[v]\}\big] =: T(v) \qquad (16.17)$$

for the error $v = u - \tilde{u}$. We show the following properties of the fixed-point operator $T : X \to X$:

(i) $T(V) \subset V$ for the closed, bounded, non-empty, and convex norm ball

$$V := \{v \in X : \|v\|_X \le \alpha\},$$

(ii) T is continuous and compact (in case (1)) or contractive on V (in case (2)), respectively.

Then, Schauder's Fixed-Point Theorem (in case (1)) or Banach's Fixed-Point Theorem (in case (2)), respectively, gives a solution $v^* \in V$ of the fixed-point equation (16.17), whence by construction $u^* := \tilde{u} + v^*$ is a solution of $\mathcal{F}(u) = 0$ satisfying (16.16).

To prove (i) and (ii), we first note that for every differentiable function $f : [0,1] \to Y$, the real-valued function $\|f\|_Y$ is differentiable almost everywhere on $[0,1]$, and $(\mathrm{d}/\mathrm{d}t)\|f\|_Y \le \|f'\|_Y$ almost everywhere on $[0,1]$. Hence, for every $v, \tilde{v} \in X$,

$$\begin{aligned} &\|\mathcal{F}(\tilde{u}+v) - \mathcal{F}(\tilde{u}+\tilde{v}) - L[v-\tilde{v}]\|_Y \qquad (16.18)\\ &= \int_0^1 \frac{\mathrm{d}}{\mathrm{d}t}\,\|\mathcal{F}(\tilde{u}+(1-t)\tilde{v}+tv) - \mathcal{F}(\tilde{u}+\tilde{v}) - tL[v-\tilde{v}]\|_Y\,\mathrm{d}t\\ &\le \int_0^1 \|\{\mathcal{F}'(\tilde{u}+(1-t)\tilde{v}+tv) - L\}[v-\tilde{v}]\|_Y\,\mathrm{d}t\\ &\le \int_0^1 \|\mathcal{F}'(\tilde{u}+(1-t)\tilde{v}+tv) - L\|_{\mathcal{B}(X,Y)}\,\mathrm{d}t\cdot\|v-\tilde{v}\|_X\\ &\le \int_0^1 g(\|(1-t)\tilde{v}+tv\|_X)\,\mathrm{d}t\cdot\|v-\tilde{v}\|_X, \end{aligned}$$

using (16.5) and (16.8) at the last step. Choosing $\tilde{v} = 0$ in (16.18) we obtain, for each $v \in X$,

$$\begin{aligned} \|\mathcal{F}(\tilde{u}+v) - \mathcal{F}(\tilde{u}) - L[v]\|_Y &\le \int_0^1 g(t\|v\|_X)\,\mathrm{d}t\;\cdot\;\|v\|_X \qquad (16.19)\\ &= \int_0^{\|v\|_X} g(s)\,\mathrm{d}s = G(\|v\|_X). \end{aligned}$$

Furthermore, (16.18) and the fact that g is non-decreasing imply, for all $v, \tilde{v} \in V$,

$$\begin{aligned} \|\mathcal{F}(\tilde{u}+v) - \mathcal{F}(\tilde{u}+\tilde{v}) - L[v-\tilde{v}]\|_Y & \qquad (16.20) \\ &\leq \int_0^1 g((1-t)\|\tilde{v}\|_X + t\|v\|_X)\,\mathrm{d}t \cdot \|v-\tilde{v}\|_X \\ &\leq g(\alpha)\|v-\tilde{v}\|_X. \end{aligned}$$

To prove (i), let $v \in V$, *i.e.*, $\|v\|_X \leq \alpha$. Now (16.17), (16.7), (16.6), (16.19), and (16.14) imply

$$\begin{aligned} \|T(v)\|_X &\leq K\|\mathcal{F}(\tilde{u}) + \{\mathcal{F}(\tilde{u}+v) - \mathcal{F}(\tilde{u}) - L[v]\}\|_Y \\ &\leq K(\delta + G(\|v\|_X)) \leq K(\delta + G(\alpha)) \leq \alpha, \end{aligned}$$

which gives $T(v) \in V$. Thus, $T(V) \subset V$.

To prove (ii), suppose first that the *compact case* applies. So (16.10), which in particular gives $L = L_0 + \mathcal{G}'(\tilde{u})$, and (16.17) imply

$$T(v) = -L^{-1}\big[\mathcal{F}(\tilde{u}) + \{\mathcal{G}(\tilde{u}+v) - \mathcal{G}(\tilde{u}) - \mathcal{G}'(\tilde{u})[v]\}\big] \quad \text{for all} \quad v \in X,$$

whence continuity and compactness of T follow from continuity and compactness of $\mathcal{G}$ and $\mathcal{G}'(\tilde{u})$, and the boundedness of L^{-1} ensured by (16.7).

If the *dual and symmetric case* applies, (16.17), (16.7), and (16.20) imply, for $v, \tilde{v} \in V$,

$$\begin{aligned} \|T(v) - T(\tilde{v})\|_X &= \|L^{-1}\{\mathcal{F}(\tilde{u}+v) - \mathcal{F}(\tilde{u}+\tilde{v}) - L[v-\tilde{v}]\}\|_X \\ &\leq K\|\mathcal{F}(\tilde{u}+v) - \mathcal{F}(\tilde{u}+\tilde{v}) - L[v-\tilde{v}]\|_Y \\ &\leq Kg(\alpha)\|v-\tilde{v}\|_X, \end{aligned}$$

whence (16.15) shows that T is contractive on V. This completes the proof of Theorem 16.1. □

In the following two subsections, the abstract approach developed in this section will be applied to the elliptic boundary value problem (16.1). This can be done in (essentially two) different ways, *i.e.*, by different choices of the Hilbert spaces X and Y, resulting in different general assumptions (*e.g.*, smoothness conditions) to be made for the 'data' of the problem and the numerical approximation $\tilde{u}$, and different conditions (16.6)–(16.8), (16.14), (16.15), as well as different 'results', *i.e.*, existence statements and error bounds (16.16).

At this point, we briefly report on some other applications of our abstract setting which cannot be discussed in more detail in this article.

For *parameter-dependent problems* (where $\mathcal{F}$ in (16.4), or f in (16.1), depends on an additional parameter λ), one is often interested in *branches* $(u_\lambda)_{\lambda \in I}$ of solutions. By additional perturbation techniques, Plum (1995) generalized the presented method to a verification method for such solution branches, as long as the parameter interval I defining the branch is compact.

Such branches may, however, contain *turning points* (where a branch 'returns' at some value λ^*) or *bifurcation points* (where several – usually two – branches cross each other). Near such points, the operator L defined in (16.5) is 'almost' singular, *i.e.*, (16.7) holds only with a very large K, or not at all, and our approach breaks down. However, there are means to overcome these problems.

In the case of (simple) turning points, Plum (1994) used the well-known method of augmenting the given equation by a *bordering* equation (as in Section 13.2): the 'new' operator $\mathcal{F}$ in (16.4) contains the 'old' one *and* the bordering functional, and the 'new' operator L is regular near the turning point if the bordering equation has been chosen appropriately.

In the case of (simple) symmetry-breaking bifurcations, Plum (1996) includes in a first step the symmetry in the spaces X and Y, which excludes the symmetry-breaking branch and regularizes the problem, whence an existence and enclosure result for the symmetric branch can be obtained. In a second step, the symmetric branch is excluded by some transformation (similar to the Lyapunov–Schmidt reduction), and defining a corresponding new operator $\mathcal{F}$ an existence and enclosure result can also be obtained for the symmetry-breaking branch.

Lahmann and Plum (2004) treated *non-self-adjoint eigenvalue problems*, again using bordering equation techniques normalizing the unknown eigenfunction. So $\mathcal{F}$ now acts on pairs (u, λ), and is defined via the eigenvalue equation and the (scalar) normalizing equation. In this way it was possible to give the first known instability proof of the Orr–Sommerfeld equation with Blasius profile, which is a fourth-order ODE eigenvalue problem on $[0, \infty)$.

Also (other) *higher-order* problems are covered by our abstract setting. Breuer, Horák, McKenna and Plum (2006) could prove the existence of 36 travelling wave solutions of a fourth-order nonlinear beam equation on the real line. Biharmonic problems (with $\Delta\Delta u$ as leading term) are currently being investigated by B. Fazekas; see also Fazekas, Plum and Wieners (2005).

16.2. Strong solutions

We first study the elliptic boundary value problem (16.1) under the additional assumptions that f and $\partial f / \partial y$ are continuous on $\bar{\Omega} \times \mathbb{R}$, that the domain $\Omega \subset \mathbb{R}^n$ (with $n \leq 3$) is *bounded* with Lipschitz boundary, and H^2-*regular* (*i.e.*, , for each $r \in L^2(\Omega)$, and that the Poisson problem $-\Delta u = r$ in Ω, $u = 0$ on $\partial\Omega$ has a unique solution $u \in H^2(\Omega) \cap H_0^1(\Omega)$).

Here and in the following, $L^2(\Omega)$ denotes the Hilbert space of all (equivalence classes of) square-integrable Lebesgue-measurable real-valued functions on Ω, endowed with the inner product

$$\langle u, v\rangle_{L^2} := \int_\Omega uv \,\mathrm{d}x,$$

and $H^k(\Omega)$ is the Sobolev space of all functions $u \in L^2(\Omega)$ with weak derivatives up to order k in $L^2(\Omega)$. $H^k(\Omega)$ is a Hilbert space with the inner product

$$\langle u, v\rangle_{H^k} := \sum_{\substack{\alpha\in\mathbb{N}_0^n \\ |\alpha|\le k}} \langle D^\alpha u, D^\alpha v\rangle_{L^2},$$

where $|\alpha| := \alpha_1+\cdots+\alpha_n$, which can also be characterized as the completion of $C^\infty(\bar\Omega)$ with respect to $\langle\cdot,\cdot\rangle_{H^k}$. Replacement of $C^\infty(\bar\Omega)$ by $C_0^\infty(\Omega)$ (with the subscript 0 indicating compact support in Ω), yields, by completion, the Sobolev space $H_0^k(\Omega)$, which incorporates the vanishing of all derivatives up to order $k-1$ on $\partial\Omega$ in a weak sense.

Note that *piecewise* C^k-smooth functions u (*e.g.*, form functions of finite element methods) belong to $H^k(\Omega)$ if and only if they are (globally) in $C^{k-1}(\overline{\Omega})$.

For example, the assumption that Ω is H^2-regular is satisfied for C^2- (or $C^{1,1}$-)*smoothly* bounded domains (Gilbarg and Trudinger 1983), and also for *convex* polygonal domains $\Omega\subset\mathbb{R}^2$ (Grisvard 1985); it is *not* satisfied, *e.g.*, for domains with re-entrant corners, such as the L-shaped domain $(-1,1)^2\setminus[0,1)^2$.

Under the assumptions made, we can choose the spaces

$$X := H^2(\Omega)\cap H_0^1(\Omega), \quad Y := L^2(\Omega), \tag{16.21}$$

and the operators

$$\mathcal{F} := L_0 + \mathcal{G}, \quad L_0[u] := -\Delta u, \quad \mathcal{G}(u) := f(\cdot, u), \tag{16.22}$$

whence indeed our problem (16.1) amounts to the abstract problem (16.4). Moreover, $L_0 : X \to Y$ is bijective by the assumed unique solvability of the Poisson problem, and clearly bounded, that is, in $\mathcal{B}(X,Y)$. Finally, $\mathcal{G} : X \to Y$ is Fréchet-differentiable with derivative

$$\mathcal{G}'(u)[v] = \frac{\partial f}{\partial y}(\cdot, u)v, \tag{16.23}$$

which follows from the fact that $\mathcal{G}$ has this derivative as an operator from $C(\bar\Omega)$ (endowed with the maximum norm $\|\cdot\|_\infty$) into itself, and that the embeddings $H^2(\Omega)\hookrightarrow C(\bar\Omega)$ and $C(\bar\Omega)\hookrightarrow L^2(\Omega)$ are bounded. In fact, $H^2(\Omega)\hookrightarrow C(\bar\Omega)$ is even a *compact* embedding by the famous Sobolev–Kondrachov–Rellich Embedding Theorem (Adams 1975) (and since $n\le 3$), which shows that $\mathcal{G}$ and $\mathcal{G}'(u)$ (for any $u \in X$) are compact. Thus, the *compact case* (see (16.10)) applies.

For the application of Theorem 16.1, we are therefore left to comment on the computation of constants δ, K and a function g satisfying (16.6)–(16.9) (in the setting (16.21), (16.22)). But first, some comments on the computation of the approximate solution $\tilde{u}$ are necessary.

16.2.1. Computation of $\tilde{u}$

Since $\tilde{u}$ is required to be in $X = H^2(\Omega) \cap H_0^1(\Omega)$, it has to satisfy the boundary condition *exactly* (in the sense of being in $H_0^1(\Omega)$), and it needs to have weak derivatives in $L^2(\Omega)$ up to order 2. If finite elements are used, this implies the need for C^1-elements (*i.e.*, *globally* C^1-smooth finite element basis functions), which is a drawback at least on a technical level. (In the alternative approach proposed in the next subsection, this drawback is avoided.)

If $\Omega = (0,a) \times (0,b)$ is a *rectangle*, there are, however, many alternatives to finite elements, for example polynomial or trigonometric polynomial basis functions. In the latter case, $\tilde{u}$ is given in the form

$$\tilde{u}(x_1,x_2) = \sum_{i=1}^{N} \sum_{j=1}^{M} \alpha_{ij} \sin\left(i\pi\frac{x_1}{a}\right) \sin\left(j\pi\frac{x_2}{b}\right), \tag{16.24}$$

with coefficients α_{ij} to be determined by some numerical procedure. Such a procedure usually consists of a Newton iteration, together with a Ritz–Galerkin or a collocation method, and a linear algebraic system solver, which possibly incorporates multigrid methods. To *start* the Newton iteration, a rough initial approximation is needed, which may be obtained by path-following methods, or by use of the numerical mountain pass algorithm proposed by Choi and McKenna (1993).

An important remark is that, no matter how $\tilde{u}$ is given or which numerical method is used, there is *no* need for any *rigorous* computation at this stage, and therefore the whole variety of numerical methods applies.

16.2.2. Defect bound δ

Computing some δ satisfying (16.6) means, due to (16.21) and (16.22), computing an upper bound for (the square root of)

$$\int_\Omega \left[-\Delta\tilde{u} + f(\cdot,\tilde{u})\right]^2 \mathrm{d}x, \tag{16.25}$$

which should be 'small' if $\tilde{u}$ is a 'good' approximate solution. In some cases this integral can be calculated in closed form, by hand or by computer algebra routines, for example if f is polynomial and $\tilde{u}$ is piecewise polynomial (as it is if finite element methods have been used to compute it), or if $f(x,\cdot)$

is polynomial and both $f(\cdot, y)$ and $\tilde{u}$ (as in (16.24)) are trigonometric polynomials. The resulting formulas have to be evaluated *rigorously*, to obtain a true upper bound for the integral in (16.25).

If closed-form integration is impossible, a *quadrature formula* should be applied, possibly piecewise, to the integral in (16.25), using, for example, the methods described in Section 12. To obtain a true upper bound for the integral, in addition a *remainder term bound* for the quadrature formula is needed, which usually requires rough $\|\cdot\|_\infty$-bounds for some higher derivatives of the integrand. These rough bounds can be obtained, for example, by subdividing Ω into (many) small boxes, and performing interval evaluations of the necessary higher derivatives over each of these boxes (which gives true supersets of the function value ranges over each of the boxes, and thus, by union, over Ω).

16.2.3. Bound K for L^{-1}

The next task is the computation of a constant K satisfying (16.7), which, due to (16.21)–(16.23), means

$$\|u\|_{H^2} \le K\|L[u]\|_{L^2} \quad \text{for all} \quad u \in H^2(\Omega) \cap H_0^1(\Omega), \tag{16.26}$$

where $L : H^2(\Omega) \cap H_0^1(\Omega) \to L^2(\Omega)$ is given by

$$L[u] = -\Delta u + cu, \quad c(x) := \frac{\partial f}{\partial y}(x, \tilde{u}(x)) \ (x \in \bar{\Omega}). \tag{16.27}$$

The first (and most crucial) step towards (16.26) is the computation of a constant K_0 such that

$$\|u\|_{L^2} \le K_0\|L[u]\|_{L^2} \quad \text{for all} \quad u \in H^2(\Omega) \cap H_0^1(\Omega). \tag{16.28}$$

Choosing some constant lower bound $\underline{c}$ for c on $\bar{\Omega}$, and using the compact embedding $H^2(\Omega) \hookrightarrow L^2(\Omega)$, we find by standard means that $(L - \underline{c})^{-1} : L^2(\Omega) \to L^2(\Omega)$ is compact, symmetric, and positive definite, and hence has a $\langle\cdot,\cdot\rangle_{L^2}$-orthonormal and complete system $(\varphi_k)_{k\in\mathbb{N}}$ of eigenfunctions $\varphi_k \in H^2(\Omega) \cap H_0^1(\Omega)$, with associated sequence $(\mu_k)_{k\in\mathbb{N}}$ of (positive) eigenvalues converging monotonically to 0. Consequently, $L[\varphi_k] = \lambda_k\varphi_k$ for $k \in \mathbb{N}$, with $\lambda_k = \mu_k^{-1} + \underline{c}$ converging monotonically to $+\infty$. Series expansion yields, for every $u \in H^2(\Omega) \cap H_0^1(\Omega)$,

$$\begin{aligned}\|L[u]\|_{L^2}^2 &= \sum_{k=1}^{\infty}\langle L[u], \varphi_k\rangle_{L^2}^2 = \sum_{k=1}^{\infty}\langle u, L[\varphi_k]\rangle_{L^2}^2 = \sum_{k=1}^{\infty}\lambda_k^2\langle u, \varphi_k\rangle_{L^2}^2 \\ &\ge \Big(\min_{j\in\mathbb{N}}\lambda_j^2\Big)\sum_{k=1}^{\infty}\langle u, \varphi_k\rangle_{L^2}^2 = \Big(\min_{j\in\mathbb{N}}\lambda_j^2\Big)\|u\|_{L^2}^2,\end{aligned}$$

which shows that (16.28) holds if (and only if) $\lambda_j \neq 0$ for all $j \in \mathbb{N}$, and

$$K_0 \geq \left(\min_{j\in\mathbb{N}} |\lambda_j|\right)^{-1}. \tag{16.29}$$

Thus, bounds for the eigenvalue(s) of L neighbouring 0 are needed to compute K_0. Such *eigenvalue bounds* can be obtained by computer-assisted means of their own. For example, *upper* bounds to $\lambda_1, \ldots, \lambda_N$ (with $N \in \mathbb{N}$ given) are easily and efficiently computed by the *Rayleigh–Ritz* method (Rektorys 1980), as follows. Let $\tilde{\varphi}_1, \ldots, \tilde{\varphi}_N \in H^2(\Omega) \cap H_0^1(\Omega)$ denote linearly independent trial functions, for example approximate eigenfunctions obtained by numerical means, and form the matrices

$$A_1 := (\langle L[\tilde{\varphi}_i], \tilde{\varphi}_j\rangle_{L^2})_{i,j=1,\ldots,N}, \quad A_0 := (\langle \tilde{\varphi}_i, \tilde{\varphi}_j\rangle_{L^2})_{i,j=1,\ldots,N}.$$

Then, with $\Lambda_1 \leq \cdots \leq \Lambda_N$ denoting the eigenvalues of the matrix eigenvalue problem

$$A_1 x = \Lambda A_0 x,$$

which can be enclosed by the methods given in Section 13.4, the Rayleigh–Ritz method gives

$$\lambda_i \leq \Lambda_i \quad \text{for} \quad i = 1, \ldots, N.$$

However, to compute K_0 via (16.29), *lower* eigenvalue bounds are also needed, which constitute a more complicated task than upper bounds. The most accurate method for this purpose was proposed by Lehmann (1963), and its range of applicability improved by Goerisch (Behnke and Goerisch 1994). Its numerical core is again (as in the Rayleigh–Ritz method) a matrix eigenvalue problem, but the accompanying analysis is more involved.

In particular, in order to compute lower bounds to the first N eigenvalues, a *rough* lower bound to the $(N+1)$st eigenvalue must already be known. This *a priori* information can usually be obtained via a *homotopy method* connecting a simple 'base problem' with known eigenvalues to the given eigenvalue problem, such that all eigenvalues increase (index-wise) along the homotopy; details are given by Plum (1997) and Breuer *et al.* (2006).

Finding a base problem for the eigenvalue problem $L[u] = \lambda u$, and a suitable homotopy connecting them, is often possible along the following lines. If Ω is a bounded rectangle (whence the eigenvalues of $-\Delta$ on $H_0^1(\Omega)$ are known), choose a constant lower bound $\underline{c}$ for c on Ω, and the coefficient homotopy

$$c_s(x) := (1-s)\underline{c} + sc(x), \quad (x \in \Omega, 0 \leq s \leq 1).$$

Then, the family of eigenvalue problems

$$-\Delta u + c_s u = \lambda^{(s)} u \ \text{ in } \ \Omega, \quad u = 0 \quad \text{on } \partial\Omega$$

connects the explicitly solvable constant-coefficient base problem ($s = 0$) to the problem $L[u] = \lambda u$ ($s = 1$), and the eigenvalues increase in s, since the Rayleigh quotient does, by Poincaré's min-max principle.

If Ω is not a rectangle (or a ball), we can first choose a rectangle Ω_0 containing Ω, and a domain deformation homotopy between Ω_0 and Ω, to enclose the (first M) eigenvalues of $-\Delta$ on $H_0^1(\Omega)$: see, *e.g.*, Plum and Wieners (2002). Then, the above coefficient homotopy is applied in a second step.

Once a constant K_0 satisfying (16.28) is known, the desired constant K (satisfying (16.26)) can relatively easily be calculated by explicit *a priori estimates*. With $\underline{c}$ again denoting a constant lower bound for c, we obtain by partial integration, for each $u \in H^2(\Omega) \cap H_0^1(\Omega)$,

$$\|u\|_{L^2}\|L[u]\|_{L^2} \geq \langle u, L[u]\rangle_{L^2} = \int_\Omega (|\nabla u|^2 + cu^2)\,\mathrm{d}x \geq \|\nabla u\|_{L^2}^2 + \underline{c}\|u\|_{L^2}^2,$$

which implies, together with (16.28), that

$$\|\nabla u\|_{L^2} \leq K_1\|L[u]\|_{L^2}, \quad \text{where } K_1 := \begin{cases} \sqrt{K_0(1-\underline{c}K_0)} & \text{if } \underline{c}K_0 \leq \frac{1}{2}, \\ \frac{1}{2\sqrt{\underline{c}}} & \text{otherwise.} \end{cases} \tag{16.30}$$

To complete the H^2-bound required in (16.26), the L^2-norm of the (Frobenius matrix norm of the) Hessian matrix u_{xx} of $u \in H^2(\Omega) \cap H_0^1(\Omega)$ is to be estimated. If Ω is convex (as we shall assume now), then

$$\|u_{xx}\|_{L^2} \leq \|\Delta u\|_{L^2} \quad \text{for all } u \in H^2(\Omega) \cap H_0^1(\Omega); \tag{16.31}$$

see, *e.g.*, Grisvard (1985) and Ladyzhenskaya and Uralt́seva (1968). For the non-convex case, see Grisvard (1985) and Plum (1992). Now, with $\bar{c}$ denoting an additional upper bound for c, we choose

$$\mu := \max\left\{0, \frac{1}{2}(\underline{c} + \bar{c})\right\},$$

and calculate

$$\|\Delta u\|_{L^2} \leq \|-\Delta u + \mu u\|_{L^2} \leq \|L[u]\|_{L^2} + \|\mu - c\|_\infty \|u\|_{L^2}.$$

Using that $\|\mu - c\|_\infty = \max\{-\underline{c}, \frac{1}{2}(\bar{c} - \underline{c})\}$, and combining with (16.28) results in

$$\|\Delta u\|_{L^2} \leq K_2\|L[u]\|_{L^2}, \quad \text{where } K_2 := 1 + K_0 \max\left\{-\underline{c}, \frac{1}{2}(\bar{c} - \underline{c})\right\}. \tag{16.32}$$

Now, (16.28), (16.30) and (16.32) give (16.26) as follows. For quantitative purposes, we use the modified inner product

$$\langle u, v\rangle_{H^2} := \gamma_0\langle u, v\rangle_{L^2} + \gamma_1\langle\nabla u, \nabla v\rangle_{L^2} + \gamma_2\langle\Delta u, \Delta v\rangle_{L^2} \tag{16.33}$$

(with positive weights $\gamma_0, \gamma_1, \gamma_2$) on X, which, due to (16.31), and to the obvious reverse inequality $\|\Delta u\|_{L^2} \leq \sqrt{n}\|u_{xx}\|_{L^2}$, is equivalent to the canonical one. Then, (16.26) obviously holds for

$$K := \sqrt{\gamma_0 K_0^2 + \gamma_1 K_1^2 + \gamma_2 K_2^2}, \tag{16.34}$$

with K_0, K_1, K_2 from (16.28), (16.30) and (16.32).

16.2.4. Local Lipschitz bound g for $\mathcal{F}'$

By (16.21), (16.22) and (16.23), condition (16.8) reads

$$\left\| \left[\frac{\partial f}{\partial y}(\cdot, \tilde{u} + u) - \frac{\partial f}{\partial y}(\cdot, \tilde{u}) \right] v \right\|_{L^2} \leq g(\|u\|_{H^2})\|v\|_{H^2} \tag{16.35}$$

$$\text{for all } \ u, v \in H^2(\Omega) \cap H_0^1(\Omega).$$

We start with a monotonically non-decreasing function $\tilde{g} : [0, \infty) \to [0, \infty)$ satisfying

$$\left| \frac{\partial f}{\partial y}(x, \tilde{u}(x) + y) - \frac{\partial f}{\partial y}(x, \tilde{u}(x)) \right| \leq \tilde{g}(|y|) \quad \text{for all } \ x \in \Omega, \ y \in \mathbb{R}, \tag{16.36}$$

and $\tilde{g}(t) \to 0$ as $t \to 0+$. In practice, such a function $\tilde{g}$ can usually be calculated by hand if a bound for $\|\tilde{u}\|_\infty$ is available, which in turn can be computed by interval evaluations of $\tilde{u}$ over small boxes (as described at the end of Section 16.2.2). Using $\tilde{g}$, the left-hand side of (16.35) can be bounded by

$$\tilde{g}(\|u\|_\infty)\|v\|_{L^2}, \tag{16.37}$$

leaving us to estimate both the norms $\|\cdot\|_{L^2}$ and $\|\cdot\|_\infty$ by $\|\cdot\|_{H^2}$. With ρ^* denoting the smallest eigenvalue of

$$-\Delta u = \rho u, \quad u \in H^2(\Omega) \cap H_0^1(\Omega),$$

we obtain by eigenfunction expansion that

$$\|\nabla u\|_{L^2}^2 = \langle u, -\Delta u \rangle_{L^2} \geq \rho^* \|u\|_{L^2}^2, \quad \|\Delta u\|_{L^2}^2 \geq (\rho^*)^2 \|u\|_{L^2}^2,$$

and thus, by (16.33),

$$\|u\|_{L^2} \leq [\gamma_0 + \gamma_1 \rho^* + \gamma_2 (\rho^*)^2]^{-\frac{1}{2}} \|u\|_{H^2} \quad \text{for all } \ u \in H^2(\Omega) \cap H_0^1(\Omega). \tag{16.38}$$

Furthermore, in Plum (1992, Corollary 1) constants C_0, C_1, C_2 are calculated which depend on Ω in a rather simple way, allowing explicit computation, such that

$$\|u\|_\infty \leq C_0\|u\|_{L^2} + C_1\|\nabla u\|_{L^2} + C_2\|u_{xx}\|_{L^2} \quad \text{for all } \ u \in H^2(\Omega) \cap H_0^1(\Omega),$$

whence by (16.31) and (16.33) we obtain

$$\|u\|_\infty \le \left[\gamma_0^{-1}C_0^2 + \gamma_1^{-1}C_1^2 + \gamma_2^{-1}C_2^2\right]^{\frac{1}{2}} \|u\|_{H^2} \quad \text{for all} \quad u \in H^2(\Omega) \cap H_0^1(\Omega). \tag{16.39}$$

Using (16.38) and (16.39) in (16.37), we find that (16.35) (and (16.9)) hold for

$$g(t) := [\gamma_0 + \gamma_1\rho^* + \gamma_2(\rho^*)^2]^{-\frac{1}{2}}\tilde{g}\left(\left[\gamma_0^{-1}C_0^2 + \gamma_1^{-1}C_1^2 + \gamma_2^{-1}C_2^2\right]^{\frac{1}{2}} t\right). \tag{16.40}$$

16.2.5. A numerical example

Consider the problem

$$\Delta u + u^2 = s\cdot\sin(\pi x_1)\sin(\pi x_2) \ \ (x = (x_1, x_2) \in \Omega := (0,1)^2), \ \ u = 0 \ \text{on} \ \partial\Omega. \tag{16.41}$$

The results reported here were established in Breuer, McKenna and Plum (2003).

Since the 1980s it had been conjectured in the PDE community that problem (16.41) has at least four solutions for sufficiently large $s > 0$.

For $s = 800$, indeed four essentially different approximate solutions could be computed by the numerical mountain pass algorithm developed by Choi and McKenna (1993), where 'essentially different' means that none of them is an elementary symmetry transform of another one. Using finite Fourier series of the form (16.24), and a Newton iteration in combination with a collocation method, the accuracy of the mountain pass solutions was improved, resulting in highly accurate approximations $\tilde{u}_1, \ldots, \tilde{u}_4$ of the form (16.24).

Then the described verification method was applied to each of these four approximations, and the corresponding four inequalities (16.14) were successfully verified, with four error bounds $\alpha_1, \ldots, \alpha_4$. Therefore, Theorem 16.1 guarantees the existence of four solutions

$$u_1, \ldots, u_4 \in H^2(\Omega) \cap H_0^1(\Omega)$$

of problem (16.41) such that

$$\|u_i - \tilde{u}_i\|_{H^2} \le \alpha_i \quad \text{for} \quad i = 1, \ldots, 4.$$

Using the embedding inequality (16.39), in addition

$$\|u_i - \tilde{u}_i\|_\infty \le \beta_i \quad \text{for} \quad i = 1, \ldots, 4 \tag{16.42}$$

for $\beta_i := [\gamma_0^{-1}C_0^2 + \gamma_1^{-1}C_1^2 + \gamma_2^{-1}C_2^2]^{\frac{1}{2}}\alpha_i$. Finally, it is easy to check on the basis of the numerical data that

$$\|S\tilde{u}_i - \tilde{u}_j\|_\infty > \beta_i + \beta_j \quad \text{for} \quad i, j = 1, \ldots, 4, \quad i \ne j$$

for each elementary (rotation or reflection) symmetry transformation S of the square Ω, so that $u_1, \ldots, u_4$ are indeed *essentially different.*

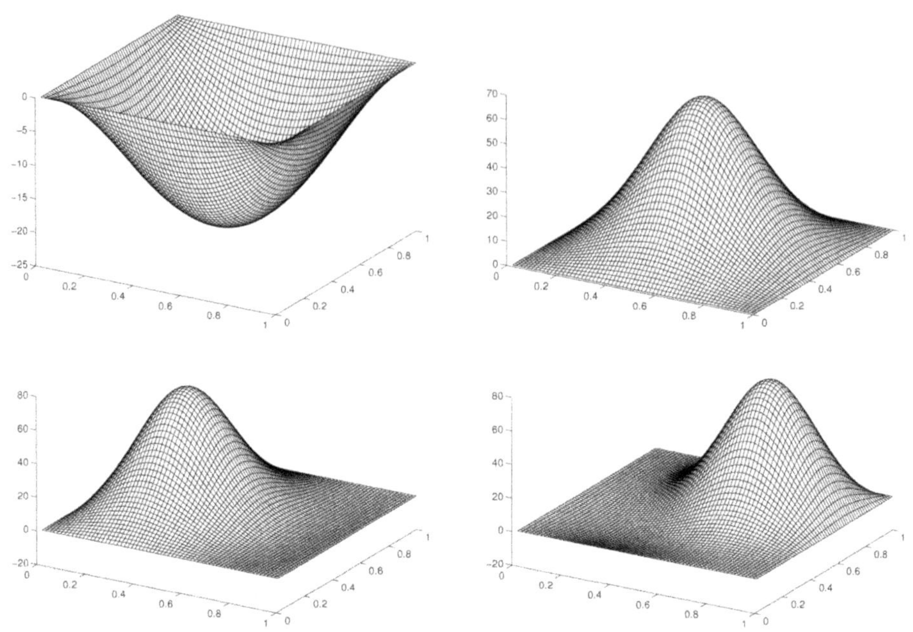

Figure 16.1. Four solutions to problem (16.41), $s = 800$.

Figure 16.1 shows plots of $\tilde{u}_1, \ldots, \tilde{u}_4$ (one might also say $u_1, \ldots, u_4$, since the error bounds β_i are much smaller than the 'optical accuracy' of the figure). The first two solutions are fully symmetric (with respect to reflection at the axes $x_1 = \frac{1}{2}, x_2 = \frac{1}{2}, x_1 = x_2, x_1 = 1 - x_2$), while the third is symmetric only with respect to $x_2 = \frac{1}{2}$, and the fourth only with respect to $x_1 = x_2$. Table 16.1 shows the defect bounds δ (see (16.6), (16.25)), the constants K satisfying (16.7) (or (16.26)), and the $\|\cdot\|_\infty$-error bounds β (see (16.42)) for the four solutions.

We wish to remark that, two years after publication of this result, Dancer and Yan (2005) gave a more general analytical proof (which we believe was stimulated by Breuer *et al.* (2003)); they even proved that the number of solutions of problem (16.41) becomes unbounded as $s \to \infty$.

16.3. Weak solutions

We will now investigate problem (16.1) under weaker assumptions on the domain $\Omega \subset \mathbb{R}^n$ and on the numerical approximation method, but stronger assumptions on the nonlinearity f, compared with the 'strong solutions' approach described in the previous subsection. Ω is now allowed to be any (bounded or unbounded) domain with Lipschitz boundary. We choose the spaces

$$X := H_0^1(\Omega), \quad Y := H^{-1}(\Omega) \tag{16.43}$$

Table 16.1. Enclosure results for problem (16.41).

Approximate solution	Defect bound δ	K (see (16.26))	Error bound β
$\tilde{u}_1$	0.0023	0.2531	$5.8222 \cdot 10^{-4}$
$\tilde{u}_2$	0.0041	4.9267	0.0228
$\tilde{u}_3$	0.0059	2.8847	0.0180
$\tilde{u}_4$	0.0151	3.1436	0.0581

for our abstract setting, where $H^{-1}(\Omega) := (H_0^1(\Omega))'$ denotes the topological dual space of $H_0^1(\Omega)$, *i.e.*, the space of all bounded linear functionals on $H_0^1(\Omega)$, where $H_0^1(\Omega)$ is endowed with the inner product

$$\langle u, v\rangle_{H_0^1} := \langle \nabla u, \nabla v\rangle_{L^2} + \sigma\langle u, v\rangle_{L^2} \tag{16.44}$$

(with some parameter $\sigma > 0$ to be chosen later), and $H^{-1}(\Omega)$ with the 'dual' inner product given by (16.11), with Φ from (16.12).

To interpret our problem (16.1) in these spaces, we first need to define Δu (for $u \in H_0^1(\Omega)$), or more generally, $\operatorname{div}\rho$ (for $\rho \in L^2(\Omega)^n$), as an element of $H^{-1}(\Omega)$. This definition simply imitates partial integration: the functional $\operatorname{div}\rho : H_0^1(\Omega) \to \mathbb{R}$ is given by

$$(\operatorname{div}\rho)[\varphi] := -\int_\Omega \rho \cdot \nabla\varphi \, \mathrm{d}x \quad \text{for all } \varphi \in H_0^1(\Omega), \tag{16.45}$$

implying in particular that

$$|(\operatorname{div}\rho)[\varphi]| \leq \|\rho\|_{L^2}\|\nabla\varphi\|_{L^2} \leq \|\rho\|_{L^2}\|\varphi\|_{H_0^1},$$

whence $\operatorname{div}\rho$ is indeed a *bounded* linear functional, and

$$\|\operatorname{div}\rho\|_{H^{-1}} \leq \|\rho\|_{L^2}. \tag{16.46}$$

Using this definition of $\Delta u (= \operatorname{div}(\nabla u))$, it is easy to check that the canonical isometric isomorphism $\Phi : H_0^1(\Omega) \to H^{-1}(\Omega)$ defined in (16.12) is now given by (note that (16.44))

$$\Phi[u] = -\Delta u + \sigma u \quad (u \in H_0^1(\Omega)), \tag{16.47}$$

where $\sigma u \in H_0^1(\Omega)$ is interpreted as an element of $H^{-1}(\Omega)$, as explained in the following.

Next, we give a meaning to a *function* being an element of $H^{-1}(\Omega)$, in order to define $f(\cdot, u)$ in (16.1) (and σu in (16.47)) in $H^{-1}(\Omega)$. For this purpose, let $\mathcal{L}$ denote the linear space consisting of all (equivalence classes of) Lebesgue-measurable functions $w : \Omega \to \mathbb{R}$ such that

$$\sup\left\{ \frac{1}{\|\varphi\|_{H_0^1}} \int_\Omega |w\varphi| \, dx : \varphi \in H_0^1(\Omega) \setminus \{0\} \right\} < \infty. \tag{16.48}$$

For each $w \in \mathcal{L}$, we can define an associated linear functional $\ell_w : H_0^1(\Omega) \to \mathbb{R}$ by

$$\ell_w[\varphi] := \int_\Omega w\varphi \, dx \quad \text{for all} \quad \varphi \in H_0^1(\Omega).$$

ℓ_w is bounded due to (16.48) and hence in $H^{-1}(\Omega)$. Identifying $w \in \mathcal{L}$ with its associated functional $\ell_w \in H^{-1}(\Omega)$, it follows that

$$\mathcal{L} \subset H^{-1}(\Omega), \tag{16.49}$$

and $\|w\|_{H^{-1}}$ is less than or equal to the left-hand side of (16.48), for every $w \in \mathcal{L}$.

To get a better impression of the functions contained in $\mathcal{L}$, recall that Sobolev's Embedding Theorem (Adams 1975, Theorem 5.4) gives $H_0^1(\Omega) \subset L^p(\Omega)$, with bounded embedding $H_0^1(\Omega) \hookrightarrow L^p(\Omega)$ (*i.e.*, there exists some constant $C_p > 0$ such that $\|u\|_{L^p} \leq C_p \|u\|_{H_0^1}$ for all $u \in H_0^1(\Omega)$), for each

$$p \in [2, \infty) \quad \text{if} \quad n = 2 \quad \text{and} \quad p \in \left[2, \frac{2n}{n-2}\right] \quad \text{if} \quad n \geq 3. \tag{16.50}$$

Here, $L^p(\Omega)$ denotes the Banach space of (equivalence classes of) Lebesgue-measurable functions $u : \Omega \to \mathbb{R}$ with finite norm

$$\|u\|_{L^p} := \left[\int_\Omega |u|^p \, dx \right]^{\frac{1}{p}}. \tag{16.51}$$

With p in the range (16.50), and p' denoting its dual number (*i.e.*, $p^{-1} + (p')^{-1} = 1$), Hölder's inequality, combined with the above embedding, yields that, for all $w \in L^{p'}(\Omega)$,

$$\int_\Omega |w\varphi| \, dx \leq \|w\|_{L^{p'}} \|\varphi\|_{L^p} \leq C_p \|w\|_{L^{p'}} \|\varphi\|_{H_0^1},$$

implying $w \in \mathcal{L}$, and $\|w\|_{H^{-1}} \leq C_p \|w\|_{L^{p'}}$. Consequently,

$$L^{p'}(\Omega) \subset \mathcal{L}, \tag{16.52}$$

and (note that (16.49)) the embedding $L^{p'}(\Omega) \hookrightarrow H^{-1}(\Omega)$ is bounded, with the same embedding constant C_p as in the 'dual' embedding $H_0^1(\Omega) \hookrightarrow L^p(\Omega)$. Note that the range (16.50) for p amounts to the range

$$p' \in (1, 2] \quad \text{if} \quad n = 2 \quad \text{and} \quad p' \in \left[\frac{2n}{n+2}, 2\right] \quad \text{if} \quad n \geq 3 \tag{16.53}$$

for the dual number p'. By (16.52), the linear span of the union of all $L^{p'}(\Omega)$, taken over p' in the range (16.53), is a subspace of $\mathcal{L}$, and this subspace is in fact all of $\mathcal{L}$, which is needed (and accessed) in practical applications.

Coming back to our problem (16.1), we now simply require that

$$f(\cdot, u) \in \mathcal{L} \quad \text{for all} \quad u \in H_0^1(\Omega), \tag{16.54}$$

in order to define the term $f(\cdot, u)$ as an element of $H^{-1}(\Omega)$. Our abstract setting requires, furthermore, that

$$\mathcal{F} : \begin{cases} H_0^1(\Omega) & \rightarrow & H^{-1}(\Omega) \\ u & \mapsto & -\Delta u + f(\cdot, u) \end{cases} \tag{16.55}$$

is Fréchet-*differentiable*. Since $\Delta : H_0^1(\Omega) \rightarrow H^{-1}(\Omega)$ is linear and bounded by (16.46), this amounts to the Fréchet-differentiability of

$$\mathcal{G} : \begin{cases} H_0^1(\Omega) & \rightarrow & H^{-1}(\Omega) \\ u & \mapsto & f(\cdot, u). \end{cases} \tag{16.56}$$

For this purpose, we require (as in the previous subsection) that $\partial f/\partial y$ is continuous on $\bar{\Omega} \times \mathbb{R}$. But in contrast to the 'strong solutions' setting, this is not sufficient here; the main reason is that $H_0^1(\Omega)$ does not embed into $C(\bar{\Omega})$. We need additional *growth restrictions* on $f(x, y)$ or $(\partial f/\partial y)(x, y)$ as $|y| \rightarrow \infty$.

An important (but not the only) admissible class consists of those functions f which satisfy

$$f(\cdot, 0) \in \mathcal{L}, \tag{16.57}$$

$$\frac{\partial f}{\partial y}(\cdot, 0) \quad \text{is a bounded function on } \Omega, \tag{16.58}$$

$$\left| \frac{\partial f}{\partial y}(x, y) - \frac{\partial f}{\partial y}(x, 0) \right| \leq c_1 |y|^{r_1} + c_2 |y|^{r_2} \quad (x \in \Omega, \ \ y \in \mathbb{R}), \tag{16.59}$$

with non-negative constants c_1, c_2, and with

$$0 < r_1 \leq r_2 < \infty \quad \text{if} \ \ n = 2, \qquad 0 < r_1 \leq r_2 \leq \frac{4}{n-2} \quad \text{if} \ \ n \geq 3. \tag{16.60}$$

(A 'small' r_1 will make condition (16.59) weak near $y = 0$, and a 'large' r_2 will make it weak for $|y| \rightarrow \infty$.)

Lemma 16.2. Let f satisfy (16.57)–(16.59), and assume the continuity of $\partial f/\partial y$. Then $\mathcal{G}$ given by (16.56) is well-defined and Fréchet-differentiable, with derivative $\mathcal{G}'(u) \in \mathcal{B}(H_0^1(\Omega), H^{-1}(\Omega))$ (for $u \in H_0^1(\Omega)$) given by

$$(\mathcal{G}'(u)[v])[\varphi] = \int_\Omega \frac{\partial f}{\partial y}(\cdot, u) v \varphi \, \mathrm{d}x \quad (v, \varphi \in H_0^1(\Omega)). \tag{16.61}$$

The proof of Lemma 16.2 is rather technical, and therefore omitted here. According to (16.45) and (16.61), we have

$$(\mathcal{F}'(u)[\varphi])[\psi] = \int_\Omega \left[\nabla\varphi \cdot \nabla\psi + \frac{\partial f}{\partial y}(\cdot, u)\varphi\psi\right] \mathrm{d}x \qquad (16.62)$$
$$= (\mathcal{F}'(u)[\psi])[\varphi] \quad (u, \varphi, \psi \in H_0^1(\Omega))$$

for the operator $\mathcal{F}$ defined in (16.55), which in particular implies condition (16.13) (for *any* $\tilde{u} \in H_0^1(\Omega)$; note that (16.5)), in the setting (16.43) and (16.55). Thus, the *dual and symmetric case* (see Section 16.2) applies.

We mention that several simplifications and extensions are possible if the domain Ω is *bounded.*

Again, we now comment on the computation of an approximate solution $\tilde{u}$, and of the terms δ, K, and g satisfying (16.6)–(16.9), needed for the application of Theorem 16.1, here in the setting (16.43) and (16.55).

16.3.1. Computation of $\tilde{u}$

By (16.43), $\tilde{u}$ needs to be in $X = H_0^1(\Omega)$ only (and no longer in $H^2(\Omega)$, as in the 'strong solutions' approach of the previous subsection). In the finite element context, this significantly increases the class of permitted elements; for example, the 'usual' linear (or quadratic) triangular elements can be used. In the case of an unbounded domain Ω, we are, furthermore, allowed to use approximations $\tilde{u}$ of the form

$$\tilde{u} = \begin{cases} \tilde{u}_0 & \text{on } \Omega_0, \\ 0 & \text{on } \Omega \setminus \Omega_0, \end{cases} \qquad (16.63)$$

with $\Omega_0 \subset \Omega$ denoting some bounded subdomain (the 'computational' domain), and $\tilde{u}_0 \in H_0^1(\Omega_0)$ some approximate solution of the differential equation (16.1) on Ω_0, subject to Dirichlet boundary conditions on $\partial\Omega_0$.

We pose the additional condition of $\tilde{u}$ being *bounded*, which on one hand is satisfied anyway for all practical numerical schemes, and on the other hand turns out to be very useful in the following.

16.3.2. Defect bound δ

By (16.43) and (16.55), condition (16.6) for the defect bound δ now amounts to

$$\|-\Delta\tilde{u} + f(\cdot, \tilde{u})\|_{H^{-1}} \leq \delta, \qquad (16.64)$$

which is a slightly more complicated task than computing an upper bound for an integral (as was needed in Section 16.2). The best general way seems to be the following. First compute an additional approximation $\rho \in H(\mathrm{div}, \Omega)$ to $\nabla\tilde{u}$. (Here, $H(\mathrm{div}, \Omega)$ denotes the space of all vector-valued functions $\tau \in L^2(\Omega)^n$ with weak derivative $\mathrm{div}\,\tau$ in $L^2(\Omega)$. Hence, obviously $H(\operatorname{div},\Omega) \supset H^1(\Omega)^n$, and ρ can be computed *e.g.*, by interpolation (or some more general projection) of $\nabla\tilde{u}$ in $H(\operatorname{div},\Omega)$, or in $H^1(\Omega)^n$. It should be noted that ρ comes 'for free' as a part of the approximation, if *mixed* finite elements are used to compute $\tilde{u}$.

Furthermore, according to the arguments before and after (16.52), applied with $p = p' = 2$,

$$\|w\|_{H^{-1}} \leq C_2\|w\|_{L^2} \quad \text{for all} \quad w \in L^2(\Omega). \tag{16.65}$$

For *explicit* calculation of C_2, we refer to the appendix in Plum (2008). By (16.46) and (16.65),

$$\begin{aligned} \|-\Delta\tilde{u} + f(\cdot,\tilde{u})\|_{H^{-1}} &\leq \|\operatorname{div}(-\nabla\tilde{u} + \rho)\|_{H^{-1}} + \|-\operatorname{div}\rho + f(\cdot,\tilde{u})\|_{H^{-1}} \\ &\leq \|\nabla\tilde{u} - \rho\|_{L^2} + C_2\|-\operatorname{div}\rho + f(\cdot,\tilde{u})\|_{L^2}, \end{aligned} \tag{16.66}$$

which reduces the computation of a defect bound δ (satisfying (16.64)) to computing bounds for two *integrals*, *i.e.*, we are back to the situation already discussed in Section 16.2.2.

There is an alternative way to compute δ if $\tilde{u}$ is of the form (16.63), with $\tilde{u}_0 \in H^2(\Omega_0) \cap H_0^1(\Omega_0)$, and with Ω_0 having a Lipschitz boundary. This situation can arise, *e.g.*, if Ω is the whole of $\mathbb{R}^n$, and the 'computational' domain Ω_0 is chosen as a 'large' rectangle, whence $\tilde{u}_0$ can be given, for instance, in the form (16.24).

Using partial integration on Ω_0, it follows that

$$\begin{aligned} &\|-\Delta\tilde{u} + f(\cdot,\tilde{u})\|_{H^{-1}} \leq \\ &\qquad C_2\Big[\|-\Delta\tilde{u}_0 + f(\cdot,\tilde{u}_0)\|^2_{L^2(\Omega_0)} + \|f(\cdot,0)\|^2_{L^2(\Omega\setminus\Omega_0)}\Big]^{\frac{1}{2}} + C_{tr}\left\|\frac{\partial\tilde{u}_0}{\partial\nu_0}\right\|_{L^2(\partial\Omega_0)}, \end{aligned} \tag{16.67}$$

with C_{tr} denoting a constant for the trace embedding $H^1(\Omega_0) \hookrightarrow L^2(\partial\Omega_0)$, the explicit computation of which is addressed in the appendix of Plum (2008), and $\partial\tilde{u}_0/\partial\nu_0$, the normal derivative on $\partial\Omega_0$.

16.3.3. Bound K for L^{-1}

According to (16.43), condition (16.7) now reads

$$\|u\|_{H_0^1} \leq K\|L[u]\|_{H^{-1}} \quad \text{for all} \quad u \in H_0^1(\Omega), \tag{16.68}$$

with L defined in (16.5), now given by (note that (16.55), (16.56))

$$L = -\Delta + \mathcal{G}'(\tilde{u}) : H_0^1(\Omega) \to H^{-1}(\Omega).$$

Under the growth conditions (16.57)–(16.60), Lemma 16.2 (or (16.61)) shows that, more concretely,

$$(L[\varphi])[\psi] = \int_\Omega \left[\nabla\varphi\cdot\nabla\psi + \frac{\partial f}{\partial y}(\cdot,\tilde{u})\varphi\psi\right] \mathrm{d}x \quad (\varphi,\psi \in H_0^1(\Omega)). \tag{16.69}$$

Making use of the isomorphism $\Phi : H_0^1(\Omega) \to H^{-1}(\Omega)$ given by (16.12) or (16.47), we obtain

$$\|L[u]\|_{H^{-1}} = \|\Phi^{-1}L[u]\|_{H_0^1} \quad (u \in H_0^1(\Omega)).$$

Since, moreover, $\Phi^{-1}L$ is $\langle\cdot,\cdot\rangle_{H_0^1}$-symmetric by (16.69) and (16.13), and defined on the whole Hilbert space $H_0^1(\Omega)$, and hence *self-adjoint*, we find that (16.68) holds for any

$$K \geq [\min\{|\lambda| : \lambda \text{ is in the spectrum of } \Phi^{-1}L\}]^{-1}, \tag{16.70}$$

provided that the min is positive (which is clearly an unavoidable condition for $\Phi^{-1}L$ being invertible with bounded inverse). Thus, in order to compute K, bounds are needed for:

(I) the essential spectrum of $\Phi^{-1}L$ (*i.e.*, accumulation points of the spectrum, and eigenvalues of infinite multiplicity), and

(II) the isolated eigenvalues of $\Phi^{-1}L$ of finite multiplicity, more precisely those neighbouring 0.

With regard to (I), if Ω is unbounded, we suppose again that $\tilde{u}$ is given in the form (16.63), with some bounded Lipschitz domain $\Omega_0 \subset \Omega$. If Ω is bounded, we may assume the same, simply choosing $\Omega_0 := \Omega$ (and $\tilde{u}_0 := \tilde{u}$).

Now define $L_0 : H_0^1(\Omega) \to H^{-1}(\Omega)$ by (16.69), but with $(\partial f/\partial y)(x, \tilde{u}(x))$ replaced by $(\partial f/\partial y)(x, 0)$. Using the Sobolev–Kondrachov–Rellich Embedding Theorem (Adams 1975), implying the compactness of the embedding $H^1(\Omega_0) \hookrightarrow L^2(\Omega_0)$, we find that $\Phi^{-1}L - \Phi^{-1}L_0 : H_0^1(\Omega) \to H_0^1(\Omega)$ is compact. Therefore, the perturbation result given in Kato (1966, IV, Theorem 5.35) shows that the essential spectra of $\Phi^{-1}L$ and $\Phi^{-1}L_0$ coincide. Thus, being left with the computation of bounds for the essential spectrum of $\Phi^{-1}L_0$, one can use Fourier transform methods, for instance, if $\Omega = \mathbb{R}^n$ and $(\partial f/\partial y)(\cdot, 0)$ is constant, or Floquet theory if $(\partial f/\partial y)(\cdot, 0)$ is periodic. Alternatively, if

$$\frac{\partial f}{\partial y}(x, 0) \geq c_0 > -\rho^* \quad (x \in \Omega), \tag{16.71}$$

with $\rho^* \in [0, \infty)$ denoting the minimal point of the spectrum of $-\Delta$ on $H_0^1(\Omega)$, we obtain by straightforward estimates of the Rayleigh quotient that the (full) spectrum of $\Phi^{-1}L_0$, and thus in particular the essential spectrum, is bounded from below by $\min\{1, (c_0 + \rho^*)/(\sigma + \rho^*)\}$.

With regard to (II), for computing bounds to eigenvalues of $\Phi^{-1}L$, the parameter σ in the H_0^1-product (16.44) is chosen such that

$$\sigma > \frac{\partial f}{\partial y}(x, \tilde{u}(x)) \quad (x \in \Omega). \tag{16.72}$$

Thus, the right-hand side of (16.72) is assumed to be bounded above. Furthermore, assume that the infimum s_0 of the essential spectrum of $\Phi^{-1}L$ is *positive*, which is true, *e.g.*, if (16.71) holds. As a particular consequence of (16.72) (and (16.47)) we obtain that $s_0 \leq 1$ and all eigenvalues of $\Phi^{-1}L$ are less than 1, and that, via the transformation $\kappa = 1/(1-\lambda)$, the eigenvalue problem $\Phi^{-1}L[u] = \lambda u$ is equivalent to

$$-\Delta u + \sigma u = \kappa\left(\sigma - \frac{\partial f}{\partial y}(\cdot, \tilde{u})\right)u \tag{16.73}$$

(to be understood as an equation in $H^{-1}(\Omega)$), which is furthermore equivalent to the eigenvalue problem for the self-adjoint operator

$$R := (I_{H_0^1(\Omega)} - \Phi^{-1}L)^{-1}.$$

Thus, *defining* the essential spectrum of problem (16.73) to be that of R, we find that it is bounded from below by $1/(1-s_0)$ if $s_0 < 1$, and is empty if $s_0 = 1$. In particular, its infimum is larger than 1, since $s_0 > 0$ by assumption.

Therefore, the computer-assisted eigenvalue enclosure methods mentioned in Section 16.2.3 (which are also applicable to eigenvalues *below* the essential spectrum of a problem like (16.73); see Zimmermann and Mertins (1995)) can be used to enclose the eigenvalue(s) of problem (16.73) neighbouring 1 (if they exist), whence by the transformation $\kappa = 1/(1-\lambda)$ enclosures for the eigenvalue(s) of $\Phi^{-1}L$ neighbouring 0 are obtained (if they exist). Also taking s_0 into account, the desired constant K can now easily be computed via (16.70). (Note that $K = s_0^{-1}$ can be chosen if no eigenvalues below the essential spectrum exist.)

16.3.4. Local Lipschitz bound g for $\mathcal{F}'$

In the setting (16.43), (16.55), condition (16.8) now reads

$$\left|\int_\Omega \left[\frac{\partial f}{\partial y}(x, \tilde{u}(x)+u(x)) - \frac{\partial f}{\partial y}(x, \tilde{u}(x))\right] v(x)\varphi(x)\,\mathrm{d}x\right| \leq g(\|u\|_{H_0^1})\|v\|_{H_0^1}\|\varphi\|_{H_0^1} \tag{16.74}$$

for all $u, v, \varphi \in H_0^1(\Omega)$. Here, we assumed that the Fréchet derivative of $\mathcal{G}$ (defined in (16.56)) is given by (16.61), which is true, under the growth conditions (16.57)–(16.60), for example, and which we assume in the following.

As in the strong solution approach treated in Section 16.2, we start with a monotonically non-decreasing function $\tilde{g} : [0, \infty) \to [0, \infty)$ satisfying

$$\left|\frac{\partial f}{\partial y}(x, \tilde{u}(x) + y) - \frac{\partial f}{\partial y}(x, \tilde{u}(x))\right| \leq \tilde{g}(|y|) \quad \text{for all} \;\; x \in \Omega,\; y \in \mathbb{R}, \tag{16.75}$$

and $\tilde{g}(t) \to 0$ as $t \to 0+$, but now we require in addition that $\tilde{g}(t^{1/r})$ is a *concave* function of t. Here, $r := r_2$ is the (larger) exponent in (16.59).

In practice, $\tilde{g}$ can often be taken to have the form

$$\tilde{g}(t) = \sum_{j=1}^{N} a_j t^{\mu_j} \quad (0 \le t < \infty),$$

where $a_1, \ldots, a_N > 0$ and $\mu_1, \ldots, \mu_N \in (0, r]$ are arranged in order to satisfy (16.75).

According to (16.60), one can find some

$$q \in (1, \infty) \quad \text{if} \quad n = 2, \qquad q \in \left[\frac{n}{2}, \infty\right) \quad \text{if} \quad n \ge 3, \tag{16.76}$$

such that qr is in the range (16.50). Since (16.76) implies that $p := 2q/(q-1)$ is also in the range (16.50), both the embeddings $H_0^1(\Omega) \hookrightarrow L^{qr}(\Omega)$ and $H_0^1(\Omega) \hookrightarrow L^p(\Omega)$ are bounded.

Using in addition the concavity of $\psi(t) := \tilde{g}(t^{1/r})$ and Jensen's inequality (Bauer 1978), one can now prove that (16.74) holds for

$$g(t) := C_2^2 \cdot \tilde{g}\big(C_{qr}(C_p/C_2)^{\frac{2}{r}} t\big) \quad (0 \le t < \infty) \tag{16.77}$$

(Plum 2008), which also satisfies (16.9) and is non-decreasing.

16.3.5. A numerical example

We consider the problem of finding non-trivial solutions to the nonlinear Schrödinger equation

$$-\Delta u + V(x)u - u^2 = 0 \quad \text{on} \quad \Omega := \mathbb{R}^2, \tag{16.78}$$

where $V(x) = A + B\sin(\pi(x_1 + x_2))\sin(\pi(x_1 - x_2))$, with real parameters A and B. The results presented here have been obtained by B. Breuer, P. J. McKenna and M. Plum (unpublished).

We are interested only in solutions which are symmetric with respect to reflection about both coordinate axes. Thus, we include these symmetries in all function spaces used, and in the numerical approximation spaces.

The particular case $A = 6$, $B = 2$ is treated. On a 'computational' domain $\Omega_0 := (-\ell, \ell) \times (-\ell, \ell)$, an approximation $\tilde{u}_0 \in H^2(\Omega_0) \cap H_0^1(\Omega_0)$ of the differential equation in (16.78) was computed, with Dirichlet boundary conditions on $\partial\Omega_0$, in a finite Fourier series form like (16.24) (with $N = M = 80$).

To find $\tilde{u}_0$, we start with a non-trivial approximate solution for Emden's equation (which is (16.78) with $A = B = 0$) on Ω_0, and perform a path-following Newton method, deforming (A, B) from $(0, 0)$ into $(6, 2)$.

In the single Newton steps, a collocation method with equidistant collocation points is used. By increasing the side length of Ω_0 in an additional path following, the approximation $\tilde{u}_0$ remains 'stable', with rapidly decreasing normal derivative $\partial\tilde{u}_0/\partial\nu_0$ (on $\partial\Omega_0$), as ℓ increases; this gives rise to some hope that a 'good' approximation $\tilde{u}$ for problem (16.78) is obtained in the form (16.63).

For $\ell = 8$, $\|\partial\tilde{u}_0/\partial\nu_0\|_{L^2(\partial\Omega_0)}$ turned out to be small enough compared with $\| -\Delta\tilde{u}_0 + V\tilde{u}_0 - \tilde{u}_0^2\|_{L^2(\Omega_0)}$, and a defect bound δ (satisfying (16.64)) is computed via (16.67) as

$$\delta = 0.7102 \cdot 10^{-2}; \tag{16.79}$$

note that, by the results mentioned in the appendix of Plum (2008),

$$C_2 = \sigma^{-\frac{1}{2}}, \quad \text{and} \quad C_{tr} = \sigma^{-\frac{1}{2}}\big[\ell^{-1} + +\sqrt{\ell^{-2} + 2\sigma}\big]^{\frac{1}{2}}.$$

Moreover, (16.72) requires $\sigma > A + B = 8$ (since $\tilde{u}$ turns out to be non-negative). Choosing $\sigma := 9$, we obtain $C_2 \leq 0.3334$ and $C_{\text{tr}} \leq 0.6968$.

Since condition (16.71) holds for $c_0 = A - B = 4$ (and $\rho^* = 0$), the arguments following (16.71) give the lower bound $s_0 := 4/9 \geq 0.4444$ for the essential spectrum of $\Phi^{-1}L$, and hence the lower bound $1/(1 - s_0) = 1.8$ for the essential spectrum of problem (16.73).

By the eigenvalue enclosure methods mentioned in Section 16.2.3, the bounds

$$\kappa_1 \leq 0.5293, \quad \kappa_2 \geq 1.1769$$

for the first two eigenvalues of problem (16.73) could be computed, which by (16.70) leads to the constant

$$K = 6.653 \tag{16.80}$$

satisfying (16.68). To compute g satisfying (16.8) or (16.74), we first note that (16.75) holds for

$$\tilde{g}(t) := 2t,$$

and (16.59) for $r_1 = r_2 = 1$, whence the additional concavity condition is satisfied. Choosing $q := 2$ yields $qr = 2$ and $p = 4$ in the arguments following (16.76), whence (16.77) gives

$$g(t) = 2C_2C_4^2 t = \frac{1}{9}t \tag{16.81}$$

since $2C_2C_4^2 = \sigma^{-1}$ by Lemma 2a) in the appendix of Plum (2008).

Using (16.79)–(16.81), it follows that (16.14) and (16.15) hold for $\alpha = 0.04811$, whence Theorem 16.1 implies the existence of a solution $u^* \in H_0^1(\mathbb{R}^2)$ to problem (16.78) with

$$\|u^* - \tilde{u}\|_{H_0^1} \leq 0.04811. \tag{16.82}$$

It is easy to check on the basis of the numerical data that $\|\tilde{u}\|_{H_0^1} > 0.04811$, whence (16.82) shows in particular that u^* is non-trivial.

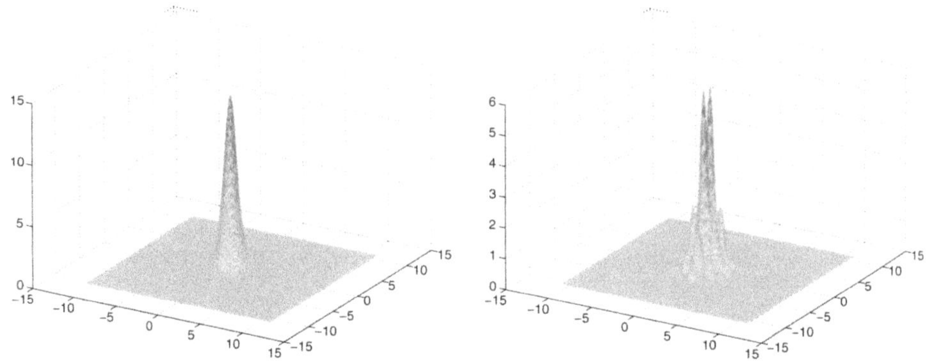

Figure 16.2. Example (16.78); $A = 6$, $B = 2$ (*left*) and $A = 6$, $B = 26$ (*right*).

We wish to remark that it would be of great interest to achieve such results also for cases where $0 < A < B$ in the potential V, because in this case V is no longer non-negative, which excludes an important class of purely analytical approaches to prove existence of a non-trivial solution.

So far, verification has not been successful for such cases, due to difficulties in the homotopy method that has to be used for our computer-assisted eigenvalue enclosures (see the brief remarks in Section 16.2.3); note that these difficulties occur on a rather 'technical' level. However, an apparently 'good' approximation $\tilde{u}$, *e.g.*, in the case $A = 6$, $B = 26$, could be computed.

Figure 16.2 shows plots of $\tilde{u}$ for the successful case $A = 6$, $B = 2$, and for the unsuccessful case $A = 6$, $B = 26$.

Acknowledgements

Many colleagues helped to improve earlier versions of this manuscript. In particular I want to express my sincere thanks to Florian Bünger, Luiz Henrique de Figueiredo, Viktor Härter, Christian Jansson, Raymond Moore, Arnold Neumaier, and Nick Trefethen for their thorough reading and for many valuable comments. Furthermore my thanks to John Pryce, Jiri Rohn, Hermann Schichl, Sergey Shary, and many others.

Moreover, my special thanks to Florian Bünger and my friend Michael Plum from Karlsruhe for their detailed advice on the penultimate section. In particular, I am indebted to Michael Plum for providing the last section of this paper. Finally, my thanks to my dear colleague and friend Professor Shin'ichi Oishi from Waseda university, Tokyo, for the opportunity to write this article in the pleasant atmosphere of his institute.

Last but not least, my thanks to the staff of *Acta Numerica*, chief amongst them Glennis Starling, for an excellent copy-editing job.

REFERENCES

J. P. Abbott and R. P. Brent (1975), 'Fast local convergence with single and multistep methods for nonlinear equations', *Austr. Math. Soc. B* **19**, 173–199.

ACRITH (1984), *IBM High-Accuracy Arithmetic Subroutine Library* (ACRITH), Release 1, IBM Deutschland GmbH, Böblingen.

R. A. Adams (1975), *Sobolev Spaces*, Academic Press, New York.

G. Alefeld, private communication.

G. Alefeld (1994), Inclusion methods for systems of nonlinear equations. In *Topics in Validated Computations* (J. Herzberger, ed.), Studies in Computational Mathematics, Elsevier, Amsterdam, pp. 7–26.

G. Alefeld and J. Herzberger (1974), *Einführung in die Intervallrechnung*, BI Wissenschaftsverlag.

G. Alefeld, V. Kreinovich and G. Mayer (1997), 'On the shape of the symmetric, persymmetric, and skew-symmetric solution set', *SIAM J. Matrix Anal. Appl.* **18**, 693–705.

G. Alefeld, V. Kreinovich and G. Mayer (2003), 'On the solution sets of particular classes of linear interval systems', *J. Comput. Appl. Math.* **152**, 1–15.

E. Anderson, Z. Bai, C. Bischof, J. Demmel, J. Dongarra, J. Du Croz, A. Greenbaum, S. Hammarling, A. McKenney, S. Ostrouchov and D. C. Sorensen (1995), *LAPACK User's Guide, Release* 2.0, 2nd edn, SIAM, Philadelphia.

M. V. A. Andrade, J. L. D. Comba and J. Stolfi (1994), Affine arithmetic. Extended abstract, presented at *INTERVAL'94*, St. Petersburg.

ARITHMOS (1986), *ARITHMOS: Benutzerhandbuch*, Siemens AG, Bibl.-Nr. U 2900-I-Z87-1 edition.

M. Aschbacher (1994), *Sporadic Groups*, Cambridge University Press.

A. Avizienis (1961), 'Signed-digit number representations for fast parallel arithmetic', *Ire Trans. Electron. Comp.* **EC-10**, 389–400.

H. Bauer (1978), *Wahrscheinlichkeitstheorie und Grundzüge der Maßtheorie*, 3rd edn, de Gruyter, Berlin.

O. Beaumont (2000), Solving interval linear systems with oblique boxes. Research report PI 1315, INRIA.

H. Behnke (1989), Die Bestimmung von Eigenwertschranken mit Hilfe von Variationsmethoden und Intervallarithmetik. Dissertation, Institut für Mathematik, TU Clausthal.

H. Behnke and F. Goerisch (1994), Inclusions for eigenvalues of selfadjoint problems. In *Topics in Validated Computations* (J. Herzberger, ed.), Studies in Computational Mathematics, Elsevier, Amsterdam, pp. 277–322.

A. Ben-Tal and A. Nemirovskii (2001), *Lectures on Modern Convex Optimization: Analysis, Algorithms, and Engineering Applications*, SIAM, Philadelphia.

F. Bernelli Zazzera, M. Vasile, M. Massari and P. Di Lizia (2004), Assessing the accuracy of interval arithmetic estimates in space flight mechanics. Final report, Ariadna id: 04/4105, Contract Number: 18851/05/NL/MV.

Y. Bertot and P. Castéran (2004), *Interactive Theorem Proving and Program Development, Coq'Art: The Calculus of Inductive Constructions*, Texts in Theoretical Computer Science, Springer.

M. Berz and K. Makino (1999), 'New methods for high-dimensional verified quadrature', *Reliable Computing* **5**, 13–22.

C. H. Bischof, A. Carle, G. Corliss and A. Griewank (1991), ADIFOR: Generating derivative codes from Fortran programs. Technical report, Mathematics and Computer Science Division, Argonne National Laboratory.

B. Borchers (1999), 'SDPLIB 1.2: A library of semidefinite programming test problems', *Optim. Methods Software* **11**, 683–690.

F. Bornemann, D. Laurie, S. Wagon and J. Waldvogel (2004), *The SIAM 100-Digit Challenge: A Study in High-Accuracy Numerical Computing*, SIAM, Philadelphia.

N. C. Börsken (1978), Komplexe Kreis-Standardfunktionen. Diplomarbeit, Freiburger Intervall-Ber. 78/2, Institut für Angewandte Mathematik, Universität Freiburg.

K. D. Braune (1987), Hochgenaue Standardfunktionen für reelle und komplexe Punkte und Intervalle in beliebigen Gleitpunktrastern. Dissertation, Universität Karlsruhe.

B. Breuer, J. Horák, P. J. McKenna and M. Plum (2006), 'A computer-assisted existence and multiplicity proof for travelling waves in a nonlinearly supported beam', *J. Diff. Equations* **224**, 60–97.

B. Breuer, P. J. McKenna and M. Plum (2003), 'Multiple solutions for a semilinear boundary value problem: A computational multiplicity proof', *J. Diff. Equations* **195**, 243–269.

B. M. Brown, D. K. R. McCormack and A. Zettl (2000), 'On a computer assisted proof of the existence of eigenvalues below the essential spectrum of the Sturm–Liouville problem', *J. Comput. Appl. Math.* **125**, 385–393.

M. W. Browne (1988), Is a math proof a proof if no one can check it? *The New York Times*, December 1988, p. 1.

J. R. Bunch, J. W. Demmel and C. F. Van Loan (1989), 'The strong stability of algorithms for solving symmetric linear systems', *SIAM J. Matrix Anal. Appl.* **10**, 494–499.

F. Bünger (2008), private communication.

O. Caprani and K. Madsen (1978), 'Iterative methods for interval inclusion of fixed points', *BIT Numer. Math.* **18**, 42–51.

F. Chatelin (1988), Analyse statistique de la qualité numérique et arithmétique de la résolution approchée d'équations par calcul sur ordinateur. Technical Report F.133, Centre Scientifique IBM–France.

Y. S. Choi and P. J. McKenna (1993), 'A mountain pass method for the numerical solutions of semilinear elliptic problems', *Nonlinear Anal. Theory Methods Appl.* **20**, 417–437.

L. Collatz (1942), 'Einschließungssatz für die charakteristischen Zahlen von Matrizen', *Math. Z.* **48**, 221–226.

G. F. Corliss and L. B. Rall (1987), 'Adaptive, self-validating numerical quadrature', *SIAM J. Sci. Statist. Comput.* **8**, 831–847.

G. Corliss, C. Faure, A. Griewank, L. Hascoët and U. Nauman (2002), *Automatic Differentiation of Algorithms: From Simulation to Optimization*, Springer.

A. Cuyt, B. Verdonk, S. Becuwe and P. Kuterna (2001), 'A remarkable example of catastrophic cancellation unraveled', *Computing* **66**, 309–320.

E. N. Dancer and S. S. Yan (2005), 'On the superlinear Lazer–McKenna conjecture', *J. Diff. Equations* **210**, 317–351.

G. Darboux (1876), 'Sur les développements en série des fonctions d'une seule variable', *J. des Mathématiques Pures et Appl.* **3**, 291–312.

M. Daumas, G. Melquiond and C. Muñoz (2005), Guaranteed proofs using interval arithmetic. In *Proc. 17th IEEE Symposium on Computer Arithmetic* (ARITH'05).

T. J. Dekker (1971), 'A floating-point technique for extending the available precision', *Numer. Math.* **18**, 224–242.

J. B. Demmel (1989), On floating point errors in Cholesky. LAPACK Working Note 14 CS-89-87, Department of Computer Science, University of Tennessee, Knoxville, TN, USA.

J. B. Demmel, B. Diament and G. Malajovich (2001), 'On the complexity of computing error bounds', *Found. Comput. Math.* **1**, 101–125.

J. B. Demmel, I. Dumitriu, O. Holtz and P. Koev (2008), Accurate and efficient expression evaluation and linear algebra. In *Acta Numerica*, Vol. 17, Cambridge University Press, pp. 87–145.

J. B. Demmel, Y. Hida, W. Kahan, X. S. Li, S. Mukherjee and E. J. Riedy (2004), Error bounds from extra precise iterative refinement. Report no. ucb/csd-04-1344, Computer Science Devision (EECS), University of California, Berkeley.

P. S. Dwyer (1951), *Linear Computations*, Wiley, New York/London.

C. Eckart and G. Young (1936), 'The approximation of one matrix by another of lower rank', *Psychometrika* **1**, 211–218.

J.-P. Eckmann, H. Koch and P. Wittwer (1984), 'A computer-assisted proof of universality for area-preserving maps', *Mem. Amer. Math. Soc.* **47**, 289.

P. Eijgenraam (1981), The solution of initial value problems using interval arithmetic.

B. Fazekas, M. Plum and C. Wieners (2005), Enclosure for biharmonic equation. In *Dagstuhl Online Seminar Proceedings 05391*. http://drops.dagstuhl.de/portal/05391/.

L. H. de Figueiredo and J. Stolfi (2004), 'Affine arithmetic: Concepts and applications', *Numer. Algorithms* **37**, 147–158.

L. V. Foster (1994), 'Gaussian elimination with partial pivoting can fail in practice', *SIAM J. Matrix Anal. Appl.* **14**, 1354–1362.

L. Fousse, G. Hanrot, V. Lefèvre, P. Pélissier and P. Zimmermann (2005), MPFR: A multiple-precision binary floating-point library with correct rounding. Research Report RR-5753, INRIA. Code and documentation available at: http://hal.inria.fr/inria-00000818.

A. Frommer (2001), Proving conjectures by use of interval arithmetic. In *Perspectives on Enclosure Methods: SCAN 2000* (U. Kulisch *et al.*, ed.), Springer.

Z. Galias and P. Zgliczynski (1998), 'Computer assisted proof of chaos in the Lorenz equations', *Physica* D **115**, 165–188.

I. Gargantini and P. Henrici (1972), 'Circular arithmetic and the determination of polynomial zeros', *Numer. Math.* **18**, 305–320.

B. Gidas, W. Ni and L. Nirenberg (1979), 'Symmetry and related problems via the maximum principle', *Comm. Math. Phys.* **68**, 209–243.

D. Gilbarg and N. S. Trudinger (1983), *Elliptic Partial Differential Equations of Second Order*, 2nd edn, Springer.

D. Goldberg (1991), 'What every computer scientist should know about floating-point arithmetic', *ACM Comput. Surv.* **23**, 5–48.

M. J. C. Gordon (2000), From LCF to HOL: A short history. In *Proof, Language, and Interaction: Essays in Honour of Robin Milner* (G. Plotkin, C. Stirling and M. Tofte, eds), MIT Press. http://www.cl.cam.ac.uk/~mjcg/papers/HolHistory.html.

D. Gorenstein, R. Lyons and R. Solomon (1994), *The Classification of the Finite Simple Groups*, Vol. 40 of *Math. Surveys Monographs*, AMS, Providence, RI.

A. Griewank (2003), A mathematical view of automatic differentiation. In *Acta Numerica*, Vol. 12, Cambridge University Press, pp. 321–398.

P. Grisvard (1985), *Elliptic Problems in Nonsmooth Domains*, Pitman, Boston.

E. Hansen (1969), The centered form. In *Topics in Interval Analysis* (E. Hansen, ed.), Oxford University Press, pp. 102–106.

E. R. Hansen and R. Smith (1967), 'Interval arithmetic in matrix computations II', *SIAM J. Numer. Anal.* **4**, 1–9.

G. Hargreaves (2002), Interval analysis in MATLAB. Master's thesis, University of Manchester. http://www.manchester.ac.uk/mims/eprints.

J. Hass, M. Hutchings and R. Schlafly (1995), 'The double bubble conjecture', *Electron. Res. Announc. Amer. Math. Soc.* **1**, 98–102.

D. J. Higham and N. J. Higham (1992*a*), 'Componentwise perturbation theory for linear systems with multiple right-hand sides', *Linear Algebra Appl.* **174**, 111–129.

D. J. Higham and N. J. Higham (1992*b*), 'Backward error and condition of structured linear systems', *SIAM J. Matrix Anal. Appl.* **13**, 162–175.

N. J. Higham (2002), *Accuracy and Stability of Numerical Algorithms*, 2nd edn, SIAM, Philadelphia.

J. Hölzl (2009), Proving real-valued inequalities by computation in Isabelle/HOL. Diplomarbeit, Fakultät für Informatik der Technischen Universität München.

IEEE 754 (2008), *ANSI/IEEE 754-2008: IEEE Standard for Floating-Point Arithmetic*, New York.

C. Jansson (1991), 'Interval linear systems with symmetric matrices, skew-symmetric matrices, and dependencies in the right hand side', *Computing* **46**, 265–274.

C. Jansson (1994), On self-validating methods for optimization problems. In *Topics in Validated Computations* (J. Herzberger, ed.), Studies in Computational Mathematics, Elsevier, Amsterdam, pp. 381–438.

C. Jansson (1997), 'Calculation of exact bounds for the solution set of linear interval systems', *Linear Algebra Appl.* **251**, 321–340.

C. Jansson (2004*a*), 'A rigorous lower bound for the optimal value of convex optimization problems', *J. Global Optim.* **28**, 121–137.

C. Jansson (2004*b*), 'Rigorous lower and upper bounds in linear programming', *SIAM J. Optim.* **14**, 914–935.

C. Jansson (2006), VSDP: A MATLAB software package for verified semidefinite programming. In *NOLTA 2006*, pp. 327–330.

C. Jansson (2009), 'On verified numerical computations in convex programming', *Japan J. Indust. Appl. Math.* **26**, 337–363.

C. Jansson and J. Rohn (1999), 'An algorithm for checking regularity of interval matrices', *SIAM J. Matrix Anal. Appl.* **20**, 756–776.

C. Jansson, D. Chaykin and C. Keil (2007), 'Rigorous error bounds for the optimal value in semidefinite programming', *SIAM J. Numer. Anal.* **46**, 180–200. http://link.aip.org/link/?SNA/46/180/1.

W. M. Kahan (1968), A more complete interval arithmetic. Lecture notes for a summer course at the University of Michigan.

Y. Kanzawa and S. Oishi (1999*a*), 'Imperfect singular solutions of nonlinear equations and a numerical method of proving their existence', *IEICE Trans. Fundamentals* **E82-A**, 1062–1069.

Y. Kanzawa and S. Oishi (1999*b*), 'Calculating bifurcation points with guaranteed accuracy', *IEICE Trans. Fundamentals* **E82-A**, 1055–1061.

T. Kato (1966), *Perturbation Theory for Linear Operators*, Springer, New York.

R. B. Kearfott (1997), 'Empirical evaluation of innovations in interval branch and bound algorithms for nonlinear systems', *SIAM J. Sci. Comput.* **18**, 574–594.

R. B. Kearfott, M. Dawande, K. Du and C. Hu (1992), 'INTLIB: A portable Fortran-77 elementary function library', *Interval Comput.* **3**, 96–105.

R. B. Kearfott, J. Dian and A. Neumaier (2000), 'Existence verification for singular zeros of complex nonlinear systems', *SIAM J. Numer. Anal.* **38**, 360–379.

R. B. Kearfott, M. T. Nakao, A. Neumaier, S. M. Rump, S. P. Shary and P. van Hentenfyck (2005) Standardized notation in interval analysis. In *Proc. XIII Baikal International School–Seminar: Optimization Methods and their Applications*, Vol. 4, Melentiev Energy Systems Institute SB RAS, Irkutsk.

C. Keil (2006), Lurupa: Rigorous error bounds in linear programming. In *Algebraic and Numerical Algorithms and Computer-assisted Proofs* (B. Buchberger, S. Oishi, M. Plum and S. M. Rump, eds), Vol. 05391 of *Dagstuhl Seminar Proceedings*, Internationales Begegnungs- und Forschungszentrum für Informatik (IBFI), Schloss Dagstuhl, Germany. http://drops.dagstuhl.de/opus/volltexte/2006/445.

C. Keil and C. Jansson (2006), 'Computational experience with rigorous error bounds for the Netlib linear programming library', *Reliable Computing* **12**, 303–321. http://www.optimization-online.org/DB_HTML/2004/12/1018.html.

R. Klatte, U. Kulisch, A. Wiethoff, C. Lawo and M. Rauch (1993), *C-XSC A C++ Class Library for Extended Scientific Computing*, Springer, Berlin.

O. Knüppel (1994), 'PROFIL/BIAS: A fast interval library', *Computing* **53**, 277–287.

O. Knüppel (1998), PROFIL/BIAS and extensions, Version 2.0. Technical report, Institut für Informatik III, Technische Universität Hamburg–Harburg.

D. E. Knuth (1969), *The Art of Computer Programming: Seminumerical Algorithms*, Vol. 2, Addison-Wesley, Reading, MA.

L. V. Kolev and V. Mladenov (1997), Use of interval slopes in implementing an interval method for global nonlinear DC circuit analysis. *Internat. J. Circuit Theory Appl.* **12**, 37–42.

W. Krämer (1987), Inverse Standardfunktionen für reelle und komplexe Intervallargumente mit *a priori* Fehlerabschätzung für beliebige Datenformate. Dissertation, Universität Karlsruhe.

W. Krämer (1991), Verified solution of eigenvalue problems with sparse matrices. In *Proc. 13th World Congress on Computation and Applied Mathematics*, pp. 32–33.

R. Krawczyk (1969*a*), 'Newton-Algorithmen zur Bestimmung von Nullstellen mit Fehlerschranken', *Computing* **4**, 187–201.

R. Krawczyk (1969*b*), 'Fehlerabschätzung reeller Eigenwerte und Eigenvektoren von Matrizen', *Computing* **4**, 281–293.

R. Krawczyk and A. Neumaier (1985), 'Interval slopes for rational functions and associated centered forms', *SIAM J. Numer. Anal.* **22**, 604–616.

V. Kreinovich, A. V. Lakeyev and S. I. Noskov (1993), 'Optimal solution of interval linear systems is intractable (NP-hard)', *Interval Comput.* **1**, 6–14.

V. Kreinovich, A. Neumaier and G. Xiang (2008), 'Towards a combination of interval and ellipsoid uncertainty', *Vych. Techn.* (*Computational Technologies*) **13**, 5–16.

U. Kulisch (1981), *Computer Arithmetic in Theory and Practice*, Academic Press.

M. La Porte and J. Vignes (1974), 'Etude statistique des erreurs dans l'arithmétique des ordinateurs: Application au contrôle des résultats d'algorithmes númeriques', *Numer. Math.* **23**, 63–72.

O. A. Ladyzhenskaya and N. N. Uralťseva (1968), *Linear and Quasilinear Elliptic Equations*, Academic Press, New York.

J. Lahmann and M. Plum (2004), 'A computer-assisted instability proof for the Orr–Sommerfeld equation with Blasius profile', *Z. Angew. Math. Mech.* **84**, 188–204.

C. W. H. Lam, L. Thiel and S. Swiercz (1989), 'The nonexistence of finite projective planes of order 10', *Canad. J. Math.* **41**, 1117–1123.

N. J. Lehmann (1963), 'Optimale Eigenwerteinschließung', *Numer. Math.* **5**, 246–272.

X. S. Li, J. W. Demmel, D. H. Bailey, G. Henry, Y. Hida, J. Iskandar, W. Kahan, S. Y. Kang, A. Kapur, M. C. Martin, B. J. Thompson, T. Tung and D. J. Yoo (2002), 'Design, implementation and testing of extended and mixed precision BLAS', *ACM Trans. Math. Software* **28**, 152–205.

E. Loh and W. Walster (2002), 'Rump's example revisited', *Reliable Computing* **8**, 245–248.

R. Lohner (1988), Einschließung der Lösung gewöhnlicher Anfangs- und Randwertaufgaben und Anordnungen. PhD thesis, University of Karlsruhe.

Maple (2009), *Release 13, Reference Manual.*

S. Markov and K. Okumura (1999), The contribution of T. Sunaga to interval analysis and reliable computing. In *Developments in Reliable Computing* (T. Csendes, ed.), Kluwer, pp. 167–188.

Mathematica (2009), *Release 7.0, Reference Manual.*

MATLAB (2004), *User's Guide, Version 7*, The MathWorks.

R. E. Moore (1962), Interval arithmetic and automatic error analysis in digital computing. Dissertation, Stanford University.

R. E. Moore (1966), *Interval Analysis*, Prentice-Hall, Englewood Cliffs.

R. E. Moore (1977), 'A test for existence of solutions for non-linear systems', *SIAM J. Numer. Anal.* **4**, 611–615.

R. E. Moore (1999), 'The dawning', *Reliable Computing* **5**, 423–424.

R. E. Moore, R. B. Kearfott and M. J. Cloud (2009), *Introduction To Interval Analysis*, Cambridge University Press.

J.-M. Muller, N. Brisebarre, F. de Dinechin, C.-P. Jeannerod, V. Lefèvre, G. Melquiond, R. Revol, D. Stehlé and S. Torres (2009), *Handbook of Floating-Point Arithmetic*, Birkhäuser, Boston.

K. Nagatou, M. T. Nakao and N. Yamamoto (1999), 'An approach to the numerical verification of solutions for nonlinear elliptic problems with local uniqueness', *Numer. Funct. Anal. Optim.* **20**, 543–565.

M. T. Nakao (1988), 'A numerical approach to the proof of existence of solutions for elliptic problems', *Japan J. Appl. Math.* **5**, 313–332.

M. T. Nakao (1993), Solving nonlinear elliptic problems with result verification using an H^{-1}type residual iteration. *Computing* (Suppl.) **9**, 161–173.

M. T. Nakao and N. Yamamoto (1995), 'Numerical verifications for solutions to elliptic equations using residual iterations with higher order finite elements', *J. Comput. Appl. Math.* **60**, 271–279.

M. T. Nakao, K. Hashimoto and Y. Watanabe (2005), 'A numerical method to verify the invertibility of linear elliptic operators with applications to nonlinear problems', *Computing* **75**, 1–14.

N. S. Nedialkov (1999), Computing rigorous bounds on the solution of an initial value problem for an ordinary differential equation. PhD dissertation, University of Toronto, Canada.

N. S. Nedialkov and K. R. Jackson (2000), ODE software that computes guaranteed bounds on the solution. In *Advances in Software Tools for Scientic Computing* (H. P. Langtangen, A. M. Bruaset and E. Quak, eds), Springer, pp. 197–224.

NETLIB (2009), Linear Programming Library. http://www.netlib.org/lp.

A. Neumaier (1974), 'Rundungsfehleranalyse einiger Verfahren zur Summation endlicher Summen', *Z. Angew. Math. Mech.* **54**, 39–51.

A. Neumaier (1984), 'New techniques for the analysis of linear interval equations', *Linear Algebra Appl.* **58**, 273–325.

A. Neumaier (1987), 'Overestimation in linear interval equations', *SIAM J. Numer. Anal.* **24**, 207–214.

A. Neumaier (1988), 'An existence test for root clusters and multiple roots', *Z. Angew. Math. Mech.* **68**, 256–257.

A. Neumaier (1989), 'Rigorous sensitivity analysis for parameter-dependent systems of equations', *J. Math. Anal. Appl.* **144**, 16–25.

A. Neumaier (1990), *Interval Methods for Systems of Equations*, Encyclopedia of Mathematics and its Applications, Cambridge University Press.

A. Neumaier (1993), 'The wrapping effect, ellipsoid arithmetic, stability and confidence regions', *Computing Supplementum* **9**, 175–190.

A. Neumaier (2001), *Introduction to Numerical Analysis*, Cambridge University Press.

A. Neumaier (2002), 'Grand challenges and scientific standards in interval analysis', *Reliable Computing* **8**, 313–320.

A. Neumaier (2003), 'Enclosing clusters of zeros of polynomials', *J. Comput. Appl. Math.* **156**, 389–401.

A. Neumaier (2004), Complete search in continuous global optimization and constraint satisfaction. In *Acta Numerica*, Vol. 13, Cambridge University Press, pp. 271–369.

A. Neumaier (2009), FMathL: Formal mathematical language. http://www.mat.univie.ac.at/~neum/FMathL.html.

A. Neumaier (2010), 'Improving interval enclosures', *Reliable Computing.* To appear.

A. Neumaier and T. Rage (1993), 'Rigorous chaos verification in discrete dynamical systems', *Physica* D **67**, 327–346.

A. Neumaier and O. Shcherbina (2004), 'Safe bounds in linear and mixed-integer programming', *Math. Program.* A **99**, 283–296.

W. Oettli and W. Prager (1964.), 'Compatibility of approximate solution of linear equations with given error bounds for coefficients and right-hand sides', *Numer. Math.* **6**, 405–409

T. Ogita, S. Oishi and Y. Ushiro (2001), 'Fast verification of solutions for sparse monotone matrix equations', *Comput. Suppl.* **15**, 175–187.

S. Oishi (1998), private communication.

S. Oishi (2000), *Numerical Methods with Guaranteed Accuracy* (in Japanese), Corona-sya.

S. Oishi and S. M. Rump (2002), 'Fast verification of solutions of matrix equations', *Numer. Math.* **90**, 755–773.

T. Okayama, T. Matsuo and M. Sugihara (2009), Error estimates with explicit constants for sinc approximation, sinc quadrature and sinc indefinite integration. Technical Report METR2009-01, The University of Tokyo.

J. B. Oliveira and L. H. de Figueiredo (2002) 'Interval computation of Viswanath's constant', *Reliable Computing* **8**, 131–138.

F. Ordóñez and R. M. Freund (2003), 'Computational experience and the explanatory value of condition measures for linear optimization', *SIAM J. Optim.* **14**, 307–333.

M. Overton (2001), *Numerical Computing with IEEE Floating Point Arithmetic*, SIAM, Philadelphia.

A. Ovseevich and F. Chernousko (1987), 'On optimal ellipsoids approximating reachable sets', *Problems of Control and Information Theory* **16**, 125–134.

M. Payne and R. Hanek (1983), 'Radian reduction for trigonometric functions', *SIGNUM Newsletter* **18**, 19–24.

K. Petras (2002), 'Self-validating integration and approximation of piecewise analytic functions', *J. Comput. Appl. Math.* **145**, 345–359.

M. Plum (1992), 'Numerical existence proofs and explicit bounds for solutions of nonlinear elliptic boundary value problems', *Computing* **49**, 25–44.

M. Plum (1994), 'Enclosures for solutions of parameter-dependent nonlinear elliptic boundary value problems: Theory and implementation on a parallel computer', *Interval Comput.* **3**, 106–121.

M. Plum (1995), 'Existence and enclosure results for continua of solutions of parameter-dependent nonlinear boundary value problems', *J. Comput. Appl. Math.* **60**, 187–200.

M. Plum (1996), Enclosures for two-point boundary value problems near bifurcation points. In *Scientific Computing and Validated Numerics: Proc. International Symposium on Scientific Computing, Computer Arithmetic and Validated Numerics, SCAN-95* (G. Alefeld *et al.*, eds), Vol. 90 of *Math. Res.*, Akademie Verlag, Berlin, pp. 265–279.

M. Plum (1997), 'Guaranteed numerical bounds for eigenvalues', In *Spectral Theory and Computational Methods of Sturm–Liouville Problems: Proc. 1996 Conference, Knoxville, TN, USA* (D. Hinton *et al.*, eds), Vol. 191 of *Lect. Notes Pure Appl. Math.*, Marcel Dekker, New York, pp. 313–332.

M. Plum (2008), 'Existence and multiplicity proofs for semilinear elliptic boundary value problems by computer assistance', *DMV Jahresbericht* **110**, 19–54.

M. Plum and C. Wieners (2002), 'New solutions of the Gelfand problem', *J. Math. Anal. Appl.* **269**, 588–606.

S. Poljak and J. Rohn (1993), 'Checking robust nonsingularity is NP-hard', *Math. Control, Signals, and Systems* **6**, 1–9.

L. B. Rall (1981), *Automatic Differentiation: Techniques and Applications*, Vol. 120 of *Lecture Notes in Computer Science*, Springer.

H. Ratschek and J. Rokne (1984), *Computer Methods for the Range of Functions*, Halsted Press, New York.

A. Rauh, E. Auer and E. P. Hofer (2006), ValEncIA-IVP: A comparison with other initial value problem solvers. In *Proc. 12th GAMM–IMACS International Symposium on Scientific Computing, Computer Arithmetic, and Validated Numerics*, SCAN, Duisburg.

K. Rektorys (1980), Variational methods in mathematics. In *Science and Engineering*, 2nd edn, Reidel, Dordrecht.

F. N. Ris (1972), Interval analysis and applications to linear algebra. PhD dissertation, Oxford University.

R. H. Risch (1969), 'The problem of integration in finite terms', *Trans. Amer. Math. Soc.* **139**, 167–189.

J. Rohn (1994), NP-hardness results for linear algebraic problems with interval data. In *Topics in Validated Computations* (J. Herzberger, ed.), Studies in Computational Mathematics, Elsevier, Amsterdam, pp. 463–471.

J. Rohn (2005), A handbook of results on interval linear problems. http://www.cs.cas.cz/rohn/handbook.

J. Rohn (2009*a*), 'Forty necessary and sufficient conditions for regularity of interval matrices: A survey', *Electron. J. Linear Algebra* **18**, 500–512.

J. Rohn (2009*b*), VERSOFT: Verification software in MATLAB/INTLAB. http://uivtx.cs.cas.cz/~rohn/matlab.

J. Rohn and V. Kreinovich (1995), 'Computing exact componentwise bounds on solutions of linear system is NP-hard', *SIAM J. Matrix Anal. Appl.* **16**, 415–420.

S. M. Rump (1980), Kleine Fehlerschranken bei Matrixproblemen. PhD thesis, Universität Karlsruhe.

S. M. Rump (1983), Solving algebraic problems with high accuracy. Habilitationsschrift, published in *A New Approach to Scientific Computation* (U. W. Kulisch and W. L. Miranker, eds), Academic Press, pp. 51–120.

S. M. Rump (1990), 'Rigorous sensitivity analysis for systems of linear and non-linear equations', *Math. Comput.* **54**, 721–736.

S. M. Rump (1994), Verification methods for dense and sparse systems of equations. In *Topics in Validated Computations* (J. Herzberger, ed.), Elsevier, Studies in Computational Mathematics, Amsterdam, pp. 63–136.

S. M. Rump (1999a), INTLAB: INTerval LABoratory. In *Developments in Reliable Computing* (T. Csendes, ed.), Kluwer, Dordrecht, pp. 77–104. http://www.ti3.tu-harburg.de/rump/intlab/index.html.

S. M. Rump (1999b), 'Fast and parallel interval arithmetic', *BIT Numer. Math.* **39**, 539–560.

S. M. Rump (2001a), Rigorous and portable standard functions. *BIT Numer. Math.* **41**, 540–562.

S. M. Rump (2001b), 'Computational error bounds for multiple or nearly multiple eigenvalues', *Linear Algebra Appl.* **324**, 209–226.

S. M. Rump (2003a), 'Ten methods to bound multiple roots of polynomials', *J. Comput. Appl. Math.* **156**, 403–432.

S. M. Rump (2003b), 'Structured perturbations I: Normwise distances', *SIAM J. Matrix Anal. Appl.* **25**, 1–30.

S. M. Rump (2003c), 'Structured perturbations II: Componentwise distances', *SIAM J. Matrix Anal. Appl.* **25**, 31–56.

S. M. Rump (2006), 'Eigenvalues, pseudospectrum and structured perturbations', *Linear Algebra Appl.* **413**, 567–593.

S. M. Rump (2009), 'Ultimately fast accurate summation', *SIAM J. Sci. Comput.* **31**, 3466–3502.

S. M. Rump and S. Graillat (2009), Verified error bounds for multiple roots of systems of nonlinear equations. To appear in *Numer. Algorithms*; published online at Numer Algor DOI 10.1007/s11075-009-9339-3.

S. M. Rump and S. Oishi (2009), Verified error bounds for multiple roots of nonlinear equations. In *Proc. International Symposium on Nonlinear Theory and its Applications: NOLTA'09*.

S. M. Rump and H. Sekigawa (2009), 'The ratio between the Toeplitz and the unstructured condition number', *Operator Theory: Advances and Applications* **199**, 397–419.

S. M. Rump and J. Zemke (2004), 'On eigenvector bounds', *BIT Numer. Math.* **43**, 823–837.

S. M. Rump, T. Ogita and S. Oishi (2008), 'Accurate floating-point summation I: Faithful rounding', *SIAM J. Sci. Comput.* **31**, 189–224.

N. V. Sahinidis and M. Tawaralani (2005), 'A polyhedral branch-and-cut approach to global optimization', *Math. Program.* B **103**, 225–249.

H. Schichl and A. Neumaier (2004), 'Exclusion regions for systems of equations', *SIAM J. Numer. Anal.* **42**, 383–408.

H. Schichl and A. Neumaier (2005), 'Interval analysis on directed acyclic graphs for global optimization', *J. Global Optim.* **33**, 541–562.

S. P. Shary (2002), 'A new technique in systems analysis under interval uncertainty and ambiguity', *Reliable Computing* **8**, 321–419.

G. W. Stewart (1990), 'Stochastic perturbation theory', *SIAM Rev.* **32**, 579–610.

T. Sunaga (1956), Geometry of numerals. Master's thesis, University of Tokyo.

T. Sunaga (1958), 'Theory of an interval algebra and its application to numerical analysis', *RAAG Memoirs* **2**, 29–46.

A. Takayasu, S. Oishi and T. Kubo (2009a), Guaranteed error estimate for solutions to two-point boundary value problem. In *Proc. International Symposium on Nonlinear Theory and its Applications: NOLTA'09*, pp. 214–217.

A. Takayasu, S. Oishi and T. Kubo (2009*b*), Guaranteed error estimate for solutions to linear two-point boundary value problems with FEM. In *Proc. Asia Simulation Conference 2009 (JSST 2009), Shiga, Japan*, pp. 1–8.

M. Tawaralani and N. V. Sahinidis (2004), 'Global optimization of mixed-integer nonlinear programs: A theoretical and computational study', *Math. Program.* **99**, 563–591.

M. J. Todd (2001), Semidefinite programming. In *Acta Numerica*, Vol. 10, Cambridge University Press, pp. 515–560.

L. N. Trefethen (2002), 'The SIAM 100-dollar, 100-digit challenge', *SIAM-NEWS* **35**, 2. http://www.siam.org/siamnews/06-02/challengedigits.pdf.

L. N. Trefethen and R. Schreiber (1990), 'Average-case stability of Gaussian elimination', *SIAM J. Matrix Anal. Appl.* **11**, 335–360.

W. Tucker (1999), 'The Lorenz attractor exists', *CR Acad. Sci., Paris, Sér. I, Math.* **328**, 1197–1202.

R. H. Tütüncü, K. C. Toh and M. J. Todd (2003), 'Solving semidefinite-quadratic-linear programs using SDPT3', *Math. Program.* B **95**, 189–217.

L. Vandenberghe and S. Boyd (1996), 'Semidefinite programming', *SIAM Review* **38**, 49–95.

J. Vignes (1978), 'New methods for evaluating the validity of the results of mathematical computations', *Math. Comp. Simul.* **XX**, 227–249.

J. Vignes (1980), *Algorithmes Numériques: Analyse et Mise en Oeuvre 2: Equations et Systèmes Non Linéaires*, Collection Langages et Algorithmes de l'Informatique, Editions Technip, Paris.

D. Viswanath (1999) 'Random Fibonacci sequences and the number 1.13198824…', *Math. Comp.* **69**, 1131–1155.

D. Viswanath and L. N. Trefethen (1998), 'Condition numbers of random triangular matrices', *SIAM J. Matrix Anal. Appl.* **19**, 564–581.

M. Warmus (1956), 'Calculus of approximations', *Bulletin de l'Academie Polonaise des Sciences* **4**, 253–259.

B. Werner and A. Spence (1984), 'The computation of symmetry-breaking bifurcation points', *SIAM J. Numer. Anal.* **21**, 388–399.

J. H. Wilkinson (1965), *The Algebraic Eigenvalue Problem*, Clarendon Press, Oxford.

S. J. Wright (1993), 'A collection of problems for which Gaussian elimination with partial pivoting is unstable', *SIAM J. Sci. Comput.* **14**, 231–238.

N. Yamanaka, T. Okayama, S. Oishi and T. Ogita (2009), 'A fast verified automatic integration algorithm using double exponential formula', *RIMS Kokyuroku* **1638**, 146–158.

R. C. Young (1931), 'The algebra of many-valued quantities', *Mathematische Annalen* **104**, 260–290.

Y.-K. Zhu, J.-H. Yong and G.-Q. Zheng (2005), 'A new distillation algorithm for floating-point summation', *SIAM J. Sci. Comput.* **26**, 2066–2078.

G. Zielke and V. Drygalla (2003), 'Genaue Lösung linearer Gleichungssysteme', *GAMM Mitt. Ges. Angew. Math. Mech.* **26**, 7–108.

S. Zimmermann and U. Mertins (1995), 'Variational bounds to eigenvalues of self-adjoint eigenvalue problems with arbitrary spectrum', *Z. Anal. Anwendungen* **14**, 327–345.

Acta Numerica (2010), pp. 451–559
doi:10.1017/S0962492910000061

Inverse problems: A Bayesian perspective

A. M. Stuart
Mathematics Institute,
University of Warwick,
Coventry CV4 7AL, UK
E-mail: a.m.stuart@warwick.ac.uk

The subject of inverse problems in differential equations is of enormous practical importance, and has also generated substantial mathematical and computational innovation. Typically some form of regularization is required to ameliorate ill-posed behaviour. In this article we review the Bayesian approach to regularization, developing a function space viewpoint on the subject. This approach allows for a full characterization of all possible solutions, and their relative probabilities, whilst simultaneously forcing significant modelling issues to be addressed in a clear and precise fashion. Although expensive to implement, this approach is starting to lie within the range of the available computational resources in many application areas. It also allows for the quantification of uncertainty and risk, something which is increasingly demanded by these applications. Furthermore, the approach is conceptually important for the understanding of simpler, computationally expedient approaches to inverse problems.

We demonstrate that, when formulated in a Bayesian fashion, a wide range of inverse problems share a common mathematical framework, and we highlight a theory of well-posedness which stems from this. The well-posedness theory provides the basis for a number of stability and approximation results which we describe. We also review a range of algorithmic approaches which are used when adopting the Bayesian approach to inverse problems. These include MCMC methods, filtering and the variational approach.

CONTENTS

1. Introduction

A significant challenge facing mathematical scientists is the development of a coherent mathematical and algorithmic framework enabling researchers to blend complex mathematical models with the (often vast) data sets which are now routinely available in many fields of engineering, science and technology. In this article we frame a range of inverse problems, mostly arising from the conjunction of differential equations and data, in the language of Bayesian statistics. In so doing our aim is twofold: (i) to highlight common mathematical structure arising from the numerous application areas where significant progress has been made by practitioners over the last few decades and thereby facilitate exchange of ideas between different application domains; (ii) to develop an abstract function space setting for the problems in order to evaluate the efficiency of existing algorithms, and to develop new algorithms. Applications are far-reaching and include fields such as the atmospheric sciences, oceanography, hydrology, geophysics, chemistry and biochemistry, materials science, systems biology, traffic flow, econometrics, image processing and signal processing.

The guiding principle underpinning the specific development of the subject of Bayesian inverse problems in this article is to *avoid discretization until the last possible moment.* This principle is enormously empowering throughout numerical analysis. For example, the first-order wave equation is not controllable to a given final state in arbitrarily small time because of finite speed of propagation. Yet every finite difference spatial discretization of the first-order wave equation gives rise to a linear system of ordinary differential equations which is controllable, in any finite time, to a given final state; asking the controllability question *before* discretization is key to understanding (Zuazua 2005). As another example consider the heat equation. If this is discretized in time by the theta method (with $\theta \in [0,1]$ and $\theta = 0$ being explicit Euler, $\theta = 1$ implicit Euler), but left undiscretized in space, the resulting algorithm on function space is only defined if $\theta \in [\frac{1}{2}, 1]$; thus it is possible to deduce that there *must* be a Courant restriction if $\theta \in [0, \frac{1}{2})$ (Richtmyer and Morton 1967) before even introducing spatial discretization. Yet another example may be found in the study of Newton methods: conceptual application of this algorithm on function space, before discretization, can yield considerable insight when applying it as an iterative method for boundary value problems in nonlinear differential equations (Deuflhard 2004). The list of problems where it is beneficial to defer discretization to the very end of the algorithmic formulation is almost endless. It is perhaps not surprising, therefore, that the same idea yields insight in the solution of inverse problems and we substantiate this idea in the Bayesian context.

The article is divided into five parts. The next section, Section 2, is devoted to a description of the basic ideas of Bayesian statistics as applied to inverse problems in the finite-dimensional setting. It also includes a pointer to the common structure that we will highlight in the remainder of the article when developing the Bayesian viewpoint in function space. Section 3 contains a range of inverse problems arising in differential equations, showing how the Bayesian approach may be applied to inverse problems for functions; in particular, we discuss the problem of recovering a field from noisy pointwise data, recovering the diffusion coefficient from a boundary value problem, given noisy pointwise observations of the solution, recovering the wave speed from noisy observations of solutions of the wave equation and recovering the initial condition of the heat equation from noisy observation of the solution at a positive time. We also describe a range of applications, involving similar but more complex models, arising in weather forecasting, oceanography, subsurface geophysics and molecular dynamics. In Section 4 we describe, and exploit, the common mathematical structure which underlies *all* of these Bayesian inverse problems for functions. In that section we prove a form of well-posedness for these inverse problems, by showing Lipschitz continuity of the posterior measure with respect to changes in the data; we also prove an approximation theorem which exploits this well-posedness to show that approximation of the forward problem (by spectral or finite element methods, for example) leads to similar approximation results for the posterior probability measure. Section 5 is devoted to a survey of the existing algorithmic tools used to solve the problems highlighted in the article. In particular, Markov chain Monte Carlo (MCMC) methods, variational methods and filtering methods are surveyed. When discussing variational methods we show, in the setting of Section 4, that posterior probability maximizers can be characterized through solution of an optimal control problem, and that this optimal control problem has a minimizer under the same conditions that lead to a well-posed Bayesian inverse problem. Section 6 contains the background probability required to read the article; the presentation in this section is necessarily terse and the reader is encouraged to follow up references in the bibliography for further detail.

A major theme of the article is thus to confront the infinite-dimensional nature of many inverse problems. This is important because, whilst all computational algorithms work on finite-dimensional approximations, these approximations are typically in spaces of very high dimension and many significant challenges stem from this fact. By formulating inverse problems in an infinite-dimensional setting we build these challenges into the fabric of the problem setting. We provide a clear concept of *the ideal solution to the inverse problem* when blending a forward mathematical model with observational data. This concept can be used to test the practical algorithms used in applications which, in many cases, use crude approximations for reasons of computational efficiency. Furthermore, it is also possible that the function space Bayesian setting will also lead to the development of improved algorithms which exploit the underlying mathematical structure common to a diverse range of applications. In particular, the theory of (Bayesian) well-posedness which we describe forms the cornerstone of many perturbation theories, including finite-dimensional approximations.

Kaipio and Somersalo (2005) provide a good introduction to the Bayesian approach to inverse problems, especially in the context of differential equations. Furthermore, Calvetti and Somersalo (2007*b*) provide a useful introduction to the Bayesian perspective in scientific computing. Another overview of the subject of inverse problems in differential equations, including a strong argument for the philosophy taken in this article, namely to formulate and study inverse problems in function space, is the book by Tarantola (2005) (see, especially, Chapter 5); however, the mathematics associated with this philosophical standpoint is not developed there to the same extent that it is in this article, and the focus is primarily on Gaussian problems. A frequentist viewpoint for inverse problems on function space is contained in the book by Ramsay and Silverman (2005); however, we adopt a different, Bayesian, perspective here, and study more involved differential equation models than those arising in Ramsay and Silverman (2005). These books indicate that the development that we undertake here is a natural one, which builds upon the existing literature.

The subject known as *data assimilation* provides a number of important applications of the material presented here. Its development has been driven, to a large extent, by practitioners working in the atmospheric and oceanographic sciences and in the geosciences, resulting in a plethora of algorithmic approaches and a number of significant algorithmic innovations. A good source for an understanding of data assimilation in the context of the atmospheric sciences, and weather prediction in particular, is the book by Kalnay (2003). A book motivated by applications in oceanography, which simultaneously highlights some of the underlying function space structure of data assimilation for linear, Gaussian problems, is that of Bennett (2002). The book by Evensen (2006) provides a good overview of many computational aspects of the subject, reflecting the author's experience in geophysical applications and related areas. The recent special edition of *Physica D* devoted to data assimilation provides a good entry point to some of the current research in this area (Ide and Jones 2007). Another application that fits the mathematical framework developed here is molecular dynamics. The problems of interest do not arise from Bayesian inverse problems, as such, but rather from conditioned diffusion processes. However, the mathematical structure has much in common with that arising in Bayesian inverse problems, and so we include a description of this problem area.

Throughout the article we use standard notation for Banach and Hilbert space norm and inner products, $\|\cdot\|, \langle\cdot,\cdot\rangle$, and the following notation for the finite-dimensional Euclidean norm and inner product: $|\cdot|, \langle\cdot,\cdot\rangle$. We also use the concept of weighted inner products and norms in any Hilbert space. For any self-adjoint positive operator $\mathcal{A}$, we define

$$\langle\cdot,\cdot\rangle_{\mathcal{A}} = \langle\mathcal{A}^{-1/2}\cdot,\mathcal{A}^{-1/2}\cdot\rangle, \quad \|\cdot\|_{\mathcal{A}} = \|\mathcal{A}^{-1/2}\cdot\|$$

in the general setting and, in finite dimensions,

$$|\cdot|_{\mathcal{A}} = |\mathcal{A}^{-1/2}\cdot|.$$

For any $a, b \in \mathcal{H}$, a Hilbert space, we define the operator $a \otimes b$ by the identity $(a \otimes b)c = \langle b, c\rangle a$ for any $c \in \mathcal{H}$. We use * to denote the adjoint of a linear operator between two Hilbert spaces. In particular, we may view $a, b \in \mathcal{H}$ as linear operators from $\mathbb{R}$ to $\mathcal{H}$ and then $a \otimes b = ab^*$.

In order to highlight the common structure arising in many of the problems in this book, we will endeavor to use the same notation repeatedly in the different contexts. A Gaussian measure will be denoted as $\mathcal{N}(m, \mathcal{C})$ with m the *mean* and $\mathcal{C}$ the *covariance operator/matrix*. The mean of the prior Gaussian measure will be m_0 and its covariance matrix/operator will be Σ_0 or $\mathcal{C}_0$ (we will drop the subscript 0 on the prior where no confusion arises in doing so). We will use the terminology *precision operator* for the (densely defined) $\mathcal{L} := \mathcal{C}^{-1}$. For inverse problems the operator mapping the unknown vector/field to the observations will be denoted by $\mathcal{G}$ and termed the *observation operator*, and the *observational noise* will be denoted by η.

We emphasize that in this article we will work for the most part with Gaussian priors. In terms of the classical theory of regularization this means that we are limiting ourselves to quadratic regularization terms, typically a Sobolev-type Hilbert space norm. We recognize that there are many applications of importance where other regularizations are natural, especially in image processing (Rudin, Osher and Fatemi 1992, Scherzer, Grasmair, Grossauer, Haltmeier and Lenzen 2009). A significant challenge is to take the material in this article and generalize it to these other settings, and there is some recent interesting work in this direction (Lassas, Saksman and Siltanen 2009).

There are other problem areas which lead to the need for computation of random functions. For example, there is a large body of work concerned with *uncertainty quantification* (DeVolder *et al.* 2002, Kennedy and O'Hagan 2001, Mohamed, Christie and Demyanov 2010, Efendiev, Datta-Gupta, Ma and Mallick 2009). In this field the input data to a differential equation is viewed as a random variable and the interest is in computing the resulting variability in the solution, as the input data varies. This is currently an active area of research in the engineering community (Spanos and Ghanem 1989, 2003). The work is thus primarily concerned with approximating measures which are the push forward, under a nonlinear map, of a Gaussian measure; in contrast, the inverse problem setting which we study here is concerned with the approximation of non-Gaussian measures whose Radon–Nikodym derivative with respect to a Gaussian measure is defined through a related nonlinear map. A rigorous numerical analysis underpinning the work of Spanos and Ghanem (1989, 2003) is an active area of research: see in particular Schwab and Todor (2006) and Todor and Schwab (2007), where the problem is viewed as an example of Schwab's more general program of tensor product approximation for high-(infinite)-dimensional problems (Gittelson and Schwab 2011). A different area where tensor products are used to form approximations of functions of many variables is computational quantum mechanics and approximation of the Schrödinger equation (Lubich 2008); this work may also be seen in the more general context of tensor product approximations in linear algebra (Kolda and Bader 2009). It would be interesting to investigate whether any of these tensor product ideas can be transferred to the approximation of probability density functions in high-dimensional spaces, as arise naturally in Bayesian inverse problems.

More generally speaking, this article is concerned with a research area which is at the interface of applied mathematics and statistics. This is a rich research interface, where there is currently significant effort. Examples include work in compressed sensing, which blends ideas from statistics, probability, approximation theory and harmonic analysis (Candès and Wakin 2008, Donoho 2006), and research aimed at efficient sampling of Gaussian random fields combining numerical linear algebra and statistics (Rue and Held 2005).

2. The Bayesian framework

2.1. Overview

This section introduces the Bayesian approach to inverse problems and outlines the common structure that we will develop in the remainder of the article. In Section 2.2 we introduce finite-dimensional inverse problems and describe the Bayesian approach to their solution, highlighting the role of observational noise which pollutes the data in many problems of practical interest. We show how to construct a formula for the posterior measure on the unknown of interest, from the data and from a prior measure incorporating structural knowledge about the problem which is present prior to the acquisition of the data. In Section 2.3 we study the effect on the posterior of small observational noise, in order to connect the Bayesian viewpoint with the classical perspective on inverse problems. We first study problems where the dimensions of the data set and the unknown match; we show that the prior measure is asymptotically irrelevant and that, in the limit of zero noise, the posterior measure converges weakly to a Dirac measure centred on the solution of the noise-free equation. We next study the special structure which arises when the mathematical model and observations are described through linear operators, and when the prior and the noise are Gaussian; this results in a Gaussian posterior measure. In this Gaussian setting we first study the limit of vanishing observational noise in the case where the dimension of the data set is greater than that of the unknown, showing that the prior is asymptotically irrelevant, and that the posterior measure approaches a Dirac concentrated on the solution of a natural least-squares problem. We then study the situation where the dimension of the data set is smaller than that of the unknown. We show that, in the limit of small observational noise, the prior remains important and we characterize this effect explicitly. Section 2.4 completes the introductory material by describing the common framework which we will illustrate and exploit in the remainder of the article when developing the Bayesian viewpoint on function space.

2.2. Linking the classical and Bayesian approaches

In applications it is frequently of interest to solve *inverse problems*: to find u, an input to a mathematical model, given y an observation of (some components of, or functions of) the solution of the model. We have an equation of the form

$$y = \mathcal{G}(u) \tag{2.1}$$

to solve for $u \in X$, given $y \in Y$, where X, Y are Banach spaces. We will refer to $\mathcal{G}$ as the *observation operator*.[1] We refer to y as *data*. It is typical of inverse problems that they are *ill-posed*: there may be no solution, or the solution may not be unique and may depend sensitively on y. One approach to the problem in this situation is to replace it by the *least-squares* optimization problem of finding, for the norm $\|\cdot\|_Y$ on Y,

$$\operatorname{argmin}_{u\in X} \frac{1}{2}\|y - \mathcal{G}(u)\|_Y^2. \tag{2.2}$$

This problem, too, may be difficult to solve as it may possess minimizing sequences $u^{(n)}$ which do not converge to a limit in X, or it may possess multiple minima and sensitive dependence on the data y. These issues can be somewhat ameliorated by solving a *regularized* minimization problem of the form, for some Banach space $(E, \|\cdot\|_E)$ contained in X, and point $m_0 \in E$,

$$\operatorname{argmin}_{u\in E} \left(\frac{1}{2}\|y - \mathcal{G}(u)\|_Y^2 + \frac{1}{2}\|u - m_0\|_E^2\right). \tag{2.3}$$

[1] This operator is often denoted with the letter $\mathcal{H}$ in the atmospheric sciences community; because we need $\mathcal{H}$ for Hilbert space later on, we use the symbol $\mathcal{G}$.

However, the choice of norms $\|\cdot\|_E, \|\cdot\|_Y$ and the point m_0 are somewhat arbitrary, without making further modelling assumptions. We will adopt a statistical approach to the inverse problems, in which these issues can be articulated and addressed in an explicit fashion. Roughly speaking, the Bayesian approach will lead to the notion of finding a *probability measure* μ^y on X, containing information about the relative probability of different states u, given the data y. For example, in the case where X, Y are both finite-dimensional, the noise polluting (2.1) is additive and Gaussian, and the prior measure is Gaussian, the posterior measure will have density π^y given by

$$\pi^y(u) \propto \exp\left(-\frac{1}{2}\|y - \mathcal{G}(u)\|_Y^2 - \frac{1}{2}\|u - m_0\|_E^2\right). \tag{2.4}$$

The properties of a measure μ^y with such a density π^y are intimately related to the minimization problem (2.3): the density is largest at minimizers. But the probabilistic approach is far richer. For example, the derivation of the probability measure μ^y will force us to confront various modelling and mathematical issues which, together, will guide the choice of norms $\|\cdot\|_E$, $\|\cdot\|_Y$ and the point m_0. Furthermore, the probabilistic approach enables us to answer questions such as: 'What is the relative probability that the unknown function u is determined by the different local minimizers of (2.3)?' 'How certain can we be that a prediction made by a mathematical model will lie in certain specified regimes?'

We now outline a probabilistic framework which will include the specific probability measure with density given by (2.4) as a special case. This framework starts from the observation that a deeper understanding of the source of data often reveals that the observations y are subject to noise and that a more appropriate model equation is often of the form

$$y = \mathcal{G}(u) + \eta, \tag{2.5}$$

where η is a mean zero random variable, whose statistical properties we might know, but whose actual value is unknown to us; we refer to η as the *observational noise*. In this context it is natural to adopt a *Bayesian* approach to the problem of determining u from y: see Section 6.6. We describe our prior beliefs about u, in terms of a probability measure μ_0, and use Bayes' formula (see (6.24)) to calculate the posterior probability measure μ^y, for u given y.

To be concrete, in the remainder of this subsection and in the next subsection we consider the case where $u \in \mathbb{R}^n, y \in \mathbb{R}^q$ and we let π_0 and π^y denote the p.d.f.s (see Section 6.1) of measures μ_0 and μ^y. We assume that $\eta \in \mathbb{R}^q$ is a random variable with density ρ. Then the probability of y given u has density

$$\rho(y|u) := \rho(y - \mathcal{G}(u)).$$

This is often referred to as the *data likelihood.* By Bayes' formula (6.24) we obtain

$$\pi^y(u) = \frac{\rho(y - \mathcal{G}(u))\pi_0(u)}{\int_{\mathbb{R}^n} \rho(y - \mathcal{G}(u))\pi_0(u)\,\mathrm{d}u}. \tag{2.6}$$

Thus

$$\pi^y(u) \propto \rho(y - \mathcal{G}(u))\pi_0(u) \tag{2.7}$$

with constant of proportionality depending only on y. Abstractly (2.7) expresses the fact that the posterior measure μ^y (with density π^y) and prior measure μ_0 (with density π_0) are related through the Radon–Nikodym derivative (see Theorem 6.2)

$$\frac{\mathrm{d}\mu^y}{\mathrm{d}\mu_0}(u) \propto \rho(y - \mathcal{G}(u)). \tag{2.8}$$

Since ρ is a density and thus non-negative, without loss of generality we may write the right-hand side as the exponential of the negative of a potential $\Phi(u; y)$, to obtain

$$\frac{\mathrm{d}\mu^y}{\mathrm{d}\mu_0}(u) \propto \exp(-\Phi(u; y)). \tag{2.9}$$

It is this form which generalizes naturally to situations where X, and possibly Y, is infinite-dimensional. We show in Section 3 that many inverse problems can be formulated in a Bayesian fashion and that the posterior measure takes this form.

In general it is hard to obtain information from a probability measure in high dimensions. One useful approach to extracting information is to find a *maximum a posteriori estimator*, or *MAP estimator*: a point u which maximizes the posterior p.d.f. π^y; such *variational* methods are surveyed in Section 5.3. Another commonly used method for interrogating a probability measure in high dimensions is *sampling*: generating a set of points $\{u_n\}_{n=1}^N$ distributed (perhaps only approximately) according to $\pi^y(u)$. In this context formula (2.7) (or (2.9) in the general setting), in which the posterior density is known only up to a constant, is useful because *MCMC methods* may be used to sample from it: MCMC methods have the advantage of sampling from a probability measure only known up to a normalizing constant; we outline these methods in Section 5.2. Time-dependent problems, where the data is acquired sequentially, also provide a class of problems where useful approximations can be developed; these *filtering* methods are outlined in Section 5.4.

We will often be interested in problems where prior μ_0 and observational noise η are Gaussian. If $\eta \sim \mathcal{N}(0, B)$ and $\mu_0 = \mathcal{N}(m_0, \Sigma_0)$, then we obtain from (2.7) the formula[2]

$$\pi^y(u) \propto \exp\Big(-\frac{1}{2}\big|B^{-1/2}(y-\mathcal{G}(u))\big|^2 - \frac{1}{2}\big|\Sigma_0^{-1/2}(u-m_0)\big|^2\Big)$$
$$= \exp\Big(-\frac{1}{2}\big|(y-\mathcal{G}(u))\big|_B^2 - \frac{1}{2}\big|(u-m_0)\big|_{\Sigma_0}^2\Big). \qquad (2.10)$$

In terms of measures this is the statement that

$$\frac{\mathrm{d}\mu^y}{\mathrm{d}\mu_0}(u) \propto \exp\Big(-\frac{1}{2}\big|(y-\mathcal{G}(u))\big|_B^2\Big). \qquad (2.11)$$

The *maximum a posteriori estimator*, or *MAP estimator*, is then

$$\operatorname{argmin}_{u\in\mathbb{R}^n}\Big(\frac{1}{2}\big|y-\mathcal{G}(u)\big|_B^2 + \frac{1}{2}\big|u-m_0\big|_{\Sigma_0}^2\Big). \qquad (2.12)$$

This is a specific instance of the regularized minimization problem (2.3). Note that in the Bayesian framework the norms $\|\cdot\|_Y, \|\cdot\|_E$ and the point m_0 all have a clear interpretation in terms of the statistics of the observational noise and the prior measure. In contrast, these norms and point are somewhat arbitrary in the classical approach.

In general the posterior probability measure (2.10) is not itself Gaussian. However, if $\mathcal{G}$ is linear then the posterior μ^y is also Gaussian. Identifying the mean and covariance (or precision) matrix can be achieved by *completing the square*, as formalized in Theorem 6.20 and Lemma 6.21 (see also Examples 6.22 and 6.23). The following simple examples illustrate this. They also show further connections between the Bayesian and classical approaches to inverse problems, a subject we develop further in the following subsection.

Example 2.1. Let $q = 1$ and $\mathcal{G}$ be linear so that

$$y = \langle g, u\rangle + \eta$$

for some $g \in \mathbb{R}^n$. Assume further that $\eta \sim \mathcal{N}(0,\gamma^2)$ and that we place a prior Gaussian measure $\mathcal{N}(0,\Sigma_0)$ on u. Then

$$\pi^y(u) \propto \exp\Big(-\frac{1}{2\gamma^2}|y-\langle g,u\rangle|^2 - \frac{1}{2}\langle u, \Sigma_0^{-1}u\rangle\Big). \qquad (2.13)$$

As the exponential of a quadratic form, this is the density of a Gaussian measure.

[2] The notation for weighted norms and inner products is defined at the end of Section 1.

From Theorem 6.20 we find that the posterior mean and covariance are given by

$$m = \frac{(\Sigma_0 g)y}{\gamma^2 + \langle g, \Sigma_0 g\rangle},$$
$$\Sigma = \Sigma_0 - \frac{(\Sigma_0 g)(\Sigma_0 g)^*}{\gamma^2 + \langle g, \Sigma_0 g\rangle}.$$

If we consider the case where observational noise disappears from the system, then we find that

$$m^+ := \lim_{\gamma\to 0} m = \frac{(\Sigma_0 g)y}{\langle g, \Sigma_0 g\rangle}, \quad \Sigma^+ := \lim_{\gamma\to 0} \Sigma = \Sigma_0 - \frac{(\Sigma_0 g)(\Sigma_0 g)^*}{\langle g, \Sigma_0 g\rangle}.$$

Notice that $\Sigma^+ g = 0$ and $\langle m^+, g\rangle = y$. This states the intuitively reasonable fact that, as the observational noise decreases, knowledge of u in the direction of g becomes certain. In directions not aligned with g, uncertainty remains, with magnitude determined by an interaction between properties of the prior and of the observation operator. Thus the prior plays a central role, even as observational noise disappears, in this example where the solution is underdetermined. ◇

Example 2.2. Assume that $q \geqslant 2$ and $n = 1$, and let $\mathcal{G}$ be nonlinear with the form

$$y = g(u + \beta u^3) + \eta,$$

where $g \in \mathbb{R}^q\backslash\{0\}$, $\beta \in \mathbb{R}$ and $\eta \sim \mathcal{N}(0, \gamma^2 I)$. Assume further that we place a Gaussian measure $\mathcal{N}(0,1)$ as a prior on u. Then

$$\pi^y(u) \propto \exp\biggl(-\frac{1}{2\gamma^2}|y - g(u + \beta u^3)|^2 - \frac{1}{2}u^2\biggr).$$

This measure is not Gaussian unless $\beta = 0$.

Consider the linear case where $\beta = 0$. The posterior measure is then Gaussian:

$$\pi^y(u) \propto \exp\biggl(-\frac{1}{2\gamma^2}|y - gu|^2 - \frac{1}{2}|u|^2\biggr).$$

By Theorem 6.20, using the identity

$$\bigl(\gamma^2 I + gg^*\bigr)^{-1} g = \bigl(\gamma^2 + |g|^2\bigr)^{-1} g,$$

we deduce that the posterior mean and covariance are given by

$$m = \frac{\langle g, y\rangle}{\gamma^2 + |g|^2},$$
$$\sigma^2 = \frac{\gamma^2}{\gamma^2 + |g|^2}.$$

In the limit where observational noise disappears, we find that

$$m^+ = \lim_{\gamma \to 0} m = \frac{\langle g, y \rangle}{|g|^2}, \quad (\sigma^+)^2 = \lim_{\gamma \to 0} \sigma^2 = 0.$$

The point m^+ is the least-squares solution of the overdetermined linear equation $y = gu$ found from the minimization problem

$$\mathrm{argmin}_{u \in \mathbb{R}} |y - gu|^2.$$

This is a minimization problem of the form (2.2). In this case, where the system is overdetermined, the prior plays no role in the limit of zero observational noise. ◇

2.3. Small noise limits of the posterior measure

We have shown that the Bayesian and classical perspectives are linked through the relationship between the posterior probability density given by (2.10) and the MAP estimator (2.12). This directly connects minimization of a regularized least-squares problem with the Bayesian perspective. Our aim now is to further the link between the Bayesian and classical approaches by considering the limit of small observational noise.

The small observational noise limit is illustrated in the two examples concluding the previous subsection. In the first, where the underlying noise-free problem is underdetermined, the prior provides information about the posterior mean, and uncertainty remains in the posterior, even as observational noise disappears; furthermore, that uncertainty is related to the choice of prior. In the second example, where the underlying noise-free problem is overdetermined, uncertainty disappears and the posterior converges to a Dirac measure centred on the least-squares solution of the limiting deterministic equation. The intuition obtained from these two examples, concerning the behaviour of the posterior measure in the small noise limit, is important. The first example suggests that in the underdetermined case the prior measure plays a role in determining the posterior measure, even as the observational noise disappears; in contrast, the second example suggests that, in the overdetermined case, the prior plays no role in the small noise limit. Many of the inverse problems for functions that we study later in this paper are underdetermined. For these problems *the prior measure plays an important role in the solution, even when observational noise is small.* A significant advantage of the Bayesian framework over classical approaches is that it makes the modelling assumptions which underly the prior both clear and explicit.

In the remainder of this subsection we demonstrate that the intuition obtained from the two examples can be substantiated on a broad class of finite-dimensional inverse problems. We first concentrate on the general case which lies between these two examples, where $q = n$ and, furthermore, equation (2.1) has a unique solution. We then restrict our attention to Gaussian problems, studying the over- and underdetermined cases in turn. We state the results first, and provide proofs at the end of the subsection. The results are stated in terms of weak convergence of probability measures, denoted by $\Rightarrow$; see the end of Section 6.1 for background on this concept. Throughout this subsection we consider the data y to be fixed, and we study the limiting behaviour of the posterior measure μ^y as the observational noise tends to zero. Other limits, where y is a random variable, depending on the observational noise, are also of interest, but we stick to the simpler setting where y is fixed, for expository purposes.

We start with the case $q = n$ and assume that equation (2.1) has a unique solution

$$u = \mathcal{F}(y) \tag{2.14}$$

for every $y \in \mathbb{R}^n$. Intuitively this unique solution should dominate the Bayesian solution to the problem (which is a probability distribution on $\mathbb{R}^n$, not a single point). We show that this is indeed the case: the probability distribution concentrates on the single point given by (2.14) as observational noise disappears.

We assume that there is a positive constant C such that, for all $y, \delta \in \mathbb{R}^n$,

$$|y - \mathcal{G}(\mathcal{F}(y) + \delta)|^2 \geqslant C \min\{1, |\delta|^2\}. \tag{2.15}$$

This condition implies that the derivative $DG(u)$ is invertible at $u = \mathcal{F}(y)$, so that the implicit function theorem holds; the condition also excludes the possibility of attaining the minimum 0 of $\frac{1}{2}|y - \mathcal{G}(u)|^2$ along a sequence $u_n \to \infty$. We then have the following.

Theorem 2.3. Assume that $k = n$, that $\mathcal{G} \in C^2(\mathbb{R}^n, \mathbb{R}^n)$ and that equation (2.1) has a unique solution given by (2.14), for every $y \in \mathbb{R}^n$. We place a Gaussian prior $\mu_0 = \mathcal{N}(m_0, \Sigma_0)$ on u and assume that the observational noise η in (2.5) is distributed as $\mathcal{N}(0, \gamma^2 I)$. Then the posterior measure μ^y, with density given by (2.10) and $B = \gamma^2 I$, satisfies $\mu^y \Rightarrow \delta_{\mathcal{F}(y)}$ as $\gamma \to 0$. $\diamond$

The preceding theorem concerns problems where the underlying equation (2.1) relating data to model is uniquely solvable. This situation rarely arises in practice, but is of course important for building links between the Bayesian and classical perspectives.

We now turn to problems which are either over- or underdetermined and, for simplicity, confine our attention to purely Gaussian problems. We again work in arbitrary finite dimensions and study the small observational noise limit and its relation to the the underlying noise-free problem (2.1). In Theorem 2.4 we show that the posterior measure converges to a Dirac measure concentrated on minimizers of the least-squares problem (2.2). Of course, when (2.1) is uniquely solvable this will lead to a Dirac measure as its solution, as in Theorem 2.3; but more generally there may be no solution to (2.1) and least-squares minimizers provide a natural generalized solution concept. In Theorem 2.5 we study the Gaussian problem in the undetermined case, showing that the posterior measure converges to a Gaussian measure whose support lies on a hyperplane embedded in the space where the unknown u lies. The structure of this Gaussian measure is determined by an interplay between the prior, the forward model and the data. In particular, *prior information remains in the small noise limit.* This illustrates the important idea that for (frequently occurring) underdetermined problems the prior plays a significant role, even when noise is small, and should therefore be treated very carefully from the perspective of mathematical modelling.

If the observational noise η is Gaussian, if the prior μ_0 is Gaussian and if $\mathcal{G}$ is a linear map, then the posterior measure μ^y is also Gaussian. This follows immediately from the fact that the logarithm of π^y given by (2.6) is quadratic in u under these assumptions. We now study the properties of this Gaussian posterior.

We assume that

$$\eta \sim \mathcal{N}(0, B), \quad \mu_0 = \mathcal{N}(m_0, \Sigma_0), \quad \mathcal{G}(u) = Au$$

and that B and Σ_0 are both invertible.

Then, since $y|u \sim \mathcal{N}(Au, B)$, Theorem 6.20 shows that the posterior measure μ^y is Gaussian $\mathcal{N}(m, \Sigma)$ with

$$m = m_0 + \Sigma_0 A^*(B + A\Sigma_0 A^*)^{-1}(y - Am_0), \tag{2.16a}$$

$$\Sigma = \Sigma_0 - \Sigma_0 A^*(B + A\Sigma_0 A^*)^{-1} A\Sigma_0. \tag{2.16b}$$

In the case where $k = n$ and A, Σ_0 are invertible, we see that, as $B \to 0$,

$$m \to A^{-1}y, \quad \Sigma \to 0.$$

From Lemma 6.5 we know that convergence of all characteristic functions implies weak convergence. Furthermore, the characteristic function of a Gaussian is determined by the mean and covariance: see Theorem 6.4. Hence, for a finite-dimensional family of Gaussians, convergence of the mean and covariance to a limit implies weak convergence to the Gaussian with that limiting mean and covariance. For this family of measures the limiting covariance is zero and thus the $B \to 0$ limit recovers a Dirac measure on the solution of the equation $Au = y$, in accordance with Theorem 2.3. It is natural to ask what happens in the limit of vanishing noise, more generally. The following two theorems provide an answer to this question.

Theorem 2.4. Assume that B and Σ_0 are both invertible. The posterior mean and covariance can be rewritten as

$$m = (A^*B^{-1}A + \Sigma_0^{-1})^{-1}(A^*B^{-1}y + \Sigma_0^{-1}m_0), \tag{2.17a}$$

$$\Sigma = (A^*B^{-1}A + \Sigma_0^{-1})^{-1}. \tag{2.17b}$$

If $\mathrm{Null}(A) = \{0\}$ and $B = \gamma^2 B_0$ then, in the limit $\gamma^2 \to 0$, $\mu^y \Rightarrow \delta_{m^+}$, where m^+ is the solution of the least-squares problem

$$m^+ = \mathrm{argmin}_{u \in \mathbb{R}^n} |B_0^{-1/2}(y - Au)|^2. \qquad \diamond$$

The preceding theorem shows that, in the overdetermined case where A^*BA is invertible, the small observational noise limit leads to a posterior which is a Dirac, centred on the solution of a least-squares problem determined by the observation operator and the relative weights on the observational noise. Uncertainty disappears, and the prior plays no role in this limit. Example 2.2 illustrates this situation.

We now assume that $y \in \mathbb{R}^q$ and $u \in \mathbb{R}^n$ with $q < n$, so that the problem is underdetermined. We assume that $\mathrm{rank}(A) = q$, so that we may write

$$A = (A_0 \;\; 0)Q^* \tag{2.18}$$

with $Q \in \mathbb{R}^{n \times n}$ an orthogonal matrix so that $Q^*Q = I$, $A_0 \in \mathbb{R}^{q \times q}$ an invertible matrix and $0 \in \mathbb{R}^{q \times (n-q)}$ a zero matrix. We also let $L_0 = \Sigma_0^{-1}$, the precision matrix for the prior, and write

$$Q^* L_0 Q = \begin{pmatrix} L_{11} & L_{12} \\ L_{12}^* & L_{22} \end{pmatrix}. \tag{2.19}$$

Here $L_{11} \in \mathbb{R}^{q \times q}$, $L_{12} \in \mathbb{R}^{q \times (n-q)}$ and $L_{22} \in \mathbb{R}^{(n-q) \times (n-q)}$; both L_{11} and L_{22} are positive definite symmetric, because Σ_0 is.

If we write

$$Q = (Q_1 \;\; Q_2) \tag{2.20}$$

with $Q_1 \in \mathbb{R}^{n \times q}$ and $Q_2 \in \mathbb{R}^{n \times (n-q)}$, then Q_1^* projects onto a q-dimensional subspace $\mathcal{O}$ and Q_2^* projects onto an $(n-q)$-dimensional subspace $\mathcal{O}^\perp$; here $\mathcal{O}$ and $\mathcal{O}^\perp$ are orthogonal.

Assume that $y = Au$ for some $u \in \mathbb{R}^n$. This identity is at the heart of the inverse problem in the small γ limit. If we define $z \in \mathbb{R}^q$ to be the unique solution of the system of equations $A_0 z = y$, then $z = Q_1^* u$. On the other hand, $Q_2^* u$ is not determined by the identity $y = Au$. Thus, intuitively we expect to determine z without uncertainty, in the limit of small noise, but for uncertainty to remain in other directions. With this in mind we define $w \in \mathbb{R}^q$ and $w' \in \mathbb{R}^{n-q}$ via the equation

$$\Sigma_0^{-1} m_0 = Q \begin{pmatrix} w \\ w' \end{pmatrix}, \tag{2.21}$$

and then set

$$z' = -L_{22}^{-1} L_{12}^* z + L_{22}^{-1} w' \in \mathbb{R}^{n-q}.$$

Theorem 2.5. Assume that B and Σ_0 are both invertible and let $B = \gamma^2 B_0$. Then, in the limit $\gamma^2 \to 0$, $\mu^y \Rightarrow \mathcal{N}(m^+, \Sigma^+)$, where

$$m^+ = Q\begin{pmatrix} z \\ z' \end{pmatrix}, \tag{2.22a}$$

$$\Sigma^+ = Q_2 L_{22}^{-1} Q_2^*. \tag{2.22b}$$

◇

We now interpret this theorem. Since $Q_2^* Q_1 = 0$, the limiting measure may be viewed as a Dirac measure, centred at z in $\mathcal{O}$, and a Gaussian measure $\mathcal{N}(z', L_{22}^{-1})$ in $\mathcal{O}^\perp$. These measures are independent, so that the theorem states that

$$\mu^y \Rightarrow \delta_z \otimes \mathcal{N}(z', L_{22}^{-1}),$$

viewed as a measure on $\mathcal{O} \oplus \mathcal{O}^\perp$. Thus, in the small observational noise limit, we determine the solution without uncertainty in $\mathcal{O}$, whilst in $\mathcal{O}^\perp$ uncertainty remains. Furthermore, the prior plays a role in the posterior measure in the limit of zero observational noise; specifically it enters the formulae for z' and L_{22}.

We finish this subsection by providing proofs of the preceding three results.

Proof of Theorem 2.3. Define $\delta := u - \mathcal{F}(y)$ and let

$$f(\delta) = -\frac{1}{2\gamma^2}|y - \mathcal{G}(\mathcal{F}(y) + \delta)|^2 - \frac{1}{2}|\Sigma_0^{-1/2}(\mathcal{F}(y) + \delta - m_0)|^2.$$

Fix $\ell \in \mathbb{R}^n$. Then, with $\mathbb{E}$ denoting expectation under μ^y,

$$\mathbb{E}\exp\bigl(\mathrm{i}\langle \ell, u\rangle\bigr) = \frac{1}{Z}\exp\bigl(\mathrm{i}\langle \ell, \mathcal{F}(y)\rangle\bigr)\int_{\mathbb{R}^n}\exp\bigl(\mathrm{i}\langle \ell, \delta\rangle + f(\delta)\bigr)\,\mathrm{d}\delta,$$

where

$$Z = \int_{\mathbb{R}^n}\exp(f(\delta))\,\mathrm{d}\delta.$$

Thus, by Lemma 6.5, it suffices to prove that, as $\gamma \to 0$,

$$\frac{1}{Z}\int_{\mathbb{R}^n}\exp\bigl(\mathrm{i}\langle \ell, \delta\rangle + f(\delta)\bigr)\,\mathrm{d}\delta \to 1.$$

Define

$$I(\ell) = \int_{\mathbb{R}^n}\exp\bigl(\mathrm{i}\langle \ell, \delta\rangle + f(\delta)\bigr)\,\mathrm{d}\delta,$$

noting that $Z = I(0)$. For $a \in (2/3, 1)$, we split $I(\ell)$ into $I(\ell) = I_1(\ell) + I_2(\ell)$ where

$$I_1(\ell) = \int_{|\delta|\leqslant\gamma^a}\exp\bigl(\mathrm{i}\langle \ell, \delta\rangle + f(\delta)\bigr)\,\mathrm{d}\delta,$$

$$I_2(\ell) = \int_{|\delta|>\gamma^a}\exp\bigl(\mathrm{i}\langle \ell, \delta\rangle + f(\delta)\bigr)\,\mathrm{d}\delta.$$

We consider $I_1(\ell)$ first so that $|\delta| \leqslant \gamma^a$. By Taylor-expanding $f(\delta)$ around $\delta = 0$, we obtain

$$f(\delta) = -\frac{1}{2\gamma^2}|B\delta|^2 - \frac{1}{2}\big|\Sigma_0^{-1/2}(\mathcal{F}(y) + \delta - m_0)\big|^2 + \mathcal{O}\Big(\frac{\delta^3}{\gamma^2}\Big),$$

where $B = D\mathcal{G}(\mathcal{F}(y))$. Thus, for $b = a \wedge (3a-2) = 3a-2$,

$$\mathrm{i}\langle \ell, \delta\rangle + f(\delta) = -\frac{1}{2}\big|\Sigma_0^{-1/2}(\mathcal{F}(y) - m_0)\big|^2 - \frac{1}{2\gamma^2}|B\delta|^2 + \mathcal{O}(\gamma^b).$$

Thus

$$I_1(\ell) = \exp\Big(-\frac{1}{2}\big|\Sigma_0^{-1/2}(\mathcal{F}(y) - m_0)\big|^2\Big)\int_{|\delta|\leqslant\gamma^a} \exp\Big(-\frac{1}{2\gamma^2}|B\delta|^2 + \mathcal{O}(\gamma^b)\Big)\,\mathrm{d}\delta.$$

It follows that

$$\begin{aligned} I_1(\ell) &= \exp\Big(-\frac{1}{2}\big|\Sigma_0^{-1/2}(\mathcal{F}(y) - m_0)\big|^2\Big)\int_{|\delta|\leqslant\gamma^a} \exp\Big(-\frac{1}{2\gamma^2}|B\delta|^2\Big) \\ &\qquad \times \big(1 + \mathcal{O}(\gamma^b)\big)\,\mathrm{d}\delta \\ &= \gamma^n \exp\Big(-\frac{1}{2}\big|\Sigma_0^{-1/2}(\mathcal{F}(y) - m_0)\big|^2\Big)\int_{|z|\leqslant\gamma^{a-1}} \exp\Big(-\frac{1}{2}|Bz|^2\Big) \\ &\qquad \times \big(1 + \mathcal{O}(\gamma^b)\big)\,\mathrm{d}z. \end{aligned}$$

We now estimate $I_2(\ell)$ and show that it is asymptotically negligible compared with $I_1(\ell)$. Note that, by (2.15),

$$\begin{aligned} f(\delta) &\leqslant -\frac{C\min\{1, |\delta|^2\}}{2\gamma^2} - \frac{1}{2}\big|\Sigma_0^{-1/2}(\mathcal{F}(y) + \delta - m_0)\big|^2 \\ &\leqslant -\frac{C\min\{1, |\delta|^2\}}{2\gamma^2}. \end{aligned}$$

Thus

$$\begin{aligned} |I_2(\ell)| &\leqslant \int_{1\geqslant|\delta|>\gamma^a} \exp\Big(-\frac{C\delta^2}{\gamma^2}\Big)\mathrm{d}\delta \\ &\qquad + \int_{|\delta|>1} \exp\Big(-\frac{C}{\gamma^2}\Big)\exp\Big(-\frac{1}{2}\big|\Sigma_0^{-1/2}(\mathcal{F}(y) + \delta - m_0)\big|^2\Big)\mathrm{d}\delta. \end{aligned}$$

Since $a < 1$, it follows that $I_2(\ell)$ is exponentially small in $\gamma \to 0$. As $I_1(\ell)$ is, to leading order, $\mathcal{O}(\gamma^n)$ and independent of ℓ, we deduce that

$$\frac{1}{Z}\int_{\mathbb{R}^n} \exp\big(\mathrm{i}\langle \ell, \delta\rangle + f(\delta)\big)\,\mathrm{d}\delta = \frac{I(\ell)}{I(0)} \to 1$$

as $\gamma \to 0$, and the result follows. $\square$

Proof of Theorem 2.4. We first note the identity

$$A^*B^{-1}(B + A\Sigma_0 A^*) = (A^*B^{-1}A + \Sigma_0^{-1})\Sigma_0 A^*,$$

which follows since Σ_0 and B are both positive definite. Since $A^*B^{-1}A+\Sigma_0^{-1}$ and $B + A\Sigma_0 A^*$ are also positive definite, we deduce that

$$(A^*B^{-1}A + \Sigma_0^{-1})^{-1}A^*B^{-1} = \Sigma_0 A^*(B + A\Sigma_0 A^*)^{-1}.$$

Thus the posterior mean may be written as

$$\begin{aligned}
m &= m_0 + (A^*B^{-1}A + \Sigma_0^{-1})^{-1}A^*B^{-1}(y - Am_0)\\
&= (A^*B^{-1}A + \Sigma_0^{-1})^{-1}(A^*B^{-1}y + A^*B^{-1}Am_0 + \Sigma_0^{-1}m_0 - A^*B^{-1}Am_0)\\
&= (A^*B^{-1}A + \Sigma_0^{-1})^{-1}(A^*B^{-1}y + \Sigma_0^{-1}m_0),
\end{aligned}$$

as required. A similar calculation establishes the desired property of the posterior covariance.

If $B = \gamma^2 B_0$ then we deduce that

$$\begin{aligned}
m &= (A^*B_0^{-1}A + \gamma^2\Sigma_0^{-1})^{-1}(A^*B_0^{-1}y + \gamma^2\Sigma_0^{-1}m_0),\\
\Sigma &= \gamma^2(A^*B_0^{-1}A + \gamma^2\Sigma_0^{-1})^{-1}.
\end{aligned}$$

Since $\mathrm{Null}(A) = \{0\}$, we deduce that there is $\alpha > 0$ such that

$$\langle \xi, A^*B_0^{-1}A\xi\rangle = |B_0^{-1/2}A\xi|^2 \geqslant \alpha|\xi|^2, \quad \forall \xi \in \mathbb{R}^n.$$

Thus $A^*B_0^{-1}A$ is invertible and it follows that, as $\gamma \to 0$, the posterior mean converges to

$$m^+ = (A^*B_0^{-1}A)^{-1}A^*B_0^{-1}y$$

and the posterior covariance converges to zero. By Lemma 6.5 we deduce the desired weak convergence of μ^y to δ_{m^+}. It remains to characterize m^+.

Since the null space of A is empty, minimizers of

$$\phi(u) := \frac{1}{2}\|B_0^{-1/2}(y - Au)\|^2$$

are unique and satisfy the normal equations

$$A^*B_0^{-1}Au = A^*B_0^{-1}y.$$

Hence m^+ solves the desired least-squares problem and the proof is complete. □

Proof of Theorem 2.5. By Lemma 6.5 we see that it suffices to prove that the mean m and covariance Σ given by the formulae in Theorem 2.4 converge to m^+ and Σ^+ given by (2.22). We start by studying the covariance matrix which, by Theorem 2.4, is given by

$$\Sigma = \left(\frac{1}{\gamma^2}A^*B_0^{-1}A + L_0\right)^{-1}.$$

Using the definition (2.18) of A, we see that

$$A^*B_0^{-1}A = Q\begin{pmatrix} A_0^*B_0^{-1}A_0 & 0 \\ 0 & 0 \end{pmatrix}Q^*.$$

Then, by (2.19) we have

$$\Sigma^{-1} = Q\begin{pmatrix} \frac{1}{\gamma^2}A_0^*B_0^{-1}A_0 + L_{11} & L_{12} \\ L_{12}^* & L_{22} \end{pmatrix}Q^*.$$

Applying the Schur complement formula for the inverse of a matrix as in Lemma 6.21, we deduce that

$$\Sigma = Q\begin{pmatrix} \gamma^2(A_0^*B_0^{-1}A_0)^{-1} & 0 \\ -\gamma^2 L_{22}^{-1}L_{12}^*(A_0^*B_0^{-1}A_0)^{-1} & L_{22}^{-1} \end{pmatrix}Q^* + \Delta, \tag{2.23}$$

where

$$\frac{1}{\gamma^2}\big(|\Delta_{11}| + |\Delta_{21}|\big) \to 0$$

as $\gamma \to 0$, and there is a constant $C > 0$ such that

$$|\Delta_{12}| + |\Delta_{22}| \leqslant C\gamma^2$$

for all γ sufficiently small. From this it follows that, as $\gamma \to 0$,

$$\Sigma \to Q\begin{pmatrix} 0 & 0 \\ 0 & L_{22}^{-1} \end{pmatrix}Q^* := \Sigma^+,$$

writing Q as in (2.20). We see that $\Sigma^+ = Q_2L_{22}^{-1}Q_2^*$, as required.

We now return to the mean. By Theorem 2.4 this is given by the formula

$$m = \Sigma(A^*B^{-1}y + \Sigma_0^{-1}m_0).$$

Using the expression $A = (A_0\ 0)Q^*$, we deduce that

$$m \to \Sigma\left(\frac{1}{\gamma^2}Q\begin{pmatrix} A_0^*B_0^{-1} \\ 0 \end{pmatrix}y + \Sigma_0^{-1}m_0\right).$$

By definition of w, w', we deduce that

$$m = \Sigma Q\begin{pmatrix} \frac{1}{\gamma^2}A_0^*B_0^{-1}y + w \\ w' \end{pmatrix}.$$

Using equation (2.23), we find that

$$m = Q\begin{pmatrix} z \\ -L_{22}^{-1}L_{12}^*z + L_{22}^{-1}w' \end{pmatrix} = Q\begin{pmatrix} z \\ z' \end{pmatrix} := m^+.$$

This completes the proof. □

2.4. Common structure

In the previous subsection we showed that, for finite-dimensional problems, Bayes' rule gives the relationship (2.6) between the prior and posterior p.d.f.s π_0 and π^y respectively. Expressed in terms of the measures μ^y and μ_0 corresponding to these densities, the formula may be written as in (2.9):

$$\frac{\mathrm{d}\mu^y}{\mathrm{d}\mu_0}(u) = \frac{1}{Z(y)} \exp(-\Phi(u;y)). \tag{2.24}$$

The normalization constant $Z(y)$ is chosen so that μ^y is a probability measure:

$$Z(y) = \int_X \exp(-\Phi(u;y))\,\mathrm{d}\mu_0(u). \tag{2.25}$$

It is this form which generalizes readily to the setting on function space where there are no densities π^y and π_0 with respect to Lebesgue measure, but where μ^y has a Radon–Nikodym derivative (see Theorem 6.2) with respect to μ_0.

In Section 3 we will describe a range of inverse problems which can be formulated in terms of finding, and characterizing the properties of, a probability measure μ^y on a separable Banach space $(X, \|\cdot\|_X)$, specified via its Radon–Nikodym derivative with respect to a reference measure μ_0 as in (2.24) and (2.25). In this subsection we highlight the common framework into which many of these problems can be placed, by studying conditions on Φ which arise naturally in a wide range of applications. This framework will then be used to develop a general theory for inverse problems in Section 4. It is important to note that, when studying inverse problems, the properties of Φ that we highlight in this section are typically determined by the *forward PDE problem*, which maps the unknown function u to the data y. In particular, probability theory does not play a direct role in verifying these properties of Φ. Probability becomes relevant when choosing the prior measure so that it charges the Banach space X, on which the desired properties of Φ hold, with full measure. We illustrate how to make such choices of prior in Section 3.

We assume that the data y is in a separable Banach space $(Y, \|\cdot\|_Y)$. When applying the framework outlined in this article we will always assume that the prior measure is Gaussian: $\mu_0 \sim \mathcal{N}(m_0, \mathcal{C}_0)$. The properties of Gaussian random measures on Banach space, and Gaussian random fields in particular, may be found in Sections 6.3, 6.4 and 6.5. The two key properties of the prior that we will use repeatedly are the tail properties of the measure as encapsulated in the Fernique Theorem (Theorem 6.9), and the ability to establish regularity properties from the covariance operator: see Theorem 6.24 and Lemmas 6.25 and 6.27. It is therefore possible to broaden the scope of this material to non-Gaussian priors, for any measures for which analogues of these two key properties hold. However, Gaussian priors do form an important class of priors for a number of reasons: they are relatively simple to define through covariance operators defined as fractional inverse powers of differential operators; they are relatively straightforward to sample from; and the Hölder and Sobolev regularity properties of functions drawn from the prior are easily understood.

The properties of Φ may be formalized through the following assumptions, which we verify on a case-by-case basis for many of the PDE inverse problems encountered in Section 3.

Assumption 2.6. The function $\Phi : X \times Y \to \mathbb{R}$ has the following properties.

(i) For every $\varepsilon > 0$ and $r > 0$ there is an $M = M(\varepsilon, r) \in \mathbb{R}$ such that, for all $u \in X$ and all $y \in Y$ with $\|y\|_Y < r$,

$$\Phi(u; y) \geqslant M - \varepsilon \|u\|_X^2.$$

(ii) For every $r > 0$ there is a $K = K(r) > 0$ such that, for all $u \in X$ and $y \in Y$ with $\max\{\|u\|_X, \|y\|_Y\} < r$,

$$\Phi(u; y) \leqslant K.$$

(iii) For every $r > 0$ there is an $L(r) > 0$ such that, for all $u_1, u_2 \in X$ and $y \in Y$ with $\max\{\|u_1\|_X, \|u_2\|_X, \|y\|_Y\} < r$,

$$|\Phi(u_1; y) - \Phi(u_2; y)| \leqslant L\|u_1 - u_2\|_X.$$

(iv) For every $\varepsilon > 0$ and $r > 0$ there is a $C = C(\varepsilon, r) \in \mathbb{R}$ such that, for all $y_1, y_2 \in Y$ with $\max\{\|y_1\|_Y, \|y_2\|_Y\} < r$, and for all $u \in X$,

$$|\Phi(u; y_1) - \Phi(u; y_2)| \leqslant \exp\big(\varepsilon \|u\|_X^2 + C\big) \|y_1 - y_2\|_Y. \qquad \diamond$$

These assumptions are, in turn, a lower bound, an upper bound and Lipschitz properties in u and in y. When Y is finite-dimensional and the observational noise is $\mathcal{N}(0, \Gamma)$, then Φ has the form

$$\begin{aligned} \Phi(u; y) &= \frac{1}{2}\big|\Gamma^{-1/2}\big(y - \mathcal{G}(u)\big)\big|^2 \\ &= \frac{1}{2}\big|\big(y - \mathcal{G}(u)\big)\big|_\Gamma^2. \end{aligned} \tag{2.26}$$

It is then natural to derive the bounds and Lipschitz properties of Φ from properties of $\mathcal{G}$.

Assumption 2.7. The function $\mathcal{G} : X \to \mathbb{R}^q$ satisfies the following.

(i) For every $\varepsilon > 0$ there is an $M = M(\varepsilon) \in \mathbb{R}$ such that, for all $u \in X$,

$$|\mathcal{G}(u)|_\Gamma \leqslant \exp\big(\varepsilon \|u\|_X^2 + M\big).$$

(ii) For every $r > 0$ there is a $K = K(r) > 0$ such that, for all $u_1, u_2 \in X$ with $\max\{\|u_1\|_X, \|u_2\|_X\} < r$,

$$|\mathcal{G}(u_1) - \mathcal{G}(u_2)|_\Gamma \leqslant K\|u_1 - u_2\|_X. \qquad \diamond$$

It is straightforward to see the following.

Lemma 2.8. Assume that $\mathcal{G} : X \to \mathbb{R}^q$ satisfies Assumption 2.7. Then $\Phi : X \times \mathbb{R}^q \to \mathbb{R}$ given by (2.26) satisfies Assumption 2.6 with $(Y, \|\cdot\|_Y) = (\mathbb{R}^q, |\cdot|_\Gamma)$. ◇

Many properties follow from these assumptions concerning the density between the posterior and the prior. Indeed, the fact that μ^y is well-defined is typically established by using the continuity properties of $\Phi(\cdot; y)$. Further properties following from these assumptions include continuity of μ^y with respect to the data y, and desirable perturbation properties of μ^y based on finite-dimensional approximation of Φ or $\mathcal{G}$. All these properties will be studied in detail in Section 4. We emphasize that many variants on the assumptions above could be used to obtain similar, but sometimes weaker, results than those appearing in this article. For example, we work with Lipschitz continuity of Φ in both arguments; similar results can be proved under the weaker assumptions of continuity in both arguments. However, since Lipschitz continuity holds for most of the applications of interest to us, we work under these assumptions.

We re-emphasize that the properties of Φ encapsulated in Assumption 2.6 are properties of the forward PDE problem, and they do not involve inverse problems and probability at all. The link to Bayesian inverse problems comes through the choice of prior measure μ_0 which, as we will see in Sections 3 and 4, should be chosen so that $\mu_0(X) = 1$; this means that functions drawn at random from the prior measure should be sufficiently regular that they lie in X with probability one, so that the properties of Φ from Assumptions 2.6 apply to it. In the function space setting, regularity of the mean function, together with the spectral properties of the covariance operator, determines the regularity of random draws. In particular, the rate of decay of the eigenvalues of the covariance operator plays a central role in determining the regularity properties. These issues are discussed in detail in Section 6.5. For simplicity we will work throughout with covariance operators which are defined through (possibly fractional) negative powers of the Laplacian, or operators that behave like the Laplacian in a sense made precise below.

To make these ideas precise, consider a second-order differential operator $\mathcal{A}$ on a bounded open set $D \subset \mathbb{R}^d$, with domain chosen so that $\mathcal{A}$ is positive definite and invertible. Let $\mathcal{H} \subset L^2(D)$. For example, $\mathcal{H}$ may be restricted to the subspace where

$$\int_D u(x)\,\mathrm{d}x = 0 \tag{2.27}$$

holds, in order to enforce positivity for an operator with Neumann or periodic boundary conditions, which would otherwise have constants in its kernel; or it may be restricted to divergence-free fields when incompressible fluid flow is being modelled.

We let $\{(\phi_k, \lambda_k)\}_{k\in\mathbb{K}}$ denote a complete orthonormal basis for $\mathcal{H}$, comprising eigenfunctions/eigenvalues of $\mathcal{A}$. Then $\mathbb{K} \subseteq \mathbb{Z}^d\backslash\{0\}$. For Laplacian-like operators we expect that the eigenvalues will grow like $|k|^2$ and that, in simple geometries, the ϕ_k will be bounded in L^∞ and the gradient of the ϕ_k will grow like $|k|$ in L^∞. We make these ideas precise below. For all infinite sums over $\mathbb{K}$ in the following we employ standard orderings.

For any $u \in \mathcal{H}$ we may write

$$u = \sum_{k\in\mathbb{K}} \langle u, \phi_k\rangle \phi_k.$$

We may then define fractional powers of $\mathcal{A}$ as follows, for any $\alpha \in \mathbb{R}$:

$$\mathcal{A}^\alpha u = \sum_{k\in\mathbb{K}} \lambda_k^\alpha \langle u, \phi_k\rangle \phi_k. \tag{2.28}$$

For any $s \in \mathbb{R}$ we define the separable Hilbert spaces $\mathcal{H}^s$ by

$$\mathcal{H}^s = \left\{u : \sum_{k\in\mathbb{K}} \lambda_k^s |\langle u, \phi_k\rangle|^2 < \infty\right\}. \tag{2.29}$$

These spaces have norm $\|\cdot\|_s$ defined by

$$\|u\|_s^2 = \sum_{k\in\mathbb{K}} \lambda_k^s |\langle u, \phi_k\rangle|^2.$$

If $s \geqslant 0$ then these spaces are contained in $\mathcal{H}$, but for $s < 0$ they are larger than $\mathcal{H}$. The following assumptions characterize a 'Laplacian-like' operator. These operators will be useful to us when constructing Gaussian priors, as they will enable us to specify regularity properties of function drawn from the prior in a transparent fashion.

Assumption 2.9. The operator $\mathcal{A}$, densely defined on a Hilbert space $\mathcal{H} \subset L^2(D;\mathbb{R}^n)$, satisfies the following properties.

(i) $\mathcal{A}$ is positive definite, self-adjoint and invertible.

(ii) The eigenfunctions/eigenvalues $\{\phi_k, \lambda_k\}_{k\in\mathbb{K}}$ of $\mathcal{A}$, indexed by $k \in \mathbb{K} \subset \mathbb{Z}^d\backslash\{0\}$, form an orthonormal basis for $\mathcal{H}$.

(iii) There exist $C^\pm > 0$ such that the eigenvalues satisfy, for all $k \in \mathbb{K}$,

$$C^- \leqslant \frac{\lambda_k}{|k|^2} \leqslant C^+.$$

(iv) There exists $C > 0$ such that

$$\sup_{k\in\mathbb{K}} \left(\|\phi_k\|_{L^\infty} + \frac{1}{|k|}\|D\phi_k\|_{L^\infty}\right) \leqslant C. \qquad \diamond$$

Note that if $\mathcal{A}$ is the Laplacian with Dirichlet or Neumann boundary conditions, then the spaces $\mathcal{H}^s$ are contained in the usual Sobolev spaces H^s. In the case of periodic boundary conditions they are identical to the Sobolev spaces $H^s_{\rm per}$. Thus the final assumption (v) above is a generalization of the following Sobolev Embedding Theorem for the Laplacian.

Theorem 2.10. (Sobolev Embedding Theorem) Assume that $\mathcal{A} := -\triangle$ is equipped with periodic, Neumann or Dirichlet boundary conditions on the unit cube. If $u \in \mathcal{H}^s$ and $s > d/2$, then $u \in C(\overline{D})$, and there is a constant $C > 0$ such that $\|u\|_{L^\infty} \leqslant C\|u\|_s$. ◇

2.5. Discussion and bibliography

An introduction to the Bayesian approach to statistical problems in general is Bernardo and Smith (1994). The approach taken to Bayesian inverse problems as outlined in Kaipio and Somersalo (2005) is to first discretize the problem and then secondly apply the Bayesian methodology to a finite-dimensional problem. This is a commonly adopted methodology. In that approach, the idea of trying to capture the limit of infinite resolution is addressed by use of statistical extrapolation techniques based on modelling the error from finite-dimensional approximation (Kaipio and Somersalo 2007*b*). The approach that is developed in this article reverses the order of these two steps: we first apply the Bayesian methodology to an infinite-dimensional problem, and then discretize. There is some literature concerning the Bayesian viewpoint for linear inverse problems on function space, including the early study by Franklin (1970), and the subsequent papers by Mandelbaum (1984), Lehtinen, Paivarinta and Somersalo (1989) and Fitzpatrick (1991); the paper by Lassas *et al.* (2009) contains a good literature review of this material, and further references. The papers of Lassas *et al.* (2009) and Lassas and Siltanen (2004) also study Bayesian inversion for linear inverse problems on function space; they introduce the notion of *discretization invariance* and investigate the question of whether it is possible to derive regularizations of families of finite-dimensional problems, in a fashion which ensures that meaningful limits are obtained; this idea also appears somewhat earlier in the data assimilation literature, for a particular PDE inverse problem, in the paper of Bennett and Budgell (1987). In the approach taken in this article, discretization invariance is guaranteed for finite-dimensional approximations of the function space Bayesian inverse problem. Furthermore, our approach is not limited to problems in which a Gaussian posterior measure appears; in contrast, existing work on discretization invariance is confined to the linear, Gaussian observational noise setting in which the posterior is Gaussian if the prior is Gaussian.

The least-squares approach to inverse problems encapsulated in (2.3) is often termed *Tikhonov regularization* (Engl, Hanke and Neubauer 1996) and, more generally, the *variational method* in the applied literature (Bennett 2002, Evensen 2006). The book by Engl *et al.* (1996) discusses regularization techniques in the Hilbert space setting and the Banach space setting is discussed in, for example, the recent papers of Kaltenbacher, Schöpfer and Schuster (2009), Neubauer (2009) and Hein (2009). As we demonstrated, regularization is closely related to finding the MAP estimator as defined in Kaipio and Somersalo (2005). As such it is clear that, from the Bayesian standpoint, regularization is intimately related to the choice of prior. Another classical regularization method for linear inverse problems is through iterative solution (Engl *et al.* 1996); this topic is related to the Bayesian approach to inverse problems in Calvetti (2007) and Calvetti and Somersalo (2005*a*).

Although we concentrate in this paper on Gaussian priors, and hence on regularization via addition of a quadratic penalization term, there is active research in the use of different regularizations (Kaltenbacher *et al.* 2009, Neubauer 2009, Hein 2009, Lassas and Siltanen 2004). In particular, the use of total variation-based regularization, and related wavelet-based regularizations, is central in image processing (Rudin *et al.* 1992, Scherzer *et al.* 2009). We will not address such regularizations in this article, but note that the development of a function space Bayesian viewpoint on such problems, along the lines developed here for Gaussian priors, is an interesting research direction (Lassas *et al.* 2009).

Theorem 2.4 concerns the small noise limit for Gaussian noise. This topic has been studied in greater detail in the papers by Engl, Hofinger and Kindermann (2005), Hofinger and Pikkarainen (2007, 2009) and Neubauer and Pikkarainen (2008), where the convergence of the posterior distribution is quantified by use of the Prokohorov and Ky Fan metrics. Gaussian problems are often amenable to closed-form analysis, as illustrated in this section, and are hence useful for illustrative purposes. Furthermore, there are many interesting applications where Gaussian structure prevails. Thus we will, on occasion, exploit Gaussianity throughout the article, for both these reasons.

The common structure underlying a wide range of Bayesian inverse problems for functions, and which we highlight in Section 2.4, is developed in Cotter, Dashti, Robinson and Stuart (2009, 2010*b*) and Cotter, Dashti and Stuart (2010*a*).

In the general framework established at the start of this section we have implicitly assumed that the observation operator $\mathcal{G}(\cdot)$ is known to us. In practice it is often approximated by some computer code $\mathcal{G}(\cdot;h)$ in which h denotes a mesh parameter, or parameter controlling missing physics. In this case (2.5) can be replaced by the equation

$$y = \mathcal{G}(u;h) + \varepsilon + \eta, \tag{2.30}$$

where $\varepsilon := \mathcal{G}(u) - \mathcal{G}(u;h)$. Whilst it is possible to lump ε and η together into one single error term, and work again with equation (2.1), this can be misleading because the observation error η and the computational model error ε are very different in character. The latter is typically not mean zero, and depends upon u; in contrast it is frequently realistic to model η as a mean zero random variable, independent of u. Attempts to model the effects of ε and η separately may be found in a number of publications, including Kaipio and Somersalo (2005, Chapter 7), Kaipio and Somersalo (2007*a*), Kaipio and Somersalo (2007*b*), Glimm, Hou, Lee, Sharp and Ye (2003), Orrell, Smith, Barkmeijer and Palmer (2001), Kennedy and O'Hagan (2001), O'Sullivan and Christie, (2006*b*, 2006*a*), Christie, Pickup, O'Sullivan and Demyanov (2008) and Christie (2010). A different approach to dealing with model error is to extend the variable u to include model terms which represent missing physics or lack of resolution in the model and to try to learn about such systematic error from the data; this approach is undertaken in Cotter, Dashti, Robinson and Stuart (2009).

3. Examples

3.1. Overview

In this section we study a variety of inverse problems arising from boundary value problems and initial-boundary value problems. Our goal is to enable application of the framework for Bayesian inverse problems on function space that is developed in Section 4, in order to justify a formula of the form (2.24) for a posterior measure μ^y on a function space, and to establish properties of the measure μ^y.

In order to carry this out it is desirable to establish that, for a wide range of problems, the common structure encapsulated in Assumptions 2.6 or 2.7 may be shown to hold. These assumptions concern properties of the *forward problem* underlying the inverse problem, and have no reference to the inverse problem, its Bayesian formulation or to probability. The link between the forward problem and the Bayesian inverse problem is provided in this section, and in the next section. In this section we show that choosing the prior measure so that $\mu_0(X) = 1$, where X is the space in which Assumptions 2.6 or 2.7 may be shown to hold, ensures that the posterior measure is well-defined; this may often be done by use of Theorem 6.31. The larger the space X, the fewer restrictions the condition $\mu_0(X) = 1$ places on the choice of prior, since it is equivalent to asking that draws from μ_0 are almost surely in the space X; the larger X is, the easier this is to satisfy. The next section is concerned with ramifications of Assumptions 2.6 or 2.7 for various stability properties of the posterior measure μ^y with respect to perturbations.

We will work in a Banach space setting and will always specify the prior measure as a Gaussian. The required background material on Gaussian measures in Banach space, and Gaussian random fields, may be found in Section 6. We also make regular use of the key Theorem 6.31, from Section 6.6, to show that the posterior is well-defined and absolutely continuous with respect to the prior. For simplicity we work with priors whose covariance operator is a fractional negative power of an operator such as the Laplacian. The reader should be aware that much greater generality than this is possible and that the simple setting for choice of priors is chosen for expository purposes. Other Gaussian priors may be chosen so long as the constraint $\mu_0(X) = 1$ is satisfied.

We start in Section 3.2 by studying the inverse problems of determining a field from *direct* pointwise observations. We use this example to illustrate our approach to identifying the Radon–Nikodym derivative between posterior and prior measures. All of the subsequent subsections in this chapter involve Bayesian inference for random fields, but in contrast to the first subsection they are based on *indirect* measurements defined through solution of a differential equation. In Section 3.3 we study the problem of finding the diffusion coefficient in a two-point boundary value problem, from observations of the solution. In Section 3.4 we consider the problem of determining the wave speed for the scalar wave equation from observations of the solution. Section 3.5 concerns the problem of recovering the initial condition for the heat equation, from observation of the entire solution at a positive time, when polluted by an additive Gaussian random field. We then describe several more involved examples arising in applications such as fluid mechanics, geophysics and molecular dynamics, all of which can be placed in the common framework developed here, but for which space precludes a full development of the details; see Sections 3.6, 3.7 and 3.8. The problems in fluid mechanics are natural extensions of the inverse problem for the initial condition of the heat equation, and those arising in subsurface geophysics generalize the inverse problem for the diffusion coefficient in a two-point boundary value problem. The problem in molecular dynamics is somewhat different, as it does not arise from a Bayesian inverse problem but rather from a conditioned diffusion process. However, the resulting mathematical structure shares much with the inverse problems and we include it for this reason. References to some of the relevant literature on these applications are given in Section 3.9.

3.2. Pointwise data for a random field

Let $D \subset \mathbb{R}^d$ be a bounded open set. Consider a field $u : D \to \mathbb{R}^n$. We view u as an element of the Hilbert space $\mathcal{H} = L^2(D)$. Assume that we are given noisy observations $\{y_k\}_{k=1}^q$ of a function $g : \mathbb{R}^n \to \mathbb{R}^\ell$ of the field at a set of points $\{x_k\}_{k=1}^q$. Thus

$$y_k = g(u(x_k)) + \eta_k, \tag{3.1}$$

where the $\{\eta_k\}_{k=1}^q$ describe the observational noise. Concatenating data, we have

$$y = \mathcal{G}(u) + \eta, \tag{3.2}$$

where $y = (y_1^*, \dots, y_q^*)^* \in \mathbb{R}^{\ell q}$ and $\eta = (\eta_1^*, \dots, \eta_q^*)^* \in \mathbb{R}^{\ell q}$. The observation operator $\mathcal{G}$ maps $X := C(\overline{D}) \subset \mathcal{H}$ into $Y := \mathbb{R}^{\ell q}$. The inverse problem is to reconstruct the field u from the data y.

We assume that the observational noise η is Gaussian $\mathcal{N}(0, \Gamma)$. We specify a prior measure μ_0 on u which is Gaussian $\mathcal{N}(m_0, \mathcal{C}_0)$ and determine the posterior measure μ^y for u given y. Since $\mathbb{P}(\mathrm{d}y|u) = \mathcal{N}(\mathcal{G}(u), \Gamma)$, informal application of Bayes' rule leads us to expect that the Radon–Nikodym derivative of μ^y with respect to μ_0 is

$$\frac{\mathrm{d}\mu^y}{\mathrm{d}\mu_0}(u) \propto \exp\biggl(-\frac{1}{2}|y - \mathcal{G}(u)|_\Gamma^2\biggr). \tag{3.3}$$

Below we deduce appropriate choices of prior measure which ensure that this measure is well-defined and does indeed determine the desired posterior distribution for u given y.

If $g : \mathbb{R}^n \to \mathbb{R}^\ell$ is linear, so that $\mathcal{G}(u) = Au$ for some linear operator $A : X \to \mathbb{R}^{\ell q}$, then the calculations in Example 6.23 show that the posterior measure μ^y is also Gaussian with $\mu^y = \mathcal{N}(m, \mathcal{C})$ where

$$m = m_0 + \mathcal{C}_0 A^*(\Gamma + A\mathcal{C}_0 A^*)^{-1}(y - Am_0), \tag{3.4a}$$

$$\mathcal{C} = \mathcal{C}_0 - \mathcal{C}_0 A^*(\Gamma + A\mathcal{C}_0 A^*)^{-1} A\mathcal{C}_0. \tag{3.4b}$$

Let $\triangle$ denote the Laplacian on D, with domain chosen so that Assumptions 2.9(i)–(iv) hold. Recall the (Sobolev-like) spaces $\mathcal{H}^s$ from (2.29). The following theorem is proved by application of Theorem 6.31, which the reader is encouraged to study before continuing in this section.

Theorem 3.1. Assume that the domain of $-\triangle$ is chosen so that Assumptions 2.9(i)–(v) hold. Let $g : \mathbb{R}^n \to \mathbb{R}^\ell$ be continuous. Assume that $\mathcal{C}_0 \propto (-\triangle)^{-\alpha}$ with $\alpha > d/2$ and assume that $m_0 \in \mathcal{H}^\alpha$. Then $\mu^y(\mathrm{d}u) = \mathbb{P}(\mathrm{d}u|y)$ is absolutely continuous with respect to $\mu_0(\mathrm{d}u) = \mathcal{N}(m_0, \mathcal{C}_0)$ with Radon–Nikodym derivative given by (3.3). Furthermore, when g is linear, so that $\mathcal{G}(u) = Au$ for some linear $A : X \to \mathbb{R}^{\ell q}$, then the posterior is Gaussian with mean and covariance given by (3.4).

Proof. The formulae for the mean and covariance of the Gaussian posterior measure $\mu^y = \mathcal{N}(m, \mathcal{C})$, which arises when g is linear, follow from Example 6.23. We now proceed to determine the posterior measure in the non-Gaussian case. Define $L_z : X \to \mathbb{R}^n$ to be the pointwise evaluation operator at $z \in D$. Notice that

$$|L_z u - L_z v| = |u(z) - v(z)| \leqslant \|u - v\|_{L^\infty}$$

so that $L_z : X \to \mathbb{R}^n$ is continuous. The function $\mathcal{G}$ is found by composing the continuous function g with the operator $L_{\cdot}$ at a finite set of points and is thus itself continuous from X into $\mathbb{R}^{\ell q}$. To apply Theorem 6.31 it suffices to show that $\mu_0(X) = 1$. Since $\mathcal{H}^\alpha$ is the Cameron–Martin space for $\mathcal{C}_0$ and since $m_0 \in \mathcal{H}^\alpha$, we deduce that $\mu_0 = \mathcal{N}(m_0, \mathcal{C}_0)$ and $\mathcal{N}(0, \mathcal{C}_0)$ are equivalent as measures, by Theorem 6.13. Thus $\mu_0(X) = 1$ since, by Lemma 6.25, draws from $\mathcal{N}(0, \mathcal{C}_0)$ are a.s. s-Hölder for all $s \in (0, \min\{1, \alpha - d/2\})$. □

In Section 2.4 we indicated that obtaining bounds and Lipschitz properties of $\mathcal{G}$ or Φ, the mappings appearing in the Radon–Nikodym derivative between μ^y and μ_0, will be important to us below. The following lemma studies this issue.

Lemma 3.2. In the setting of Theorem 3.1 assume, in addition, that $g \in C^1(\mathbb{R}^n, \mathbb{R}^\ell)$ and that g is polynomially bounded. Then $\mathcal{G}$ satisfies Assumption 2.7 with $X = C(\overline{D})$ and $Y = \mathbb{R}^{\ell q}$. Furthermore, if Dg is polynomially bounded then $K(r)$ is polynomially bounded.

Proof. Since g is polynomially bounded and $\mathcal{G}$ is found by pointwise evaluation at a finite number of points, it follows that

$$|\mathcal{G}(u)| \leqslant p(\|u\|_X)$$

for some polynomial $p : \mathbb{R} \to \mathbb{R}$. The bound (i) of Assumption 2.7 follows. By the mean-value theorem (Taylor theorem with remainder) we have that

$$|\mathcal{G}(u) - \mathcal{G}(v)|_\Gamma \leqslant \max_{1 \leqslant k \leqslant K} \left| \int_0^1 Dg\big(su(x_k) + (1-s)v(x_k)\big)\, \mathrm{d}s \big(u(x_k)) - v(x_k)\big) \right|.$$

Thus, for all u, v satisfying $\max\{\|u\|_X, \|v\|_X\} < r$,

$$|\mathcal{G}(u) - \mathcal{G}(v)|_\Gamma \leqslant K(r)\|u - v\|_X.$$

Furthermore, K may be bounded polynomially if Dg is bounded polynomially. The result follows. □

3.3. Inverse problem for a diffusion coefficient

The previous example illustrated the formulation of an inverse problem for a function, using the Bayesian framework. However, the observations of the function comprised direct measurements of the function at points in its domain D. We now consider a problem where the measurements are more indirect, and are defined through the solution of a differential equation.

We consider the inverse problem of determining the diffusion coefficient from observations of the solution of the two-point boundary value problem

$$-\frac{\mathrm{d}}{\mathrm{d}x}\left(k(x)\frac{\mathrm{d}p}{\mathrm{d}x}\right) = 0, \tag{3.5a}$$

$$p(0) = p^-, \quad p(1) = p^+. \tag{3.5b}$$

We assume that $p^+ > p^- > 0$ and we make observations of $\{p(x_k)\}_{k=1}^q$, at a set of points $0 < x_1 < \cdots < x_q < 1$ subject to Gaussian measurement error. We write the observations as

$$y_j = p(x_j) + \eta_j, \quad j = 1, \ldots, q \tag{3.6}$$

and, for simplicity, assume that the η_k form an i.i.d. sequence with $\eta_1 \sim \mathcal{N}(0, \gamma^2)$. Our interest is in determining the diffusion coefficient k from y. To ensure that k is strictly positive on $[0,1]$, we introduce $u(x) = \ln(k(x))$ and view $u \in L^2((0,1))$ as the basic unknown function.

The forward problem (3.5) for p given u is amenable to considerable explicit analysis, and we now use this to write down a formula for the observation operator $\mathcal{G}$ and to study its properties. We first define $J_x : L^\infty((0,1)) \to \mathbb{R}$ by

$$J_x(w) = \int_0^x \exp(-w(z))\,\mathrm{d}z. \tag{3.7}$$

The solution of (3.5) may be written as

$$p(x) = (p^+ - p^-)\frac{J_x(u)}{J_1(u)} + p^- \tag{3.8}$$

and is monotonic increasing; furthermore, $p(x)$ is unchanged under $u(x) \to u(x) + \lambda$ for any $\lambda \in \mathbb{R}$. The observation operator is then given by the formula

$$\mathcal{G}(u) = \big(p(x_1), \ldots, p(x_q)\big)^*. \tag{3.9}$$

Lemma 3.3. The observation operator $\mathcal{G} : C([0,1]) \to \mathbb{R}^q$ is Lipschitz and satisfies the bound

$$|\mathcal{G}(u)| \leqslant \sqrt{q}p^+. \tag{3.10}$$

Indeed, $\mathcal{G}$ satisfies Assumption 2.7 with $X = C([0,1])$ and $K(\cdot)$ exponentially bounded: there are $a, b > 0$ such that $K(r) \leqslant a\exp(br)$.

Proof. The fact that $\mathcal{G}$ is defined on $C([0,1])$ follows from the explicit solution given in equation (3.8). The bound on $\mathcal{G}$ follows from the monotonicity of the solution. For the Lipschitz property it suffices to consider the case $q = 1$ and, without loss of generality, take $x_1 = 1/2$. Note that then

$$\begin{aligned}\frac{|\mathcal{G}(u) - \mathcal{G}(v)|}{p^+ - p^-} &= \frac{1}{J_1(u)J_1(v)}\big|J_{\frac12}(u)J_1(v) - J_{\frac12}(v)J_1(u)\big| \\ &= \frac{1}{J_1(u)J_1(v)}\big|J_{\frac12}(u)\big(J_1(v) - J_1(u)\big) + J_1(u)\big(J_{\frac12}(u) - J_{\frac12}(v)\big)\big| \\ &\leqslant J_1(v)^{-1}|J_1(v) - J_1(u)| + J_1(v)^{-1}\big|J_{\frac12}(u) - J_{\frac12}(v)\big|.\end{aligned}$$

But

$$J_1(v)^{-1} \leqslant \exp\big(\|v\|_\infty\big)$$

and

$$|J_x(u) - J_x(v)| \leqslant x \exp\big(\max\{\|u\|_\infty, \|v\|_\infty\}\big) \|u - v\|_\infty.$$

Thus we deduce that

$$|\mathcal{G}(u) - \mathcal{G}(v)| \leqslant \frac{3}{2}(p^+ - p^-) \exp\big(\|v\|_\infty + \max\{\|u\|_\infty, \|v\|_\infty\}\big) \|u - v\|_\infty. \quad \square$$

We place a Gaussian prior measure $\mu_0 \sim \mathcal{N}(u_0, \mathcal{C}_0)$ on u. We say that k is *log-normal.* Since changing u by an arbitrary additive constant does not change the solution of (3.5), we cannot expect to determine any information about the value of this constant from the data. Thus we must build our assumptions about this constant into the prior. To do this we assume that u integrates to zero on $(0, 1)$ and define the prior measure μ_0 on the space

$$\mathcal{H} = \left\{ u \in L^2((0,1)) \,\Big|\, \int_0^1 u(x)\,\mathrm{d}x = 0 \right\}. \tag{3.11}$$

We define $A = -\mathrm{d}^2/\mathrm{d}x^2$ to be a densely defined operator on $\mathcal{H}$ with

$$D(A) = \left\{ u \in H^2_{\mathrm{per}}((0,1)) \,\Big|\, \int_0^1 u(x)\,\mathrm{d}x = 0 \right\}.$$

Then A is positive definite self-adjoint and, for any $\beta > 0$ and $\alpha > 1/2$ (which ensures that the covariance operator is trace-class), we may define the Gaussian measure $\mathcal{N}(m_0, \beta A^{-\alpha})$ on $\mathcal{H}$.

We have

$$y = \mathcal{G}(u) + \eta,$$

where $y = (y_1, \dots, y_q)^* \in \mathbb{R}^q$ and $\eta \in \mathbb{R}^q$ is distributed as $\mathcal{N}(0, \gamma^2 I)$. The probability of y given u (the data likelihood) is

$$\mathbb{P}(y|u) \propto \exp\left(-\frac{1}{2\gamma^2}|y - \mathcal{G}(u)|^2\right).$$

We wish to find $\mu^y(\mathrm{d}u) = \mathbb{P}(\mathrm{d}u|y)$. Informal use of Bayes' rule suggests that

$$\frac{\mathrm{d}\mu^y}{\mathrm{d}\mu_0}(u) \propto \exp\left(-\frac{1}{2\gamma^2}|y - \mathcal{G}(u)|^2\right). \tag{3.12}$$

We now justify this formula. Since $\mathcal{G}(\cdot)$ is Lipschitz on $X := C([0,1])$, by Lemma 3.3, the basic idea underlying the justification of (3.12) in the next theorem is to choose α so that $\mu_0(X) = 1$, so that we may apply Theorem 6.31.

Theorem 3.4. Consider the Bayesian inverse problem for $u(x) = \ln(k(x))$ subject to observation in the form (3.6), with p solving (3.5), and prior measure $\mu_0 = \mathcal{N}(m_0, \mathcal{C}_0)$ with $m_0 \in \mathcal{H}^\alpha$ and $\mathcal{C}_0 = \beta A^{-\alpha}$. If $\beta > 0$ and $\alpha > 1/2$ then $\mu^y(\mathrm{d}u) = \mathbb{P}(\mathrm{d}u|y)$ is absolutely continuous with respect to $\mu_0(\mathrm{d}u)$ with Radon–Nikodym derivative given by (3.12), with $\mathcal{G}$ defined in (3.9).

Proof. We apply Theorem 6.31. The function $\mathcal{G}$ is continuous from X into $\mathbb{R}^q$. Hence it suffices to show that $\mu_0(X) = 1$. Since $\mathcal{H}^\alpha$ is the Cameron–Martin space for $\mathcal{C}_0$ and since $m_0 \in \mathcal{H}^\alpha$, we deduce that $\mu_0 = \mathcal{N}(m_0, \mathcal{C}_0)$ and $\mathcal{N}(0, \mathcal{C}_0)$ are equivalent as measures, by Theorem 6.13. Thus $\mu_0(X) = 1$ since, by Lemma 6.25, draws from $\mathcal{N}(0, \mathcal{C}_0)$ are a.s. s-Hölder for all $s \in (0, \min\{1, \alpha - 1/2\})$. □

3.4. *Wave speed for the wave equation*

Consider the equation

$$\frac{\partial v}{\partial t} + c(x)\frac{\partial v}{\partial x} = 0, \quad (x,t) \in \mathbb{R} \times (0, \infty) \tag{3.13a}$$

$$v = f, \quad (x,t) \in \mathbb{R} \times \{0\} \tag{3.13b}$$

We assume that the wave speed $c(x)$ is known to be a positive, one-periodic function, and that we are interested in the inverse problem of determining c given the observations

$$y_j = v(1, t_j) + \eta_j, \quad j = 1, \ldots, q. \tag{3.14}$$

We assume that the observational noise $\{\eta_j\}_{j=1}^q$ is mean zero Gaussian. Since c is positive, we write $c = \exp(u)$ and view the inverse problem as being the determination of u. We thus concatenate the data and write

$$y = \mathcal{G}(u) + \eta,$$

where $\eta \sim \mathcal{N}(0, \Gamma)$ and $\mathcal{G} : X \to \mathbb{R}^q$ where $X = C^1(\mathbb{S})$; here $\mathbb{S}$ denotes the unit circle $[0, 1)$ with end points identified to enforce periodicity. We equip X with the norm

$$\|u\|_X = \sup_{x\in\mathbb{S}} |u(x)| + \sup_{x\in\mathbb{S}} \left|\frac{\mathrm{d}u}{\mathrm{d}x}(x)\right|.$$

Note that we may also view u as a function in $X_{\text{per}} := C^1_{\text{per}}(\mathbb{R})$, the space of 1-periodic C^1 functions on $\mathbb{R}$. Before defining the inverse problem precisely, we study the properties of the forward operator $\mathcal{G}$.

Lemma 3.5. Assume that $f \in C^1(\mathbb{R}; \mathbb{R})$ and f is polynomially bounded: there are constants $K > 0$ and $p \in \mathbb{Z}^+$ such that

$$|f(x)| \leqslant K\big(1 + |x|^p\big).$$

Then $\mathcal{G} : X \to \mathbb{R}^q$ satisfies the following conditions.

- There is a constant $C > 0$:

$$|\mathcal{G}(u)| \leqslant C\big(1 + \exp(p\|u\|_X)\big).$$

- For all $u, w \in X : \|u\|_X, \|w\|_X < r$ there exists $L = L(r)$:

$$|\mathcal{G}(u) - \mathcal{G}(w)| \leqslant L\|u - w\|_\infty.$$

Proof. It suffices to consider the case $q = 1$ and take $t_1 = 1$ for simplicity. Let $\Psi(\cdot; t, u) : \mathbb{R} \to \mathbb{R}$ denote the one-parameter group given by the solution operator for the equation

$$\frac{\mathrm{d}x}{\mathrm{d}t} = -\exp(u(x)), \tag{3.15}$$

where we view u as an element of X_{per} in order to define the solution of this equation. Then v solving (3.13) with $c = \exp(u)$ is given by the formula

$$v(x,t) = f\big(\Psi(x; t, u)\big).$$

Thus

$$\mathcal{G}(u) = v(1,1) = f\big(\Psi(1; 1, u)\big) \tag{3.16}$$

and

$$|\mathcal{G}(u)| = |v(1,1)| \leqslant K\big(1 + |\Psi(1; 1, u)|^p\big).$$

But the solution of (3.15) subject to the condition $x(0) = 1$ satisfies

$$\begin{aligned} |x(1)| &\leqslant 1 + \int_0^1 \exp\big(u(x(s))\big)\,\mathrm{d}s \\ &\leqslant 1 + \exp(\|u\|_X). \end{aligned}$$

Hence

$$\Psi(1; 1, u) \leqslant 1 + \exp(\|u\|_X), \tag{3.17}$$

and the first result follows.

For the second result let $x(t) = \Psi(1; t, u)$ and $y(t) = \Psi(1; t, w)$ so that, by (3.15),

$$x(t) = 1 - \int_0^t \exp\big(u(x(s))\big)\,\mathrm{d}s,$$

$$y(t) = 1 - \int_0^t \exp\big(u(y(s))\big)\,\mathrm{d}s + \int_0^t \exp\big(u(y(s))\big)\,\mathrm{d}s - \int_0^t \exp\big(w(y(s))\big)\mathrm{d}s.$$

Thus, using (3.17),

$$|x(t) - y(t)| \leqslant C(\|u\|_X, \|w\|_X)\bigg(\int_0^1 |x(s) - y(s)|\,\mathrm{d}s + \|u - w\|_\infty\bigg).$$

Application of the Gronwall inequality gives

$$\sup_{t \in [0,1]} |x(t) - y(t)| \leqslant C(\|u\|_X, \|w\|_X)\|u - w\|_\infty.$$

Thus

$$|\Psi(1; 1, u) - \Psi(1, 1, w)| \leqslant C(\|u\|_X, \|w\|_X)\|u - w\|_\infty.$$

Hence, using (3.16), the fact that f is C^1 and the bound (3.17), we deduce that

$$\begin{aligned}|\mathcal{G}(u)-\mathcal{G}(w)| &= \big|f\big(\Psi(1;1,u)\big)-f\big(\Psi(1;1,w)\big)\big|\\ &\leqslant L(r)\|u-w\|_\infty\\ &\leqslant L(r)\|u-w\|_X.\end{aligned}$$

□

We wish to find $\mu^y(\mathrm{d}u)=\mathbb{P}(\mathrm{d}u|y)$. Informal use of Bayes' rule gives us

$$\frac{\mathrm{d}\mu^y}{\mathrm{d}\mu_0}(u)\propto\exp\biggl(-\frac{1}{2}|y-\mathcal{G}(u)|_\Gamma^2\biggr). \tag{3.18}$$

We now justify this formula by choice of prior and by application of Theorem 6.31.

We place a prior measure μ_0 on the space X by assuming that $u\sim\mu_0$ is Gaussian and that

$$u'(x)=\frac{\mathrm{d}u}{\mathrm{d}x}(x)\sim\mathcal{N}(0,\beta\mathcal{A}^{-\alpha}),$$

where $\mathcal{A}=-\mathrm{d}^2/\mathrm{d}x^2$ is a densely defined operator on $\mathcal{H}=L^2(\mathbb{S})$ with

$$D(\mathcal{A})=\biggl\{u\in H^2(\mathbb{S})\biggm| \int_0^1 u(x)\,\mathrm{d}x=0\biggr\}.$$

If $\beta>0$ and $\alpha>1/2$ then u' is almost surely a continuous function, by Lemma 6.25. Defining

$$u(x)=u_0+\int_0^x u'(s)\,\mathrm{d}s,$$

where $u_0\sim\mathcal{N}(0,\sigma^2)$, determines the distribution of u completely. Furthermore, for $\beta>0$ and $\alpha>1/2$ we have that u drawn from this measure is in X with probability 1: $\mu_0(X)=1$. Hence we deduce the following result.

Theorem 3.6. Consider the Bayesian inverse problem for $u(x)=\ln(c(x))$ subject to observation in the form (3.14), with v solving (3.13), and prior measure μ_0 as constructed immediately preceding this theorem. If $\beta>0$ and $\alpha>1/2$ then $\mu^y(\mathrm{d}u)=\mathbb{P}(\mathrm{d}u|y)$ is absolutely continuous with respect to $\mu_0(\mathrm{d}u)$, with Radon–Nikodym derivative given by (3.18).

Proof. To apply Theorem 6.31 it suffices to show that $\mu_0(X)=1$, since the function $\mathcal{G}$ is continuous from X into $\mathbb{R}^q$. The fact that $\mu_0(X)=1$ is established immediately prior to this theorem. □

3.5. Initial condition for the heat equation

We now study an inverse problem where the data y is a function, and is hence infinite-dimensional, in contrast to preceding examples where the data has been finite-dimensional. We assume that our observation is the solution of the heat equation at some fixed positive time $T > 0$, with an added Gaussian random field as observational noise, and that we wish to determine the initial condition.

To be concrete we consider the heat equation on a bounded open set $D \subset \mathbb{R}^d$, with Dirichlet boundary conditions, and written as an ODE in Hilbert space $\mathcal{H} = L^2(D)$:

$$\frac{\mathrm{d}v}{\mathrm{d}t} + Av = 0, \quad v(0) = u. \tag{3.19}$$

Here $A = -\triangle$ with $D(A) = H_0^1(D) \bigcap H^2(D)$. We assume sufficient regularity conditions on D and its boundary ∂D to ensure that the operator A is positive and self-adjoint on $\mathcal{H}$, and is the generator of an analytic semigroup. We define the (Sobolev-like) spaces $\mathcal{H}^s$ as in (2.29).

Assume that we observe the solution v at time T, subject to error which has the form of a Gaussian random field, and that we wish to recover the initial condition u. This problem is classically ill-posed, because the heat equation is smoothing, and inversion of this operator is not continuous on $\mathcal{H}$. Nonetheless, we will construct a well-defined Bayesian inverse problem.

We place a prior measure on u, which is a Gaussian $\mu_0 \sim \mathcal{N}(m_0, \mathcal{C}_0)$ with $\mathcal{C}_0 = \beta A^{-\alpha}$, for some $\beta > 0$ and $\alpha > d/2$; consequently $u \in \mathcal{H}$ μ_0-a.s. by Lemma 6.27. Our aim is to determine conditions on α, and on m_0, which ensure that the Bayesian inverse problem is well-defined. In particular, we would like conditions under which the posterior measure is equivalent (as a measure, see Section 6) to the prior measure.

We model the observation y as

$$y = \mathrm{e}^{-AT} u + \eta, \tag{3.20}$$

where $\eta \sim \mathcal{N}(0, \mathcal{C}_1)$ is independent of u. The observation operator $\mathcal{G} : \mathcal{H} \to \mathcal{H}$ is given by $\mathcal{G}(u) = \mathrm{e}^{-AT} u$ and, in fact, $\mathcal{G} : \mathcal{H} \to \mathcal{H}^\ell$ for any $\ell > 0$. If we assume that $\mathcal{C}_1 = \delta A^{-\gamma}$ for some $\gamma > d/2$ and $\delta > 0$, we then have that, almost surely, $\eta \in \mathcal{H}$ by Lemma 6.27.

Consider the Gaussian random variable $(u, y) \in \mathcal{H} \times \mathcal{H}$. We have

$$\mathbb{E}(u, y) := (\overline{u}, \overline{y}) = (m_0, \mathrm{e}^{-AT} m_0).$$

Straightforward calculation shows that

$$\begin{aligned} \mathbb{E}(u - \overline{u}) \otimes (u - \overline{u}) &= \mathcal{C}_0, \\ \mathbb{E}(u - \overline{u}) \otimes (y - \overline{y}) &= \mathcal{C}_0 \, \mathrm{e}^{-AT}, \\ \mathbb{E}(y - \overline{y}) \otimes (y - \overline{y}) &= \mathrm{e}^{-AT} \mathcal{C}_0 \, \mathrm{e}^{-AT} + \mathcal{C}_1. \end{aligned}$$

By Theorem 6.20 we find that the posterior measure μ^y for u given y is also

Gaussian, with mean

$$m = m_0 + \frac{\beta}{\delta} e^{-AT} A^{\gamma-\alpha}\left(I + \frac{\beta}{\delta} e^{-2AT} A^{\gamma-\alpha}\right)^{-1} (y - e^{-AT} m_0) \tag{3.21}$$

and covariance operator

$$\mathcal{C} = \mathcal{C}_0 (I + e^{-2AT} \mathcal{C}_0 \mathcal{C}_1^{-1})^{-1} \tag{3.22}$$

$$= \beta A^{-\alpha}\left(I + \frac{\beta}{\delta} e^{-2AT} A^{\gamma-\alpha}\right)^{-1}. \tag{3.23}$$

We now show that the posterior (Gaussian) measure on $\mathcal{H}$ is indeed equivalent to the prior. We will assume $\alpha > d/2$ since this ensures that samples from the prior are continuous functions, by Lemma 6.25.

Theorem 3.7. Consider an initial condition for the heat equation (3.19) with prior Gaussian measure $\mu_0 \sim \mathcal{N}(m_0, \beta A^{-\alpha})$, $m_0 \in \mathcal{H}^\alpha$, $\beta > 0$ and $\alpha > d/2$. If an observation is given in the form (3.20) then the posterior measure μ^y is also Gaussian, with mean and variance determined by (3.21) and (3.23). Furthermore, μ^y and the prior measure μ_0 are equivalent Gaussian measures.

Proof. Let $\{\phi_k, \lambda_k\}_{k\in\mathbb{K}}$, $\mathbb{K} = \mathbb{Z}^d \backslash \{0\}$, denote the eigenvalues of A and define $\kappa := \frac{\beta}{\delta} \sup_{k\in\mathbb{K}} e^{-2\lambda_k T} \lambda_k^{\gamma-\alpha}$ which is finite since $T > 0$ and A generates an analytic semigroup. Furthermore, the operator

$$K = \left(I + \frac{\beta}{\delta} e^{-2AT} A^{\gamma-\alpha}\right)^{-1}$$

is diagonalized in the same basis as A, and is a bounded and invertible linear operator with all eigenvalues lying in $[(1+\kappa)^{-1}, 1]$. Now, from (3.22), for any $h \in \mathcal{H}$,

$$\frac{1}{1+\kappa}\langle h, \mathcal{C}_0 h\rangle \leqslant \langle h, \mathcal{C} h\rangle = \langle h, \mathcal{C}_0 K h\rangle \leqslant \langle h, \mathcal{C}_0 h\rangle.$$

Thus, by Lemma 6.15, we deduce that condition (i) of Theorem 6.13 is satisfied, with $E = \mathcal{H}^\alpha = \operatorname{Im}(\mathcal{C}_0^{1/2})$.

From (3.21) we deduce that

$$m - m_0 = \frac{\beta}{\delta} e^{-AT} A^{\gamma-\alpha} K (y - e^{-AT} m_0).$$

Since A generates an analytic semigroup and since K is bounded, we deduce that $m - m_0 \in \mathcal{H}^r$ for any $r \in \mathbb{R}$. Hence condition (ii) of Theorem 6.13 is satisfied. To check the remaining condition (iii), define

$$\begin{aligned}\mathcal{T} &= \mathcal{C}_0^{-1/2}\mathcal{C}\mathcal{C}_0^{-1/2} - I \\ &= -\frac{\beta}{\delta}\left(I + \frac{\beta}{\delta}\mathrm{e}^{-2AT}A^{\gamma-\alpha}\right)^{-1} A^{\gamma-\alpha}\mathrm{e}^{-2AT}.\end{aligned}$$

The operator T is clearly Hilbert–Schmidt because its eigenvalues μ_k satisfy

$$|\mu_k| \leqslant \frac{\beta}{\delta}\lambda_k^{\gamma-\alpha}\mathrm{e}^{-2\lambda_k T}$$

and hence decay exponentially fast. This establishes (iii) of Theorem 6.13 and the proof is complete. □

The preceding theorem uses the Gaussian structure of the posterior measure explicitly. To link the presentation to the other examples in this section, it is natural to ask whether a similar result can be obtained less directly.

We define $\Phi : \mathcal{H} \times \mathcal{H} \to \mathbb{R}$ by

$$\Phi(u; y) = \frac{1}{2}\|\mathrm{e}^{-AT}u\|_{\mathcal{C}_1}^2 - \langle \mathrm{e}^{-AT}u, y\rangle_{\mathcal{C}_1},$$

and use this function to derive Bayes' formula for the measure $\mu^y(\mathrm{d}u) = \mathbb{P}(\mathrm{d}u|y)$. We will show that $\mu^y(\mathrm{d}u)$ is absolutely continuous with respect to the prior $\mu_0(\mathrm{d}u)$ with density

$$\frac{\mathrm{d}\mu^y}{\mathrm{d}\mu_0}(u) \propto \exp\bigl(-\Phi(u; y)\bigr). \tag{3.24}$$

Remark 3.8. It would be tempting to define a potential

$$\begin{aligned}\Psi(u; y) &= \frac{1}{2}\|y - \mathcal{G}(u)\|_{\mathcal{C}_1}^2 \\ &= \frac{1}{2}\|y - \mathrm{e}^{-AT}u\|_{\mathcal{C}_1}^2\end{aligned}$$

in analogy with the examples in the two previous sections: this Ψ is a least-squares functional measuring model/data mismatch. However, this quantity is almost surely infinite, with respect to the random variable y, since draws from a Gaussian measure in infinite dimensions do not lie in the corresponding Cameron–Martin space $\mathrm{Im}(\mathcal{C}_1^{1/2})$: see Lemma 6.10. This undesirable property of Ψ stems directly from the fact that the data y is a function rather than a finite-dimensional vector. To avoid the problem we work with $\Phi(u; y)$ which, informally, may be viewed as being given by the identity

$$\Phi(u; y) = \Psi(u; y) - \frac{1}{2}\|y\|_{\mathcal{C}_1}^2.$$

Thus we 'subtract off' the infinite part of Ψ. Since Bayes' formula in the form (3.24) only gives the density up to a y-dependent constant, we see intuitively why this subtraction of a term involving y is reasonable. The issues outlined in this remark arise quite generally when the data y is infinite-dimensional and the observational noise η is Gaussian. ◇

The form of Φ arising in this problem, and the fact that the data is infinite-dimensional, precludes us from using Theorem 6.31 to establish that (3.24) is correct; however, the method of proof is very similar to that used to prove Theorem 6.31.

Before proving that (3.24) is correct, we state and prove some properties of the potential Φ.

Lemma 3.9. The function $\Phi : \mathcal{H} \times \mathcal{H} \to \mathbb{R}$ satisfies Assumptions 2.6 with $X = Y = \mathcal{H}$ and $L(r)$ linearly bounded.

Proof. We may write

$$\Phi(u; y) = \frac{1}{2}\|\mathcal{C}_1^{-1/2}\mathrm{e}^{-AT}u\|^2 - \langle \mathcal{C}_1^{-1/2}\mathrm{e}^{-\frac{1}{2}AT}u, \mathcal{C}_1^{-1/2}\mathrm{e}^{-\frac{1}{2}AT}y\rangle.$$

Since $\mathcal{C}_1^{-1} = \delta A^{\gamma}$ we deduce that $\mathcal{K}_\lambda := \mathcal{C}_1^{-1/2}\mathrm{e}^{-\lambda AT}$ is a compact operator on $\mathcal{H}$ for any $\lambda > 0$. By the Cauchy–Schwarz inequality we have, for any $a > 0$,

$$\Phi(u; y) \geqslant -\frac{a^2}{2}\|\mathcal{C}_1^{-1/2}\mathrm{e}^{-\frac{1}{2}AT}u\|^2 - \frac{1}{2a^2}\|\mathcal{C}_1^{-1/2}\mathrm{e}^{-\frac{1}{2}AT}y\|^2.$$

By the compactness of $\mathcal{K}_{1/2}$ and by choosing a arbitrarily small, we deduce that Assumption 2.6(i) holds. Assumption 2.6(ii) holds by a similar Cauchy–Schwarz argument. Since Φ is quadratic in u, and using the compactness of $\mathcal{K}_{1/2}$ and $\mathcal{K}_1$, we see that

$$\begin{aligned}
|\Phi(u_1; y) - \Phi(u_2; y)| &\leqslant C\big(\|\mathcal{K}_1 u_1\| + \|\mathcal{K}_1 u_2\| + \|\mathcal{K}_{1/2} y\|\big)\|\mathcal{K}_{1/2}(u_1 - u_2)\| \\
&\leqslant C\big(\|u_1\| + \|u_2\| + \|y\|\big)\|\mathrm{e}^{-\frac{1}{4}AT}(u_1 - u_2)\| \qquad (3.25)\\
&\leqslant C\big(\|u_1\| + \|u_2\| + \|y\|\big)\|u_1 - u_2\|, \qquad (3.26)
\end{aligned}$$

and similarly

$$|\Phi(u; y_1) - \Phi(u; y_2)| \leqslant C\|u\|\|y_1 - y_2\|,$$

so that Assumptions 2.6(iii) and (iv) hold. □

Theorem 3.10. Consider the inverse problem for the initial condition u in (3.19), subject to observation in the form (3.20) and with prior Gaussian measure $\mu_0 = \mathcal{N}(m_0, \beta A^{-\alpha})$. If $m_0 \in \mathcal{H}^\alpha, \beta > 0$ and $\alpha > d/2$, then the posterior measure $\mu^y(\mathrm{d}u) = \mathbb{P}(\mathrm{d}u|y)$ and the prior $\mu_0(\mathrm{d}u)$ are equivalent with Radon–Nikodym derivative given by (3.24).

Proof. Recall that $\mathcal{C}_1 = \delta A^{-\gamma}$ and that $\mathcal{C}_0 = \beta A^{-\alpha}$. Define the measures

$$\begin{aligned}\mathbb{Q}_0(\mathrm{d}y) &= \mathcal{N}(0, \mathcal{C}_1),\\ \mathbb{Q}(\mathrm{d}y|u) &= \mathcal{N}(\mathrm{e}^{-AT}u, \mathcal{C}_1),\\ \mu_0(\mathrm{d}u) &= \mathcal{N}(m_0, \mathcal{C}_0),\end{aligned}$$

and then define

$$\begin{aligned}\nu_0(\mathrm{d}y, \mathrm{d}u) &= \mathbb{Q}_0(\mathrm{d}y) \otimes \mu_0(\mathrm{d}u),\\ \nu(\mathrm{d}y, \mathrm{d}u) &= \mathbb{Q}(\mathrm{d}y|u)\mu_0(\mathrm{d}u).\end{aligned}$$

By Theorem 6.14 we deduce that

$$\frac{\mathrm{d}\mathbb{Q}}{\mathrm{d}\mathbb{Q}_0}(y|u) = \exp\biggl(-\frac{1}{2}\|\mathrm{e}^{-AT}u\|_{\mathcal{C}_1}^2 + \langle \mathrm{e}^{-AT}u, y\rangle_{\mathcal{C}_1}\biggr).$$

The measure ν is well-defined because the function $\Phi(\cdot; y) : \mathcal{H} \to \mathbb{R}$ is continuous and hence μ_0-measurable if $\mu_0(\mathcal{H}) = 1$. This last fact follows from Lemma 6.27, which shows that draws from μ_0 are almost surely in $\mathcal{H}$. Hence

$$\frac{\mathrm{d}\nu}{\mathrm{d}\nu_0}(y, u) = \exp\biggl(-\frac{1}{2}\|\mathrm{e}^{-AT}u\|_{\mathcal{C}_1}^2 + \langle \mathrm{e}^{-AT}u, y\rangle_{\mathcal{C}_1}\biggr).$$

By applying Theorem 6.29, noting that under ν_0 the random variables y and u are independent with $u \sim \mu_0$, we deduce that

$$\frac{\mathrm{d}\mu^y}{\mathrm{d}\mu_0}(u) \propto \exp\biggl(-\frac{1}{2}\|\mathrm{e}^{-AT}u\|_{\mathcal{C}_1}^2 + \langle \mathrm{e}^{-AT}u, y\rangle_{\mathcal{C}_1}\biggr),$$

with constant of proportionality independent of u. □

3.6. Fluid mechanics

The preceding four subsections provide a range of examples where somewhat explicit calculations, using the solution of various forward linear PDE problems, establish that the associated inverse problems may be placed in the general framework that we outlined in Section 2.4 and will study further in Section 4. However, it is by no means necessary to have explicit solutions of the forward problem to use the framework developed in this article, and the examples of this subsection, and the two subsections which follow it, illustrate this.

Fluid mechanics provides an interesting range of applications where the technology of inverse problems is relevant. We outline examples of such problems and sketch their formulation as Bayesian inverse problems for functions. We also show that these problems may be formulated to satisfy Assumptions 2.7. Unlike the previous three sections, however, we do not provide full details; we refer to other works for these, in the bibliography subsection.

In *weather forecasting* a variety of instruments are used to measure the velocity of the air in the atmosphere. Examples include weather balloons, data from commercial and military aircraft, as well as special-purpose aircraft, and satellites. An important inverse problem is to determine the global velocity field, and possibly other fields, from the *Eulerian data* comprising the various noisy measurements described above.

As a concrete, and simplified, model of this situation we consider the linearized shallow water equations on a two-dimensional torus. The equations are a coupled pair of PDEs for the two-dimensional velocity field v and a scalar height field h, with the form

$$\frac{\partial v}{\partial t} = Sv - \nabla h, \quad (x,t) \in \mathbb{T}^2 \times [0,\infty), \tag{3.27a}$$

$$\frac{\partial h}{\partial t} = -\nabla \cdot v, \quad (x,t) \in \mathbb{T}^2 \times [0,\infty). \tag{3.27b}$$

The two-dimensional unit torus $\mathbb{T}^2$ is shorthand for the unit square with periodic boundary conditions imposed. The skew matrix S is given by

$$S = \begin{pmatrix} 0 & 1 \\ -1 & 0 \end{pmatrix}$$

and the term involving it arises from the Coriolis effect.

The objective is to find the initial velocity and height fields $(v(0), h(0)) = (u, p) \in \mathcal{H}$, where

$$\mathcal{H} := \left\{ u \in L^2(\mathbb{T}^2; \mathbb{R}^3) \Big| \int_{\mathbb{T}^2} u \, \mathrm{d}x \right\}.$$

We assume that we are given noisy observations of the velocity field at positions $\{x_j\}_{j=1}^{\mathrm{J}}$ and times $\{t_k\}_{k=1}^{\mathrm{K}}$, all positive. Concatenating data, we write

$$y = \mathcal{G}(u,p) + \eta. \tag{3.28}$$

Here $\mathcal{G}$ maps a dense subset of $\mathcal{H}$ into $\mathbb{R}^{2\mathrm{JK}}$ and is the *observation operator*. Because the PDE (3.27) is linear, so too is $\mathcal{G}$. We assume that $\eta \sim \mathcal{N}(0,\Gamma)$ is independent of u and we consider the Bayesian inverse problem of finding the posterior measure $\mu^y(\mathrm{d}u) = \mathbb{P}(\mathrm{d}u|y)$ from the prior μ_0. We let $A = -\triangle$ on $\mathbb{T}^2$ with domain

$$D(A) = \left\{ H^2(\mathbb{T}^2) \Big| \int_{\mathbb{T}^2} u \, \mathrm{d}x = 0 \right\}$$

and define the prior through fractional powers of A.

Theorem 3.11. Consider an initial condition for the shallow water equations (3.27) with prior Gaussian measure $\mu_0 = \mathcal{N}(m_0, \beta A^{-\alpha})$ with $m_0 \in \mathcal{H}^\alpha$, $\beta > 0$ and $\alpha > 2$. If a noisy observation is made in the form (3.28), then the posterior measure μ^y is also Gaussian, and is absolutely continuous with respect to the prior measure μ_0, with Radon–Nikodym derivative

$$\frac{\mathrm{d}\mu^y}{\mathrm{d}\mu_0}(u,p) \propto \exp\left(-\frac{1}{2}|y-\mathcal{G}(u,p)|_\Gamma^2\right), \tag{3.29}$$

where $\mathcal{G}$ is given by (3.28). Furthermore, the observation operator $\mathcal{G}$ satisfies Assumptions 2.7 with $X = \mathcal{H}^s$ and K globally bounded, for any $s > 1$. ◇

In *oceanography* a commonly used method of gathering data about ocean currents, temperature, salinity and so forth is through the use of Lagrangian instruments which are transported by the fluid velocity field and transmit positional information using GPS. An important inverse problem is to determine the velocity field in the ocean from the Lagrangian data comprising the GPS information about the position of the instruments. As an idealized model consider the incompressible Stokes ($\iota = 0$) or Navier–Stokes ($\iota = 1$) equations written in the form:

$$\frac{\partial v}{\partial t} + \iota v.\nabla v = \nu\Delta v - \nabla p + f, \quad (x,t) \in \mathbb{T}^2 \times [0,\infty), \tag{3.30a}$$

$$\nabla \cdot v = 0, \qquad (x,t) \in \mathbb{T}^2 \times [0,\infty), \tag{3.30b}$$

$$v = u, \qquad (x,t) \in \mathbb{T}^2 \times \{0\}. \tag{3.30c}$$

As in the preceding example we impose periodic boundary conditions, here on the velocity field v and the pressure p. We assume that f has zero average over D, noting that this implies the same for $v(x,t)$, provided that $u(x) = v(x,0)$ has zero initial average. We define the Stokes operator A and Leray projector P in the standard fashion, together with the Sobolev spaces $\mathcal{H}^s = D(A^{s/2})$ as in (2.29). The equations (3.30) can be written as an ODE in the Hilbert space $\mathcal{H}$:

$$\frac{\mathrm{d}v}{\mathrm{d}t} + \iota B(v,v) + \nu Au = \psi, \tag{3.31}$$

where $\psi = Pf$ and $B(v,v)$ represents the projection, under P, of the nonlinear convective term.

We assume that we are given noisy observations of Lagrangian tracers with position z solving the integral equation

$$z_j(t) = z_{j,0} + \int_0^t v(z_j(s),s)\,\mathrm{d}s. \tag{3.32}$$

These equations have a unique solution if $u \in \mathcal{H}$ and $\psi \in L^2((0,T);\mathcal{H})$.

For simplicity assume that we observe all the tracers z at the same set of times $\{t_k\}_{k=1}^{\mathrm{K}}$, and that the initial particle tracer positions $z_{j,0}$ are known to us:

$$y_{j,k} = z_j(t_k) + \eta_{j,k}, \quad j = 1,\ldots,\mathrm{J} \text{ and } k = 1,\ldots,\mathrm{K}, \tag{3.33}$$

where the $\eta_{j,k}$ are zero mean Gaussian random variables. The times $\{t_k\}$ are assumed to be positive. Concatenating data, we may write

$$y = \mathcal{G}(u) + \eta, \tag{3.34}$$

with $y = (y_{1,1}^*, \dots, y_{J,K}^*)^*$ and $\eta \sim \mathcal{N}(0, \Gamma)$ for some covariance matrix Γ capturing the correlations present in the noise. The function $\mathcal{G}$ maps a dense subspace of $\mathcal{H}$ into $\mathbb{R}^{2JK}$. The objective is to find the initial velocity field u, given y. We start by stating a result concerning the observation operator.

Lemma 3.12. Assume that $\psi \in C([0,T]; \mathcal{H}^\gamma)$ for some $\gamma > 0$. Then $\mathcal{G}$ given by (3.34) satisfies Assumptions 2.7 with $X = \mathcal{H}^\ell$ for any $\ell > 0$. ◇

These properties of the observation operator $\mathcal{G}$ lead to the following result.

Theorem 3.13. Let $\mu_0 = \mathcal{N}(m_0, \beta A^{-\alpha})$ denote a prior Gaussian measure on μ_0. If $m_0 \in \mathcal{H}^\alpha$, $\beta > 0$ and $\alpha > 1$ then the measure $\mu^y(\mathrm{d}u) = \mathbb{P}(\mathrm{d}u|y)$ is absolutely continuous with respect to μ_0, with Radon–Nikodym derivative given by

$$\frac{\mathrm{d}\mu^y}{\mathrm{d}\mu_0}(u) \propto \exp\left(-\frac{1}{2}|y - \mathcal{G}(u)|_\Gamma^2\right), \tag{3.35}$$

with $\mathcal{G}$ defined by (3.34). ◇

Notice that the required lower bound on the exponent α in the preceding theorem is lower than that appearing in Theorem 3.11. This is because the (Navier–)Stokes equation is smoothing, and hence less regularity is required on the initial condition in order to define the observation operator $\mathcal{G}$ than for the linearized shallow water equations.

3.7. Subsurface geophysics

Determining the permeability of subsurface rock is enormously important in a range of different applications. Among these applications are the prediction of transport of radioactive waste from underground waste repositories, and the optimization of oil recovery from underground fields. We give an overview of some inverse problems arising in this area. As in the previous subsection we do not give full details, leaving these to the cited literature in the bibliography subsection.

The permeability tensor K is a central component of *Darcy's law*, which relates the velocity field v to the gradient of the pressure p in porous media flow:

$$v = -K\nabla p. \tag{3.36}$$

In general K is a tensor field. However, the problem is often simplified by assuming that $K = kI$, where k is a scalar field and I the identity tensor; we make this simplification.

In many subsurface flow problems it is reasonable to model the velocity field as incompressible. Combining this constraint with Darcy's law (3.36) shows that the pressure p is governed by the PDE

$$\nabla \cdot (-k\nabla p) = 0, \quad x \in D, \tag{3.37a}$$

$$p = h, \quad x \in \partial D. \tag{3.37b}$$

This model is a widely used simplified model in groundwater flow modelling. The inverse problem is to find the permeability k from observations of the pressure at points in the interior of D; this information can be found by measuring the height of the water table. For simplicity we work in two or three dimensions d and assume that $D \subset \mathbb{R}^d$ is bounded and open. As in Section 3.3 it is physically and mathematically important that k be positive, in order that the elliptic equation for the pressure is well-posed. Hence we write $k = \exp(u)$ and consider the problem of determining u.

We assume that we observe

$$y_j = p(x_j) + \eta_j, \quad j = 1, \ldots, \mathrm{J}, \tag{3.38}$$

and note that this may be written as

$$y = \mathcal{G}(u) + \eta \tag{3.39}$$

for some implicitly defined function $\mathcal{G}$. We assume that $\eta \sim \mathcal{N}(0, \Gamma)$ is independent of u. Before formulating the Bayesian inverse problem, we state the following result concerning the forward problem.

Lemma 3.14. Assume that the boundary of D, ∂D, is C^1-regular and that the boundary data h may be extended to a function $h \in W^{1,2r}(D)$ with $r > d/2$. The function $\mathcal{G}$ satisfies Assumptions 2.7 with $X = C(\overline{D})$. ◇

We define the prior Gaussian measure through fractional powers of the Laplacian $\mathcal{A} = -\triangle$ with

$$D(\mathcal{A}) = \left\{ u \in H^2(D) \middle| \nabla u \cdot n = 0, \int_D u(x)\,\mathrm{d}x = 0 \right\}.$$

Here n denotes the unit outward normal on the boundary of D.

Theorem 3.15. Let the assumptions of Lemma 3.14 hold and let $\mu_0 = \mathcal{N}(0, \beta \mathcal{A}^{-\alpha})$ denote a prior Gaussian measure on μ_0. If $\beta > 0$ and $\alpha > d - 1/2$, then the measure $\mu^y(\mathrm{d}u) = \mathbb{P}(\mathrm{d}u|y)$ is absolutely continuous with respect to μ_0, with Radon–Nikodym derivative

$$\frac{\mathrm{d}\mu^y}{\mathrm{d}\mu_0}(x) \propto \exp\left(-\frac{1}{2}|y - \mathcal{G}(u)|_\Gamma^2 \right) \tag{3.40}$$

and $\mathcal{G}$ given by (3.39). ◇

Once the posterior measure on u is known it can be used to quantify uncertainty in predictions made concerning the Lagrangian transport of radioactive particles under the velocity field v given by (3.36). In particular, the push forward of the measure μ^y onto v, and hence onto particle trajectories z obeying

$$\frac{\mathrm{d}z}{\mathrm{d}t} = v(z),$$

will define a measure on the possible spread of radioactive contaminants, enabling risk assessment to be undertaken.

The oil industry routinely confronts an inverse problem similar to but more complex than that arising in the nuclear waste industry. Again, uncertainty quantification is important as it enables more effective decision making concerned with the substantial investment of resources required to extract oil from increasingly complex environments. The primary difference between the simple model we have described for nuclear waste management and that which we are about to describe for oil extraction arises because the subsurface fluid flow for oil extraction is multiphase (gas, water, oil) and significant on much shorter time scales than in the nuclear waste management scenario. We study a simplified two-phase problem, for oil and water alone. The physical model contains two unknown scalar fields, the water saturation S (volume fraction of water in an oil–water mixture) and pressure p, and is posed in a bounded open set $D \subset \mathbb{R}^d$. Darcy's law now takes the form

$$v = -\lambda(S)k\nabla p. \tag{3.41}$$

Mass conservation and transport, respectively, give the equations

$$\begin{aligned} -\nabla \cdot (\lambda(S)k\nabla p) &= h_1, \qquad (x,t) \in D \times [0,\infty), \\ \frac{\partial S}{\partial t} + v \cdot \nabla f(S) &= \eta \triangle S, \quad (x,t) \in D \times [0,\infty), \qquad (3.42) \\ p &= h_2, \qquad (x,t) \in \partial D \times [0,\infty). \qquad (3.43) \end{aligned}$$

The flux function f is known (typically the Buckley–Leverett form is used) and the source/boundary terms h_1, h_2 are also both assumed known. The scalar η is the (also assumed known) diffusivity of the multiphase flow, typically very small. Initial conditions for S are specified on D at time $t = 0$. There are additional boundary conditions on S which we now describe. We partition $\partial D = \partial D^{\text{out}} \bigcup \partial D^{\text{in}}$. We think of pumping water in on the boundary ∂D^{in}, so that $S = 1$ there, and specify a Robin boundary condition on ∂D^{out}, determining the flux of fluid in terms of S the water saturation.

We assume that we have access to noisy measurements of the *fractional flow* $F(t)$, which quantifies the fraction of oil produced on a subset ∂D^{meas}

of the outflow boundary ∂D^{out}. This measurement is via the function

$$F(t) = 1 - \frac{\int_{\partial D^{\text{meas}}} f(S) v_n \, \mathrm{d}l}{\int_{\partial D^{\text{meas}}} v_n \, \mathrm{d}l},$$

where v_n is the component of the velocity v which is normal to the boundary and $\mathrm{d}l$ denotes integration along the boundary. Assume that we make measurements of F at times $\{t_k\}_{k=1}^{\mathrm{K}}$, polluted by Gaussian noise. Then the data are as follows:

$$y_k = F(t_k) + \eta_k, \quad k = 1, \dots, \mathrm{K},$$

where the η_k are zero mean Gaussian random variables. Concatenating data, we may write

$$y = \mathcal{G}(u) + \eta$$

where, as before, $k(x) = \exp(u(x))$. We assume that $\eta \sim \mathcal{N}(0, \Gamma)$ for some covariance matrix Γ encapsulating measurement errors. The prior μ_0 is a Gaussian measure on u, specified as in the previous section. We once again anticipate that

$$\frac{\mathrm{d}\mu}{\mathrm{d}\mu_0}(x) \propto \exp\left(-\frac{1}{2}|y - \mathcal{G}(u)|_B^2\right). \tag{3.44}$$

This is similar to the nuclear waste problem, but the observation operator $\mathcal{G}$ is now more complicated. However, similar analyses of the properties of the forward problem, and the resulting Bayesian inverse problem, can be undertaken.

3.8. Molecular dynamics

Consider a molecule described by the positions x of N atoms moving in $\mathbb{R}^d$, with $d = 1, 2$ or 3. If we assume that the particles interact according to a potential $V : \mathbb{R}^d \to \mathbb{R}$ and are subject to thermal activation, then, in the *over-damped* limit where the inertial relaxation time is fast, we obtain the *Brownian dynamics* model for the position of x:

$$\frac{\mathrm{d}x}{\mathrm{d}t} = -\nabla V(x) + \sqrt{\frac{2}{\beta}} \frac{\mathrm{d}W}{\mathrm{d}t}. \tag{3.45}$$

Here W is a standard $\mathbb{R}^{Nd}$-valued Brownian motion and β the inverse temperature. One of the key challenges in molecular dynamics is to understand how molecules rearrange themselves to change from one configuration to another: in some applications this may represent a chemical reaction, and in others a conformational change such as seen in biomolecules. When the temperature is small ($\beta \gg 1$), the solutions of (3.45) spend most of their time near the minima of the potential V. Transitions between different minima of the potential are rare events. Simply solving the SDE starting from one of the minima will be a computationally infeasible way of generating sample paths which jump between minima, since the time to make a transition is exponentially small in β. Instead we may condition on this rare event occurring. This may be viewed as an inverse problem to determine the control W which drives the system from one configuration to another. However, we will work directly with the functions x which result from this control, as these constitute the more physically interesting quantity. Because the Brownian motion W is a random function, this leads naturally to the question of determining the probability measure on functions x undergoing the desired transition between configurations. The desired transition can be defined by conditioning the dynamics given by (3.45) to satisfy the boundary conditions

$$x(0) = x^-, \quad x(T) = x^+. \tag{3.46}$$

We view x as an element of $L^2((0,T);\mathbb{R}^{Nd})$ and denote the Nd-dimensional *Brownian bridge measure* arising from (3.45) and (3.46) in the case $V \equiv 0$ by μ_0. We also define μ to be the desired bridge diffusion measure arising from the given V. We may view both μ_0 and μ as measures on $L^2((0,T);\mathbb{R}^{Nd})$; the measure μ_0 is Gaussian but, unless V is quadratic, the measure μ is not. We now proceed to determine the Radon–Nikodym derivative of μ with respect to the Gaussian bridge diffusion μ_0.

Theorem 3.16. Assume $V \in C^2(\mathbb{R}^{Nd};\mathbb{R})$ and that the stochastic initial value problem, found from (3.45) and (3.46) without the condition $x(T) = x^+$, has solutions which do not explode almost surely on $t \in [0,T]$. Then the measure μ defined by the bridge diffusion problem (3.45) and (3.46) is absolutely continuous with respect to the Brownian bridge measure μ_0 found from (3.45) and (3.46) in the case $V \equiv 0$. Furthermore, the Radon–Nikodym derivative is given by

$$\frac{\mathrm{d}\mu}{\mathrm{d}\mu_0}(x) \propto \exp(-\Phi(x)), \tag{3.47}$$

where the potential Φ is defined by

$$\Phi(x) = \frac{\beta}{2}\int_0^T G(x(t))\,\mathrm{d}t, \tag{3.48a}$$

$$G(x) = \frac{1}{2}\|\nabla V(x)\|^2 - \frac{1}{\beta}\Delta V(x). \tag{3.48b}$$

◇

In addition, we find that a large class of problems leads to the common structure of Section 2.4. There is no explicit data $y \in Y$ in this problem, but we can let $y \in \mathbb{R}^p$ denote the parameters appearing in the potential V, and hence in G. (Note that β is not such a parameter, as it appears in G but not in V; more fundamentally it appears in μ_0 and so is not simply a parameter in the potential Φ). We thus write $V(x;y)$ and $G(x;y)$.

Lemma 3.17. Consider the function Φ defined by (3.48a) and (3.48b) with $V : \mathbb{R}^{Nd} \times \mathbb{R}^p \to \mathbb{R}$. Assume that for any $\varepsilon, r > 0$ there exists $M = M(\varepsilon, r) \in \mathbb{R}$ such that, for all $\|y\| < r$,

$$G(x; y) \geqslant -\varepsilon^2 |x|^2 + M;$$

assume also that $G \in C^1(\mathbb{R}^{Nd} \times \mathbb{R}^p, \mathbb{R})$ with derivative $D_y G(x; y)$ which is polynomially bounded in x. Then Φ satisfies Assumptions 2.6 with $X = H^1((0,T))$. ◇

3.9. Discussion and bibliography

We started this section by studying the problem of determining a field from observation. This is intimately related to the study of interpolation of data by splines, a subject comprehensively developed and reviewed in Wahba (1990). The link between spline interpolation and inverse problems using Gaussian fields is surveyed in Gu (2002).

The inverse problem for the diffusion coefficient in Section 3.3 is a one-dimensional analogue of the inverse problems arising in the geophysics community, which we outline in Section 3.7; these problems, which arise in the study of groundwater flow and are hence of interest to the burial of (radioactive nuclear and other) waste, are discussed in Zimmerman *et al.* (1998). A related inverse problem for the diffusion coefficient of an elliptic PDE is that arising in electrical impedence tomography; this widely studied inverse problem requires recovery of the diffusion coefficient from measurements of the boundary flux. It is of central importance in the medical sciences, and also has a rich mathematical structure; see Borcea (2002) and Uhlmann (2009) for reviews.

Inverse problems for the heat equation, the subject of Section 3.5, are widely studied. See, for example, the cited literature in Beck, Blackwell and Clair (2005) and Engl *et al.* (1996). An early formulation of this problem in a Bayesian framework appears in Franklin (1970).

We study applications to fluid dynamics in Section 3.6: the subject known as data assimilation. Kalnay (2003) and Bennett (2002) survey inverse problems in fluid mechanics from the perspective of weather prediction and oceanography respectively; see also Apte, Jones, Stuart and Voss (2008*b*), Lorenc (1986), Ide, Kuznetsov and Jones (2002), Kuznetsov, Ide and Jones (2003), Nichols (2003*a*) and Nodet (2006) for representative examples, some closely related to the specific model problems that we study in this article. Theorem 3.11, arising in our study of Eulerian observations and integration into a wave equation model, is proved in Dashti, Pillai and Stuart (2010*b*). Lemma 3.12 and Theorem 3.13, arising in the study of Lagrangian observations, are proved in Cotter *et al.* (2009) (Navier–Stokes case) and Cotter *et al.* (2010*a*) (Stokes case). A major question facing the research community in data assimilation for fluid mechanics applications is to determine whether future increase in available computer resources is used to increase resolution of the computational models, or to improve estimates of uncertainty. (The question is discussed, in the context of climate modelling, in Palmer *et al.* (2009).) The framework developed in Section 3.6 allows for a systematic treatment of uncertainty, as quantified by the variability in the posterior measure; furthermore, the framework may be extended to make inference not only about the initial condition but also about forcing to the model, thereby enabling model error to be uncovered in a systematic fashion. In this context we define model error to be an error term in the dynamical model equations, as in Hagelberg, Bennett and Jones (1996). Note, however, that in practical data assimilation, model errors are sometimes combined with the observation errors (Cohn 1997). Further discussion of model error for problems arising in the atmospheric sciences may be found in the papers of Nichols (2003*b*) and Fang *et al.* (2009*b*). In Cotter *et al.* (2009) we discuss both Eulerian and Lagrangian data assimilation with and without model error, with fluid flow model given by the Navier–Stokes equations (3.30) with $\iota = 1$.

The subject of minimal regularity required to define Lagrangian trajectories (3.32) in a Navier–Stokes velocity field is covered in Chemin and Lerner (1995) and Dashti and Robinson (2009). This theory is easily extended to cover the case of the Stokes equations.

The systematic treatment of Lagrangian data assimilation is developed in the sequence of papers by Ide *et al.* (2002), Kuznetsov *et al.* (2003), Salman, Kuznetsov, Jones and Ide (2006) and Salman, Ide and Jones (2008) with recent application in Vernieres, Ide and Jones (2010). Although the subject had been treated in an applied context, these were the first papers to develop a clear dynamical systems framework in which the coupled (skew-product) dynamical system for the fluid and the Lagrangian particles was introduced as the fundamental object of study.

The papers by Pimentel, Haines and Nichols (2008*a*, 2008*b*), Bell, Martin and Nichols (2004), Huddleston, Bell, Martin and Nichols (2004) and Martin, Bell and Nichols (2002) describe a variety of applications of ideas from data assimilation to problems in oceanography. The paper by Wlasak, Nichols and Roulstone (2006) discusses data assimilation in the atmospheric sciences, using a potential vorticity formulation. In Bannister, Katz, Cullen, Lawless and Nichols (2008), forecast errors are studied for data assimilation problems in fluid flow. The paper by Alekseev and Navon (2001) uses a wavelet-based approach to study the inverse problem of determining inflow fluid properties from outflow measurements.

Some of the earliest work concerning the statistical formulation of inverse problems was motivated by geophysical applications (Backus 1970*a*, 1970*b*, 1970*c*), such as those introduced in Section 3.7. The interpolation of a random field, observed at a finite set of points, is outlined in Gu (2008) and is often referred to as 'kriging' (Cressie 1993). Overviews of issues arising in oil reservoir simulation may be found in Farmer (2005, 2007). The mathematical statement of the oil reservoir simulation problem as outlined here is formulated in Ma, Al-Harbi, Datta-Gupta and Efendiev (2008) and further discussion of numerical methods is undertaken in Dostert, Efendiev, Hou and Luo (2006). Lemma 3.14 and Theorem 3.15, concerning the elliptic inverse problem for subsurface flow, are proved in Dashti, Harris and Stuart (2010*a*).

The formulation of problems from molecular dynamics in terms of probability measures on time-dependent functions has a long history. On the mathematical side this is intimately related to the theory of rare events (Freidlin and Wentzell 1984) and an overview of some of the sampling techniques used for this problem may be found in Bolhuis, Chandler, Dellago and Geissler (2002). The particular formulation of the problem that we undertake here, in which the length of the transition T is specified *a priori*, can be found in Dashti *et al.* (2010*b*); see also Reznikoff and Vanden Eijnden (2005), Hairer, Stuart, Voss and Wiberg (2005) and Hairer, Stuart and Voss (2007). A generalization to second-order Newtonian dynamics models, in place of the over-damped Brownian dynamics model (3.45) may be found in Hairer, Stuart and Voss (2010*a*).

4. Common structure

4.1. Overview

It is natural to view the posterior measure μ^y given by (2.24) as *the ideal solution* to the problem of combining a mathematical model with data y. However, obtaining a formula such as this is only the beginning: we are confronted with the formidable task of extracting information from this formula. At a high level this entire section is devoted to the question of the *stability* of measures μ^y to perturbations of various kinds, under Assumptions 2.6 or 2.7. These stability results help to create firm foundations for the *algorithms* designed to obtain information from the measure μ^y; these algorithms are summarized in the next section.

In this section, then, we study the well-posedness of problems with respect to parameters, or data, entering the definition of the measure: we show Lipschitz properties of the posterior measure with respect to changes in the data. We also study the related issue of approximating the measure, in particular the approximation by measures defined over a finite-dimensional space. Section 4.2 concerns well-posedness in the setting where the data is in the form of a function: it is infinite-dimensional. In practical applications the data will always be finite, but when the data is very dense it is a useful abstraction to consider the data as being a function, and so this situation is conceptually important. However, when the data is sparse it is best viewed as finite, as a number of mathematical simplifications follow from this. The well-posedness of the posterior measure in this finite data situation is studied in Section 4.3. In Section 4.4 we study the effect of approximating the potential Φ and the effect of this approximation on the measure μ^y given by (2.24).

A key idea throughout this section is the use of metrics to study distances between probability measures. This topic is discussed in Section 6.7 and, in particular, the Hellinger metric which we use throughout this section is introduced. The primary message concerning the Hellinger metric is this: consider two measures which are absolutely continuous with respect to a common Gaussian reference measure and which are distance ε apart in the Hellinger metric. Then the expectations of polynomially bounded functions under these two measures are also $\mathcal{O}(\varepsilon)$ apart. In particular, the mean and covariance operator are $\mathcal{O}(\varepsilon)$ apart.

4.2. Well-posedness

The probability measure of interest is typically defined through a density with respect to a Gaussian reference measure $\mu_0 = \mathcal{N}(0, \mathcal{C})$ on a Hilbert space $\mathcal{H}$ which, by shift of origin, we have taken to have mean zero. We assume that, for some separable Banach space X, we have $\mu_0(X) = 1$. We let $\{\phi_k, \gamma_k\}_{k=1}^{\infty}$ denote the eigenfunctions and eigenvalues of $\mathcal{C}$.

As in our previous developments, μ^y denotes the measure of interest, with y denoting parameters, or data, entering its definition. As in (2.24) we assume that

$$\frac{\mathrm{d}\mu^y}{\mathrm{d}\mu_0}(u) = \frac{1}{Z(y)} \exp\big(-\Phi(u; y)\big). \tag{4.1}$$

Recall that $\Phi(u; y)$ is the *potential* and that the *normalization constant* $Z(y)$ is chosen so that μ^y is a probability measure:

$$Z(y) = \int_H \exp\big(-\Phi(u; y)\big)\, \mathrm{d}\mu_0(u). \tag{4.2}$$

Both for this integral, and for others below, we observe that if $\mu_0(X) = 1$ we may write

$$Z(y) = \int_X \exp\big(-\Phi(u; y)\big)\, \mathrm{d}\mu_0(u),$$

and hence use properties of $\Phi(\cdot; y)$ which hold on X.

In the preceding section we showed that a number of inverse problems give rise to a probability measure μ^y of the form (4.1), where $\Phi : X \times Y \to \mathbb{R}$ satisfies Assumptions 2.6. The data (or parameters) y is (are) assumed to lie in a Banach space $(Y, \|\cdot\|_Y)$. We allow for the case where Y is infinite-dimensional and the data is in the form of a function. The four Assumptions 2.6(i)–(iv) play different roles, indicated by the following two theorems. The third assumption is important for showing that the posterior probability measure is well-defined, whilst the fourth is important for showing continuity with respect to data. The first and second assumptions lead to bounds on the normalization constant Z from above and below, respectively.

Theorem 4.1. *Let Φ satisfy Assumptions 2.6(i), (ii) and (iii) and assume that μ_0 is a Gaussian measure satisfying $\mu_0(X) = 1$. Then μ^y given by (4.1) is a well-defined probability measure on H.*

Proof. Assumption 2.6(ii) may be used to show that Z is bounded below, as shown in the proof of Theorem 4.2 below. Under Assumption 2.6(iii) it follows that Φ is μ_0-measurable, and hence the measure μ^y is well-defined by (4.1). By Assumption 2.6(i) we have that, for $\|y\|_Y < r$ and all ε sufficiently small,

$$\begin{aligned} Z(y) &= \int_X \exp\bigl(-\Phi(u;y)\bigr)\, \mathrm{d}\mu_0(u) \\ &\leqslant \int_X \exp\bigl(\varepsilon\|u\|_X^2 - M(\varepsilon,r)\bigr)\, \mathrm{d}\mu_0(u) \\ &\leqslant C\exp\bigl(-M(\varepsilon,r)\bigr) < \infty, \end{aligned}$$

since μ_0 is a Gaussian probability measure and we may choose ε sufficiently small so that the Fernique Theorem (Theorem 6.9) applies. Thus the measure is normalizable and the proof is complete. □

This proof directly shows that the posterior measure is a well-defined probability measure, without recourse to a conditioning argument. The conditioning argument used in Theorem 6.31 provides the additional fact that $\mu^y(\mathrm{d}u) = \mathbb{P}(\mathrm{d}u|y)$.

Now we study continuity properties of the measure μ^y with respect to $y \in Y$, under Assumptions 2.6(i), (ii) and (iv). This establishes the robustness of many of the problems introduced in the preceding section to changes in data.

Theorem 4.2. *Let Φ satisfy Assumptions 2.6(i), (ii) and (iv). Assume also that μ_0 is a Gaussian measure satisfying $\mu_0(X) = 1$ and that the measure $\mu^y \ll \mu_0$ with Radon–Nikodym derivative given by (4.1), for each $y \in Y$. Then μ^y is Lipschitz in the data y, with respect to the Hellinger distance: if μ^y and $\mu^{y'}$ are two measures corresponding to data y and y' then there exists $C = C(r) > 0$ such that, for all y, y' with $\max\{\|y\|_Y, \|y'\|_Y\} < r$,*

$$d_{\mathrm{Hell}}(\mu^y, \mu^{y'}) \leqslant C\|y - y'\|_Y.$$

Consequently the expectation of any polynomially bounded function $f : X \to E$ is continuous in y. In particular the mean and, in the case where X is a Hilbert space, the covariance operator, are continuous in y.

Proof. Throughout the proof, all integrals are over X, unless specified otherwise. The constant C may depend on ε and r and changes from occurrence to occurrence. Let $Z = Z(y)$ and $Z' = Z(y')$ denote the normalization constants for μ^y and $\mu^{y'}$ so that

$$
\begin{aligned}
Z &= \int \exp(-\Phi(u;y))\, \mathrm{d}\mu_0(u),\\
Z' &= \int \exp(-\Phi(u;y'))\, \mathrm{d}\mu_0(u).
\end{aligned}
$$

Using Assumption 2.6(ii) gives, for any $r > 0$,

$$
Z \geqslant \int_{\{\|u\|_X \leqslant r\}} \exp(-K(r))\, \mathrm{d}\mu_0(u) = \exp(-K(r))\mu_0\{\|u\|_X \leqslant r\}.
$$

This lower bound is positive because μ_0 has full measure on X and is Gaussian, so that all balls in X have positive probability. We have an analogous lower bound for $|Z'|$.

Using Assumptions 2.6(i) and (iv) and using the Fernique Theorem, for μ_0,

$$
\begin{aligned}
|Z - Z'| &\leqslant \left(\int \exp(\varepsilon\|u\|_X^2 - M) \exp(\varepsilon\|u\|_X^2 + C)\, \mathrm{d}\mu_0(u) \right) \|y - y'\|_Y\\
&\leqslant C\|y - y'\|_Y.
\end{aligned}
$$

From the definition of Hellinger distance, we have

$$
\begin{aligned}
2d_{\mathrm{Hell}}(\mu^y, \mu^{y'})^2 &= \int \left(Z^{-1/2} \exp\left(-\frac{1}{2}\Phi(u;y) \right) \right.\\
&\qquad \left. - (Z')^{-1/2} \exp\left(-\frac{1}{2}\Phi(u;y') \right) \right)^2 \mathrm{d}\mu_0(u)\\
&\leqslant I_1 + I_2
\end{aligned}
$$

where

$$
\begin{aligned}
I_1 &= \frac{2}{Z} \int \left(\exp\left(-\frac{1}{2}\Phi(u;y) \right) - \exp\left(-\frac{1}{2}\Phi(u;y') \right) \right)^2 \mathrm{d}\mu_0(u),\\
I_2 &= 2|Z^{-1/2} - (Z')^{-1/2}|^2 \int \exp(-\Phi(u;y'))\, \mathrm{d}\mu_0(u).
\end{aligned}
$$

Now, again using Assumptions 2.6(i) and (iv) and the Fernique Theorem,

$$
\begin{aligned}
\frac{Z}{2} I_1 &\leqslant \int \frac{1}{4} \exp(\varepsilon\|u\|_X^2 - M) \exp(2\varepsilon\|u\|_X^2 + 2C)\, \|y - y'\|_Y^2\, \mathrm{d}\mu_0(u)\\
&\leqslant C\|y - y'\|_Y^2.
\end{aligned}
$$

A similar use of the Fernique Theorem and Assumption 2.6(i) shows that the integral in I_2 is finite. Also, using the bounds on Z, Z' from below,

$$\begin{aligned}|Z^{-1/2} - (Z')^{-1/2}|^2 &\leqslant C(Z^{-3} \vee (Z')^{-3})|Z - Z'|^2 \\ &\leqslant C\|y - y'\|_Y^2.\end{aligned}$$

Combining gives the desired continuity result in the Hellinger metric.

Finally all moments of u in X are finite under μ^y and $\mu^{y'}$ because the change of measure from Gaussian μ_0 involves a term which may be bounded by use of Assumption 2.6(i). The Fernique Theorem may then be applied. The desired result concerning the continuity of moments follows from Lemma 6.37. □

Example 4.3. An example in which the data is a function is given in Section 3.5, where we study the inverse problem of determining the initial condition for the heat equation, given noisy observation of the solution at a positive time; in Lemma 3.9 we establish that Assumptions 2.6 hold in this case. ◇

4.3. *Well-posedness: finite data*

For Bayesian inverse problems in which a finite number of observations are made, the potential Φ has the form

$$\Phi(u; y) = \frac{1}{2}|y - \mathcal{G}(u)|_\Gamma^2, \tag{4.3}$$

where $y \in \mathbb{R}^q$ is the data, $\mathcal{G} : X \to \mathbb{R}^q$ is the observation operator and $|\cdot|_\Gamma$ is a covariance weighted norm on $\mathbb{R}^q$. In this case it is natural to express conditions on the potential Φ in terms of $\mathcal{G}$. Recall that this is undertaken in Assumptions 2.7. By Lemma 2.8 we know that Assumptions 2.7 imply Assumptions 2.6 for Φ given by (4.3). The following corollary of Theorem 4.2 is hence automatic.

Corollary 4.4. Assume that $\Phi : X \times \mathbb{R}^q \to \mathbb{R}$ is given by (4.3) and let $\mathcal{G}$ satisfy Assumptions 2.7. Assume also that μ_0 is a Gaussian measure satisfying $\mu_0(X) = 1$. Then the measure μ^y given by (4.1) is a well-defined probability measure and is Lipschitz in the data y, with respect to the Hellinger distance: if μ^y and $\mu^{y'}$ are two measures corresponding to data y and y', then there is $C = C(r) > 0$ such that, for all y, y' with $\max\{|y|_\Gamma, |y'|_\Gamma\} < r$,

$$d_{\text{Hell}}(\mu^y, \mu^{y'}) \leqslant C|y - y'|_\Gamma.$$

Consequently the expectation of any polynomially bounded function $f : X \to E$ is continuous in y. In particular the mean and, in the case where X is a Hilbert space, the covariance operator are continuous in y. ◇

Example 4.5. The first example of a problem with the structure of Assumptions 2.7 may be found in the discussion of finite-dimensional inverse problems in Section 2.2, and formula (2.8) in the case where ρ is a Gaussian density; if, for example, $\mathcal{G}$ is differentiable and polynomially bounded, then Assumptions 2.7 hold: see Example 2.2 for an explicit illustration. All the examples in Section 3, with the exception of the heat equation example, for which the data is infinite, and the oil reservoir problem, for which the appropriate analysis and choice of X has not yet been carried out, fit the framework of Corollary 4.4. ◇

4.4. Approximation of measures in the Hellinger metric

To implement algorithms designed to sample the posterior measure μ^y given by (4.1), we need to make finite-dimensional approximations. We study this issue here. Since the dependence on y is not relevant in this section, we study measures μ given by

$$\frac{\mathrm{d}\mu}{\mathrm{d}\mu_0}(u) = \frac{1}{Z}\exp\bigl(-\Phi(u)\bigr), \tag{4.4}$$

where the normalization constant Z is given by

$$Z = \int_X \exp\bigl(-\Phi(u)\bigr)\,\mathrm{d}\mu_0(u). \tag{4.5}$$

We approximate μ by approximating Φ. In particular, we define μ^N by

$$\frac{\mathrm{d}\mu^N}{\mathrm{d}\mu_0}(u) = \frac{1}{Z^N}\exp\bigl(-\Phi^N(u)\bigr), \tag{4.6}$$

where

$$Z^N = \int_X \exp\bigl(-\Phi^N(u)\bigr)\,\mathrm{d}\mu_0(u). \tag{4.7}$$

Our interest is in translating approximation results for Φ (determined by the forward problem) into approximation results for μ (which describes the inverse problem).

The following theorem proves such a result, bounding the Hellinger distance, and hence the total variation distance, between measures μ and μ^N in terms of the error in approximating Φ.

Theorem 4.6. Assume that the measures μ and μ^N are both absolutely continuous with respect to μ_0, satisfying $\mu_0(X) = 1$, with Radon–Nikodym derivatives given by (4.4) and (4.6) and that Φ and Φ^N satisfy Assumptions 2.6(i) and (ii) with constants uniform in N. Assume also that for any $\varepsilon > 0$ there exists $K = K(\varepsilon) > 0$ such that

$$|\Phi(u) - \Phi^N(u)| \leqslant K\exp\bigl(\varepsilon\|u\|_X^2\bigr)\psi(N), \tag{4.8}$$

where $\psi(N) \to 0$ as $N \to \infty$. Then the measures μ and μ^N are close with respect to the Hellinger distance: there is a constant C, independent of N, and such that

$$d_{\mathrm{Hell}}(\mu, \mu^N) \leqslant C\psi(N). \tag{4.9}$$

Consequently the expectation under μ and μ^N of any polynomially bounded function $f : X \to E$ are $\mathcal{O}(\psi(N))$ close. In particular, the mean and, in the case where X is a Hilbert space, the covariance operator are $\mathcal{O}(\psi(N))$ close.

Proof. Throughout the proof, all integrals are over X. The constant C changes from occurrence to occurrence. The normalization constants Z and Z^N satisfy lower bounds which are identical to that proved for Z in the course of establishing Theorem 4.2.

From Assumption 2.6(i) and (4.8), using the fact that μ_0 is a Gaussian probability measure so that the Fernique Theorem 6.9 applies,

$$\begin{aligned}|Z - Z^N| &\leqslant \int K\psi(N)\exp(\varepsilon\|u\|_X^2 - M)\exp(\varepsilon\|u\|_X^2)\,\mathrm{d}\mu_0(u)\\ &\leqslant C\psi(N).\end{aligned}$$

From the definition of Hellinger distance, we have

$$\begin{aligned}2d_{\mathrm{Hell}}(\mu, \mu^N)^2 &= \int \Bigl(Z^{-1/2}\exp\Bigl(-\frac{1}{2}\Phi(u)\Bigr)\\ &\qquad - (Z^N)^{-1/2}\exp\Bigl(-\frac{1}{2}\Phi^N(u)\Bigr)\Bigr)^2\,\mathrm{d}\mu_0(u)\\ &\leqslant I_1 + I_2,\end{aligned}$$

where

$$\begin{aligned}I_1 &= \frac{2}{Z}\int \Bigl(\exp\Bigl(-\frac{1}{2}\Phi(u)\Bigr) - \exp\Bigl(-\frac{1}{2}\Phi^N(u)\Bigr)\Bigr)^2\,\mathrm{d}\mu_0(u),\\ I_2 &= 2\bigl|Z^{-1/2} - (Z^N)^{-1/2}\bigr|^2\int \exp(-\Phi^N(u))\,\mathrm{d}\mu_0(u).\end{aligned}$$

Now, again using Assumption 2.6(i) and equation (4.8), together with the Fernique Theorem,

$$\begin{aligned}\frac{Z}{2}I_1 &\leqslant \int K^2\exp(3\varepsilon\|u\|_X^2 - M)\psi(N)^2\,\mathrm{d}\mu_0(u)\\ &\leqslant C\psi(N)^2.\end{aligned}$$

A similar use of the Fernique Theorem and Assumption 2.6(i) shows that the integral in I_2 is finite. Thus, using the bounds on Z, Z^N from below,

$$\begin{aligned}\bigl|Z^{-1/2} - (Z^N)^{-1/2}\bigr|^2 &\leqslant C(Z^{-3} \vee (Z^N)^{-3})|Z - Z^N|^2\\ &\leqslant C\psi(N)^2.\end{aligned}$$

Combining gives the desired continuity result in the Hellinger metric.

Finally, all moments of u in X are finite under μ and μ^N because the change of measure from Gaussian μ_0 involves a term which may be controlled by the Fernique Theorem. The desired results follow from Lemma 6.37. □

Example 4.7. Consider the inverse problem for the heat equation, from Section 3.5, in the case where $D = (0,1)$. Approximate the Bayesian inverse problem by use of a spectral approximation of the forward map $\mathrm{e}^{-AT} : \mathcal{H} \to \mathcal{H}$. Let P^N denote the orthogonal projection in $\mathcal{H}$ onto the first N eigenfunctions of A. Then, for any $T > 0$ and $r \geqslant 0$,

$$\|\mathrm{e}^{-AT} - \mathrm{e}^{-AT} P^N\|_{\mathcal{L}(\mathcal{H},\mathcal{H}^r)} = \mathcal{O}(\exp(-cN^2)).$$

From (3.25) we have the Lipschitz property that

$$|\Phi(u) - \Phi(v)| \leqslant C\big(\|u\| + \|v\| + \|y\|\big)\|\mathrm{e}^{-\frac{1}{4}AT}(u-v)\|.$$

If we define $\Phi^N(u) = \Phi(P^N u)$, then the two preceding estimates combine to give, for some $C, c > 0$ and independent of (u, y),

$$|\Phi(u) - \Phi^N(u)| \leqslant C\big(\|u\| + \|y\|\big)\|u\| \exp(-cN^2).$$

Thus (4.8) holds and Theorem 4.6 shows that the posterior measure is perturbed by a quantity with order of magnitude $\mathcal{O}(\exp(-cN^2))$ in the Hellinger metric. ◇

Remark 4.8. Approximation may come from two sources: (i) from representing the target function u in a finite-dimensional basis; and (ii) from approximating the forward model, and hence the potential Φ, by a numerical method such as a finite element or spectral method. In general these two sources of approximation error are distinct and must be treated separately. An important issue is to balance the two sources of error to optimize workload. In the case where u is a subset of, or the entire, initial condition for a dynamical system and $\mathcal{G}$ is defined through composition of some function with the solution operator, then (i) and (ii) will overlap if a spectral approximation is employed for (ii), using the finite-dimensional basis from (i). This is the situation in the preceding example. ◇

For Bayesian inverse problems with finite data, the potential Φ has the form given in (4.3), where $y \in \mathbb{R}^q$ is the data, $\mathcal{G} : X \to \mathbb{R}^q$ is the observation operator and $|\cdot|_\Gamma$ is a covariance weighted norm on $\mathbb{R}^q$. If $\mathcal{G}^N$ is an approximation to $\mathcal{G}$ and we define

$$\Phi^N := \frac{1}{2}|y - \mathcal{G}^N(u)|_\Gamma^2, \tag{4.10}$$

then we may define an approximation μ^N to μ as in (4.6). The following corollary relating μ and μ^N is useful.

Corollary 4.9. Assume that the measures μ and μ^N are both absolutely continuous with respect to μ_0, with Radon–Nikodym derivatives given by (4.4), (4.3) and (4.6), (4.10) respectively. Assume also that $\mathcal{G}$ is approximated by a function $\mathcal{G}^N$ with the property that, for any $\varepsilon > 0$, there is $K' = K'(\varepsilon) > 0$ such that

$$|\mathcal{G}(u) - \mathcal{G}^N(u)| \leqslant K' \exp\bigl(\varepsilon \|u\|_X^2\bigr)\psi(N), \tag{4.11}$$

where $\psi(N) \to 0$ as $N \to \infty$. If $\mathcal{G}$ and $\mathcal{G}^N$ satisfy Assumption 2.7(i) uniformly in N, then there is a constant C, independent of N, and such that

$$d_{\text{Hell}}(\mu, \mu^N) \leqslant C\psi(N). \tag{4.12}$$

Consequently the expectation under μ and μ^N of any polynomially bounded function $f : X \to E$ is $\mathcal{O}(\psi(N))$ close. In particular, the mean and, in the case where X is a Hilbert space, the covariance operator are $\mathcal{O}(\psi(N))$ close.

Proof. We simply show that the conditions of Theorem 4.6 hold. That (i) and (ii) of Assumptions 2.6 hold follows as in the proof of Lemma 2.8. Also (4.8) holds since (for some $K : \mathbb{R}^+ \to \mathbb{R}^+$ defined in the course of the following chain of inequalities)

$$\begin{aligned}
|\Phi(u) - \Phi^N(u)| &\leqslant \frac{1}{2}|2y - \mathcal{G}(u) - \mathcal{G}^N(u)|_\Gamma |\mathcal{G}(u) - \mathcal{G}^N(u)|_\Gamma \\
&\leqslant \bigl(|y|_\Gamma + \exp\bigl(\varepsilon\|u\|_X^2 + M(\varepsilon)\bigr)\bigr) \times K'(\varepsilon)\exp\bigl(\varepsilon\|u\|_X^2\bigr)\psi(N) \\
&\leqslant K(2\varepsilon)\exp\bigl(2\varepsilon\|u\|_X^2\bigr)\psi(N),
\end{aligned}$$

as required. □

A notable fact concerning Theorem 4.6 is that the rate of convergence attained in the solution of the forward problem, encapsulated in approximation of the function Φ by Φ^N, is transferred into the rate of convergence of the related inverse problem for measure μ given by (4.4) and its approximation by μ^N. Key to achieving this transfer of rates of convergence is the dependence of the constant in the forward error bound (4.8) on u. In particular it is necessary that this constant is integrable by use of the Fernique Theorem. In some applications it is not possible to obtain such dependence. Then convergence results can sometimes still be obtained, but at weaker rates. We state a theorem applicable in this situation.

Theorem 4.10. Assume that the measures μ and μ^N are both absolutely continuous with respect to μ_0, satisfying $\mu_0(X) = 1$, with Radon–Nikodym derivatives given by (4.4) and (4.6), and that Φ and Φ^N satisfy Assumptions 2.6(i) and (ii) with constants uniform in N. Assume also that for any $R > 0$ there is a $K = K(R) > 0$ such that, for all u with $\|u\|_X \leqslant R$,

$$|\Phi(u) - \Phi^N(u)| \leqslant K\psi(N), \tag{4.13}$$

where $\psi(N) \to 0$ as $N \to \infty$. Then the measures μ and μ^N are close with respect to the Hellinger distance:

$$d_{\text{Hell}}(\mu, \mu^N) \to 0 \tag{4.14}$$

as $N \to \infty$. Consequently the expectation of any polynomially bounded function $f : X \to E$ under μ^N converges to the corresponding expectation under μ as $N \to \infty$. In particular, the mean and, in the case where X is a Hilbert space, the covariance operator converge. ◇

4.5. Discussion and bibliography

The idea of placing a number of inverse problems within a common Bayesian framework, and studying general properties in this abstract setting, is developed in Cotter *et al.* (2009). That paper contains Theorems 4.1 and 4.2 under Assumptions 2.6 in the case where (i) is satisfied trivially because Φ is bounded from below by a constant; note that this case occurs whenever the data is finite-dimensional. Generalizing the theorems to allow for (i) as stated here was undertaken in Hairer, Stuart and Voss (2010*b*), in the context of signal processing for stochastic differential equations.

Theorem 4.2 is a form of well-posedness. Recall that, in the approximation of forward problems in differential equations, well-posedness and a local approximation property form the key concepts that underpin the equivalence theorems of Dahlquist (Hairer, Nørsett and Wanner 1993, Hairer and Wanner 1996), Lax (Richtmyer and Morton 1967) and Sanz-Serna and Palencia (Sanz-Serna and Palencia 1985). It is also natural that the well-posedness that we have exhibited for inverse problems should, when combined with forward approximation, give rise to approximation results for the inverse problem. This is the basic idea underlying Theorem 4.6. That result, Corollary 4.9 and Theorem 4.10 are all stated and proved in Cotter *et al.* (2010*a*).

The underlying well-posedness of properly formulated Bayesian inverse problems has a variety of twists and turns which we do not elaborate fully here. The interested reader should consult Dashti *et al.* (2010*b*).

5. Algorithms

5.1. Overview

We have demonstrated that a wide range of inverse problems for functions u given data y give rise to a posterior measure μ^y with the form (2.24). This formula encapsulates neatly the ideal information that we have about a function, formed from conjunction of model and data. Furthermore, for many applications, the potential Φ satisfies Assumptions 2.6. From this we have shown in Section 4 that the formula (2.24) indeed leads to a well-defined posterior μ^y and that this measure enjoys nice robustness properties with respect to changes in the data or approximation of the forward problem. However, we have not yet addressed the issue of how to obtain information from the formula (2.24) for the posterior measure. We devote this section to an overview of the computational issues which arise in this context.

If the prior measure is Gaussian and the potential $\Phi(\cdot; y)$ is quadratic, then the posterior is also Gaussian. This situation arises, for example, in the inverse problem for the heat equation described in Section 3.5. The measure μ^y is then characterized by a function (the mean) and an operator (the covariance) and formulae can be obtained for these quantities by completing the square using Theorem 6.20: see the developments for the heat equation, or Example 6.23, for an illustration of this.

However, in general there is no explicit way of characterizing the measure μ^y as can be done in the Gaussian case. Thus approximations and computational tools are required to extract information from the formula (2.24). One approach to this problem is to employ sampling techniques which (approximately) generate sample functions according to the probability distribution implied by (2.24). Among the most powerful generic tools for sampling are the *Markov chain Monte Carlo* (MCMC) methods, which we review in Section 5.2. However, whilst these methods can be very effective when tuned carefully to the particular problem at hand, they are undeniably costly and, for many applications, impracticable at current levels of computer resources. For this reason we also devote two subsections to *variational* and *filtering* methods, which are widely used in practice because of their computational expedience. When viewed in terms of their relation to (2.24) these methods constitute approximations. Furthermore, these approximations are, in many cases, not well understood. In the near future we see the main role of MCMC methods as providing controlled approximations to the true posterior measure μ^y, against which variational and filtering methodologies can be tested, on well-designed model problems. In the longer term, as computational power and algorithmic innovation grows, we also anticipate increasing use of MCMC methods in their own right to approximate (2.24).

From a Bayesian perspective, the variational methods of Section 5.3 start from the premise that variability in the posterior measure is small and that most of the information resides in a single peak of the probability distribution, which can be found by optimization techniques. We view this problem from the standpoint of optimal control, showing that a minimizer exists whenever the common framework of Section 2.4 applies; we also review algorithmic practice in the area. Section 5.4 describes the widely used filtering methods which approximate the posterior measure arising in time-dependent data assimilation problems by a sequence of probability measures in time, updated sequentially. The importance of this class of algorithms stems from the fact that, in many applications, solutions are required on-line, with updates required as more data is acquired; thus sequential updating of the posterior measure at the current time is natural. Furthermore, sequential updates are computationally efficient as they reduce the dimensionality of the desired posterior measure, breaking a correlated measure at a sequence of times into a sequence of conditionally independent measures at each time, provided there is an underlying Markovian structure. We conclude, in Section 5.5, with references to the literature concerning algorithms.

When discussing MCMC methods and variational methods, the dependence of the potential Φ appearing in (2.24) will not be relevant and we will consider the problem for the posterior measure written in the form

$$\frac{\mathrm{d}\mu}{\mathrm{d}\mu_0}(u) = \frac{1}{Z}\exp\big(-\Phi(u)\big), \tag{5.1}$$

with normalization constant

$$Z = \int_X \exp\big(-\Phi(u)\big)\,\mathrm{d}\mu_0(u). \tag{5.2}$$

We refer to μ as the *target distribution*. For the study of both MCMC and variational methods, we will also find it useful to define

$$I(u) = \frac{1}{2}\|u\|_C^2 + \Phi(u). \tag{5.3}$$

This is, of course, a form of regularized least-squares functional as introduced in Section 2.

5.2. Markov chain Monte Carlo

The basic idea of MCMC methods is simple: design a Markov chain with the property that a single sequence of output from the chain $\{u_n\}_{n=0}^{\infty}$ is distributed according to μ given by (5.1). This is a *very* broad algorithmic prescription and allows for significant innovation in the design of methods tuned to the particular structure of the desired target distribution. We will focus on a particular class of MCMC methods known as *Metropolis–Hastings* (MH) methods.

The key ingredient of these methods is a probability measure on X, parametrized by $u \in X$: a Markov transition kernel $q(u, dv)$. This kernel is used to propose moves from the current state of the Markov chain u_n to a new point distributed as $q(u_n, \cdot)$. This proposed point is then accepted or rejected according to a criterion which uses the target distribution μ. The resulting Markov chain has the desired property of preserving the target distribution. Key to the success of the method is the choice of q. We now give details of how the method is constructed.

Given $q(u,\cdot)$ and the target μ we define a new measure on $X \times X$ defined by

$$\nu(\mathrm{d}u, \mathrm{d}v) = q(u, dv)\mu(\mathrm{d}u).$$

We define the same measure, with the roles of u and v reversed, by

$$\nu^\top(\mathrm{d}u, \mathrm{d}v) = q(v, \mathrm{d}u)\mu(\mathrm{d}v).$$

Provided that $\nu^\top$ is absolutely continuous with respect to ν, we may define

$$\alpha(u,v) = \min\left\{1, \frac{\mathrm{d}\nu^\top}{\mathrm{d}\nu}(u,v)\right\}.$$

Now define a random variable $\gamma(u,v)$, independent of the probability space underlying the transition kernel q, with the property that

$$\gamma(u,v) = \begin{cases} 1 & \text{with probability } \alpha(u,v), \\ 0 & \text{otherwise.} \end{cases} \tag{5.4}$$

We now create a random Markovian sequence $\{u_n\}_{n=0}^\infty$ as follows. Given a proposal $v_n \sim q(u_n,\cdot)$, we set

$$u_{n+1} = \gamma(u_n, v_n)v_n + (1 - \gamma(u_n, v_n))u_n. \tag{5.5}$$

If we choose the randomness in the proposal v_n and the binary random variable $\gamma(u_n, v_n)$ independently of each other for each n, and independently of their values for different n, then this construction gives rise to a Markov chain with the desired property.

Theorem 5.1. Under the given assumptions, the Markov chain defined by (5.5) is invariant for μ: if $u_0 \sim \mu$ then $u_n \sim \mu$ for all $n \geqslant 0$. Furthermore, if the resulting Markov chain is ergodic then, for any continuous bounded function $f : X \to \mathbb{R}$, any $M \geqslant 0$, and for u_0 μ-a.s.,

$$\frac{1}{N}\sum_{n=1}^{N} f(u_{n+M}) \to \int_X f(u)\mu(\mathrm{d}u) \quad \text{as } N \to \infty. \tag{5.6}$$

◇

In words, this theorem states that the empirical distribution of the Markov chain converges weakly to that of the target measure μ. However, this nice abstract development has not addressed the question of actually constructing an MH method. If $X = \mathbb{R}^n$ and the target measures have positive density with respect to Lebesgue measure, then this is straightforward: any choice of kernel $q(u, dv)$ will suffice, provided it too has positive density with respect to Lebesgue measure, for every u. It then follows that $\nu^\top \ll \nu$. From this wide range of admissible proposal distributions, the primary design choice is to identify proposals which lead to low correlation in the resulting Markov chain, as this increases efficiency.

Example 5.2. A widely used proposal kernel is simply that of a *random walk*; for example, if $\mu_0 = \mathcal{N}(0, \mathcal{C})$ it is natural to propose

$$v = u + \sqrt{2\delta}\xi, \tag{5.7}$$

where $\xi \sim \mathcal{N}(0, \mathcal{C})$. A straightforward calculation shows that

$$\alpha(u, v) = \min\{1, \exp(I(u) - I(v))\}$$

where I is given by (5.3). Thus, if the proposed state corresponds to a lower value of the regularized least-squares functional I, then the proposal is automatically accepted; otherwise it will be accepted with a probability depending on $I(u) - I(v)$.

The parameter δ is a scalar which controls the size of the move. Large values lead to proposals which are hence unlikely to be accepted, leading to high correlation in the Markov chain. On the other hand small moves do not move very far, again leading to high correlation in the Markov chain. Identifying appropriate values of δ between these extremes is key to making effective algorithms. More complex proposals use additional information about $D\Phi$ in an attempt to move into regions of high probability (low Φ). $\diamond$

In infinite dimensions things are not so straightforward: a random walk will not typically deliver the required condition $\nu^{\top} \ll \nu$. For example, if $\mu_0 = \mathcal{N}(0, \mathcal{C})$ and X is infinite-dimensional, then the proposal (5.7) will not satisfy this constraint. However, a little thought shows that appropriate modifications are possible.

Example 5.3. The random walk can be modified to obtain the desired absolute continuity of $\nu^{\top}$ with respect to ν. The proposal

$$v = (1 - 2\delta)^{1/2} u + \sqrt{2\delta}\xi, \tag{5.8}$$

where $\xi \sim \mathcal{N}(0, \mathcal{C})$, will satisfy the desired condition for any $\delta \in \mathbb{R}$. The acceptance probability is

$$\alpha(u, v) = \min\{1, \exp(\Phi(u) - \Phi(v))\}.$$

Thus, if the proposed state corresponds to a lower value of Φ than does the current state, it will automatically be accepted.

The proposal in (5.8) should be viewed as an appropriate analogue of the random walk proposal in infinite-dimensional problems. Intuition as to why this proposal works in the infinite-dimensional setting can be obtained by observing that, if $u \sim \mathcal{N}(0, \mathcal{C})$ and v is constructed using (5.8), then $v \sim \mathcal{N}(0, \mathcal{C})$; thus the proposal preserves the underlying reference measure (prior) μ_0. In contrast, the proposal (5.7) does not: if $u \sim \mathcal{N}(0, \mathcal{C})$ then $v \sim \sqrt{(1 + 2\delta)}\mathcal{N}(0, \mathcal{C})$.

Note that the choice $\delta = 1/2$ in (5.8) yields an *independence sampler* where proposals v are made from the prior measure μ_0, independently of the current state of the Markov chain u. As in finite dimensions, improved proposals can be found by including information about $D\Phi$ in the proposal. ◇

In computational practice, of course, we always implement a sampling method in finite dimensions. The error incurred by doing so may be quantified by use of Theorem 4.6. It is natural to ask whether there is any value in deriving MH methods on function space, especially since this appears harder than doing so in finite dimensions. The answer, of course, is 'yes'. Any MH method in finite dimensions which does not correspond to a well-defined limiting MH method in the function space (infinite-dimensional) limit will degenerate as the dimension of the space increases. This effect can be quantified and compared with what happens when proposals defined on function space are used. In conclusion, then, the function space viewpoint on MCMC methods is a useful one which leads to improved algorithms, and an understanding of the shortcomings of existing algorithms.

5.3. Variational methods

Variational methods attempt to answer the following question: 'How do we find the most likely function u under the posterior measure μ given by (5.1)?' To understand this consider first the case where $X = \mathbb{R}^n$ and $\mu_0 = \mathcal{N}(0, \mathcal{C})$ is a Gaussian prior. Then μ has density with respect to Lebesgue measure and the negative logarithm of this density is given by (5.3).[3] Thus the Lebesgue density of μ is maximized by minimizing I over $\mathbb{R}^n$. Another way of looking at this is as follows: if $\overline{u}$ is such a minimizer then the probability of a small ball of radius ε and centred at u will be maximized, asymptotically as $\varepsilon \to 0$, by choosing $u = \overline{u}$.

If X is an infinite-dimensional Hilbert space then there is no Lebesgue measure on X, and we cannot directly maximize the density. However, we may again consider the probability of small balls at $u \in X$, of radius ε. We may then ask how u should be chosen to maximize the probability of the ball, asymptotically as $\varepsilon \to 0$. Again taking $\mu_0 = \mathcal{N}(0, \mathcal{C})$ this question leads to the conclusion that u should be chosen as a global minimizer of I given by (5.3) over the Cameron–Martin space E with inner product $\langle \cdot, \cdot \rangle_{\mathcal{C}}$ and norm $\| \cdot \|_{\mathcal{C}}$.

Recall that Φ measures model/data mismatch, in the context of applications to inverse problems. In the case where y is finite-dimensional it has the form (4.3). It is thus natural to minimize Φ directly, as in (2.2). However, when X is infinite-dimensional, this typically leads to minimizing sequences

[3] Recall that for economy of notation we drop explicit reference to the y dependence of Φ in this subsection, as it plays no role.

which do not converge in any reasonable topology. The addition of the quadratic penalization in E may be viewed as a *Tikhonov regularization* to overcome this problem. Minimization of I is thus a regularized *nonlinear least-squares problem* as in (2.3). Of course this optimization approach can be written down directly, with no reference to probability. The beauty of the Bayesian approach is that it provides a rational basis for the choice of norms underlying the objective functional Φ, as well as the choice of norm in the regularization term proportional to $\|u\|_{\mathcal{C}}^2$. Furthermore, the Bayesian viewpoint gives an interpretation of the resulting optimization problem as a probability maximizer. And finally the framework of Section 2.4, which leads to well-posed posterior measures, also leads directly to an existence theory for probability maximizers. We now describe this theory.

Theorem 5.4. Let Assumptions 2.6(i) and (iii) hold, and assume that $\mu_0(X) = 1$. Then there exists $\overline{u} \in E$ such that

$$I(\overline{u}) = \overline{I} := \inf\{I(u) : u \in E\}.$$

Furthermore, if $\{u_n\}$ is a minimizing sequence satisfying $I(u_n) \to I(\overline{u})$, then there is a subsequence $\{u_{n'}\}$ that converges strongly to $\overline{u}$ in E.

Proof. First we show that I is weakly lower semicontinuous on E. Let $u_n \rightharpoonup \overline{u}$ in E. By the compact embedding of E in X, which follows from Theorem 6.11 since $\mu_0(X) = 1$, we deduce that $u_n \to \overline{u}$, strongly in X. By the Lipschitz continuity of Φ in X (Assumption 2.6(iii)) we deduce that $\Phi(u_n) \to \Phi(\overline{u})$. Thus Φ is weakly continuous on E. The functional $J(u) := \frac{1}{2}\|u\|_{\mathcal{C}}^2$ is weakly lower semicontinuous on E. Hence $I(u) = J(u) + \Phi(u)$ is weakly lower semicontinuous on E.

Now we show that I is coercive on E. Again using the fact that E is compactly embedded in X, we deduce that there is a $K > 0$ such that

$$\|u\|_X^2 \leqslant K\|u\|_{\mathcal{C}}^2.$$

Hence, by Assumption 2.6(i), it follows that, for any $\varepsilon > 0$, there is an $M(\varepsilon) \in \mathbb{R}$ such that

$$\left(\frac{1}{2} - K\varepsilon\right)\|u\|_{\mathcal{C}}^2 + M(\varepsilon) \leqslant I(u).$$

By choosing ε sufficiently small, we deduce that there is an $M \in \mathbb{R}$ such that, for all $u \in E$,

$$\frac{1}{4}\|u\|_{\mathcal{C}}^2 + M \leqslant I(u). \tag{5.9}$$

This establishes coercivity.

Consider a minimizing sequence. For any $\delta > 0$ there is an $N_1 = N_1(\delta)$:

$$M \leqslant \overline{I} \leqslant I(u_n) \leqslant \overline{I} + \delta, \quad \forall n \geqslant N_1.$$

Using (5.9) we deduce that the sequence $\{u_n\}$ is bounded in E and, since E is a Hilbert space, there exists $\overline{u} \in E$ such that (possibly along a subsequence) $u_n \rightharpoonup \overline{u}$ in E. From the weak lower semicontinuity of I it follows that, for any $\delta > 0$,

$$\overline{I} \leqslant I(\overline{u}) \leqslant \overline{I} + \delta.$$

Since δ is arbitrary the first result follows.

Now consider the subsequence $u_n \rightharpoonup \overline{u}$. Then there is an $N_2 = N_2(\delta) > 0$ such that, for $n, \ell \geqslant N_2$,

$$\begin{aligned}
\frac{1}{4}\|u_n - u_\ell\|_C^2 &= \frac{1}{2}\|u_n\|_C^2 + \frac{1}{2}\|u_\ell\|_C^2 - \|\frac{1}{2}(u_n + u_\ell)\|_C^2 \\
&= I(u_n) + I(u_\ell) - 2I\left(\frac{1}{2}(u_n + u_\ell)\right) - \Phi(u_n) \\
&\quad - \Phi(u_\ell) + 2\Phi\left(\frac{1}{2}(u_n + u_\ell)\right) \\
&\leqslant 2(\overline{I} + \delta) - 2\overline{I} - \Phi(u_n) - \Phi(u_\ell) + 2\Phi\left(\frac{1}{2}(u_n + u_\ell)\right) \\
&\leqslant 2\delta - \Phi(u_n) - \Phi(u_\ell) + 2\Phi\left(\frac{1}{2}(u_n + u_\ell)\right).
\end{aligned}$$

But u_n, u_ℓ and $\frac{1}{2}(u_n + u_\ell)$ all converge strongly to $\overline{u}$ in X. Thus, by continuity of Φ, we deduce that, for all $n, \ell \geqslant N_3(\delta)$,

$$\frac{1}{4}\|u_n - u_\ell\|_C^2 \leqslant 3\delta.$$

Hence the sequence is Cauchy in E and converges strongly, and the proof is complete. □

Example 5.5. Recall the inverse problem for the diffusion coefficient of the one-dimensional elliptic problem described in Section 3.3. The objective is to find $u(x)$ appearing in

$$\begin{aligned}
-\frac{\mathrm{d}}{\mathrm{d}x}\left(\exp(u(x))\frac{\mathrm{d}p}{\mathrm{d}x}\right) &= 0, \\
p(0) = p^- \quad p(1) &= p^+,
\end{aligned}$$

where $p^+ > p^-$. The observations are

$$y_k = p(x_k) + \eta_k, \quad k = 1, \dots, q$$

written succinctly as

$$y = \mathcal{G}(u) + \eta,$$

where $\eta \in \mathbb{R}^q$ is distributed as $\mathcal{N}(0, \gamma^2 I)$. The function $\mathcal{G}$ is Lipschitz in the space of continuous functions $X = C([0,1])$ by Lemma 3.3.

Recall that changing u by an arbitrary additive constant does not change the solution of (3.5), and so we assume that u integrates to zero on $(0,1)$. We define

$$\mathcal{H} = \left\{ u \in L^2((0,1)) \middle| \int_0^1 u(x)\,\mathrm{d}x = 0 \right\}.$$

We take $\mathcal{A} = -\mathrm{d}^2/\mathrm{d}x^2$ with

$$D(\mathcal{A}) = \left\{ u \in H^2_{\mathrm{per}}((0,1)) \middle| \int_0^1 u(x)\,\mathrm{d}x = 0 \right\}.$$

Then $\mathcal{A}$ is positive definite self-adjoint, and we may define the prior Gaussian measure $\mu_0 = \mathcal{N}(0, \mathcal{A}^{-1})$ on $\mathcal{H}$. By Lemma 6.25 we deduce that $\mu_0(X) = 1$. The Cameron–Martin space

$$E = \mathrm{Im}(\mathcal{A}^{-1/2}) = \left\{ u \in H^1_{\mathrm{per}}((0,1)) \middle| \int_0^1 u(x)\,\mathrm{d}x = 0 \right\}$$

is compactly embedded into $C([0,1])$ by Theorem 2.10; this is also a consequence of the general theory of Gaussian measures since $\mu_0(X) = 1$. By the Lipschitz continuity of $\mathcal{G}$ in X and Theorem 5.4, we deduce that

$$I(u) := \frac{1}{2}\|u\|^2_{H^1_{\mathrm{per}}} + \frac{1}{2\gamma^2}|y - \mathcal{G}(u)|^2$$

attains its infimum at $\overline{u} \in E$. ◇

In summary, the function space Bayesian viewpoint on inverse problems is instructive in developing an understanding of variational methods. In particular it implicitly guides choice of the regularization that will lead to a well-posed minimization problem.

5.4. *Filtering*

There are two key ideas underlying filtering: the first is to build up knowledge about the posterior sequentially, and hence perhaps more efficiently; the second is to break up the unknown u and build up knowledge about its constituent parts sequentially, hence reducing the computational dimension of each sampling problem. Thus the first idea relies on decomposing the *data* sequentially, whilst the second relies on decomposing the *unknown* sequentially.

The first basic idea is to build up information about μ^y sequentially as the size of the data set increases. For simplicity assume that the data is finite-dimensional and can be written as $y = \{y_j\}_{j=1}^J$. Assume also that each data point y_j is found from a mapping $\mathcal{G}_j : X \to \mathbb{R}^\ell$ and subject to independent Gaussian observational noises $\eta_j \sim \mathcal{N}(0, \Gamma_j)$ so that

$$y_j = \mathcal{G}_j(u) + \eta_j. \tag{5.10}$$

Thus the data is in $\mathbb{R}^q$ for $q = \ell\mathrm{J}$. The posterior measure has the form

$$\frac{\mathrm{d}\mu^y}{\mathrm{d}\mu_0}(u) \propto \exp\biggl(-\frac{1}{2}\sum_{j=1}^{\mathrm{J}} |y_j - \mathcal{G}_j(u)|^2_{\Gamma_j}\biggr). \tag{5.11}$$

Now let μ_i^y denote the posterior distribution given only the data $y = \{y_j\}_{j=1}^i$. Then

$$\frac{\mathrm{d}\mu_i^y}{\mathrm{d}\mu_0}(u) \propto \exp\biggl(-\frac{1}{2}\sum_{j=1}^{i} |y_j - \mathcal{G}_j(u)|^2_{\Gamma_j}\biggr). \tag{5.12}$$

Furthermore, setting $\mu_0^y = \mu_0$, we have

$$\frac{\mathrm{d}\mu_{i+1}^y}{\mathrm{d}\mu_i^y}(u) \propto \exp\biggl(-\frac{1}{2}|y_{i+1} - \mathcal{G}_{i+1}(u)|^2_{\Gamma_j}\biggr). \tag{5.13}$$

Compare formulae (5.11) and (5.13). When J is large, it is intuitive that μ_{i+1}^y is closer to μ_i^y than $\mu^y = \mu_{\mathrm{J}}^y$ is to μ_0. This suggests that formula (5.13) may be used as the basis for obtaining μ_{i+1}^y from μ_i^y, and thereby to approach $\mu^y = \mu_{\mathrm{J}}^y$ by iterating this over i. In summary, the first key idea enables us to build up our approximation to μ^y incrementally over an ordered set of data.

The second key idea involves additional structure. Imagine that we have $y_j = y(t_j)$ for some set of times

$$0 \leqslant t_1 < t_2 < \cdots < t_{\mathrm{J}} < \infty.$$

Assume furthermore that u is also time-dependent and can be decomposed as $u = \{u_j\}_{j=1}^{\mathrm{J}}$, where $u_j = u(t_j)$, and that (5.10) simplifies to

$$y_j = \mathcal{G}_j(u_j) + \eta_j. \tag{5.14}$$

Then it is reasonable to seek to find the conditional measures

$$\nu_{i|1:i}(\mathrm{d}u_i) := \mathbb{P}(\mathrm{d}u_i | \{y_j\}_{j=1}^i). \tag{5.15}$$

Notice that each of these measures lives on a smaller space than does μ^y and this dimension reduction is an important feature of the methodology. Assuming that the sequence $u = \{u_j\}_{j=1}^{\mathrm{J}}$ is governed by a Markovian evolution, the measure (5.15) uniquely determines the measure

$$\nu_{i+1|1:i}(\mathrm{d}u_{i+1}) := \mathbb{P}(\mathrm{d}u_{i+1} | \{y_j\}_{j=1}^i).$$

Incorporating the $(i+1)$st data point, we find that

$$\frac{\mathrm{d}\nu_{i+1|1:i+1}}{\mathrm{d}\nu_{i+1|1:i}}(u_{i+1}) \propto \exp\biggl(-\frac{1}{2}|y_{i+1} - \mathcal{G}_{i+1}(u_{i+1})|^2_{\Gamma_j}\biggr). \tag{5.16}$$

Thus we have a way of building the measures given by (5.15) incrementally in i.

Clearly, by definition, $\nu_{J|1:J}(du_J)$ agrees with the marginal distribution of $\mu^y(du)$ on the coordinate $u_J = u(t_J)$; however, the distribution of $\nu_{i|1:i}(du_i)$ for $i < J$ does not agree with the marginal distribution of $\mu^y(du)$ on coordinate $u_i = u(t_i)$. Thus the algorithm is potentially very powerful at updating the current state of the system given data up to that time; but it fails to update previous states of the system, given data that subsequently becomes available. We discuss the implications of this in Section 5.5.

5.5. *Discussion and bibliography*

We outline the methods described in this section, highlight some relevant related literature, and discuss inter-relations between the methodologies. A number of aspects concerning computational methods for inverse problems, both classical and statistical, are reviewed in Vogel (2002). An important conceptual algorithmic distinction to make in time-dependent data assimilation problems is between *forecasting* methods, which are typically used online to make predictions as data is acquired sequentially, and *hindcasting* methods which are used offline to obtain improved understanding (this is also called *reanalysis*) and, for example, may be used for the purposes of parameter estimation to obtain improved models. MCMC methods are natural for hindcasting and reanalysis; filtering is natural in the forecasting context. Filtering methods update the estimate of the state based only on data from the past, whereas the full posterior measure estimates the state at any given time based on both past and future observations; methods based on this full posterior measure are known as *smoothing methods* and include MCMC methods based on the posterior and variational methods which maximize the posterior probability.

The development of MCMC methods was initiated with the paper by Metropolis, Rosenbluth, Teller and Teller (1953), in which a symmetric random walk proposal was used to determine thermodynamic properties, such as the equation of state, from a microscopic statistical model. Hastings (1970) demonstrated that the idea could be generalized to quite general families of proposals, providing the seed for the study of these methods in the statistics community (Gelfand and Smith 1990, Smith and Roberts 1993, Bernardo and Smith 1994). The paper of Tierney (1998) provides the infinite-dimensional framework for MH methods that we outline here; in particular, Theorem 5.1 follows from the work in that paper. Ergodic theorems, such as the convergence of time averages as in (5.6), can in many cases be proved for much wider classes of functions than continuous bounded functions. The general methodology is described in Meyn and Tweedie (1993) and an application to MH methods is given in Roberts and Tweedie (1996).

The degeneration of many MH methods on state spaces of finite but growing dimension is a well-known phenomenon to many practitioners. An analysis and quantification of this effect was first undertaken in Roberts,

Gelman and Gilks (1997), where random walk proposals were studied for an i.i.d. target, and subsequently in Roberts and Rosenthal (1998, 2001), Beskos and Stuart (2009) and Beskos, Roberts and Stuart (2009) for other target distributions and proposals; see Beskos and Stuart (2010) for an overview. The idea of using proposals designed to work in the infinite-dimensional context to overcome this degeneration is developed in Stuart, Voss and Wiberg (2004) and Beskos, Roberts, Stuart and Voss (2008) in the context of sampling conditioned diffusions, and is described more generally in Beskos and Stuart (2009), Beskos *et al.* (2009), Beskos and Stuart (2010) and Cotter, Dashti, Robinson and Stuart (2010*b*).

The use of MCMC methods for sampling the posterior distribution arising in the Bayesian approach to inverse problems is highlighted in Kaipio and Somersalo (2000, 2005), Calvetti and Somersalo (2006) and Calvetti, Kuceyeski and Somersalo (2008). When sampling complex high-dimensional posterior distributions, such as those that arise from finite-dimensional approximation of measures μ^y given by (2.24), can be extremely computationally challenging. It is, however, starting to become feasible; recent examples of work in this direction include Calvetti and Somersalo (2006), Dostert *et al.* (2006), Kaipio and Somersalo (2000), Heino, Tunyan, Calvetti and Somersalo (2007) and Calvetti, Hakula, Pursiainen and Somersalo (2009). In Cotter *et al.* (2010*b*) inverse problems such as those in Section 3.6 are studied by means of the MH technology stemming from the proposal (5.8). Examples of application of MCMC techniques to the statistical solution of inverse problems arising in oceanography, hydrology and geophysics may be found in Efendiev *et al.* (2009), Cui, Fox, Nicholls and O'Sullivan (2010), McKeague, Nicholls, Speer and Herbei (2005), Herbei, McKeague and Speer (2008), McLaughlin and Townley (1996), Michalak and Kitanidis (2003) and Mosegaard and Tarantola (1995). The paper by Herbei and McKeague (2009) studies the geometric ergodicity properties of the resulting Markov chains, employing the framework developed in Meyn and Tweedie (1993).

The idea of using proposals more general than (5.7), and in particular proposals that use derivative information concerning Φ, is studied in Roberts and Tweedie (1996). A key concept here is the *Langevin equation*: a stochastic differential equation for which μ is an invariant measure. Discretizing this equation, which involves the derivative of Φ, is the basis for good proposals. This is related to the fact that, for small discretization parameter, the proposals nearly inherit this invariance under μ. Applying this idea in the infinite-dimensional context is described in Apte, Hairer, Stuart and Voss (2007) and Beskos and Stuart (2009), based on the idea of Langevin equations in infinite dimensions (Hairer *et al.* 2005, Hairer *et al.* 2007, Hairer, Stuart and Voss 2009).

Characterizing the centres of small balls with maximum probability has been an object of interest in the theory of stochastic differential equations for some time. See Ikeda and Watanabe (1989) and Dürr and Bach (1978) for the simplest setting, and Zeitouni and Dembo (1987) for a generalization to signal processing problems. Our main Theorem 5.4 concerning the existence of probability maximizers provides a nice link between Bayesian inverse problems and optimal control. The key ingredients are continuity of the forward mapping from the unknown function to the data, in the absence of observational noise, in a space X, and choice of a prior measure which has the properties that draws from it are almost surely in X: $\mu_0(X) = 1$; this then guarantees that the Tikhonov regularization, which is in the Cameron–Martin norm for the prior measure, is sufficient to prove existence of a minimizer for the variational method.

The idea concluding the proof of the first part of Theorem 5.4 is standard in the theory of calculus of variations: see Dacarogna (1989, Chapter 3, Theorem 1.1). The strong convergence argument generalizes an argument from Kinderlehrer and Stampacchia (1980, Theorem II.2.1). The PhD thesis of Nodet (2005) contains a specific instance of Theorem 5.4, for a model of Lagrangian data assimilation in oceanography, and motivated the approach that we take here; related work is undertaken in White (1993) for Burgers' equation. An alternative approach to the existence of minimizers is to study the Euler–Lagrange equations. The paper of Hagelberg *et al.* (1996) studies existence by this approach for a minimization problem closely related to the MAP estimator. The paper studies the equations of fluid mechanics, formulated in terms vorticity–streamfunction variables. Their approach has the disadvantage of requiring a derivative to define the Euler–Lagrange equations, a short time interval to obtain existence of a solution, and also requires further second-derivative information to distinguish between minimizers and saddle points. However, it does form the basis of a numerical approach to find the MAP estimator. For linear differential equations subject to Gaussian noise there is a beautiful explicit construction of the MAP estimator, using the Euler–Lagrange equations, known as the *representer method*. This method is described in Bennett (2002).

Variational methods in image processing are reviewed in Scherzer *et al.* (2009) and the Bayesian approach to this field is exemplified by Calvetti and Somersalo (2005*b*, 2007*a*, 2008) and, implicitly, in Ellerbroek and Vogel (2009). Variational methods are known in the atmospheric and oceanographic literature as *4DVAR* methods (Derber 1989, Courtier and Talagrand 1987, Talagrand and Courtier 1987, Courtier 1997) and, as we have shown, they are linked to probability maximizers. In the presence of model error the method is known as *weak constraint 4DVAR* (Zupanski 1997). There are also variational methods for sequential problems which update the probability maximizer at a sequence of times; this methodology is known as *3DVAR* (Courtier *et al.* 1998) and is closely related to filtering. Indeed, although filtering and variational methods may be viewed as competing methodologies, they are, in fact, not distinct methodologies, and hybrid methods are sought which combine the advantages of both; see Kalnay, Li, Miyoshi, Yang and Ballabrera-Poy (2007), for example.

Although we strongly advocate the function space viewpoint on variational methods, a great deal of work is carried out by first discretizing the problem and then defining the variational problem. Some representative papers which take this approach for large-scale applications arising in fluid mechanics include Bennett and Miller (1990), Bennett and Chua (1994), Eknes and Evensen (1997), Chua and Bennett (2001), Yu and O'Brien (1991), Watkinson, Lawless, Nichols and Roulstone (2007), Gratton, Lawless and Nichols (2007), Johnson, Hoskins, Nichols and Ballard (2006), Lawless and Nichols (2006), Johnson, Hoskins and Nichols (2005), Lawless, Gratton and Nichols (2005*b*, 2005*a*), Stanton, Lawless, Nichols and Roulstone (2005) and Wlasak and Nichols (1998). The paper of Griffith and Nichols (1998) contains an overview of adjoint methods, used in the solution of data assimilation problems with model error, primarily in the context of variational methods. A discussion of variational methods for the Lorenz equations, and references to the extensive literature in this area, may be found in Evensen (2006).

The regularized nonlinear least-squares or Tikhonov approach to inverse problems is widely studied, including in the infinite-dimensional setting of Hilbert spaces – see the book by Engl *et al.* (1996) and the references therein – and Banach spaces – see the papers by Kaltenbacher *et al.* (2009), Neubauer (2009) and Hein (2009) and the references therein. Although we have concentrated on Bayesian priors, and hence on regularization via addition of a quadratic penalization term, there is active research in the use of different regularizations (Kaltenbacher *et al.* 2009, Neubauer 2009, Hein 2009, Lassas and Siltanen 2004). In particular, the use of total variation-based regularization, and related wavelet-based regularizations, is central in image processing (Rudin *et al.* 1992).

Solving the very high-dimensional optimization problems which arise from discretizing the minimization problem (5.3) is extremely challenging and, as with filtering methods, ideas from model reduction (Antoulas, Soresen and Gugerrin 2001) are frequently used to obtain faster algorithms. Some applications of model reduction techniques, mainly to data assimilation problems arising in fluid mechanics, may be found in Lawless, Nichols, Boess and Bunse-Gerstner (2008*a*, 2008*b*), Griffith and Nichols (1998, 2000), Akella and Navon (2009), Fang *et al.* (2009*a*, 2009*b*) and the references therein. Another approach to dealing with the high-dimensional problems that arise in data assimilation is to use ideas from machine learning (Mitchell *et al.* 1990) to try to find good quality low-dimensional approximations to the posterior measure; see, for example, Shen *et al.* (2008*b*), Shen, Cornford, Archambeau and Opper (2010), Vrettas, Cornford and Shen (2009), Shen,

Archambeau, Cornford and Opper (2008*a*), Archambeau, Opper, Shen, Cornford and Shawe-Taylor (2008) and Archambeau, Cornford, Opper and Shawe-Taylor (2007).

There are some applications where the objective functional may not be differentiable. This can arise for two primary reasons. Firstly the PDE model itself may have discontinuous features arising from switches, or shock-like solutions; and secondly the method of observing the PDE may have switches at certain threshold values of the physical parameters. In this case it is of interest to find computational algorithms to identify MAP estimators which do not require derivatives of the objective functional; see Zupanski, Navon and Zupanski (2008).

An overview of the algorithmic aspects of particle filtering, for non-Gaussian problems, is contained in the edited volume by Doucet and Gordon (2001) and a more mathematical treatment of the subject may be found in Bain and Crisan (2009). An introduction to filtering in continuous time, and a derivation of the Kalman–Bucy filter in particular, which exploits the Gaussian structure of linear problems with additive Gaussian noise, is undertaken in Oksendal (2003). It should be emphasized that these methods are all developed primarily in the context of low-dimensional problems. In practice filtering in high-dimensional systems is extremely hard. This is because the iterative formulae (5.13) and (5.16) do not express the density of the target measure with respect to an easily understood Gaussian measure, as happens in (2.24). To overcome this issue, particle approximations of the reference measures are used, corresponding to approximation by Dirac masses; thus the algorithms build up sequential approximations based on Dirac masses. In high dimensions this can be extremely computationally demanding and various forms of approximation are employed to deal with the curse of dimensionality. See Bengtsson, Bickel and Li (2008) and Bickel, Li and Bengtsson (2008) for discussion of the fundamental difficulties arising in high-dimensional filtering, and Snyder, Bengtsson, Bickel and Anderson (2008) for a development of these ideas in the context of applications. A review of some recent mathematical developments in the subject of high-dimensional filtering, especially in the context of the modelling or turbulent atmospheric flows, may be found in Majda, Harlim and Gershgorin (2010). A review of filtering from the perspective of geophysical applications, may be found in Van Leeuwen (2009). A widely used approach is that based on the ensemble Kalman filter (Burgers, Van Leeuwen and Evensen 1998, Evensen and Van Leeuwen 2000, Evensen 2006), which uses an ensemble of particles to propagate the dynamics, but incorporates data using a Gaussian approximation which is hard to justify in general; see also Berliner (2001) and Ott *et al.* (2004). Further approaches based on the use of ensembles to approximate error covariance propagation may be found in Chorin and Krause (2004) and Livings, Dance and Nichols (2008). The paper of Bengtsson, Snyder and Nychka (2003) describes a generalization of the ensemble Kalman filter, based on mixtures of Gaussians, motivated by the high-dimensional systems arising in fluid dynamics data assimilation problems. The paper of Bennett and Budgell (1987) studies the use of filtering techniques in high dimensions, motivated by oceanographic data assimilation, and contains a study of the question of how to define families of finite-dimensional filters which converge to a function-space-valued limit as the finite-dimensional computation is refined; it is thus related to the concept of discretization invariance referred to in Section 2.5. However, the methodology for proving limiting behaviour in Bennett and Budgell (1987), based on Fourier analysis, is useful only for linear Gaussian problems; in contrast, the approach developed here, namely formulation of the inverse problem on function space, gives rise to algorithms which are robust under discretization even in the non-Gaussian case.

In Apte *et al.* (2007) and Apte, Jones and Stuart (2008*a*), studies of the ideal solution obtained from applying MCMC methods to the posterior (2.24) are compared with ensemble Kalman filter methods. The context is a Lagrangian data assimilation problem driven by a low-dimensional truncation of the linearized shallow water equations (3.27) and the results demonstrate pitfalls in the ensemble Kalman filter approach. An unambiguous and mathematically well-defined definition of the *ideal solution*, as given by (2.24), plays an important role in underpinning such computational studies.

A study of particle filters for Lagrangian data assimilation is undertaken in Spiller, Budhiraja, Ide and Jones (2008), and another application of filtering to oceanographic problems can be found in Brasseur *et al.* (2005). Recent contributions to the study of filtering in the context of the high-dimensional systems of interest in geophysical applications include Bergemann and Reich (2010), Cui *et al.* (2010), Chorin and Krause (2004), Chorin and Tu (2009, 2010), Majda and Grote (2007), Majda and Gershgorin (2008), Majda and Harlim (2010) and Van Leeuwen (2001, 2003). A comparison of various filtering methods, for the Kuramoto–Sivashinsky equation, may be found in Jardak, Navon and Zupanski (2010). In the paper of Pikkarainen (2006), filtering is studied in the case where the the state space for the dynamical variable is infinite-dimensional, and modelled by an SPDE. An attempt is made to keep track of the error made when approximating the infinite-dimensional system by a finite-dimensional one. In this regard, a useful approximation is introduced in Huttunen and Pikkarainen (2007), building on ideas in Kaipio and Somersalo (2007*a*). Parameter estimation in the context of filtering can be problematic, and smoothing should ideally be used when parameters are also to be estimated. However, there is some activity to try and make parameter estimation feasible in online scenarios; see Hurzeler and Künsch (2001) for a general discussion and Vossepoel and Van Leeuwen (2007) for an application.

We conclude this bibliography by highlighting an important question confronting many applied disciplines for which data assimilation is important. It is typically the case that models in fields such as climate prediction, oceanography, oil reservoir simulation and weather prediction are not fully resolved and various subgrid-scale models are used to compensate for this fact. This then raises the question: 'Should future increased computer resources be invested in further model resolution, or in more detailed study of uncertainty?' In the language of this section a stark version of this question is as follows: 'Should we employ only variational methods which identify probability maximizers, but do not quantify risk, investing future computer power in resolving the function space limit more fully? Or should we use MCMC methods, which quantify risk and uncertainty very precisely, but whose implementation is very costly and will preclude further model resolution?' This is a hard question. An excellent discussion in the context of climate models may be found in Palmer *et al.* (2009).

6. Probability

6.1. Overview

This section contains an overview of the probabilistic ideas used throughout the article. The presentation is necessarily terse and the reader is referred to the bibliography subsection at the end for references to material containing the complete details. Section 6.2 describes a number of basic definitions from the theory of probability that we will use throughout the article. In Section 6.3 we introduce Gaussian measures on Banach spaces and describe the central ideas of the Cameron–Martin space and the Fernique Theorem. Section 6.4 describes some explicit calculations concerning Gaussian measures on Hilbert space. In particular, we discuss the Karhunen–Loève expansion and conditioned Gaussian measures. The Karhunen–Loève expansion is a basic tool for constructing random draws from a Gaussian measure on Hilbert space, and for analysing the regularity properties of such random draws. Conditioned measures are key to the Bayesian approach to inverse problems and the Gaussian setting provides useful examples which help to build intuition. In Section 6.5 we introduce random fields and, in the Gaussian case, show how these may be viewed as Gaussian measures on vector fields. The key idea that we use from this subsection is to relate the properties of the covariance operator to sample function regularity. In Section 6.6 we describe Bayesian probability and a version of Bayes' theorem appropriate for function space. This will underpin the approach to inverse problems that we take in this article. We conclude, in Section 6.7, with a discussion of metrics on probability measures, and describe properties of the Hellinger metric in particular. This will enable us to measure distance between pairs of probability measures, and is a key ingredient in the definition of well-posed posterior measures described in this article.

In this section, and indeed throughout the article, we will use the following notational conventions. The measure μ_0 will denote a prior measure, and π_0 its density with respect to Lebesgue measure when the state space is $\mathbb{R}^n$. Likewise the measure μ^y will denote a posterior measure, given data y, and π^y its density with respect to Lebesgue measure when the state space is $\mathbb{R}^n$; occasionally we will drop the y dependence and write μ and π. Given a density $\rho(u, y)$ on a pair of jointly distributed random variables, we will write $\rho(u|y)$ (resp. $\rho(y|u)$) for the density of the random variable u (resp. y), given a single observation of y (resp. u). We also write $\rho(u)$ for the marginal density found by integrating out y, and similarly $\rho(y)$ for the marginal density found by integrating out u. We will use similar conventions for other densities, and the densities arising from conditioning and marginalization.

6.2. Basic concepts

A measure (resp. probability) space is a triplet $(\Omega, \mathcal{F}, \mu)$, where Ω is the sample space, $\mathcal{F}$ the σ-algebra of events and μ the measure (resp. probability measure). In this article we will primarily be concerned with situations in which Ω is a separable Banach space $(X, \|\cdot\|_X)$ and $\mathcal{F}$ is the Borel σ-algebra $\mathcal{B}(X)$ generated by the open sets, in the strong topology. We are interested in *Radon measures* on X which are characterized by the property

$$\mu(A) = \sup\{\mu(B) \mid B \subset A,\ B \,\text{compact}\}, \quad A \in \mathcal{B}(X).$$

We use $\mathbb{E}$ and $\mathbb{P}$ to denote expectation and probability, respectively, and $\mathbb{E}(\cdot|\cdot)$ and $\mathbb{P}(\cdot|\cdot)$ for conditional expectation and probability; on occasion we will use the notation $\mathbb{E}^\mu$ or $\mathbb{P}^\mu$ if we wish to indicate that the expectation (or probability) in question is with respect to a particular measure μ. We use $\sim$ as shorthand for *is distributed as*; thus $x \sim \mu$ means that x is drawn from a probability measure μ. A real-valued measurable function on the measure space $(\Omega, \mathcal{F}, \mu)$ is one for which the pre-image of every Borel set in $\mathbb{R}$ is in $\mathcal{F}$ (is μ-measurable).

A function $m \in X$ is called *the mean* of μ on Banach space X if, for all $\ell \in X^*$, where X^* denotes the dual space of linear functionals on X,

$$\ell(m) = \int_X \ell(x)\mu(\mathrm{d}x).$$

If $m = 0$ the measure is called *centred*. In the Hilbert space setting we have that, for $x \sim \mu$, $m = \mathbb{E}x$. A linear operator $K : X^* \to X$ is called the covariance operator if, for all $k, \ell \in X^*$,

$$k(K\ell) = \int_X k(x-m)\ell(x-m)\mu(\mathrm{d}x).$$

In the Hilbert space setting where $X = X^*$, the covariance operator is characterized by the identity

$$\langle k, K\ell\rangle = \mathbb{E}\langle k, (x-m)\rangle\langle (x-m), \ell\rangle, \tag{6.1}$$

for $x \sim \mu$ and for all $k, \ell \in X$. Thus

$$K = \mathbb{E}(x-m) \otimes (x-m). \tag{6.2}$$

If μ and ν are two measures on the same measure space, then μ is *absolutely continuous* with respect to ν if $\nu(A) = 0$ implies $\mu(A) = 0$. This is sometimes written $\mu \ll \nu$. The two measures are *equivalent* if $\mu \ll \nu$ and $\nu \ll \mu$. If the measures are supported on disjoint sets then they are *mutually singular* or *singular*.

A family of measures $\mu^{(n)}$ on Banach space X is said to *converge weakly* to measure μ on X if

$$\int_X f(x)\mu^{(n)}(\mathrm{d}x) \to \int_X f(x)\mu(\mathrm{d}x)$$

for all continuous bounded $f : E \to \mathbb{R}$. We write $\mu^{(n)} \Rightarrow \mu$.[4]

The characteristic function of a probability distribution μ on a separable Banach space X is, for $\ell \in X^*$,

$$\varphi_\mu(\ell) = \mathbb{E}\exp(\mathrm{i}\ell(x)).$$

Theorem 6.1. If μ and ν are two Radon measures on a separable Banach space X and if $\varphi_\mu(\ell) = \varphi_\nu(\ell)$ for all $\ell \in X^*$, then $\mu = \nu$. $\diamond$

The following *Radon–Nikodym Theorem* plays an important role in this article.

Theorem 6.2. (Radon–Nikodym Theorem) Let μ and ν be two measures on the same measure space $(\Omega, \mathcal{F})$. If $\mu \ll \nu$ and ν is σ-finite then there exists ν-measurable function $f : \Omega \to [0, \infty]$ such that, for all ν-measurable sets $A \in \mathcal{F}$,

$$\mu(A) = \int_A f(x)\,\mathrm{d}\nu(x). \qquad \diamond$$

The function f is known as the *Radon–Nikodym derivative* of μ with respect to ν. The derivative is written as

$$\frac{\mathrm{d}\mu}{\mathrm{d}\nu}(x) = f(x). \tag{6.3}$$

We will sometimes simply refer to $f = \mathrm{d}\mu/\mathrm{d}\nu$ as the *density* of μ with

[4] This should not be confused with weak convergence of functions.

respect to ν. If μ is also a probability measure then

$$1 = \mu(\Omega) = \int_{\Omega} f(x)\,\mathrm{d}\nu(x).$$

Thus, if ν is a probability measure, $\mathbb{E}^{\nu} f(x) = 1$.

We give an example which illustrates a key idea underlying the material we develop in this section. We work in finite dimensions but highlight what can be transferred to probability measures on a Banach space.

Example 6.3. For a probability measure μ on $\mathbb{R}^d$ which is absolutely continuous with respect to Lebesgue measure λ, we use the shorthand p.d.f. for the probability density function, or *density*, ρ defined so that

$$\mu(A) = \int_A \rho(x)\,\mathrm{d}x \tag{6.4}$$

for $A \in \mathcal{F}$, where $\mathcal{F}$ is the sigma algebra generated by the open sets in $\mathbb{R}^d$. Strictly speaking this is the p.d.f. *with respect to Lebesgue measure*, as we integrate the density against Lebesgue measure to find the probability of a set A. Note that

$$\frac{\mathrm{d}\mu}{\mathrm{d}\lambda}(x) = \rho(x).$$

It is also possible to find the density of μ with respect to a Gaussian measure. To illustrate this, let $\mu_0 = \mathcal{N}(0, I)$ denote a standard unit Gaussian in $\mathbb{R}^d$. Then

$$\mu_0(\mathrm{d}x) = (2\pi)^{-d/2} \exp\left(-\frac{1}{2}|x|^2\right)\mathrm{d}x.$$

Thus the density of μ with respect to μ_0 is

$$\rho_{\mathrm{g}}(x) = (2\pi)^{d/2} \exp\left(\frac{1}{2}|x|^2\right)\rho(x).$$

We then have the identities

$$\mu(A) = \int_A \rho_{\mathrm{g}}(x)\mu_0(\mathrm{d}x) \tag{6.5}$$

and

$$\frac{\mathrm{d}\mu}{\mathrm{d}\mu_0}(x) = \rho_{\mathrm{g}}(x).$$

It turns out that, in the infinite-dimensional setting, the formulation (6.5) generalizes much more readily than does (6.4). This is because infinite-dimensional Gaussian measure is well-defined, and because many measures have a density (Radon–Nikodym derivative) with respect to an infinite-dimensional Gaussian measure. In contrast, infinite-dimensional Lebesgue measure does not exist. $\diamond$

We conclude this subsection with two definitions of operators, both important for definitions associated with Gaussian measures on Hilbert space. Let $\{\phi_k\}_{k=1}^{\infty}$ denote an orthonormal basis for a separable Hilbert space $\mathcal{H}$. A linear operator $A : \mathcal{H} \to \mathcal{H}$ is *trace-class* or *nuclear* if

$$\mathrm{Tr}(A) := \sum_{k=1}^{\infty} \langle A\phi_k, \phi_k \rangle < \infty. \tag{6.6}$$

The sum is independent of the choice of basis. The operator A is *Hilbert–Schmidt* if

$$\sum_{k=1}^{\infty} \|A\phi_k\|^2 < \infty. \tag{6.7}$$

If A is self-adjoint and we choose the $\{\phi_k\}$ to be the eigenfunctions of A, then the sum in (6.6) is simply the sum of the eigenvalues of A. A weaker condition is that the eigenvalues are square-summable, which is (6.7).

6.3. Gaussian measures

We will primarily employ Gaussian measures in the Hilbert space setting. However, they can also be defined on Banach spaces and, on occasion, we will employ this level of generality. Indeed, when studying Gaussian random fields in Section 6.5, we will show that, for a Gaussian measure μ on a Hilbert space $\mathcal{H}$, there is often a Banach space X which is continuously embedded in $\mathcal{H}$ and has the property that $\mu(X) = 1$. We would then like to define the measure μ on the Banach space X. We thus develop Gaussian measure theory on separable Banach spaces here.

Having defined Gaussian measure, we describe its characteristic function and we state the Fernique Theorem, which exploits tail properties of Gaussian measure. We follow this with definition and discussion of the Cameron–Martin space. We then describe the basic tools required to study the absolute continuity of two Gaussian measures.

A measure μ on $(X, \mathcal{B}(X))$ is *Gaussian* if, for any $\ell \in X^*$, $\ell(x) \sim \mathcal{N}(m_\ell, \sigma_\ell^2)$ for some $m_\ell \in \mathbb{R}, \sigma_\ell \in \mathbb{R}$. Note that $\sigma_\ell = 0$ is allowed, so that the induced measure on $\ell(x)$ may be a Dirac mass at m_ℓ. Note also that it is expected that $m_\ell = \ell(m)$, where m is the mean defined above, and $\sigma_\ell^2 = \ell(K\ell)$, where K is the covariance operator. The mean m and covariance operator K are indeed well-defined by this definition of covariance operator.

Theorem 6.4. A Gaussian measure on $(X, \mathcal{B}(X))$ has a mean m and covariance operator K. Further, the characteristic function of the measure is

$$\varphi(\ell) = \exp\left(\mathrm{i}\ell(m) - \frac{1}{2}\ell(K\ell) \right). \qquad \diamond$$

Hence, by Theorem 6.1 we see that the mean and covariance completely characterize the Gaussian measure, and so we are justified in denoting it by $\mathcal{N}(m, K)$. The following lemma demonstrates an important role for characteristic functions in studying weak convergence.

Lemma 6.5. Consider a family of probability measures $\mu^{(n)}$. Assume that, for all $\ell \in X^*$,

$$\varphi_{\mu^{(n)}}(\ell) \to \exp\Bigl(\mathrm{i}\ell(m^+) - \frac{1}{2}\ell(K^+\ell)\Bigr).$$

Then $\mu^{(n)} \Rightarrow \mathcal{N}(m^+, K^+)$. ◇

In the Hilbert space setting we refer to the inverse of the covariance operator $\mathcal{C}$ as the *precision operator* and denote it by $\mathcal{L}$. It is natural to ask what conditions an operator must satisfy in order to be a covariance operator. Good intuition can be obtained by thinking of the precision operator as a (possibly) fractional differential operator of sufficiently high order. To pursue this issue a little further we confine ourselves to the Hilbert space setting. The following theorem provides a precise answer to the question concerning properties of the covariance operator.

Theorem 6.6. If $\mathcal{N}(0, \mathcal{C})$ is a Gaussian measure on a Hilbert space $\mathcal{H}$, then $\mathcal{C}$ is a self-adjoint, positive semi-definite trace-class operator on $\mathcal{H}$. Furthermore, for any integer p, there is a constant $C = C_p \geqslant 0$ such that, for $x \sim \mathcal{N}(0, \mathcal{C})$,

$$\mathbb{E}\|x\|^{2p} \leqslant C_p(\mathrm{Tr}(\mathcal{C}))^p.$$

Conversely, if $m \in \mathcal{H}$, and $\mathcal{C}$ is a self-adjoint, positive semi-definite, trace-class linear operator on a Hilbert space $\mathcal{H}$, then there is a Gaussian measure $\mu = \mathcal{N}(m, \mathcal{C})$ on $\mathcal{H}$. ◇

Example 6.7. Unit Brownian bridge on $J = (0, 1)$ may be viewed as a Gaussian process on $L^2(J)$ with precision operator $\mathcal{L} = -\mathrm{d}^2/\mathrm{d}x^2$ and $D(\mathcal{L}) = H^2(J) \bigcap H_0^1(J)$. Thus the eigenvalues of $\mathcal{C}$ are $\gamma_k = (k^2\pi^2)^{-1}$ and are summable. ◇

If $x \sim \mathcal{N}(0, \mathcal{C})$, then $\mathbb{E}\|x\|^2 = \mathrm{Tr}(\mathcal{C})$. Combining this fact with the previous theorem we have the following generalization of the well-known property concerning the moments of finite-dimensional Gaussian measures.

Corollary 6.8. If $\mathcal{N}(0, \mathcal{C})$ is a Gaussian measure on a Hilbert space $\mathcal{H}$ then, for any positive integer p, there exists $C_p \geqslant 0$ such that $\mathbb{E}\|x\|^{2p} \leqslant C_p(\mathbb{E}\|x\|^2)^p$. ◇

In fact, as in finite dimensions, the exponentials of certain quadratic functionals are bounded for Gaussian measures. This is the Fernique Theorem, which we state in the Banach space context.

Theorem 6.9. (Fernique Theorem) If $\mu = \mathcal{N}(0, K)$ is a Gaussian measure on Banach space X, so that $\mu(X) = 1$, then there exists $\alpha > 0$ such that

$$\int_X \exp\big(\alpha \|x\|_X^2\big)\mu(\mathrm{d}x) < \infty. \qquad \diamond$$

We define the *Cameron–Martin space* E associated with a Gaussian measure $\mu = \mathcal{N}(0, K)$ on Banach space X to be the intersection of all linear spaces of full measure under μ.[5]

Lemma 6.10. Let E be the Cameron–Martin space of Gaussian measure $\mu = \mathcal{N}(0, K)$ on Banach space X. In infinite dimensions it is necessarily the case that $\mu(E) = 0$. Furthermore, E can be endowed with a Hilbert space structure. Indeed, for Gaussian measures $\mathcal{N}(0, \mathcal{C})$ on the Hilbert space $(\mathcal{H}, \langle\cdot,\cdot\rangle)$, the Cameron–Martin space is the Hilbert space $E := \mathrm{Im}(\mathcal{C}^{1/2})$ with inner product

$$\langle\cdot,\cdot\rangle_{\mathcal{C}} = \langle \mathcal{C}^{-1/2}\cdot, \mathcal{C}^{-1/2}\cdot\rangle. \qquad \diamond$$

Note that the covariance operator $\mathcal{C}$ of a Gaussian probability measure on a Hilbert space $\mathcal{H}$ is necessarily compact because $\mathcal{C}$ is trace-class, so that the eigenvalues of $\mathcal{C}^{1/2}$ decay at least algebraically. Thus the Cameron–Martin space $\mathrm{Im}(\mathcal{C}^{1/2})$ is compactly embedded in $\mathcal{H}$. In fact we have the following more general result.

Theorem 6.11. The Cameron–Martin space E associated with a Gaussian measure $\mu = \mathcal{N}(0, K)$ on Banach space X is compactly embedded in all separable Banach spaces X' with full measure $(\mu(X') = 1)$ under μ. $\diamond$

Example 6.12. Consider a probability measure ν on $\mathbb{R}^2$ which is a product measure of the form $\delta_0 \otimes \mathcal{N}(0, 1)$. Introduce coordinates (x_1, x_2) so that the marginal on x_1 is δ_0 and the marginal on x_2 is $\mathcal{N}(0, 1)$. The intersection of all linear spaces with full measure is the subset of $\mathbb{R}^2$ defined by the line

$$E = \{(x_1, x_2) \in \mathbb{R}^2 : x_1 = 0\}.$$

Note, furthermore, that this subset is characterized by the property that the measures $\nu(\cdot)$ and $\nu(a + \cdot)$ are equivalent (as measures) if and only if $a \in E$. Thus, for this example, the Cameron–Martin space defines the space of allowable shifts, under which equivalence of the measures holds. $\diamond$

We now generalize the last observation in the preceding example: we show that the Cameron–Martin space characterizes precisely those shifts in the mean of a Gaussian measure which preserve equivalence.

[5] In most developments of the subject this characterization is given after a more abstract definition of the Cameron–Martin space. However, for our purposes this level of abstraction is not needed.

Theorem 6.13. Two Gaussian measures $\mu_i = \mathcal{N}(m_i, \mathcal{C}_i)$, $i = 1, 2$, on a Hilbert space $\mathcal{H}$ are either singular or equivalent. They are equivalent if and only if the following three conditions hold:

(i) $\mathrm{Im}(\mathcal{C}_1^{1/2}) = \mathrm{Im}(\mathcal{C}_2^{1/2}) := E$,

(ii) $m_1 - m_2 \in E$,

(iii) the operator $T := \big(\mathcal{C}_1^{-1/2}\mathcal{C}_2^{1/2}\big)\big(\mathcal{C}_1^{-1/2}\mathcal{C}_2^{1/2}\big)^* - I$ is Hilbert–Schmidt in $\overline{E}$. ◇

In particular, choosing $\mathcal{C}_1 = \mathcal{C}_2$ we see that shifts in the mean give rise to equivalent Gaussian measures if and only if the shifts lie in the Cameron–Martin space E. It is of interest to characterize the Radon–Nikodym derivative arising from such shifts in the mean.

Theorem 6.14. Consider two measures $\mu_i = \mathcal{N}(m_i, \mathcal{C})$, $i = 1, 2$, on Hilbert space $\mathcal{H}$, where $\mathcal{C}$ has eigenbasis $\{\phi_k, \lambda_k\}_{k=1}^{\infty}$. Denote the Cameron–Martin space by E. If $m_1 - m_2 \in E$, then the Radon–Nikodym derivative is given by

$$\frac{\mathrm{d}\mu_1}{\mathrm{d}\mu_2}(x) = \exp\bigg(\langle m_1 - m_2, x - m_2\rangle_{\mathcal{C}} - \frac{1}{2}\|m_1 - m_2\|_{\mathcal{C}}^2\bigg). \qquad ◇$$

Since $m_1 - m_2 \in \mathrm{Im}(\mathcal{C}^{1/2})$, the quadratic form $\|m_1 - m_2\|_{\mathcal{C}}^2$ is defined; the random variable $x \to \langle m_1 - m_2, x - m_2\rangle_{\mathcal{C}}$ is defined via a limiting procedure as follows. By the the Karhunen–Loève expansion (6.9) below, we have the representation of $x \sim \mathcal{N}(0, \mathcal{C})$ as

$$x - m_2 = \sum_{k=1}^{\infty} \sqrt{\lambda_k}\omega_k\phi_k,$$

where $\omega = \{\omega_k\}_{k=1}^{\infty} \in \Omega$ is an i.i.d. sequence of $\mathcal{N}(0, 1)$ random variables. Then $\langle m_1 - m_2, x - m_2\rangle_{\mathcal{C}}$ is defined as the $L^2(\Omega; \mathcal{H})$ limit in n of the series

$$\sum_{k=1}^{n} \frac{1}{\sqrt{\lambda_k}}\langle m_1 - m_2, \phi_k\rangle\omega_k.$$

In establishing the first of the conditions in Theorem 6.13, the following lemma is often useful.

Lemma 6.15. For any two positive definite, self-adjoint, bounded linear operators $\mathcal{C}_i$ on a Hilbert space $\mathcal{H}$, $i = 1, 2$, the condition $\mathrm{Im}(\mathcal{C}_1^{1/2}) \subset \mathrm{Im}(\mathcal{C}_2^{1/2})$ holds if and only if there exists a constant $K > 0$ such that

$$\langle h, \mathcal{C}_1 h\rangle \leqslant K\langle h, \mathcal{C}_2 h\rangle, \quad \forall h \in \mathcal{H}. \qquad ◇$$

Example 6.16. Consider two Gaussian measures μ_i on $\mathcal{H} = L^2(J), J = (0, 1)$ both with precision operator $\mathcal{L} = -\mathrm{d}^2/\mathrm{d}x^2$ and the domain of $\mathcal{L}$ being $H_0^1(J) \cap H^2(J)$. (Informally $-\mathcal{L}$ is the Laplacian on J with homogeneous

Dirichlet boundary conditions.) The mean of μ_1 is a function $m \in \mathcal{H}$ and the mean of μ_2 is 0. Thus $\mu_1 \sim \mathcal{N}(m, \mathcal{C})$ and $\mu_2 \sim \mathcal{N}(0, \mathcal{C})$, where $\mathcal{C} = \mathcal{L}^{-1}$. Here $\mathcal{C}_1 = \mathcal{C}_2 = \mathcal{C}$ and $T = 0$, so that (i) and (iii) in Theorem 6.13 are satisfied with $E = \mathrm{Im}\,(\mathcal{C}^{1/2}) = H_0^1(J)$. It follows that the measures are equivalent if and only if $m \in E$. If this condition is satisfied then, from Theorem 6.14, the Radon–Nikodym derivative between the two measures is given by

$$\frac{\mathrm{d}\mu_1}{\mathrm{d}\mu_2}(x) = \exp\left(\langle m, x\rangle_{H_0^1} - \frac{1}{2}\|m\|_{H_0^1}^2\right). \qquad \diamond$$

Example 6.17. Consider two mean zero Gaussian measures μ_i on $\mathcal{H} = L^2(J)$, $J = (0,1)$ with norm $\|\cdot\|$ and precision operators $\mathcal{L}_1 = -\mathrm{d}^2/\mathrm{d}x^2 + I$ and $\mathcal{L}_2 = -\mathrm{d}^2/\mathrm{d}x^2$ respectively, both with domain $H_0^1(J) \cap H^2(J)$.

The operators $\mathcal{L}_1, \mathcal{L}_2$ share the same eigenfunctions,

$$\phi_k(x) = \sqrt{2}\sin(k\pi x),$$

and have eigenvalues

$$\lambda_k(1) = \lambda_k(2) + 1, \quad \lambda_k(2) = k^2\pi^2,$$

respectively. Thus $\mu_1 \sim \mathcal{N}(0, \mathcal{C}_1)$ and $\mu_2 \sim \mathcal{N}(0, \mathcal{C}_2)$ where, in the basis of eigenfunctions, $\mathcal{C}_1$ and $\mathcal{C}_2$ are diagonal with eigenvalues

$$\frac{1}{k^2\pi^2 + 1}, \quad \frac{1}{k^2\pi^2},$$

respectively. We have, for $h_k = \langle h, \phi_k\rangle$,

$$\frac{\pi^2}{\pi^2 + 1} \leqslant \frac{\langle h, \mathcal{C}_1 h\rangle}{\langle h, \mathcal{C}_2 h\rangle} = \frac{\sum_{k\in\mathbb{Z}^+}(1 + k^2\pi^2)^{-1}h_k^2}{\sum_{k\in\mathbb{Z}^+}(k\pi)^{-2}h_k^2} \leqslant 1.$$

Thus, by Lemma 6.15, Theorem 6.13(i) is satisfied. Part (ii) holds trivially. Notice that

$$T = \mathcal{C}_1^{-1/2}\mathcal{C}_2\mathcal{C}_1^{-1/2} - I$$

is diagonalized in the same basis as the $\mathcal{C}_i$, and has eigenvalues

$$\frac{1}{k^2\pi^2}.$$

These are square-summable, and so part (iii) of Theorem 6.13 holds and the two measures are absolutely continuous with respect to one another. $\diamond$

A Hilbert space $(X, \langle\cdot,\cdot\rangle_X)$ of functions $f : D \subset \mathbb{R}^d \to \mathbb{R}$ is called a *reproducing kernel Hilbert space*, RKHS for short, if pointwise evaluation is a continuous linear functional in the Hilbert space. If $f(y) = \langle f, r_y\rangle_X$, then r_y is called the *representer* of the RKHS.

Example 6.18. Let $J = (0,1)$. Note that $\mathcal{H} = L^2(J;\mathbb{R})$ is not an RKHS. Consider $X = H^1(J;\mathbb{R})$ equipped with the inner product

$$(a,b) = a(0)b(0) + \int_0^1 a'(x)b'(x)\,\mathrm{d}x. \tag{6.8}$$

If $r_{\mathrm{y}}(x) = 1 + x \wedge y$ then $f(y) = (f, r_{\mathrm{y}})$. Notice that $r_{\mathrm{y}} \in X$. Thus, by the Cauchy–Schwarz inequality,

$$\begin{aligned} |f(y) - g(y)| &\leqslant |(f-g, r_{\mathrm{y}})| \\ &\leqslant \|f-g\|_X \|r_{\mathrm{y}}\|_X, \end{aligned}$$

demonstrating that pointwise evaluation is a continuous linear functional on X. Notice, furthermore, that the expression $f(y) = (f, r_{\mathrm{y}})$ is an explicit statement of the Riesz Representation Theorem. ◇

In the literature there is often an overlap of terminology surrounding the RKHS and the Cameron–Martin space. This is related to the fact that the representer of an RKHS can often be viewed as the covariance function (see Section 6.5 below) of a covariance operator associated to a Gaussian measure on $L^2(D;\mathbb{R})$.

6.4. Explicit calculations with Gaussian measures

In this section we confine our attention to Gaussian measures on Hilbert space. We provide a number of explicit formulae that are helpful throughout the article, and which also help to build intuition about measures on infinite-dimensional spaces.

We can construct random draws from a Gaussian measure on Hilbert space $\mathcal{H}$ as follows, using the *Karhunen–Loève expansion*.

Theorem 6.19. Let $\mathcal{C}$ be a self-adjoint, positive semi-definite, nuclear operator in a Hilbert space $\mathcal{H}$ and let $m \in \mathcal{H}$. Let $\{\phi_k, \gamma_k\}_{k=1}^{\infty}$ be an orthonormal set of eigenvectors/eigenvalues for $\mathcal{C}$ ordered so that

$$\gamma_1 \geqslant \gamma_2 \geqslant \cdots .$$

Take $\{\xi_k\}_{k=1}^{\infty}$ to be an i.i.d. sequence with $\xi_1 \sim \mathcal{N}(0,1)$. Then the random variable $x \in \mathcal{H}$ given by the *Karhunen–Loève expansion*

$$x = m + \sum_{k=1}^{\infty} \sqrt{\gamma_k}\,\xi_k \phi_k \tag{6.9}$$

is distributed according to $\mu = \mathcal{N}(m, \mathcal{C})$. ◇

In applications the eigenvalues and eigenvectors of $\mathcal{C}$ will often be indexed over a different countable set, say $\mathbb{K}$. In this context certain calculations are cleaner if we write the Karhunen–Loève expansion (6.9) in the form

$$x = m + \sum_{k\in\mathbb{K}} \sqrt{\gamma_k}\xi_k\phi_k. \tag{6.10}$$

Here the $\{\xi_k\}_{k\in\mathbb{K}}$ are an i.i.d. set of random variables all distributed as $\mathcal{N}(0,1)$. Of course, the order of summation does, in general, matter; whenever we use (6.10), however, the ordering will not be material to the outcome and will streamline the calculations to use (6.10).

The next theorem concerns conditioning of Gaussian measures.

Theorem 6.20. Let $\mathcal{H} = \mathcal{H}_1 \oplus \mathcal{H}_2$ be a separable Hilbert space with projectors $\Pi_i\colon \mathcal{H} \to \mathcal{H}_i$. Let $(x_1, x_2) \in \mathcal{H}_1 \oplus \mathcal{H}_2$ be an $\mathcal{H}$-valued Gaussian random variable with mean $m = (m_1, m_2)$ and positive definite covariance operator $\mathcal{C}$. Define

$$\mathcal{C}_{ij} = \mathbb{E}(x_i - m_i) \otimes (x_j - m_j).$$

Then the conditional distribution of x_1 given x_2 is Gaussian with mean

$$m' = m_1 + \mathcal{C}_{12}\mathcal{C}_{22}^{-1}(x_2 - m_2) \tag{6.11}$$

and covariance operator

$$\mathcal{C}' = \mathcal{C}_{11} - \mathcal{C}_{12}\mathcal{C}_{22}^{-1}\mathcal{C}_{21}. \tag{6.12}$$

◇

To understand this theorem it is useful to consider the following finite-dimensional result concerning block matrix inversion.

Lemma 6.21. Consider a positive definite matrix C with the block form

$$C = \begin{pmatrix} C_{11} & C_{12} \\ C_{12}^* & C_{22} \end{pmatrix}.$$

Then C_{22} is positive definite symmetric and the *Schur complement* S defined by $S = C_{11} - C_{12}C_{22}^{-1}C_{12}^*$ is positive definite symmetric. Furthermore,

$$C^{-1} = \begin{pmatrix} S^{-1} & -S^{-1}C_{12}C_{22}^{-1} \\ -C_{22}^{-1}C_{12}^*S^{-1} & C_{22}^{-1} + C_{22}^{-1}C_{12}^*S^{-1}C_{12}C_{22}^{-1} \end{pmatrix}.$$

Now let (x, y) be jointly Gaussian with distribution $\mathcal{N}(m, C)$ and $m = (m_1^*, m_2^*)^*$. Then the conditional distribution of x given y is Gaussian with mean m' and covariance matrix C' given by

$$m' = m_1 + C_{12}C_{22}^{-1}(y - m_2),$$

$$C' = C_{11} - C_{12}C_{22}^{-1}C_{12}^*.$$

◇

Example 6.22. Consider a random variable u with Gaussian prior probability distribution $\mathcal{N}(0,1)$, and hence associated p.d.f.

$$\pi_0(u) \propto \exp\left(-\frac{1}{2}u^2\right).$$

Let y be the random variable $y = u + \xi$, where $\xi \sim \mathcal{N}(0,\sigma^2)$ is independent of u. Then the likelihood of y given u has p.d.f. proportional to

$$\exp\left(-\frac{1}{2\sigma^2}|y-u|^2\right).$$

The joint probability of (u,y) thus has p.d.f. proportional to

$$\exp\left(-\frac{1}{2\sigma^2}|y-u|^2 - \frac{1}{2}|u|^2\right).$$

Since

$$\frac{1}{2\sigma^2}|y-u|^2 + \frac{1}{2}|u|^2 = \left(\frac{\sigma^2+1}{2\sigma^2}\right)\left|u - \frac{1}{\sigma^2+1}y\right|^2 + c_y,$$

where c_y is independent of u, we see that the $u|y$ is a Gaussian $\mathcal{N}(m,\gamma^2)$ with

$$m = \frac{1}{\sigma^2+1}y, \quad \gamma^2 = \frac{\sigma^2}{\sigma^2+1}.$$

This technique for deriving the mean and covariance of a Gaussian measure is often termed *completing the square*; it may be rigorously justified by Theorem 6.20 as follows. First we observe that $m_1 = m_2 = 0$, that $\mathcal{C}_{11} = \mathcal{C}_{12} = \mathcal{C}_{21} = 1$ and that $\mathcal{C}_{22} = 1 + \sigma^2$. The formulae (6.11) and (6.12) then give identical results to those found by completing the square. ◇

We now study an infinite-dimensional version of the previous example.

Example 6.23. Consider a random variable u on a Hilbert space $\mathcal{H}$ distributed according to a measure $\mu_0 \sim \mathcal{N}(m_0, \mathcal{C}_0)$. We assume that $m_0 \in \mathrm{Im}(\mathcal{C}_0^{1/2})$. Assume that $y \in \mathbb{R}^m$ is also Gaussian and is given by

$$y = Au + \eta,$$

where $A : X \to \mathbb{R}^m$ is linear and continuous on a Banach space $X \subseteq \mathcal{H}$ with $\mu_0(X) = 1$. The adjoint of A, denoted A^*, is hence the operator from $\mathbb{R}^m \to X^*$ defined by the identity

$$\langle Au, v\rangle = (A^*v)(u),$$

which holds for all $v \in \mathbb{R}^m$, $u \in X$, and where $A^*v \in X^*$ is a linear functional on X. We also assume that $\eta \sim \mathcal{N}(0,\Gamma)$ is independent of u and that Γ is positive definite. Thus $y|u$ is Gaussian with density proportional to $\exp\left(-\frac{1}{2}|y - Au|_\Gamma^2\right)$. We would like to characterize the Gaussian measure μ^y

for $u|y$. Let $\mu^y = \mathcal{N}(m, \mathcal{C})$. To calculate $\mathcal{C}$ and m we first use the idea of completing the square, simply computing formally as if the Hilbert space for u were finite-dimensional and had a density with respect to Lebesgue measure; we will then justify the resulting formulae after the fact by means of Theorem 6.20. The formal Lebesgue density for $u|y$ is proportional to

$$\exp\left(-\frac{1}{2}|y - Au|_\Gamma^2 - \frac{1}{2}\|u - m_0\|_{\mathcal{C}_0}^2\right).$$

But

$$\frac{1}{2}|y - Au|_\Gamma^2 + \frac{1}{2}\|u - m_0\|_{\mathcal{C}_0}^2 = \frac{1}{2}\|u - m\|_{\mathcal{C}}^2 + c_y$$

with c_y independent of u, and hence completing the square gives

$$\mathcal{C}^{-1} = A^*\Gamma^{-1}A + \mathcal{C}_0^{-1}, \tag{6.13a}$$

$$m = \mathcal{C}(A^*\Gamma^{-1}y + \mathcal{C}_0^{-1}m_0). \tag{6.13b}$$

We now justify this informal calculation.

The pair (u, y) is jointly Gaussian with $\mathbb{E}u = m_0$ and $\mathbb{E}y = Am_0$. We define $\overline{u} = u - m_0$ and $\overline{y} = y - Am_0$. Note that $\overline{y} = A\overline{u} + \eta$. The pair (u, y) has covariance operator with components

$$\begin{aligned}
\mathcal{C}_{11} &= \mathbb{E}\overline{u}\overline{u}^* = \mathcal{C}_0, \\
\mathcal{C}_{22} &= \mathbb{E}\overline{y}\overline{y}^* = A\mathcal{C}_0A^* + \Gamma, \\
\mathcal{C}_{21} &= \mathbb{E}\overline{y}\overline{u}^* = A\mathcal{C}_0.
\end{aligned}$$

Thus, by Theorem 6.20, we deduce that the mean m and covariance operator $\mathcal{C}$ for u conditional on y are given, respectively, by

$$m = m_0 + \mathcal{C}_0A^*(\Gamma + A\mathcal{C}_0A^*)^{-1}(y - Am_0) \tag{6.14}$$

and

$$\mathcal{C} = \mathcal{C}_0 - \mathcal{C}_0A^*(\Gamma + A\mathcal{C}_0A^*)^{-1}A\mathcal{C}_0. \tag{6.15}$$

We now demonstrate that the formulae (6.14) and (6.15) agree with (6.13). To check agreement with the formula for the inverse of $\mathcal{C}$ found by completing the square, we show that the product is indeed the identity. Note that

$$\begin{aligned}
&\big(\mathcal{C}_0 - \mathcal{C}_0A^*(\Gamma + A\mathcal{C}_0A^*)^{-1}A\mathcal{C}_0\big)\big(\mathcal{C}_0^{-1} + A^*\Gamma^{-1}A\big) \\
&\quad = \big(I - \mathcal{C}_0A^*(\Gamma + A\mathcal{C}_0A^*)^{-1}A\big)\big(I + \mathcal{C}_0A^*\Gamma^{-1}A\big) \\
&\quad = I + \mathcal{C}_0A^*\Gamma^{-1}A - \mathcal{C}_0A^*(\Gamma + A\mathcal{C}_0A^*)^{-1}(A + A\mathcal{C}_0A^*\Gamma^{-1}A) \\
&\quad = I + \mathcal{C}_0A^*\Gamma^{-1}A - \mathcal{C}_0A^*\Gamma^{-1}A \\
&\quad = I.
\end{aligned}$$

To check agreement with the two formulae for the mean, we proceed as follows. We have

$$\Gamma^{-1} - (\Gamma + A\mathcal{C}_0A^*)^{-1}A\mathcal{C}_0A^*\Gamma^{-1} = (\Gamma + A\mathcal{C}_0A^*)^{-1}. \tag{6.16}$$

The formula for the mean derived by completing the square gives

$$\begin{aligned} m &= \mathcal{C}\big((\mathcal{C}^{-1} - A^*\Gamma^{-1}A)m_0 + A^*\Gamma^{-1}y\big) \\ &= m_0 + \mathcal{C}A^*\Gamma^{-1}(y - Am_0). \end{aligned}$$

To get agreement with the formula (6.14) it suffices to show that

$$\mathcal{C}A^*\Gamma^{-1} = \mathcal{C}_0A^*(\Gamma + A\mathcal{C}_0A^*)^{-1}.$$

By (6.15) and (6.16),

$$\begin{aligned} \mathcal{C}A^*\Gamma^{-1} &= \mathcal{C}_0A^*\Gamma^{-1} - \mathcal{C}_0A^*(\Gamma + A\mathcal{C}_0A^*)^{-1}A\mathcal{C}_0A^*\Gamma^{-1} \\ &= \mathcal{C}_0A^*(\Gamma + A\mathcal{C}_0A^*)^{-1}, \end{aligned}$$

and we are done. ◇

6.5. Gaussian random fields

Our aim in this subsection is to construct, and study the properties of, Gaussian random functions. We first consider the basic construction of random functions, then Gaussian random functions, following this by a study of the regularity properties of Gaussian random functions.

Let $(\Omega, \mathcal{F}, \mathbb{P})$ be a probability space and $D \subseteq \mathbb{R}^d$ an open set. A *random field* on D is a measurable mapping $u : D \times \Omega \to \mathbb{R}^n$. Thus, for any $x \in D$, $u(x;\cdot)$ is an $\mathbb{R}^n$-valued random variable; on the other hand, for any $\omega \in \Omega$, $u(\cdot;\omega) : D \to \mathbb{R}^n$ is a vector field. In the construction of random fields it is commonplace to first construct the *finite-dimensional distributions.* These are found by choosing any integer $q \geqslant 1$, and any set of points $\{x_k\}_{k=1}^q$ in D, and then considering the random vector $(u(x_1;\cdot)^*, \dots, u(x_q;\cdot)^*)^* \in \mathbb{R}^{nq}$. From the finite-dimensional distributions of this collection of random vectors we would like to be able to make sense of the probability measure μ on X, a Banach space, via the formula

$$\mu(A) = \mathbb{P}(u(\cdot;\omega) \in A), \quad A \in \mathcal{B}(X), \tag{6.17}$$

where ω is taken from a common probability space on which the random element $u \in X$ is defined. It is thus necessary to study the joint distribution of a set of q $\mathbb{R}^n$-valued random variables, all on a common probability space. Such $\mathbb{R}^{nq}$-valued random variables are, of course, only defined up to a set of zero measure. It is desirable that all such finite-dimensional distributions are defined on a common subset $\Omega_0 \subset \Omega$ with full measure, so that u may be viewed as a function $u : D \times \Omega_0 \to \mathbb{R}^n$; such a choice of random field is termed a *modification.* In future developments, statements about almost sure (regularity) properties of a random field should be interpreted as statements concerning the existence of a modification possessing the stated almost sure regularity property. We will often simply write $u(x)$, suppressing the explicit dependence on the probability space.

A *Gaussian random field* is one where, for any integer $q \geqslant 1$, and any set of points $\{x_k\}_{k=1}^q$ in D, the random vector $(u(x_1;\cdot)^*, \ldots, u(x_q;\cdot)^*)^* \in \mathbb{R}^{nq}$ is a Gaussian random vector. The *mean function* of a Gaussian random field is $m(x) = \mathbb{E}u(x)$. The *covariance function* is $c(x, y) = \mathbb{E}(u(x) - m(x))(u(y) - m(y))^*$. For Gaussian random fields this function, together with the mean function, completely specify the joint probability distribution for $(u(x_1;\cdot)^*, \ldots, u(x_q)^*)^* \in \mathbb{R}^{nq}$. Furthermore, if we view the Gaussian random field as a Gaussian measure on $L^2(D; \mathbb{R}^n)$, then the covariance operator can be constructed from the covariance function as follows. Without loss of generality we consider the mean zero case; the more general case follows by shift of origin. Since the field has mean zero we have, from (6.1),

$$\begin{aligned}\langle h_1, \mathcal{C}h_2\rangle &= \mathbb{E}\langle h_1, u\rangle\langle u, h_2\rangle \\ &= \mathbb{E}\int_D\int_D h_1(x)^*(u(x)u(y)^*)h_2(y)\,\mathrm{d}y\,\mathrm{d}x \\ &= \mathbb{E}\int_D h_1(x)^*\Big(\int_D (u(x)u(y)^*)h_2(y)\,\mathrm{d}y\Big)\mathrm{d}x \\ &= \int_D h_1(x)^*\Big(\int_D c(x,y)h_2(y)\,\mathrm{d}y\Big)\mathrm{d}x\end{aligned}$$

and we deduce that

$$(\mathcal{C}\phi)(x) = \int_D c(x, y)\phi(y)\,\mathrm{d}y. \tag{6.18}$$

Thus the covariance operator of a Gaussian random field is an integral operator with kernel given by the covariance function.

If we view the Gaussian random field as a measure on the space $X = C(\overline{D}; \mathbb{R}^n)$, then the covariance operator $K : X^* \to X$ may also be written as an integral operator as follows. For simplicity we consider the case $n = 1$. We note that $\ell = \ell_\mu \in X^*$ may be identified with a signed measure μ on D. Then similar arguments to those used in the Hilbert space case show that

$$(K\ell_\mu)(x) = \int_D c(x, y)\mu(\mathrm{d}y). \tag{6.19}$$

This may be extended to the case of random fields taking values in $\mathbb{R}^n$.

A mean zero Gaussian random field is termed *stationary* if $c(x, y) = s(x - y)$ for some matrix-valued function s, so that shifting the field by a fixed random vector does not change the statistics. It is *isotropic* if, in addition, $s(x - y) = \iota(|x - y|)$, for some matrix-valued function ι.

An important general question concerning random fields is to find criteria to establish their regularity, expressed in terms of the covariance function or operator. An important tool in this context is the *Kolmogorov Continuity Theorem*, which follows below. This theorem expresses sample function regularity in terms of the covariance function of the random field. Another key tool in establishing regularity is the Karhunen–Loève expansion (6.10), which expresses a random draw from a Gaussian measure in terms of the eigenfunctions and eigenvalues of the covariance operator and may be used to express sample function regularity in terms of the decay of the eigenvalues of the covariance operator. Both these approaches to sample function regularity, one working from the covariance functions, and one from eigenvalues of the covariance operator, are useful in practice when considering Bayesian inverse problems for functions; this is because prior Gaussian measures may be specified via either the covariance function or the covariance operator.

Theorem 6.24. (Kolmogorov Continuity Theorem) Consider an $\mathbb{R}^n$-valued random field u on a bounded open set $D \subset \mathbb{R}^d$. Assume that there are constants $K, \varepsilon > 0$ and $\delta \geqslant 1$ such that

$$\mathbb{E}|u(x) - u(y)|^\delta \leqslant K|x - y|^{2d+\varepsilon}.$$

Then u is almost surely Hölder-continuous on D with any exponent smaller than $\min\{1, \varepsilon/\delta\}$. ◇

In this article we mainly work with priors specified through the covariance operator on a simple geometry, as this makes the exposition more straightforward. Specifically, we consider covariance operators constructed as fractional powers of operators $\mathcal{A}$ whose leading-order behaviour is like that of the Laplacian on a rectangle. Precisely, we will assume that Assumptions 2.9 hold.

By using the Kolmogorov Continuity Theorem we can now prove the following.

Lemma 6.25. Let $\mathcal{A}$ satisfy Assumptions 2.9(i)–(iv). Consider a Gaussian measure $\mu = \mathcal{N}(0, \mathcal{C})$ with $\mathcal{C} = \mathcal{A}^{-\alpha}$ with $\alpha > d/2$. Then $u \sim \mu$ is almost surely s-Hölder-continuous for any exponent $s < \min\{1, \alpha - d/2\}$.

Proof. The Karhunen–Loève expansion (6.10) shows that

$$u(x) = \sum_{k \in \mathbb{K}} \frac{1}{|\lambda_k|^{\alpha/2}} \xi_k \phi_k(x).$$

Thus, for any $\iota > 0$ and for C a (possibly changing) constant independent of t, x and ξ,

$$\begin{aligned}
\mathbb{E}|u(x+h) - u(x)|^2 &\leqslant C \sum_{k \in \mathbb{K}} \frac{1}{|k|^{2\alpha}} |\phi_k(x+h) - \phi_k(x)|^2 \\
&\leqslant C \sum_{k \in \mathbb{K}} \frac{1}{|k|^{2\alpha}} \min\{|k|^2|h|^2, 1\} \\
&\leqslant C \int_{|k| \geqslant 1} \frac{1}{|k|^{2\alpha}} \min\{|k|^2|h|^2, 1\} \, \mathrm{d}k
\end{aligned}$$

$$\begin{aligned}
&\leqslant C\int_{1\leqslant|k|\leqslant|h|^{-\iota}} |k|^{2(1-\alpha)}|h|^2\,\mathrm{d}k + C\int_{|k|\geqslant|h|^{-\iota}} |k|^{-2\alpha}\,\mathrm{d}k\\
&\leqslant C|h|^2\int_1^{|h|^{-\iota}} r^{d-1+2(1-\alpha)}\,\mathrm{d}r + C\int_{|h|^{-\iota}}^{\infty} r^{d-1-2\alpha}\,\mathrm{d}r\\
&= C\big(|h|^{2-\iota(d+2(1-\alpha))} + |h|^{-\iota(d-2\alpha)}\big).
\end{aligned}$$

Making the optimal choice $\iota = 1$ gives

$$\mathbb{E}|u(x+h)-u(x)|^2 \leqslant C|h|^{2\alpha-d}.$$

Thus, by Corollary 6.8 with $\mathcal{H} = \mathbb{R}^n$,

$$\mathbb{E}|u(x)-u(y)|^{2p} \leqslant C|x-y|^{(2\alpha-d)p}$$

for any $p \in \mathbb{N}$. Choosing the exponents $\delta = 2p$ and $\varepsilon = (2\alpha-d)p - 2d$ and letting $p \to \infty$, we deduce from Theorem 6.24 that the function is s-Hölder with any exponent s as specified. □

Example 6.26. Assume that a Gaussian random field with measure μ has the property that, for $X = C(\overline{D};\mathbb{R}^n)$, $\mu(X) = 1$. Then the Cameron–Martin space for this measure, denoted by $(E, \langle\cdot,\cdot\rangle_E)$, is compactly embedded in X, by Theorem 6.11, and hence there is a constant $C > 0$ such that

$$\|\cdot\|_X \leqslant C\|\cdot\|_E.$$

Thus pointwise evaluation is a continuous linear functional on the Cameron–Martin space so that this space may be viewed as an RKHS.

As an example consider the Gaussian measure $\mathcal{N}(0, \beta\mathcal{A}^{-\alpha})$ on $\mathcal{H}$, with $\mathcal{A}$ satisfying Assumptions 2.9(i)–(iv). Then $\mu(X) = 1$ for $\alpha > d/2$ by Lemma 6.25. The Cameron–Martin space is just $\mathcal{H}^\alpha$. This shows that the space $\mathcal{H}^\alpha$ is compactly embedded in the space of continuous functions, for $\alpha > d/2$. (Of course, a related fact follows more directly from the Sobolev Embedding Theorem, Theorem 2.10.) ◇

We now turn to Sobolev regularity, again using the Karhunen–Loève expansion. Recall the Sobolev-like spaces (2.29) defining $\mathcal{H}^s = D(\mathcal{A}^{s/2})$.

Lemma 6.27. Consider a Gaussian measure $\mu = \mathcal{N}(0, \mathcal{A}^{-\alpha})$, where $\mathcal{A}$ satisfies Assumptions 2.9(i)–(iii) and $\alpha > d/2$. Then $u \sim \mu$ is in $\mathcal{H}^s$ almost surely for any $s \in [0, \alpha - d/2)$.

Proof. The Karhunen–Loève expansion (6.10) shows that

$$u = \sum_{k\in\mathbb{K}} \sqrt{\gamma_k}\,\xi_k\phi_k,$$

with $\{\xi_k\}$ an i.i.d. $\mathcal{N}(0,1)$-sequence and $\gamma_k = \lambda_k^{-\alpha}$. Thus

$$\mathbb{E}\|u\|_s^2 = \sum_{k\in\mathbb{K}} \gamma_k\lambda_k^s.$$

If the sum is finite then $\mathbb{E}\|u\|_s^2 < \infty$ and $u \in \mathcal{H}^s$ μ-a.s. We have

$$\sum_{k\in\mathbb{K}} \gamma_k \lambda_k^s = \sum_{k\in\mathbb{K}} \lambda_k^{s-\alpha}.$$

Since the eigenvalues λ_k of $\mathcal{A}$ grow like $|k|^2$, we deduce that this sum is finite if and only if $\alpha > s + d/2$, by comparison with an integral. □

It is interesting that the Hölder and Sobolev exponents predicted by Lemmas 6.25 and Lemma 6.27 agree for $d/2 < \alpha < d/2+1$. The proof of Hölder regularity uses Gaussianity in a fundamental way to obtain this property. In particular, in the proof of Lemma 6.25, we use the fact that the second moment of Gaussians can be used to bound arbitrarily high moments. Note that using the Sobolev Embedding Theorem, together with Lemma 6.27, to determine Hölder properties does not, of course, give results which are as sharp as those obtained from Lemma 6.25. For example, using Lemma 6.27 and Theorem 2.10 shows that choosing $\alpha > d$ ensures that u is almost surely continuous. On the other hand Lemma 6.25 shows that choosing $\alpha = d$ ensures that u is almost surely Hölder-continuous with any exponent less than $d/2$; in particular, u is almost surely continuous.

Example 6.28. Consider the case $d = 2, n = 1$ and $D = [0,1]^2$. Define the Gaussian random field through the measure $\mu = \mathcal{N}(0, (-\triangle)^{-\alpha})$, where $\triangle$ is the Laplacian with domain $H_0^1(D) \cap H^2(D)$. Then Assumptions 2.9 are satisfied by $-\triangle$. By Lemma 6.27 it follows that choosing $\alpha > 1$ suffices to ensure that draws from μ are almost surely in $L^2(D)$. Then, by Lemma 6.25, it follows that, in fact, draws from μ are almost surely in $C(D)$. ◇

In many applications in this article we will be interested in constructing a probability measure μ on a Hilbert space $\mathcal{H}$ which is absolutely continuous with respect to a given reference Gaussian measure μ_0. We can then write, via Theorem 6.2,

$$\frac{\mathrm{d}\mu}{\mathrm{d}\mu_0}(x) \propto \exp\bigl(-\Phi(x)\bigr). \tag{6.20}$$

The Theorem 6.14 provides an explicit example of this structure when μ and μ_0 are both Gaussian. For expression (6.20) to make sense we require that the potential $\Phi : \mathcal{H} \mapsto \mathbb{R}$ is μ_0-measurable. Implicit in the statement of Theorem 6.14 is just such a measurability property of the logarithm of the density between the two Gaussian measures. We return to the structure (6.20) again, in the case where μ is not necessarily Gaussian, in the next subsection.

6.6. Bayesian probability

Bayesian probability forms the underpinnings of the approach to inverse problems taken in this article. In this subsection we first discuss the general concept of conditioned measures. We then turn to Bayesian probability in the finite-dimensional case, and finally generalize Bayes' theorem to the function space setting. The following theorem is of central importance.

Theorem 6.29. Let μ, ν be probability measures on $S \times T$, where $(S, \mathcal{A})$ and $(T, \mathcal{B})$ are measurable spaces. Let (x, y), with $x \in S$ and $y \in T$, denote an element of $S \times T$. Assume that $\mu \ll \nu$ and that μ has Radon–Nikodym derivative ϕ with respect to ν. Assume further that the conditional distribution of $x|y$ under ν, denoted by $\nu^y(\mathrm{d}x)$, exists. Then the conditional distribution of $x|y$ under μ, denoted $\mu^y(\mathrm{d}x)$, exists and $\mu^y \ll \nu^y$. The Radon–Nikodym derivative is given by

$$\frac{\mathrm{d}\mu^y}{\mathrm{d}\nu^y}(x) = \begin{cases} \frac{1}{c(y)}\phi(x, y) & \text{if } c(y) > 0, \text{ and} \\ 1 & \text{else,} \end{cases} \tag{6.21}$$

with $c(y) = \int_S \phi(x, y)\, \mathrm{d}\nu^y(x)$ for all $y \in T$. ◇

Given a probability triplet $(\Omega, \mathcal{F}, \mathbb{P})$ and two sets $A, B \in \mathcal{F}$ with $\mathbb{P}(A) > 0, \mathbb{P}(B) > 0$, we define the probabilities of A given B and B given A by

$$\mathbb{P}(A|B) = \frac{1}{\mathbb{P}(B)}\mathbb{P}(A \cap B),$$
$$\mathbb{P}(B|A) = \frac{1}{\mathbb{P}(A)}\mathbb{P}(A \cap B).$$

Combining gives Bayes' formula:

$$\mathbb{P}(A|B) = \frac{1}{\mathbb{P}(B)}\mathbb{P}(B|A)\mathbb{P}(A). \tag{6.22}$$

If $(u, y) \in \mathbb{R}^d \times \mathbb{R}^\ell$ is a jointly distributed pair of random variables with Lebesgue density $\rho(u, y)$, then the infinitesimal version of the preceding formula tells us that

$$\rho(u|y) \propto \rho(y|u)\rho(u), \tag{6.23}$$

and where the normalization constant depends only on y. Thus

$$\rho(u|y) = \frac{\rho(y|u)\rho(u)}{\int_{\mathbb{R}^d} \rho(y|u)\rho(u)\, \mathrm{d}u}. \tag{6.24}$$

This gives an expression for the probability of a random variable u, given a single observation of a random variable y, which requires knowledge of only the *prior* (unconditioned) probability density $\rho(u)$ and the conditional probability density $\rho(y|u)$ of y given u. Both these expressions are readily available in many modelling scenarios, as we demonstrate in Section 3. This observation is the starting point for the Bayesian approach to probability. Furthermore, there is a wide range of sampling methods which are designed to sample probability measures known only up to a multiplicative constant (see Section 5), and knowledge of the normalization constant is not required in this context: the formula (6.23) may be used directly to implement these algorithms. Recall that in the general Bayesian framework introduced in Section 1 we refer to the observation y as *data* and to $\rho(y|u)$ as the *likelihood* of the data.

Example 6.30. Consider Example 6.22. The random variable (u, y) is distributed according to a measure $\mu_0(u, y)$, which has density with respect to Lebesgue measure given by

$$\pi_0(u,y) = \frac{1}{2\pi\sigma}\exp\left(-\frac{1}{2}u^2 - \frac{1}{2\sigma^2}|y-u|^2\right).$$

By completing the square we showed that the posterior probability measure for u given y is $\mu_0(u|y)$ with density

$$\pi_0(u|y) = \sqrt{\left(\frac{1+\sigma^2}{2\pi\sigma^2}\right)}\exp\left(-\left(\frac{\sigma^2+1}{2\sigma^2}\right)\left|u - \frac{1}{\sigma^2+1}y\right|^2\right).$$

This result also follows from (6.23), which shows that

$$\pi_0(u|y) = \frac{\pi(u,y)}{\int_{\mathbb{R}^d}\pi(u,y)\,\mathrm{d}u}.$$

Now consider a random variable (u, y) distributed according to measure $\mu(u, y)$, which has density $\rho(u, y)$ with respect to $\mu_0(u, y)$. We assume that $\rho > 0$ everywhere on $\mathbb{R}^d \times \mathbb{R}^\ell$. By Theorem 6.29 the random variable found by conditioning u from μ on y has density

$$\rho(u|y) = \frac{\rho(u,y)}{\int_{\mathbb{R}^d}\rho(u,y)\pi_0(u|y)\,\mathrm{d}u}$$

with respect to $\pi_0(u|y)$. ◇

The expression (6.23) may be rewritten to give an expression for the *ratio* of the posterior and prior p.d.f.s:

$$\frac{\rho(u|y)}{\rho(u)} \propto \rho(y|u), \tag{6.25}$$

with constant of proportionality which depends only on y, and not on u. Stated in this way, the formula has a natural generalization to infinite dimensions, as we now explain.

Let u be a random variable distributed according to measure μ_0 on a separable Banach space $(X, \|\cdot\|)$. We assume that the *data* $y \in \mathbb{R}^m$ is given in terms of the *observation operator* $\mathcal{G}$ by the formula $y = \mathcal{G}(u) + \eta$, where $\eta \in \mathbb{R}^m$ is independent of u and has density ρ with respect to Lebesgue measure;

for simplicity we assume that the support of ρ is $\mathbb{R}^m$. Define $\Phi(u;y)$ to be any function which differs from $-\log(\rho(y-\mathcal{G}(u)))$ by an additive function of y only. Hence it follows that

$$\frac{\rho(y-\mathcal{G}(u))}{\rho(y)} \propto \exp\bigl(-\Phi(u;y)\bigr),$$

with constant of proportionality independent of u. Use of Bayes' rule in the form (6.25) suggests that the probability measure for u given y, denoted $\mu^y(\mathrm{d}u)$, has Radon–Nikodym derivative with respect to μ_0 given by

$$\frac{\mathrm{d}\mu^y}{\mathrm{d}\mu_0}(u) \propto \exp\bigl(-\Phi(u;y)\bigr). \tag{6.26}$$

We refer to such an argument as *informal application of Bayes' rule.* We now justify the formula rigorously.

Theorem 6.31. Assume that $\mathcal{G}: X \to \mathbb{R}^m$ is continuous, that ρ has support equal to $\mathbb{R}^m$ and that $\mu_0(X)=1$. Then $u|y$ is distributed according to the measure $\mu^y(\mathrm{d}u)$, which is absolutely continuous with respect to $\mu_0(\mathrm{d}u)$ and has Radon–Nikodym derivative given by (6.26).

Proof. Throughout the proof $C(y)$ denotes a constant depending on y, but not on u, and possibly changing between occurrences. Let $\mathbb{Q}_0(\mathrm{d}y)=\rho(y)\,\mathrm{d}y$ and $\mathbb{Q}(\mathrm{d}y|u)=\rho(y-\mathcal{G}(u))\,\mathrm{d}y$. By construction,

$$\frac{\mathrm{d}\mathbb{Q}}{\mathrm{d}\mathbb{Q}_0}(y|u) = C(y)\exp\bigl(-\Phi(u;y)\bigr),$$

with constant of proportionality independent of u. Now define

$$\begin{aligned}\nu_0(\mathrm{d}y,\mathrm{d}u) &= \mathbb{Q}_0(\mathrm{d}y)\otimes\mu_0(\mathrm{d}u),\\ \nu(\mathrm{d}y,\mathrm{d}u) &= \mathbb{Q}(\mathrm{d}y|u)\mu_0(\mathrm{d}u).\end{aligned}$$

Note that ν_0 is a product measure under which u and y are independent random variables. Since $\mathcal{G}: X\to\mathbb{R}^m$ is continuous we deduce that $\Phi: X\to\mathbb{R}$ is continuous and hence, since $\mu_0(X)=1$, μ_0-measurable. Thus ν is well-defined and is absolutely continuous with respect to ν_0 with Radon–Nikodym derivative

$$\frac{\mathrm{d}\nu}{\mathrm{d}\nu_0}(y,u) = C(y)\exp\bigl(-\Phi(u;y)\bigr);$$

again the constant of proportionality depends only on y. Note that

$$\int_X \exp\bigl(-\Phi(u;y)\bigr)\mu_0(\mathrm{d}u) = C(y)\int_X \rho(y-\mathcal{G}(u))\mu_0(\mathrm{d}u) > 0$$

since $\rho>0$ everywhere on $\mathbb{R}^m$ and since $\mathcal{G}: X\to\mathbb{R}^m$ is continuous. By Theorem 6.29 we have the desired result, since $\nu_0(\mathrm{d}u|y)=\mu_0(\mathrm{d}u)$. □

Remark 6.32. Finally we remark that, if μ^y is absolutely continuous with respect to μ_0 then any property which holds almost surely under μ_0 will also hold almost surely under μ^y. The next example illustrates how useful this fact is. ◇

Example 6.33. Let μ_0 denote the Gaussian random field constructed in Example 6.28, with $\alpha > 1$ so that draws from μ_0 are almost surely continuous. Now imagine that we observe y, the $L^2(D)$-norm of u drawn from μ_0, subject to noise η:

$$y = \|u\|^2 + \eta.$$

We assume that $\eta \sim \mathcal{N}(0, \gamma^2)$, independently of u. The $L^2(D)$-norm is a continuous function on $X = C(D)$ and $\mu_0(X) = 1$; hence evaluation of the $L^2(D)$-norm is μ_0-measurable, and the measure $\mu^y(\mathrm{d}u) = \mathbb{P}(\mathrm{d}u|y)$ is absolutely continuous with respect to μ_0, with Radon–Nikodym derivative given by

$$\frac{\mathrm{d}\mu^y}{\mathrm{d}\mu_0}(u) \propto \exp\left(-\frac{1}{2\gamma^2}\left|y - \|u\|^2\right|^2\right).$$

Note that the probability measure μ^y is not Gaussian. Nonetheless, any function drawn from μ^y is almost surely in $C(D)$. ◇

6.7. Metrics on measures

In Section 4 it will be important to estimate the distance between two probability measures and thus we will be interested in metrics which measure distance between probability measures.

In this section we introduce two useful metrics on measures: the *total variation distance* and the *Hellinger distance.* We discuss the relationships between the metrics and indicate how they may be used to estimate differences between expectations of random variables under two different measures.

Assume that we have two probability measures μ and μ', both absolutely continuous with respect to the same reference measure ν. The following definitions give two concepts of distance between μ and μ'.

Definition 6.34. The *total variation distance* between μ and μ' is

$$d_{\mathrm{TV}}(\mu, \mu') = \frac{1}{2}\int\left|\frac{\mathrm{d}\mu}{\mathrm{d}\nu} - \frac{\mathrm{d}\mu'}{\mathrm{d}\nu}\right|\mathrm{d}\nu.$$ ◇

In particular, if μ' is absolutely continuous with respect to μ, then

$$d_{\mathrm{TV}}(\mu, \mu') = \frac{1}{2}\int\left|1 - \frac{\mathrm{d}\mu'}{\mathrm{d}\mu}\right|\mathrm{d}\mu. \tag{6.27}$$

Definition 6.35. The *Hellinger distance* between μ and μ' is

$$d_{\mathrm{Hell}}(\mu,\mu') = \sqrt{\left(\frac{1}{2}\int\left(\sqrt{\frac{\mathrm{d}\mu}{\mathrm{d}\nu}} - \sqrt{\frac{\mathrm{d}\mu'}{\mathrm{d}\nu}}\right)^2 \mathrm{d}\nu\right)}. \qquad \diamond$$

In particular, if μ' is absolutely continuous with respect to μ, then

$$d_{\mathrm{Hell}}(\mu,\mu') = \sqrt{\left(\frac{1}{2}\int\left(1 - \sqrt{\frac{\mathrm{d}\mu'}{\mathrm{d}\mu}}\right)^2 \mathrm{d}\mu\right)}. \tag{6.28}$$

The total variation distance as defined is invariant under the choice of ν in that it is unchanged if a different reference measure, with respect to which μ and μ' are absolutely continuous, is used. Furthermore, it follows from the definition that

$$0 \leqslant d_{\mathrm{TV}}(\mu,\mu') \leqslant 1.$$

The Hellinger distance is also unchanged if a different reference measure, with respect to which μ and μ' are absolutely continuous, is used. Furthermore, it follows from the definition that

$$0 \leqslant d_{\mathrm{Hell}}(\mu,\mu') \leqslant 1.$$

The Hellinger and total variation distances are related as follows:

Lemma 6.36. Assume that two probability measures μ and μ' are both absolutely continuous with respect to a measure ν. Then

$$\frac{1}{\sqrt{2}} d_{\mathrm{TV}}(\mu,\mu') \leqslant d_{\mathrm{Hell}}(\mu,\mu') \leqslant d_{\mathrm{TV}}(\mu,\mu')^{1/2}. \qquad \diamond$$

The Hellinger distance is particularly useful for estimating the difference between expectation values of functions of random variables under different measures. This idea is encapsulated in the following lemma.

Lemma 6.37. Assume that two probability measures μ and μ' on a Banach space $(X, \|\cdot\|_X)$ are both absolutely continuous with respect to a measure ν. Assume also that $f : X \to E$, where $(E, \|\cdot\|)$ is a Banach space, has second moments with respect to both μ and μ'. Then

$$\|\mathbb{E}^{\mu} f - \mathbb{E}^{\mu'} f\| \leqslant 2\big(\mathbb{E}^{\mu}\|f\|^2 + \mathbb{E}^{\mu'}\|f\|^2\big)^{1/2} d_{\mathrm{Hell}}(\mu,\mu').$$

Furthermore, if $(E, \langle\cdot,\cdot\rangle, \|\cdot\|)$ is a Hilbert space and $f : X \to E$ has fourth moments, then

$$\|\mathbb{E}^{\mu} f\otimes f - \mathbb{E}^{\mu'} f\otimes f\| \leqslant 2\big(\mathbb{E}^{\mu}\|f\|^4 + \mathbb{E}^{\mu'}\|f\|^4\big)^{1/2} d_{\mathrm{Hell}}(\mu,\mu'). \qquad \diamond$$

Remark 6.38. Note, in particular, that choosing $X = E$, and with f chosen to be the identity mapping, we deduce that the differences in mean and covariance operators under two measures are bounded above by the Hellinger distance between the two measures. $\diamond$

6.8. Discussion and bibliography

For a general classical introduction to probability theory see Breiman (1992), and for a modern treatment of the subject see Grimmett and Stirzaker (2001). For a concise, modern (and more advanced) treatment of the subject see Williams (1991). The text by Chorin and Hald (2006) provides an overview of tools from probability and stochastic processes aimed at applied mathematicians.

The discussion of Gaussian measures in a Hilbert space, and proofs of Lemma 6.15 and Theorems 6.6, 6.2, 6.13 and 6.14 may be found in Da Prato and Zabczyk (1992). Theorem 6.14 is also proved in Bogachev (1998). The lecture notes by Hairer (2009) are also a good source, and contain a proof of Theorem 6.1 as well as the Fernique Theorem. Bogachev (1998), Hairer (2009) and Lifshits (1995) all discuss Gaussian measures in the Banach space setting. In particular, Theorem 6.4 is proved in Lifshits (1995), and Hairer (2009) has a nice exposition of the Fernique Theorem.

The Karhunen–Loève expansion is described in Loève (1977, 1978) and a modern treatment of Gaussian random fields is contained in Adler (1990). Recent work exploiting the Karhunen–Loève expansion to approximate solutions of differential equations with random coefficients may be found in Schwab and Todor (2006) and Todor and Schwab (2007).

Theorem 6.29 is proved in Dudley (2002, Section 10.2). For a general discussion of Bayes' rule in finite dimensions see, for example, Bickel and Doksum (2001). The approach to Bayes' rule in infinite dimensions that we adopt in Theorem 6.31 was used to study a specific problem arising in signal processing in Hairer *et al.* (2007). The topic of metrics on probability measures, and further references to the literature, may be found in Gibbs and Su (2002). Note that the choice of normalization constants in the definitions of the total variation and Hellinger metrics differs in the literature.

Acknowledgements

The material in this article is developed in greater detail in the lecture notes of Dashti *et al.* (2010*b*). These notes are freely available for download from: http://www.warwick.ac.uk/~masdr/inverse.html. The author is grateful to his co-authors Masoumeh Dashti and Natesh Pillai for their input into this article. The author also thanks Sergios Agapiou, Andrew Duncan, Stephen Harris, Sebastian Reich and Sebastian Vollmer for numerous comments which improved the presentation, and to Daniella Calvetti and Erkki Somersalo for useful pointers to relevant literature. Finally, the author is grateful to have received financial support from the Engineering and Physical Sciences Research Council (UK), the European Research Council and from the US Office of Naval Research during the writing of this article. This funded research has helped shape much of the material presented.

REFERENCES

R. J. Adler (1990), *An Introduction to Continuity, Extrema, and Related Topics for General Gaussian Processes*, Vol. 12 of *Institute of Mathematical Statistics Lecture Notes: Monograph Series*, Institute of Mathematical Statistics, Hayward, CA.

S. Akella and I. Navon (2009), 'Different approaches to model error formulation in 4D-Var: A study with high resolution advection schemes', *Tellus* **61A**, 112–128.

A. Alekseev and I. Navon (2001), 'The analysis of an ill-posed problem using multiscale resolution and second order adjoint techniques', *Comput. Meth. Appl. Mech. Engrg* **190**, 1937–1953.

A. Antoulas, D. Soresen and S. Gugerrin (2001), *A Survey of Model Reduction Methods for Large Scale Dynamical Systems*, AMS.

A. Apte, M. Hairer, A. Stuart and J. Voss (2007), 'Sampling the posterior: An approach to non-Gaussian data assimilation', *Physica D* **230**, 50–64.

A. Apte, C. Jones and A. Stuart (2008*a*), 'A Bayesian approach to Lagrangian data assimilation', *Tellus* **60**, 336–347.

A. Apte, C. Jones, A. Stuart and J. Voss (2008*b*), 'Data assimilation: Mathematical and statistical perspectives', *Internat. J. Numer. Methods Fluids* **56**, 1033–1046.

C. Archambeau, D. Cornford, M. Opper and J. Shawe-Taylor (2007), Gaussian process approximations of stochastic differential equations. In *JMLR Workshop and Conference Proceedings 1: Gaussian Processes in Practice* (N. Lawrence, ed.), The MIT Press, pp. 1–16.

C. Archambeau, M. Opper, Y. Shen, D. Cornford and J. Shawe-Taylor (2008), Variational inference for diffusion processes. In *Advances in Neural Information Processing Systems 20* (J. Platt, D. Koller, Y. Singer and S. Roweis, eds), The MIT Press, Cambridge, MA, pp. 17–24.

G. Backus (1970*a*), 'Inference from inadequate and inaccurate data I', *Proc. Nat. Acad. Sci.* **65**, 1–7.

G. Backus (1970*b*), 'Inference from inadequate and inaccurate data II', *Proc. Nat. Acad. Sci.* **65**, 281–287.

G. Backus (1970*c*), 'Inference from inadequate and inaccurate data III', *Proc. Nat. Acad. Sci.* **67**, 282–289.

A. Bain and D. Crisan (2009), *Fundamentals of Stochastic Filtering*, Springer.

R. Bannister, D. Katz, M. Cullen, A. Lawless and N. Nichols (2008), 'Modelling of forecast errors in geophysical fluid flows', *Internat. J. Numer. Methods Fluids* **56**, 1147–1153.

J. Beck, B. Blackwell and C. Clair (2005), *Inverse Heat Conduction: Ill-Posed Problems*, Wiley.

M. Bell, M. Martin and N. Nichols (2004), 'Assimilation of data into an ocean model with systematic errors near the equator', *Quart. J. Royal Met. Soc.* **130**, 873–894.

T. Bengtsson, P. Bickel and B. Li (2008), 'Curse of dimensionality revisited: The collapse of importance sampling in very large scale systems', *IMS Collections: Probability and Statistics: Essays in Honor of David Freedman* **2**, 316–334.

T. Bengtsson, C. Snyder and D. Nychka (2003), 'Toward a nonlinear ensemble filter for high-dimensional systems', *J. Geophys. Res.* **108**, 8775.

A. Bennett (2002), *Inverse Modeling of the Ocean and Atmosphere*, Cambridge University Press.

A. Bennett and W. Budgell (1987), 'Ocean data assimilation and the Kalman filter: Spatial regularity', *J. Phys. Oceanography* **17**, 1583–1601.

A. Bennett and B. Chua (1994), 'Open ocean modelling as an inverse problem', *Monthly Weather Review* **122**, 1326–1336.

A. Bennett and R. Miller (1990), 'Weighting initial conditions in variational assimilation schemes', *Monthly Weather Review* **119**, 1098–1102.

K. Bergemann and S. Reich (2010), 'A localization technique for ensemble transform Kalman filters', *Quart. J. Royal Met. Soc.* To appear.

L. Berliner (2001), 'Monte Carlo based ensemble forecasting', *Statist. Comput.* **11**, 269–275.

J. Bernardo and A. Smith (1994), *Bayesian Theory*, Wiley.

A. Beskos and A. Stuart (2009), MCMC methods for sampling function space. In *Invited Lectures: Sixth International Congress on Industrial and Applied Mathematics, ICIAM07* (R. Jeltsch and G. Wanner, eds), European Mathematical Society, pp. 337–364.

A. Beskos and A. M. Stuart (2010), Computational complexity of Metropolis–Hastings methods in high dimensions. In *Monte Carlo and Quasi-Monte Carlo Methods 2008* (P. L'Ecuyer and A. B. Owen, eds), Springer, pp. 61–72.

A. Beskos, G. O. Roberts and A. M. Stuart (2009), 'Optimal scalings for local Metropolis–Hastings chains on non-product targets in high dimensions', *Ann. Appl. Probab.* **19**, 863–898.

A. Beskos, G. O. Roberts, A. M. Stuart and J. Voss (2008), 'MCMC methods for diffusion bridges', *Stochastic Dynamics* **8**, 319–350.

P. Bickel and K. Doksum (2001), *Mathematical Statistics*, Prentice-Hall.

P. Bickel, B. Li and T. Bengtsson (2008), 'Sharp failure rates for the bootstrap particle filter in high dimensions', *IMS Collections: Pushing the Limits of Contemporary Statistics* **3**, 318–329.

V. Bogachev (1998), *Gaussian Measures*, AMS.

P. Bolhuis, D. Chandler, D. Dellago and P. Geissler (2002), 'Transition path sampling: Throwing ropes over rough mountain passes', *Ann. Rev. Phys. Chem.* **53**, 291–318.

L. Borcea (2002), 'Electrical impedence tomography', *Inverse Problems* **18**, R99–R136.

P. Brasseur, P. Bahurel, L. Bertino, F. Birol, J.-M. Brankart, N. Ferry, S. Losa, E. Remy, J. Schroeter, S. Skachko, C.-E. Testut, B. Tranchat, P. Van Leeuwen and J. Verron (2005), 'Data assimilation for marine monitoring and prediction: The Mercator operational assimilation systems and the Mersea developments', *Quart. J. Royal Met. Soc.* **131**, 3561–3582.

L. Breiman (1992), *Probability*, Vol. 7 of *Classics in Applied Mathematics*, SIAM, Philadelphia, PA. Corrected reprint of the 1968 original.

G. Burgers, P. Van Leeuwen and G. Evensen (1998), 'On the analysis scheme in the ensemble Kalman filter', *Monthly Weather Review* **126**, 1719–1724.

D. Calvetti (2007), 'Preconditioned iterative methods for linear discrete ill-posed problems from a Bayesian inversion perspective', *J. Comput. Appl. Math.* **198**, 378–395.

D. Calvetti and E. Somersalo (2005*a*), 'Priorconditioners for linear systems', *Inverse Problems* **21**, 1397–1418.

D. Calvetti and E. Somersalo (2005*b*), 'Statistical elimination of boundary artefacts in image deblurring', *Inverse Problems* **21**, 1697–1714.

D. Calvetti and E. Somersalo (2006), 'Large-scale statistical parameter estimation in complex systems with an application to metabolic models', *Multiscale Modeling and Simulation* **5**, 1333–1366.

D. Calvetti and E. Somersalo (2007*a*), 'Gaussian hypermodel to recover blocky objects', *Inverse Problems* **23**, 733–754.

D. Calvetti and E. Somersalo (2007*b*), *Introduction to Bayesian Scientific Computing*, Vol. 2 of *Surveys and Tutorials in the Applied Mathematical Sciences*, Springer.

D. Calvetti and E. Somersalo (2008), 'Hypermodels in the Bayesian imaging framework', *Inverse Problems* **24**, #034013.

D. Calvetti, H. Hakula, S. Pursiainen and E. Somersalo (2009), 'Conditionally Gaussian hypermodels for cerebral source location', *SIAM J. Imag. Sci.* **2**, 879–909.

D. Calvetti, A. Kuceyeski and E. Somersalo (2008), 'Sampling based analysis of a spatially distributed model for liver metabolism at steady state', *Multiscale Modeling and Simulation* **7**, 407–431.

E. Candès and M. Wakin (2008), 'An introduction to compressive sampling', *IEEE Signal Processing Magazine*, March 2008, 21–30.

J.-Y. Chemin and N. Lerner (1995), 'Flot de champs de veceurs non lipschitziens et équations de Navier–Stokes', *J. Diff. Equations* **121**, 314–328.

A. Chorin and O. Hald (2006), *Stochastic Tools in Mathematics and Science*, Vol. 1 of *Surveys and Tutorials in the Applied Mathematical Sciences*, Springer, New York.

A. Chorin and P. Krause (2004), 'Dimensional reduction for a Bayesian filter', *Proc. Nat. Acad. Sci.* **101**, 15013–15017.

A. Chorin and X. Tu (2009), 'Implicit sampling for particle filters', *Proc. Nat. Acad. Sci.* **106**, 17249–17254.

A. Chorin and X. Tu (2010), 'Interpolation and iteration for nonlinear filters', *Math. Model. Numer. Anal.* To appear.

M. Christie (2010), Solution error modelling and inverse problems. In *Simplicity, Complexity and Modelling*, Wiley, New York, to appear.

M. Christie, G. Pickup, A. O'Sullivan and V. Demyanov (2008), Use of solution error models in history matching. In *Proc. European Conference on the Mathematics of Oil Recovery XI*, European Association of Geoscientists and Engineers.

B. Chua and A. Bennett (2001), 'An inverse ocean modelling system', *Ocean. Meteor.* **3**, 137–165.

S. Cohn (1997), 'An introduction to estimation theory', *J. Met. Soc. Japan* **75**, 257–288.

S. Cotter, M. Dashti, J. Robinson and A. Stuart (2009), 'Bayesian inverse problems for functions and applications to fluid mechanics', *Inverse Problems* **25**, #115008.

S. Cotter, M. Dashti and A. Stuart (2010*a*), 'Approximation of Bayesian inverse problems', *SIAM J. Numer. Anal.* To appear.

S. Cotter, M. Dashti, J. Robinson and A. Stuart (2010*b*). In preparation.

P. Courtier (1997), 'Dual formulation of variational assimilation', *Quart. J. Royal Met. Soc.* **123**, 2449–2461.

P. Courtier and O. Talagrand (1987), 'Variational assimilation of meteorological observations with the adjoint vorticity equation II: Numerical results', *Quart. J. Royal Met. Soc.* **113**, 1329–1347.

P. Courtier, E. Anderson, W. Heckley, J. Pailleux, D. Vasiljevic, M. Hamrud, A. Hollingworth, F. Rabier and M. Fisher (1998), 'The ECMWF implementation of three-dimensional variational assimilation (3D-Var)', *Quart. J. Royal Met. Soc.* **124**, 1783–1808.

N. Cressie (1993), *Statistics for Spatial Data*, Wiley.

T. Cui, C. Fox, G. Nicholls and M. O'Sullivan (2010), 'Using MCMC sampling to calibrate a computer model of a geothermal field'. Submitted.

G. Da Prato and J. Zabczyk (1992), *Stochastic Equations in Infinite Dimensions*, Vol. 44 of *Encyclopedia of Mathematics and its Applications*, Cambridge University Press.

B. Dacarogna (1989), *Direct Methods in the Calculus of Variations*, Springer, New York.

M. Dashti and J. Robinson (2009), 'Uniqueness of the particle trajectories of the weak solutions of the two-dimensional Navier–Stokes equations', *Nonlinearity* **22**, 735–746.

M. Dashti, S. Harris and A. M. Stuart (2010*a*), Bayesian approach to an elliptic inverse problem. In preparation.

M. Dashti, N. Pillai and A. Stuart (2010*b*), *Bayesian Inverse Problems in Differential Equations*. Lecture notes, available from:
http://www.warwick.ac.uk/~masdr/inverse.html.

J. Derber (1989), 'A variational continuous assimilation technique', *Monthly Weather Review* **117**, 2437–2446.

P. Deuflhard (2004), *Newton Methods for Nonlinear Problems: Affine Invariance and Adaptive Algorithms*, Springer.

B. DeVolder, J. Glimm, J. Grove, Y. Kang, Y. Lee, K. Pao, D. Sharp and K. Ye (2002), 'Uncertainty quantification for multiscale simulations', *J. Fluids Engrg* **124**, 29–42.

D. Donoho (2006), 'Compressed sensing', *IEEE Trans. Inform. Theory* **52**, 1289–1306.

P. Dostert, Y. Efendiev, T. Hou and W. Luo (2006), 'Coarse-grain Langevin algorithms for dynamic data integration', *J. Comput. Phys.* **217**, 123–142.

N. Doucet, A. de Frietas and N. Gordon (2001), *Sequential Monte Carlo in Practice*, Springer.

R. Dudley (2002), *Real Analysis and Probability*, Cambridge University Press, Cambridge.

D. Dürr and A. Bach (1978), 'The Onsager–Machlup function as Lagrangian for the most probable path of a diffusion process', *Comm. Math. Phys.* **160**, 153–170.

Y. Efendiev, A. Datta-Gupta, X. Ma and B. Mallick (2009), 'Efficient sampling techniques for uncertainty quantification in history matching using nonlinear error models and ensemble level upscaling techniques', *Water Resources Res.* **45**, #W11414.

M. Eknes and G. Evensen (1997), 'Parameter estimation solving a weak constraint variational formulation for an Ekman model', *J. Geophys. Res.* **12**, 479–491.

B. Ellerbroek and C. Vogel (2009), 'Inverse problems in astronomical adaptive optics', *Inverse Problems* **25**, #063001.

H. Engl, M. Hanke and A. Neubauer (1996), *Regularization of Inverse Problems*, Kluwer.

H. Engl, A. Hofinger and S. Kindermann (2005), 'Convergence rates in the Prokhorov metric for assessing uncertainty in ill-posed problems', *Inverse Problems* **21**, 399–412.

G. Evensen (2006), *Data Assimilation: The Ensemble Kalman Filter*, Springer.

G. Evensen and P. Van Leeuwen (2000), 'An ensemble Kalman smoother for nonlinear dynamics', *Monthly Weather Review* **128**, 1852–1867.

F. Fang, C. Pain, I. Navon, M. Piggott, G. Gorman, P. Allison and A. Goddard (2009*a*), 'Reduced order modelling of an adaptive mesh ocean model', *Internat. J. Numer. Methods Fluids* **59**, 827–851.

F. Fang, C. Pain, I. Navon, M. Piggott, G. Gorman, P. Farrell, P. Allison and A. Goddard (2009*b*), 'A POD reduced-order 4D-Var adaptive mesh ocean modelling approach', *Internat. J. Numer. Methods Fluids* **60**, 709–732.

C. Farmer (2005), Geological modelling and reservoir simulation. In *Mathematical Methods and Modeling in Hydrocarbon Exploration and Production* (A. Iske and T. Randen, eds), Springer, Heidelberg, pp. 119–212.

C. Farmer (2007), Bayesian field theory applied to scattered data interpolation and inverse problems. In *Algorithms for Approximation* (A. Iske and J. Levesley, eds), Springer, pp. 147–166.

B. Fitzpatrick (1991), 'Bayesian analysis in inverse problems', *Inverse Problems* **7**, 675–702.

J. Franklin (1970), 'Well-posed stochastic extensions of ill-posed linear problems', *J. Math. Anal. Appl.* **31**, 682–716.

M. Freidlin and A. Wentzell (1984), *Random Perturbations of Dynamical Systems*, Springer, New York.

A. Gelfand and A. Smith (1990), 'Sampling-based approaches to calculating marginal densities', *J. Amer. Statist. Soc.* **85**, 398–409.

A. Gibbs and F. Su (2002), 'On choosing and bounding probability metrics', *Internat. Statist. Review* **70**, 419–435.

C. Gittelson and C. Schwab (2011), Sparse tensor discretizations of high-dimensional PDEs. To appear in *Acta Numerica*, Vol. 20.

J. Glimm, S. Hou, Y. Lee, D. Sharp and K. Ye (2003), 'Solution error models for uncertainty quantification', *Contemporary Mathematics* **327**, 115–140.

S. Gratton, A. Lawless and N. Nichols (2007), 'Approximate Gauss–Newton methods for nonlinear least squares problems', *SIAM J. Optimization* **18**, 106–132.

A. Griffith and N. Nichols (1998), Adjoint methods for treating model error in data assimilation. In *Numerical Methods for Fluid Dynamics VI*, ICFD, Oxford, pp. 335–344.

A. Griffith and N. Nichols (2000), 'Adjoint techniques in data assimilation for treating systematic model error', *J. Flow, Turbulence and Combustion* **65**, 469–488.

G. Grimmett and D. Stirzaker (2001), *Probability and Random Processes*, Oxford University Press, New York.

C. Gu (2002), *Smoothing Spline ANOVA Models*, Springer.

C. Gu (2008), 'Smoothing noisy data via regularization', *Inverse Problems* **24**, #034002.

C. Hagelberg, A. Bennett and D. Jones (1996), 'Local existence results for the generalized inverse of the vorticity equation in the plane', *Inverse Problems* **12**, 437–454.

E. Hairer and G. Wanner (1996), *Solving Ordinary Differential Equations II*, Vol. 14 of *Springer Series in Computational Mathematics*, Springer, Berlin.

E. Hairer, S. P. Nørsett and G. Wanner (1993), *Solving Ordinary Differential Equations I*, Vol. 8 of *Springer Series in Computational Mathematics*, Springer, Berlin.

M. Hairer (2009), *Introduction to Stochastic PDEs*. Lecture notes.

M. Hairer, A. M. Stuart and J. Voss (2007), 'Analysis of SPDEs arising in path sampling II: The nonlinear case', *Ann. Appl. Probab.* **17**, 1657–1706.

M. Hairer, A. M. Stuart and J. Voss (2009), Sampling conditioned diffusions. In *Trends in Stochastic Analysis*, Vol. 353 of *London Mathematical Society Lecture Notes*, Cambridge University Press, pp. 159–186.

M. Hairer, A. Stuart and J. Voss (2010*a*), 'Sampling conditioned hypoelliptic diffusions'. Submitted.

M. Hairer, A. Stuart and J. Voss (2010*b*), Signal processing problems on function space: Bayesian formulation, stochastic PDEs and effective MCMC methods. In *Oxford Handbook of Nonlinear Filtering* (D. Crisan and B. Rozovsky, eds), Oxford University Press, to appear.

M. Hairer, A. Stuart, J. Voss and P. Wiberg (2005), 'Analysis of SPDEs arising in path sampling I: The Gaussian case', *Comm. Math. Sci.* **3**, 587–603.

W. K. Hastings (1970), 'Monte Carlo sampling methods using Markov chains and their applications', *Biometrika* **57**, 97–109.

T. Hein (2009), 'On Tikhonov regularization in Banach spaces: Optimal convergence rate results', *Applicable Analysis* **88**, 653–667.

J. Heino, K. Tunyan, D. Calvetti and E. Somersalo (2007), 'Bayesian flux balance analysis applied to a skeletal muscle metabolic model', *J. Theor. Biol.* **248**, 91–110.

R. Herbei and I. McKeague (2009), 'Geometric ergodicity of hybrid samplers for ill-posed inverse problems', *Scand. J. Statist.* **36**, 839–853.

R. Herbei, I. McKeague and K. Speer (2008), 'Gyres and jets: Inversion of tracer data for ocean circulation structure', *J. Phys. Oceanography* **38**, 1180–1202.

A. Hofinger and H. Pikkarainen (2007), 'Convergence rates for the Bayesian approach to linear inverse problems', *Inverse Problems* **23**, 2469–2484.

A. Hofinger and H. Pikkarainen (2009), 'Convergence rates for linear inverse problems in the presence of an additive normal noise', *Stoch. Anal. Appl.* **27**, 240–257.

M. Huddleston, M. Bell, M. Martin and N. Nichols (2004), 'Assessment of wind stress errors using bias corrected ocean data assimilation', *Quart. J. Royal Met. Soc.* **130**, 853–872.

M. Hurzeler and H. Künsch (2001), Approximating and maximizing the likelihood for a general state space model. In *Sequential Monte Carlo Methods in Practice* (A. Doucet, N. de Freitas and N. Gordon, eds), Springer, pp. 159–175.

J. Huttunen and H. Pikkarainen (2007), 'Discretization error in dynamical inverse problems: One-dimensional model case', *J. Inverse and Ill-posed Problems* **15**, 365–386.

K. Ide and C. Jones (2007), 'Data assimilation', *Physica D* **230**, vii–viii.

K. Ide, L. Kuznetsov and C. Jones (2002), 'Lagrangian data assimilation for point-vortex system', *J. Turbulence* **3**, 53.

N. Ikeda and S. Watanabe (1989), *Stochastic Differential Equations and Diffusion Processes*, second edn, North-Holland, Amsterdam.

M. Jardak, I. Navon and M. Zupanski (2010), 'Comparison of sequential data assimilation methods for the Kuramoto–Sivashinsky equation', *Internat. J. Numer. Methods Fluids* **62**, 374–402.

C. Johnson, B. Hoskins and N. Nichols (2005), 'A singular vector perspective of 4DVAR: Filtering and interpolation', *Quart. J. Royal Met. Soc.* **131**, 1–20.

C. Johnson, B. Hoskins, N. Nichols and S. Ballard (2006), 'A singular vector perspective of 4DVAR: The spatial structure and evolution of baroclinic weather systems', *Monthly Weather Review* **134**, 3436–3455.

J. Kaipio and E. Somersalo (2000), 'Statistical inversion and Monte Carlo sampling methods in electrical impedance tomography', *Inverse Problems* **16**, 1487–1522.

J. Kaipio and E. Somersalo (2005), *Statistical and Computational Inverse problems*, Vol. 160 of *Applied Mathematical Sciences*, Springer.

J. Kaipio and E. Somersalo (2007*a*), 'Approximation errors in nonstationary inverse problems', *Inverse Problems and Imaging* **1**, 77–93.

J. Kaipio and E. Somersalo (2007*b*), 'Statistical inverse problems: Discretization, model reduction and inverse crimes', *J. Comput. Appl. Math.* **198**, 493–504.

E. Kalnay (2003), *Atmospheric Modeling, Data Assimilation and Predictability*, Cambridge University Press.

E. Kalnay, H. Li, S. Miyoshi, S. Yang and J. Ballabrera-Poy (2007), '4D-Var or ensemble Kalman filter?', *Tellus* **59**, 758–773.

B. Kaltenbacher, F. Schöpfer and T. Schuster (2009), 'Iterative methods for nonlinear ill-posed problems in Banach spaces: Convergence and applications to parameter identification problems', *Inverse Problems* **25**, #065003.

M. Kennedy and A. O'Hagan (2001), 'Bayesian calibration of computer models', *J. Royal Statist. Soc.* **63B**, 425–464.

D. Kinderlehrer and G. Stampacchia (1980), *An Introduction to Variational Inequalities and their Applications*, SIAM.

T. Kolda and B. Bader (2009), 'Tensor decompositions and applications', *SIAM Review* **51**, 455–500.

L. Kuznetsov, K. Ide and C. Jones (2003), 'A method for assimilation of Lagrangian data', *Monthly Weather Review* **131**, 2247–2260.

M. Lassas and S. Siltanen (2004), 'Can one use total variation prior for edge-preserving Bayesian inversion?', *Inverse Problems* **20**, 1537–1563.

M. Lassas, E. Saksman and S. Siltanen (2009), 'Discretization-invariant Bayesian inversion and Besov space priors', *Inverse Problems and Imaging* **3**, 87–122.

A. Lawless and N. Nichols (2006), 'Inner loop stopping criteria for incremental four-dimensional variational data assimilation', *Monthly Weather Review* **134**, 3425–3435.

A. Lawless, S. Gratton and N. Nichols (2005*a*), 'Approximate iterative methods for variational data assimilation', *Internat. J. Numer. Methods Fluids* **47**, 1129–1135.

A. Lawless, S. Gratton and N. Nichols (2005*b*), 'An investigation of incremental 4D-Var using non-tangent linear models', *Quart. J. Royal Met. Soc.* **131**, 459–476.

A. Lawless, N. Nichols, C. Boess and A. Bunse-Gerstner (2008*a*), 'Approximate Gauss–Newton methods for optimal state estimation using reduced order models', *Internat. J. Numer. Methods Fluids* **56**, 1367–1373.

A. Lawless, N. Nichols, C. Boess and A. Bunse-Gerstner (2008*b*), 'Using model reduction methods within incremental four-dimensional variational data assimilation', *Monthly Weather Review* **136**, 1511–1522.

M. Lehtinen, L. Paivarinta and E. Somersalo (1989), 'Linear inverse problems for generalized random variables', *Inverse Problems* **5**, 599–612.

M. Lifshits (1995), *Gaussian Random Functions*, Vol. 322 of *Mathematics and its Applications*, Kluwer, Dordrecht.

D. Livings, S. Dance and N. Nichols (2008), 'Unbiased ensemble square root filters', *Physica D: Nonlinear Phenomena* **237**, 1021–1028.

M. Loève (1977), *Probability Theory I*, fourth edn, Vol. 45 of *Graduate Texts in Mathematics*, Springer, New York.

M. Loève (1978), *Probability Theory II*, fourth edn, Vol. 46 of *Graduate Texts in Mathematics*, Springer, New York.

A. Lorenc (1986), 'Analysis methods for numerical weather prediction', *Quart. J. Royal Met. Soc.* **112**, 1177–1194.

C. Lubich (2008), *From Quantum to Classical Molecular Dynamics: Reduced Models and Numerical Analysis*, European Mathematical Society.

X. Ma, M. Al-Harbi, A. Datta-Gupta and Y. Efendiev (2008), 'Multistage sampling approach to quantifying uncertainty during history matching geological models', *Soc. Petr. Engrg J.* **13**, 77–87.

A. Majda and B. Gershgorin (2008), 'A nonlinear test model for filtering slow–fast systems', *Comm. Math. Sci.* **6**, 611–649.

A. Majda and M. Grote (2007), 'Explicit off-line criteria for stable accurate filtering of strongly unstable spatially extended systems', *Proc. Nat. Acad. Sci.* **104**, 1124–1129.

A. Majda and J. Harlim (2010), 'Catastrophic filter divergence in filtering nonlinear dissipative systems', *Comm. Math. Sci.* **8**, 27–43.

A. Majda, J. Harlim and B. Gershgorin (2010), 'Mathematical strategies for filtering turbulent dynamical systems', *Disc. Cont. Dyn. Sys.* To appear.

A. Mandelbaum (1984), 'Linear estimators and measurable linear transformations on a Hilbert space', *Probab. Theory Rel. Fields* **65**, 385–397.

M. Martin, M. Bell and N. Nichols (2002), 'Estimation of systematic error in an equatorial ocean model using data assimilation', *Internat. J. Numer. Methods Fluids* **40**, 435–444.

I. McKeague, G. Nicholls, K. Speer and R. Herbei (2005), 'Statistical inversion of south Atlantic circulation in an abyssal neutral density layer', *J. Marine Res.* **63**, 683–704.

D. McLaughlin and L. Townley (1996), 'A reassessment of the groundwater inverse problem', *Water Resources Res.* **32**, 1131–1161.

N. Metropolis, R. Rosenbluth, M. Teller and E. Teller (1953), 'Equations of state calculations by fast computing machines', *J. Chem. Phys.* **21**, 1087–1092.

S. P. Meyn and R. L. Tweedie (1993), *Markov Chains and Stochastic Stability*, Communications and Control Engineering Series, Springer, London.

A. Michalak and P. Kitanidis (2003), 'A method for enforcing parameter nonnegativity in Bayesian inverse problems with an application to contaminant source identification', *Water Resources Res.* **39**, 1033.

T. Mitchell, B. Buchanan, G. DeJong, T. Dietterich, P. Rosenbloom and A. Waibel (1990), 'Machine learning', *Annual Review of Computer Science* **4**, 417–433.

L. Mohamed, M. Christie and V. Demyanov (2010), 'Comparison of stochastic sampling algorithms for uncertainty quantification', *Soc. Petr. Engrg J.* To appear. http://dx.doi.org/10.2118/119139-PA

K. Mosegaard and A. Tarantola (1995), 'Monte Carlo sampling of solutions to inverse problems', *J. Geophys. Research* **100**, 431–447.

A. Neubauer (2009), 'On enhanced convergence rates for Tikhonov regularization of nonlinear ill-posed problems in Banach spaces', *Inverse Problems* **25**, #065009.

A. Neubauer and H. Pikkarainen (2008), 'Convergence results for the Bayesian inversion theory', *J. Inverse and Ill-Posed Problems* **16**, 601–613.

N. Nichols (2003*a*), Data assimilation: Aims and basic concepts. In *Data Assimilation for the Earth System* (R. Swinbank, V. Shutyaev and W. A. Lahoz, eds), Kluwer Academic, pp. 9–20.

N. Nichols (2003*b*), Treating model error in 3-D and 4-D data assimilation. In *Data Assimilation for the Earth System* (R. Swinbank, V. Shutyaev and W. A. Lahoz, eds), Kluwer Academic, pp. 127–135.

M. Nodet (2005), Mathematical modeling and assimilation of Lagrangian data in oceanography. PhD thesis, University of Nice.

M. Nodet (2006), 'Variational assimilation of Lagrangian data in oceanography', *Inverse Problems* **22**, 245–263.

B. Oksendal (2003), *Stochastic Differential Equations: An Introduction with Applications*, sixth edn, Universitext, Springer.

D. Orrell, L. Smith, J. Barkmeijer and T. Palmer (2001), 'Model error in weather forecasting', *Non. Proc. in Geo.* **8**, 357–371.

A. O'Sullivan and M. Christie (2006*a*), 'Error models for reducing history match bias', *Comput. Geosci.* **10**, 405–405.

A. O'Sullivan and M. Christie (2006*b*), 'Simulation error models for improved reservoir prediction', *Reliability Engineering and System Safety* **91**, 1382–1389.

E. Ott, B. Hunt, I. Szunyogh, A. Zimin, E. Kostelich, M. Corazza, E. Kalnay, D. Patil and J. Yorke (2004), 'A local ensemble Kalman filter for atmospheric data assimilation', *Tellus A* **56**, 273–277.

T. Palmer, F. Doblas-Reyes, A. Weisheimer, G. Shutts, J. Berner and J. Murphy (2009), 'Towards the probabilistic earth-system model', *J. Climate* **70**, 419–435.

H. Pikkarainen (2006), 'State estimation approach to nonstationary inverse problems: Discretization error and filtering problem', *Inverse Problems* **22**, 365–379.

S. Pimentel, K. Haines and N. Nichols (2008*a*), 'The assimilation of satellite derived sea surface temperatures into a diurnal cycle model', *J. Geophys. Research: Oceans* **113**, #C09013.

S. Pimentel, K. Haines and N. Nichols (2008*b*), 'Modelling the diurnal variability of sea surface temperatures', *J. Geophys. Research: Oceans* **113**, #C11004.

J. Ramsay and B. Silverman (2005), *Functional Data Analysis*, Springer.

M. Reznikoff and E. Vanden Eijnden (2005), 'Invariant measures of SPDEs and conditioned diffusions', *CR Acad. Sci. Paris* **340**, 305–308.

D. Richtmyer and K. Morton (1967), *Difference Methods for Initial Value Problems*, Wiley.

G. Roberts and J. Rosenthal (1998), 'Optimal scaling of discrete approximations to Langevin diffusions', *J. Royal Statist. Soc. B* **60**, 255–268.

G. Roberts and J. Rosenthal (2001), 'Optimal scaling for various Metropolis–Hastings algorithms', *Statistical Science* **16**, 351–367.

G. Roberts and R. Tweedie (1996), 'Exponential convergence of Langevin distributions and their discrete approximations', *Bernoulli* **2**, 341–363.

G. Roberts, A. Gelman and W. Gilks (1997), 'Weak convergence and optimal scaling of random walk Metropolis algorithms', *Ann. Appl. Probab.* **7**, 110–120.

L. Rudin, S. Osher and E. Fatemi (1992), 'Nonlinear total variation based noise removal algorithms', *Physica D* **60**, 259–268.

H. Rue and L. Held (2005), *Gaussian Markov Random Fields: Theory and Applications*, Chapman & Hall.

H. Salman, K. Ide and C. Jones (2008), 'Using flow geometry for drifter deployment in Lagrangian data assimilation', *Tellus* **60**, 321–335.

H. Salman, L. Kuznetsov, C. Jones and K. Ide (2006), 'A method for assimilating Lagrangian data into a shallow-water equation ocean model', *Monthly Weather Review* **134**, 1081–1101.

J. M. Sanz-Serna and C. Palencia (1985), 'A general equivalence theorem in the theory of discretization methods', *Math. Comp.* **45**, 143–152.

O. Scherzer, M. Grasmair, H. Grossauer, M. Haltmeier and F. Lenzen (2009), *Variational Methods in Imaging*, Springer.

C. Schwab and R. Todor (2006), 'Karhunen–Loeve approximation of random fields in domains by generalized fast multipole methods', *J. Comput. Phys.* **217**, 100–122.

Y. Shen, C. Archambeau, D. Cornford and M. Opper (2008*a*), Variational Markov chain Monte Carlo for inference in partially observed nonlinear diffusions. In *Proceedings of the Workshop on Inference and Estimation in Probabilistic*

Time-Series Models (D. Barber, A. T. Cemgil and S. Chiappa, eds), Isaac Newton Institute for Mathematical Sciences, Cambridge, pp. 67–78.

Y. Shen, C. Archambeau, D. Cornford, M. Opper, J. Shawe-Taylor and R. Barillec (2008*b*), 'A comparison of variational and Markov chain Monte Carlo methods for inference in partially observed stochastic dynamic systems', *J. Signal Processing Systems*. In press (published online).

Y. Shen, D. Cornford, C. Archambeau and M. Opper (2010), 'Variational Markov chain Monte Carlo for Bayesian inference in partially observed non-linear diffusions', *Comput. Statist.* Submitted.

A. Smith and G. Roberts (1993), 'Bayesian computation via the Gibbs sampler and related Markov chain Monte Carlo methods', *J. Royal Statist. Soc. B* **55**, 3–23.

T. Snyder, T. Bengtsson, P. Bickel and J. Anderson (2008), 'Obstacles to high-dimensional particle filtering', *Monthly Weather Review* **136**, 4629–4640.

P. Spanos and R. Ghanem (1989), 'Stochastic finite element expansion for random media', *J. Engrg Mech.* **115**, 1035–1053.

P. Spanos and R. Ghanem (2003), *Stochastic Finite Elements: A Spectral Approach*, Dover.

E. Spiller, A. Budhiraja, K. Ide and C. Jones (2008), 'Modified particle filter methods for assimilating Lagrangian data into a point-vortex model', *Physica D* **237**, 1498–1506.

L. Stanton, A. Lawless, N. Nichols and I. Roulstone (2005), 'Variational data assimilation for Hamiltonian problems', *Internat. J. Numer. Methods Fluids* **47**, 1361–1367.

A. Stuart, J. Voss and P. Wiberg (2004), 'Conditional path sampling of SDEs and the Langevin MCMC method', *Comm. Math. Sci* **2**, 685–697.

P. Talagrand and O. Courtier (1987), 'Variational assimilation of meteorological observations with the adjoint vorticity equation I: Theory', *Quart. J. Royal Met. Soc.* **113**, 1311–1328.

A. Tarantola (2005), *Inverse Problem Theory*, SIAM.

L. Tierney (1998), 'A note on Metropolis–Hastings kernels for general state spaces', *Ann. Appl. Probab.* **8**, 1–9.

R. Todor and C. Schwab (2007), 'Convergence rates for sparse chaos approximations of elliptic problems with stochastic coefficients', *IMA J. Numer. Anal.* **27**, 232–261.

G. Uhlmann (2009), Visibility and invisibility. In *Invited Lectures, Sixth International Congress on Industrial and Applied Mathematics, ICIAM07* (R. Jeltsch and G. Wanner, eds), European Mathematical Society, pp. 381–408.

P. Van Leeuwen (2001), 'An ensemble smoother with error estimates', *Monthly Weather Review* **129**, 709–728.

P. Van Leeuwen (2003), 'A variance minimizing filter for large-scale applications', *Monthly Weather Review* **131**, 2071–2084.

P. Van Leeuwen (2009), 'Particle filtering in geophysical systems', *Monthly Weather Review* **137**, 4089–4114.

G. Vernieres, K. Ide and C. Jones (2010), 'Lagrangian data assimilation, an application to the Gulf of Mexico', *Physica D.* Submitted.

C. Vogel (2002), *Computational Methods for Inverse Problems*, SIAM.

F. Vossepoel and P. Van Leeuwen (2007), 'Parameter estimation using a particle method: Inferring mixing coefficients from sea-level observations', *Monthly Weather Review* **135**, 1006–1020.

M. Vrettas, D. Cornford and Y. Shen (2009), A variational basis function approximation for diffusion processes. In *Proceedings of the 17th European Symposium on Artificial Neural Networks*, D-side publications, Evere, Belgium, pp. 497–502.

G. Wahba (1990), *Spline Models for Observational Data*, SIAM.

L. Watkinson, A. Lawless, N. Nichols and I. Roulstone (2007), 'Weak constraints in four dimensional variational data assimilation', *Meteorologische Zeitschrift* **16**, 767–776.

L. White (1993), 'A study of uniqueness for the initialization problem for Burgers' equation', *J. Math. Anal. Appl.* **172**, 412–431.

D. Williams (1991), *Probability with Martingales*, Cambridge University Press, Cambridge.

M. Wlasak and N. Nichols (1998), Application of variational data assimilation to the Lorenz equations using the adjoint method. In *Numerical Methods for Fluid Dynamics VI*, ICFD, Oxford, pp. 555–562.

M. Wlasak, N. Nichols and I. Roulstone (2006), 'Use of potential vorticity for incremental data assimilation', *Quart. J. Royal Met. Soc.* **132**, 2867–2886.

L. Yu and J. O'Brien (1991), 'Variational estimation of the wind stress drag coefficient and the oceanic eddy viscosity profile', *J. Phys. Ocean.* **21**, 1361–1364.

O. Zeitouni and A. Dembo (1987), 'A maximum *a posteriori* estimator for trajectories of diffusion processes', *Stochastics* **20**, 221–246.

D. Zimmerman, G. de Marsily, C. Gotway, M. Marietta, C. Axness, R. Beauheim, R. Bras, J. Carrera, G. Dagan, P. Davies, D. Gallegos, A. Galli, J. Gomez-Hernandez, P. Grindrod, A. Gutjahr, P. Kitanidis, A. Lavenue, D. McLaughlin, S. Neuman, B. RamaRao, C. Ravenne and Y. Rubin (1998), 'A comparison of seven geostatistically based inverse approaches to estimate transmissivities for modeling advective transport by groundwater flow', *Water Resources Res.* **6**, 1373–1413.

E. Zuazua (2005), 'Propagation, observation, control and numerical approximation of waves approximated by finite difference method', *SIAM Review* **47**, 197–243.

O. Zupanski (1997), 'A general weak constraint applicable to operational 4DVAR data assimilation systems', *Monthly Weather Review* **125**, 2274–2292.

M. Zupanski, I. Navon and D. Zupanski (2008), 'The maximum likelihood ensemble filter as a non-differentiable minimization algorithm', *Quart. J. Royal Met. Soc.* **134**, 1039–1050.

Acta Numerica (2010), pp. 561–598
doi:10.1017/S0962492910000073

Kepler, Newton and numerical analysis

G. Wanner
University of Geneva,
Section de Mathématiques,
CP 64, CH-1211 Genève 4,
Switzerland
E-mail: Gerhard.Wanner@unige.ch

Numerical methods are usually constructed for solving mathematical problems such as differential equations or optimization problems. In this contribution we discuss the fact that numerical methods, applied inversely, were also important in *establishing* these models. We show in detail the discovery of the laws of planetary motion by Kepler and Newton, which stood at the beginning of modern science. The 400th anniversary of the publication of Kepler's laws (1609) is a good occasion for this investigation.

CONTENTS

1. Origins of numerical analysis

We start with an overview of the origins of numerical methods for ODEs. The problem we choose is the movement of a body with negligible mass in the gravitational field of two fixed bodies with masses A and B, positions $x_1 = 0$ for the first, $x_1 = a$ for the second, and $x_2 = 0$ for both. The so-called *Newton's equations* for this problem are

$$\frac{\mathrm{dd}x_1}{\mathrm{d}t^2} = -\frac{Ax_1}{v^3} - \frac{B(x_1-a)}{u^3}, \qquad \frac{\mathrm{dd}x_2}{\mathrm{d}t^2} = -\frac{Ax_2}{v^3} - \frac{Bx_2}{u^3}, \tag{1.1}$$

$$v = \sqrt{x_1^2 + x_2^2}, \quad u = \sqrt{(x_1-a)^2 + x_2^2}, \quad A = 2, \quad B = 1, \quad a = 1. \tag{1.2}$$

This is one of the problems for which Euler managed, by an incredible *tour de force*, to find analytical formulas for the solutions in the two-dimensional case (Euler E301 1766) and the three-dimensional case (Euler E328 1767). Here we use this problem to demonstrate numerical methods for the chosen initial position $x_{10} = 1.47$, $\dot{x}_{10} = -0.81$, $x_{20} = 0.8$, $\dot{x}_{20} = 0$.

1.1. *Euler's method*

$$b' = b + A(a' - a)$$
$$b'' = b' + A'(a'' - a')$$
$$b''' = b'' + A''(a''' - a'')$$

Ipsius	valores successivi
x	a ; a' ; a'' ; a''' ; a^{IV} ; .... $'x$; x
y	b ; b' ; b'' ; b''' ; b^{IV} ; .... $'y$; y
V	A ; A' ; A'' ; A''' ; A^{IV} ; .... 'V ; V.

Figure 1.1. First publication of Euler's method in E342, written for the differential equation $\frac{\mathrm{d}y}{\mathrm{d}x} = V(x, y)$ with initial values $y(a) = b$.

In his monumental three-volume treatise *Institutiones Calculi Integralis* (Euler E342 1768, E366 1769, E385 1770), after having explained analytic formulas for integrals, so-called Riemann sums for integrals without analytic solutions, and analytic formulas for ordinary differential equations, Euler reached the problem of finding approximate solutions of differential equations without analytic solutions by an extension of the idea of Riemann sums (Euler E342 1768, Pars II, §650). This led to Euler's method, whose first publication, written for the differential equation $\frac{\mathrm{d}y}{\mathrm{d}x} = V(x, y)$ with initial values $y(a) = b$, is reproduced in Figure 1.1.

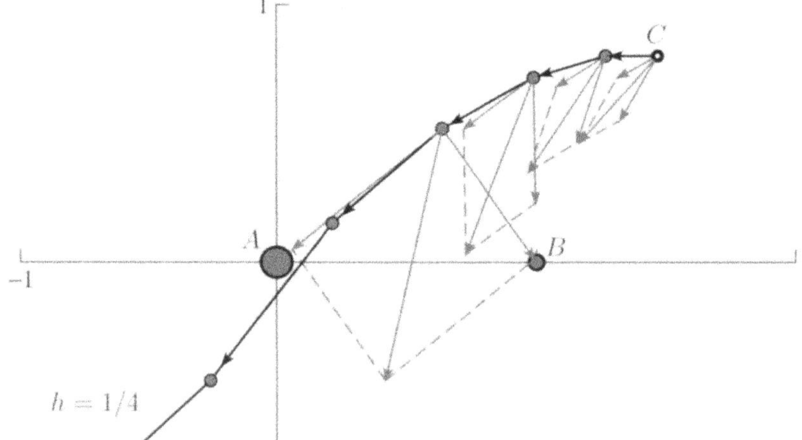

Figure 1.2. The problem with two fixed mass points and solution using Euler's method for $h = 1/4$.

For an equation of the type (1.1), *i.e.*,

$$\frac{\mathrm{dd}x}{\mathrm{d}t^2} = F(t,x) \qquad \text{or} \qquad \frac{\mathrm{d}x}{\mathrm{d}t} = v, \quad \frac{\mathrm{d}v}{\mathrm{d}t} = F(t,x), \tag{1.3}$$

this method becomes

$$\begin{gathered} x_{n+1} = x_n + hv_n, \quad v_{n+1} = v_n + hF(t_n, x_n), \qquad h = \Delta t, \\ x = a + \omega\,;\; y = b + c\omega\,;\; p = c + F\omega \end{gathered} \tag{1.4}$$

(Euler E366, Liber I, Pars II, Sectio I, Caput XII, Problema 137, §1082); Euler's x, y, p, ω, a, b, c are our $t, x, v, h, t_0, x_0, v_0$, respectively. The *modus operandi* of this method is illustrated in Figure 1.2: the solution point moves from the initial point x_0 with constant initial velocity v_0 during the time interval h to x_1 (first formula of (1.4)). The initial velocity is updated by using the force $F(t_0, x_0)$ evaluated at the initial point (second formula of (1.4)) to obtain v_1, which is then used to move the solution point during the second time interval to x_2, and so on.

Euler did not, perhaps, require much genius to design this primitive-looking method, but he needed great genius *to understand that it works*, despite all these accumulated truncation errors and their propagation. He demonstrated the convergence of the solution, as $h \to 0$, for several examples. We demonstrate the convergence of the solution for problem (1.1) graphically in Figure 1.3. A formal convergence proof, using what was later called a *Lipschitz condition*, was given by Cauchy in 1824.

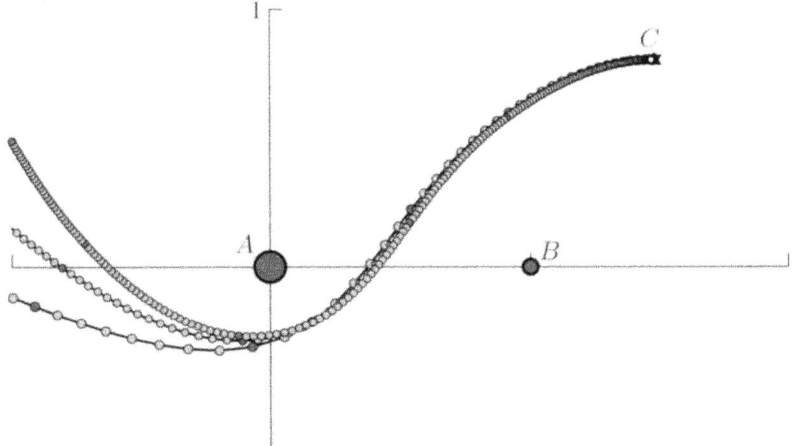

Figure 1.3. The solution of the problem with two fixed mass points using Euler's method for $h = 1/32$, $h = 1/64$ and $h = 1/128$.

$$y = b + \frac{(x-a)\,db}{da} + \frac{(x-a)^2\,ddb}{1\cdot 2\,da^2} + \frac{(x-a)^3\,d^3b}{1\cdot 2\cdot 3\,da^3} + \text{etc.}$$

$$\frac{ddy}{dx^2} = \left(\frac{dV}{dx}\right) + V\left(\frac{dV}{dy}\right)$$

$$\frac{d^3y}{dx^3} = \left(\frac{ddV}{dx^2}\right) + \left(\frac{dV}{dx}\right)\left(\frac{dV}{dy}\right) + 2V\left(\frac{ddV}{dxdy}\right) + V\left(\frac{dV}{dy}\right)^2 + VV\left(\frac{ddV}{dy^2}\right).$$

Figure 1.4. Euler's third-order Taylor method.

1.2. Euler's higher-order methods and implicit methods

A couple of paragraphs later (in §656 of E342), Euler realizes that one can very much improve ('magis perficere') the above method ('methodum praecedentem'), by adding, at each step, to the term $y_{n+1} = y_n + hV(x_n, y_n)$ higher-order terms of the Taylor series

$$y_{n+1} = y_n + hy_n' + \frac{h^2}{1\cdot 2}y_n'' + \frac{h^3}{1\cdot 2\cdot 3}y_n''' + \cdots,$$

computing the higher derivatives of $y' = V(x,y)$ by implicit differentiation $y'' = \frac{\partial V}{\partial x} + \frac{\partial V}{\partial y}\cdot y'$, *etc.* (see Figure 1.4).

It can be observed graphically in Figure 1.5 that this same method, applied with order 3 (up to this order, Euler wrote the derivatives explicitly) to our problem with two fixed mass points, *really* improves the solution, especially if the step-size h becomes small.

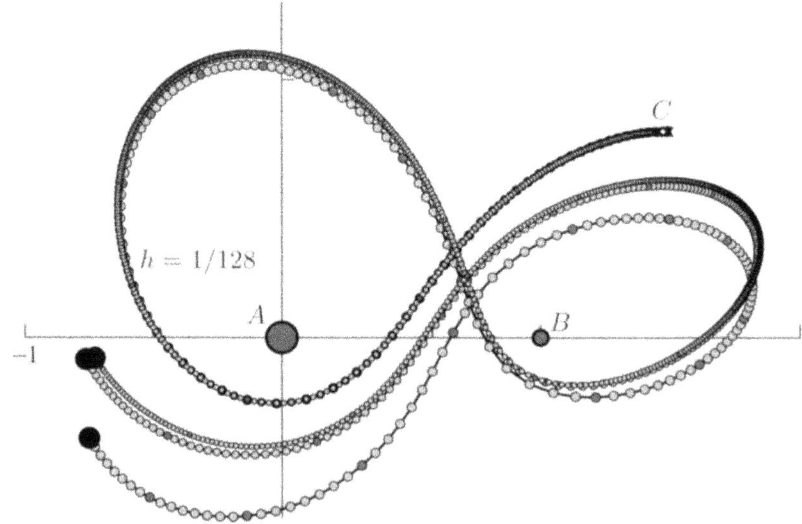

Figure 1.5. The solutions using Euler's third-order method for $h = 1/32$, $h = 1/64$ and $h = 1/128$.

$$y = b + \frac{(x-a)dy}{dx} - \frac{(x-a)^2 ddy}{1.2\, dx^2} + \frac{(x-a)^3 d^3y}{1.2.3\, dx^3} - \frac{(x-a)^4 d^4y}{1.2.3.4\, dx^4} + \text{etc.}$$

Figure 1.6. Euler's third-order implicit 'Obreschkoff' method.

But what represents the formula reproduced in Figure 1.6, with its curious sign changes, which Euler had written half a page earlier? Well, we can observe that the initial value b and the solution value y have been interchanged. *This is the implicit Euler method for order* 1 *and represents the so-called Hermite–Obreschkoff methods of the first column of the Padé table for higher orders*; see Hairer, Nørsett and Wanner (1993, Section II.13). Euler realizes that the application of such a formula requires the solution of an 'aequatio algebraica'.

1.3. Recursive computation of Taylor coefficients

At the beginning of the era of *automatic electronic calculating machines*, many authors had discovered independently the possibility of computing, in a more or less general setting, the Taylor coefficients of the solutions of differential equations in a *differentiation-free* recursive algorithm.[1] We demonstrate the method using the example $y' = x^2 + y^2$. Set

$$y(x_0+h) = y_0 + hy_1 + h^2y_2 + h^3y_3 + h^4y_4 + \cdots, \quad x^2 = x_0^2 + 2x_0h + h^2,$$

develop

$$\begin{aligned} y' = {} & y_1 + 2y_2h + 3y_3h^2 + 4y_4h^3 + \cdots \\ = {} & x_0^2 + 2x_0h + h^2 \\ & + y_0^2 + 2y_1y_0h + 2y_0y_2h^2 + 2y_0y_3h^3 \\ & \qquad + y_1^2h^2 + 2y_1y_2h^3 + \cdots \end{aligned} \tag{1.5}$$

and compare coefficients. In each column y_{n+1}, say, appears above terms of lower index, which allows us to compute them recursively via

$$y_1 = x_0^2 + y_0^2, \; 2y_2 = 2x_0 + 2y_1y_0, \; 3y_3 = 1 + 2y_0y_2 + y_1^2, \; 4y_4 = 2y_0y_3 + 2y_1y_2, \tag{1.6}$$

and so on. This can be turned into the following *general algorithm*.

Suppose that f of $y' = f(x, y)$ is composed of elementary functions, such as $r = pq$ or $r = p^c$ with c a constant. Then each of these functions is replaced by a formula computing the ith term of the Taylor coefficient of r

[1] This author, in his very first book, written in 1968, compiled the following list: J. F. Steffensen 1956, A. Gibbons 1960, W. Gautschi 1966, R. E. Moore 1966, I. Mennig 1964, Miller-Hurst 1958, E. Rabe 1961, E. Fehlberg 1964, Deprit-Zahar 1966, Leavitt 1966, Richtmyer 1957, Shanks 1964, Chiarella-Reichel 1968, A. J. Strecok 1968.

$\alpha = aa + bb;\ 2\beta = 2ab + 2a;\ 3\gamma = 2\beta b + \alpha\alpha + 1$
$4\delta = 2\gamma b + 2a\beta;\ 5\varepsilon = 2\delta b + 2a\gamma + \beta\beta;$
$6\zeta = 2\varepsilon b + 2a\delta + 2\beta\gamma$ etc.

Figure 1.7. Euler's recursive differentiation-free computation of the Taylor coefficients of the solutions of $y' = x^2 + y^2$.

from coefficients of lower order of r, or of the same order of p and q:

$$r = pq : \Rightarrow \quad r_i = \sum_{j=0}^{i} p_j q_{i-j}, \qquad i = 0, 1, 2, \ldots,$$

$$r = p^c : \Rightarrow \ r_0 = p_0^c,\ r_i = \frac{1}{ip_0}\left(\sum_{j=0}^{i-1}\big(ci - (c+1)j\big) r_j p_{i-j}\right), \qquad i = 1, 2, \ldots.$$

Thus the following scheme computes all derivatives recursively:

$$\begin{array}{llll}
x_0, y_0 & \mapsto \cdots & p_0, q_0 \rightarrow r_0 & \cdots \mapsto \ f_0 \\
\cdots & & \cdots & \cdots \\
y_1 = f_0 & \mapsto \cdots & p_1, q_1 \rightarrow r_1 & \cdots \mapsto \ f_1 \\
\cdots & & \cdots & \cdots \\
y_2 = \frac{1}{2} f_1 & \mapsto \cdots & p_2, q_2 \rightarrow r_2 & \cdots \mapsto \ f_2 \\
\cdots & & \cdots & \cdots \\
y_3 = \frac{1}{3} f_2 & \mapsto \cdots & p_3, q_3 \rightarrow r_3 & \cdots \mapsto \ f_3 \\
\cdots & & \cdots & \cdots \\
y_4 = \frac{1}{4} f_3 & etc., & &
\end{array} \tag{1.7}$$

to any order. And what is the surprise? If we compare Figure 1.7, which is copied from §663 of E342, with the formulas of (1.6), we see that *Euler invented this too.* ('Uti haec methodus simplicior ...')

1.4. The 'symplectic' Euler method

There is another way of improving Euler's method, not with the heavy machinery of higher derivatives, but with a small and clever modification of the formulas. We saw in Figure 1.2 that with Euler's method (1.4) the velocity updates $v_1 - v_0$ are computed from the force F at the point x_0, which lies just outside the interval of action of these two velocities. Hence *we obtain a much more symmetric situation if we update the velocity with the force evaluated at the point* x_1, and have in general

$$x_{n+1} = x_n + hv_n, \quad v_{n+1} = v_n + hF(t_{n+1}, x_{n+1}), \qquad h = \Delta t. \tag{1.8}$$

This modification is free of charge: it just requires us to exchange two lines of the code, but leads to a significant increase of the performance, as can be seen in Figure 1.8.

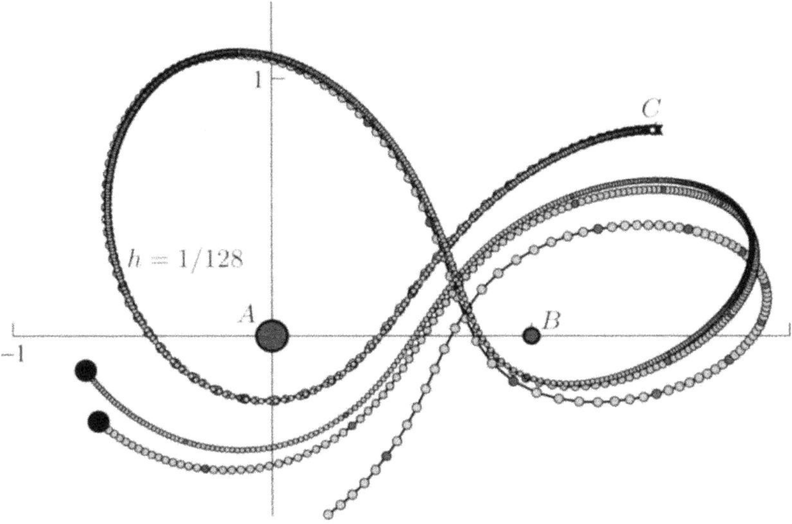

Figure 1.8. The solutions with the symplectic Euler method for $h = 1/32$, $h = 1/64$ and $h = 1/128$.

This method, for reasons which are explained in Hairer, Lubich and Wanner (2006, Section VI.3), bears the name *symplectic Euler method*, and was designed by De Vogelaere (1956) for computations in nuclear physics. This very first paper on symplectic integration was so ahead of current research interests that it never found a publisher.

Now, after all we have seen above, we might ask if there is yet another miracle, and whether this method, too, can be found somewhere in Euler's work? No, this miracle has not taken place, but another miracle has: *this method appears in Newton's Principia Mathematica, from 1687*, and will play an important role in the discussions below. Did you say 1687, the reader may ask, a century earlier than Euler? Yes, but it was then used *inversely*, as we will see in the following sections.

2. 'Inverse' numerical methods

How are the differential equations for modelling phenomena of nature justified? We will see the surprising fact that numerical methods for ODEs are *older* than ODEs themselves, but were used *inversely* to *establish these ODEs*. Here are some examples.

2.1. The tractrix of Leibniz

During the years 1674/75 young Leibniz was visiting Paris and received from Christiaan Huygens his first introduction to 'modern' mathematics. During this visit, the famous architect and medical doctor Claude Perrault stated the following challenge: Which curve is described by a silver pocket watch ('horologio portabili suae thecae argentae'), when it is pulled across the table, where the end of the chain moves on a straight line (see (2.1) *left*)? Today's students are told that 'trivially' this curve must have a tangent of constant length a and that therefore we have to solve

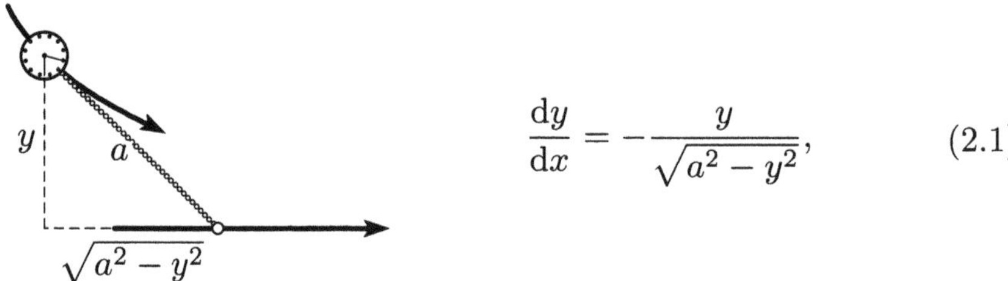

$$\frac{\mathrm{d}y}{\mathrm{d}x} = -\frac{y}{\sqrt{a^2-y^2}}, \tag{2.1}$$

which can be solved by methods revealed by Euler (E342 1768), namely, separation of the variables and then an integral containing a square root of a quadratic polynomial.

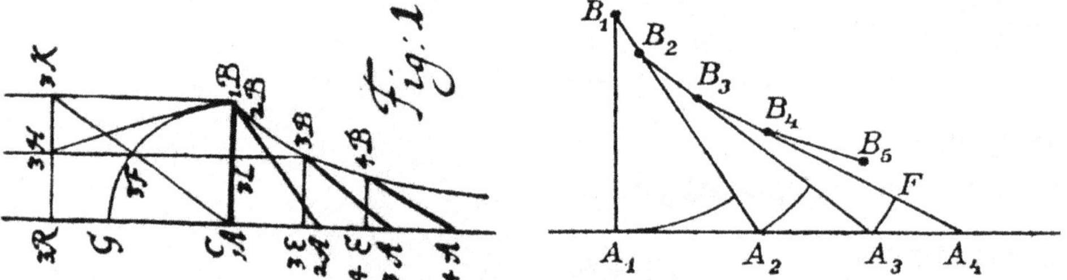

Figure 2.1. Drawings by Leibniz (*left*) and Kowalewski (*right*).

But for Leibniz who, as one of the inventors of differential calculus, was certainly not a stupid man, this conclusion was not so easy (see Figure 2.1, from Leibniz (1693); the drawing by Kowalewski, in his translation of 1908, is even clearer): physical intuition tells us that pulling the watch with a finger along a line is the same as rotating the chain, which is assumed to be without weight, by a small angle to the right, then pulling in the direction of the chain, until the imposed line is reached again, and so on. The polygon $B_1, B_2, B_3 \ldots$, which we obtain in this way, is precisely the result of the implicit Euler method applied to equation (2.1). We see that *this differential equation is justified through the inverse use of the implicit Euler method.*

A brilliant history of the tractrix, including a detailed description from the earliest publications up to later developments in the 20th century, has recently been published by Tournès (2009). In particular, it contains on page 11 an autograph drawing by Jacob Bernoulli (dated before 1692) explaining the properties of the curve using the *explicit* Euler method.

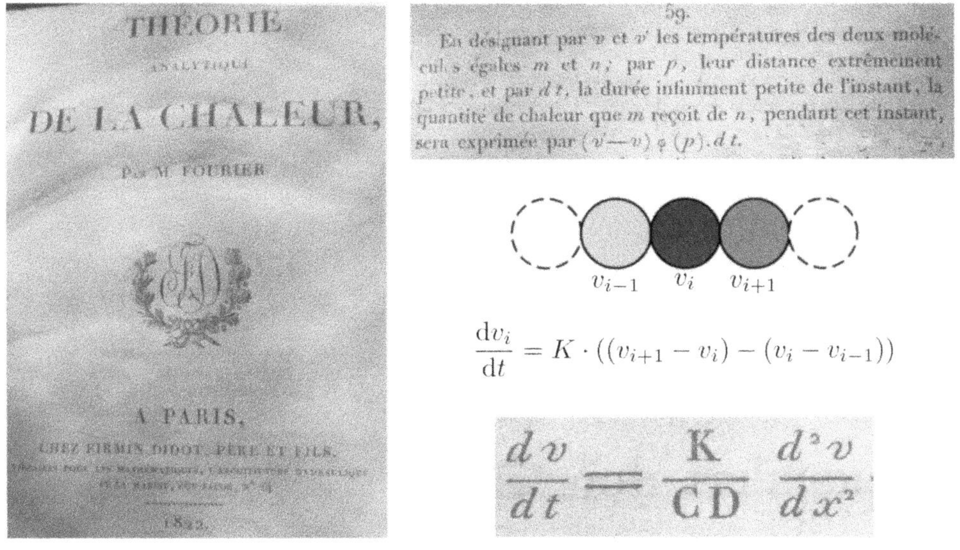

Figure 2.2. The discovery of the heat equation by Fourier (1822).

2.2. Fourier's heat equation

One of the most important books of science in the nineteenth century, *La Théorie Analytique de la Chaleur* by Fourier (1822), originates from a manuscript sent to the French Academy of Sciences in 1807, which was rejected due to 'lack of rigour' and could only be published after Lagrange's death. It is here that Fourier discovers the parabolic partial differential equation which governs the transport of heat, and whose solution led to the important concepts of separation of variables, eigenvalue problems, Fourier series and Fourier transform.

We see in Figure 2.2 the first publication of this equation, and in words (not formulas) Fourier's motivation for justifying it: in §59 he applies Newton's observation that the quantity of heat passing from one soup pot to another is proportional to the difference of the temperatures, on the molecular level. Therefore, if v_i is the temperature of the ith molecule (Fourier used the symbols v and v'), the heat it receives from the right-hand neighbour during an instance 'infiniment petite' of time is $\mathrm{d}v_i = K \cdot (v_{i+1} - v_i) \cdot \mathrm{d}t$; similarly the heat received from the left-hand neighbour will be $\mathrm{d}v_i = K \cdot (v_{i-1} - v_i) \cdot \mathrm{d}t$. The sum of these two terms is a second-order finite difference of the values v_{i-1}, v_i, v_{i+1}; when the distance of the molecules tends to zero, this expression will become the second partial derivative with respect to space. The heat equation is thus justified *by the method of lines applied inversely.*

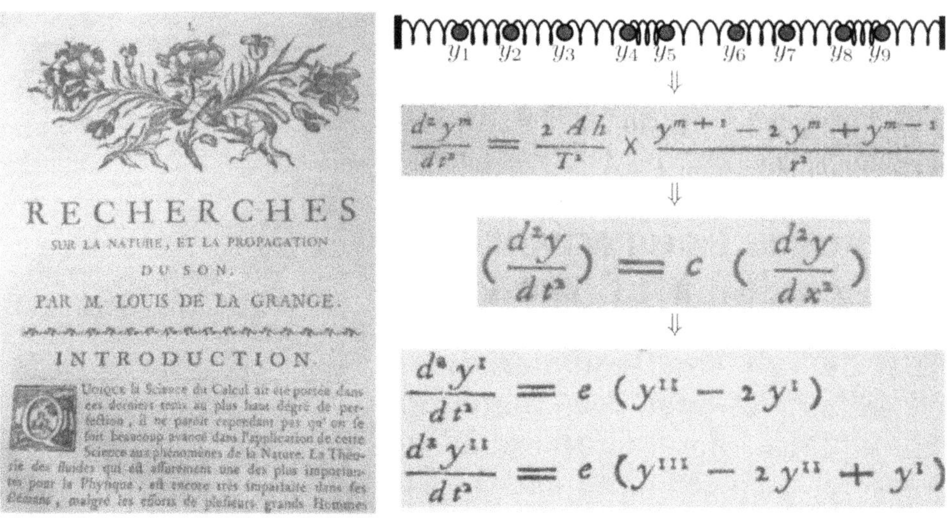

Figure 2.3. The discovery of the equation of sound by Lagrange (1759).

2.3. Lagrange's theory of sound

In Lagrange's paper of 1759, one of the first papers with which the young rising scientific star filled the newly founded *Miscellanea Taurinensia*, the above ideas are explained even more clearly: the air is thought to consist of a sequence of molecules, tied together with elastic forces. If we denote by y_i the displacement of the ith molecule, then the elastic forces acting from the left-hand and the right-hand neighbour are, following Hooke's law, proportional to the differences $y_{i-1} - y_i$ to the left, and $y_{i+1} - y_i$ to the right. The sum of these two forces is again a second finite difference and must be proportional to the acceleration $\frac{\mathrm{d}^2 y_i}{\mathrm{d}t^2}$. This gives the first formula in Figure 2.3 (*right*). The *inverse* use of the method of lines then leads to the partial differential equation for sound (second formula). In order to solve it, Lagrange proceeds *back to the method of lines* (third formula) which represents a linear system of ODEs with constant coefficients, which Lagrange solved using the 'now-so-familiar formulas' of Euler. After lengthy calculations, Lagrange finally stood at the door of Fourier series but did not open it; for more details see Hairer *et al.* (1993, pp. 28–29).

2.4. Euler's equations for variational calculus

We all learned in analysis lectures that Euler found his famous differential equation for a general variational problem,

$$J = \int_a^b Z\,\mathrm{d}x = \min! \text{ vel } \max!, \tag{2.2}$$

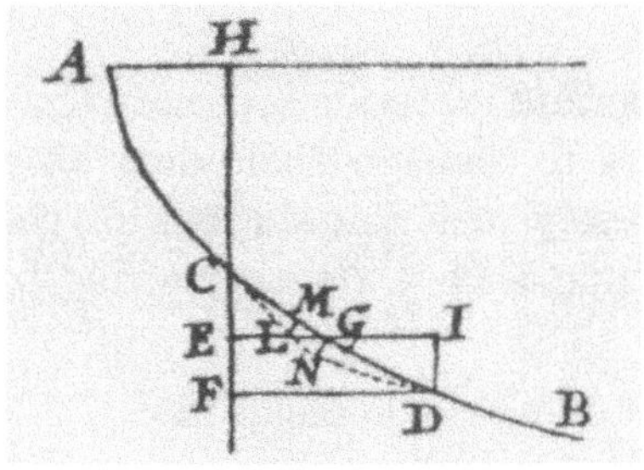

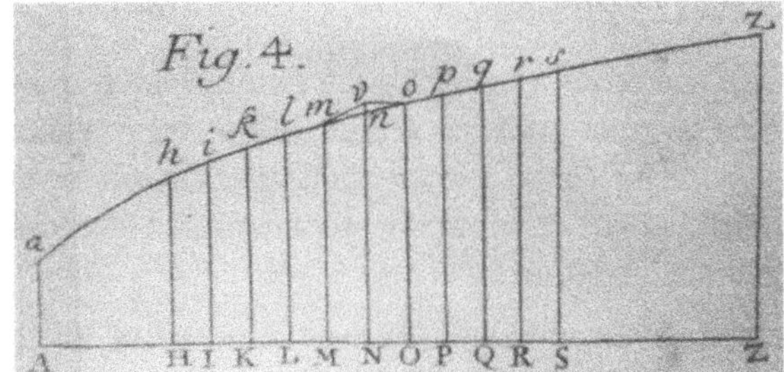

Figure 2.4. The 'variation' of one single y-value in Jacob Bernoulli's solution of the brachistochrone problem (*left*) and in Euler's derivation of his general formula (*right*).

where $Z = Z(x, y, p)$ is an arbitrarily given function of x, y and $p = \frac{\mathrm{d}y}{\mathrm{d}x}$, in his masterpiece, Euler E65 (1744), and that the proof usually given is due to Lagrange in 1755. This leads naturally to the question: How did Euler himself discover this equation, which enabled him to solve many dozens of such problems of all kinds? We will see that Euler's original route, displayed in Figure 2.4 (*right*), is quite elegant.

The solution $y(x)$ is represented by a discrete sequence $y, y', y'', y''', \ldots$ (in the figure $h, i, k, l, m, \ldots$; today we would say 'piecewise linear finite elements'), and the integral J in (2.2) by a 'finite Riemann sum',

$$J = Z\,dx + Z'\,dx + Z''\,dx + Z'''\,dx + \&c. \tag{2.3}$$

This J must be minimal for each choice of the values $h, i, k, l, m, \ldots$. We therefore differentiate it with respect to any of these (in Figure 2.4 Euler moves n to ν). We write (2.3) more explicitly, merging Euler's notation with ours, and replacing the derivative p by a finite divided difference

$$J = Z\left(x, y, \frac{y' - y}{\mathrm{d}x}\right)\mathrm{d}x + Z\left(x', y', \frac{y'' - y'}{\mathrm{d}x}\right)\mathrm{d}x + \cdots, \tag{2.4}$$

differentiate with respect to y', which appears in three places, set this derivative to zero, and obtain

$$(P + N'dx - P') = 0, \qquad \text{where } N = \frac{\partial Z}{\partial y},\ P = \frac{\partial Z}{\partial p}. \tag{2.5}$$

This condition, which must hold everywhere, is precisely the differential equation

$$N - \frac{dP}{dx} = 0 \qquad \text{or} \qquad \frac{\partial Z}{\partial y} - \frac{\mathrm{d}}{\mathrm{d}x}\left(\frac{\partial Z}{\partial p}\right) = 0, \tag{2.6}$$

discretized by the *implicit Euler method.* Hence, again, the *inverse* use of this method, and all other numerical procedures applied before, establishes equation (2.6), the famous *Euler's equation of variational calculus.*

2.5. Conclusion

We see from all the above examples that numerical methods are more than just simple recipes which allow lazy scientists to consign their duty to a computer, but they constitute, applied inversely, the foundations of the application of analysis to science. We conclude with a quotation due to Bertrand Russell:[2]

> Although this may seem a paradox, all exact science is dominated by the idea of approximation.
>
> B. Russell, *The Scientific Outlook* (1931)

3. The origin of Kepler's laws

> Astronomy is older than physics. In fact, it got physics started by showing the beautiful simplicity of the motion of the stars and planets, the understanding of which was the *beginning* of physics.
>
> R. Feynman (1963); published in *Six Easy Pieces* (1994), p. 59

> [Gravity] was one of the first great laws to be discovered and it has an interesting history. You may say, 'Yes, but then it is old hat, I would like to hear something about a more modern science'. More recent perhaps, but not more modern. ... I do not feel at all bad about telling you about the Law of Gravitation because in describing its history and methods, the character of its discovery, its quality, I am being completely modern.
>
> R. Feynman, *Messenger Lectures* (1964);
> published in *The Character of Physical Law* (1967)

We now go back to the very beginning of science (see the first quotation from Feynman), or rather, to the beginning of *modern science* (second quotation, from a public lecture in which Feynman, soon to receive the Nobel Prize, spoke, not about recent discoveries in quantum electrodynamics, but, to the astonishment of everybody, about Kepler's and Newton's laws).

3.1. Ptolemy (∼ AD 150)

The first great observer of the sky, with thousands of precise measurements of the position of the stars and planets, was Ptolemy, who lived around AD 150. His fundamental work, originally called the *Great Collection*, or, in Greek, ἡ μεγάλη σύνταξις, was given by Islamic astronomers the Arabic definite article 'al', to become the multilingual conglomerate *Almagest.* It was translated in the second half of the fifteenth century by Peurbach and

[2] Communicated to the author by Jan Lacki, Geneva.

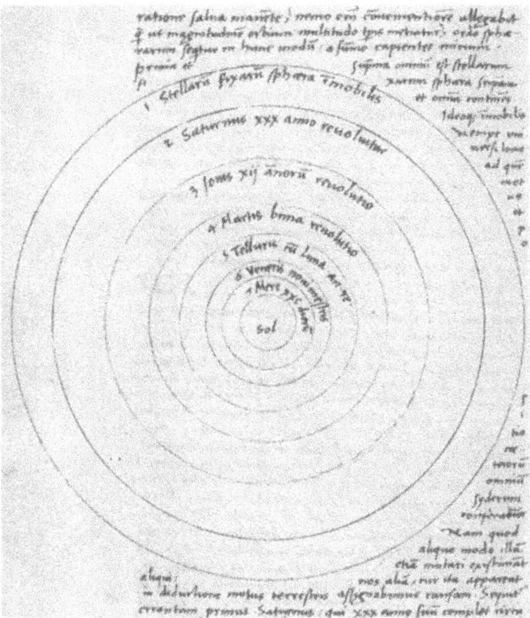

Figure 3.1. *Left*: Frontispiece of Regiomontanus' translation of Ptolemy's *Almagest*, published in 1496; we see the Earth in the centre, with the stars rotating around attached to a solid machinery. *Right*: A page of Copernicus' manuscript *De Revolutionibus*, representing the heliocentric system, published just before his death in 1543. Reproduced, with permission, from the copy of *De Revolutionibus*, call number BJ10000, in the Library of the Jagiellonian University, Kraków, Poland.

Regiomontanus and published in 1496 in Venice as one of the very first scientific books ever printed (see Figure 3.1, *left*). This book had a big influence on the astronomers of that time, in particular a young Polish fellow studying in Bologna, Nicolaus Copernicus. Ptolemy's conclusion after all his measurements is explained in the central drawing of Figure 3.2 (*right*): the Earth is in the centre, the Sun moves around the Earth, and the planets move on *epicycles*, circles whose centres move around the Earth.

3.2. Nicolaus Copernicus (1473–1543)

Kepler (1609, p. 131) wrote: 'I start by explaining these things in the Copernican setting, where they are easiest to understand.' Hence, the Copernican system is explained in the first image of Kepler's drawing in Figure 3.2 (*right*): the Sun is in the centre and all planets, including the Earth, rotate around the Sun on circles which are, however, in an *eccentric* position. This model, which makes no distinction between the Earth and the other planets, is the simplest of all, and therefore an attractive choice. An

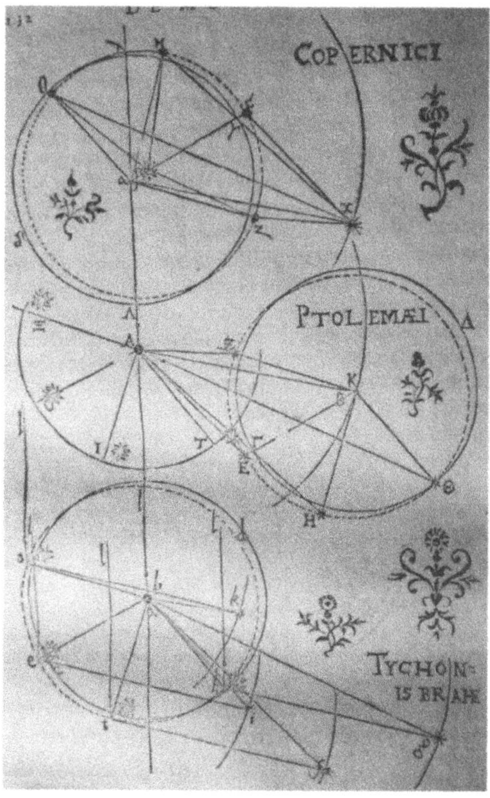

Figure 3.2. *Left*: Tycho Brahe (1546–1601). *Right*: Explanation of the three systems of Copernicus (*above*), Ptolemy (*centre*) and Tycho Brahe (*below*) in Kepler's *Astronomia Nova* (1609).

outermost circle ('sphaera immobilis') is then reserved for the fixed stars ('stellarii fixarii'; see Figure 3.1, *right*). Copernicus, who had obtained this result after life-long effort and observations, hesitated to make it public, but this was finally achieved with the help of the German mathematician G. J. Rheticus, just before Copernicus' death in 1543.

3.3. Tycho Brahe (1546–1601)

Tycho Brahe was a Danish nobleman who became impassioned by astronomy after a solar eclipse of 1560, the supernova of 1572 and the comet of 1577. The imperfection of existing astronomical predictions, based on Ptolemy's calculations, motivated him to become 'a second Ptolemy', by building the huge astronomical observatory Uraniborg on the island of Hven, where, over many years, he made thousands of astronomical measurements with unequalled precision (accurate to 1 arc minute). After problems with the new king, he left Denmark in 1597 with all his tables and went to Prague.

Tycho did not accept the Copernican system, because if the Earth were moving around the Sun, then the fixed stars must show a parallax, which, however, could not be observed.[3] Therefore, Tycho's model of the universe was as follows (see the third drawing in Figure 3.2, *right*): *the Sun rotates around the Earth on an eccentric circle, and the planets rotate around the Sun, also on eccentric circles.* The sphere with the fixed stars does not move.

3.4. The Ptolemy–Copernicus–Brahe model for planetary motion

As long as we are only interested in the *relative* position of a planet with respect to the Sun, *i.e.*, in the geometry of such an orbit, and not its physical quantities such as forces and accelerations, it does not matter which body is moving and which is not (since we are in Denmark, we may say 'to move or not to move, that is the question ...'). In this case, all three above models are equivalent and state the following (see Figure 3.3).

(1) The planets move around the Sun on circular orbits. The centre B of this circle is called the *Mean Sun* and its distance from the Sun S is governed by the *eccentricity.* This law describing the geometric form of the orbit is called in the old literature 'inequalitas prima'.

(2) The speed of the planet on this orbit, which has been observed being faster close to the Sun and slower away from it, is governed by a *punctum aequans* C, which is at the same distance from B as S, in the opposite position. Seen from this *punctum aequans*, the planets move at constant angular speed. This law is called 'inequalitas secunda'.

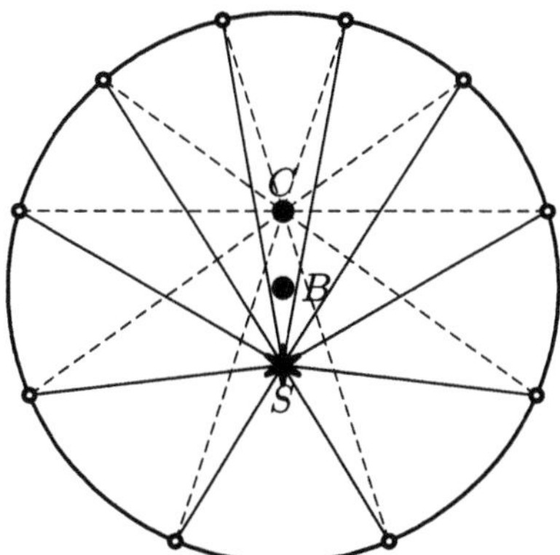

Figure 3.3. The Ptolemy–Copernicus–Brahe model for planetary motion: S the Sun, B the Mean Sun, C the *punctum aequans.*

[3] Because of the tremendous distance of the fixed stars, their parallaxes, which were finally discovered more than 200 years later (by Bessel in 1838), required more than a hundred times greater precision ($<$ 0.3 arc seconds).

In numerical calculations lasting years, Tycho tried to adjust the three parameters (radius, eccentricity and inclination of the line CS) of each of the planetary orbits for all his observations.

This worked fine for all planets **except for Mars!**

When young Johannes Kepler, who had fled Graz because of religious persecution, arrived in Prague, Tycho gave him the data of Mars to study. It is suspected that Tycho, who found the young man too self-assured and ambitious, wanted to cool him down with an apparently impossible task.

3.5. Kepler's Astronomia Nova

During the year 1605 Kepler finally unveiled the secret of the orbit of Mars, four years after the sudden death of Tycho. However, owing to differences with Tycho's family, lack of money and the tremendous size of Kepler's manuscript, the book could not be published until 1609. In the same year, Galileo pointed his first telescope towards the sky. Both events, 400 years ago, can be said to mark the birth of modern astronomy, and thus of all modern science.

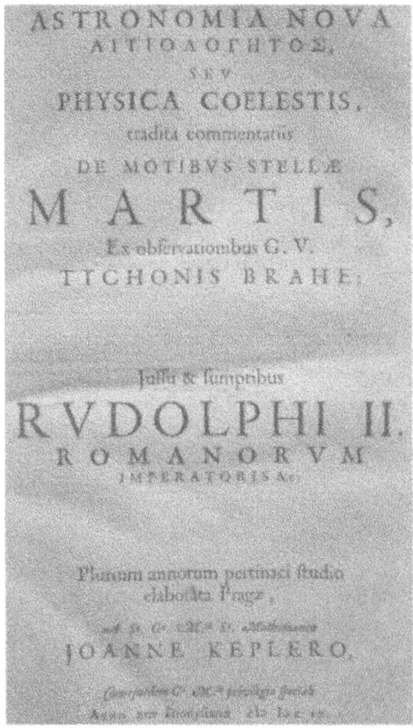
ASTRONOMIA NOVA
ΑΙΤΙΟΛΟΓΗΤΟΣ,
SEV
PHYSICA COELESTIS,
tradita commentariis
DE MOTIBVS STELLÆ
MARTIS,
Ex observationibus G. V.
TYCHONIS BRAHE:
Jussu & sumptibus
RVDOLPHI II.
ROMANORVM
IMPERATORIS &c:
Plurium annorum pertinaci studio
elaborata Pragæ,
JOANNE KEPLERO,

Figure 3.4. Frontispiece and synopsis of Kepler's *Astronomia Nova*.

We see in Figure 3.4 (*left*) the frontispiece of this work. Following the title, Kepler writes 'Αἰτιολογητός seu Physica Coelestis', which expresses that Kepler is also interested in the *physical reasons* for the movement of the planets, and not only in their geometry. Then comes 'of the movement of the star Mars from the observations of Tycho Brahe'. We see that Tycho's name is printed in precisely the same size as 'the mathematician Johannes Kepler' below. We also read 'after many years of pertinacious study in Prague'. The name of Emperor Rudolf II, who paid for all this, is in the same huge letters as the planet Mars.

After a long hymn of praise for the emperor, several poems and an obligatory page by Tycho's son-in-law F. G. Tengnagel praising Tycho, the book starts with an introduction, which sounds very 'modern', about the difficulty of writing (and reading) a mathematical book (see Figure 3.5).

The book, with its 340 quarto pages, arranged in 70 chapters, is indeed very difficult to read. To help the reader, Kepler included a 'Synopsis Totius Operis' on a huge sheet, which must be (carefully) unfolded from the book, listing all the 70 chapters in the form of a binary tree (see Figure 3.4, *right*). We see that the book consists of five parts; parts three (discovery of Kepler's Second Law) and four (Kepler's First Law) are the central core.

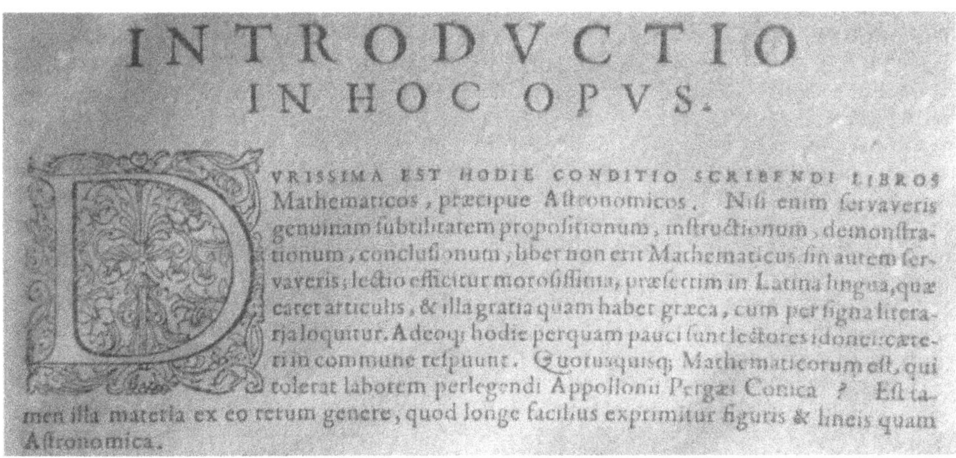

INTRODVCTIO
IN HOC OPVS.

DVRISSIMA EST HODIE CONDITIO SCRIBENDI LIBROS Mathematicos, præcipue Astronomicos. Nisi enim servaveris genuinam subtilitatem propositionum, instructionum, demonstrationum, conclusionum; liber non erit Mathematicus: sin autem servaveris; lectio efficitur morosissima, præsertim in Latina lingua, quæ caret articulis, & illa gratia quam habet græca, cum per signa literaria loquitur. Adeoq; hodie perquam pauci sunt lectores idonei: cæteri in commune respuunt. Quotusquisq; Mathematicorum est, qui tolerat laborem perlegendi Appollonii Pergæi Conica? Est tamen illa materia ex eo rerum genere, quod longe facilius exprimitur figuris & lineis quam Astronomica.

Figure 3.5. The 'very modern' introduction of Kepler's *Astronomia Nova*: 'Today the requirements for writing mathematical books, especially astronomical ones, are very hard. If you do not preserve the original details of the propositions, instructions, demonstrations and conclusions, the book will not be mathematical. If, however, you do preserve them, the book will be very boring to read.'

3.6. Kepler's Pars Secunda: 'Ad imitationem veterum'

> If you find these calculations tedious ('pertaesum'), then have pity on me ('jure mei te miserat'): I did them at least 70 times, losing a lot of time ('ad minimum septuagies ivi cum plurima temporis jactura').
>
> Kepler, *Astronomia Nova* (1609), p. 95

In part two, Kepler tries to obtain the best possible results 'by imitating the Ancients' (the 'Ancients' are Ptolemy, Copernicus and Tycho Brahe, 25 years older than him). His main observation is that there is no convincing reason for the assumption that, in Figure 3.3, the *punctum aequans* C is at the same distance from B as the Sun S. Therefore he allows an arbitrary position for C, which increases the number of free parameters to 4, hence these constants must be determined from 4 observations. This leads to very tedious calculations (see quotation), which Kepler solved by an iterative scheme similar to the regula falsi (see Figure 3.6). As a result, Kepler obtained a very precise model, the Hypothesis Vicaria, for the orbit of the Earth. The importance of this is the fact that only through knowing the distance of the Earth at any moment can the distances of the other planets be computed accurately, by measuring all the angles and using the sine theorem of trigonometry.

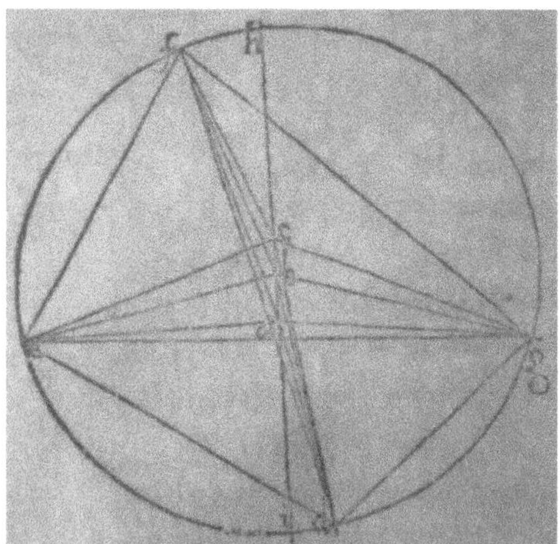

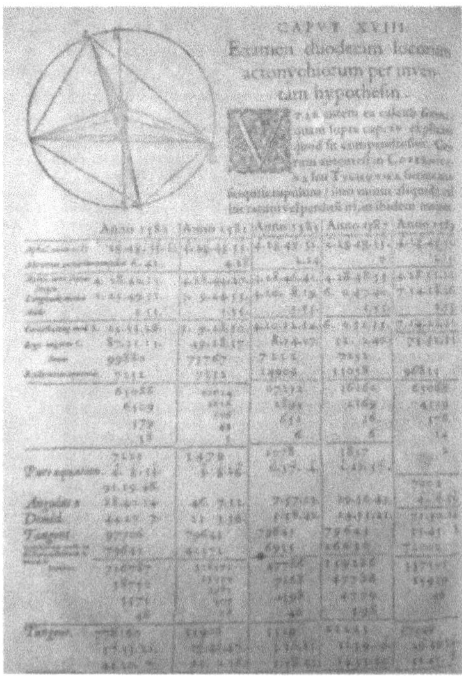

CAPVT XVIII.
Examen duodecim locorum acronychiorum per inventam hypothesin.

Figure 3.6. *Above*: Kepler's improved model leading to the *Hypothesis Vicaria*. *Right:* An example of the recursive computation of its parameters.

3.7. Kepler's Pars Tertia: 'Ex propria sententia' (his own opinion).

In his third part (Chapters 22–40), Kepler wants to rid himself more and more of all this geometric thinking, with the *punctum aequans* C as a kind of machine pushing the planets, and wants to accept only truly physical reasons, as announced in his title:

ΑΙΤΙΟΛΟΓΗΤΟΣ PHYSICA COELESTIS,

These discussions fill Chapters 32–39. But if we throw away the *punctum aequans*, we have to replace it by something else. What could that be? The only really fixed object *is the Sun*, in the centre of the Universe ('I myself am of the opinion of Copernicus: I admit that the Earth is one of the planets', p. 170). After long deliberations about all possible reasons, forces, magnetic forces, the light coming from the Sun, wind from an ether, he finally concludes that planets must have a soul, which looks to the Sun and 'wishes' to move, seeing its diameter inversely proportional to the distance ('Is it so, Kepler, that you attribute two eyes to each planet? Not at all.' p. 191). 'You see thus, reader, with reflection and spirit' (p. 191), that *the speed of the planets is inversely proportional to the distance* (see Figure 3.7, *left*).

However, the arc length of a curve is a nasty mathematical expression: it involves Pythagoras' theorem and an uncomfortable square root. So, finally, inspired by the ideas of a great mathematical god, Archimedes,

diſtantias omnes ineſſe. Nam memineram, ſic olim & ARCHIMEDEM, cum circumferentiæ proportionem ad diametrum quæreret, circulum in infinita triangula diſſecuiſſe: nam hæc vis occulta eſt ejus demon-

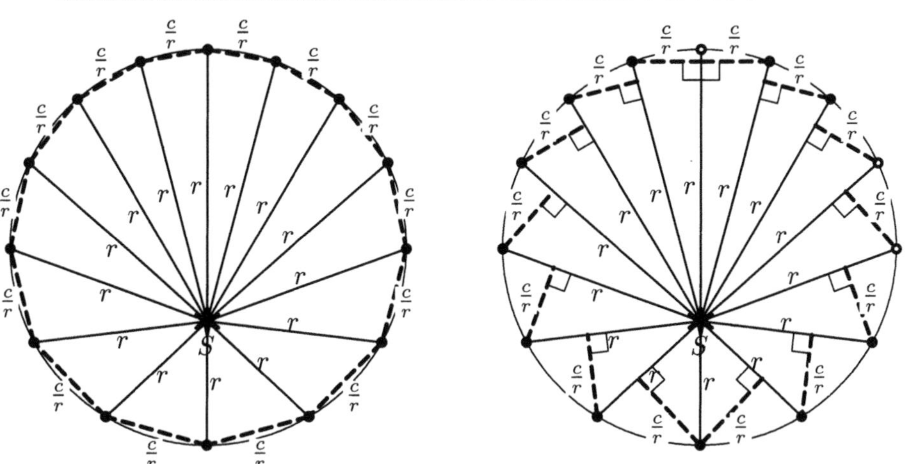

Figure 3.7. *Left*: Kepler's model from Chapter 39 with the speed inversely proportional to the distance. *Right*: Kepler's 'methodus imperfecta' from Chapter 40: equal times correspond to equal areas.

he decides to replace the hypothenuse by the leg (the orthogonal distance of the rays) and arrive at the 'methodus imperfecta', saying that the heights of the triangles are inversely proportional to the distance, *i.e.*, that *the areas of the triangles for equal times are all the same* (see Figure 3.7, *right*). In this way, *Kepler's Second Law* was finally discovered, and corresponded perfectly with the observations of Tycho.

3.8. Kepler's Pars Quarta: the elliptic orbits

The most difficult part was yet to come: the definite renunciation of all Greek and medieval thinking on circular orbits. This nearly hopeless struggle fills Chapters 41–55, until the epiphany in Chapter 56 (p. 267):

> plane nihil dictum eſſe, itaque futilem fuiſſe meum de Marte triumphum; forte fortuito incido in ſecantem anguli 5.18. quæ eſt menſura æquationis Opticæ maximæ. Quem cum viderem eſſe 100429, hic quaſi e ſomno expergefactus, & novam lucem intuitus, ſic cœpi ratio-

> As I reflected ... that my triumph over Mars had been futile, I fell by chance on the observation that the secant of the angle $5°18'$ is 1.00429, which was the error of the measure of the maximal point. I awoke as if from sleep, & a new light broke on me.

This decisive discovery is explained in Figure 3.8 (*right*). If the orbit of Mars were a circle of radius 1, then the eccentricity $e = OS$ is such that the angle OBS would be $5°18'$, where B is the point with the largest elongation from the axis SOC. Therefore the distance BS would be $1/\cos 5°18' = 1.00429$. *But Tycho measured* 1 *for this distance.* Therefore, we should move the point B to the point B', whose distance $B'S$ is that of BO; in other words, we have to replace, once again, the hypothenuse BS by the leg BO ('Hence, what brought us to despair in Chapter 39, now changes here into an argument to attain the truth,' p. 267). Kepler applied the same recipe to other points: move the point P to the position P' so that the length $P'S$ is that of the leg PR, which is

$$P'S = PR = 1 + e\cos u, \tag{3.1}$$

because the angle u, called the *eccentric anomaly*, reappears as angle SOR, so $OR = e\cos u$. Kepler finally concluded that 'these distances are confirmed by very numerous and very sure measurements' (Chapter 56, end).

In Chapters 58 and 59, Kepler finally achieves the proof that the orbit expressed by formula (3.1) represents an ellipse, with the Sun in one focus. The theory of conics had been developed to high perfection by Apollonius ($\sim$ 250 BC) and completed by Pappus ($\sim$ AD 300–350), but then forgotten for more than 1000 years. One rediscovery, albeit in a very rudimentary form, had been by Kepler himself in 1604, so Kepler claimed his proof to be

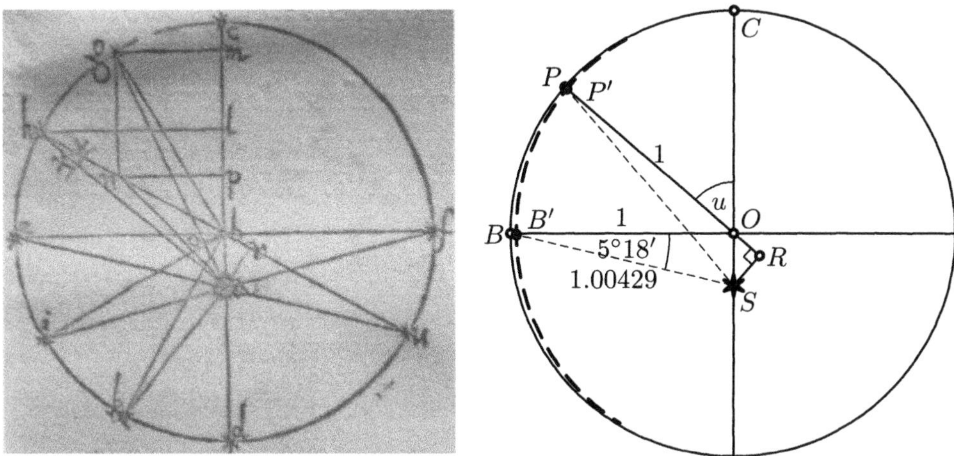

Figure 3.8. The discovery of Kepler's First Law (Chapter 56, p. 267): Kepler's drawing (*left*), modern drawing (*right*).

'although very sure, lacking art and not geometric' (p. 293). He concluded that the reader should consult the conics of Apollonius, 'which requires very strong meditations and reflections about these matters' (p. 295). In fact, formula (3.1) just expresses the result of Pappus that the distance PS is in a constant ratio $e : 1$ with the distance of P to a directrix.

In Chapter 60, finally, Kepler establishes for the area $P'SC$, which by his Second Law is proportional to the elapsed time t, the expression

$$\mathrm{Const} \cdot t = \frac{B'O}{2} \cdot (u + e \sin u)$$

('composed of two portions of area, a sector and a triangle', p. 299). The solution of this equation allows one to find u for any given time t, 'but I believe myself unable to solve it, and whoever shows the way would be for me a great Apollonius' (p. 300).

3.9. Kepler's Third Law

Work on the *Astronomia Nova*, with all its 'artless' numerical calculations and hazardous conclusions, was for Kepler something of an interruption to his true vocation, which was to unveil the harmonies of God's creations with the help of the beauty of mathematics, in particular geometry and music.[4]

[4] 'Geometria enim, ... Deo coaeterna, inque Mente divina relucens, exempla Deo suppeditavit, ... exornandi Mundi, ut is fieret Optimus et Pulcherrimus, denique Creatoris simillimus.' (p. 13; see also *Gesammelte Werke*, Vol. 6, pp. 104 and 489; 'Geometry, eternal as God, and shining out of the Divine Mind, has supplied God with the models for shaping the World, making it the Best and most Beautiful, hence similar to the Creator.')

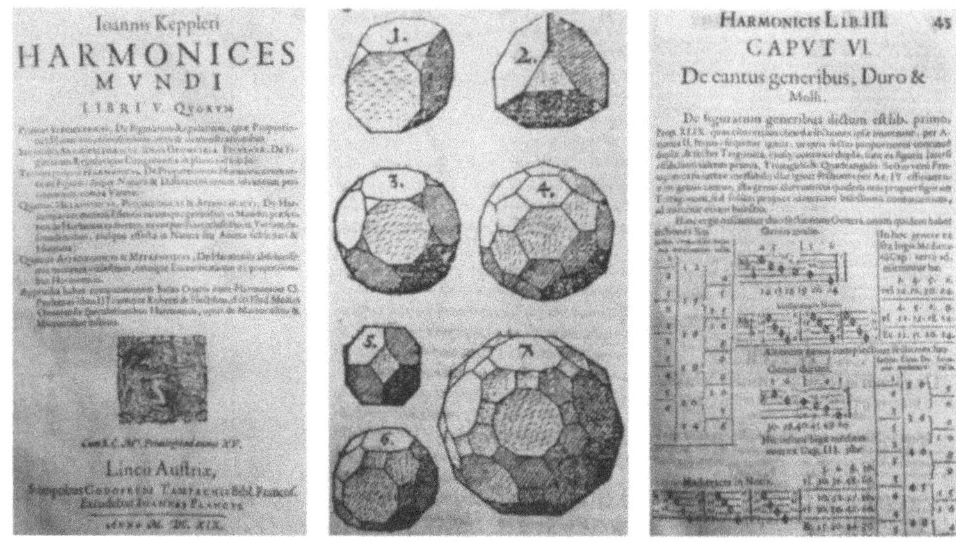
Ioannis Kepleri
HARMONICES
MVNDI
LIBRI V. QVORVM

Lincii Austriæ,

HARMONICES LIB. III. 45
CAPVT VI.
De cantus generibus, Duro & Molli.

Researching harmonies ... from geometry ... and music ...

Figure 3.9. Kepler's *Harmonices Mundi* (1619).

This research which he had begun in 1599 led finally to the *Harmonices Mundi* (Kepler 1619), in five books and an appendix (see Figure 3.9).

Although the books contain many beautiful results in mathematics, in particular the first rediscovery of the complete list of the 13 so-called *Archimedean solids* (some of which are shown in Figure 3.9, *centre*), the applications of all this to interesting results in astronomy led constantly to failure. Finally, the most famous result of this book was again discovered by vulgar numerical calculations and announced towards the end (Kepler 1619, Liber V, Caput 3, §8, p. 189) as follows:

inter principia, primò crederem. Sed res est certissima exactissimaque, quòd *proportio quæ est inter binorum quorumcunque Planetarum tempora periodica, sit præcisè sesquialtera proportionis* mediarum distanciarum, id

'It is extremely certain and extremely exact that the ratio of the time period for two planets is one and a half of the ratio of the mean distances.' In this way Kepler expresses the fact that

$$T_1/T_2 = (a_1/a_2)^{3/2}.$$

3.10. Galileo Galilei (1564–1642)

> ... io grandemente dubito che Aristotele non sperimentasse ...
> (I have strong doubts if Aristotle did make any experiments ...)
>
> Galileo's *Discorsi* (1638), First Day

> De subiecto vetustissimo novissimam promovemus scientiam.
> (We present an entirely new science on a very old subject.)
>
> Galileo's *Discorsi* (1638), Third Day

Another important influence on modern science came from Galileo in Italy. A first manuscript by Galileo on mechanics, finished around 1629, was translated into French and published in 1634 by Marin Mersenne. He soon ran into problems with the Roman Curia, and his masterly publication of 1638, *Discorsi e Dimostrazioni Matematiche*, written in the form of discussions between three persons of different scientific level in six days, was smuggled out of Italy and published in the Netherlands.

In 1609, the same year in which Kepler's *Astronomia Nova* appeared, Galileo first directed his telescope towards the sky. His discovery of the satellites of Jupiter and the phases of Venus left no doubts about the Copernican system. At the same time, he tried to understand mechanics by experimenting with ropes, heavy stones and beams down on Earth (see Figure 3.10). While the first two 'days' of the *Discorsi* were written in popular Italian, for the third 'day' Galileo turned to serious Latin (see quotation), and laid the fundamental principles of an 'entirely new' mechanics, such as mass, forces and acceleration, as we know it today.

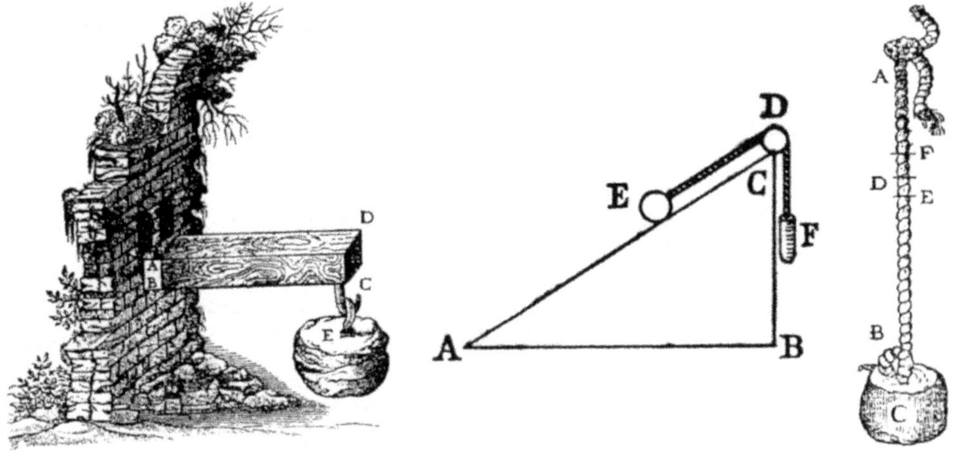

Figure 3.10. Illustrations from Galileo's *Discorsi* of 1638 (*left* and *right*), and his *Mechanics* of 1629 (*centre*).

3.11. Conclusion

We conclude this section with a quotation from Kepler (1609, p. 95):

> EXISTENT acuti Geometræ VIETÆ ſimiles, qui magnum aliquid eſſe putabunt demonſtrare hujus METHODI ἀτεχνίαν. Id enim & PTOLEMÆO & COPERNICO & REGIOMONTANO objectum in hoc negocio a VIETA. Eant igitur & ſchema Geometrice ipſi ſolvant, & erunt mihi magni Apollines. MIHI ſufficit ad quatuor vel quinque con-

> There might exist ingenious geometers, like Viète, who think they are doing something great in showing this [numerical] method to be *artless*.[5] Indeed, Viète made this very criticism of Ptolemy, Copernicus & Regiomontanus in his work. Let them step forward, then, & solve the scheme geometrically. They will be great Apollos for me.

This *great Apollo* will lead us to the next section.

4. Newton's discovery of the Law of Gravitation

> This sudden change of emphasis has been provoked by a visit from Edmund Halley (1656–1742), which probably took place in August [1684].
>
> S. Mandelbrote, *Footprints of the Lion* (2001), p. 88

During the half-century which separated Kepler's works from Newton's studies in Cambridge, Kepler's laws slowly became known and accepted, not through the books we have cited above, but through the *Rudolphine Tables* (Kepler 1627), a huge compendium of more than 300 pages of tables for the positions of stars and planets, which Kepler computed with the help of his laws, and which became the universal tool for generations of astronomers – and astrologers. Also, Galileo's principles of mechanics became known, particularly in Cambridge through the lectures of Isaac Barrow, who had visited Paris and Florence in 1655/56.

It thus became a natural challenge to understand the principles governing the movement of the planets from a mechanical point of view. Newton, who was in priority dispute with Robert Hooke of London over this discovery, declared later that he had discovered all this during the plague years 1665/66, but no written evidence about this claim could be found in his manuscripts.

The first manuscript clearly showing Newton's ideas is one dating to 1684, initiated by a visit to Cambridge by Edmund Halley, bringing news from

[5] In fact, this is still a very 'modern' opinion.

London (see quotation). All the illustrations produced below are reproduced from this manuscript.[6] This manuscript (and others) later led to the epoch-making *Principia* (Newton 1687).

4.1. Proof of Kepler's Second Law

> ... what Newton writes is correct, clear, and short; in earlier works the brilliant diamonds of discovery lie concealed in an opaque matrix of wordy special cases, laborious details, metaphysics, confusion, and error, while Newton follows a vein of pure gold.
>
> C. Truesdell, *Essays in the History of Mechanics* (1968), p. 88

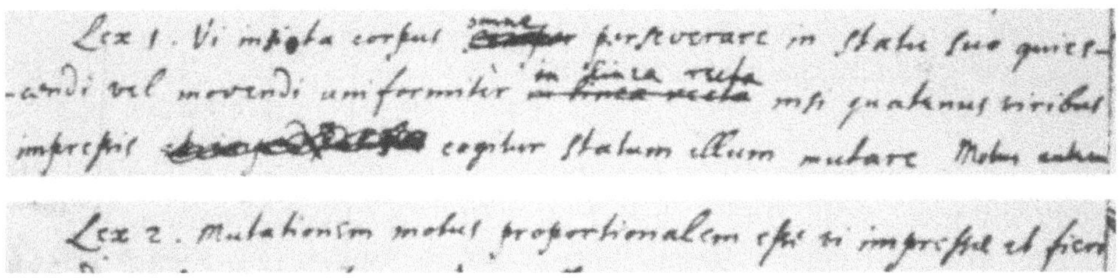

Figure 4.1. Newton's Lex 1 and Lex 2 from the manuscript Add. 3965[7a].

The first principles of motion fill many pages of axioms, theorems and discussions of the third 'day' of Galileo's *Discorsi* and many pages of beautiful prose in Barrow's lectures; for example: 'The following Axiom of *Aristotle* concerning Motion is famous ...: *He that is ignorant of Motion, must necessarily know nothing of Nature,*' (Barrow 1670, p. 2), or 'You know the very trite Saying of St. *Austin,*[7] *If no one asks me, I know; but if any Person should require me to tell him, I cannot,*' (Barrow 1670, p. 4). But Newton subsumes everything into three short and precise laws, the first two of which are as follows (see Figure 4.1).

Lex 1. Without force a body remains in uniform motion on a straight line.

Lex 2. The change of motion is proportional to the motive force impressed.

These laws were also expressed independently by Huygens (1673).

The basic idea is now displayed in Figure 4.2. Instead of thinking of the body moving on a curve $ABCDEF\ldots$ under the continuous influence of a force F acting from the Sun, we let it move under Lex 1 from A to B during a time interval Δt without force, and replace the forces acting by one force impulse of size $\Delta t \cdot F$ at the end of this step. If this force is directed towards the Sun, by Lex 2 the velocity which was in direction AB will change to the

[6] With courtesy of Cambridge University Library.

[7] St. Augustine.

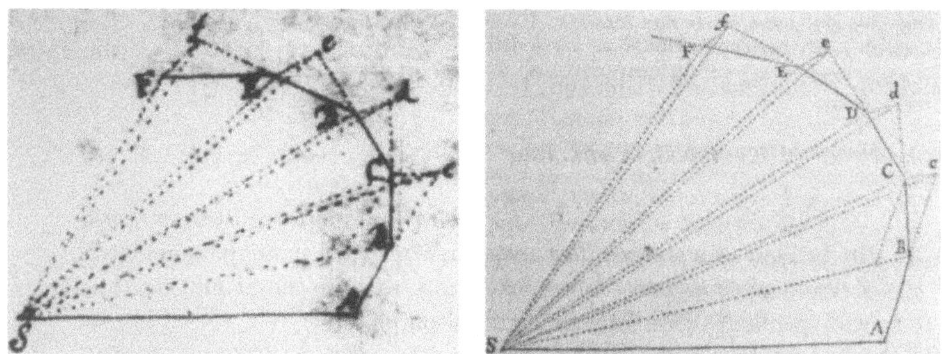

Figure 4.2. Newton's drawing of the symplectic Euler method for the proof of Kepler's Second Law; manuscript Add. 3965[7a] (*left*), publication in the *Principia* (*right*).

direction AV such that BVS are aligned. So for the next time interval, the body will move from B to C such that $ABCV$ is a parallelogram.

The proof of Kepler's Second Law is now as follows (see Figure 4.3). If the force impulse at B had not occurred, the body would have continued under Lex 1 until c such that $AB = Bc$. The triangles SAB and SBc have the same altitudes and the same bases. Hence, by Eucl. I.41, they have the same areas. Next, since the triangles ABV and BcC are the same, cC will be parallel to BV, which, by hypothesis, is in the direction of BS. This means that the triangles SBc and SBC again have the same bases (which is SB) and the same altitudes, and thus the same areas. We conclude that ABS has the same area as BCS, and if we continue like this, all triangles ABS, BCS, CDS, DES, *etc.*, will have the same areas.

Remark. If we recall the explanations of Section 1.4 above, we see that this proof of Newton uses *precisely the symplectic Euler method*, because the force impulse uses the force evaluated at the *end* of the interval. Here, this method is used inversely; the properties of the numerical solution are pulled

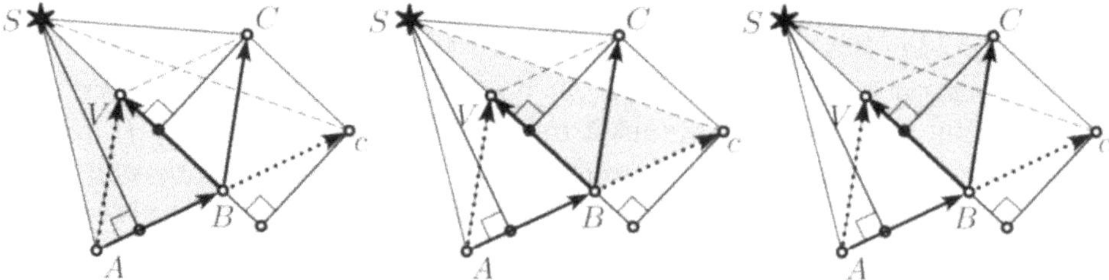

Figure 4.3. Eucl. I.41: All the triangles have the same area; Newton's proof of Kepler's Second Law.

back to the continuous problem, and then constitute Kepler's Second Law. This became the 'Theorema 1' of the *Principia* and Kepler would surely have been very happy with this superbly elegant proof.

4.2. Newton's discovery of the Law of Gravitation from Kepler's First and Second Laws

> ... one of the most far-reaching generalizations of the human mind. While we are admiring the human mind, we should take some time off to stand in awe of a *nature* that could follow with such completeness and generality such an elegantly simple principle as the law of gravitation.
>
> R. Feynman (1963); published in *Six Easy Pieces* (1994), p. 89

The next, and greater, challenge is to find out the *quantity* of this force. Two lemmas will give the answer.

Lemma 1. For a fixed time interval Δt, the force impulse is proportional to the distance RQ, where R is on the tangent and Q is on the orbit (see Figure 4.4, *right*).

Proof. Newton's motivation is shown in Figure 4.4 (*left*). Let our body move, under the continuous force, from A to D. If there were no force, it would move to B on the tangent; if it had no initial velocity, it would move to C. By a principle of superposition of forces and movements (which requires another lemma), we see that $BD = AC$, which is proportional, for a fixed Δt and by Lex 2, to the acting force. Huygens (1673) illustrated the same result with the picture shown in Figure 4.4 (*centre*).

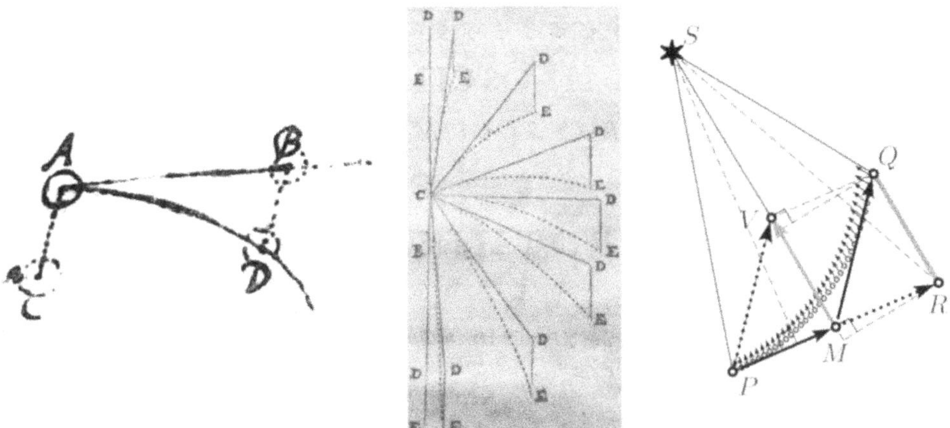

Figure 4.4. Proof of Lemma 1: Newton's manuscript Add. 3965[6], from 1684 (*left*); Huygens, 1673 (*centre*); the Störmer–Verlet method (*right*).

A third way of seeing this result is to place the force impulse $\Delta t \cdot F$ in *the middle* M of the time interval (Figure 4.4, *right*). Our intuitive intelligence, and more numerical analysis, tell us that the resulting point Q is very close to the orbit, and by parallel lines we see that the acting force, which is proportional to MV by Lex 2, is also proportional to RQ.

This last method bears the name *Störmer–Verlet.* It is also symplectic; many further properties are explained in Hairer, Lubich and Wanner (2003). □

Lemma 2. Let an ellipse be given with focus S. Then there is a constant such that, for every P on the ellipse, Q on the ellipse close to P, PR a tangent, RQS and PTS aligned and T the orthogonal projection of Q to PS (see Figure 4.5), we have

$$RQ \approx \text{Const} \cdot QT^2. \tag{4.1}$$

Proof. Half of Newton's proof is displayed in Figure 4.6, where Lemma 2 is called 'Prob. 3'. Most readers, perhaps, would like some more explanations. The result is easier to see for circular motion (see Figure 4.7), where the formula $AT \cdot TP = QT^2$ is known in geometry as the 'Theorem of the Altitudes', or, for experts, Eucl. II.14. Then, if Q moves towards P,

$$RQ \to TP = \frac{QT^2}{AT} \approx \frac{QT^2}{2a}. \tag{4.2}$$

In the *general case* we draw the 'diametri conjugata' GP and DK. The latter is parallel to the tangent, to which we draw the parallel QXV (see Figure 4.6 as well as Figure 4.8). Now formula (4.2) has to be scaled by the halved lengths of these diameters, for which we use the letters c and d respectively, and we instead obtain

$$VP \approx \frac{c}{2d^2} \cdot QV^2. \tag{4.3}$$

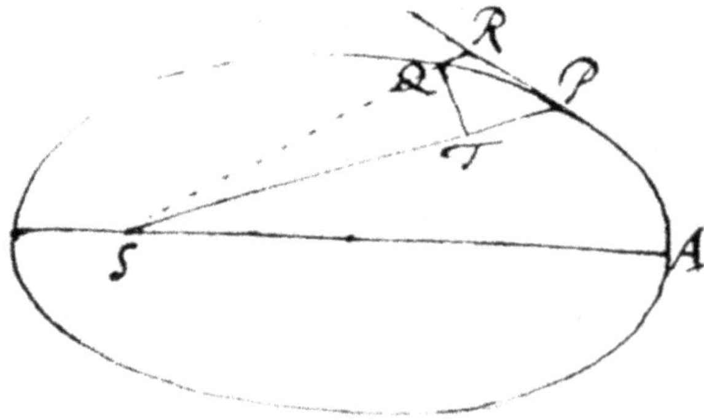

Figure 4.5. Newton's drawing for Lemma 2.

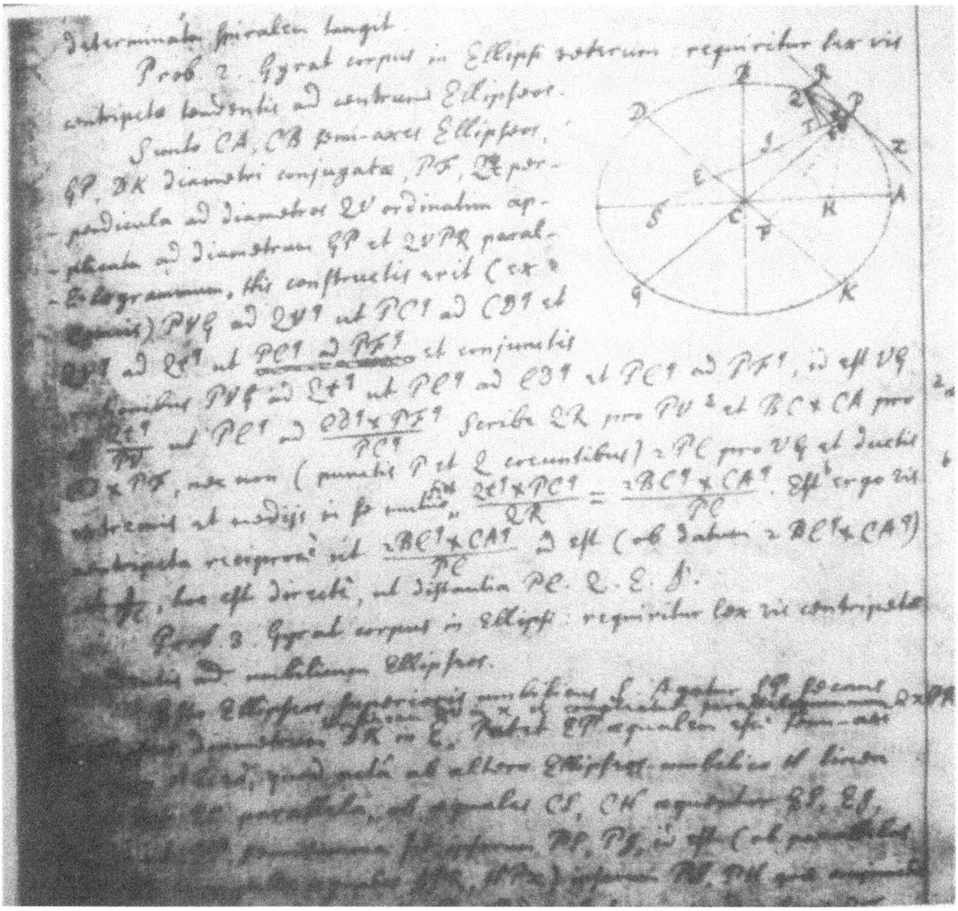

Figure 4.6. Proof of Lemma 2 in Newton's autograph Add. 3965[6] (1684).

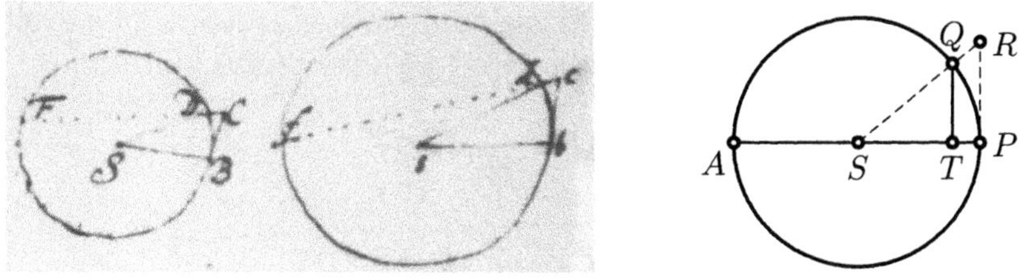

Figure 4.7. The proof of Lemma 2 for circular motion.

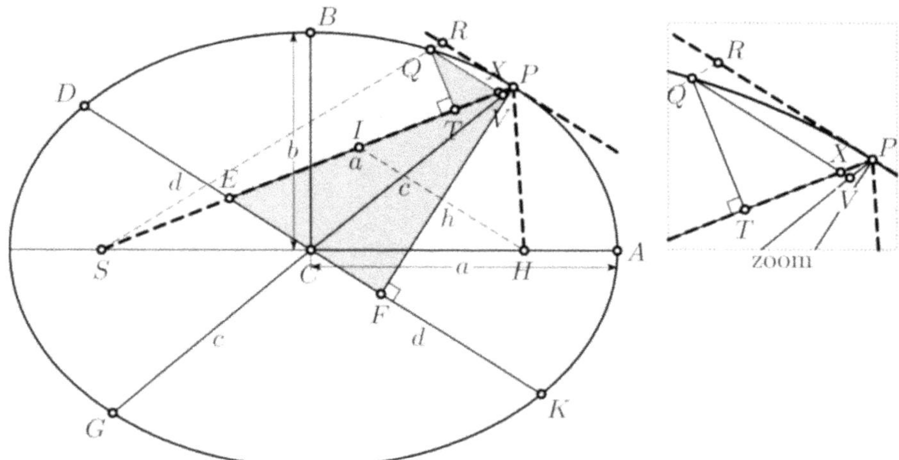

Figure 4.8. Modern illustration of Newton's proof of Lemma 2.

The next result we need is that $EP = a$, the major semi-axis. This follows from known theorems of Apollonius, *i.e.*, that $SP + PH = 2a$, $SE = EI$ and $IP = PH$, but for Newton it was not an easy task (see the beginning of the proof of 'Prob. 3' in the lower part of Figure 4.6). This allows us to compute VP from $XP \approx RQ$ and $QV \approx QX$ from QT, by similar triangles $XVP \sim ECP$ as well as $QTX \sim PFE$, giving

$$VP \approx \frac{c}{a} \cdot RQ \qquad \text{and} \qquad QV \approx \frac{a}{h} \cdot QT. \tag{4.4}$$

These two formulas inserted into (4.3) lead to

$$RQ \approx \frac{a^3}{2h^2d^2} \cdot QT^2. \tag{4.5}$$

Our last difficulty is in understanding that $hd = ab$. This is another theorem of Apollonius (Apoll. VII.31) which Newton had to rediscover, and which states that parallelograms based on conjugate diameters of an ellipse all have the same area. Inserting this into (4.5), we finally obtain

$$RQ \approx \frac{a}{2b^2} \cdot QT^2, \tag{4.6}$$

the desired result. □

Theorem 1. (Proposition XI of the *Principia*) A body P, orbiting according to Kepler's First and Second Laws, is moving under the effect of a centripetal force, directed to the centre S, inversely proportional to the square of the distance.

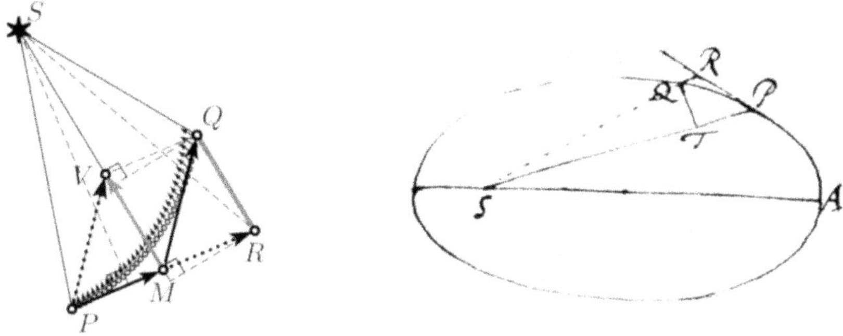

Force proportional to RQ (Lemma 1) RQ proportional to QT^2 (Lemma 2)

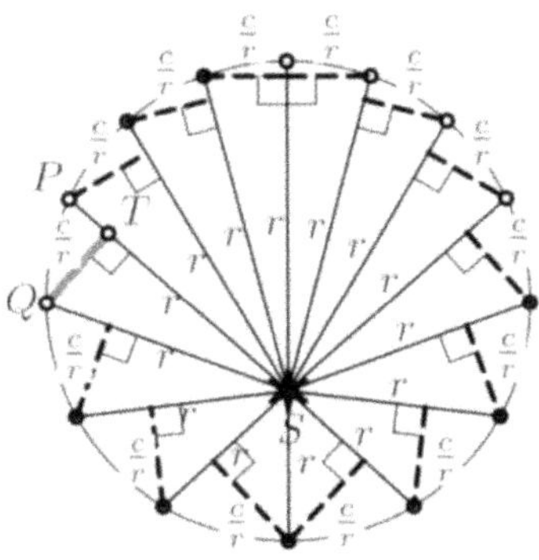

QT proportional to $\frac{1}{r}$ (Kepler 2)

Figure 4.9. The three steps of the proof of the Law of Gravitation.

The *proof* is in three steps, as shown in Figure 4.9, by combining Lemma 1, Lemma 2 and Kepler's Second Law, which together lead to:

$$\text{The force is proportional to } \frac{1}{r^2}.$$

Another century later we arrive at Euler E112 (1749, p. 103), where the so-called 'Newton's equations'

$$\text{I. } \frac{2ddx}{dt^2} = \frac{X}{M}; \quad \text{II. } \frac{2ddy}{dt^2} = \frac{Y}{M}; \quad \text{III. } \frac{2ddz}{dt^2} = \frac{Z}{M}$$

are published for the first time, and for which Newton's ideas are *inverse numerical methods.*

> While physicists call these 'Newton's equations', they occur nowhere in the work of Newton or of anyone else prior to 1747 ... such is the universal ignorance of the true history of mechanics.
>
> C. Truesdell, *Essays in the History of Mechanics* (1968), p. 117

5. Richard Feynman's elegant proof

> Pour voir présentement que cette courbe ABC ... est toûjours une Section Conique, ainsi que M. Newton l'a supposé ... sans le démontrer; il y faut bien plus d'adresse:
> (To see that the curve ABC ... is always a conic section, which Newton supposed it to be, without proof, requires much more ability.)
>
> Johann Bernoulli (1710); in Radelet-de Grave and Villaggio (2007)

> ... no calculus required, no differential equations, no conservation laws, no dynamics, no angular momentum, no constants of integration. This is Feynman at his best: reducing something seemingly big, complicated, and difficult to something small, simple, and easy.
>
> B. Beckman (2006)

As we have just seen, in the *Principia* of 1687 Newton proved that a body moving around the Sun according to Kepler's First and Second Laws possesses the centrifugal force prescribed by the inverse square law. The reciprocal question, however, still remains open: Is *every* movement under a central inverse square force always an ellipse (or a conic)? Johann Bernoulli, in his usual sarcastic style, did not attribute to Newton enough ability ('adresse') to answer this question (see quotation). For other authors such as Arnol'd (1989), this result is clear from the uniqueness of the solutions. In any case, those who have seen one of the usual proofs in calculus know that the computations are not easy.

Fortunately, there *is* an elegant idea, in a 'lost lecture' of Richard Feynman, which D. L. and J. R. Goodstein (1996) discovered under inches of dust in Feynman's papers.

Feynman's idea. We see in Figure 5.1 the movement of our body moving under the inverse squares law represented *with constant time steps* Δt, so that the force impulses towards the Sun are proportional to $1/r^2$. Let us now modify this picture (see Figure 5.2), not with constant time steps, but with *constant angles* $\Delta\phi$. Numerical analysts would say that we use another step-size control. Now the areas of the triangles, which are all similar, *become proportional to* r^2 (this is Eucl. VI.19). Now, by Kepler's Second Law, whose proof is the same as for Lemma 1 above, all time steps Δt become proportional to r^2. As a consequence, all force impulses $\Delta t \cdot F$ are:

(1) of constant length, and

(2) under an angle which changes constantly by $\Delta\phi$.

We now draw the velocities $\dot{P}$ in the velocity plane, called the *hodograph*. Because of Newton's Lex 2 and the two properties above, this hodograph

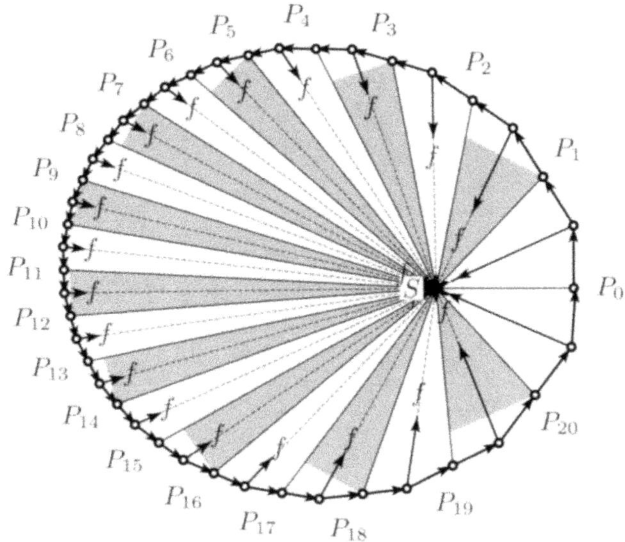

Figure 5.1. Planetary motion in constant time intervals.

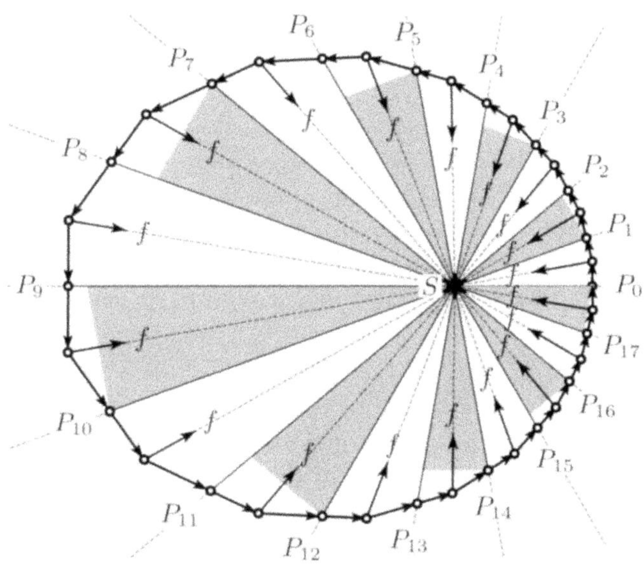

Figure 5.2. Planetary motion in constant angles.

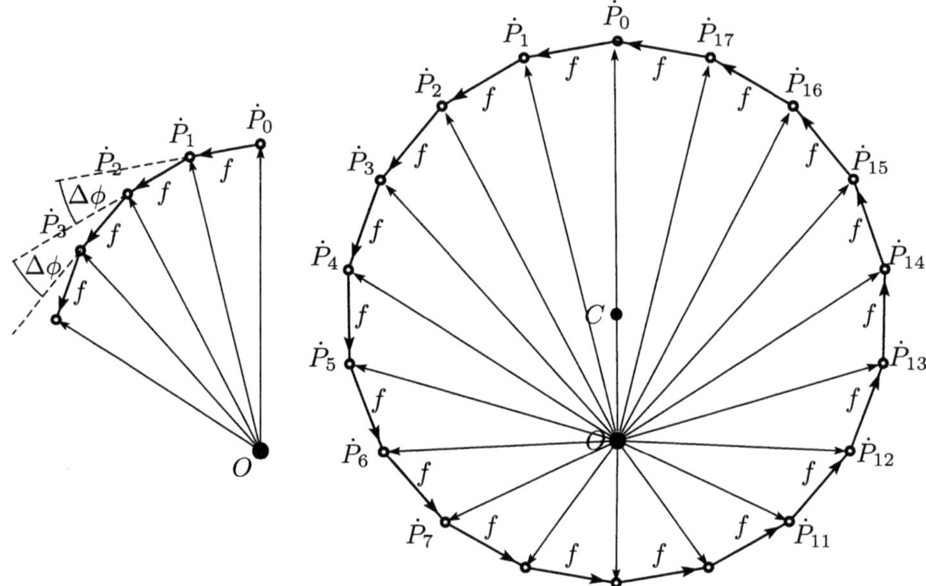

Figure 5.3. The hodograph for inverse square motion is a circle.

behaves like a dog which is dragged across the floor, pulled through the same distances each time, but in different directions. So the dog, starting from the velocity $\dot{P}_0$, which is in an upward direction with maximal speed, is first dragged to the left, later downwards, to $\dot{P}_1$, then to $\dot{P}_2$ and so on, describing the arc of a circle (Figure 5.3). The origin O, corresponding to speed 0, may lie *in* the circle (this corresponds to elliptic movement), or *on* the circle (parabolic), or *outside* the circle (hyperbolic movement).

Our last problem is to find a connection between the circle of Figure 5.3 and the orbit in Figure 5.2. Concerning this question, Feynman said 'I took a long time to find that,' and Beckman (2006) said 'Take a deep breath and look at the following.'

We know from geometry that if we have an ellipse with foci, say, C and O, then for every point P on the ellipse the sum of the distances satisfies $CP+PO=2a$. This means, too, that the distance of P from O is the same as its distance from the circle centred at C with radius $2a$.

Well, let us have a second look at Figure 5.3: we see a circle centred at C and a point O inside this circle. This leads to the idea of considering the curve of points P having the same distance from these two objects (see Figure 5.4). We know, as we just saw, that it is an ellipse. It is also known from geometry (Apoll. III.48), that the tangent at any point, say $\widetilde{P}_3$, reflects the ray $O\widetilde{P}_3$ to C, or, equivalently, that this tangent is *orthogonal* to $O\dot{P}_3$. On the contrary, in Figure 5.2 the tangent of the orbit is *parallel* to $O\dot{P}_3$. Furthermore, the angles in C, respectively S, move with the same constant

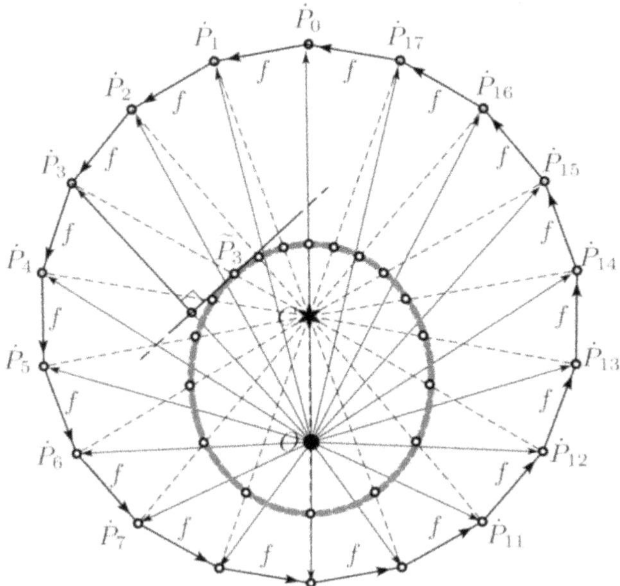

Figure 5.4. The hodograph with the ellipse of the same distance.

speed in both curves. Consequently, *the orbit of Figure 5.2 is proportional to the ellipse of Figure 5.4, rotated by* 90°. This concludes the proof.

Feynman comments on this beautiful proof as follows (lecture of March 13, 1964, 35th minute):

> It is not easy to use the geometrical method to discover things. It is very difficult, but the elegance of the demonstrations after the discoveries are made is really very great. The power of the analytic method is that it is much easier to discover things and to prove things. But not in any degree of elegance. It's a lot of dirty paper, with x's and y's and crossed out cancellations and so on. [laughter]
>
> Published in D. L. and J. R. Goodstein, *Feynman's Lost Lecture* (1996)

What about rigour? Many mathematicians would not consider the above to constitute rigorous proofs but merely nice illustrations. For example, the footnotes in D. T. Whiteside's edition of Newton's *Mathematical Papers* (2008), trying to render Newton's proofs rigorous, with all sorts of curved movements and estimations, are three times as long as Newton's original text. We numerical analysts know, however, that all the above formulas and pictures represent numerical methods, which are known to converge for $h \to 0$ to the corresponding solution, and for which properties of the numerical solution carry over to corresponding properties of the continuous model. So there is no need to announce all the time, as did, for example, Beckman (2006), that 'the argument is water tight'.

Acknowledgements

It is my pleasure to express my thanks first of all to Ernst Hairer, from whom and with whom I had the privilege to learn numerical analysis over many decades, further to Christian Lubich, our co-author of the latest book of our trilogy on the numerical analysis of differential equations, to Philippe Henry, my co-author of a book in progress on Euler, to Alexander Ostermann, my co-author of a book in progress on geometry, to Alexei Shadrin, Cambridge, and Christian Aebi and Bernard Gisin, Geneva, for important hints on the literature, and to Glennis Starling, *Acta Numerica*'s copy-editor, for excellent and extremely helpful support.

I also thank the Bibliothèque de Genève, as well as the Bibliothèque Georges de Rham in Geneva, and Cambridge University Library for permission to reproduce all the beautiful photographs from the original works in their possession. All illustrations from printed books in the Bibliothèque de Genève (BGE) are reproduced, with permission, from the specimen kept under catalogue numbers Ka337 (Euler E342), Rb1 (Leibniz), La163 (Fourier), Ra406 (Lagrange), Ka368 (Euler E65), Kb12 (*Almagest*), Kb31 (Brahe), Ka123 (Kepler, *Astronomia Nova*), Kb39 (Kepler, *Harmonices Mundi*), Kc216 (Galileo), Kb127 (Newton, *Principia*), Kc182 (Huygens) and Ra3 (Euler E112).

REFERENCES

V. I. Arnol'd (1989), *Huygens and Barrow, Newton and Hooke: Pioneers in Mathematical Analysis and Catastrophe Theory from Evolvements to Quasicrystals*, Nauka Moskva (1989). English translation: Birkhäuser (1990).

I. Barrow (1670), *Geometrical Lectures: Explaining the Generation, Nature and Properties of Curve Lines*, translated from the Latin edition (published 1670), revised, corrected and amended by Isaac Newton, edited by Edmund Stone, published in London (1735).

B. Beckman (2006), 'Feynman says: Newton implies Kepler, no calculus needed!' *J. Symbolic Geom.* **1**, 57–72.

N. Copernicus (1543), *De Revolutionibus Orbium Coelestium*, published in Nuremberg. Original manuscript preserved in the Jagiellonian University Library, Kraków, Poland.

R. De Vogelaere (1956), Methods of integration which preserve the contact transformation property of the Hamiltonian equations. Report no. 4, Department of Mathematics, University of Notre Dame, IN.

L. Euler (E65 1744), *Methodus Inveniendi Lineas Curvas Maximi Minimive Proprietate Gaudentes Sive Solutio Problematis Isoperimetrici Latissimo Sensu Accepti*, published in Lausanne and Geneva. Reprinted in *Opera Omnia*, Ser. I, Vol. XXIV.

L. Euler (E112 1749), *Recherches sur le Mouvement des Corps Célestes en Général*, Vol. 3 of *Mem. de l'Acad. des Sciences de Berlin*, pp. 93–143. Reprinted in *Opera Omnia*, Ser. II, Vol. XXV, pp. 1–44.

L. Euler (E301 1766), *De Motu Corporis ad Duo Centra Virium Fixa Attracti*, Vol. 10 of *Novi Comm. Acad. Scient. Petropolitanae*, pp. 207–242. Reprinted in *Opera Omnia*, Ser. II, Vol. VI, pp. 209–246.

L. Euler (E328 1767), *De Motu Corporis ad Duo Centra Virium Fixa Attracti*, Vol. 11 of *Novi Comm. Acad. Scient. Petropolitanae*, pp. 152–184. Reprinted in *Opera Omnia*, Ser. II, Vol. VI, pp. 247–273.

L. Euler (E342 1768), *Institutiones Calculi Integralis*, Vol. I, St. Petersburg. Reprinted in *Opera Omnia*, Ser. I, Vol. XI.

L. Euler (E366 1769), *Institutiones Calculi Integralis*, Vol. II, St. Petersburg. Reprinted in *Opera Omnia*, Ser. I, Vol. XII.

L. Euler (E385 1770), *Institutiones Calculi Integralis*, Vol. III, St. Petersburg. Reprinted in *Opera Omnia*, Ser. I, Vol. XIII.

R. Feynman (1967), *The Character of Physical Law*, MIT Press.

R. Feynman (1994), *Six Easy Pieces: Essentials of Physics, Explained by its Most Brilliant Teacher*, Perseus Books.

J. B. J. Fourier (1822): *La Théorie Analytique de la Chaleur*, Paris. A manuscript from 1807, *Sur la Propagation de la Chaleur*, had its publication refused 'due to lack of rigour'.

G. Galilei (1629), *Les Méchaniques de Galilée, Mathématicien & Ingénieur du Duc de Florence, Traduites de l'Italien par le P. Marin Mersenne*, 1629 manuscript published in Paris (1634). Critical edition by B. Rochod, Paris (1966).

G. Galilei (1638), *Discorsi e Dimostrazioni Matematiche, Intorno à Due Nuove Scienze, Attenti Alla Mecanica & i Movimenti Locali, del Signor Galileo Galilei Linceo*, Elsevier, Leiden. Critical edition by E. Giusti, Giulio Einaudi Editore, Torino (1990). German translation by A. von Oettingen, Ostwald's Klassiker, Leipzig (1890/91).

D. L. Goodstein and J. R. Goodstein (1996), *Feynman's Lost Lecture: The Motion of Planets Around the Sun*, Norton, New York.

E. Hairer, C. Lubich and G. Wanner (2003), Geometric numerical integration illustrated by the Störmer–Verlet method. In Vol. 12 of *Acta Numerica*, Cambridge University Press, pp. 399–450.

E. Hairer, C. Lubich and G. Wanner (2006), *Geometric Numerical Integration: Structure-Preserving Algorithms for Ordinary Differential Equations*, 2nd edn, Springer, Berlin.

E. Hairer, S. P. Nørsett and G. Wanner (1993), *Solving Ordinary Differential Equations I: Nonstiff Problems*, 2nd edn, Springer, Heidelberg.

C. Huygens (1673), *Horologium Oscillatorium, Sive de Motu Pendulorum ad Horologia Aptato Demonstrationes Geometricae*, published in Paris.

J. Kepler (1604), *Ad Vitellionem Paralipomena, Quibus Astronomiae Pars Optica Traditur, Potissimum de Artificiosa Observatione et Aestimatione Diametrorum Deliquiorumque, Solis & Lunae cum Exemplis Insignium Eclipsium.* Reprinted in *Gesammelte Werke*, Vol. 2.

J. Kepler (1609), *Astronomia Nova: Αἰτιολογητός seu Physica Coelestis, Traditia Commentariis de Motibus Stellae Martis, ex Observationibus G. V. Tychonis Brahe*, published in Prague. Reprinted in *Gesammelte Werke*, Vol. 3, edited by M. Caspar (1937). French translation by J. Peyroux (1979).

J. Kepler (1619), *Harmonices Mundi, Libri V*, published in Linz. Reprinted in *Gesammelte Werke*, Vol. 6, edited by M. Caspar (1940).

J. Kepler (1627), *Tabulae Rudolphinae*, published in Ulm. Reprinted in *Gesammelte Werke* vol. 10, edited by F. Hammer (1969).

J. L. de Lagrange (1759), 'Recherches sur la nature et la propagation du son', *Miscellanea Taurinensia*, Vol. I. *Oeuvres*, Vol. 1, pp. 39–148.

G. W. Leibniz (1693), *Supplementum Geometriae Dimensoriae seu Generalissima Omnium Tetragonismorum Effectio per Motum: Similiterque Multiplex Constructio Linea ex Data Tangentium Conditione*, published in *Acta Eruditorum*, Leipzig, pp. 385–392. German translation by G. Kowalewski, *Leibniz über die Analysis des Unendlichen*, No. 162 of *Ostwalds Klassiker* (1908), pp. 24–34.

S. Mandelbrote (2001), *Footprints of the Lion*, Cambridge University Library.

I. Newton (1684), several manuscripts preparing the *Principia*, preserved in Cambridge University Library, in particular manuscript Add. 3965^{7a} (1684).

I. Newton (1687), *Philosophiae Naturalis Principia Mathematica*, published in London. Russian translation with commentaries by A. N. Krylov (1936); reprinted (1989).

I. Newton (1713), *Philosophiae Naturalis Principia Mathematica*, 2nd edn.

C. Ptolemy ($\sim$ AD 150), ἡ μεγάλη σύνταξις (= Great Collection = *Almagest* = *Al* μεγίστη). Latin translation by G. Peurbach and J. Regiomontanus, *Epitoma Almagesti per Magistrum Georgium de Peurbach et eius Discipulum Magistrum Jo. de Künigsperg ...*, completed 1462, printed 1496.

P. Radelet-de Grave and P. Villaggio, eds (2007), *Die Werke von Johann I und Nicolaus II Bernoulli*, Birkhäuser.

B. Russell (1931), *The Scientific Outlook*, Allen & Unwin.

D. Tournès (2009), *La Construction Tractionnelle des Équations Différentielles*, Collection Sciences dans l'Histoire, Albert Blanchard, Paris.

C. Truesdell (1968), *Essays in the History of Mechanics*, Springer.

D. T. Whiteside, ed. (2008), *The Mathematical Papers of Isaac Newton*, Cambridge University Press.